Chemistry of Nanocarbons

Chemistry of Nanocarbons

Edited by

TAKESHI AKASAKA

Center for Tsukuba Advanced Research Alliance
University of Tsukuba, Tsukuba, Japan

FRED WUDL

Department of Chemistry and Biochemistry
University of California, Santa Barbara, USA

SHIGERU NAGASE

Department of Theoretical and Computational Molecular Science
Institute for Molecular Science, Myodaiji, Japan

WILEY

A John Wiley and Sons, Ltd, Publication

This edition first published 2010

Registered office
John Wiley & Sons Ltd, The Atrium, Southern Gate, Chichester, West Sussex, PO19 8SQ, United Kingdom

For details of our global editorial offices, for customer services and for information about how to apply for permission to reuse the copyright material in this book please see our website at www.wiley.com.

Library of Congress Cataloging-in-Publication Data

Akasaka, Takeshi.
Chemistry of nanocarbons / Takeshi Akasaka, Fred Wudl, Shigeru Nagase.
p. cm.
Includes bibliographical references and index.
ISBN 978-0-470-72195-7 (cloth)
1. Nanotubes. 2. Fullerenes. 3. Nanodiamonds. I. Wudl, Fred. II. Nagase, Shigeru. III. Title.
TA418.9.N35A427 2010
620′.5–dc22

2010004437

A catalogue record for this book is available from the British Library.

ISBN 978-0-470-72195-7 (HB)

Set in 10/12pt, Times Roman by Thomson Digital, Noida, India
Printed and bound in Singapore by Markono Print Media Pte Ltd.

The manufacturer's authorized representative according to the EU General Product Safety Regulation is Wiley-VCH GmbH, Boschstr. 12, 69469 Weinheim, Germany, e-mail: Product_Safety@wiley.com.

We dedicate this monograph to the memory of R. Smalley and to the original discoverers Harry Kroto and Sumio Iijima

Contents

Preface

The first time I heard about the possibility of the existence of the molecule we now call buckminsterfullerene was at a lecture given by the late Prof. Orville Chapman in the mid 1980s, followed by the first disclosure by Kroto *et al.* in their *Nature* paper of 1985. In 1990, while visiting Robert Haddon at the AT&T Bell laboratories, I learnt that it had actually been synthesized, not by chemists but by physicists, referring, of course, to a preprint by W. Kraetschmer *et al*'s now famous 1990 *Nature* paper that was floating around the Labs. Since then, buckminsterfullerene has spawned an entire field of endeavor and this book tries to capture the most salient features of the novel molecular allotropes of carbon.

The chapters within this volume present the most up-to-date research on chemical aspects of nanometer sized forms of carbon. It therefore emphasizes the chemistry aspects of fullerenes, nanotubes and nanohorns. All modern chemical aspects are mentioned, including noncovalent interactions, supramolecular assembly, dendrimers, nanocomposites, chirality, nanodevices, host-guest interactions, endohedral fullerenes, magnetic resonance imaging, nanodiamond particles and graphene. The reader will be exposed to the most recent potential and actual applications of these remarkable allotropes of carbon in molecular electronics as well as medicine. The authors of the nineteen chapters are the current principal exponents of nano allotropes of carbon.

The subjects of this book would not be possible without the pioneering work of (in alphabetical order) Curl, Huffman, Iijima, Kraetschmer, Kroto and Smalley, and it is hoped that the book's contents will contribute to the lasting memory of these scientists.

Acknowledgements

T. Akasaka, F. Wudl and S. Nagase gratefully acknowledge the support they received from their respective institutions during the process of this book's edition. We also thank the chapter authors for their prompt cooperation and help to produce this book that we believe will be an invaluable source of information to future researchers in the field.

Contributors

Akasaka, Takeshi, Center for Tsukuba Advanced Research Alliance, University of Tsukuba, Tsukuba, Japan

Ananta, Jeyarama S., Department of Chemistry & Smalley Institute of Nanoscale Science and Technology, Rice University, Houston, TX, USA

Backes, Claudia, Institute of Advanced Materials and Processes, University of Erlangen, Fuerth, Germany

Balch, Alan L., Department of Chemistry, University of California, Davis, CA, USA

Campidelli, Stéphane, CEA, IRAMIS, Laboratoire d'Electronique Moléculaire, Gif sur Yvette, France

Chaur, Manuel N., Department of Chemistry, Clemson University, Clemson, SC, USA

Chen, Zhongfang, Department of Chemistry, Institute for Functional Nanomaterials, University of Puerto Rico, PR, USA

Diederich, François, Laboratory of Organic Chemistry, ETH Zurich, Zurich, Switzerland

Dorn, Harry C., Department of Chemistry, Virginia Polytechnic Institute & State University, Blacksburg, VA, USA

Echegoyen, Luis, Department of Chemistry, Clemson University, Clemson, SC, USA

Fujigaya, Tsuyohiko, Department of Applied Chemistry, Graduate School of Engineering, Kyushu University, Fukuoka, Japan

Fukuzumi, Shunichi, Department of Material and Life Science, Graduate School of Engineering, Osaka University, Japan

Gao, Xingfa, Department of Theoretical and Computational Molecular Science, Institute for Molecular Science, Myodaiji, Okazaki, Japan

Gibson, Harry W., Department of Chemistry, Virginia Polytechnic Institute & State University, Blacksburg, VA, USA

Guldi, Dirk M., Department of Chemistry and Pharmacy & Interdisciplinary Center for Molecular Materials (ICMM), Erlangen, Germany

Herranz, Mª Ángeles, Departamento de Química Orgánica, Universidad Complutense, Madrid, Spain

Hirsch, Andreas, Institute of Organic Chemistry II, University of Erlangen, Erlangen, Germany

Iijima, S., Nanotube Research Center, Meijo University, Japan

Illescas, Beatriz M., Departamento de Química Orgánica, Universidad Complutense, Madrid, Spain

Imahori, Hiroshi, Institute for Integrated Cell-Material Sciences (iCeMS), Kyoto University, Kyoto, Japan

Jiang, De-en, Chemical Science Division, Oak Ridge National Laboratory, Oak Ridge, TN, USA

Kawase, Takeshi, Graduate School of Engineering, University of Hyogo, Hyogo, Japan

Komatsu, Koichi, Department of Environmental and Biotechnological Frontier Engineering, Fukui University of Technology, Fukui, Japan

Kraszewska, Agnieszka, Laboratory of Organic Chemistry, ETH Zurich, Zurich, Switzerland

Lu, Jing, Mesoscopic Physics Laboratory, Department of Physics, Peking University, Beijing, People's Republic of China

Maeda, Yutaka, Department of Chemistry, Tokyo Gakugei University, Koganei, Tokyo, Japan

Martin, Juan-José Cid, Laboratoire de Chimie des Matériaux Moléculaires, Université de Strasbourg et CNRS (UMR 7509), Strasbourg, France

Martín, Nazario, Departamento de Química Orgánica, Universidad Complutense, Madrid, Spain

Matsuo, Yutaka, Nakamura Functional Carbon Cluster Project, ERATO, Japan Science and Technology Agency and Department of Chemistry, The University of Tokyo, Tokyo, Japan

Murata, Michihisa, Institute for Chemical Research, Kyoto University, Kyoto, Japan

Murata, Yasujiro, Institute for Chemical Research, Kyoto University, Kyoto, Japan

Nagase, Shigeru, Institute for Molecular Science, Myodaiji, Okazaki, Japan

Nakamura, Eiichi, Nakamura Functional Carbon Cluster Project, ERATO, Japan Science and Technology Agency and Department of Chemistry, The University of Tokyo, Tokyo, Japan

Nakashima, Naotoshi, Department of Applied Chemistry, Graduate School of Engineering, Kyushu University, Fukuoka, Japan

Nierengarten, Jean-François, Laboratoire de Chimie des Matériaux Moléculaires, Université de Strasbourg et CNRS (UMR 7509), Strasbourg, France

Okazaki, T., Nanotube Research Center, Meijo University, Japan

Olmstead, Marilyn M., Department of Chemistry, University of California, Davis, CA, USA

Ortiz, Angy, Department of Chemistry, Clemson University, Clemson, SC, USA

Ōsawa, Eiji, Nanocarbon Research Institute, Ltd., Asama Research Extension Centre, Faculty of Textile Science and Technology, Shinshu University, Ueda, Nagano, Japan

Palkar, Amit J., ConocoPhillips Company, Ponca City, Oklahoma, USA

Pérez, Emilio M., Departamento de Química Orgánica, Universidad Complutense, Madrid, Spain

Pinzon, Julio R., Department of Chemistry, Clemson University, Clemson, SC, USA

Prato, Maurizio, INSTM, Università di Trieste,Trieste, Italy

Thilgen, Carlo, Laboratory of Organic Chemistry, ETH Zurich, Zurich, Switzerland

Tsuchiya, Takahiro, Center for Tsukuba Advanced Research Alliance, University of Tsukuba, Tsukuba, Japan

Wilson, Lon J., Department of Chemistry & Smalley Institute of Nanoscale Science and Technology, Rice University, Houston, TX, USA

Yudasaka, M., Nanotube Research Center, Meijo University, Japan

Abbreviations

ACCVD	alcohol catalytic chemical vapor deposition
AFM	antiferromagnetic
AFM	atomic force microscopy
AGNRs	armchair-edged graphene nanoribbons
AMI	Austin model 1
AMOs	antibonding molecular orbitals
ArcNTs	AP-grade single-walled carbon nanotubes
ATRP	atom transfer radical polymerization
BET	Brunauer, Emmett, and Teller
BIGCHAP	*N,N*-bis(3-D-gluconamidopropyl) cholamide
BMOs	bonding molecular orbitals
BODA	*bis-o*-diynyl arene
BSA	bovine serum albumin
BWF	Breit–Wigner–Fano
CAs	circumacenes
CAN	ammonium cerium(IV) nitrate
CaNCN	calcium cyanamide
CAPTEAR	chemically adjusting plasma temperature, energy, and reactivity
CD	circular dichroism
CIP	Cahn, Ingold, Prelog
CNG	carbon nanographene
CNOs	carbon nano onions
CNs	carbon nanotubes
CNTs	carbon nanotubes
COOH	carboxylic acid
CPE	constant potential electrolysis
CPPAs	cyclic [n]paraphenyleneacetylenes
CSCNTs	cup-stacked carbon nanotubes
CSP	chiral stationary phases
CT	charge transfer
CV	cyclic voltammetry
CVD	chemical vapor deposition
DABCO	1,4-diazabicyclo[2.2.2]octane
DBU	1,8-diazabicyclo[5.4.0]undec-7-ene
DFT	density functional theory

DFT	discrete Fourier transform
DFT-GGA	density functional theory-generalized gradient-corrected approximation
DGU	density gradient ultracentrifugation
DLS	dynamic light scattering
DMA	dimethylacetamide
DMA	9,10-dimethylanthracene
DMAc	dimethylacetamide
DMAP	dimethylaminopyridine
DMF	dimethylformamide
DMRG	density matrix renormalization group
DMSO	dimethylsulfoxide
DN	detonation nanodiamond
DNA	deoxyribonucleic acid
DOS	density of states
DPV	differential pulse voltammetry
dsDNA	double-strand DNA
DTAB	dodecyltrimethylammonium bromide
DWNT	double wall carbon nanotube
ECF	extracellular fluid space
EMAPS	electromagnetically accelerated plasma spraying
EMFs	endohedral metallofullerenes
EPR	electron paramagnetic resonance
ES	electrostatic
exTTFs	π-extended tetrathiafulvalenes
FAD	flavine adenine dinucleotide cofactor
FET	field-effect transistors
FFF	field flow fractionation
FM	ferromagnetic
FTIR	Fourier transform infrared spectroscopy
GBL	γ-butyrolactone
G/D	graphite/defect
GGA	generalized-gradient approximation
GIAO	gauge-including atomic orbital
GlcNAc	*N*-acetyl-D-glucosamine
GNR	graphene nanoribbon
GOx	glucose oxidase
GPC	gel permeation chromatography
HEM	high energy mode
HiPco	high-pressure CO conversion
HMQC	hetero multiple bond correlation
HOMO	highest occupied molecular orbital

HOPG	highly oriented pyrolitic graphite
HPHT	high pressure high temperature
HPLC	high performance liquid chromatography
HRTEM	high-resolution transmission electron microscope
HSVM	high-speed vibration milling
HTAB	hexadecyltrimethylammonium bromide
IEC	ion exchange chromatography
IPCE	internal photon-to-current efficiency
IPR	isolated pentagon rule
IR	infrared
ITO	indium tin oxide
IUPAC	International Union of Pure and Applied Chemistry
LB	Langmuir-Blodgett
LCAO	linear combination of atomic orbitals
LDA	local density approximation
LDS	lithium, dodecyl sulfate
LPC	lysophosphatidylcholine
LPG	lysophosphatidylglycerol
LUMO	lowest unoccupied molecular orbital
MALDI-TOF-MS	matrix assisted laser desorption ionization time-of-flight mass spectrometry
MCPBA	*m*-chloroperbenzoic acid
MEM	maximum entropy method
MeOH	methanol
MNDO	modified neglect of differential overlap
MPWB1K	hybrid meta DFT method for kinetics
MRA	magnetic resonance angiography
MRI	magnetic resonance imaging
MWCNTs	multi-walled carbon nanotubes
MWNTs	multi-walled carbon nanotubes
NFE	nearly free electron
NHE	normal hydrogen electrode
NICS	nucleus independent chemical shifts
NIR	near-IR
NM	nonmagnetic
NMP	*N*-methyl-2-pyrrolidone
NMR	nuclear magnetic resonance
NMRD	nuclear magnetic relaxation dispersion
NSB	nonspecific binding
NW	nanowire

OC	*o*-carboxymethyl chitosan
ODA	octadecylamine
ODCB	*o*-dichlorobenzene
OITB	orbital interactions through bonds
OPV	oligophenylenevinylene
PABS	poly(*m*-aminobenzenesulfonic acid)
PAH	polycyclic aromatic hydrocarbons
PAMAM	poly(amido amine)
PAmPV	poly{(5-alkoxy-m-phenylenevinylene)-co-[(2,5-dioctyloxy-p-phenylene)-vinylene]}
PArcNTs	oxidized single-walled carbon nanotubes
PAs	periacenes
PBS	phosphate buffered saline
PCBM	methanofullerene phenyl-C_{61}-butyric acid methyl ester
PDDA	poly(diallyl dimethylammonium) chloride
PEO	polyethylene oxide
PEO-PDEM	poly(ethylene oxide)-*b*-poly[2-(*N*,*N*-dimethylamino)-ethyl methacrylate]
PEO-PDMS-PEO	poly(ethyleneoxide)-*b*-poly(dimethylsiloxane)-*b*-poly-(ethylene oxide)
PEO-PPO	poly(ethylene oxide)-*b*-poly(propylene oxide)
PEO-PPO-PEO	poly(ethylene oxide)-*b*-poly(propylene oxide)-*b*-poly-(ethylene oxide)
PDMS	poly(dimethylsiloxane)
PFG-NMR	pulsed-field gradient nuclear magnetic resonance
PFH-A	poly[(9,9-dihexylfluorenyl-2,7-diyl)-*co*-(9,19-anthracence)]
PFO	poly(9,9-dioctylfluorenyl-2,7-diyl
PFO-BT	poly[9,9-dioctylfluorenyl-2,7-diyl)-co-1,4-benzo-{2,1′-3}-thiadiazole)]
PhCN	benzonitrile
PL	photoluminescence
PLE	photoluminescence excitation
PLV	pulsed-laser vaporization
PMMA	poly(methylmethacrylate)
PMMA-PEO	poly(methylmethacrylate)-*b*-poly(ethylene oxide)
PmPV	poly-*m*-phenylenevinylene
PNIPAM	poly(N-isopropylacrylamide)
POAV	p-orbital axis vector analysis
PPEs	poly(aryleneethynylene)s
PPV	*p*-phenylenevinylene
PS-P4VP	polystyrene-*b*-poly(4-vinylpyridine)
PS-PBA	polystyrene-*b*-poly(*tert*-butyl acrylate)
PS-PBD-PS	polystyrene-*b*-polybutadiene-*b*-polystyrene
PS-PEO	polystyrene-*b*-poly(ethylene oxide)
PS-PI	polystyrene-*b*-polyisoprene

PS-PMAA	polystyrene-*b*-poly(methacrylic acid)
PS-PSCI	polystyrene-*b*-poly[sodium(2-sulfamate-3-carboxylate)isoprene]
PSA	prostate specific antigen
PSS^{n-}	poly(sodium 4-styrenesulfonate)
PTCDA	perylene tetracarboxylic dianhydride
$PVBTA^{n+}$	poly((vinylbenzyl)trimethylammonium chloride)
PVP	poly(4-vinylpyridine)
PZC	point of zero charge
QCM	quartz crystal microbalance
RBM	radial breathing mode
RDX	cyclotrimethylenetrinitramine
RNA	ribonucleic acid
SAM	self-assembled monolayers
SANS	small angle neutron scattering
SBM	Solomon-Bloembergen-Morgan
SC	sodium cholate
SCC-DFTB	self-consistent charges density functional theory of tight binding
SCCNT	stacked-cup carbon nanotubes
SDBS	sodium dodecyl benzene sulfonate
SDC	sodium deoxycholate
SDS	sodium dodecyl sulfate
SEC	size exclusion chromatography
SEM	scanning electron microscopy
SGC	sodium glycocholate
SiPc	silicon-phthalocyanine
SNBD	single-nano buckydiamond
SpA	staphylococcal protein A
ssDNA	single-strand DNA
STC	sodium taurocholate
STDC	sodium taurodeoxycholate
SWCNTs	single-walled carbon nanotubes
SWNHox	hole-opened single-walled nanohorns
SWNHs	single-walled nanohorns
SWNTs	single-walled carbon nanotubes
SWNs	single-walled carbon nanotubes
TDAE	tetrakis(dimethylamino)ethylene
TEM	transmission electron microscopic
TFA	trifluoroacetic acid
TGA	thermogravimetric analysis
THF	tetrahydrofuran
THPP	5,10,15,20-tetrakis(hexadecyloxyphenyl)-21*H*,23*H*-porphyrin
TMPD	*N*,*N*,*N*′,*N*′-tetramethyl-*p*-phenylenediamine

TMWCNTs	thin multi-walled carbon nanotubes
TNT	trinitrotoluene
TNTs	trimetallic nitride template endohedral fullerenes
TTAP	tetradecyl trimethyl ammonium bromide
TTF	tetrathiafulvalene
UDD	ultra-dispersed diamond
US	ultra-short
UV-vis	ultraviolet-visible
VDW	Van der Waals
VTMWCNTs	very thin multi-walled carbon nanotubes
VT-NMR	variable temperature nuclear magnetic resonance
XPS	X-ray photoelectron spectrum
XRD	X-ray diffraction
ZGNR	zigzag-edged graphene nanoribbon
ZINDO	*Zerner* Intermediate Neglect of Differential Overlap
ZnNc	zinc naphthalocyanine
ZnP	zinc tetraphenylporphyrin
ZnPP	zinc protoporphyrin

1

Noncovalent Functionalization of Carbon Nanotubes

Claudia Backes[a,b] and Andreas Hirsch[b]

[a]*Institute of Advanced Materials and Processes, Fuerth, Germany*
[b]*Department of Chemistry and Pharmacy, Erlangen, Germany*

1.1 Introduction

Within the past decades extensive research has shed light into the structure, reactivity and properties of carbon nanotubes (CNTs) [1–3]. This new carbon allotrope is theoretically constructed by rolling up a graphene sheet into a cylinder with the hexagonal rings joining seamlessly. Commonly, carbon nanotubes are classified into single-walled carbon nanotubes (SWCNTs) which consist of one cylinder and multi-walled carbon nanotubes (MWCNTs) comprising an array of tubes being concentrically nested. Depending on the roll-up vector which defines the arrangement of the hexagonal rings along the tubular surface, single-walled carbon nanotubes exhibit different physical and electronic properties, e.g. they either possess metallic or semiconducting character.

Apart from their outstanding electronic properties providing the foundation for multiple applications as nanowires, field-effect transistors and electronic devices [3–5], carbon nanotubes surmount any other substance class in their mechanical properties. The exceptionally high tensile modulus (640 GPa) and tensile strength ($\approx$100 GPa) together with the high aspect ratio (300–1000) make nanotubes an ideal candidate for reinforcing fibers and polymers [6, 7].

However, in order to tap the full potential of nanotubes in electronics, photonics, as sensors or in composite materials, two major obstacles have to be overcome, e.g. separation

Chemistry of Nanocarbons Edited by Takeshi Akasaka, Fred Wudl and Shigeru Nagase

according to diameter and/or chirality on the one hand and uniform dispersability in a solvent or matrix on the other hand. Responding to a growing interest, progress in the diameter control during carbon nanotube production has been achieved [8]. However, up to now, the as-produced material contains nanotubes of differing lengths, diameters and chiralities, therefore including semiconducting and metallic nanotubes. This inhomogeneity still forms the bottleneck for nanotube-based technological progress. Furthermore, the strong intertube van der Waals interactions of 0.5 eV/μm, which render nanotubes virtually insoluble in common organic solvents and water also constrict any application [9, 10].

Among the efforts to increase processability of this unique material, chemical and especially noncovalent functionalization represents a cornerstone, as in its nondestructive meaning it does not alter the intrinsic properties of CNTs. Furthermore, tailoring of the surface properties of carbon nanotubes is accessible, boosting solubility in a variety of solvents and increasing matrix interactions, as will be summarized within this chapter.

1.2 Overview of Functionalization Methods

In CNT functionalization chemistry four main approaches are at hand, as outlined by Figure 1.1. The covalent attachment of groups onto the nanotube scaffold includes defect functionalization on the one hand and direct sidewall functionalization on the other hand. Defect functionalization is associated with chemical transformation of defects already present or induced. Contrarily, direct sidewall functionalization is directly related to rehybridization of the sp^2 carbon atoms of the CNT framework into sp^3 carbon atoms. Even though no further defects like kinks and holes in the nanotube walls are created hereby, both covalent methods disrupt the sp^2 carbon network of the nanotube resulting in mostly undesirable alterations of the physical and chemical properties of this extraordinary material. Certainly, covalent functionalization is a prosperous field of research for modifying surface properties of carbon nanotubes, as it allows attachment of a large variety of groups, which is nicely documented in a series of review articles related to this topic [11–18].

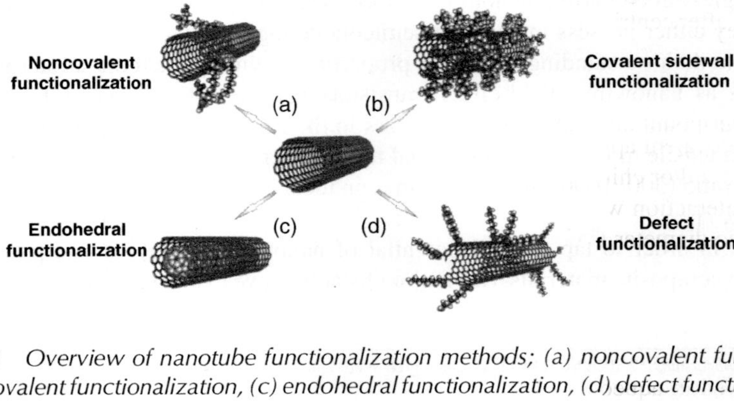

Figure 1.1 *Overview of nanotube functionalization methods; (a) noncovalent functionalization, (b) covalent functionalization, (c) endohedral functionalization, (d) defect functionalization*

Opposed to covalent functionalization, endohedral functionalization [17, 19] is concerned with filling the inner cavities of carbon nanotubes to store guest molecules like fullerenes or small proteins inside the nanotube. However, this approach only slightly influences the surface properties and therefore represents a special case in nanotube functionalization.

Noncovalent exohedral functionalization takes advantage of the supramolecular approach and involves adsorption of various inorganic and organic molecules onto the sidewall of a nanotube via noncovalent interactions including π-π-stacking, van der Waals or charge transfer interactions. The great potential of this method has been realized within the past years [6, 20–25], as it is completely nondestructive and preserves the intrinsic structure of the tubular network without changing the configuration of the carbon atoms.

The following chapter shall mainly deal with concepts and progresses towards monodisperse carbon nanotubes by the noncovalent approach, as this functionalization route may pave the road for many nanotube-based applications due to its nondestructive nature and scalability.

1.3 The Noncovalent Approach

1.3.1 Dispersability of Carbon Nanotubes

1.3.1.1 Aim and Prospects

Noncovalent functionalization mainly targets an enhancement in dispersability and solubility of pristine carbon nanotubes. In the following, the term dispersion refers to homogeneously distributed nanotubes in a colloidal state also including small bundles, while solution implies that the nanotubes appear individualized, e.g. the carbon nanotube bundles are exfoliated. The aim and prospects of noncovalent carbon nanotube functionalization are illustrated by Figure 1.2: The introduction of a bifunctional molecule to the sidewall of a nanotube not only serves the purpose of yielding stable nanotube dispersions and solutions (Figure 1.2a), the interaction with a polymer matrix can also be increased (Figure 1.2b). The use of molecules selectively adsorbing on the nanotube sidewall may also be exploited for nanotube purification, as merely the nanotubes are solubilized, while the impurities consisting of amorphous carbon and metal catalyst particles can be removed as precipitate after centrifugation (Figure 1.2c). This concept is especially advantageous over oxidative purification techniques, as it preserves the intrinsic nanotube structure. The major obstacle here is the necessity to completely individualize the nanotubes, as the impurities are often trapped inside the bundles.

The noncovalent approach also opens the door to the separation of SWCNTs according to diameter and/or chirality either by density gradient ultracentrifugation (Figure 1.2d) or selective interaction with bridged mediators acting as nanotweezers to extract nanotubes of a specific diameter (Figure 1.2e).

1.3.1.2 Aqueous Dispersions

The most widespread noncovalent functionalization method to obtain nanotubes dispersed and dissolved in aqueous media is their encapsulation in surfactant micelles. In general,

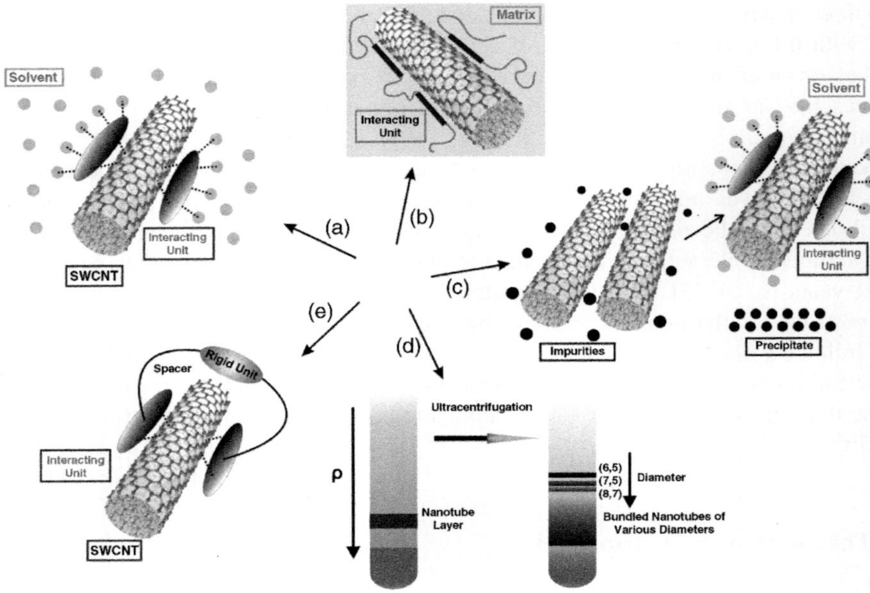

Figure 1.2 Aim and prospects of noncovalent functionalization: (a) solubility tuning, (b) composite reinforcement, (c) purification, (d) separation of different tube species by density gradient ultracentrifugation, (e) separation of different tube species by selective interaction

surfactants can be described as molecules with a hydrophilic region usually referred to as polar head group and a hydrophobic region denoted as tail. Due to this amphiphilicity they tend to adsorb at interfaces and self-accumulate into supramolecular structures.

Three adsorption mechanisms of surfactants onto SWCNTs have been proposed as depicted by Figure 1.3. In analogy to the epitaxial adsorption of surfactants on graphite, specific self-alignment as cylindrical micelles (Figure 1.3a) or hemimicelles (Figure 1.3b) has been suggested [26, 27]. More recently, structureless random adsorption has been favored [28]. In this case, no preferred arrangement of the head and tail groups stabilizes the dispersion/solution (Figure 1.3c).

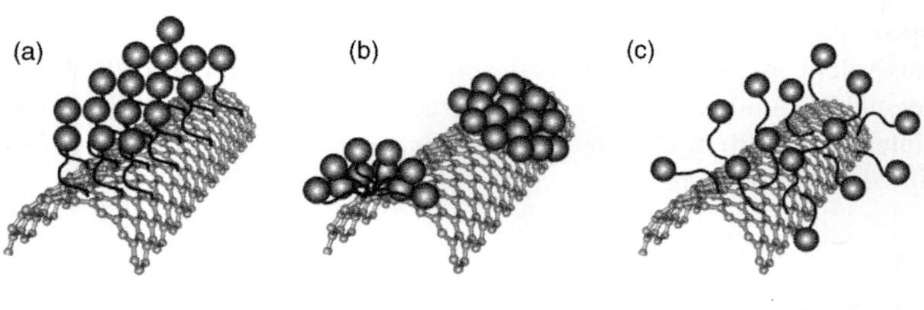

Figure 1.3 Adsorption of surfactants: (a) SWCNTs encapsulated in a cylindrical micelle by aligned adsorption of the amphiphiles, (b) hemimicellar adsorption, (c) random adsorption

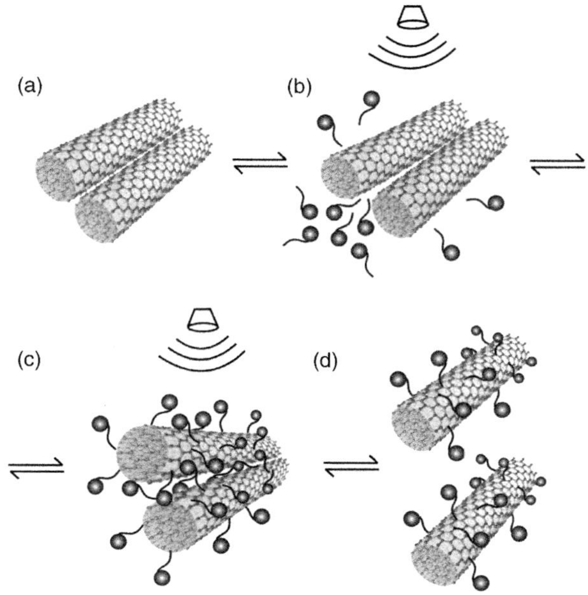

Figure 1.4 *Mechanism of nanotube exfoliation from bundles with the aid of a surfactant and ultrasonication according to the unzippering mechanism*

According to the unzippering mechanism proposed by Strano *et al.* [29], nanotubes are isolated from bundles by ultrasonication in the presence of a surfactant (Figure 1.4). During the first step, the energy input by ultrasonication provides high local sheer, resulting in dangling ends in the nanotube bundles (Figure 1.4b) which become adsorption sites for surfactants preventing the loosened tubes from reaggregation (Figure 1.4c). Due to the relative movement of the partly individualized nanotube relative to the bundle, the surfactant continuously progresses along the nanotube length resulting in the isolation of the individual tube (Figure 1.4d). Hereby, an equilibrium is established between free individuals and bundled aggregates limiting the concentration of stably individualized SWCNTs.

The exfoliation of SWCNTs is specifically important for their characterization. A cornerstone in nanotube characterization has been laid by O'Connell *et al.* who have first reported on the observation of nanotube fluorescence directly across the bandgap of semiconducting SWCNTs [26]. They have revealed that photoemission is only observed when nanotubes are exfoliated, as aggregation otherwise quenches the fluorescence by interaction with metallic nanotubes. Based on this finding, Bachilo *et al.* have used spectrofluorimetry to assign the optical transitions to specific (n,m)-nanotubes for determining the detailed composition of a bulk sample of individualized SWCNTs [30]. Exfoliation in both cases was achieved by the anionic detergent sodium dodecyl sulfate (SDS) nicely underlining the importance of nanotube solubilization by surfactants.

Surfactants are usually classified according to the nature of the head charge, e.g. anionic, cationic, nonionic and zwitterionic. A variety of nanotube surfactants (Figure 1.5) has been investigated including the most common detergents sodium dodecyl benzene sulfonate

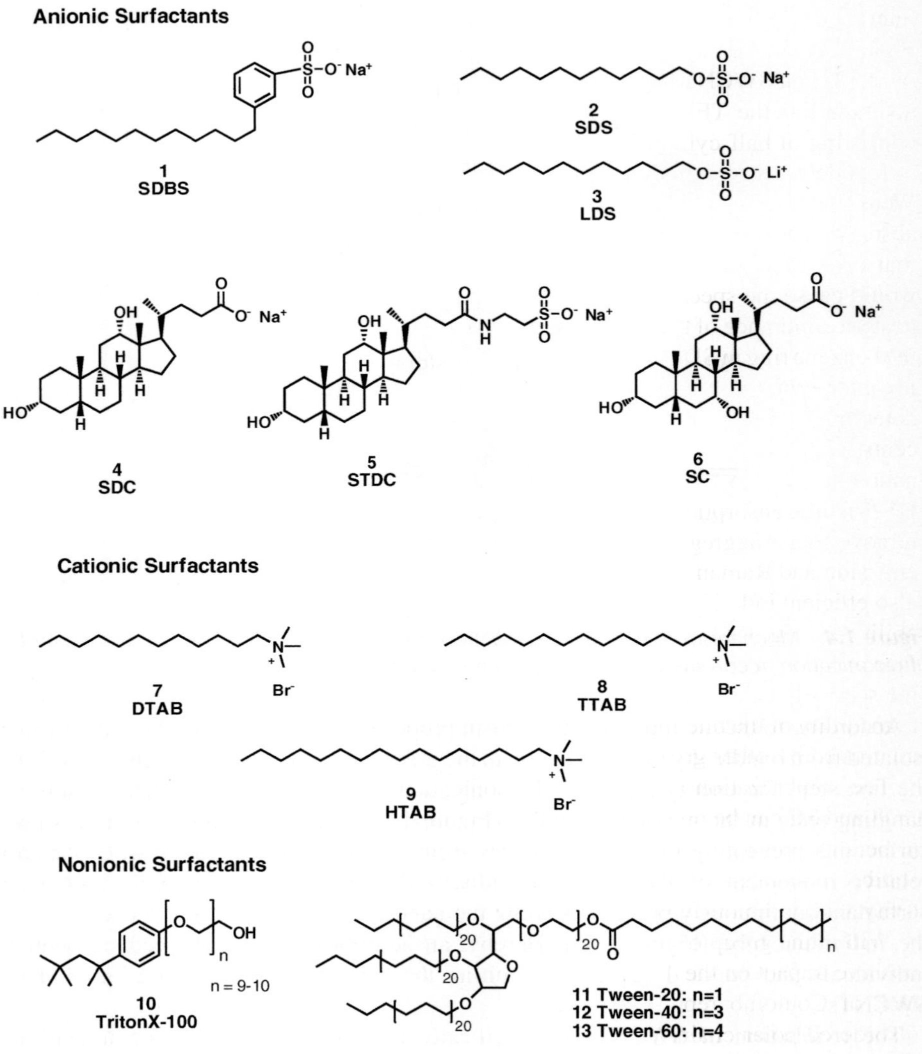

Figure 1.5 *Structure of the most common SWCNT detergents*

(SDBS **1**), sodium dodecyl sulfate (SDS **2**), lithium, dodecyl sulfate (LDS **3**), the bile salts sodium deoxycholate (SDC **4**), sodium taurodeoxycholate (STDC **5**) and sodium cholate (SC **6**) as representatives of anionic surfactants, dodecyltrimethylammonium bromide (DTAB **7**), tetradecyl trimethyl ammonium bromide (TTAP **8**), hexadecyltrimethylammonium bromide (HTAB **9**) among cationic surfactants and TritonX-100 **10**, Tween-20 **11**, Tween-40 **12** and Tween-60 **13** as nonionic surfactants.

A comparative study of SDS, SDBS and TritonX-100 has shown that SDBS and TritonX-100 are more effective than SDS in dispersing SWCNTs [31]. It has been suggested that the aromatic rings in SDBS and TritonX-100 have a positive effect due to additional π-stacking interactions with the nanotube sidewall when adsorbed in a hemimicellar fashion (compare Figure 1.3b). The transmission electron microscopic (TEM) investigations of Mioskowski *et al.* [27] on SWCNTs dispersed in SDS, however, support a cylindrical adsorption of the detergent, as the TEM images have revealed that SDS forms supramolecular structures consisting of half-cylinders.

In a detailed study on the dispersion of arc discharge SWCNTs by various surfactants, Wenseleers *et al.* [32] have observed no spectral shift in the characteristic nanotube absorption bands in aqueous solutions of SDBS and TritonX-100. They have concluded that π-stacking of the benzene rings is therefore not likely, as interaction with the nanotubes would cause the spectral features to be red-shifted. They have assigned their observation to steric hindrance of the rather bulky substituents in TritonX-100 and SDBS. Additionally, the benzene ring in SDBS is located at the polar end of the detergent rendering it unfavorable for interacting with the nanotube. Furthermore, their analysis has revealed that bile salt detergents (SDC, STDC and SC) are highly effective in dispersing SWCNTs before centrifugation. In order to probe the stability of the dispersion and solubility of the nanotubes, the dispersed nanotubes have been characterized by Raman, nIR emission and UV/Vis/nIR absorption spectroscopy after ultracentrifugation which serves the purpose to remove coarse aggregates and nanotube bundles. In the case of the bile salts, the absorption, emission and Raman intensity is high, yielding well resolved features strongly supporting also efficient individualization. This superior dispersion behavior has been ascribed to the ability of the bile salts to stack into ordered layers due to the hydrophobic and hydrophilic face of the apolar half of the molecule: depending on the position of the hydroxyl group in the semi-rigid cholesterol unit the molecule has a more polar and a more apolar side. The high ability of the bile salts to form stable nanotube dispersions and solutions has also been confirmed by other groups [33–35].

The solubilization and individualization of SWCNTs in aqueous solutions of SDBS, SDS and SC can be probed by photoluminescence mapping [36], as fluorescence is known to be quenched in nanotube bundles. It has been indicated that SDS and SC preferentially solubilize smaller-diameter nanotubes, while SDBS shows no significant diameter selectivity within the range of $d = 0.83–0.97$ nm.

Within the group of the nonionic surfactants an increase in the molecular weight has a positive impact on the dispersability of nanotubes [37]. This behavior is traced back to the lack of Coulomb repulsion in the head groups resulting in the long and/or branched disordered polar chains (usually poly(ethylene glycol)) to be the key factor in nanotube dispersability.

As outlined in the section above, the results especially concerning the adsorption mechanism partly appear contradictory, nicely demonstrating that the dispersability and solubility of carbon nanotubes is highly sensitive to the environment and the dispersion parameters, e.g. ultrasonication power and time [38], centrifugation or precipitation conditions, temperature, etc. Furthermore, it has been shown that the concentration of the nanotube [39], as well as the detergent [40, 41] has a tremendous impact on both, the quality of the dispersion and the general dispersion behavior. The composition of the pristine nanotube material (amount of impurities, diameter distribution, etc.) constitutes a further impediment towards comparability [38].

In a systematic study, the effect of purification, sonication time and surfactant concentration on the dispersability of SWCNTs in an aqueous solution of SDBS has also been investigated [42]. It has been revealed that the purification method has an impact on the surface properties of the nanotube, e.g. the point of zero charge (PZC). However, the introduced positive or negative charges on the nanotube, being dependent on the pH, only influence the interaction with the negatively charged SDBS molecules at pH values far from the PZC, indicating that the nanotube-detergent interactions are hydrophobic in nature. Further adsorption studies have shown that, at saturation, the detergent molecules cover the nanotubes as monolayer with the tails oriented vertically on the surface. This indicates that the nanotubes are rather dispersed by adsorption of the SDBS molecules than by enclosing the SWCNTs in cylindrical micelles. It has also been pointed out that the sonication time plays a key role in nanotube dispersion and dissolution, as dispersion remained ineffective without the aid of sonication. Finally, the investigations have unveiled that nanotubes can be dispersed in an aqueous solution of SDBS below the critical micelle concentration (cmc) of SDBS further underlining that the formation of micelles is not a requirement for suspendability. The dispersability of the nanotubes reaches a maximum at [SDBS] = 2.5 mM (0.87 wt%) under the experimental conditions chosen.

Based on preliminary research on the zeta potential of aqueous nanotube dispersions [43], Coleman and coworkers [44] have been able to relate the zeta potential of detergent coated SWCNTs to the quality of the dispersion. The zeta potential in colloidal science can be defined as the electrical potential in the vicinity of the surface of the colloid dispersed, e.g. the nanotube. By a detailed atomic force microscopy (AFM) analysis on nanotubes dispersed in SDBS, SDS, LDS, TTAP, SC and fairy liquid (a common kitchen surfactant) they have quantified the quality of the dispersion by four parameters: the saturation value (at low concentration) of the root-mean-square bundle diameter, the maximum value of the total number of dispersed objects per unit volume of dispersion, the saturation value (at low concentration) of the number fraction of individual tubes and the maximum value of the number of individual nanotubes per unit volume of dispersion. They have included the four parameters in a metric, allowing quantification of the quality of the dispersion. The dispersion quality metric scales very well with the zeta potential decreasing in the following order: SDS > LDS > SDBS > TTAB > SC > fairy liquid. Dispersion and solubilization of nanotubes by ionic surfactants imparts an effective charge on the nanotube, stabilizing the nanotubes from reaggregation due to electrostatic repulsion. Thus, it is reasonable that higher zeta potentials are related to an increased stability of the dispersion. This means that the number of adsorbed surfactant molecules per unit area of tube surface should be maximized. In a preliminary study, White *et al.* [43] have demonstrated that the zeta potential of a dispersion of nanotubes in SDS augments with increasing the SDS concentration (for a fixed nanotube concentration). Furthermore they have shown that the zeta potential is increased when reducing the chain length of the detergent. Both observations are consistent with the zeta potential scaling with the total charge in the vicinity of the nanotube. Typical values of the zeta potential in the study range from –20 mV for the SC solution to –72 mV for the LDS dispersion (for a nanotube concentration of 0.065 g/l) [44]. It has clearly been outlined that the dispersion quality can presumably be significantly improved by using surfactants coating the nanotubes to give hybrids with magnitudes of the zeta potential of 100 mV and higher.

Additionally to the readily available detergents, bifunctional polycyclic aromatic compounds are also excellent candidates for the dispersion/dissolution of nanotubes. In principal, a strong and specific interaction with the nanotube can be ensured *via*

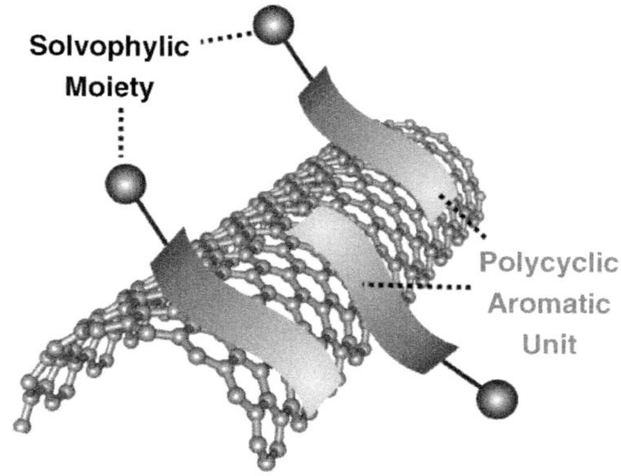

Figure 1.6 *Concept of nanotube dispersion by polycyclic aromatic compounds equipped with a solvophylic moiety*

π-π-stacking which is, in many cases, favorable over the nonspecific hydrophobic interaction being exploited by detergents. Water solubility is provided by solvophylic moieties covalently attached to the aromatic backbone of the dispersing agent (Figure 1.6).

Pyrene derivatives, especially trimethyl-(2-oxo-2-pyrene-1-yl-ethyl)-ammonium bromide **14** are prominent examples nicely underlining the effectiveness of this concept. It has been demonstrated that **14** is capable of dispersing and individualizing both as-produced and purified SWCNTs under mild dispersion conditions [45, 46]. Photoluminescence measurements have revealed that a significant red-shift of the nanotube spectral features occurs, being indicative for the π-π-stacking interaction. Furthermore, semiconducting nanotubes in the diameter range of 0.89–1.00 nm are preferentially individualized. TEM results have indicated that purification of the raw nanotube material occurs upon dissolution in **14**, as fewer catalyst particles are observed compared to nanotubes dispersed in an aqueous solution of HTAB. Since the finding of **14** being an excellent nanotube solubilizer, the pyrene moiety has widely been applied as noncovalent anchoring group, for instance, for the immobilization of fullerenes, proteins, porphyrins and metal nanoparticles (ref [25] and references therein).

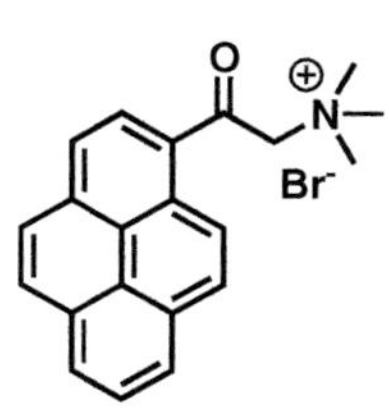

14

The dispersion of SWCNTs by noncovalent functionalization with ionic pyrene and naphthalene derivatives has been explored [47]. The nondestructive nature of the interaction has been confirmed by UV/Vis/nIR absorption, emission and Raman spectroscopy, as well as by X-ray photoelectron spectrum (XPS). Presumably, charge transfer from the adsorbate to the nanotube takes place, as a shift to higher binding energies in the XPS C1s core level spectra has been observed. The presence of a free amino group, especially in the case of the naphthalene derivatives, plays a key role in the dispersion process due to an increased interaction with the nanotube. Thus, specific interactions between the adsorbate's substituents, e.g. by charge transfer and cation-π interactions additionally to the π-π-stacking interaction has been unveiled as important, especially in the case of rather small aromatic molecules. Furthermore, the XPS measurements have shown that the ionic surface charge density on the nanotubes in the composites is almost constant indicating that electrostatic repulsion between the adsorbate molecules is the limiting factor for noncovalent functionalization of SWCNTs with water soluble polycyclic aromatic compounds.

Furthermore, water soluble perylene bisimide derivatives are highly effective in individualizing SWCNTs [48]. The perylene derivative **15** represents a novel class of SWCNT surfactants, as it can be regarded as three-component molecule bearing a 2G-Newkome dendrimer as solvophylic moiety, a perylene bisimide unit for interacting with the nanotube surface *via* π-π-stacking interactions and an aliphatic tail responsible for the highly amphyphilic nature. In fact, it has previously been shown by cryo-TEM that **15** forms regular micelles with a diameter of approximately 16 nm in buffered aqueous media (pH = 7.2) [49].

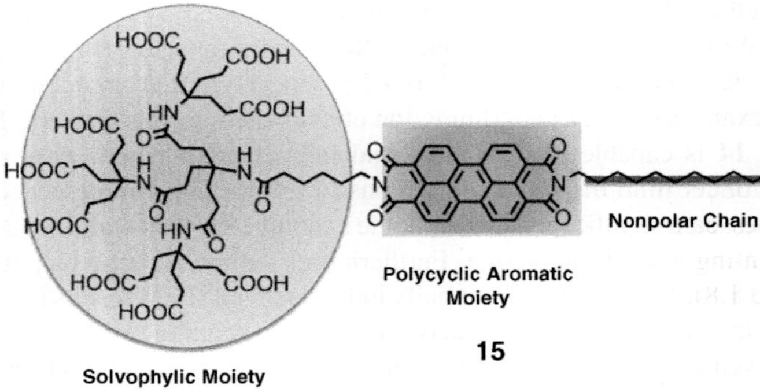

15

After sonicating SWCNTs immersed in a buffered aqueous solution of **15**, with a concentration as low as 0.01 wt%, stable dispersions are formed. After centrifugation (25 000 g), the population of individual SWCNTs is much higher compared to nanotubes dispersed in a solution of SDBS, under the same experimental conditions, as demonstrated by statistical AFM analysis. Adsorption of the perylene unit of **15** has been indicated by the red-shift of the characteristic absorption and emission features of the nanotubes. This has further been supported by the fluorescence quenching of the perylene unit. Cryo-TEM imaging has also underlined the high degree of individualization and revealed that less catalyst particles are present when nanotubes are dispersed with the perylene bisimide derivative **15** compared to a dispersion of nanotubes in SDBS (Figure 1.7).

Additional to pyrene **14** and perylene **15**, porphyrin derivatives represent a third class of polycyclic aromatic surfactants to aid the dispersion of nanotubes in water or organic solvents (see Section 1.3.1.3). Porphyrin derivatives are highly efficient in constructing SWCNT-nanohybrids. However, only the water-soluble porphyrin derivative **16** (*meso*-(tetrakis-4-sulfonatophenyl)porphyrin) will be mentioned in this section. It has been demonstrated by fluorescence and absorption spectroscopy that the free base of **16** is

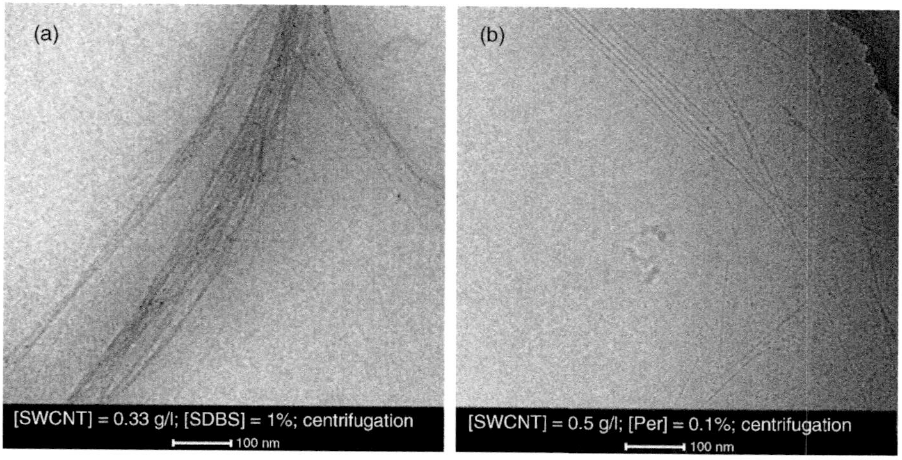

Figure 1.7 *Representative cryo-TEM images of SWCNTs dispersed in an aqueous solution of (a) SDBS and (b) perylene **15**. Reprinted with permission from reference [48]*

responsible for dispersing SWCNTs in water [50]. The stabilizing interaction upon adsorption of the porphyrin to the nanotube sidewall renders protonation to the diacid form more difficult. At pH = 5, the nucleation of *J*-aggregates being unstable in solution cause the nanotube porphyrin complex to precipitate. Furthermore, the porphyrin functionalized nanotubes can be precisely aligned on poly(dimethylsiloxane) (PDMS) stamps by combing. Printing then allows transfer of the nanotubes to a silicon surface as imaged by AFM (Figure 1.8).

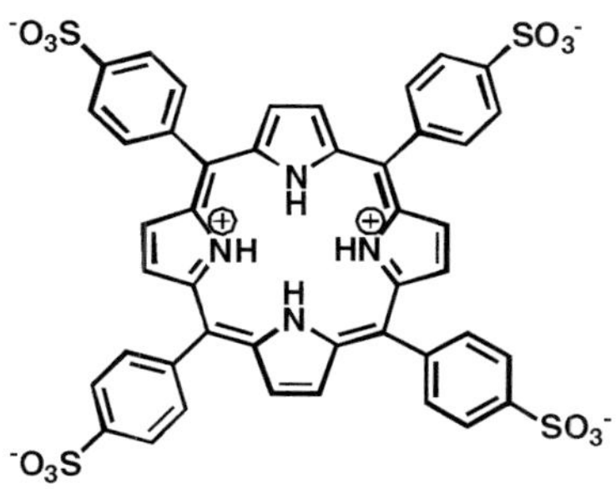

16

1.3.1.3 Dispersion in Organic Solvents

Only limited research has thus far focused on the dispersion of nanotubes in organic solvents compared to water-based systems. Since carbon nanotubes are hydrophobic, they are expected to be wetted by organic solvents as opposed to aqueous media. However, pristine CNTs are colloidally dispersed only in a limited number of solvents, e.g. *o*-dichlorobenzene (ODCB) [51–55], *N*-methyl-2-pyrrolidone (NMP) [56–60], *N,N*-dimethylformamide

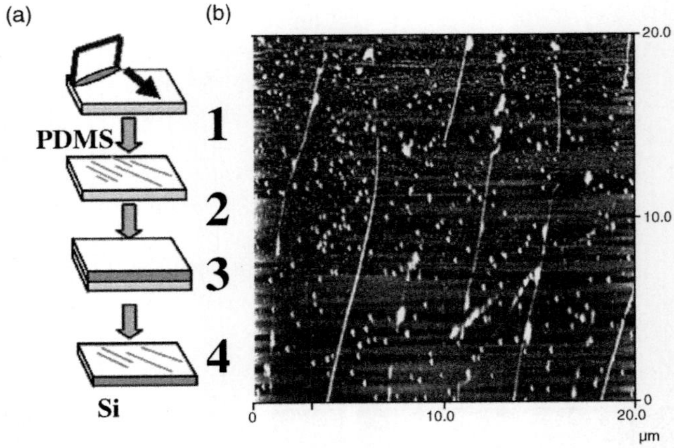

Figure 1.8 *(a) Alignment of porphyrin functionalized SWCNTs onto PDMS stamps by combing followed by transfer printing of the aligned nanotubes onto a silicon substrate, (b) AFM image of aligned SWCNTs. Reprinted with permission from reference [50]*

(DMF) [56–61] and *N,N*-dimethylacetamide (DMA) [57, 60]. Even though dispersion in such solvents is convenient, it is important to note that the stability of the dispersion is usually poor being accompanied by the formation of nanotube aggregates within hours or days.

Dispersability of carbon nanotubes in ODCB has been the topic of discussion, as ODCB was found to degrade upon sonication which is commonly used in carbon nanotube processing. In 2003 Niyogi *et al.* [53] have pointed out that the sonochemical decomposition and polymerization of ODCB results in additional stabilization of the nanotube dispersion. The dispersion stability has been found to be drastically reduced when adding ethanol which may act as radical quencher in inhibiting the polymerization of ODCB. Interestingly, if ODCB is allowed to polymerize sonochemically prior to the addition of nanotubes, the SWCNTs are not efficiently dispersed indicating that dispersion of nanotubes in ODCB upon sonication follows a more complex mechanism. This has further been supported by the observation that nanotubes are irreversibly damaged upon extended sonication in ODCB. Two years later Geckeler and coworkers [54] have demonstrated that by-products of sonochemical degradation of ODCB such as sonopolymers can be removed by ultracentrifugation (325 000 g). However, small oligomeric species are still present in the supernatant solution. After ultracentrifugation, they have found that nanotubes are highly exfoliated containing 85 % of individual nanotubes as shown by statistical AFM analysis. In a recent study, Moonoosawmy *et al.* have revealed that the electronic band structure of SWCNTs is disrupted by sonication in chlorinated solvents such as ODCB, dichloromethane, chloroform and 1,2-dichloroethane due to p-type doping [62]. Chlorinated solvents are sonochemically decomposed to form species like hydrogen chloride and chlorine gas. These, in turn react with residual iron catalyst often present in the SWCNT pristine material to form iron chlorides being identified as p-dopant by XPS. The doping behavior is characterized by a loss of intensity in the shoulder of the Raman G band, an increase in relative intensity of the G band, as well as an upward shift of the D* band. Furthermore, it has been recently demonstrated that nanotube dispersions in chlorinated aromatic solvents such as ODCB produced by mild sonication exhibit are highly light scattering, interfering with the acquisition of conventional absorption spectroscopic measurements [61].

Additionally to ODCB, HiPco SWCNTs can be dispersed and exfoliated in NMP without additional dispersants by diluting stock solutions [63]. The number fraction of individual nanotubes approaches 70 % at a concentration of 0.004 g/l as revealed by statistical AFM analysis, while the number density of individual nanotubes has a maximum at a concentration of 0.010 g/l. The presence of an equilibrium bundle number density has been proposed so that the dispersions self-arrange themselves and always remain close to the dilute/semidilute boundary. Optical absorption and emission, as well as Raman investigations have confirmed the presence of individualized SWCNTs at all nanotube concentrations and have underlined the conclusions drawn from the AFM analysis. The dispersions are stable against aggregation and sedimentation for at least two weeks as shown by absorption spectroscopy and AFM [63, 64].

Recently, it has been pointed out that γ-butyrolactone (GBL), often referred to as liquid ecstasy, is a suitable solvent for the dispersion and solubilization of SWCNTs [65]. In contrast to NMP, the dispersions show an anisotropic, liquid crystalline behavior at nanotube concentrations above 0.105 g/l as revealed by absorption spectroscopy, crossed polarized microscopy and scanning electron microscopy (SEM). The aligned liquid crystalline phase (Figure 1.9) can be removed by mild centrifugation. The upper limit of the pure isotropic phase has been detected to be at a nanotube concentration of 0.004 g/l. At intermediate concentrations, the dispersion can be regarded as biphasic. As shown by sedimentation and AFM measurements, the isotropic dispersions obtained after centrifugation are stable against aggregation. Since the degree of individualization is increased

Figure 1.9 *SEM image of SWCNT anisotropic phase after centrifugation of SWCNTs in GBL. Shown in the bottom left corner is a magnified region depicting aligned (gold-coated) bundles with diameters of the order of 100 nm. Reprinted with permission from reference [65]*

in dispersions of low nanotube concentrations, the presence of an equilibrium characterized by a maximum number density of bundles has been suggested, similarily to the NMP dispersions described above. The maximum fraction of individual nanotubes, approaching 40% at a concentration of 0.6 mg/l, is however lower than for the NMP dispersions.

Before the discovery that GBL is a suitable solvent for nanotubes, it was widely recognized that the required characteristics for a nanotube dispersing solvent are large solvent-nanotube interactions relative to the nanotube-nanotube and solvent-solvent interactions in combination with the absence of ordering at the nanotube-solvent-nanotube interface. Ausman *et al.* [56], as well as Landi *et al.* [57] and Furtado *et al.* [59] have subsequently pointed out that nanotube dispersing solvents are characterized by a high electron pair donicity suggesting that a weak charge transfer from the nitrogen electron lone pair in NMP or DMF to the nanotube results in an increased solvent-nanotube interaction. Furthermore Landi *et al.* [57] have proposed that alkyl groups attached to the carbonyl group of the amide solvents stabilize the double bond character in the amide and thus the dipole moment resulting in a stronger solvent-nanotube interaction. However, both criteria are fulfilled by dimethylsulfoxide (DMSO) which is not a suitable nanotube dispersing solvent, while GBL matches none of the criteria. Based on these data, Coleman and coworkers have followed a different approach to shed light into the dispersion of nanotubes by organic solvents [66]. They have asserted that the ideal situation for the dispersion of nanotubes would be to find a true solvent where the free energy of mixing is negative, i.e. the solution is thermodynamically stable. They have pinpointed that nanotube dissolution is prohibited by the small entropy of mixing due to the large molecular weight and high rigidity of the nanotubes on the one hand and the positive enthalpy of mixing due to the strong mutual attractions between the nanotubes on the other hand. Thus, the goal is to find solvents leading to an enthalpy of mixing close to zero resulting in a slightly negative free energy of mixing. This would lead to spontaneous exfoliation of nanotubes without the aid of ultrasonication. They have clearly been able to demonstrate by optical absorption spectroscopy and AFM that this is the case upon diluting SWCNT-NMP dispersions, as a dynamic equilibrium, characterized by a significant population of nonfunctionalized individual nanotubes and small bundles, is formed. The exfoliation process is therefore concentration dependent and can be accelerated by sonication; however, sonication is not a prerequisite for the dissolution of nanotubes. In general, they have pointed out that nanotube dispersability is maximized in solvents for which the surface energy matches that of graphitic surfaces, finally answering the fundamental question of nanotube solubility in organic solvents and thus providing the cornerstone for solution based experimental and processing procedures in a two component system consisting of nanotubes and solvent only.

Similar to the concept of dispersion of nanotubes in aqueous media by the addition of dispersants, the dispersability of nanotubes in organic solvents can also be increased by designed additives. However, one main driving force for dispersion in aqueous media, namely the hydrophobic effect, cannot be exploited in this case. Nonetheless, some examples exist, e.g. porphyrins which have shown to also successfully disperse nanotubes in nonaqueous media. The first report appeared in 2003 revealing that zinc protoporphyrin IX (ZnPP; **17**) is capable of dispersing and also individualizing SWCNTs in DMF as shown by AFM and absorption spectroscopy [67]. The filtrated supernatant solution after centrifugation is redispersable in DMF supporting strong noncovalent interactions of porphyrin **17** with the nanotube sidewall.

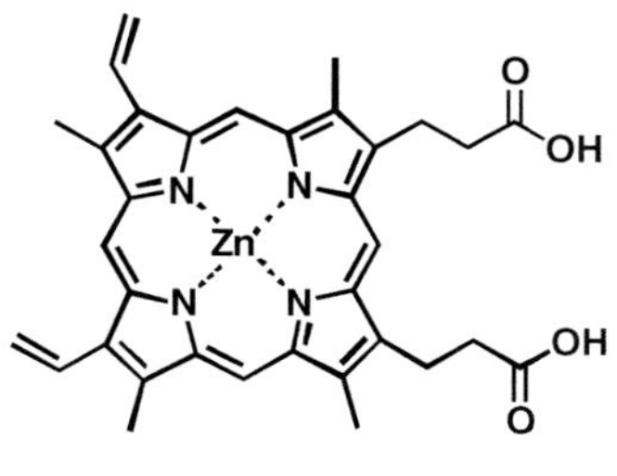

17

The dispersion of SWCNTs in toluene can be increased by the aid of small dye molecules such as terphenyl and anthracene [68]. The π-π-stacking interaction is indicated by the fluorescence quenching of the dye molecules in the nanotube composites. Interestingly, a Raman spectroscopic investigation has revealed the presence of vibrations in the composites that could not be detected in the starting materials, e.g. the nanotubes and the small dye molecules alone. Possibly, the modes arise from intrinsically IR active vibrations becoming Raman active in the composite.

Further investigations on dispersability of nanotubes in organic solvents include the use of tripodal porphyrin hosts to yield stable dispersions of SWCNTs in DMF also containing individualized nanotubes [69]. The same system has previously been shown to bind to C_{60} in a toluene solution to give supramolecular complexes with interesting 3D packing in the crystalline phase. Investigations by microscopic (TEM, SEM, AFM) and spectroscopic (Raman and emission) techniques were combined with density functional theory gas phase modeling to predict a model for the geometry adopted by the preorganized host in the presence of the nanotube guests.

In a supramolecular dispersion approach, a mixture of barbituric acid and triaminopyrimidine has been used for the solubilization of SWCNTs in DMF by sonication and a mechanochemical high-speed vibration milling technique [70]. Since neither barbituric acid nor triaminopyrimidine alone are capable of significantly increasing nanotube dispersability, the formation of a hydrogen-bonding network is responsible for multipoint interactions with the nanotube surface. Similarily, dispersion and precipitation of CoMoCAT SWCNTs can be controlled by using a copper complexed 2,2′-bipyridine derivative bearing two cholesteryl groups [71]. The copper complex shows a reversible sol-gel phase transition by changing the redox state of the Cu^{I}/Cu^{II} complexes. It has been revealed that the Cu^{II} complex is highly efficient in dispersing SWCNTs in chloroform attributed to the expansion of the π-conjugated system in the planar complex (Figure 1.10a). However, upon reduction of the copper Cu^{II} to Cu^{I} the nanotubes are precipitated due to the conformational change to the tetrahedral structure (Figure 1.10b). The precipitation can be reversed by oxidation with O_2.

Furthermore, oligomeric thiophene derivatives act as surfactants and dispersants for SWCNTs in NMP [72]. By systematic variations of the number of the head groups, the regioregularity of the head groups and the head to tail ratio, the structural design of the

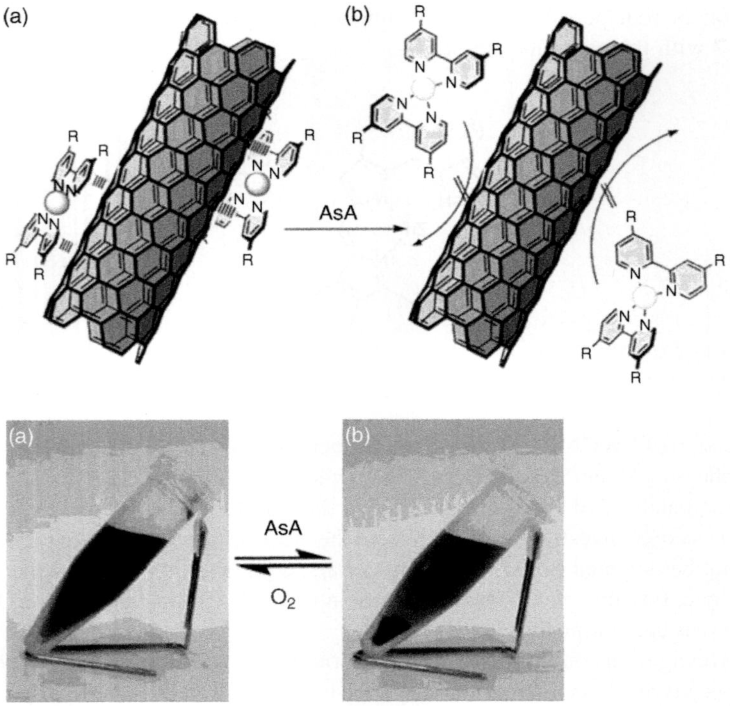

Figure 1.10 Schematic representation of the redox induced conformational change of a copper bipyridyl complex on a SWCNT (top) and the corresponding dispersions of SWCNT in chloroform: (a) the planar Cu^{II} complex is highly efficient in dispersing SWCNTs; (b) upon reduction by ascorbic acid (AsA) the conformation changes to tetrahedral structure which results in the precipitation of the nanotubes. Reprinted with permission from reference [71]

dispersants has been emphasized. The dispersability is improved by increasing the number of head groups in the oligomers. Regioregularity is also found to have an impact on the dispersion behavior of the nanotubes. Raman spectroscopy and XPS furthermore have indicated that a charge transfer from the SWCNT to the strongly electronegative sulfur atom in the thiophene head group is responsible for the strong adsorption of the dispersant on the nanotube sidewall which is responsible for the formation of high quality dispersions at dispersant concentrations as low as 0.1 g/l.

By relying solely on π-π-stacking interactions of an extended diazapentacene derivative **18** with the SWCNT sidewall, it has been demonstrated that stable nanotube dispersions in THF are formed in the presence of **18** as confirmed by AFM and optical spectroscopy [73]. Most interestingly, no solvophobic forces are exploited in this case as indicated by transient absorption measurements so that nanotube dispersions are formed by a noncovalent dispersion approach without alteration of the electronic properties of the SWCNT. This result also suggests that the shifts of nanotube absorption and emission features upon solubilization with aromatic dispersants are widely influenced by solvatochromic effects rather than by the π-π-stacking interaction itself.

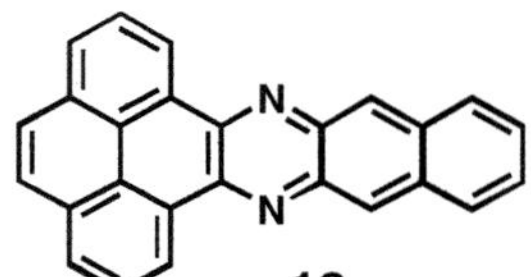

18

Just recently, a new concept for the exfoliation of SWCNTs in NMP and THF has been introduced in which a perylene dye intercalant is combined with a functionalized peylene derivative dispersant [74]. Owing to its flatness and aromaticity, 3,4,9,10-perylene tetracarboxylic dianhydride (PTCDA **19**) has been chosen as additive for the intercalation between the nanotube. The dispersion is then stabilized by a perylene bisimide derivative with polyethylene oxide (PEO) attached as tail group **20** to increase the solubility of the system. As evidenced by TEM and UV/Vis/nIR absorption spectroscopy, the nanotubes are well individualized at a weight ratio of SWCNT to additive of 1 at low concentrations of additives (0.1 g/l).

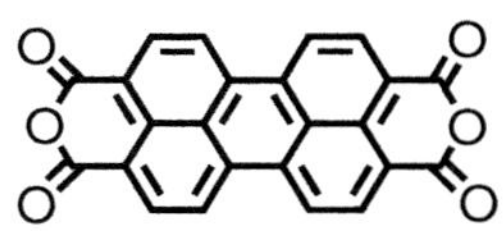

PTCDA
19

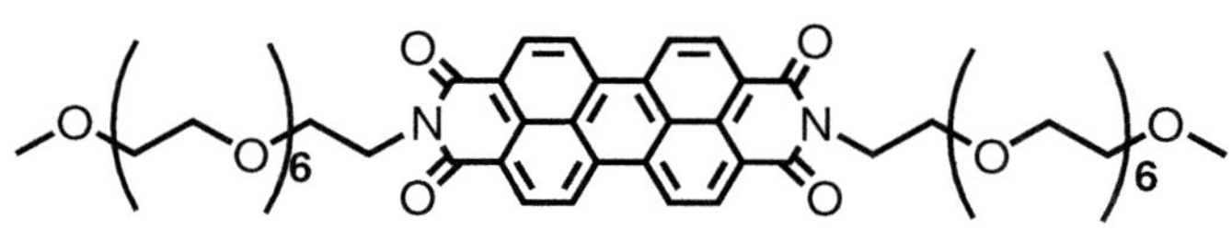

20

As outlined in the section above, the prosperous field of nanotube dispersion in organic solvents has born possibilities to overcome the high mutual attraction between the nanotubes either by designed additives or new dispersion concepts. Apart from the monomeric and oligomeric additives described so far, it has also been recognized that polymeric substances are promising in exfoliating SWCNTs as outlined below.

1.3.1.4 Stabilization of Dispersions by Natural and Synthetic Polymers

Designed synthetic, as well as natural polymers such as DNA, peptides and proteins, carbohydrates and lipids partially exhibit excellent nanotube exfoliation capabilities, as will be summarized in the following paragraphs. In general, it is believed that polymers may wrap around a nanotube due to van der Waals and possibly π-π-stacking interactions as depicted by Figure 1.11.

Bioapplications of nanotubes have been predicted and explored ever since their discovery [75–78]. Thus, a combination of SWCNTs and biological molecules is highly desirable in many chemical and biological areas from the viewpoints of both fundamentals and applications. Among natural polymers, the most prominent candidate that exhibits superior SWCNT dissolution is single (ss) and double stranded (ds) DNA. The first publications on

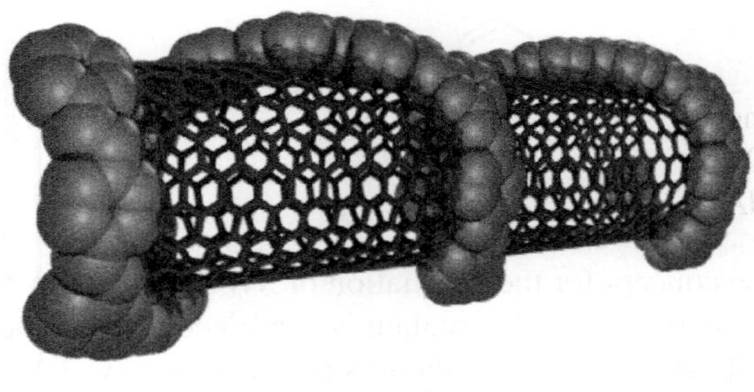

Figure 1.11 *Schematic representation of polymer wrapping on the sidewall of a SWCNT*

DNA-assisted dispersion of SWCNTs appeared in 2003 by Nakashima *et al.* for dsDNA [79], as well as by Zheng *et al.* [80] for ssDNA and short dsDNA. Both groups have revealed that DNA is highly capable of dispersing and individualizing SWCNTs in aqueous media. Simulation studies have shown that the nature of nanotube solubilization by DNA is based on nonspecific DNA-SWCNT interactions due to the nucleic acid base stacking to the nanotube sidewall with the hydrophilic sugar-phosphate backbone pointing towards the exterior, thus guaranteeing water solubility (Figure 1.12). The mode of interaction could thereby be helical wrapping and/or simple surface adsorption. The base stacking mechanism has been supported by the observation that polyadenine and cytosine strands, known to strongly self-stack in solution, exhibit lower exfoliation power than poly guanines and thymines [81]. In their first report, Zheng *et al.* [80] have also indicated that DNA based nanotube dispersions may be applied in the separation of metallic and semiconducting SWCNTs *via* ion-exchange chromatography, as shall be discussed later. To elucidate the nature of DNA interaction with the nanotube, the dispersions have been characterized in detail by fluorescence and Raman spectroscopic investigations revealing a strong diameter dependence of the DNA adsorption or wrapping, respectively [82, 83].

Interestingly, nanotube-DNA dispersions at high concentrations of nanotubes created by simple solvent evaporation have been reported to form a water-based nematic phase of unfunctionalized and freely dispersed SWCNTs [84]. This approach may be versatile in the construction of aligned nanotubes in macroscopic materials.

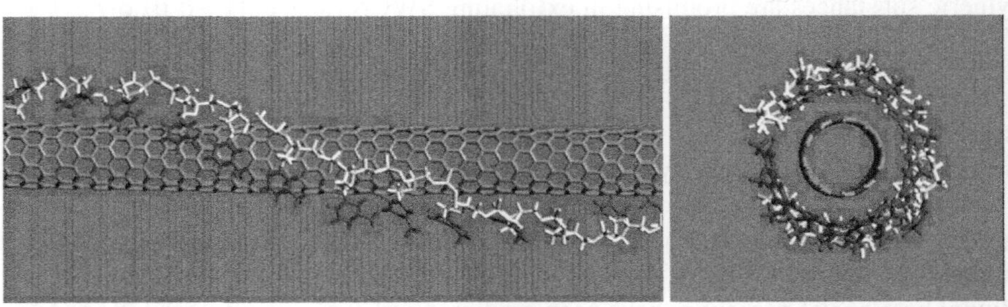

Figure 1.12 *Schematic representation of the helical wrapping binding model of a (10,10)-SWCNT by a polyt(T) DNA sequence. The bases orient to stack with the nanotube framework thus extending away from the sugar-phosphate backbone. Reprinted with permission from reference [80]*

In contrast to the study of Zheng *et al.* [81] having been carried out with nonnatural DNA with an optimal length of less than 150 bases preferably being constructed of guanine and thymine bases, a more recent investigation has focused on the use of long genomic DNA with more than 100 bases of completely random sequence [85]. It has been pointed out that the ability of ssDNA to form tight helices around SWCNTs with distinct periodic pitches is responsible for the dispersion of nanotubes. However, removal of the complementary ssDNA strands is a prerequisite for the wrapping mechanism. When following the same procedure with short 50-base oligomers with random base sequence, the dispersion capability is significantly reduced presumably due to the different folding characteristics of short ssDNA opposed to long genomic DNA.

Furthermore, it has been outlined that natural salmon testes DNA is indeed a powerful dispersing agent, as this additive is capable of exfoliating SWCNTs in water spontaneously, e.g. without the need of ultrasonication and ultracentrifugation by merely diluting a stock dispersion as revealed by statistical AFM analysis and photoluminescence spectroscopy [86]. At lower nanotube concentration, the amount of individualized nanotubes increases. The maximum number fraction of individual nanotubes reaches 83% at a nanotube concentration of 0.027 g/l.

In general, DNA-nanotube conjugates which combine the unique properties of SWCNTs with the sequence-specific pairing interaction and conformational flexibility of DNA have been extensively pursued [87–89] for their promising prospects in a number of applications, such as nanoscale devices, nanotube separation, biosensors, electronic sequencing and therapeutic delivery.

The second class of nanotube dispersants for biological applications has been presented by Dieckmann and coworkers who have designed an amphiphilic helical peptide they denoted as **nano-1** [90]. They have constructed their artificial peptide on the basis of the preliminary works of Wang *et al.* [91] who used phage display to identify several peptides with a high affinity for carbon nanotubes. An analysis of the peptide conformations has suggested that the binding sequence is flexible and folds into a structure matching the geometry of the nanotube [91]. Figure 1.13 schematically illustrates the structure of the peptide and its interaction with a SWCNT. When folded into an α-helix, as proposed on the basis of CD spectroscopy, the hydrophobic valine and phenylalanine residues in positions *a* and *d*, respectively, create an apolar surface of the peptide suitable for interacting with the nanotube sidewall. The introduction of polar residues in positions *e* and *g* generate favorable helix-helix interactions, while the oppositely charged residues in positions *b* and *f* provide favorable interactions between the helices from different peptides. Specifically the latter factor can be easily influenced by changing the solution's ionic strength resulting in controlled solubility characteristics, modulated by influencing peptide-peptide interactions. Nanotubes dispersed and exfoliated by **nano-1** can then be assembled into ordered fiber-like hierarchical structures [92], presumably by end-to-end connections [93].

Selective individualization of SWCNTs according to diameter has been achieved by reversible cyclization of artificial peptides [94]. After wrapping the peptides of specific lengths around the nanotube, a head to tail covalent bond formation has been induced between the thiol moieties of the peptide termini. Enrichment of certain diameters after the solubilization process has been monitored by absorption and Raman spectroscopy, as well as AFM. Moreover, peptide cross linking by the formation of amide bonds between amino acid side chains increases the stability of the nanotube dispersions and facilitates

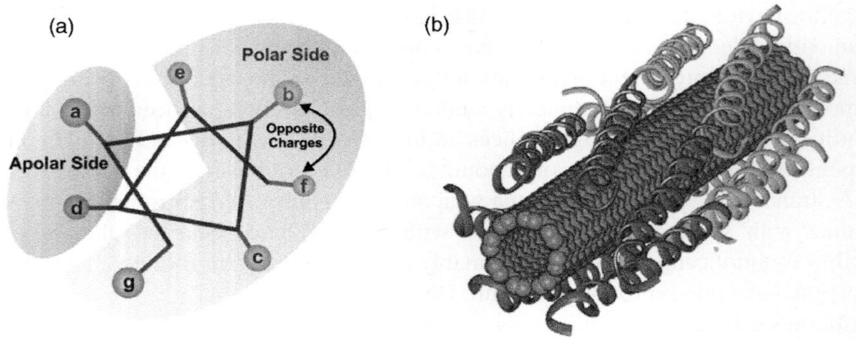

Figure 1.13 *(a) Schematic representation of the designed helical peptide **nano-1**. The residues in positions a and d are hydrophobic in nature thus creating an apolar side of the peptide presumably interacting with the nanotube; (b) model of the peptide wrapping of **nano-1** on the nanotube with head to tail alignment of helices in two adjacent layers. Reprinted with permission from reference [90]*

self-assembly into fiber-like structures [95]. Further investigations have demonstrated the importance of the aromatic content in the apolar side of the designed peptide, as it has been shown that the degree of nanotube exfoliation increases with an increasing amount of aromatic moieties indicating that π-π-stacking interaction plays an important role [96, 97]. Nanotubes are also spontaneously debundled by the artificial peptide **nano-1** similar to DNA-based dispersions [98]. However, the number fraction of individual nanotubes even surmounts that of the DNA-based dispersions, as the maximum reaches 95 %.

Apart from the designed peptides constructed by Dieckmann's group, other peptides have been studied with respect to nanotube adsorption and solubilization. A series of branched anionic and cationic amphiphilic peptides has also been discovered to efficiently solubilize SWCNTs in aqueous media as demonstrated by the aid of TEM and optical absorption spectroscopy [99]. A multifunctional peptide has further been used to disperse nanotubes and to direct the precipitation of silica and titania onto the nanotube sidewall at room temperature [100]. Highly exfoliated SWCNTs in water have also been obtained by noncovalent functionalization with designed peptides combining a combinatorial library sequence to bind to nanotubes with a rationally designed section to yield controllable solubility characteristics as evidenced by optical absorption and emission spectroscopy, as well as cryo-TEM imaging [101]. In contrast to previous works on nanotube solubilization by the aid of peptides having focused on maximizing the interaction of the peptide with the nanotubes, recent investigations have probed the fluorescence properties of SWCNTs dispersed in various custom-designed peptides [102]. It has been revealed that self-assembling properties of the peptide onto the nanotube scaffold are beneficial for the degree of dispersion on the one hand and for the preservation of the SWCNT emission features on the other hand. The brightest nanotube emission has been found for peptides that uniformly coat the nanotube which has been attributed to nanotubes templated self-assembly of the peptide dispersants.

Similarly to natural and artificial peptides, some proteins also interact with nanotubes and can thus be considered as promising candidates for nanotube solubilization. In an attempt to

Figure 1.14 *TEM images of streptavidin immobilized on MWCNTs: (a) stochastic binding of streptavidin molecules on a MWCNT with a diameter smaller than 15 nm; (b) helical organization of streptavidin molecules on a carbon nanotube with a suitable diameter of 16 nm. The bar represents 50 nm. Reprinted with permission from reference [103]*

trace crystallization of proteins by electron microscopy, it was found that upon incubation of streptavidin in the presence of MWCNTs, the nanotubes are almost completely covered by the protein molecules under ideal conditions [103]. In some instances, the nanotubes showed lateral striations regularly spaced at 6.4 nm along with perpendicular striations suggesting that the streptavidin molecules were organized in a square lattice along the nanotube backbone (Figure 1.14).

In order to elucidate the adsorption mechanism of proteins on the nanotube sidewall, the structure and function of enzymes was probed [104], as the catalytic activity of enzymes requires the near complete retention of their native structure. The structure and therefore function of the enzymes is strongly influenced by the hydrophobic, nanoscale environment of a SWCNT, however with varying extend. As revealed by IR and circular dichroism (CD) spectroscopy, as well as AFM, α-chymotrypsin unfolds upon adsorption on the nanotube leading to a loss of its native activity. In contrast, soybean peroxidase retains 30 % of its activity due to the preservation of its three-dimensional shape underlining the complexity of the adsorption process of proteins on nanotubes.

It has further been shown that proteins do not only adsorb onto the nanotube backbone, but are also capable of acting as nanotube solubilizers in water as demonstrated by UV/Vis/nIR absorption, Raman spectroscopy and AFM [105]. Removal of the unbound proteins by dialysis leads to flocculation of the nanotubes indicating the presence of an adsorption-desorption equilibrium. Due to the rich functionality of proteins with respect to functional groups and biorecognition abilities, a protein based dispersion approach is highly versatile for the preparation of self-assembled nanostructures and nanobioconjugates.

Among nanotube biosurfactants, lysozyme, a well-studied antibactierial cationic protein, holds great promise for applications of nanotubes as optical pH sensors and in biomedical research, as nanotube exfoliation by lysozyme has been revealed to be highly pH sensitive [106]. Thus, the aggregation state of the nanotube can be reversibly tuned by varying the pH: the SWCNTs are highly debundled below a pH of 8 and above a pH of 11, while being aggregated in the pH range of 8–11. Furthermore, the secondary structure of the protein remains largely intact as indicated by CD spectroscopy.

Since the structural properties of proteins are highly complex compared to the well-defined characteristics of detergents, it is reasonable that the task of exploring the protein adsorption mechanism onto the nanotube backbone is tedious to resolve. One approach towards this topic was presented by a cryo-TEM investigation of the nanotube dispersion by bovine serum albumin (BSA) labeled with gold nanoparticles (GNP) to yield a high density contrast [107]. The TEM analysis has unveiled that the majority of the BSA-GNP complexes are distributed at distances of 20–80 nm from each other along the individually dispersed nanotube. Based on the cryo-TEM study in combination with AFM and CD techniques, it has been proposed that the BSA molecules are adsorbed on the SWCNT with their hydrophobic domains resulting in partial unfolding. Thus the majority of the ionized residues interact with the solvent.

In analogy to monomeric carbohydrates that have been introduced as nanotube surfactants as outlined in Section 1.3.1.2, oligomeric and polymeric carbohydrates also induce nanotube exfoliation to achieve solubilization in aqueous media. The first report was presented by Star *et al.* who have shown that SWCNTs are effectively solubilized by common starch, provided starch is activitated towards complexation by wrapping itself helically around small molecules [108]. This is nicely reflected by the observation that SWCNTs are insoluble in an aqueous solution of starch, while being dispersed and individualized in an aqueous solution of a starch-iodine complex due to the preorganization of the amylose in starch into a helical conformation by iodine. The solubilization process is reversible at high temperatures and preferred for nanotubes compared to amorphous carbon and catalyst particle impurities. Since addition of glucosidase to the starched nanotubes results in precipitation of the SWCNTs, starch wrapping gives access to a completely nondestructive purification route.

SWCNTs are also solubilized by the aid of amylose in a DMSO-water mixture [109]. In an optimal procedure, SWCNTs are presonicated in water to partly exfoliate the nanotube bundles, followed by addition of amylose in a DMSO-water mixture to maximize cooperative interactions between the nanotubes and amylose leading to solubilization. The ideal solvent condition is 10–20 % DMSO, in which amylose is characterized by an interrupted loose helix indicating that the helical organization of amylose is not a prerequisite for nanotube solubilization.

Among carbohydrates, β-1,3-glucans such as single-chain schizophyllan and curdlan, also effectively disperse as grown and cut SWCNTs [110, 111]. In the case of cut SWCNTs, mainly nanotube bundles are dispersed, while exfoliation occurs for as-grown nanotubes. Upon adsorption of the β-1,3-glucans, a right-handed helical superstructure is formed on the nanotube backbone with two carbohydrate chains twining one nanotube.

Sodium carboxymethylcellulose, an etherified derivative of cellulose, is a promising candidate for increasing nanotube processability, as it does not merely highly exfoliate nanotubes in solution, but also retains the individualized state upon film formation.

Significantly, the nanotubes in the films tend to align as demonstrated by considerable dichroism in their absorption spectra. These homogeneous thin films of high quality constitute a further step towards the development of nanotube based optical devices with wavelength tuning capability.

Another example of a carbohydrate nanotube dispersing additive is presented in chitosan and its derivatives [33, 112–116]. The great value of chitosan and its derivatives lies in the biocompability and their use as biosensors. The dispersion state of the nanotubes can be controlled by applying pH changes as stimulus: chitosan disperses nanotubes in acidic aqueous media, while an inverse dispersion behavior is observed for N-succinyl chitosan, as nanotube precipitate below a pH of 4.66. Carboxymethyl chitosan allows dispersion below a pH value of 6.7 and above 7.3 and 2-hydroxypropyltrimethylammonium chloride chitosan allows nanotube dispersion in the pH range of 2-12 [115]. Furthermore, *o*-carboxymethyl chitosan (OC) and OC modified by poly(ethyleneglycol) at the COOH terminus are effectively exfoliating SWCNTs in neutral pH solutions completing the picture of pH sensitive dispersion of nanotubes [116].

Among natural polymers, gum arabic, a natural, highly branched polysaccharide with a small amount of arabinogalactan-protein complexes, has also been unveiled as excellent additive for nanotube dispersion and exfoliation as shown by cryo-TEM and HRTEM [117].

Owing to their amphiphilic nature, lipids can be used for the solubilization of carbon nanotubes. Based on solubility and TEM investigations, a half-cylindrical binding mode has been suggested for lysophospholipids such as lysophosphatidylcholine LPC 18:0 and lysophosphatidylglycerol LPG 16:0 [118]. In the TEM study, a periodic wrapping in the lipid phase has been observed with the size and regularity of the striations being dependent on the polarity of the lysophospholipid as denoted by Figure 1.15.

After performing molecular dynamics simulations, the same authors have concluded that the adsorption of the lipids on the SWCNT varies from the proposed hemimicellar adsorption mechanism, as this organization requires the lipid micelles to break from the middle and to reassemble in tandem onto the nanotube backbone. They have pointed out that the lipids are organized into 'crests' consisting of several lipid layers shifted along the tube axis and packed in parallel and antiparallel directions that wrap the nanotube spirally [119].

Recently, it has been discovered that SWCNTs are also debundled by natural polyelectrolytes like sodium lignosulfonate, humic acid, fulvic acid and tannic acid [120]. This does not appear to be surprising, as natural polyelectrolytes are generally amphiphilic in nature comprising a mixture of amorphous, polydispersed organic polyelectrolytes of mixed aliphatic and aromatic constituents in which the aromatic moieties are known to interact *via* π-π-stacking interactions, as demonstrated for the solubilization of C_{60}-fullerene [121]. Most remarkably, SWCNTs are exfoliated at polyelectrolyte concentrations as low as 0.15 g/l as demonstrated by various spectroscopic and microscopic techniques.

In general, dispersion of CNTs in synthetic polymers is highly desirable, as it enhances the intrinsic properties of the polymer due to the outstanding electronic and mechanical properties ascribed to the nanotubes. To fully exploit the potential of the reinforcing procedure, nanotubes need to be individualized in order to be homogeneously distributed in the polymer matrix. Commonly, exfoliation is aided by the presence of aromatic moieties in the polymer (Fig. 1.16). The reports on dispersion of nanotubes in polymers are numerous and merely some can be considered here.

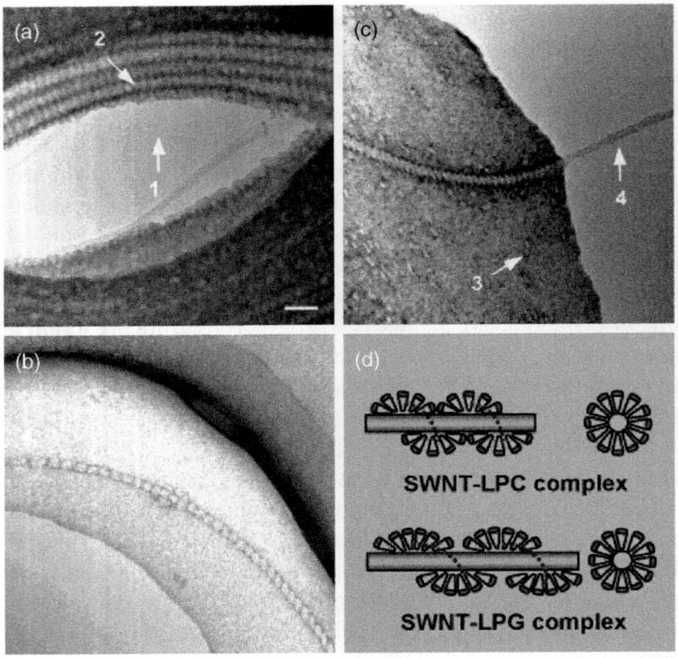

Figure 1.15 TEM images of SWCNT-LPC (a and c) and SWCNT-LPG (b) complexes. Numbers in a and c correspond to (1) an isolated SWCNT in the vacuum phase, (2) an LPC striation on an SWCNT/SWCNT bundle, (3) possibly an LPC micelle on the substrate in the lipid phase, and (4) an uncoated SWCNT bundle in the vacuum phase. Note the less organized and wider striations of SWCNT-LPG complexes in (b), as compared to those in (a) and (c) for SWCNT-LPC. Scale bar: 20 nm (same for a–c). (d) Hypothesized microscopic binding modes of LPC and LPG with SWCNTs. The lysophospholipids are shown as truncated triangles, their headgroups are shown in black, and the SWCNTs as gray bar. The left section of (d) illustrates the proposed lipid spiral wrapping along the tube axis, while the right section shows their possible binding along the circumference of the tubes. Reprinted with permission from reference [118]

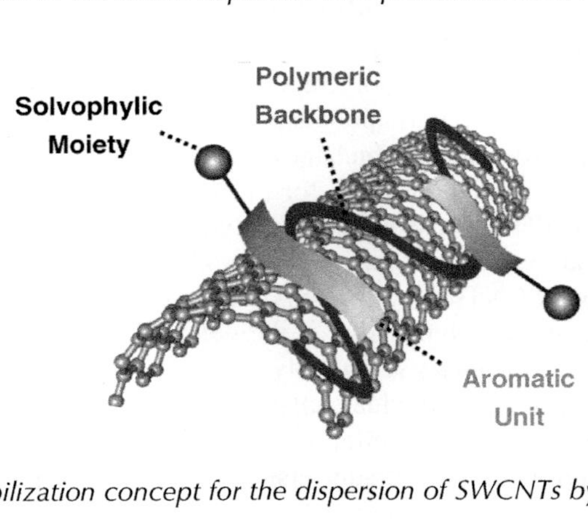

Figure 1.16 Solubilization concept for the dispersion of SWCNTs by aromatic polymers

Early reports have focused on nanotube composites with PPV (*p*-phenylenevinylene) [122], as PPVs exhibit interesting optoelectronic properties and may be applied as light-emitting semiconductor in organic light emitting devices [123]. Especially the structural analogue PmPV (poly-*m*-phenylenevinylene-*co*-2,5-dioctoxy-*p*-phenylenevinylene) can be used as additive to form stable nanotube dispersions in organic solvents such as chloroform. The SWCNT/PmPV hybrids increase the conductivity of undoped PmPV by eight times while maintaining the luminescence properties [124–126] which allowed production of photovoltaic devices [126] and the application as electron transport layers in organic light emitting devices [127]. For further details on this subject the reader is referred to several review articles concerned with this topic [6, 11, 12, 25, 128].

By using polymers carrying polar side chains such as PVP (polyvinylpyrrolidone), PSS (polystyrenesulfonate) or bovine serum albumin, stable aqueous SWCNT dispersions are obtained [129]. The wrapping of the polymer around the nanotube has been shown to be robust not being dependent on the presence of free bulk polymer giving access to stable nanotube dispersions with concentrations up to 1.4 g/l. The wrapping and solubilization of the SWCNTs associated with the SWCNT-polymer interaction can be reversed by addition of an organic solvent such as THF. The addition of PVP also stabilizes nanotube dispersion in NMP allowing higher concentrations of nanotubes to be homogeneously distributed [130, 131]. Furthermore, especially oxidatively purified SWCNTs are efficiently dispersed in an aqueous solution of diamine-terminated oligomeric poly(ethylene glycol) [132].

Noncovalent functionalization of SWCNTs by designed conducting polymers essentially being based on PAmPV (poly{(5-alkoxy-m-phenylenevinylene)-co-[(2,5-dioctyloxy-p-phenylene)-vinylene]}) disperse nanotube bundles in organic solvents [133]. The formation of pseudorotaxanes has been achieved by PAmPV derivatives bearing tethers and rings, respectively, to yield threaded complexes which might be of interest for the development of molecular actuators and switches.

In contrast to polymer wrapping, SWCNTs are also dispersed in organic solvents by conjugated PPEs (poly(aryleneethynylene)s) which cannot wrap around the nanotube due to their rigid backbone [134]. Thus, dispersion has been attributed to π-π-stacking interactions. Advantageously, various neutral and ionic groups can be introduced onto the nanotube surface. The versatility of the π-π-stacking approach is also reflected in the dispersion of both SWCNTs and MWCNTs by a series of polymers containing a pyrene moiety to increase the interaction with the nanotube surface [135–140].

A very efficient route towards nanotube polymer composites based on polyimide has been presented by *in situ* polymerization in the presence of nanotubes [141–144]. In the case of a polyimide derivative equipped with sulfonic acid groups, exfoliated nanotubes in organic solvents have been obtained as revealed by fluorescence spectroscopy, as well as microscopic investigations [144].

As already previously indicated, amphiphilicity of the dispersing agent is beneficial. Accordingly, numerous block copolymers have been presented for nanotube dissolution [145–157]. A highly interesting nanotube dispersing blockpolymer is polystyrene-*block*-poly(4-vinylpyridine), as SWCNTs can be exfoliated both in polar and apolar solvents [158]. In the case of dissolution in toluene and other apolar solvents, the polystyrene block is exposed to the solution phase, while the poly(vinylpyridine) forms a micellar shell incorporating the nanotube when polar solvents are used as revealed by transmission electron microscopy.

As summarized above, systematic research has shed light into the dispersion and exfoliation of SWCNTs. However, dissolution of the nanotubes merely constitutes one step towards the realization of nanotube-based applications. An obstacle even greater to overcome is the polydisperse nature of the raw material. Thus, the following section is devoted to the separation of SWCNTs according to diameter and/or chirality.

1.3.2 The Role of Noncovalent Functionalization in Nanotube Separation

Even though the extraordinary potential of carbon nanotubes as new super materials especially in CNT-based electronics had been recognized soon after their discovery, integration of millions of nanotubes in functional circuits can thus far be merely considered a vision, as nanotube samples with defined electronic classification, or preferably single chirality with defined length are a prerequisite. Despite recent progresses in the field of controlled nanotube production [8], it is reasonable that this goal may not be achieved by controlled synthesis alone by considering the following aspects: SWCNTs are grown from metal catalyst particles widely defining their diameter. However, the futility of the exact chirality control by predefined catalyst particle size during synthesis can be imagined by recognizing that the diameter difference between a (10,10) metallic and a [9, 11] semiconducting SWCNTs is merely 0.03 Å. Furthermore, the high temperatures during nanotube production presumably induce thermal vibrations allowing variations in the SWCNT diameters even for identically sized catalyst particles. Thus, the development of post-synthetic separation techniques such as chromatography, electrophoresis and density gradient ultracentrifugation is deemed necessary and has received considerable attention in nanotube research [20, 159, 160]. Since exfoliation of the nanotube bundles is a precondition for efficient separation, noncovalent functionalization is highly versatile, as shall be discussed in the following sections on a variety of examples.

1.3.2.1 Selective Carbon Nanotube Interaction

Additionally to the design of separation techniques, focus has been laid on the exploration of selective interaction of various molecules with SWCNTs according to electronic type, diameter and/or chirality, as differences between (n,m)-SWCNTs can thus be amplified aiding the separation process. Therefore, they shall be discussed first.

Based on preliminary works revealing that adsorption of linear alkylamines induces significant changes in the electrical conductance of oxidized semiconducting SWCNTs, while retaining the conductance behavior of metallic SWCNTs [161, 162], a separation method according to electronic type with the aid of octadecylamine has been proposed [163, 164]. The method relies on additional stabilization of oxidized semiconducting SWCNTs in THF by amines as opposed to their metallic counterparts allowing for the precipitation of the metallic SWCNTs. Most importantly, SWCNTs dispersed in THF/(octyl)amine solutions show the characteristic photoluminescence signals of individualized (semiconducting) nanotubes [165]. A multilaser Raman analysis of the enrichment process has shown that larger diameter metallic SWCNTs (above 1 nm) can be detected along with semiconducting nanotubes in the supernatant rendering separation less effective for laser ablation nanotubes with a higher average diameter as opposed to HiPco SWCNTs [164]. This behavior is related to amine assisted dedoping of the oxidized nanotubes and therefore the redox characteristics of the nanotubes. The reduction removes physisorbed counterions being accompanied with an increased organization of ODA in the case of the nanotube species remaining in the supernatant [166]. Conversely, enrichment of metallic SWCNTs in the supernatant has been achieved by a dispersion-centrifugation experiment of pristine SWCNTs in THF by the aid of propylamine and isopropylamine unveiling that amines preferentially interact with metallic SWCNTs, as long as they are not carboxy-functionalized [167, 168]. After five iterative dispersion-centrifugation steps, metallic nanotubes have been estimated to be enriched from 41% in the as produced mixture to 72% in the THF/propylamine supernatant solution.

Selective interaction of SWCNTs produced by different techniques with fluorene based polymers has also been under investigation [169–172]. Hereby, different polymers have been shown to discriminate between nanotube species either by diameter or chiral angle. Upon dispersion of the nanotubes in toluene/polymer solutions, significant alterations of the nanotube photoluminescence features have been observed reflecting selective interaction depending on the polymer structure. Based on the optical properties of the nanotube dispersions, it has been concluded that PFO (poly(9,9-dioctylfluorenyl-2,7-diyl **21**) [169–172], and PFH-A (poly[(9,9-dihexylfluorenyl-2,7-diyl)-*co*-(9,19-anthracence)] **22**) [170] preferentially individualize SWCNTs with high chiral angle (> 24.5°), while PFO-BT (poly[9,9-dioctylfluorenyl-2,7-diyl)-co-1,4-benzo-{2,1′-3}-thiadiazole)] **23**) [169, 170, 172] exhibits diameter selective exfoliation in the range of 1.02–1.06 nm. However, the selective interaction is dependent on the solvent used for the dispersion. It has furthermore been revealed that overall solubilization follows the opposite trend to selective interaction, as more flexible conformations of the polymers allow more nanotubes to be dispersed, while at the same time reducing the selectivity [172]. Additionally, PFO **21** has been used as extracting agent for semiconducting SWCNTs in toluene assisted by ultracentrifugation allowing the fabrication of improved nanotube based field-effect transistors (FET) [173].

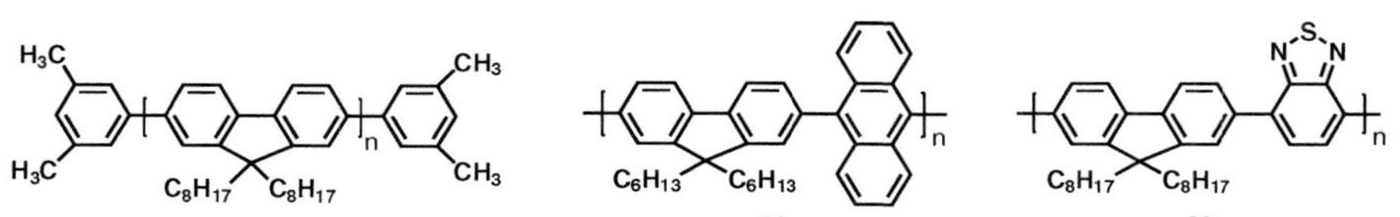

21
PFO
22
PFH-A
23
PFO-BT

Among biomolecules, a variety of additives have been suggested to promote selective interaction with specific SWCNTs. For example, SWCNTs can be threaded by large-ring cyclodextrins providing water solubility on the one hand and partial discrimination with respect to diameter on the other hand [174]. In a similar approach, diameter sorting has been achieved by reversible cyclization of designed peptides [94]. Artificial peptides containing thiol groups on the N and C terminus are capable of solubilizing and encircling SWCNTs within a certain diameter range (depending on the size of the peptide) by controlled formation of disulfide bonds on the termini with the advantage of avoiding dissociation of the peptide by the introduction of a covalent bond. Furthermore, individually suspended SWCNTs in aqueous media by adsorption of phosphatidylcholine are enriched in smaller diameter nanotubes as indicated by Raman spectroscopy [175].

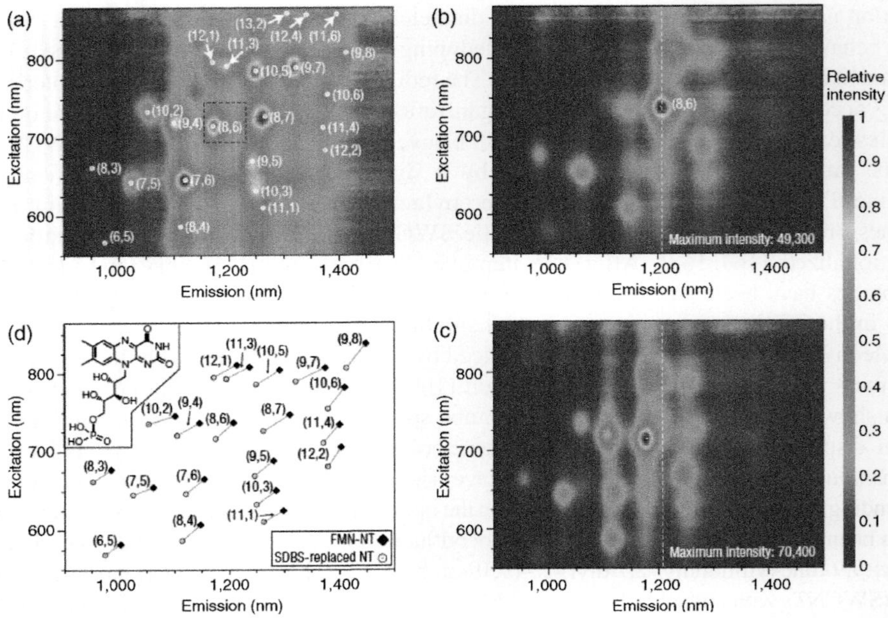

Figure 1.17 *Photoluminescence emission maps of HiPco-SWCNTs dispersed in (a) SDBS, (b) flavin mononucleotide and (c) flavin mononucleotide after replacement with SDBS; (d) Plot of transitions for flavin dispersed SWCNTs (diamonds) and SDBS dispersed SWCNTs (circles). The inset in (d) represents the chemical structure of flavin mononucleotide. Reprinted with permission from reference [176]*

Recently, the interaction of the common redox cofactor flavin mononucleotide with SWCNTs has been investigated [176, 177]. Due to cooperative hydrogen bonding between adjacent flavin moieties adsorbed on the SWCNT *via* π-π-stacking interaction, a helical ribbon organizes around the nanotube backbone, as demonstrated by HRTEM. Most interestingly, a strong chirality dependency in the interaction has been unveiled by replacing the flavin dispersant with SDBS. The replacement could be mapped by fluorescence spectroscopy, as π-π-stacking of the flavin induces a red-shift of the nanotube photoluminescence features, as demonstrated by Figure 1.16. The PLE map after complete replacement of the flavin from the nanotube sidewall (Figure 1.17c) reveals an enrichment of the (8,6)-SWCNT indicating that the (8,6)-SWCNT exhibits a profound affinity for the flavin helix. This strongly selective interaction can be exploited for enrichment of the (8,6) nanotubes (85 % enrichment value). For this purpose, an appropriate amount of SDBS was added to nanotubes dispersed in a solution of flavin mononucleotide to yield replacement of the flavin moieties on all chiralities except for the (8,6) nanotube. Since it has previously been reported that SDS suspended SWCNTs can be precipitated out of solution by addition of NaCl [178], all nanotubes being enclosed in SDBS micelles could be flocculated by the addition of NaCl to yield a nanotube sample highly enriched in a single chirality to remain in suspension.

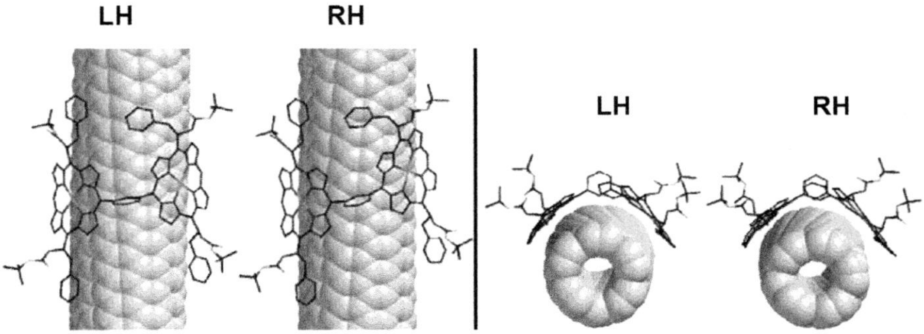

Figure 1.18 Schematic representation of the separation of left-handed (LH) and right-handed (RH) SWCNTs with a chiral diporphyrin derivative. Reprinted with permission from reference [182]

Another approach to achieve diameter selective enrichment of SWCNTs is presented in the concept of selective dispersion of the nanotube material by nanotweezers consisting of an anchor moiety shaped like a folded ribbon which adsorbs onto the nanotube sidewall and furthermore a solvophyilc moiety, as schematically depicted in Figure 1.2. This concept has been realized in the separation according to diameter in toluene by noncovalent functionalization with a pentacene-based molecular tweezer [179]. This nanotweezer principle has been expanded to the extraction of optically pure SWCNTs with the aid of chiral diporphyrins [180–182]. Figure 1.18 schematically illustrates complexation of a chiral diporphyrin preferentially solubilizing SWCNTs of a specific chirality. Thus, left and right-handed mirror images of chiral nanotubes can be separated. The resulting nanotube suspensions are distinguishable by circular dichroism after removal of the chiral extracting agents. The composition and the optical purity of the nanotubes can be tuned by varying the bridging moiety of the dispersing additives.

Furthermore, enrichment of metallic and semiconducting SWCNTs can be achieved on the foundation of the stronger interaction of bromine with metallic SWCNTs [183]. For this purpose, cut nanotubes dispersed in Triton X-100 have been treated with bromine which resulted in the formation of charge transfer complexes, preferentially with the metallic species. Due to the increased density of the complexes, the metallic nanotubes can be separated from the semiconducting counterparts by centrifugation.

Selective noncovalent functionalization of semiconducting nanotubes has been realized by porphyrin chemistry involving 5,10,15,20-tetrakis(hexadecyloxyphenyl)-21*H*,23*H*-porphyrin (THPP) [184]. Upon redispersing noncovalently functionalized SWCNTs in THF, it has been shown that the suspended nanotubes are enriched in semiconducting species, while the precipitate is enchriched in metallic SWCNTs after repeated extractions.

The concept of solubilizing SWCNTs by π-π-stacking additives has already been outlined as versatile approach for noncovalent functionalization of nanotubes. It is reasonable to assume that chirality recognition of nanotubes is possible, when a large enough aromatic moiety is chosen, as the π-π-stacking interaction may lead to preferred orientations along the nanotube backbone to ensure maximum p-orbital overlap. The fertility of this concept has recently been indicated by fluorescence spectroscopic investigations on the selective interaction of two aromatic amphiphiles with a pentacenic moiety and a quaterrylene moiety, respectively [185]. Upon adsorption of the pentacene-based amphiphile, the fluorescence of nanotubes with small to medium helicity angles is quenched, while it is retained for the quaterrylene-based additive.

Even though separation of nanotubes according to diameter and chirality by the supramolecular approaches described here is highly desirable, it suffers from the drawback that the nanotube complexes are not readily separated, as selective interaction does not necessarily result in enrichment of certain nanotube species without the additional use of well established separation techniques which shall be summarized in the following.

1.3.2.2 Chromatography

Chromatographic techniques are well established in chemistry and biology to separate materials on the molecular scale. Thus, it is not surprising that evaluating the suitability of chromatographic techniques for the separation and purification of nanotubes has commenced soon after their discovery. Focus has been laid on size exclusion chromatography (SEC), gel permeation chromatography (GPC), field flow fractionation (FFF) and ion exchange chromatography (IEC).

Size exclusion chromatography has shown to be effective in purification and length sorting of both MWCNT and SWCNT material. A prerequisite for SEC to be effective is the dispersion of nanotubes either by covalent [192] or noncovalent functionalization methods, e.g. encapsulation in surfactant micelles [186–189], or DNA wrapping [190, 191]. Length separation of SWCNTs has shown to be highly effective, when nanotubes are cut prior to injection in the SEC column [193]. Furthermore, purification and length sorting on oxidatively shortened SWCNTs was achieved without the addition of dispersing additives [194]. SEC can also be used to remove the unbound dispersing additives, e.g. DNA, yielding information about the stability of the noncovalently functionalized nanotubes in the absence of bulk dispersing agent [195].

In addition to SEC, length separation of zwitterionic functionalized SWCNTs has been achieved by GPC [196]. It has also been revealed that the purification efficiency of oxidatively shortened SWCNTs in THF could be improved by GPC [197]. Field flow fractionation, a separation technique where a field is applied to the mixtures flow allowing fractionation due to different mobilities of the various components in the electrical field, has also been successful in purification and length sorting of shortened SWCNTs [198–200].

Among chromatographic techniques, ion exchange chromatography is most promising, as it allows for the separation of SWCNTs according to diameter and/or electronic type. As previously described, SWCNTs can be efficiently individualized by DNA. After ion-exchange chromatography, nanotubes are separated by electronic type and diameter, as revealed by optical absorption, fluorescence and Raman spectroscopy [80, 83]. The differences in the optical properties are reflected by the distinguishable color of the fractions. Further investigations demonstrated that the sorting quality is strongly dependent on the DNA sequence [81], as the effective charge density of the DNA-SWCNT hybrid governs the separation process [201]. A major obstacle in the IEC separation of DNA-wrapped SWCNTs is presented in the broad length distribution of the SWCNTs, as separation is achieved by differential movement of the nanotubes stimulated by an external field. This problem has been overcome by combining SEC to obtain length separation with

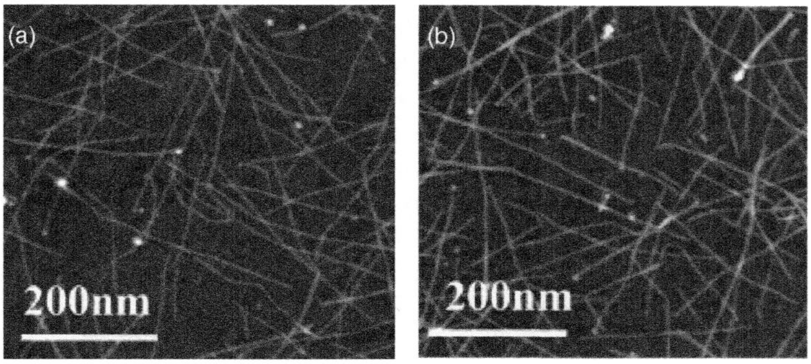

Figure 1.19 *AFM images of DNA-wrapped SWCNTs sorted first by length in a SEC column, followed by diameter sorting by IEC. The AFM images are taken from different fractions after the IEC with an average diameter of 1.20 ± 0.11 nm (a) and 1.37 ± 0.14 nm (b). Reprinted with permission from reference [203]*

IEC to refine the sorting by diameter and/or chirality [202, 203]. Thus, nanotubes with defined lengths and diameters are accessible, as exemplarily depicted by the atomic force micrographs in Figure 1.19. The versatility of this approach is particularly reflected by the successful separation of two nanotube species with the same diameter, but different chirality, namely the (9,1) and the (6,5) SWCNTs.

1.3.2.3 Electrophoresis

Since carbon nanotubes are similar in dimension to biomolecules, attempts to adopt separation techniques from life sciences such as electrophoresis have been undertaken. Electrophoretic separation approaches can be classified in conventional direct current (dc) electrophoresis on the one hand which is based on sorting nanotubes according to their different mobilities through a gel, capillary or solution upon applying an external electrical field and alternating current (ac) dielectrophoresis on the other hand which exploits the different polarizabilities of metallic and semiconducting SWCNTs. Apart from separation of nanotubes, dc electrophoresis [204–206] and ac dielectrophoresis [207–214] has been applied to align and deposit nanotubes in a controlled fashion – a crucial aspect for the fabrication of nanotube-based electronic devices.

In direct current electrophoretic separation, the mobility of the objects in the electrical field is regarded as a main driving force for separation. However, the charge-density differences between the nanotubes of different geometry dispersed in a surfactant solution is also expected to influence the movement in the external field. Since the total charge on the nanotube is defined by the surface area, the charge density differences are diameter dependent so that dc electrophoretic separation is theoretically capable of sorting SWCNTs by diameter. Thus far, separation of nanotubes according to diameter has not yet been realized by dc electrophoresis, even though length separation, separation of bundled and individualized SWCNTs, as well as purification of nanotubes dispersed in an aqueous solution has been achieved by capillary electrophoresis [215, 216]. The reproducibility of the experiments could be improved by adding small amounts of hydroxypropyl methyl cellulose during the dispersion process allowing to precisely evaluate nanotube size distributions which is valuable for the optimization of nanotube synthesis [217]. Similar results have been obtained when nanotubes are dispersed in the ionic liquid 1-butyl-3-methylimidazolium tetrafluoroborate prior to encapsulation in SDS micelles [218]. In analogy to capillary electrophoresis, length sorting of SWCNTs dispersed in sodium cholate [219] or RNA and DNA [220], as well as purification [221] is permitted by gel electrophoresis. Interestingly, the separation by length occurs alongside with some diameter selection, as the scission process during ultrasonication is diameter dependent, e.g. smaller diameter nanotubes tend to be cut to a higher degree.

Among electrophoretic techniques, alternating current dielectrophoresis holds most promise, as nanotube separation by electronic type may be accomplished. Based on the preliminary work on the deposition of nanotube bundles containing at least one metallic tube [222], Krupke and coworkers have been able to demonstrate that metallic nanotubes are selectively deposited between the electrodes in ac field, when a nanotube dispersion with a high degree of individualization is dropped onto the [223]. Upon subjecting nanotubes to an external electric field, a dipole moment is induced resulting in a translational motion along the field gradient which depends both on the dielectric constant of the nanotubes ε_p and solvent medium ε_S. The static dielectric constant for semiconducting HiPco nanotubes has been calculated to be less than 5, while that of metallic SWCNTs has been estimated to be around 1000 [224]. Since the dielectric constant of the solvent, in this case an aqueous solution of sodium dodecyl sulfate, is around 80, semiconducting nanotubes exhibit a negative dielectrophoretic force, e.g. they move towards the low electric field region, while the electrophoretic force in the case of metallic nanotubes is positive so that they move towards the high field region. Upon investigation of the deposited material by incident-light dark-field microscopy it has been shown that the Rayleigh scattered light is polarized perpendicular to the electrode revealing the alignment of the nanotubes.

The characterization by Raman spectroscopy indicates that up to 80 % of the deposited nanotubes are metallic. However, the Raman characterization of the deposited material has been questioned, as the resonant conditions may change due to bundling upon deposition [225–228]. Meanwhile, the enrichment of semiconducting nanotubes in the leftover suspension has been ascertained by repeated dielectrophoretic filtering of the metallic species, strongly supporting the proof of principle experiments [229].

Further experiments on ac dielectrophoresis on SWCNTS have demonstrated that the electrophoretic mobility of sidewall functionalized SWCNTs is strongly decreased suggesting that the dielectric function of the functionalized material is strongly altered [227]. Subsequent work has furthermore been able to show that the sorting efficiency is increased by increasing the frequency of the electrical field [230, 231] even though numerical calculations have suggested the use of low frequency fields [232], or by using a surfactant system composed of anionic (SDS) and cationic (HTAB) additives, as the surface charge on the semiconducting nanotubes is neutralized by this procedure [233]. It has also been demonstrated that metallic and semiconducting SWCNTs can be simultaneously separated and assembled in a multigap nanoelectrode setup to yield a sequential metallic-semiconducting-metallic multiarray structure [234].

Even though dielectrophoretic separation is highly promising in sorting SWCNTs by electronic properties, the method suffers from the disadvantage of limited throughput. In any case, dielectric spectroscopy on SWCNT suspensions allows rapid and accurate determination of both the dielectric properties of the SWCNTs, as well as the proportions of metallic and semiconducting nanotubes [235]. Nonetheless, several attempts have focused on upscaling the original setup by using larger electrodes [236] or making use of dielectrophoretic field flow-fractionation [237]. Furthermore, by the use of a radio frequency dielectrophoresis setup, nanotube films with a thickness of 100 nm can be constructed [238]. The carbon network at very large electrical fields is composed of aligned metallic and randomly oriented semiconducting SWCNTs as revealed by polarization dependent absorption measurements. The deposition of the semiconducting SWCNTs has been explained by a refined model which takes into account the longitudinal and transversal polarizability of the nanotubes.

1.3.2.4 Density Gradient Ultracentrifugation

The last postsynthetic separation technique to be discussed in this chapter is density gradient ultracentrifugation (DGU). The method exploits subtle differences in the buoyant density of the material to be separated. In principle, the sample is loaded into an aqueous solution with a known density gradient established by a gradient medium such as iodixanol, nycodenz or sucrose. Upon applying a centrifugal force, the species travel towards their respective isopycnic points, e.g. the position where their density is equal to that of the gradient. The spatially separated bands can then be fractionated.

If differences in the buoyant density were merely related to the diameter of the nanotube, larger diameter SWCNTs would have smaller density than smaller diameter nanotubes. However, as described by a hydrodynamic model [239], the thickness and hydration of the surfactant coating, as well as the eventual filling of the nanotubes with water [240], strongly alters the buoyant density. The choice of the surfactant is a crucial aspect for the nanotube sorting criteria by DGU. When a surfactant is chosen which uniformly coats all nanotubes equally, the sorting is related to the diameter. In this case, the density increases with increasing diameter (Figure 1.20a). If nanotubes are dispersed in a surfactant, or a combination of surfactants that exhibits preferences for some (n,m)-species, separation by properties beyond geometrical aspects, e.g. sorting by electronic structure may be achieved.

In a first report, enrichment of DNA-wrapped HiPco and CoMoCAT SWCNTs by diameter has been described [241]. Further works focusing on SWCNTs dispersed in conventional detergent solutions have demonstrated the versatility of the DGU approach [242–251]. Multiple possibilities for sorting of CoMoCAT, laser ablation and arc discharge SWCNTs according to diameter or electronic properties exist up to now, so that merely a few shall be summarized here. For example, sodium cholate encapsulated CoMoCAT nanotubes can be sorted by diameter as evidenced by the evolution of visibly colored bands as illustrated in Figure 1.20b [242]. Due to the high optical purity of the fractions, investigations concerning the photoluminescence quantum yields could be carried out revealing that the quantum yields exceed 1 % which is by a factor of 5 higher than previously reported for aqueous nanotube dispersions [243, 244]. Electronic-type separation has been achieved by using surfactant mixtures of SC and SDS. Most remarkably the position of the metallic and semiconducting fractions, respectively can be tailored by changing the ratio of the two surfactants, e.g. semiconducting nanotubes have a lower density when sodium cholate is the main surfactant [242, 252], while metallic nanotubes

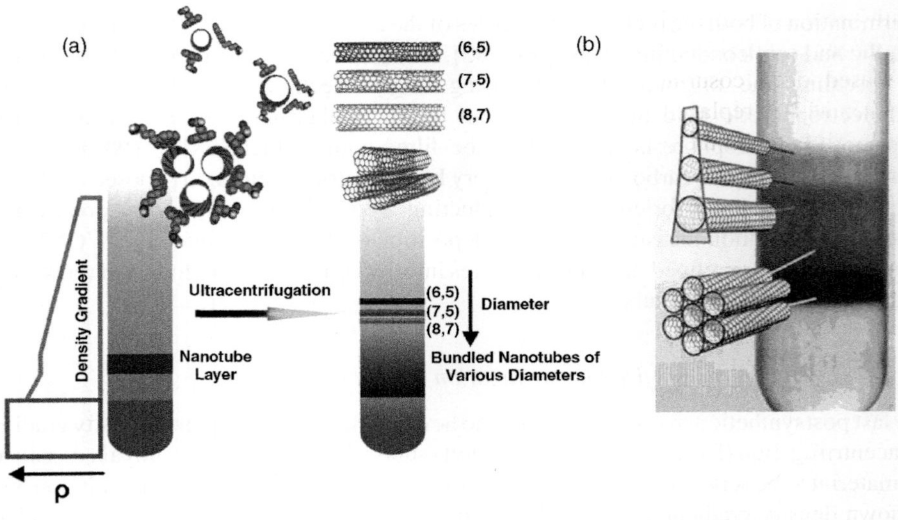

Figure 1.20 Principle of density gradient ultracentrifugation on SNWTs. (a) Prior to ultracentrifugation surfactant encapsulated nanotubes are injected into an approximately linear density gradient. In the centrifugal field, the nanotubes move to the respective isopycninc points in the centrifuge vial resulting in separation according to diameter (or electronic properties). (b) The separation process is evidenced by the formation of colored bands. Reprinted with permission from reference [242]

show a lower density with SDS as main surfactant and SC as cosurfactant [245, 249, 250, 253, 254]. The metallic fractions have been applied to the preparation of colored semi-transparent conductive coatings with colors varying between cyan, magenta and yellow, depending on the composition of the starting material [245, 249, 250, 253, 254]. The semiconducting counterparts allowed the fabrication of thin film nanotube transistors [252, 255]. Furthermore an assignment of the (n,m) indices is possible by the use of an aberration corrected transmission electron microscopic study [249].

As has been mentioned above, the sorting of SWCNTs according to electronic type in DGU has been attributed to inequivalent binding of two surfactants as a function of nanotube polarizability and therefore electronic type. This principle has been confirmed by electronic type sorting of narrowly distributed (n,m) SWCNTs by iterative centrifugation steps in a SDS-SC surfactant mixture without the aid of a density gradient [246].

The density gradient medium usually chosen for SWCNT-DGU is iodixanol which has the disadvantage of being equipped with iodine atoms potentially acting as electron acceptor. Furthmore, iodixanol is an expensive reagent and a rather large molecule causing problems when removing the gradient medium from the nanotube sample. Consequently, other gradient media deserve some attention. Recently, it has been demonstrated that electronic-type sorting also occurs in sucrose as gradient medium when temperature and surfactant concentration is adjusted [248].

Even though significant progress in DGU-based separations has been achieved, the sorting of HiPco SWCNTs has shown to be difficult due to the wide diameter distribution with small average diameters. Recently, two variations of the commonly used procedure have been reported which are potentially capable of overcoming this obstacle. One approach is based on a cosurfactant replacement DGU, where the initial perylene derivative surfactant **2** is replaced by SDS during the centrifugation procedure [256]. The second approach exploits the higher packing density of SDS on the nanotubes with increasing electrolyte concentration [251]. For this purpose, varying amounts of NaCl have been added prior to DGU which resulted in separation of HiPco SWCNTs according to electronic type, as SDS presumably preferentially adsorbs on metallic SWCNTs.

Finally, it is worthwhile mentioning that DGU on functionalized SWCNTs has revealed that the density of the covalently functionalized nanotubes is altered allowing separation of functionalized from nonfunctionalized SWCNTs which is an important step for precise reaction control [257]. Length sorting of nanotubes in DGU has also been reported by exploiting the transient motion regime, as opposed to the equilibrium regime which is approached for diameter and electronic-type sorting [258].

1.4 Conclusion

As indicated above, recent years have been affected by tangible progress in nanotube research, especially concerning their dispersion and separation according to diameter and/or chirality. This is nicely reflected by the number of publications in the vast field of nanotube research. Soon after the discovery of this novel super material by Ijima in 1991 [259], the number of publications about nanotube purification, production, functionalization, separation and application has risen exponentially reaching a number of approximately 10400 in 2008. Obviously the climax of this plot has not yet been reached and the progression resembles that of Moore's law.

It seems that the polydispersability problem which has been the major challenge so far will unambiguously be solved by the noncovalent approach in combination with selective growth and clever refinement techniques such as chromatography, electrophoresis and density gradient centrifugation. However, one should keep in mind that other obstacles still need to be overcome including the issue of scalability, alignment, process compability and economic aspects. Attention should furthermore be drawn towards establishing a standard for the precise determination of SWCNT purity. Thus, we are still somehow at the beginning so that following and participating in nanotube research will become even more exciting as time progresses.

References

[1] Dresselhaus MS, Dresselhaus G, Avouris P., *Carbon Nanotubes: Synthesis, Structure, Properties, and Applications*. Berlin: Springer; 2001.
[2] Ouyang M, Huang J-L, Lieber CM. Fundamental electronic properties and applications of single-walled carbon nanotubes. *Acc Chem Res*. 2002;35(12):1018–1025.
[3] Tanaka K, Yamabe T, Fukui K. *The Science and Technology of Carbon Nanotubes*. Oxford: Elsevier; 1999.

[4] Baughman RH, Zakhidov AA, de Heer WA. Carbon nanotubes-the route toward applications. *Science*. 2002;297(5582):787–792.
[5] Avouris P. Molecular electronics with carbon nanotubes. *Acc Chem Res*. 2002;35(12): 1026–1034.
[6] Vaisman L, Wagner HD, Marom G. The role of surfactants in dispersion of carbon nanotubes. *Adv Colloid Interface Sci*. 2006;128–130: 37–46.
[7] Grossiord N, Loos J, Regev O, Koning CE. Toolbox for dispersing carbon nanotubes into polymers to get conductive nanocomposites. *Chem Mater*. 2006;18(5):1089–1099.
[8] Grobert N. Carbon nanotubes – becoming clean. *Mater Today*. 2007;10(1–2):28–35.
[9] Girifalco LA, Hodak M, Lee RS. Carbon nanotubes, buckyballs, ropes, and a universal graphitic potential. *Phys Rev B*. 2000;62(19):13104–13110.
[10] Nuriel S, Liu L, Barber AH, Wagner HD. Direct measurement of multiwall nanotube surface tension. *Chem Phys Lett*. 2005;404(4–6):263–266.
[11] Tasis D, Tagmatarchis N, Bianco A, Prato M. Chemistry of carbon nanotubes. *Chem Rev*. 2006;106(3):1105–1136.
[12] Hirsch A, Vostrowsky O. Functionalization of carbon nanotubes. *Top Curr Chem*. 2005;245 (Functional Molecular Nanostructures):193–237.
[13] Banerjee S, Hemraj-Benny T, Wong SS. Covalent surface chemistry of single-walled carbon nanotubes. *Adv Mater*. 2005;17(1):17–29.
[14] Tasis D, Tagmatarchis N, Georgakilas V, Prato M. Soluble carbon nanotubes. *Chem Eur J*. 2003;9(17):4000–4008.
[15] Lin T, Bajpai V, Ji T, Dai L. Chemistry of carbon nanotubes. *Aust J Chem*. 2003;56(7): 635–651.
[16] Niyogi S, Hamon MA, Hu H, Zhao B, Bhowmik P, Sen R, *et al.* Chemistry of single-walled carbon nanotubes. *Acc Chem Res*. 2002;35(12):1105–1113.
[17] Hirsch A. Functionalization of single-walled carbon nanotubes. *Angew Chem Int Ed*. 2002;41(11):1853–1859.
[18] Bahr JL, Tour JM. Covalent chemistry of single-wall carbon nanotubes. *J Mater Chem*. 2002;12(7):1952–1958.
[19] Britz DA, Khlobystov AN. Noncovalent interactions of molecules with single walled carbon nanotubes. *Chem Soc Rev*. 2006;35(7):637–659.
[20] Hersam MC. Progress towards monodisperse single-walled carbon nanotubes. *Nat Nanotechnol*. 2008;3(7):387–394.
[21] Ke PC. Fiddling the string of carbon nanotubes with amphiphiles. *Physical Chemistry Chemistry Physics*. 2007;9 439–447.
[22] Nakashima N. Solubilization of single-walled carbon nanotubes with condensed aromatic compounds. *Sci Technol Adv Mater*. 2006;7(7):609–616.
[23] Murakami H, Nakashima N. Soluble carbon nanotubes and their applications. *J Nanosci Nanotechnol*. 2006;6(1):16–27.
[24] Nakashima N. Soluble carbon nanotubes: Fundamentals and applications. *Int J Nanosci*. 2005;4(1):119–137.
[25] Fujigaya T, Nakashima N. Methodology for homogeneous dispersion of single-walled carbon nanotubes by physical modification. *Polymer Journal* (Tokyo, Japan). 2008;40(7): 577–589.
[26] O'Connell MJ, Bachilo SM, Huffman CB, Moore VC, Strano MS, Haroz EH, *et al.* Band gap fluorescence from individual single-walled carbon nanotubes. *Science*. 2002;297(5581): 593–596.
[27] Richard C, Balavoine F, Schultz P, Ebbesen TW, Mioskowski C. Supramolecular self-assembly of lipid derivatives on carbon nanotubes. *Science*. 2003;300(5620):775–778.
[28] Yurekli K, Mitchell CA, Krishnamoorti R. Small-angle neutron scattering from surfactant-assisted aqueous dispersions of carbon nanotubes. *J Am Chem Soc*. 2004;126(32):9902–9903.
[29] Strano MS, Moore VC, Miller MK, Allen MJ, Haroz EH, Kittrell C, *et al.* The role of surfactant adsorption during ultrasonication in the dispersion of single-walled carbon nanotubes. *J Nanosci Nanotechnol*. 2003;3(1/2):81–86.

[30] Bachilo SM, Strano MS, Kittrell C, Hauge RH, Smalley RE, Weisman RB. Structure-assigned optical spectra of single-walled carbon nanotubes. *Science*. 2002;298(5602):2361–2366.
[31] Islam MF, Rojas E, Bergey DM, Johnson AT, Yodh AG. High weight fraction surfactant solubilization of single-wall carbon nanotubes in water. *Nano Lett*. 2003;3(2):269–273.
[32] Wenseleers W, Vlasov II, Goovaerts E, Obraztsova ED, Lobach AS, Bouwen A. Efficient isolation and solubilization of pristine single-walled nanotubes in bile salt micelles. *Adv Funct Mater*. 2004;14(11):1105–1112.
[33] Haggenmueller R, Rahatekar SS, Fagan JA, Chun J, Becker ML, Naik RR, *et al.* Comparison of the quality of aqueous dispersions of single wall carbon nanotubes using surfactants and biomolecules. *Langmuir*. 2008;24(9):5070–5078.
[34] Ishibashi A, Nakashima N. Strong chemical structure dependence for individual dissolution of single-walled carbon nanotubes in aqueous micelles of biosurfactants. *Bull Chem Soc Jpn*. 2006;79(2):357–359.
[35] Ishibashi A, Nakashima N. Individual dissolution of single-walled carbon nanotubes in aqueous solutions of steroid or sugar compounds and their Raman and near-IR spectral properties. *Chem Eur J*. 2006;12(29):7595–7602.
[36] Okazaki T, Saito T, Matsuura K, Ohshima S, Yumura M, Iijima S. Photoluminescence Mapping of 'As-Grown' Single-Walled Carbon Nanotubes: A Comparison with Micelle-Encapsulated Nanotube Solutions. *Nano Lett*. 2005;5(12):2618–2623.
[37] Moore VC, Strano MS, Haroz EH, Hauge RH, Smalley RE, Schmidt J, *et al.* Individually suspended single-walled carbon nanotubes in various surfactants. *Nano Lett*. 2003;3(10): 1379–1382.
[38] Grossiord N, Regev O, Loos J, Meuldijk J, Koning CE. Time-dependent study of the exfoliation process of carbon nanotubes in aqueous dispersions by using uv-visible spectroscopy. *Anal Chem*. 2005;77(16):5135–5139.
[39] Priya BR, Byrne HJ. Investigation of sodium dodecyl benzene sulfonate assisted dispersion and debundling of single-wall carbon nanotubes. *J Phys Chem C*. 2008;112(2):332–337.
[40] Shin J-Y, Premkumar T, Geckeler KE, Dispersion of single-walled carbon nanotubes by using surfactants: are the type and concentration important? *Chem Eur J*. 2008;14(20): 6044–6048.
[41] Utsumi S, Kanamaru M, Honda H, Kanoh H, Tanaka H, Ohkubo T, *et al.* RBM band shift-evidenced dispersion mechanism of single-wall carbon nanotube bundles with NaDDBS. *J Colloid Interface Sci*. 2007;308(1):276–284.
[42] Matarredona O, Rhoads H, Li Z, Harwell JH, Balzano L, Resasco DE. Dispersion of single-walled carbon nanotubes in aqueous solutions of the anionic surfactant NaDDBS. *J Phys Chem B*. 2003;107(48):13357–13367.
[43] White B, Banerjee S, O'Brien S, Turro NJ, Herman IP. Zeta-potential measurements of surfactant-wrapped individual single-walled carbon nanotubes. *J Phys Chem C*. 2007; 111(37):13684–13690.
[44] Sun Z, Nicolosi V, Rickard D, Bergin SD, Aherne D, Coleman JN. Quantitative evaluation of surfactant-stabilized single-walled carbon nanotubes: dispersion quality and its correlation with zeta potential. *J Phys Chem C*. 2008;112(29):10692–10699.
[45] Nakashima N, Tomonari Y, Murakami H. Water-soluble single-walled carbon nanotubes via noncovalent sidewall-functionalization with a pyrene-carrying ammonium ion. *Chem Lett*. 2002(6):638–639.
[46] Tomonari Y, Murakami H, Nakashima N. Solubilizaton of single-walled carbon nanotubes by using polycyclic aromatic ammonium amphiphiles in water – strategy for the design of high-performance solubilizers. *Chem Eur J*. 2006;12(15):4027–4034.
[47] Paloniemi H, Aeaeritalo T, Laiho T, Liuke H, Kocharova N, Haapakka K, *et al.* Water-soluble full-length single-wall carbon nanotube polyelectrolytes: preparation and characterization. *J Phys Chem B*. 2005;109(18):8634–8642.
[48] Backes C, Schmidt CD, Hauke F, Boettcher C, Hirsch A. High population of individualized SWCNTs through the adsorption of water-soluble perylenes. *J Am Chem Soc*. 2009;131(6): 2172–2184.

[49] Schmidt CD, Bottcher C, Hirsch A. Synthesis and aggregation properties of water-soluble Newkome-dendronized perylenetetracarboxdiimides. *Eur J Org Chem*. 2007(33): 5497–5505.

[50] Chen J, Collier CP. Noncovalent functionalization of single-walled carbon nanotubes with water-soluble porphyrins. *J Phys Chem B*. 2005;109(16):7605–7609.

[51] Diehl MR, Yaliraki SN, Beckman RA, Barahona M, Heath JR. Self-assembled, deterministic carbon nanotube wiring networks. *Angew Chem Int Ed*. 2002;41(2):353–356.

[52] Bahr JL, Mickelson ET, Bronikowski MJ, Smalley RE, Tour JM. Dissolution of small diameter single-wall carbon nanotubes in organic solvents?. *Chem Commun*. 2001(2):193–194.

[53] Niyogi S, Hamon MA, Perea DE, Kang CB, Zhao B, Pal SK, *et al.* Ultrasonic dispersions of single-walled carbon nanotubes. *J Phys Chem B*. 2003;107(34):8799–8804.

[54] Kim DS, Nepal D, Geckeler KE, *Individualization of single-walled carbon nanotubes: Is the solvent important? Small*. 2005;1(11):1117–1124.

[55] Fagan SB, Souza Filho AG, Lima JOG, Mendes Filho J, Ferreira OP, Mazali IO, *et al.* 1,2-dichlorobenzene interacting with carbon nanotubes. *Nano Lett*. 2004;4(7):1285–1288.

[56] Ausman KD, Piner R, Lourie O, Ruoff RS, Korobov M. Organic solvent dispersions of single-walled carbon nanotubes: toward solutions of pristine nanotubes. *J Phys Chem B*. 2000; 104(38):8911–8915.

[57] Landi BJ, Ruf HJ, Worman JJ, Raffaelle RP. Effects of alkyl amide solvents on the dispersion of single-wall carbon nanotubes. *J Phys Chem B*. 2004;108(44):17089–17095.

[58] Liu J, Casavant MJ, Cox M, Walters DA, Boul P, Lu W, *et al.* Controlled deposition of individual single-walled carbon nanotubes on chemically functionalized templates. *Chem Phys Lett*. 1999;303(1,2):125–129.

[59] Furtado CA, Kim UJ, Gutierrez HR, Pan L, Dickey EC, Eklund PC. Debundling and dissolution of single-walled carbon nanotubes in amide solvents. *J Am Chem Soc*. 2004;126(19): 6095–6105.

[60] Wang J, Blau WJ. Solvent effect on optical limiting properties of single-walled carbon nanotube dispersions. *J Phys Chem C*. 2008;112(7):2298–2303.

[61] Cheng Q, Debnath S, Gregan E, Byrne HJ. Effect of solvent solubility parameters on the dispersion of single-walled carbon nanotubes. *J Phys Chem C*. 2008;112(51): 20154–20158.

[62] Moonoosawmy KR, Kruse P. To dope or not to dope: the effect of sonicating single-wall carbon nanotubes in common laboratory solvents on their electronic structure. *J Am Chem Soc*. 2008;130(40):13417–13424.

[63] Giordani S, Bergin SD, Nicolosi V, Lebedkin S, Kappes MM, Blau WJ, *et al.* Debundling of single-walled nanotubes by dilution: observation of large populations of individual nanotubes in amide solvent dispersions. *J Phys Chem B*. 2006;110(32):15708–15718.

[64] Giordani S, Bergin S, Nicolosi V, Lebedkin S, Blau WJ, Coleman JN. Fabrication of stable dispersions containing up to 70% individual carbon nanotubes in a common organic solvent. *Phys Stat Sol (B)*. 2006;243(13):3058–3062.

[65] Bergin SD, Nicolosi V, Giordani S, de Gromard A, Carpenter L, Blau WJ, *et al.* Exfoliation in ecstasy: liquid crystal formation and concentration-dependent debundling observed for single-wall nanotubes dispersed in the liquid drug g-butyrolactone. *Nanotechnology*. 2007;18(45): 455705/1-/10.

[66] Bergin SD, Nicolosi V, Streich PV, Giordani S, Sun Z, Windle AH, *et al.* Towards solutions of single-walled carbon nanotubes in common solvents. *Adv Mater*. 2008;20(10):1876–1881.

[67] Murakami H, Nomura T, Nakashima N. Noncovalent porphyrin-functionalized single-walled carbon nanotubes in solution and the formation of porphyrin-nanotube nanocomposites. *Chem Phys Lett*. 2003;378(5,6):481–485.

[68] Hedderman TG, Keogh SM, Chambers G, Byrne HJ. Solubilization of SWNTs with organic dye molecules. *J Phys Chem B*. 2004;108(49):18860–18865.

[69] Pascu SI, Kuganathan N, Tong LH, Jacobs RMJ, Barnard PJ, Chu BT, *et al.* Interactions between tripodal porphyrin hosts and single walled carbon nanotubes: an experimental and theoretical (DFT) account. *J Mater Chem*. 2008;18(24):2781–2788.

[70] Ikeda A, Tanaka Y, Nobusawa K, Kikuchi J-I. Solubilization of single-walled carbon nanotubes by supramolecular complexes of barbituric acid and triaminopyrimidines. *Langmuir*. 2007;23(22):10913–10915.
[71] Nobusawa K, Ikeda A, Kikuchi J-i Kawano S-i Fujita N, Shinkai S. Reversible solubilization and precipitation of carbon nanotubes through oxidation-reduction reactions of a solubilizing agent. *Angew Chem Int Ed*. 2008;47(24):4577–4580.
[72] Kim KK, Yoon S-M, Choi J-Y, Lee J, Kim B-K, Kim JM, *et al.* Design of dispersants for the dispersion of carbon nanotubes in an organic solvent. *Adv Funct Mater*. 2007;17(11): 1775–1783.
[73] Mateo-Alonso A, Ehli C, Chen KH, Guldi DM, Prato M. Dispersion of single-walled carbon nanotubes with an extended diazapentacene derivative. *J Phys Chem A*. 2007;111(49): 12669–12673.
[74] Lee J-H, Yoon S-M, Kim KK, Cha I-S, Park YJ, Choi J-Y, *et al.* Exfoliation of single-walled carbon nanotubes induced by the structural effect of perylene derivatives and their optoelectronic properties. *J Phys Chem C*. 2008;112(39):15267–15273.
[75] Tsang SC, Davis JJ, Green MLH, Hill HAO, Leung YC, Sadler PJ. Immobilization of small proteins in carbon nanotubes: high-resolution transmission electron microscopy study and catalytic activity. *J Chem Soc, Chem Commun*. 1995(24):2579.
[76] Davis JJ, Green MLH, Hill HAO, Leung YC, Sadler PJ, Sloan J, *et al.* The immobilization of proteins in carbon nanotubes. *Inorg Chim Acta*. 272(1,2):1998; 261–266.
[77] Wong SS, Joselevich E, Woolley AT, Cheung CL, Lieber CM. Covalently functionalized nanotubes as nanometer-sized probes in chemistry and biology. *Nature*. 1998;394(6688): 52–55.
[78] Baughman RH, Cui C, Zakhidov AA, Iqbal Z, Barisci JN, Spinks GM, *et al.* Carbon nanotube actuators. *Science*. 1999;284(5418):1340–1344.
[79] Nakashima N, Okuzono S, Murakami H, Nakai T, Yoshikawa K. DNA dissolves single-walled carbon nanotubes in water. *Chem Lett*. 2003;32(5):456–457.
[80] Zheng M, Jagota A, Semke ED, Diner BA, McLean RS, Lustig SR, *et al.* DNA-assisted dispersion and separation of carbon nanotubes. *Nat Mater*. 2003;2(5):338–342.
[81] Zheng M, Jagota A, Strano MS, Santos AP, Barone P, Chou SG, *et al.* Structure-based carbon nanotube sorting by sequence-dependent DNA assembly. *Science*. 2003;302(5650): 1545–1548.
[82] Chou SG, Ribeiro HB, Barros EB, Santos AP, Nezich D, Samsonidze GG, *et al.* Optical characterization of DNA-wrapped carbon nanotube hybrids. *Chem Phys Lett*. 2004;397(4–6): 296–301.
[83] Strano MS, Zheng M, Jagota A, Onoa GB, Heller DA, Barone PW, *et al.* Understanding the nature of the DNA-assisted separation of single-walled carbon nanotubes using fluorescence and Raman spectroscopy. *Nano Lett*. 2004;4(4):543–550.
[84] Badaire S, Zakri C, Maugey M, Derre A, Barisci JN, Wallace G, *et al.* Liquid crystals of DNA-stabilized carbon nanotubes. *Adv Mater*. 2005;17(13):1673–1676.
[85] Gigliotti B, Sakizzie B, Bethune DS, Shelby RM, Cha JN. Sequence-independent helical wrapping of single-walled carbon nanotubes by long genomic DNA. *Nano Lett*. 2006;6(2): 159–164.
[86] Cathcart H, Quinn S, Nicolosi V, Kelly JM, Blau WJ, Coleman JN. Spontaneous debundling of single-walled carbon nanotubes in DNA-based dispersions. *J Phys Chem C*. 2007;111(1): 66–74.
[87] Xu Y, Pehrsson PE, Chen L, Zhang R, Zhao W. Double-Stranded DNA Single-walled carbon nanotube hybrids for optical hydrogen peroxide and glucose sensing. *J Phys Chem C*. 2007;111(24):8638–8643.
[88] Tu X, Pehrsson PE, Zhao W. Redox reaction of DNA-encased HiPco carbon nanotubes with hydrogen peroxide: a near infrared optical sensitivity and kinetics study. *J Phys Chem C*. 2007;111(46):17227–17231.
[89] Li X, Peng Y, Qu X. Carbon nanotubes selective destabilization of duplex and triplex DNA and inducing B-A transition in solution. *Nucleic Acids Res*. 2006;34(13):3670–3676.

[90] Dieckmann GR, Dalton AB, Johnson PA, Razal J, Chen J, Giordano GM, *et al.* Controlled assembly of carbon nanotubes by designed amphiphilic peptide helices. *J Am Chem Soc.* 2003;125(7):1770–1777.
[91] Wang S, Humphreys ES, Chung S-Y, Delduco DF, Lustig SR, Wang H, *et al.* Peptides with selective affinity for carbon nanotubes. *Nat Mater.* 2003;2(3):196–200.
[92] Dalton AB, Ortiz-Acevedo A, Zorbas V, Brunner E, Sampson WM, Collins S, *et al.* Hierarchical self-assembly of peptide-coated carbon nanotubes. *Adv Funct Mater.* 2004;14(12):1147–1151.
[93] Zorbas V, Ortiz-Acevedo A, Dalton AB, Yoshida MM, Dieckmann GR, Draper RK, *et al.* Preparation and characterization of individual peptide-wrapped single-walled carbon nanotubes. *J Am Chem Soc.* 2004;126(23):7222–7227.
[94] Ortiz-Acevedo A, Xie H, Zorbas V, Sampson WM, Dalton AB, Baughman RH, *et al.* Diameter-selective solubilization of single-walled carbon nanotubes by reversible cyclic peptides. *J Am Chem Soc.* 2005;127(26):9512–9517.
[95] Xie H, Ortiz-Acevedo A, Zorbas V, Baughman RH, Draper RK, Musselman IH, *et al.* Peptide cross-linking modulated stability and assembly of peptide-wrapped single-walled carbon nanotubes. *J Mater Chem.* 2005;15(17):1734–1741.
[96] Zorbas V, Smith AL, Xie H, Ortiz-Acevedo A, Dalton AB, Dieckmann GR, *et al.* Importance of aromatic content for peptide/single-walled carbon nanotube interactions. *J Am Chem Soc.* 2005;127(35):12323–12328.
[97] Xie H, Becraft EJ, Baughman RH, Dalton AB, Dieckmann GR. Ranking the affinity of aromatic residues for carbon nanotubes by using designed surfactant peptides. *J Pept Sci.* 2008;14(2): 139–151.
[98] Nicolosi V, Cathcart H, Dalton AR, Aherne D, Dieckmann GR, Coleman JN. Spontaneous exfoliation of single-walled carbon nanotubes dispersed using a designed amphiphilic peptide. *Biomacromolecules.* 2008;9(2):598–602.
[99] Arnold MS, Guler MO, Hersam MC, Stupp SI. Encapsulation of carbon nanotubes by self-assembling peptide amphiphiles. *Langmuir.* 2005;21(10):4705–4709.
[100] Pender MJ, Sowards LA, Hartgerink JD, Stone MO, Naik RR. Peptide-mediated formation of single-wall carbon nanotube composites. *Nano Lett.* 2006;6(1):40–44.
[101] Witus LS, Rocha J-DR Yuwono VM, Paramonov SE, Weisman RB, Hartgerink JD. Peptides that non-covalently functionalize single-walled carbon nanotubes to give controlled solubility characteristics. *J Mater Chem.* 2007;17(19):1909–1915.
[102] Tsyboulski DA, Bakota EL, Witus LS, Rocha J-DR Hartgerink JD, Weisman RB. Self-assembling peptide coatings designed for highly luminescent suspension of single-walled carbon nanotubes. *J Am Chem Soc.* 2008;130(50):17134–17140.
[103] Balavoine F, Schultz P, Richard C, Mallouh V, Ebbesen TW, Mioskowski C. Helical crystallization of proteins on carbon nanotubes: a first step towards the development of new biosensors. *Angew Chem Int Ed.* 38(13/14):1999; 1912–1915.
[104] Karajanagi SS, Vertegel AA, Kane RS, Dordick JS. Structure and function of enzymes adsorbed onto single-walled carbon nanotubes. *Langmuir.* 2004;20(26):11594–11599.
[105] Karajanagi SS, Yang H, Asuri P, Sellitto E, Dordick JS, Kane RS. Protein-assisted solubilization of single-walled carbon nanotubes. *Langmuir.* 2006;22(4):1392–1395.
[106] Nepal D. Geckeler KE. pH-sensitive dispersion and debundling of single-walled carbon nanotubes: lysozyme as a tool. *Small.* 2006;2(3):406–412.
[107] Goldberg-Oppenheimer P, Regev O. Exploring a nanotube dispersion mechanism with gold-labeled proteins via cryo-TEM imaging. *Small.* 2007;3(11):1894–1899.
[108] Star A, Steuerman DW, Heath JR, Stoddart JF. Starched carbon nanotubes. *Angew Chem Int Ed.* 2002;41(14):2508–2512.
[109] Kim O-K, Je J, Baldwin JW, Kooi S, Pehrsson PE, Buckley LJ. Solubilization of single-wall carbon nanotubes by supramolecular encapsulation of helical amylose. *J Am Chem Soc.* 2003;125(15):4426–4427.
[110] Numata M, Asai M, Kaneko K, Bae A-H, Hasegawa T, Sakurai K, *et al.* Inclusion of cut and as-grown single-walled carbon nanotubes in the helical superstructure of schizophyllan and curdlan (b-1,3-Glucans). *J Am Chem Soc.* 2005;127(16):5875–5884.

[111] Numata M, Asai M, Kaneko K, Hasegawa T, Fujita N, Kitada Y, *et al.* Curdlan and schizophyllan (b-1,3-glucans) can entrap single-wall carbon nanotubes in their helical super-structure. *Chem Lett.* 2004;33(3):232–233.
[112] Takahashi T, Luculescu CR, Uchida K, Ishii T, Yajima H. Dispersion behavior and spectroscopic properties of single-walled carbon nanotubes in chitosan acidic aqueous solutions. *Chem Lett.* 2005;34(11):1516–1517.
[113] Wang S-F, Shen L, Zhang W-D, Tong Y-J. Preparation and mechanical properties of chitosan/carbon nanotubes composites. biomacromolecules. 2005;6(6):3067–3072.
[114] Yang H, Wang SC, Mercier P, Akins DL. Diameter-selective dispersion of single-walled carbon nanotubes using a water-soluble, biocompatible polymer. *Chem Commun.* 2006(13): 1425–1427.
[115] Zhang J, Wang Q, Wang L, Wang A. Manipulated dispersion of carbon nanotubes with derivatives of chitosan. *Carbon.* 2007;45(9):1917–1920.
[116] Yan LY, Poon YF, Chan-Park MB, Chen Y, Zhang Q. Individually dispersing single-walled carbon nanotubes with novel neutral pH Water-soluble chitosan derivatives. *J Phys Chem C.* 2008;112(20):7579–7587.
[117] Bandyopadhyaya R, Nativ-Roth E, Regev O, Yerushalmi-Rozen R. Stabilization of individual carbon nanotubes in aqueous solutions. *Nano Lett.* 2002;2(1):25–28.
[118] Wu Y, Hudson JS, Lu Q, Moore JM, Mount AS, Rao AM, *et al.* Coating single-walled carbon nanotubes with phospholipids. *J Phys Chem B.* 2006;110(6):2475–2478.
[119] Qiao R, Ke PC. Lipid-carbon nanotube self-assembly in aqueous solution. *J Am Chem Soc.* 2006;128(42):13656–13657.
[120] Liu Y, Gao L, Zheng S, Wang Y, Sun J, Kajiura H, *et al.* Debundling of single-walled carbon nanotubes by using natural polyelectrolytes. *Nanotechnology.* 2007;18(36): 365702/1-/6.
[121] Terashima M, Nagao S. Solubilization of [60]fullerene in water by aquatic humic substances. *Chem Lett.* 2007;36(2):302–303.
[122] Romero DB, Carrard M, De Heer W, Zuppiroli L. A carbon nanotube/organic semiconducting polymer heterojunction. *Adv Mater.* 1996;8(11):899–902.
[123] Burroughes JH, Bradley DDC, Brown AR, Marks RN, Mackay K, Friend RH, *et al.* Light-emitting diodes based on conjugated polymers. *Nature.* 1990;347(6293):539–541.
[124] Curran SA, Ajayan PM, Blau WJ, Carroll DL, Coleman JN, Dalton AB, *et al.* A composite from poly(m-phenylenevinylene-co-2,5-dioctoxy-p-phenylenevinylene) and carbon nanotubes. A novel material for molecular optoelectronics. *Adv Mater.* 1998;10(14):1091–1093.
[125] Coleman JN, Curran S, Dalton AB, Davey AP, McCarthy B, Blau W, *et al.* Percolation-dominated conductivity in a conjugated-polymer-carbon-nanotube composite. *Phys Rev B.* 1998;58(12):R7492–R7500.
[126] Star A, Stoddart JF, Steuerman D, Diehl M, Boukai A, Wong EW, *et al.* Preparation and properties of polymer-wrapped single-walled carbon nanotubes. *Angew Chem Int Ed.* 2001;40(9):1721–1725.
[127] Coleman JN, Dalton AB, Curran S, Rubio A, Davey AP, Drury A, *et al.* Phase separation of carbon nanotubes and turbostratic graphite using a functional organic polymer. *Adv Mater.* 2000;12(3):213–216.
[128] Nakashima N, Fujigaya T. Fundamentals and applications of soluble carbon nanotubes. *Chem Lett.* 2007;36(6):692–697.
[129] O'Connell MJ, Boul P, Ericson LM, Huffman C, Wang Y, Haroz E, *et al.* Reversible water-solubilization of single-walled carbon nanotubes by polymer wrapping. *Chem Phys Lett.* 2001;342(3,4):265–271.
[130] Hasan T, Scardaci V, Tan P, Rozhin AG, Milne WI, Ferrari AC. Stabilization and 'de-bundling' of single-wall carbon nanotube dispersions in N-Methyl-2-pyrrolidone (NMP) by polyvinyl-pyrrolidone (PVP). *J Phys Chem C.* 2007;111(34):12594–12602.
[131] Hasan T, Tan PH, Bonaccorso F, Rozhin AG, Scardaci V, Milne WI, *et al.* Polymer-assisted isolation of single wall carbon nanotubes in organic solvents for optical quality nanotube-polymer composites. *J Phys Chem C.* 2009.

[132] Huang W, Fernando S, Allard LF, Sun Y-P. Solubilization of single-walled carbon nanotubes with diamine-terminated oligomeric poly(ethylene glycol) in different functionalization reactions. *Nano Lett*. 2003;3(4):565–568.
[133] Star A, Liu Y, Grant K, Ridvan L, Stoddart JF, Steuerman DW, *et al.* Noncovalent side-wall functionalization of single-walled carbon nanotubes. *Macromolecules*. 2003;36(3):553–560.
[134] Chen J, Liu H, Weimer WA, Halls MD, Waldeck DH, Walker GC. Noncovalent engineering of carbon nanotube surfaces by rigid, functional conjugated polymers. *J Am Chem Soc*. 2002;124(31):9034–9035.
[135] Lou X, Daussin R, Cuenot S, Duwez A-S, Pagnoulle C, Detrembleur C, *et al.* Synthesis of pyrene-containing polymers and noncovalent sidewall functionalization of multiwalled carbon nanotubes. *Chem Mater*. 2004;16(21):4005–4011.
[136] Petrov P, Stassin F, Pagnoulle C, Jerome R. Noncovalent functionalization of multi-walled carbon nanotubes by pyrene containing polymers. *Chem Commun*. 2003(23):2904–2905.
[137] Nakashima N, Okuzono S, Tomonari Y, Murakami H. Solubilization of carbon nanotubes with a pyrene-carrying polymer in water. *Trans Mater Res Soc Jpn*. 2004;29(2):525–528.
[138] Yuan WZ, Sun JZ, Dong Y, Haeussler M, Yang F, Xu HP, *et al.* Wrapping carbon nanotubes in pyrene-containing poly(phenylacetylene) chains: solubility, stability, light emission, and surface photovoltaic properties. *Macromolecules*. 2006;39(23):8011–8020.
[139] Bahun GJ, Wang C, Adronov A. Solubilizing single-walled carbon nanotubes with pyrene-functionalized block copolymers. *J Polym Sci, Part A*. 2006;44(6):1941–1951.
[140] Wang D, Ji W-X, Li Z-C, Chen L. A Biomimetic 'Polysoap' for single-walled carbon nanotube dispersion. *J Am Chem Soc*. 2006;128(20):6556–6557.
[141] Park C, Ounaies Z, Watson KA, Crooks RE, Smith J, Lowther SE, *et al.* Dispersion of single wall carbon nanotubes by in-situ polymerization under sonication. *Chem Phys Lett*. 2002; 364(3,4):303–308.
[142] Lillehei PT, Park C, Rouse JH, Siochi EJ. Imaging carbon nanotubes in high performance polymer composites via magnetic force microscopy. *Nano Lett*. 2002;2(8):827–829.
[143] Ounaies Z, Park C, Wise KE, Siochi EJ, Harrison JS. Electrical properties of single wall carbon nanotube reinforced polyimide composites. *Compos Sci Technol*. 2003;63(11): 1637–1646.
[144] Shigeta M, Komatsu M, Nakashima N. Individual solubilization of single-walled carbon nanotubes using totally aromatic polyimide. *Chem Phys Lett*. 2006;418(1–3):115–118.
[145] Kang Y, Taton TA. Micelle-encapsulated carbon nanotubes: a route to nanotube composites. *J Am Chem Soc*. 2003;125(19):5650–5651.
[146] Nativ-Roth E, Shvartzman-Cohen R, Bounioux C, Florent M, Zhang D, Szleifer I, *et al.* Physical adsorption of block copolymers to SWNT and MWNT: a nonwrapping mechanism. *Macromolecules*. 2007;40(10):3676–3685.
[147] Lu L, Zhou Z, Zhang Y, Wang S, Zhang Y. Reinforcement of styrene-butadiene-styrene triblock copolymer by multi-walled carbon nanotubes via melt mixing. *Carbon*. 2007;45(13): 2621–2627.
[148] Gilmore KJ, Moulton SE, Wallace GG. Incorporation of carbon nanotubes into the biomedical polymer poly(styrene-beta -isobutylene-beta -styrene). *Carbon*. 2007;45(2):402–410.
[149] Cotiuga I, Picchioni F, Agarwal US, Wouters D, Loos J, Lemstra PJ. Block-copolymer-assisted solubilization of carbon nanotubes and exfoliation monitoring through viscosity. *Macromol Rapid Commun*. 2006;27(13):1073–1078.
[150] Shvartzman-Cohen R, Levi-Kalisman Y, Nativ-Roth E, Yerushalmi-Rozen R. Generic approach for dispersing single-walled carbon nanotubes: the strength of a weak interaction. *Langmuir*. 2004;20(15):6085–6088.
[151] Slusarenko NN, Heurtefeu B, Maugey M, Zakri C, Poulin P, Lecommandoux S. Diblock copolymer stabilization of multi-wall carbon nanotubes in organic solvents and their use in composites. [Erratum to document cited in CA147:167044]. *Carbon*. 2007;45(4):903.
[152] Mountrichas G, Pispas S, Tagmatarchis N. Aqueous carbon-nanotube-amphiphilic-block-copolymer nanoensembles: towards realization of charge-transfer processes with semiconductor quantum dots. *Small*. 2007;3(3):404–407.

[153] Mountrichas G, Tagmatarchis N, Pispas S. Synthesis and Solution Behavior of Carbon Nanotubes Decorated with Amphiphilic Block Polyelectrolytes. *J Phys Chem B*. 2007; 111(29):8369–8372.
[154] Sinani VA, Gheith MK, Yaroslavov AA, Rakhnyanskaya AA, Sun K, Mamedov AA, *et al.* Aqueous dispersions of single-wall and multiwall carbon nanotubes with designed amphiphilic polycations. *J Am Chem Soc*. 2005;127(10):3463–3472.
[155] Sung J-w Park JM, Choi UH, Huh J, Jung B, Min BG, *et al.* Control of SWNT dispersion by block copolymers with a side-chain polarity modifier. *Macromol Rapid Commun*. 2007;28(2):176–182.
[156] Park C, Lee S, Lee JH, Lim J, Lee SC, Park M, *et al.* Controlled assembly of carbon nanotubes encapsulated with amphiphilic block copolymer. *Carbon*. 2007;45(10):2072–2078.
[157] Zou J, Liu L, Chen H, Khondaker SI, McCullough RD, Huo Q, *et al.* Dispersion of pristine carbon nanotubes using conjugated block copolymers. *Adv Mater*. 2008;20(11): 2055–2060.
[158] Shin H-i Min BG, Jeong W, Park C. Amphiphilic block copolymer micelles: New dispersant for single wall carbon nanotubes. *Macromol Rapid Commun*. 2005;26(18):1451–1457.
[159] Krupke R, Hennrich F. Separation techniques for carbon nanotubes. *Adv Eng Mater*. 2005; 7(3):111–116.
[160] Banerjee S, Hemraj-Benny T, Wong SS. Routes towards separating metallic and semiconducting nanotubes. *J Nanosci Nanotechnol*. 2005;5(6):841–855.
[161] Basiuk EV, Basiuk VA, Banuelos J-G, Saniger-Blesa J-M, Pokrovskiy VA, Gromovoy TY, *et al.* Interaction of oxidized single-walled carbon nanotubes with vaporous aliphatic amines. *J Phys Chem B*. 2002;106(7):1588–1597.
[162] Kong J, Dai H. Full and modulated chemical gating of individual carbon nanotubes by organic amine compounds. *J Phys Chem B*. 2001;105(15):2890–2893.
[163] Chattopadhyay D, Galeska I, Papadimitrakopoulos F. A Route for bulk separation of semiconducting from metallic single-wall carbon nanotubes. *J Am Chem Soc*. 2003;125(11): 3370–3375.
[164] Samsonidze GG, Chou SG, Santos AP, Brar VW, Dresselhaus G, Dresselhaus MS, *et al.* Quantitative evaluation of the octadecylamine-assisted bulk separation of semiconducting and metallic single-wall carbon nanotubes by resonance Raman spectroscopy. *Appl Phys Lett*. 2004;85(6):1006–1008.
[165] Maeda Y, Kimura S, Hirashima Y, Kanda M, Lian Y, Wakahara T, *et al.* Dispersion of single-walled carbon nanotube bundles in nonaqueous solution. *J Phys Chem B*. 2004;108(48): 18395–18397.
[166] Kim SN, Luo Z, Papadimitrakopoulos F. Diameter and metallicity dependent redox influences on the separation of single-wall carbon nanotubes. *Nano Lett*. 2005;5(12): 2500–2504.
[167] Maeda Y, Kanda M, Hashimoto M, Hasegawa T, Kimura S-I, Lian Y, *et al.* Dispersion and separation of small-diameter single-walled carbon nanotubes. *J Am Chem Soc*. 2006;128(37):12239–12242.
[168] Maeda Y, Kimura S, Kanda M, Hirashima Y, Hasegawa T, Wakahara T, *et al.* Large-scale separation of metallic and semiconducting single-walled carbon nanotubes. *J Am Chem Soc*. 2005;127(29):10287–10290.
[169] Nish A, Hwang J-Y, Doig J, Nicholas RJ. Highly selective dispersion of single-walled carbon nanotubes using aromatic polymers. *Nat Nanotechnol*. 2007;2(10):640–646.
[170] Chen F, Wang B, Chen Y, Li L-J. Toward the extraction of single species of single-walled carbon nanotubes using fluorene-based polymers. *Nano Lett*. 2007;7(10):3013–3017.
[171] Hennrich F, Lebedkin S, Kappes MM. Improving separation techniques for single-walled carbon nanotubes: towards monodisperse samples. *Phys Stat Sol (B)*. 2008;245(10): 1951–1953.
[172] Hwang J-Y, Nish A, Doig J, Douven S, Chen C-W, Chen L-C, *et al.* Polymer structure and solvent effects on the selective dispersion of single-walled carbon nanotubes. *J Am Chem Soc*. 2008;130(11):3543–3553.

[173] Izard N, Kazaoui S, Hata K, Okazaki T, Saito T, Iijima S, *et al.* Semiconductor-enriched single wall carbon nanotube networks applied to field effect transistors. *Appl Phys Lett.* 2008;92(24):243112/1-/3.

[174] Dodziuk H, Ejchart A, Anczewski W, Ueda H, Krinichnaya E, Dolgonos G, *et al.* Water solubilization, determination of the number of different types of single-wall carbon nanotubes and their partial separation with respect to diameters by complexation with h-cyclodextrin. *Chem Commun.* 2003(8):986–987.

[175] Tasis D, Papagelis K, Douroumis D, Smith JR, Bouropoulos N, Fatouros DG. Diameter-selective solubilization of carbon nanotubes by lipid micelles. *J Nanosci Nanotechnol.* 2008;8(1):420–423.

[176] Ju S-Y, Doll J, Sharma I, Papadimitrakopoulos F. Selection of carbon nanotubes with specific chiralities using helical assemblies of flavin mononucleotide. *Nat Nanotechnol.* 2008; 3(6):356–362.

[177] Lin CS, Zhang RQ, Niehaus TA, Frauenheim T. Geometric and electronic structures of carbon nanotubes adsorbed with flavin adenine dinucleotide: a theoretical study. *J Phys Chem C.* 2007;111(11):4069–4073.

[178] Niyogi S, Boukhalfa S, Chikkannanavar SB, McDonald TJ, Heben MJ, Doorn SK. Selective aggregation of single-walled carbon nanotubes via salt addition. *J Am Chem Soc.* 2007;129(7): 1898–1899.

[179] Tromp RM, Afzali A, Freitag M, Mitzi DB, Chen Z. Novel Strategy for diameter-selective separation and functionalization of single-wall carbon nanotubes. *Nano Lett.* 2008;8(2): 469–472.

[180] Peng X, Komatsu N, Kimura T, Osuka A. Simultaneous enrichments of optical purity and (n,m) abundance of SWNTs through extraction with 3,6-carbazolylene-bridged chiral diporphyrin nanotweezers. *ACS Nano.* 2008;2(10):2045–2050.

[181] Peng X, Komatsu N, Kimura T, Osuka A. Improved optical enrichment of SWNTs through extraction with chiral nanotweezers of 2,6-pyridylene-bridged diporphyrins. *J Am Chem Soc.* 2007;129(51):15947–15953.

[182] Peng X, Komatsu N, Bhattacharya S, Shimawaki T, Aonuma S, Kimura T, *et al.* Optically active single-walled carbon nanotubes. *Nat Nanotechnol.* 2007;2(6):361–365.

[183] Chen Z, Du X, Du M-H, Rancken CD, Cheng H-P, Rinzler AG. Bulk separative enrichment in metallic or semiconducting single-walled carbon nanotubes. *Nano Lett.* 2003;3(9): 1245–1249.

[184] Li H, Zhou B, Lin Y, Gu L, Wang W, Fernando KAS, *et al.* Selective interactions of porphyrins with semiconducting single-walled carbon nanotubes. *J Am Chem Soc.* 2004;126(4): 1014–1015.

[185] Marquis R, Greco C, Sadokierska I, Lebedkin S, Kappes MM, Michel T, *et al.* Supramolecular discrimination of carbon nanotubes according to their helicity. *Nano Lett.* 2008;8(7): 1830–1835.

[186] Duesberg GS, Muster J, Krstic V, Burghard M, Roth S. Chromatographic size separation of single-wall carbon nanotubes. *Appl Phys A.* 1998;67(1):117–119.

[187] Duesberg GS, Burghard M, Muster J, Philipp G, Roth S. Separation of carbon nanotubes by size exclusion chromatography. *Chem Commun.* 1998(3):435–436.

[188] Duesberg GS, Blau W, Byrne HJ, Muster J, Burghard M, Roth S. Chromatography of carbon nanotubes. *Synth Met.* 1999;103(1–3):2484–2485.

[189] Arnold K, Hennrich F, Krupke R, Lebedkin S, Kappes MM. Length separation studies of single walled carbon nanotube dispersions. *Phys Stat Sol (B).* 2006;243(13): 3073–3076.

[190] Huang X, McLean RS, Zheng M. High-resolution length sorting and purification of DNA-wrapped carbon nanotubes by size-exclusion chromatography. *Anal Chem.* 2005;77(19): 6225–6228.

[191] Bauer BJ, Fagan JA, Hobbie EK, Chun J, Bajpai V. Chromatographic fractionation of SWNT/DNA dispersions with on-line multi-angle light scattering. *J Phys Chem C.* 2008;112(6):1842–1850.

[192] Niyogi S, Hu H, Hamon MA, Bhowmik P, Zhao B, Rozenzhak SM, *et al.* Chromatographic purification of soluble single-walled carbon nanotubes (s-SWNTs). *J Am Chem Soc.* 2001;123(4):733–734.

[193] Farkas E, Elizabeth Anderson M, Chen Z, Rinzler AG. Length sorting cut single wall carbon nanotubes by high performance liquid chromatography. *Chem Phys Lett.* 2002;363(1,2): 111–116.

[194] Yang Y, Xie L, Chen Z, Liu M, Zhu T, Liu Z. Purification and length separation of single-walled carbon nanotubes using chromatographic method. *Synth Met.* 2005;155(3):455–460.

[195] Noguchi Y, Fujigaya T, Niidome Y, Nakashima N. Single-walled carbon nanotubes/DNA hybrids in water are highly stable. *Chem Phys Lett.* 2008;455(4–6):249–251.

[196] Chattopadhyay D, Lastella S, Kim S, Papadimitrakopoulos F. Length separation of zwitterion-functionalized single wall carbon nanotubes by GPC. *J Am Chem Soc.* 2002;124(5):728–729.

[197] Zhao B, Hu H, Niyogi S, Itkis ME, Hamon MA, Bhowmik P, *et al.* Chromatographic purification and properties of soluble single-walled carbon nanotubes. *J Am Chem Soc.* 2001;123(47):11673–11677.

[198] Liu J, Rinzler AG, Dai H, Hafner JH, Bradley RK, Boul PJ, *et al.* Fullerene pipes. *Science.* 1998;280(5367):1253–1256.

[199] Chen B, Selegue JP. Separation and characterization of single-walled and multiwalled carbon nanotubes by using flow field-flow fractionation. *Anal Chem.* 2002;74(18):4774–4780.

[200] Moon MH, Kang D, Jung J, Kim J. Separation of carbon nanotubes by frit inlet asymmetrical flow field-flow fractionation. *J Sep Sci.* 2004;27(9):710–717.

[201] Lustig SR, Jagota A, Khripin C, Zheng M. Theory of structure-based carbon nanotube separations by ion-exchange chromatography of DNA/CNT hybrids. *J Phys Chem B.* 2005;109(7):2559–2566.

[202] Zheng M, Semke ED. Enrichment of single chirality carbon nanotubes. *J Am Chem Soc.* 2007;129(19):6084–6085.

[203] Zhang L, Zaric S, Tu X, Wang X, Zhao W, Dai H. Assessment of chemically separated carbon nanotubes for nanoelectronics. *J Am Chem Soc.* 2008;130(8):2686–2691.

[204] Gao B, Yue GZ, Qiu Q, Cheng Y, Shimoda H, Fleming L, *et al.* Fabrication and electron field emission properties of carbon nanotube films by electrophoretic deposition. *Adv Mater.* 2001;13(23):1770–1773.

[205] Kamat PV, Thomas KG, Barazzouk S, Girishkumar G, Vinodgopal K, Meisel D. Self-assembled linear bundles of single wall carbon nanotubes and their alignment and deposition as a film in a d.c. field. *J Am Chem Soc.* 2004;126(34):10757–10762.

[206] Chen Z, Yang Y, Wu Z, Luo G, Xie L, Liu Z, *et al.* Electric-field-enhanced assembly of single-walled carbon nanotubes on a solid surface. *J Phys Chem B.* 2005;109(12):5473–5477.

[207] Yamamoto K, Akita S, Nakayama Y. Orientation and purification of carbon nanotubes using ac electrophoresis. *J Phys D: Appl Phys.* 1998;31(8):L34–L40.

[208] Chen XQ, Saito T, Yamada H, Matsushige K. Aligning single-wall carbon nanotubes with an alternating-current electric field. *Appl Phys Lett.* 2001;78(23):3714–3716.

[209] Nagahara LA, Amlani I, Lewenstein J, Tsui RK. Directed placement of suspended carbon nanotubes for nanometer-scale assembly. *Appl Phys Lett.* 2002;80(20):3826–3828.

[210] Krupke R, Hennrich F, Weber HB, Beckmann D, Hampe O, Malik S, *et al.* Contacting single bundles of carbon nanotubes with alternating electric fields. *Appl Phys A.* 2003;76(3):397–400.

[211] Suehiro J, Zhou G, Hara M. Fabrication of a carbon nanotube-based gas sensor using dielectrophoresis and its application for ammonia detection by impedance spectroscopy. *J Phys D: Appl Phys.* 2003;36(21):L109–L114.

[212] Li J, Zhang Q, Yang D, Tian J. Fabrication of carbon nanotube field effect transistors by AC dielectrophoresis method. *Carbon.* 2004;42(11):2263–2267.

[213] Lee SW, Lee DS, Yu HY, Campbell EEB, Park YW. Production of individual suspended single-walled carbon nanotubes using the ac electrophoresis technique. *Appl Phys A.* 2004;78(3): 283–286.

[214] Chen Z, Yang Y, Chen F, Qing Q, Wu Z, Liu Z. Controllable interconnection of single-walled carbon nanotubes under ac electric field. *J Phys Chem B.* 2005;109(23):11420–11423.

[215] Doorn SK, Fields RE III, Hu H, Hamon MA, Haddon RC, Selegue JP, *et al.* High resolution capillary electrophoresis of carbon nanotubes. *J Am Chem Soc.* 2002;124(12):3169–3174.

[216] Doorn SK, Strano MS, O'Connell MJ, Haroz EH, Rialon KL, Hauge RH, *et al.* Capillary electrophoresis separations of bundled and individual carbon nanotubes. *J Phys Chem B.* 2003;107(25):6063–6069.

[217] Suarez B, Simonet BM, Cardenas S, Valcarcel M. Separation of carbon nanotubes in aqueous medium by capillary electrophoresis. *Journal of Chromatography, A.* 1128(1–2):2006; 282–289.

[218] Lopez-Pastor M, Dominguez-Vidal A, Ayora-Canada MJ, Simonet BM, Lendl B, Valcarcel M. Separation of single-walled carbon nanotubes by use of ionic liquid-aided capillary electrophoresis. *Anal Chem.* 2008;80(8):2672–2679.

[219] Heller DA, Mayrhofer RM, Baik S, Grinkova YV, Usrey ML, Strano MS. Concomitant length and diameter separation of single-walled carbon nanotubes. *J Am Chem Soc.* 2004;126(44): 14567–14573.

[220] Vetcher AA, Srinivasan S, Vetcher IA, Abramov SM, Kozlov M, Baughman RH, *et al.* Fractionation of SWNT/nucleic acid complexes by agarose gel electrophoresis. *Nanotechnology.* 2006;17(16):4263.

[221] Xu X, Ray R, Gu Y, Ploehn HJ, Gearheart L, Raker K, *et al.* Electrophoretic analysis and purification of fluorescent single-walled carbon nanotube fragments. *J Am Chem Soc.* 2004;126(40):12736–12737.

[222] Krupke R, Hennrich F, Weber HB, Kappes MM, von Loehneysen H. Simultaneous deposition of metallic bundles of single-walled carbon nanotubes using Ac-dielectrophoresis. *Nano Lett.* 2003;3(8):1019–1023.

[223] Krupke R, Hennrich F, Lohneysen Hilbert v Kappes M. Separation of metallic from semiconducting single-walled carbon nanotubes. *Science.* 2003;301(5631):344–347.

[224] Benedict LX, Louie SG, Cohen ML. Static polarizabilities of single-wall carbon molecules. *Phys Rev B: Condens Matter.* 1995;52(11):8541–8549.

[225] Nair N, Strano MS. Using the selective functionalization of metallic single-walled carbon nanotubes to control dielectrophoretic mobility. *Reply. J Phys Chem B.* 2005;109(35): 17016–17018.

[226] Krupke R, Hennrich F. Using the selective functionalization of metallic single-walled carbon nanotubes to control dielectrophoretic mobility. *Comment. J Phys Chem B.* 2005;109(35): 17014–17015.

[227] Baik S, Usrey M, Rotkina L, Strano M. Using the selective functionalization of metallic single-walled carbon nanotubes to control dielectrophoretic mobility. *J Phys Chem B.* 2004;108(40): 15560–15564.

[228] Ericson LM, Pehrsson PE. Aggregation effects on the raman spectroscopy of dielectrophoretically deposited single-walled carbon nanotubes. *J Phys Chem B.* 2005;109(43): 20276–20280.

[229] Lee DS, Kim DW, Kim HS, Lee SW, Jhang SH, Park YW, *et al.* Extraction of semiconducting CNTs by repeated dielectrophoretic filtering. *Appl Phys A.* 2004;80(1):5–8.

[230] Krupke R, Hennrich F, Kappes MM, von Loehneysen H. Surface conductance induced dielectrophoresis of semiconducting single-walled carbon nanotubes. *Nano Lett.* 2004;4(8): 1395–1399.

[231] Krupke R, Linden S, Rapp M, Hennrich F. Thin films of metallic carbon nanotubes prepared by dielectrophoresis. *Adv Mater.* 2006;18(11):1468–1470.

[232] Dimaki M, Boggild P. Dielectrophoresis of carbon nanotubes using microelectrodes: a numerical study. *Nanotechnology.* 2004;15(8):1095–1102.

[233] Kim Y, Hong S, Jung S, Strano MS, Choi J, Baik S. Dielectrophoresis of surface conductance modulated single-walled carbon nanotubes using catanionic surfactants. *J Phys Chem B.* 2006;110(4):1541–1545.

[234] Chen Z, Wu Z, Tong L, Pan H, Liu Z. Simultaneous dielectrophoretic separation and assembly of single-walled carbon nanotubes on multigap nanoelectrodes and their thermal sensing properties. *Anal Chem.* 2006;78(23):8069–8075.

[235] Mureau N, Mendoza E, Silva SRP, Hoettges KF, Hughes MP. In situ and real time determination of metallic and semiconducting single-walled carbon nanotubes in suspension via dielectrophoresis. *Appl Phys Lett*. 2006;88(24):243109/1-/3.

[236] Lutz T, Donovan KJ. Macroscopic scale separation of metallic and semiconducting nanotubes by dielectrophoresis. *Carbon*. 2005;43(12):2508–2513.

[237] Peng H, Alvarez NT, Kittrell C, Hauge RH, Schmidt HK. Dielectrophoresis field flow fractionation of single-walled carbon nanotubes. *J Am Chem Soc*. 2006;128(26):8396–8397.

[238] Blatt S. Hennrich F, von Loehneysen H, Kappes MM. Vijayaraghavan A, Krupke R. Influence of structural and dielectric anisotropy on the dielectrophoresis of single-walled carbon nanotubes. *Nano Lett*. 2007;7(7):1960–1966.

[239] Nair N, Kim W-J, Braatz RD, Strano MS. Dynamics of surfactant-suspended single-walled carbon nanotubes in a centrifugal field. *Langmuir*. 2008;24(5):1790–1795.

[240] Hennrich F, Arnold K, Lebedkin S, Quintilla A, Wenzel W, Kappes MM. Diameter sorting of carbon nanotubes by gradient centrifugation: role of endohedral water. *Physica Status Solidi B*. 2007;244(11):3896–3900.

[241] Arnold MS, Stupp SI, Hersam MC. Enrichment of single-walled carbon nanotubes by diameter in density gradients. *Nano Lett*. 2005;5(4):713–718.

[242] Arnold MS, Green AA, Hulvat JF, Stupp SI, Hersam MC. Sorting carbon nanotubes by electronic structure using density differentiation. *Nat Nanotechnol*. 2006;1(1):60–65.

[243] Crochet J, Clemens M, Hertel T. Quantum yield heterogeneities of aqueous single-wall carbon nanotube suspensions. *J Am Chem Soc*. 2007;129(26):8058–8059.

[244] Crochet J, Clemens M, Hertel T. Optical properties of structurally sorted single-wall carbon nanotube ensembles. *Phys Stat Sol (B)*. 2007;244(11):3964–3968.

[245] Miyata Y, Yanagi K, Maniwa Y, Kataura H. Highly stabilized conductivity of metallic single wall carbon nanotube thin films. *J Phys Chem C*. 2008;112(10):3591–3596.

[246] Wei L, Wang B, Goh TH, Li L-J, Yang Y, Chan-Park MB, *et al.* Selective enrichment of (6,5) and (8,3) single-walled carbon nanotubes via cosurfactant extraction from narrow (n,m) distribution samples. *J Phys Chem B*. 2008;112(10):2771–2774.

[247] Yanagi K, Miyata Y, Kataura H. Optical and conductive characteristics of metallic single-wall carbon nanotubes with three basic colors; cyan, magenta, and yellow. *Applied Physics Express*. 2008;1(3):034003/1-/3.

[248] Yanagi K, Iitsuka T, Fujii S, Kataura H. Separations of metallic and semiconducting carbon nanotubes by using sucrose as a gradient medium. *J Phys Chem C*. 2008;112(48): 18889–18894.

[249] Sato Y, Yanagi K, Miyata Y, Suenaga K, Kataura H, Iijima S. Chiral-angle distribution for separated single-walled carbon nanotubes. *Nano Lett*. 2008;8(10):3151–3154.

[250] Miyata Y, Yanagi K, Maniwa Y, Kataura H. Optical properties of metallic and semiconducting single-wall carbon nanotubes. *Phys Stat Sol (B)*. 2008;245(10):2233–2238.

[251] Niyogi S, Densmore CG, Doorn SK. Electrolyte tuning of surfactant interfacial behavior for enhanced density-based separations of single-walled carbon nanotubes. *J Am Chem Soc*. 2009.

[252] Lee CW, Weng C-H, Wei L, Chen Y, Chan-Park MB, Tsai C-H, *et al.* Toward high-performance solution-processed carbon nanotube network transistors by removing nanotube bundles. *J Phys Chem C*. 2008;112(32):12089–12091.

[253] Green AA, Hersam MC. Colored semitransparent conductive coatings consisting of monodisperse metallic single-walled carbon nanotubes. *Nano Lett*. 2008;8(5):1417–1422.

[254] Miyata Y, Yanagi K, Maniwa Y, Kataura H. Optical evaluation of the metal-to-semiconductor ratio of single-wall carbon nanotubes. *J Phys Chem C*. 2008;112(34):13187–13191.

[255] Engel M, Small JP, Steiner M, Freitag M, Green AA, Hersam MC, *et al.* Thin film nanotube transistors based on self-assembled, aligned, semiconducting carbon nanotube arrays. *ACS Nano*. 2008;2(12):2445–2452.

[256] Backes C, Schmidt C, Hauke F, Hirsch A. Fractioning HiPco and CoMoCAT SWCNTs via density gradient ultracentrifugation by the aid of a novel perylene bisimide derivative surfactant. *Chem Commun*. 2009; 10.1039/b818141a.

[257] Kim W-J, Nair N, Lee CY, Strano MS. Covalent functionalization of single-walled carbon nanotubes alters their densities allowing electronic and other types of separation. *J Phys Chem C*. 2008;112(19):7326–7331.
[258] Fagan JA, Becker ML, Chun J, Hobbie EK. Length fractionation of carbon nanotubes using centrifugation. *Adv Mater*. 2008;20(9):1609–1613.
[259] Iijima S. Helical microtubules of graphitic carbon. *Nature*. 1991;354(6348):56–58.

2

Supramolecular Assembly of Fullerenes and Carbon Nanotubes Hybrids

Mª Ángeles Herranz, Beatriz M. Illescas, Emilio M. Pérez and Nazario Martín

Departamento de Química Orgánica, Facultad de Ciencias Químicas, Universidad Complutense, Madrid, Spain
IMDEA-Nanociencia, Facultad de Ciencias Módulo C-IX. Ciudad Universitaria de Cantoblanco, Madrid, Spain

2.1 Introduction

A major goal in the field of *Supramolecular Chemistry* is the search for new artificial photosynthetic models, in which biomimetic principles can be used to construct different molecular electronic devices [1]. Among the nanomaterials that have exerted a profound impact on the preparation of these architectures, fullerenes and carbon nanotubes (CNT) stand out owing to their extraordinary physicochemical properties [2].

The development of reliable and reproducible methodologies to integrate fullerenes and CNTs into functional structures such as donor acceptor hybrids, able to transform sunlight into electrical or chemical energy, has emerged as an area of intense research. To this end, we have extensively explored the combination of tetrathiafulvalene (TTF) [3]

Chemistry of Nanocarbons Edited by Takeshi Akasaka, Fred Wudl and Shigeru Nagase

and its π–extended analogues (exTTFs) – in which the 1,3 dithiole rings are covalently connected to a π-conjugated core – to construct photo- and electroactive donor-acceptor dyads and triads with C_{60} [4]. TTF is well known for its electron donor ability, which has been exploited besides in the preparation of the donor-acceptor nanohybrids mentioned above, in a breadth of molecular devices, including organic field-effect transistors [5], cation sensors and bistable molecular shuttles and catenanes [6]. Analogously, exTTFs have mainly been exploited as electron donor fragments. In contrast, the possibility of furthermore utilizing their distorted curved shape in the construction of molecular receptors for fullerene had by and large been overlooked. The main focus of this chapter is to present recent progress toward the preparation of supramolecular TTF or exTTF ensembles with fullerenes and carbon nanotubes particularly concentrating on this aspect. Guided by the structures based on TTF or exTTF that have been combined with fullerenes or CNTs, we will summarize the recent work carried out by ourselves and others under the three following subtopics: (i) hydrogen bonded C_{60}•Donor ensembles, (ii) concave exTTF derivatives as recognizing motifs for fullerenes, and (iii) noncovalent functionalization of carbon nanotubes.

2.2 Hydrogen Bonded C_{60}•Donor Ensembles

Hydrogen bonds, with binding energies ranging between 4 and 120 KJ·mol^{-1}, constitute a versatile supramolecular methodology, mainly because of their high degree of specificity and directionality [7]. Although one single bond is characteristically too weak to guarantee stable architectures, the use of multiple H-bonds can overcome this drawback. Alternatively, the combination of H-bonds with additional supramolecular interactions, such as hydrophobic or electrostatic forces, can get to high values of binding constants. Whereas association constants of $\sim$10 M^{-1} are obtained for a simple D•A array built on one H-donor and one H-acceptor, triple H-bonding motifs (i.e. DAD, etc.) can reach K_a values as large as $\sim$10^2–10^3 M^{-1} (Figure 2.1) [8]. Even higher association constants of $\sim$10^5 M^{-1} are achieved in self-complementary quadruple H-bonding motifs (Figure 2.1d) [9]. If, remarkably, the presence of multiple H-bonds combines with attractive secondary interactions, high K_a values can be accomplished. This is the case for 2-ureido-4-pyrimidinones (UP), with K_a values $\sim$10^7 M^{-1} (Figure 2.1e) [10].

By using the binding motif represented in Figure 2.1a, in collaboration with Mendoza's group, we have reported a series of H-bonded C_{60}•TTF ensembles (**1a–d**) [11]. In these cases, the combination of complementary DD•AA H-bonds with electrostatic interactions through guanidinium and carboxylate ion pairs holds the fullerene and TTF units together (Figure 2.2). Two different spacers of different lengths (i.e. phenyl *vs.* biphenyl) as well as two functional groups (i.e. ester *vs.* amide) have been used in order to modulate the molecular architectures.

In these supramolecular dyads, the flexible nature of the spacer results in through-space electron transfer processes. The lifetime obtained for the radical ion pair states, i.e. $C_{60}^{\bullet-}$•$TTF^{\bullet+}$, are in the range of hundred of nanoseconds, thus being several orders of magnitude higher than those reported for covalently linked C_{60}-TTF dyads [4].

A set of noncovalently associated C_{60}-porphyrin ensembles (**2•3**) was prepared by using a two-point amidinium-carboxylate binding motif, which is again particularly

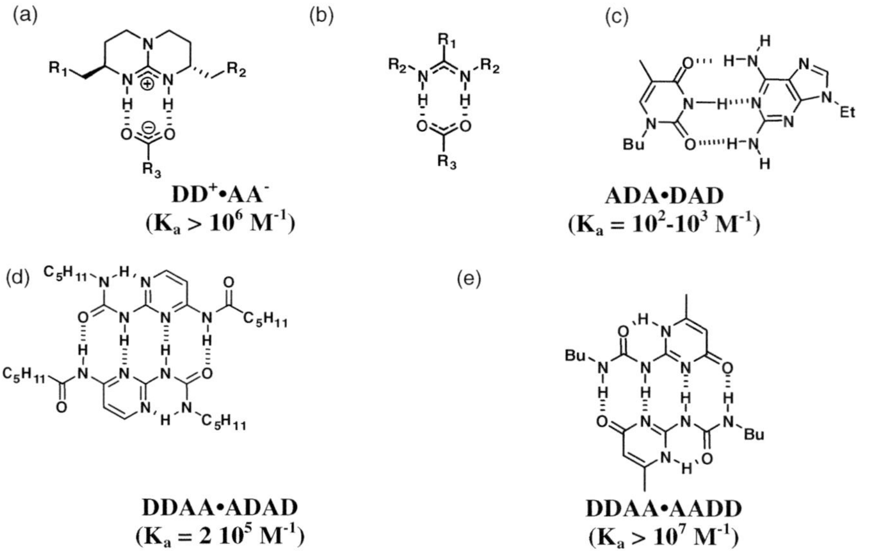

Figure 2.1 *H-bonding motifs with high K_a values [measured in toluene/DMSO 99/1 for (a) and (b) and in $CHCl_3$ for (c)–(e)]*

stable as a result of the synergy of hydrogen bonds and electrostatic interactions (Figure 2.3) [12]. This amidinium-carboxylate ion pairing diminishes other possible bonding modes, thus favouring the linearity of the donor-acceptor pair and ensuring an optimal pathway for the motion of the charges. The association constants for these pairs were evaluated by fluorescence spectroscopy and reach values up to 10^7

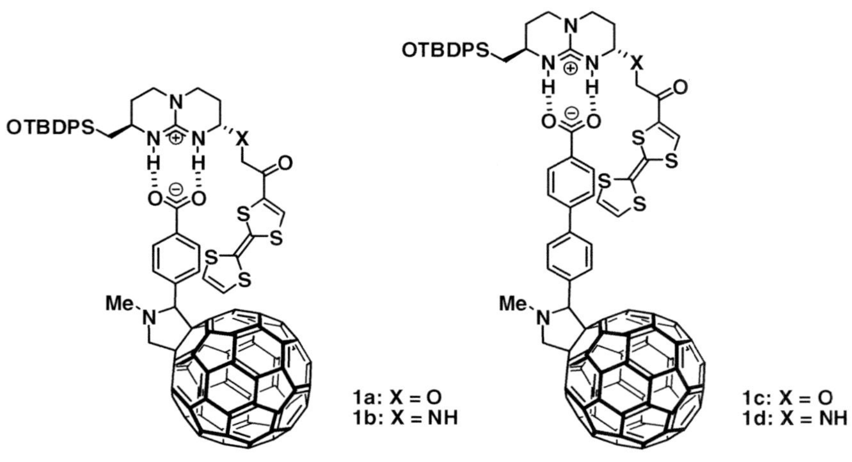

Figure 2.2 *Structure of H-bonded C_{60}•TTF dyads*

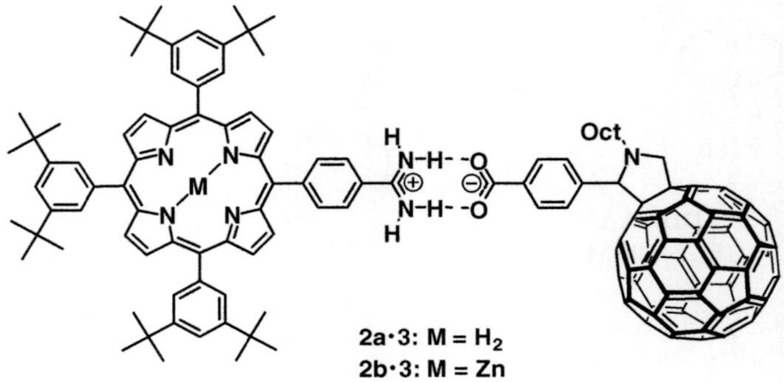

Figure 2.3 *Amidinium-carboxylate interfaced porphyrin-C_{60} ensembles*

in toluene or 10^5 in THF. This strong binding gives rise to an exceptionally strong electronic coupling between both electroactive elements ($36\,cm^{-1}$ for **2b•3**), which in turn facilitates the formation of long-lived $C_{60}^{\bullet -}$•$P^{\bullet +}$ radical pairs with a lifetime of ~1 μs in THF.

It is well established that ammonium-crown ether interaction is relatively weak, with a maximun strength of $\sim 10^3\,M^{-1}$ [13]. However, the introduction of additional recognition elements can increase dramatically the stability of the complexes. Thus, for example, an unusual high value of the association constant was observed for the complex formed by a porphyrin-crown ether receptor and a C_{60}-based ammonium host [14]. This additional stabilization was attributed to the intramolecular π–stacking of the two chromophores and gave rise to a K_a value of $3.75 \cdot 10^5\,M^{-1}$, two orders of magnitude higher than those reported for other complexes based on ammonium-crown ether noncovalent interactions [15]. In this sense, the concave aromatic surface of exTTF introduces an additional recognizing motif which should favor the self-assembly of complemantary ammonium salt-crown ether interaction.

Our first approach to this subject was to synthesize a supramolecular triad in which a crown ether receptor endowed with two exTTF moieties (**4**) formed a pseudo-rotaxane with a fullerene-based secondary ammonium salt (**4•5**, Figure 2.4) [16]. The complexation between this macrocycle and the fullerene host was investigated by ^1H NMR (nuclear magnetic resonance) binding titration in $CDCl_3/CD_3CN$, and a $K_a \sim 50\,M^{-1}$ was obtained. In complementary work, the fullerene-macrocycle complexation was tested in fluorescence experiments. The decrease in intensity of the fullerene fluorescence upon addition of increasing amounts of macrocycle, allowed to determine a K_a value of $55 \pm 5\,M^{-1}$, in excellent agreement with the data based on NMR experiments.

Interestingly, macrocycle **4** shows two closely-spaced two-electron oxidation waves, probably due to the flexibility of the polyether chains, which originates weak interactions between the two exTTF donor moieties.

The weak interaction observed in this case is explained by the big and hindered cavity of the crown ether **4**. Therefore, in a next step we decided to synthesize a series of exTTF-based

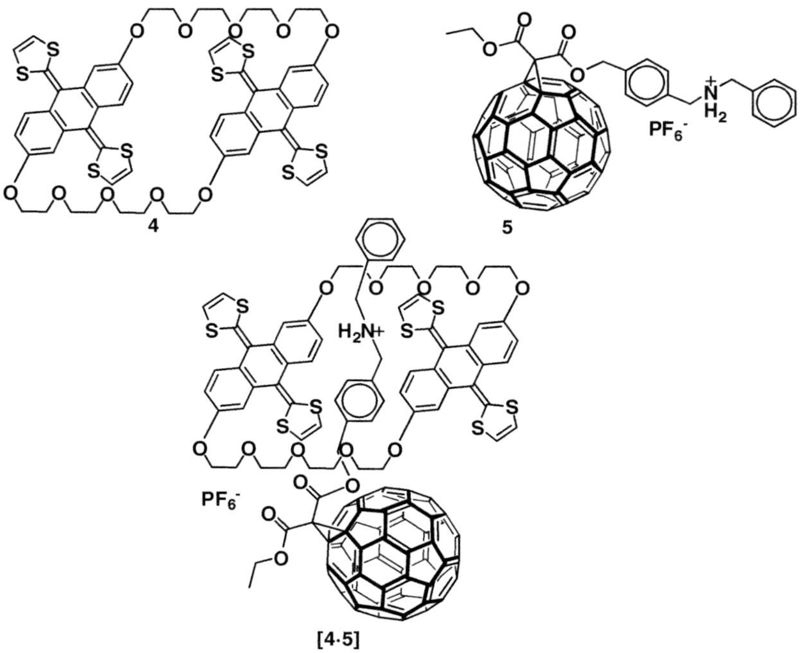

Figure 2.4 *Chemical structure of macrocycle **4**, ammonium salt **5** and its supramolecular complex*

secondary ammonium salts and study their self-assembly with fullerene C_{60} endowed with a DB24C8 crown ether appendage.

For the design of these salts we considered different aspects: (i) a rigid or a flexible spacer between the exTTF and the ammonium group; (ii) one or two recognition sites and (iii) different donor ability of the exTTF moiety. Bearing these considerations in mind, compounds **6–9** were prepared, and complexation experiments with DB24C8-fullerene derivative **10** were carried out by ^{1}H NMR titration and fluorescence studies (Figure 2.5) [17].

K_a values from $8.6 \times 10^2\,M^{-1}$ (for **[9·10]**) to $1.4 \times 10^4\,M^{-1}$ (for **[7·10$_2$]**) were obtained for the different supramolecular assemblies. The enhancement of the constant in **[7·10$_2$]** can be considered mainly associated to the presence of two binding sites. The low variation observed for the K_a values, together with the finding that complexation does not strongly influence the redox properties of the components, as shown by cyclic voltammetry studies, suggest that the electroactive units are not spatially close enough to allow measurable electronic interactions between them.

The length and flexibility of the spacers between the complementary ammonium-crown ether bonding motifs must be crucial to allow the intramolecular interaction between the fullerene sphere and the π concave surface of exTTF. Therefore, we prepared a new exTTF-crown ether derivative, **11**, and studied its supramolecular

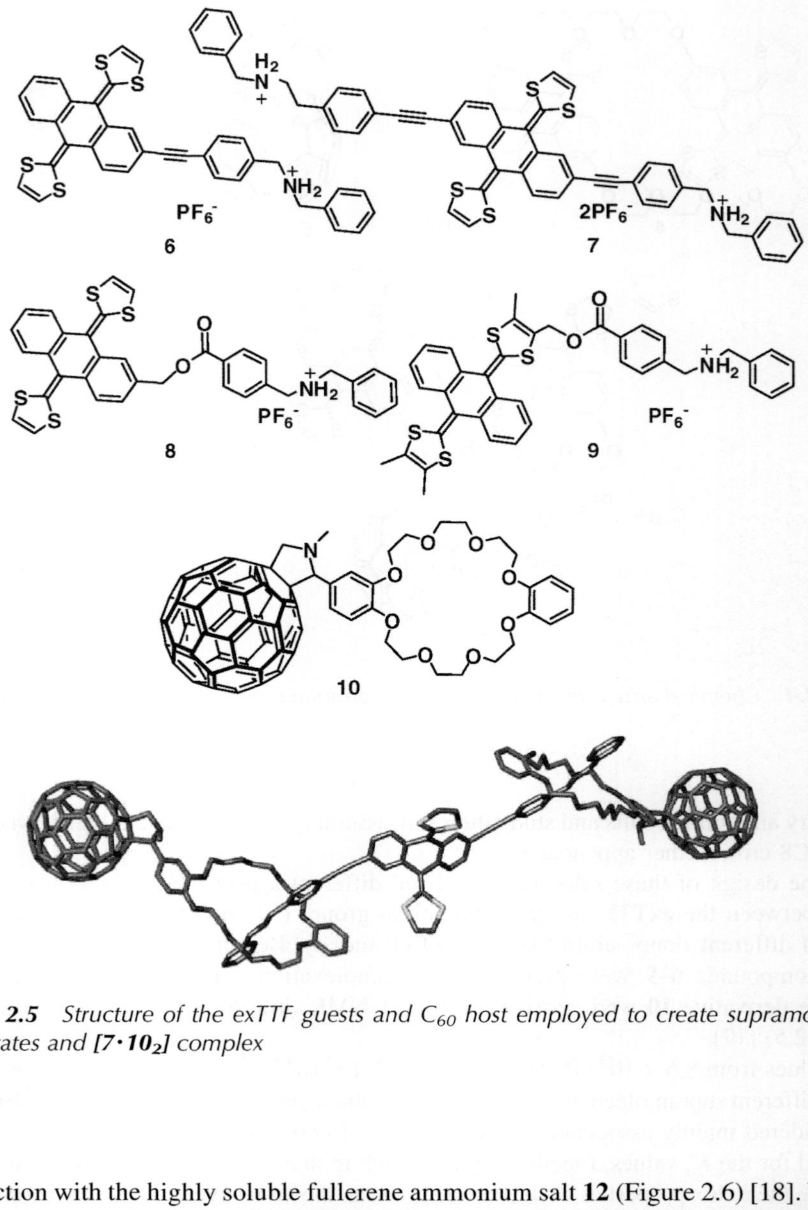

Figure 2.5 *Structure of the exTTF guests and C_{60} host employed to create supramolecular conjugates and [7·**10**$_2$] complex*

interaction with the highly soluble fullerene ammonium salt **12** (Figure 2.6) [18]. UV-vis and fluorescence titrations evidenced the formation of the supramolecular complex and a binding constant of $(1.58 \pm 0.82) \times 10^6\,M^{-1}$ in chlorobenzene was obtained. Upon complexation, an anodic shift of ~100 mV was now observed for the oxidation potentials of exTTF by cyclic voltammetry, thus accounting for the high K_a value obtained. Time resolved transient absorption spectroscopy experiments revealed the photoinduced generation of a charge separated state with a short lifetime (9.3 ps in chlorobenzene).

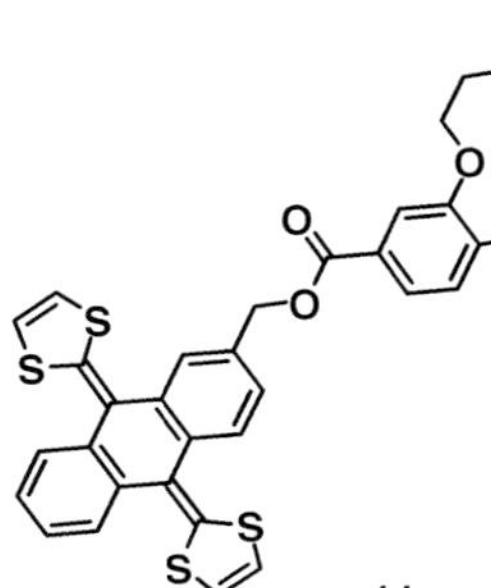

11

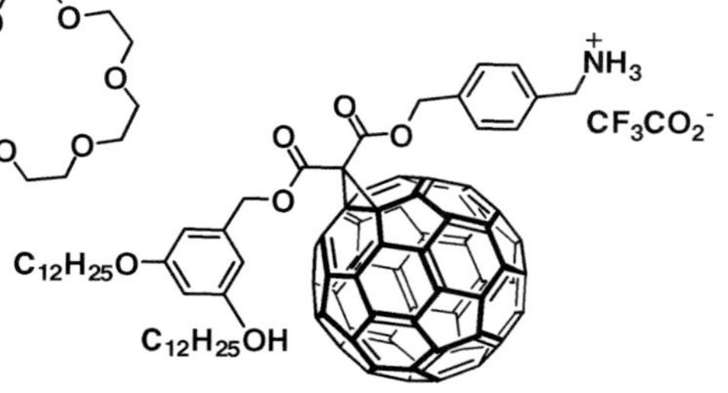

12

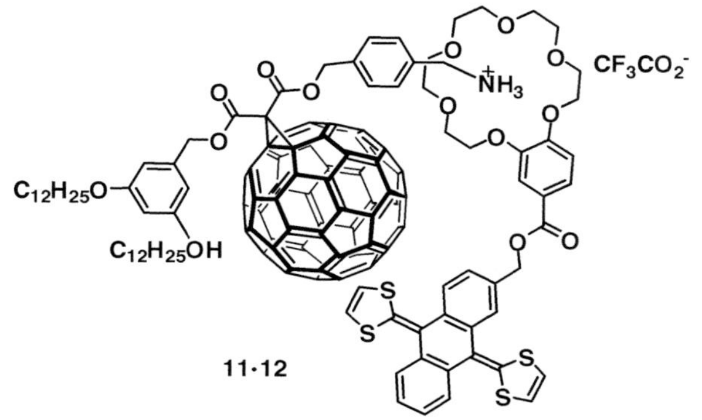

11·12

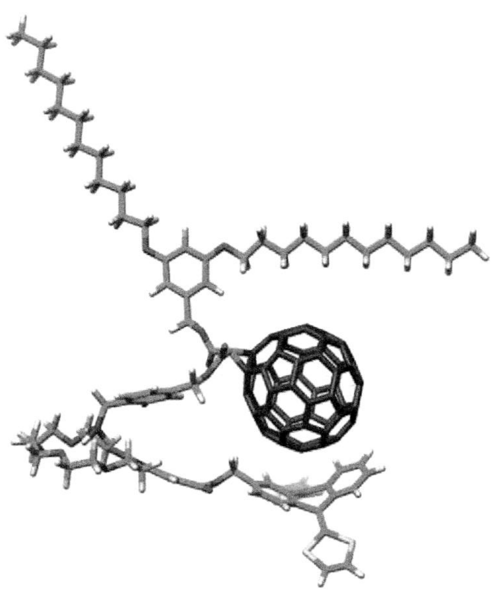

[11· 12]

Figure 2.6 *exTTF-crown ether **11**, ammonium salt **12** and molecular model showing its supramolecular complex*

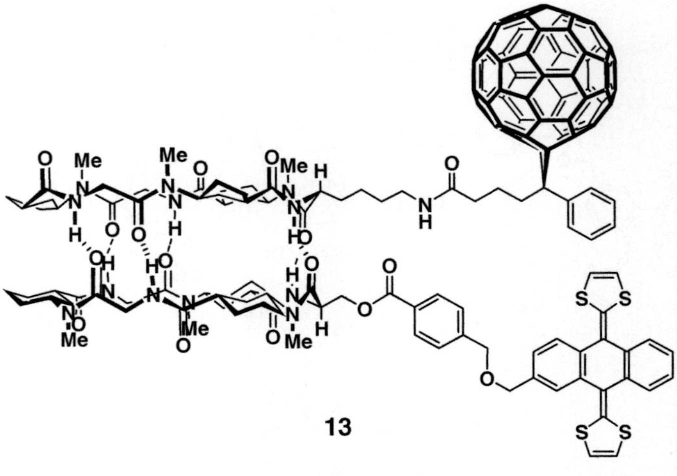

13

Figure 2.7 *Structure of the bioinspired cyclopeptidic heterodimer* ***13***

Therefore, the cooperativity between π-π and H-bonding interactions generates a highly stable supramolecular electron donor acceptor hybrid facilitated by the close proximity of the exTTF unit to the fullerene core, which allows the noncovalent interaction between the benzene concave rings of the exTTF unit and the fullerene convex sphere.

Similar cooperative forces were explored in the preparation of the bioinspired cyclopeptidic heterodimers **13** built on β-sheet-like hydrogen-bonding networks (Figure 2.7). The equilibrium mixture of the three **13** species obtained -differing in the relative positions of their C_{60} and exTTF moieties- exhibits a remarkable association constant, of at least $10^6\,M^{-1}$, as extracted from fluorescence spectroscopy [19]. In addition, steady-state and time-resolved spectroscopies evidenced an electron transfer process from the exTTF to the photoexcited C_{60} that results in the generation of a radical ion pair state stabilized for up to 1 μs before recombining to the ground state. The structure of **13** in principle allows its extension to form a nanotubular self-organized material for electronic and photonic applications, while also serving as a valuable artificial model for natural photosynthetic reaction centers.

2.3 Concave exTTF Derivatives as Recognizing Motifs for Fullerene

Following the example mentioned above, we noticed that the shape complementarity between the concave aromatic face of exTTF and the convex exterior of fullerenes should lead to large and positive noncovalent interactions (see Figure 2.8). In fact, theoretical calculations (DFT) predicted binding energies up to 7.00 kcal mol^{-1} between a single unit of exTTF and C_{60} [20]. However, no experimental evidence of association was found in either UV-vis or NMR titrations. We thus turned our view to a tweezer-like design, in which two exTTF units would serve as recognizing units, and an isophthalate [21] or a terephthalate [22]

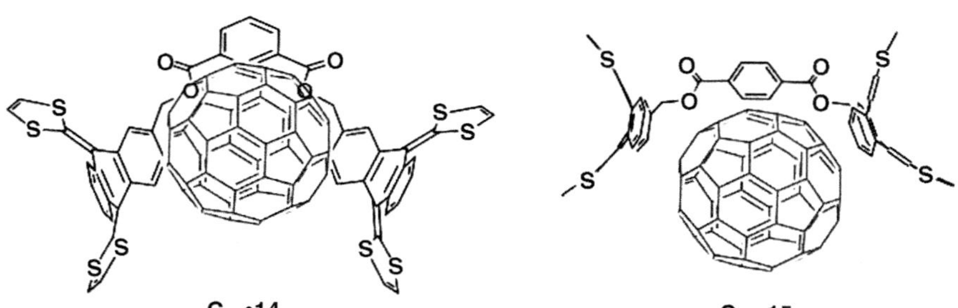

C60•14 **C60•15**

Figure 2.8 *exTTF-based receptors **14** and **15** forming pincer-like complexes with C_{60}*

diester would act as a spacer. Receptors **14** and **15** (Figure 2.8) were synthesized in excellent yields from easily available exTTF methyl alcohol and commercially available isophthaloyl or terephthaloyl dichloride. We were pleased to observe that the electronic absorption spectra of receptor **14** showed significant changes upon addition of fullerene. A decrease in the absorption band characteristic of exTTF ($\lambda_{max} = 434$ nm) is accompanied by the appearance of what seems to be a charge-transfer band ($\lambda_{max} = 482$ nm). Fitting of these spectral changes to a 1 : 1 binding isotherm afforded a binding constant of $3.0 \times 10^3\ M^{-1}$ in chlorobenzene at room temperature ($2.9 \times 10^4\ M^{-1}$ according to fluorescence titrations). Similar binding constants were obtained for the case of receptor **15**. The considerable stability of the **14/15**·C_{60} complexes, given the lack of preorganization of the receptors, demonstrates the validity of exTTF as a building block for fullerene receptors. We were surprised to find that the complexation behavior of receptor **14** towards C_{60} in $CHCl_3/CS_2$ mixtures was rather different. Although the spectral changes are analogous to those found in chlorobenzene, the binding isotherm turned out to be sigmoidal in shape. This is generally regarded as indicative of cooperative binding events. Indeed, the binding isotherm fitted very well to the Hill equation, to yield a Hill coefficient of 2.7 ± 0.3 and an apparent binding constant of $3.6 \times 10^3\ M^{-1}$. Although it is often considered a direct indication of the number of available binding sites on the receptor, the Hill coefficient is best thought of as an interaction coefficient reflecting the extent of cooperativity, with a maximum value equal to the number of binding sites [23]. Thus, a value of $n_H > 2$ rules out the formation of the expected pincerlike **14**·C_{60} complex since it features two binding sites only. As 1 : 1 stoichiometry was experimentally found by continuous variation plots, this strongly suggests the formation of a supramolecular tetramer in which two units of C_{60} are sandwiched between two molecules of receptor **14** [24].

The most significant feature of these **14/15**·C_{60} complexes is the unique combination of supramolecular and electronic complementarity. In fact, in our group we had previously shown that photoinduced electron transfer (PET) from the electron donor exTTF to the acceptor fullerene readily took place in a variety of covalently linked dyads [4]. Considering that the ultimate goal of this novel recognition motif should be its utilization to organize electroactive materials in the solid state, we decided to investigate if through space intracomplex PET was possible in these associates. Indeed, upon excitation of the charge transfer band at 484 nm of mixtures of either **14** or **15** with C_{60} in benzonitrile, in transient absorption measurements the spectral features clearly reveal the immediate (i.e. $> 10^{12}\ s^{-1}$)

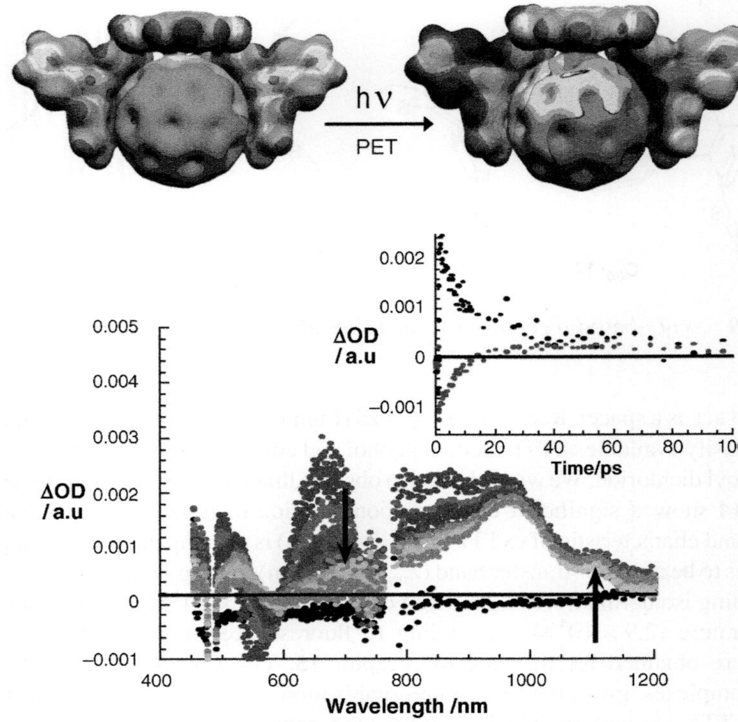

Figure 2.9 Top: B3LYP/6-31G** electrostatic potential maps calculated for MTW•C_{60} in the ground electronic state and in the photoinduced charge-separated state. Bottom: differential absorption spectra (visible and near-infrared) obtained upon femtosecond flash photolysis (484 nm) of a MTW•C_{60} mixture (MTW: 2.5×10^{-5}; C_{60}: 2.5×10^{-3}) in benzonitrile. Inset – time-absorption profile at 670 nm (open circles) and 580 nm (filled circles), reflecting the charge separation and charge recombination dynamics

formation of a fully $C_{60}^{\bullet -}$/*ex*TTF$^{\bullet +}$ charge-separated state (see Figure 2.9). In particular, we observed a transient centered at 668 nm to the one-electron oxidized radical cation of the exTTF of the receptor and the radical anion of C_{60}, which shows up in the near-infrared, at approximately 1100 nm. The charge separated state lifetimes, as determined from the 668 and 1100 nm decays, were found to be very short, for instance, in benzonitrile we determined lifetimes of 12.7 ps for **14** and 9.6 ps for **15**.

The relatively high association constant of receptors **14** and **15** towards C_{60} despite their inherent lack of preorganization, got us interested in the specific contribution to the overall stabilization of the complex arising from the concave shape of the recognizing unit. In this regard, the group of Kawase had coined the term 'concave–convex interactions' to refer to the increase in noncovalent interactions between curved aromatic hosts and guests, and suggested these might play a distinct role in the stabilization of the complexes [25, 26]. In order to get an insight into whether these concave–convex interactions did really contribute to stabilize our complexes, and if so, to what extent, we designed and synthesized

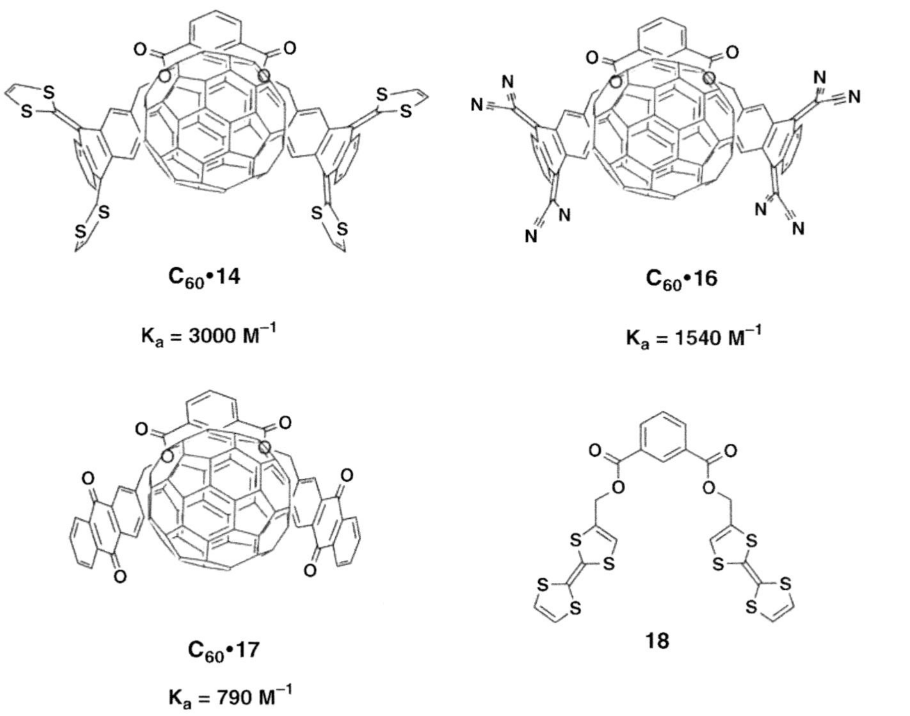

Figure 2.10 *Structure of the C_{60}·**14**, C_{60}·**16** y C_{60}·**17** complexes, with their corresponding binding constants determined by ^{1}H NMR titrations and chemical structure of receptor **18***

a collection of structurally-related receptors **16-18** (Figure 2.10) [27]. These, together with **14**, provided a full collection of receptors in which the size, shape and the electronic character of the recognizing motifs were selectively modified. The binding constants of all receptors towards C_{60} were investigated through ^{1}H NMR titrations in $CDCl_3$.

Receptor **14** featured five aromatic rings –two per recognizing unit plus the isophthalic spacer–, a large and concave van der Waals surface and an electron-rich character. Unsurprisingly, **14** is the strongest binder for C_{60}, with a $K_a = (3.00 \pm 0.12) \times 10^3\,M^{-1}$. Receptor **16** utilizes 11,11,12,12-tetracyano-9,10-anthraquinodimethane (TCAQ) as the recognizing element. Thus, as compared to **14**, it presents equal number of aromatic rings and surface available for recognition, with close to identical curvature, but electron-poor character. The change in electronic nature results in a decrease of K_a to $(1.54 \pm 0.15) \times 10^3\,M^{-1}$. A similar drop-off in the association constant is observed when moving from **16** to **17**. In this case, the surface available for van der Waals interactions is similar to that of **14** and **16**, but **17** lacks both the concave-convex and the electronic complementarity. This results in a binding constant of $(0.79 \pm 0.05) \times 10^3\,M^{-1}$. Finally, no sign of association with C_{60} was observed in either the ^{1}H NMR or the electronic absorption spectra of receptor **18**, which is decorated with the electron rich, small and nonaromatic tetrathiafulvalene (TTF) unit.

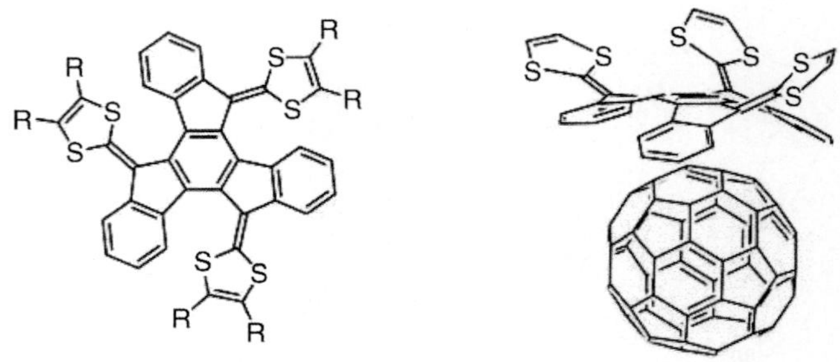

19a: R = H
19b: R = SMe
19c: R = $(SCH_2)_2$

C_{60}•**19a**

Figure 2.11 Chemical structure of truxTTFs **19a-c** and of the **19a**·C_{60} complex

Comparison of the binding constants of **14** and **16** towards C_{60} suggested an important contribution of coulombic interactions, in agreement with preceding reports. However, the fact that **18** did not show any sign of complexation towards C_{60} denoted that this contribution is not quantitatively comparable to those of π–π and van der Waals forces. Remarkably, we observed for the first time that concave–convex complementarity does play its own role, as illustrated by the cases of receptors **16** and **17**. In spite of the more electron-poor character of **16** when compared to **17**, its binding constant towards C_{60} is larger, which is necessarily related to concave shape of the TCAQ recognizing units.

An alternative strategy to the molecular tweezer is the design of a donor molecule bearing more than two dithioles and two benzene rings. With this in mind, we designed and synthesized a new family of TTF derivatives, truxene-TTFs, **19a-c** [28]. As shown in Figure 2.11, truxene-TTFs feature three 1,3-dithiole rings connected to a truxene core. To accommodate the dithioles, the truxene moiety breaks down its planar structure and adopts an all-*cis* sphere-like geometry with the three dithiole rings protruding outside (Figure 2.11). The concave shape adopted by the truxene core perfectly mirrors the convex surface of fullerenes, indicating that van der Waals and concave-convex π–π interactions between them should be maximized. Indeed, the association of trux-TTF and fullerenes in solution was investigated by ^{1}H NMR titrations with C_{60} and C_{70} as guests affording binding constants of $(1.2 \pm 0.3) \times 10^3\ M^{-1}$ and $(8.0 \pm 1.5) \times 10^3\ M^{-1}$ for C_{60} and C_{70} in $CDCl_3/CS_2$, respectively. DFT (MPWB1K/6-31G** level) calculations provided satisfactory explanation for this difference in binding constant, which arises from the increase in surface from C_{60} to C_{70}.

The combination of supramolecular and electronic reciprocity between exTTF and C_{60} suggested that this novel host-guest system would be a good candidate to be utilized in the self-organization of electroactive materials. With this in mind, we designed **20** and **21** (Figure 2.12) as monomers for the construction of redox-amphoteric supramolecular polymers [29] and dendrimers [30] through π–π and van der Waals interactions.

In fact, a systematic collection of experiments, including variable concentration and VT-NMR, PFG-NMR, MALDI-TOF-MS, Dynamic Light Scattering, and AFM demonstrated that **20** forms linear multimeric supramolecular aggregates, while **21** forms arborescent and dynamically polydisperse supramolecular aggregates, both in solution, gas and solid phase.

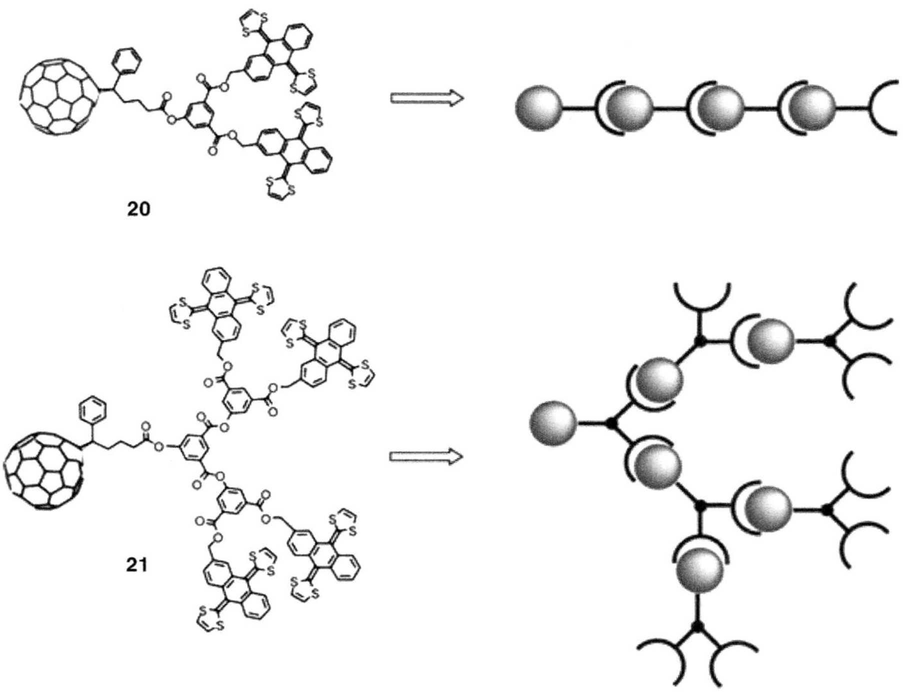

Figure 2.12 *Structures of the supramolecular monomers **20** and **21** and schematic representation of their self-association to form linear and dendritic supramolecular architectures, respectively*

The electronic characterization of $(\mathbf{20})_n$ and $(\mathbf{21})_n$ was carried out by means of Cyclic Voltammetry (CV), Differential Pulse Voltammetry (DPV) and UV-vis, and showed that there is electronic communication between electroactive fragments in the ground state.

Besides this noncovalent assemblies, we decided to investigate the possibility of synthesizing large covalently linked dendrimes decorated on their periphery with multiple units of receptor **14** [31]. We were able to synthesize dendrimers from 2nd up to 4th generation (Figure 2.13). We were glad to observe that several units of C_{60} were associated by the exTTF rich exterior. Furthermore, UV-vis titration experiments demonstrated the complexation of C_{60} to occur in a positive cooperative manner [31].

Apparently, these new systems are of interest for the construction of optoelectronic devices in which donor and acceptors are ensembled supramolecularly in an organized manner.

2.4 Noncovalent Functionalization of Carbon Nanotubes

The noncovalent modification of carbon nanotubes (CNTs) with different donor units has also been explored in the preparation of multifunctional hybrids for a wide range of applications [32].

CNTs are characterized by outstanding and unprecedented electronic and mechanical properties: they represent the ultimate carbon fiber, with the highest thermal conductivity [33]

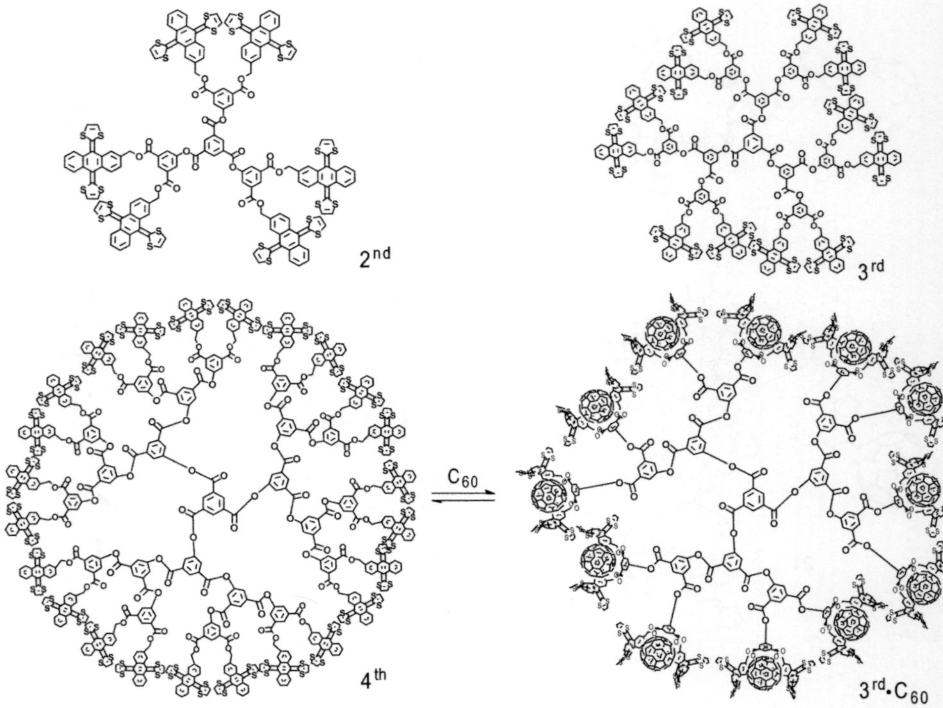

Figure 2.13 *Schematic representation of the chemical structure of the 2nd to 4th generation of dendrimers obtained from receptor* ***14*** *and of the idealized structure of 4th generation dendrimer associating* C_{60}

and the highest tensile strength of any material [34]. In particular, single-walled carbon nanotubes (SWCNTs) have Young's moduli of around 1 TPa and are thus up to 100 times stronger than steel. However, the insolubility [35] of SWCNTs in most organic solvents and the difficulties of handling these carbon nanostructures [36] has restricted their applications to a considerable extent. To improve upon the solubility of CNTs, the controlled defect and sidewall functionalization has been pursued in the past years and demonstrated that the formation of covalent linkages can drastically enhance the solubility of these species in various solvents at the same time that guarantees their structural integrity [32, 37]. However, it also alters the intrinsic physical properties of CNTs because of a modification of the sp^2-carbon framework [32]. An alternative strategy for the preservation of the intrinsic electronic and mechanical properties of CNTs consists in the noncovalent modification of CNTs [38]. In this approach, hydrophobic, van der Waals and electrostatic forces are primarily involved and require the physical adsorption of suitable molecules onto the sidewalls of the CNTs. Noncovalent functionalization has been achieved by polymer wrapping [37k], adsorption of surfactants or small aromatic molecules [39], and interaction with porphyrins [40] or biomolecules such as DNA and peptides [41]. A special case is the endohedral functionalization of CNTs by filling their inner surface with atoms or small molecules (peapods) [42].

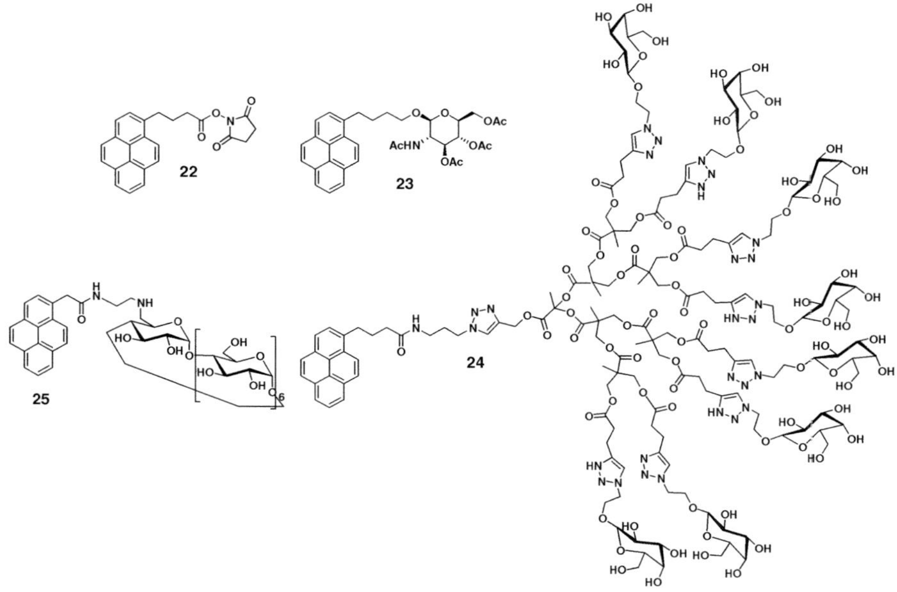

Figure 2.14 *Water-soluble pyrene derivatives used for the formation of different aggregates with CNTs*

The anchoring of small aromatic molecules, in particular pyrene derivatives, to the sidewalls of CNTs by means of π–π stacking interactions, has resulted specially useful in the noncovalent modification of CNTs. Dai *et al.* [43] reported a general and attractive approach to the supramolecular functionalization of SWCNTs sidewalls and the subsequent immobilization of biological molecules onto SWCNTs with a high degree of control and specificity. They found that the bifunctional molecule, *N*-succinimidyl-1-pyrenebutanoate **22** (Figure 2.14), is adsorbed irreversibly onto the hydrophobic surface of SWCNTs in either DMF or MeOH. The anchored *N*-succinimidyl-1-pyrenebutanoate molecules on the surface of the SWCNTs are highly resistant to desorption in aqueous solution, which has lead to a further functionalization of SWCNTs with succinimidyl ester groups that are reactive to nucleophilic substitution by primary and secondary amines of some proteins, such as ferritin, streptavidin, and biotinyl-3,6-dioxaoctanediamine.

The combination of pyrenes with bioactive monosaccharides, such as *N*-acetyl-D-glucosamine (GlcNAc) (**23**), served to prepare glycosylated CNTs that are able to biocompatibly interface with living cells and detect the dynamic secretion of biomolecules of them [44]. Pyrene-based glycodendrimers (**24**) have also been prepared considering this approach, and demonstrated to function as homogeneous bioactive coatings for SWCNTs that also mitigate their citotoxicity [45]. In recent examples, Stoddart *et al.* [46] have fabricated pyrenecyclodextrine-decorated SWCNT field effect transistor (FET) devices considering the pyrene-modified β-cyclodextrin derivative **25**. In the presence of certain organic molecules, the transistor characteristics of the pyrene cyclodextrin-decorated SWCNT/FET device shift toward negative gate voltage due to the molecular recognition by the cyclodextrin torus. When a ruthenium complex with an adamantyl tether is used as the

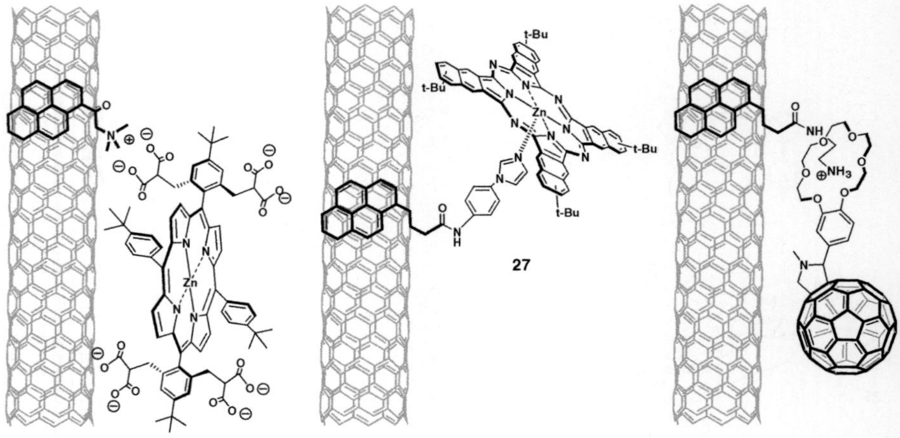

Figure 2.15 *Donor-acceptor systems prepared by considering pyrene-π–π interactions followed by complementary electrostatics (**26**), axial coordination (**27**), or crown ether-alkyl ammonium ion interactions (**28**)*

sensing guest, the SWCNT/FET device can indeed serve as a tuneable photosensor to detect luminescent molecules [46b].

The noncovalent association of SWCNTs with pyrene and different electron-donor derivatives leads to novel electron donor-acceptor nanohybrids, which, upon photoexcitation, undergo fast electron transfer, followed by the generation of long-lived charge-separated species [32]. In particular, pyrenes bearing positive or negative charges [47], nitrogenated bases [48], or alkyl ammonium ions [49], through π–π interactions followed by assembling the electron/energy donor molecules by complementary electrostatics, axial coordination, or crown ether-alkyl ammonium ion interactions, respectively, has resulted in stable donor-acceptor systems with maximum preservation of the electronic and mechanical properties of CNTs (Figure 2.15).

With the CNT surface covered with positively or negatively charged ionic head groups, van der Waals and electrostatic interactions were utilized to complex oppositely charged electron donors. Water soluble porphyrins (i.e. octapyridinium ZnP/H_2P salts or octacarboxylate ZnP/H_2P salts) have been used to form SWCNT-(π–π interaction)-pyrene$^+$-electrostatic-ZnP/H_2P, **26** or SWCNT-(π–π interaction)-pyrene$^-$-electrostatic-ZnP/H_2P electron-donor acceptor nanohybrids [47]. Photoexcitation of all the resulting nanohybrids with visible light revealed the formation of long-lived radical ion pairs, with lifetimes in the range of microseconds. The better delocalization of electrons in MWCNTs enhanced the stability of the radical ion pairs formed ($5.8 \pm 0.2\,\mu s$) when compared to the analogous SWCNT systems ($0.4 \pm 0.05\,\mu s$).

The imidazole ligand of soluble imidazol-pyrene-SWCNTs aggregates, served to anchor donor entities to the SWCNTS surface, such as zinc tetraphenylporphyrin (ZnP) and zinc naphthalocyanine (ZnNc) (**27**), by axial coordination of the nitrogen to the metallic center of the macrocycle. Utilization of ZnNc as the electron donor in these supramolecular structures

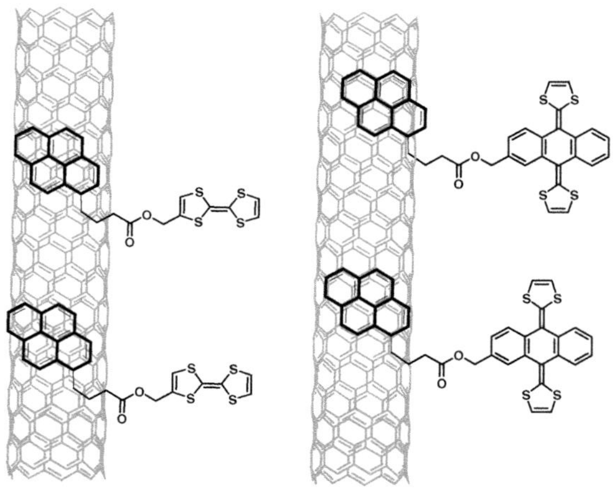

Figure 2.16 SWCNT•pyrene-TTF (**29**) and SWCNT•pyrene-exTTF (**30**) hybrids

enabled to observe the donor cation radical ($ZnNc^{\bullet+}$) which acts as the direct evidence for the photo-induced electron transfer within these systems [48].

As already mentioned, self-assembly by using an ammonium ion-crown ether interaction is regarded as one of the most powerful method among the noncovalent methodologies reported to date. The efficient selectivity of the 18-crown-6 moiety towards ammonium cations and ease of formation of the size-fit complex even in polar solvents, enabled to built porphyrin and fullerene based donor-acceptor supramolecular systems with CNTs [49]. In the SWCNTs-$PyrNH_3^+$-crown-C_{60} nanohybrids **28**, free energy calculations suggested the possibility of electron transfer from the CNT to the singlet excited fullerene, resulting in the formation of a $SWCNTs^{\bullet+}$-$PyrNH_3^+$-crown-$C_{60}^{\bullet-}$ charge separated state [49b]. Transient absorption spectroscopy confirmed the electron transfer as the quenching mechanism affording a lifetime for the radical ion-pair over 100 ns, which suggests a further charge stabilization due to the supramolecular assembly.

Considering this background, we have used two powerful electron-donor moieties, TTF and exTTF to decorate SWCNTs, following a variety of covalent [50] and noncovalent approaches. Our strategy for the supramolecular modification of CNTs involved the preparation of bifunctional systems – pyrene-TTF (**29**) or pyrene-exTTF (**30**) (Figure 2.16) – where the use of pyrene is particularly crucial to achieve surface immobilization onto CNTs through directed π–π interactions. The synthesis of these new molecules was based on the covalent linkage of both units through a flexible and medium-length chain which favours a facile interaction with the SWCNTs surface and, where the pyrene fragment functions exclusively as a template that guarantees the immobilization of the electron donor onto the CNT surface.

Stable dispersions of CNT•pyrene-TTF [51] or CNT•pyrene-exTTF [52] were obtained by using a mixture of 2 mg of the corresponding CNT and 1 mg of **29** or **30**. After a process that involves stirring, sonication and centrifugation, the aggregates were obtained as solids that were redissolved in THF for characterization considering different analytical, spectroscopic and microscopic techniques.

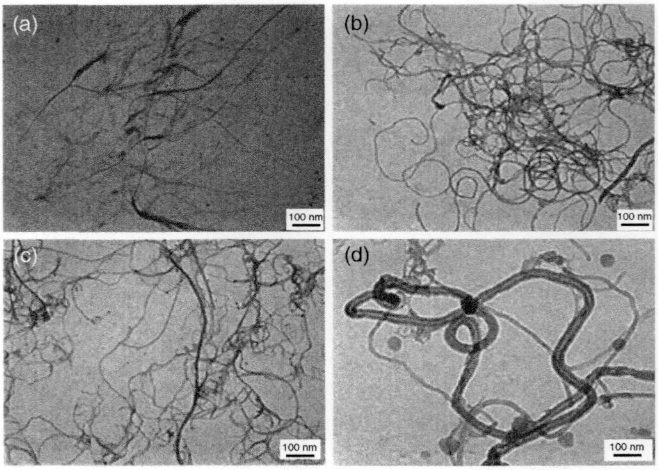

Figure 2.17 *TEM images of (a) SWCNT•pyrene-TTF **30**, (b) VTMWCNT•pyrene- TTF **30**, (c) TMWCNT•pyrene-TTF **30** and (d) MWCNT•pyrene-TTF **30** on TEM grids*

The presence of CNTs in all the studied samples was corroborated by means of transmission electron microscopy (TEM) and atomic force microscopy (AFM). Representative images of the aggregates formed by **29** with SWCNTs, multi-walled carbon nanotubes (MWCNTs), very thin multi-walled carbon nanotubes (VTMWCNTs) and thin multi-walled carbon nanotubes (TMWCNTs) (Figure 2.17), reveal high aspect-ratio objects that appear throughout the scanned region – typically objects from 500 nm to several micrometers long. In some cases aggregation is still evident. Common to all the CNT•pyrene-TTF samples is a good dispersion in the solvent and a marked degree of debundling, especially for SWCNTs and DWCNTs.

From AFM was corroborated a variable length scale between several hundred nanometres and several micrometres depending of the type of tube employed to form the supramolecular aggregate. The height of the tubes, on the other hand, ranges from 1.5 nm SWCNTs to 20 nm (MWCNTs) matching the diameters of individual SWCNTs, DWCNTs, VTMCWNTs, TMWCNTs or MWCNTs, respectively [51].

Further support for the successful immobilization of **29** and **30** onto the surface of SWCNTs was obtained from thermogravimetric analysis (TGA), where a loss weight of about a 6–7 % was obtained for SWCNT•pyrene-TTF **29** and SWCNT•pyrene-exTTF **30** samples, which corresponds to a ratio of a single pyrene -TTF or pyrene-exTTF molecule per 750 carbon atoms of SWCNTs [52]. Electrochemical investigations resulted to be particularly relevant for the case of SWCNT•pyrene-exTTF nanohybrids: the oxidation processes corresponding to the free pyrene-exTTF **30** molecules ($E^1_{ox} = +170$ mV, $E^2_{ox} = +1035$ mV) where observed together with the presence of pyrene-exTTF **30** molecules noncovalently bonded to the SWNTs structure ($E^1_{ox} = +170$ mV, $E^2_{ox} = +960$ mV) (Figure 2.18). The stabilization of the pyrene radical cation –when interacting with SWCNTs – evidence the strong π–π interactions with this planar and aromatic structure [47c].

The photophysical properties of these supramolecular assemblies were investigated by steady-state and time-resolved fluorescence as well as femtosecond transient absorption

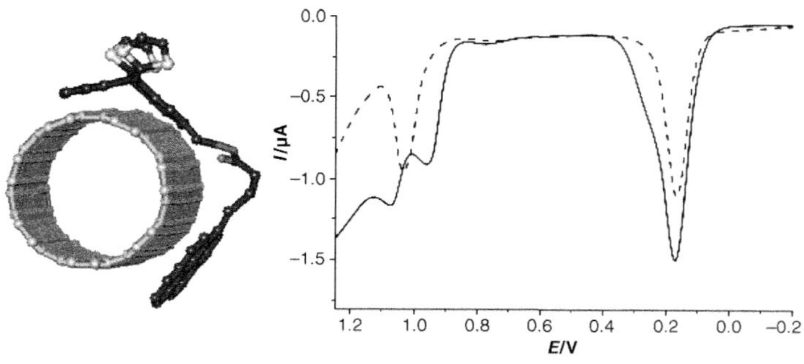

Figure 2.18 *SWVs obtained for the pyrene-exTTF molecule **30** (dashed line) and a SWCNT•pyrene-exTTF mixture (solid line)*

spectroscopy. Because of the close proximity of the TTF or exTTF to the electron acceptor, a very rapid intrahybrid electron transfer affords a photogenerated radical ion pair, whose lifetime is only a few nanoseconds for the case of SWCNTs. Important differences are observed when the CNT•pyrene-TTF series are considered: charge injection into the conduction band of CNTs afforded stable radical ion pair states only for MWCNTs, while the lifetimes observed for SWCNTs are much shorter, as the rate constant decay for the radical ion pair state indicates ($>3 \times 10^{11}\ s^{-1}$) [51]. The presence of a large number of concentric tubes, providing different acceptor levels in MWCNTs, could be a rational explanation for this additional stabilization of the transient radical ion pairs.

Our next step will comprise the interface of these donor–acceptor structures with suitable electron conducting layers to construct highly organized layer-by-layer composites for application in photovoltaics.

2.5 Summary and Outlook

In this chapter, we have provided a general picture of the research carried out pairing C_{60} or carbon nanotube species with the potent electron-donor TTF and, more specifically, with its *p*-quinonoid congener (exTTF) that has resulted in an outstanding family of photo and electroactive conjugates of interest for applications in research areas such as artificial photosynthesis and photovoltaics. In particular, the unique combination of supramolecular and electronic reciprocity between the receptors based on TTF-type curved aromatic systems and fullerenes could result very valuable for the development of self-assembled nanometric optoelectronic devices.

The same basic principles of supramolecular organization can be also applied to carbon nanotubes. Although considerably less studied, the CNT-based TTF ensembles reveal that these new carbon allotropes are as efficient as the parent fullerenes in electron transfer events. Indeed, π–π interactions between the concave hydrocarbon skeleton of exTTF and the convex surface of SWCNTs adds further strength and stability to the SWCNT•pyrene-exTTF nanohybrid. These results pave the way toward the use of CNTs as appealing and promising materials for photovoltaic applications in the near future.

Acknowledgements

We want to express our gratitude to Prof. Dirk M. Guldi for the photophysical studies shown in this chapter. The authors wish also to express their gratitude to the MEC of Spain (projects CTQ2008-00795 and Consolider-Ingenio 2010 CSD2007-0010 Nanociencia Molecular) and the CAM (project P-PPQ-000225-0505) for generous financial support.

References

[1] (a) W. Fritzsche and M. Köhler, *Nanotechnology: An Introduction to Nanostructuring Techniques*; Wiley: Weinheim, Germany, 2004; (b) C. Rao, *Chemistry of Nanomaterials: Synthesis, Properties and Applications*; Wiley: Weinheim, Germany, 2004; (c) E. Wolf, *Nanophysics and Nanotechnology: An Introduction to Modern Concepts in Nanoscience*; Wiley: Weinheim, Germany, 2004; (d) G. Cao, *Nanostructures and Nanomaterials Synthesis, Properties & Applications*; Imperial College: London, 2004.

[2] (a) *Carbon Nanotubes and Related Structures: New Materials for the Twenty-First Century*; P. Harris, Ed.; Cambridge University Press: Cambridge, 2001; (b) *Fullerenes: From Synthesis to Optoelectronic Properties*, Ed. D.M. Guldi and N. Martín, Kluver Academic Publishers: Dordrecht, The Netherlands, 2002; (c) A. Hirsch, *The Chemistry of Fullerenes*, Wiley-VCH: Weinheim, Germany, 2005; (d) *Fullerenes. Principles and Applications*, F. Langa and J.-F. Nierengarten, Eds. Royal Society of Chemistry: Cambridge, United Kingdom, 2007.

[3] J.L. Segura and N. Martín, New concepts in tetrathiafulvalene chemistry, *Angew. Chem. Int. Ed.*, **40**, 1372–1409 (2001).

[4] N. Martín, L. Sánchez, M.A. Herranz, B. Illescas and D.M. Guldi, Electronic communication in tetrathiafulvalene (TTF)/C_{60} systems: toward molecular solar energy conversion materials? *Acc. Chem. Res. Ed.*, **40**, 1015–1024 (2007).

[5] M. Mas-Torrent and C. Rovira, Tetrathiafulvalene derivatives for organic field effect transistors, *J. Mater. Chem.*, **16**, 433–436 (2006).

[6] E.R. Kay, D.A. Leigh and F. Zerbetto, Molecular motors and mechanical machines, *Angew. Chem Int. Ed.*, **46**, 72–191 (2007).

[7] (a) R.P. Sijbesma and E.W. Meijer, Quadruple hydrogen bonded systems, *Chem. Commun.*, 5–16 (2003); (b) G.A. Jeffrey, *An Introduction to Hydrogen Bonding*, Oxford University Press: Oxford, 1997; (c) L. Sánchez, N. Martín and D.M. Guldi, Hydrogen-bonding motifs in fullerene chemistry, *Angew. Chem. Int. Ed.*, **44**, 5374–5382 (2005).

[8] (a) G.M. Whitesides, E.E. Simanek, J.P. Mathias, C.T. Seto, D.N. Chin, M. Mammen and D.M. Gordon, Noncovalent synthesis: using physical-organic chemistry to make aggregates, *Acc. Chem. Res.*, **28**, 37–44 (1995); (b) F.H. Beijer, R.P. Sijbesma, J.A.J.M. Vekemans, E.W. Meijer, H. Kooijman and A.L. Spek, Hydrogen-bonded complexes of diaminopyridines and diaminotriazines: opposite effect of acylation on complex stabilities, *J. Org. Chem.*, **61**, 6371–6380 (1996); (c) S.C. Zimmermann and P.S. Corbin, Heteroaromatic modules for self-assembly using multiple hydrogen bonds, *Struct. Bonding*, **96**, 63–94 (2000).

[9] F.H. Beijer, H. Kooijman, A.L. Spek, R.P. Sijbesma and E.W. Meijer, Self-complementarity achieved through quadruple hydrogen bonding, *Angew. Chem. Int. Ed.*, **37**, 75–78 (1998).

[10] S.H.M. Söntjens, R.P. Sijbesma, M.H.P. van Genderen and E.W. Meijer, Stability and lifetime of quadruply hydrogen bonded 2-ureido-4[1H]-pyrimidinone dimers, *J. Am. Chem. Soc.*, **122**, 7487–7493 (2000).

[11] M. Segura, L. Sánchez, J. de Mendoza, N. Martín and D.M. Guldi, Hydrogen bonding interfaces in fullerene•TTF ensembles, *J. Am. Chem. Soc.*, **125**, 15093–15100 (2003).

[12] L. Sánchez, M. Sierra, N. Martín, A.J. Myles, T.J. Dale, J. Rebek, Jr., W. Seitz and D.M. Guldi, Exceptionally strong electronic communication through hydrogen bonds in porphyrin-C_{60} pairs, *Angew. Chem. Int. Ed.*, **45**, 4637–4641 (2006).

[13] J.-M. Lehn, *Supramolecular Chemistry, Concepts and Perspectives*, VCH, Weinheim: Germany, 1995.

[14] (a) N. Solladié, M.E. Walther, M. Gross, T.M. Figueira Duarte, C. Bourgogne and J.-F. Nierengarten, A supramolecular cup-and-ball C_{60}–porphyrin conjugate system, *Chem. Commun.*, 2412–2413 (2003); (b) N. Solladié, M.E. Walther, H. Herschbach, E. Leize, A. Van Dorsselaer, T.M. Figueira Duarte and J.-F. Nierengarten, Supramolecular complexes obtained from porphyrin–crown ether conjugates and a fullerene derivative bearing an ammonium unit, *Tetrahedron*, **62**, 1979–1987 (2006).

[15] M. Gutiérrez-Nava, H. Nierengarten, P. Massou, A. van Dorsselaer and J.-F. Nierengarten, A supramolecular oligophenylenevinylene–C_{60} conjugate, *Tetrahedron Lett.*, **44**, 3043–3046 (2003).

[16] M.C. Díaz, B.M. Illescas, N. Martín, J. Fraser Stoddart, M.A. Canales, J. Jiménez-Barbero, G. Sarova and D.M. Guldi, Supramolecular pseudo-rotaxane type complexes from π-extended TTF dimer crown ether and C_{60}, *Tetrahedron*, **62**, 1998–2002 (2006).

[17] B.M. Illescas, J. Santos, M.C. Díaz, N. Martín, C.M. Atienza and D.M. Guldi, Supramolecular threaded complexes from fullerene-crown ether and extended TTF derivatives, *Eur. J. Org. Chem.*, 5027–5037 (2007).

[18] J. Santos, B. Grimm, B.M. Illescas, D.M. Guldi and N. Martín, Cooperativity between – and H-bonding interactions – a supramolecular complex formed by C_{60} and exTTF, *Chem. Commun.*, 5993–5995 (2008).

[19] R.J. Brea, L. Castedo, J.R. Granja, M.A. Herranz, L. Sánchez, N. Martín, W. Seitz and D.M. Guldi, Electron transfer in Me-blocked heterodimeric α,β–peptide nanotubular donor–acceptor hybrids, *Proc. Natl. Acad. Sci. USA*, **104**, 5291–5294 (2007).

[20] Unpublished results in collaboration with E. Ortí (University of Valencia). Calculations carried out at the BH & H/6-31+G** level; $\Delta G_{\text{binding}} = 9.47$ kcal/mol which is reduced to 7.00 kcal/mol after BSSE correction.

[21] E.M. Pérez, L. Sánchez, G. Fernández and N. Martín, exTTF as a building block for fullerene receptors. Unexpected solvent-dependent positive homotropic cooperativity, *J. Am. Chem. Soc.*, **128**, 7172–7173 (2006).

[22] S.S. Gayathri, M. Wielopolski, E.M. Pérez, G. Fernández, L. Sánchez, R. Viruela, E. Ortí, D.M. Guldi and N. Martín, Discrete supramolecular donor-acceptor complexes, *Angew. Chem. Int. Ed.*, **48**, 815–819. (2009).

[23] K.A. Connors, *Binding Constants. The Measurement of Molecular Complex Stability*, John Wiley & Sons, Inc., New York, 1987.

[24] The applicability of the Hill equation to self-assembling systems has recently been under, discussion. In a careful analysis, G. Ercolani (*J. Am. Chem. Soc.*, **125**, 16097–16103 (2003)) argued against its use except in the case of intermolecular binding of a monovalent ligand to a multivalent receptor. Our proposed binding mode is close to this case, since both binding events are intermolecular – at least to a first approximation.

[25] T. Kawase and H. Kurata, Ball-, bowl-, and belt-shaped conjugated systems and their complexing abilities: exploration of the concave–convex π–π interaction, *Chem. Rev.*, **106**, 5250–5273 (2006).

[26] E.M. Pérez and N. Martín, Curves ahead: molecular receptors for fullerenes based on concave–convex complementarity, *Chem. Soc. Rev.*, **37**, 1512–1519 (2008).

[27] E.M. Pérez, A.L. Capodilupo, G. Fernández, L. Sánchez, P.M. Viruela, R. Viruela, E. Ortí, M. Bietti and N. Martín, Weighting noncovalent forces in the molecular recognition of C_{60}. Relevance of concave convex complementarity, *Chem. Commun.*, 4567–4569 (2008).

[28] E.M. Pérez, M. Sierra, L. Sánchez, M.R. Torres, R. Viruela, P.M. Viruela, E. Ortí and N. Martín, Concave tetrathiafulvalene-type donors as supramolecular partners for fullerenes, *Angew. Chem., Int. Ed.*, **46**, 1847–1851 (2007).

[29] G. Fernández, E.M. Pérez, L. Sánchez and N. Martín, Self-organization of electroactive materials: a head-to-tail donor-acceptor supramolecular polymer, *Angew. Chem., Int. Ed.*, **47**, 1094–1097 (2008).

[30] G. Fernández, E.M. Pérez, L. Sánchez and N. Martín, An electroactive dynamically polydisperse supramolecular dendrimer, *J. Am. Chem. Soc.*, **130**, 2410–2411 (2008).

[31] G. Fernández, L. Sánchez, E.M. Pérez and N. Martín, Large exTTF-based dendrimers. self-assembly and peripheral cooperative multiencapsulation of C_{60}, *J. Am. Chem. Soc.*, **130**, 10674–10683 (2008).

[32] (a) D.M. Guldi, G.M.A. Rahman, V. Sgobba and C. Ehli, Multifunctional molecular carbon materials – from fullerenes to carbon nanotubes, *Chem. Soc. Rev.*, **35**, 471–487 (2006); (b) D.M. Guldi, Nanometer scale carbon structures for charge-transfer systems and photovoltaic applications, *Phys. Chem. Chem. Phys.*, 1400–1420 (2007); (c) V. Sgobba and D.M. Guldi, Carbon nanotubes – electronic/electrochemical properties and application for nanoelectronics and photonics, *Chem. Soc. Rev.*, **38**, 165–184 (2009).

[33] J. Hone, B. Batlogg, Z. Benes, A.T. Johnson and J.E. Fischer, Quantized phonon spectrum of single-wall carbon nanotubes, *Science*, **289**, 1730–1734 (2000).

[34] M.F. Yu, B.S. Files, S. Arepalli and R.S. Ruoff, Tensile loading of ropes of single wall carbon nanotubes and their mechanical properties, *Phys. Rev. Lett.*, **84**, 5552–5555 (2000).

[35] M. Ouyang, J.-L. Huang and C.M. Lieber, Fundamental electronic properties and applications of single-walled carbon nanotubes, *Acc. Chem. Res.*, **35**, 1018–1025 (2002).

[36] For a review on the different nanoforms of carbon, please see: J.L. Delgado, M.A. Herranz and N. Martín, The nanoforms of carbon, *J. Mater. Chem.*, **18**, 1417–1426 (2008).

[37] (a) A. Hirsch, Functionalization of single-walled carbon nanotubes, *Angew. Chem. Int. Ed.*, **41**, 1853–1859 (2002); (b) J.L. Barh and J.M. Tour, Covalent chemistry of single-wall carbon nanotubes, *J. Mater. Chem.*, **12**, 1952–1958 (2002); (c) S. Nigoyi, M.A. Hamon, H. Hu, B. Zhao, P. Bhomwik, R. Sen, M.E. Itkis and R.C. Haddon, Chemistry of single-walled carbon nanotubes, *Acc. Chem. Res.*, **35**, 1105–1113 (2002); (d) Y.-P. Sun, K. Fu, Y. Lin and W. Huang, Functionalized carbon nanotubes: Properties and applications, *Acc. Chem. Res.*, **35**, 1096–1104 (2002); (e) S. Banerjee, M.G.C. Kahn and S.S. Wong, Rational chemical strategies for carbon nanotube functionalization, *Chem. Eur. J.*, **9**, 1898–1908 (2003); (f) D. Tasis, N. Tagmatarchis, V. Georgakilas and M. Prato, Soluble carbon nanotubes, *Chem. Eur. J.*, **9**, 4000–4008 (2003); (g) C.A. Dyke and J.M. Tour, Overcoming the insolubility of carbon nanotubes through high degrees of sidewall functionalization, *Chem. Eur. J.*, **10**, 812–817 (2004); (h) S. Banerjee, T. Hemraj-Benny and S.S. Wong, Covalent surface chemistry of single-walled carbon nanotubes, *Adv. Mater.*, **17**, 17–29 (2005); (i) D.M. Guldi, G.M.A. Rahman, F. Zerbetto and M. Prato, Carbon nanotubes in electron donor-acceptor nanocomposites, *Acc. Chem. Res.*, **38**, 871–878 (2005); (j) A. Hirsch and O. Vostrowsky, Functionalization of Carbon Nanotubes, *Top. Curr. Chem.*, **245**, 193–237 (2005); (k) D. Tasis, N. Tagmatarchis and A. Bianco, Chemistry of carbon nanotubes, *Chem. Rev.*, **106**, 1105–1136 (2006).

[38] Y.-L. Zhao, J.F. Stoddart, Noncovalent functionalization of single-walled carbon nanotubes, *Acc. Chem. Res.*, **42**, 1161–1171 (2009).

[39] (a) J.M. Simmons, I. In, V.E. Campbell, T.J. Mark, F. Léonard, P. Gopalan and M.A. Eriksson, Optically modulated conduction in chromophore-functionalized single-wall carbon nanotubes, *Phys. Rev. Lett.*, **98**, 086802(4), (2007); (b) C. Backes, C.D. Schmidt, F. Hauke, C. Böttcher and A. Hirsch, High population of individualized SWCNTs through the adsorption of water-soluble perylenes, *J. Am. Chem. Soc.*, **131**, 2172–2184 (2009); (c) R. Marquis, C. Greco, I. Sadokierska, S. Lebedkin, M.M. Kappes, T. Michel, L. Alvarez, J.-L. Sauvajol, S. Meunier and C. Mioskowski, Supramolecular discrimination of carbon nanotubes according to their helicity, *Nano Lett.*, **8**, 1830–1835 (2008); (d) C. Ehli, C. Oelsner, D.M. Guldi, A. Mateo-Alonso, M. Prato, C. Schmidt, C. Backes, F. Hauke and A. Hirsch, Manipulating single-wall carbon nanotubes by chemical doping and charge transfer with perylene dyes, *Nature Chem.*, **1**, 243–249 (2009).

[40] (a) W. Wang, K.A.S. Fernando, Y. Lin, M.J. Meziani, L.M. Veca, L. Cao, P. Zhang, M.M. Kimani and Y.-P. Sun, Metallic single-walled carbon nanotubes for conductive nanocomposites, *J. Am. Chem. Soc.*, **130**, 1415–1419 (2008); (b) W. Tu, J. Lei and H. Ju, Functionalization of carbon nanotubes with water-insoluble porphyrin in ionic liquid: direct electrochemistry and highly sensitive amperometric biosensing for trichloroacetic acid, *Chem. Eur., J.*, **15**, 779–784 (2009); (c) K. Saito, V. Troiani, H. Qiu, N. Solladié, T. Sakata, H. Mori, M. Ohama and S. Fukuzumi, Nondestructive formation of supramolecular nanohybrids of single-walled carbon nanotubes with flexible porphyrinic polypeptides, *J. Phys. Chem. C*, **111**, 1194–1199 (2007).

[41] M. Prato, K. Kostarelos and A. Bianco, Functionalized carbon nanotubes in drug design and discovery, *Acc. Chem. Res.*, **41**, 60–68 (2008).

[42] D.A. Britz and A.N. Khlobystov, Noncovalent interactions of molecules with single walled carbon nanotubes, *Chem. Soc. Rev.*, **35**, 637–659 (2006).

[43] R.J. Chen, Y. Zhang, D. Wang and H. Dai, A self-healing oxygen-evolving catalyst, *J. Am. Chem. Soc.*, **123**, 3838–3839 (2001).

[44] H.G. Sudiya, J. Ma, X. Dong, S. Ng, L.J. Li, X.-W. Lu and P. Chen, Interfacing glycosylated carbon-nanotube-network devices with living cells to detect dynamic secretion of biomolecules, *Angew. Chem. Int. Ed.*, **48**, 2723–2726. (2009).

[45] P. Wu, X. Chen, N. Hu, U.C. Tam, O. Blixt, A. Zettl and C.R. Bertozzi, Biocompatible carbon nanotubes generated by functionalization with glycodendrimers, *Angew. Chem. Int. Ed.*, **47**, 5100–5103 (2008).

[46] (a) Y.-L. Zhao, L. Hu, J.F. Stoddart and G. Grüner, Pyrenecyclodextrin-decorated single-walled carbon nanotube field-effect transistors as chemical sensors, *Adv. Mat.*, **20**, 1910–1915 (2008); (b) Y.-L. Zhao, L. Hu, G. Grüner and J.F. Stoddart, A tunable photosensor, *J. Am. Chem. Soc.*, **130**, 16996–17003 (2008).

[47] (a) D.M. Guldi, G.M.A. Rahman, N. Jux, D. Balbinot, N. Tagmatarchis and M. Prato, Multiwalled carbon nanotubes in donor–acceptor nanohybrids – towards long-lived electron transfer products, *Chem. Commun.*, 2038–2040 (2005); (b) D.M. Guldi, G.M.A. Rahman, N. Jux, D. Balbinot, U. Hartnagel, N. Tagmatarchis and M. Prato, Functional single-wall carbon nanotube nanohybrids associating SWNTs with water-soluble enzyme model systems, *J. Am. Chem. Soc.*, **127**, 9830–9838 (2005); (c) C. Ehli, G.M.A. Rahman, N. Jux, D. Balbinot, D.M. Guldi, F. Paolucci, M. Marcaccio, D. Paolucci, M. Melle-Franco, F. Zerbetto, S. Campidelli and M. Prato, Interactions in single wall carbon nanotubes/pyrene/porphyrin nanohybrids, *J. Am. Chem. Soc.*, **128**, 11222–11231 (2006).

[48] R. Chitta, A.S.D. Sandanayaka, A.L. Schumacher, L. D'Souza, Y. Araki, O. Ito and F. D'Souza, Donor–acceptor nanohybrids of zinc naphthalocyanine or zinc porphyrin noncovalently linked to single-wall carbon nanotubes for photoinduced electron transfer, *J. Phys. Chem. C*, **111**, 6947–6955 (2007).

[49] (a) F. D'Souza, R. Chitta, A.S.D. Sandanayaka, N.K. Subbaiyan, L. D'Souza, Y. Araki and O. Ito, Self-assembled single-walled carbon nanotube:zinc-porphyrin hybrids through ammonium ion-crown ether interaction: construction and electron transfer, *Chem. Eur. J.*, **13**, 8277–8284 (2007); (b) F. D'Souza, R. Chitta, A.S.D. Sandanayaka, N.K. Subbaiyan, L. D'Souza, Y. Araki and O. Ito, Supramolecular carbon nanotube-fullerene donor–acceptor hybrids for photoinduced electron transfer, *J. Am. Chem. Soc.*, **129**, 15865–15871 (2007); (c) R. Chitta and F. D'Souza, Self-assembled tetrapyrrole-fullerene and tetrapyrrole-carbon nanotube donor-acceptor hybrids for light induced electron transfer applications: a review, *J. Mater. Chem.*, **18**, 1440–1471 (2008).

[50] M.A. Herranz, N. Martín, S. Campidelli, M. Prato, G. Brehm and D.M. Guldi, Control over electron transfer in tetrathiafulvalene-modified single-walled carbon nanotubes, *Angew. Chem. Int. Ed.*, **45**, 4478–4482 (2006).

[51] C. Ehli, D.M. Guldi, M.A. Herranz, N. Martín, S. Campidelli and M. Prato, Pyrene-tetrathiafulvalene supramolecular assembly with different types of carbon nanotubes, *J. Mater. Chem.*, **18**, 1498–1503 (2008).

[52] M.A. Herranz, C. Ehli, S. Campidelli, M. Gutiérrez, G.L. Hug, K. Ohkubo, S. Fukuzumi, M. Prato, N. Martín and D.M. Guldi, Spectroscopic characterization of photolytically generated radical ion pairs in single-wall carbon nanotubes bearing surface-immobilized tetrathiafulvalenes, *J. Am. Chem. Soc.*, **130**, 66–73 (2008).

3

Properties of Fullerene-Containing Dendrimers

Juan-José Cid Martin and Jean-François Nierengarten

Laboratoire de Chimie des Matériaux Moléculaires, Université de Strasbourg et CNRS (UMR 7509), Ecole Européenne de Chimie, Polymères et Matériaux (ECPM), Strasbourg Cedex 2, France

3.1 Introduction

Dendrimers have attracted increased attention among various scientific communities in the last twenty years [1–2]. This interest is mainly related to the capability of dendritic architectures to generate specific properties, as a result of their unique molecular structures [1–2]. For example, a dendritic framework can surround active core molecules, thus creating specific site-isolated microenvironments capable of affecting the properties of the core itself [3–5]. The multiplication of functional groups at the periphery of a dendritic structure also provides several advantages. For example, the dendrimer surface can be used as a platform for amplification of substrate binding or as an antenna for light-harvesting [6–9]. Furthermore dendrimers can be used as traps for small molecules or ions with the aim of releasing them where needed (e.g. in biological tissues) [10] or improving their properties (e.g. luminescence) [11]. Among the large number of molecular subunits used for dendrimer chemistry, C_{60} has proven to be a versatile building block and fullerene-functionalized dendrimers, i.e. *fullerodendrimers* [12], have generated significant research activities in recent years [13–18]. In particular, the peculiar physical properties of fullerene derivatives make fullerodendrimers attractive candidates for a variety of interesting features in supramolecular chemistry and materials science [15]. In this section, recent developments on the

Chemistry of Nanocarbons Edited by Takeshi Akasaka, Fred Wudl and Shigeru Nagase

molecular engineering of fullerene-containing dendrimers will be presented. The aim of this chapter is not to show an exhaustive review on such systems but to present significant examples to illustrate the current state-of-the-art of fullerene chemistry for the development of new functional dendrimers. In particular, specific features resulting from the dendritic structures will be highlighted.

3.2 Dendrimers with a Fullerene Core

C_{60} itself is a convenient core for dendrimer chemistry [13] and the functionalization of C_{60} with a controlled number of dendrons dramatically improves the solubility of the fullerenes [13]. Furthermore, variable degrees of addition within the fullerene core are possible and its almost spherical shape leads to globular systems even with low-generation dendrons [19, 20]. On the other hand, specific advantages are brought about by the encapsulation of a fullerene moiety in the middle of a dendritic structure [12]. The shielding effect resulting from the presence of the surrounding shell has been found useful to obtain amphiphilic derivatives with good spreading characteristics [21, 22], or to prepare fullerene-containing liquid crystalline materials [23]. In the present section, the use of the fullerene sphere as a photoactive core unit will be emphasized.

3.2.1 A Fullerene Core to Probe Dendritic Shielding Effects

Dendrimers with a fullerene core appear to be appealing candidates to demonstrate the shielding effects resulting from the presence of the surrounding dendritic shell. Effectively, the lifetime of the first triplet excited state of fullerene derivatives is sensitive to the solvent [24]. Therefore, lifetime measurements in different solvents can be used to evaluate the degree of isolation of the central C_{60} moiety from external contacts. With this idea in mind, two series of fullerodendrimers have been prepared (Figure 3.1) [25, 26]. In the design of these compounds, it was decided to attach poly(aryl ether) dendritic branches terminated with peripheral triethyleneglycol chains to obtain derivatives soluble in a wide range of solvents [24–26]. The synthetic approach to prepare compounds **1–4** relies upon the 1,3-dipolar cycloaddition [27] of the dendritic azomethine ylides generated in situ from the corresponding aldehydes and *N*-methylglycine. Dendrimers **5–8** [26] have been obtained by taking advantage of the versatile regioselective reaction developed in the group of Diederich [28], which led to macrocyclic bis-adducts of C_{60} by a cyclization reaction at the C sphere with bis-malonate derivatives in a double Bingel cyclopropanation [29].

The photophysical properties of **1–8** have been studied in different solvents (PhMe, CH_2Cl_2 and CH_3CN). The lifetimes of the lowest triplet excited states are summarized in Table 3.1.

For both series of dendrimers interesting trends can be obtained from the analysis of triplet lifetimes in air-equilibrated solutions (Table 3.1) [24, 25]. A steady increase of lifetimes is found by increasing the dendrimers size in all solvents, suggesting that the dendritic wedges are able to shield, at least partially, the fullerene core from external contacts with the solvent and from quenchers such as molecular oxygen. For compounds **1–4**, the increase is particularly marked in polar CH_3CN, where a better shielding of the fullerene chromophore is expected as a consequence of a tighter contact between the

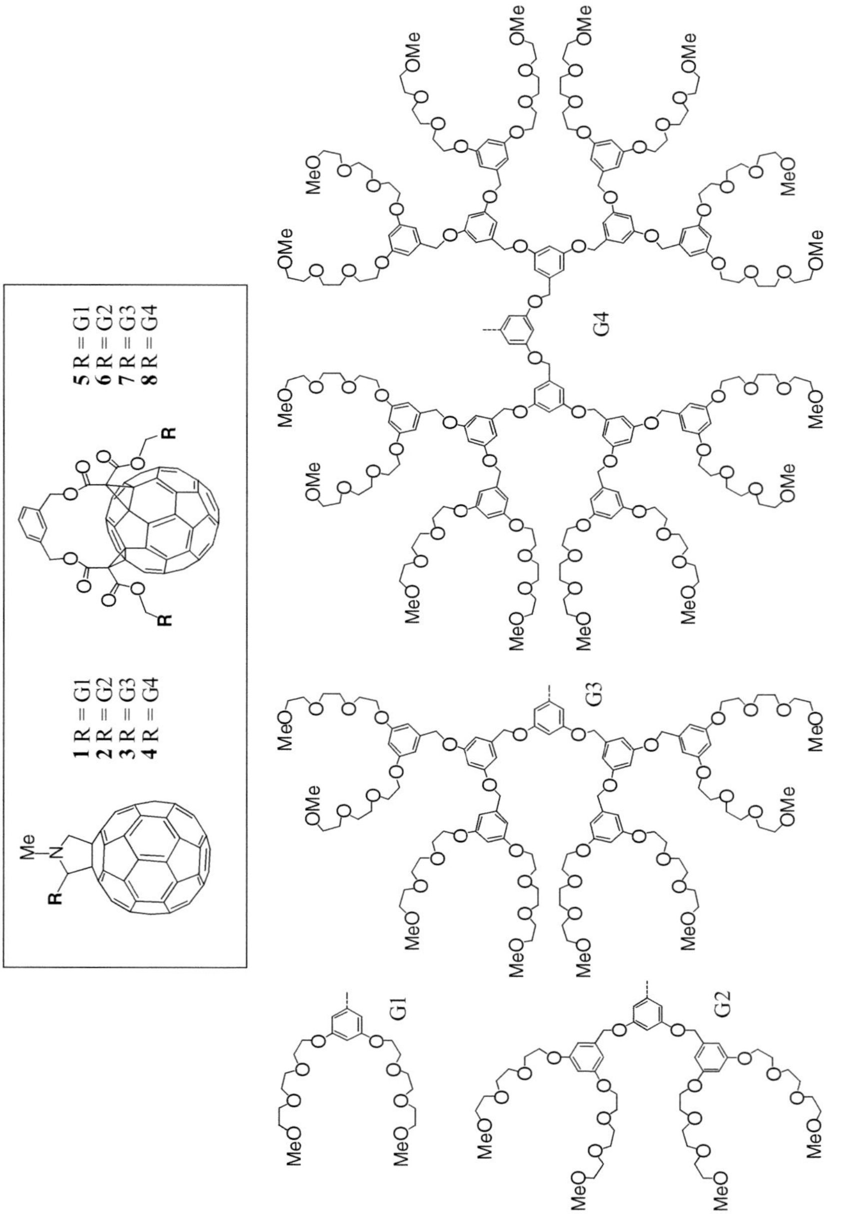

Figure 3.1 *Fullerodendrimers **1–8***

Table 3.1 *Life time of the first triplet excited state of **1–8** in air equilibrated solutions determined by transient absorption at room temperature*

Compound	τ (ns) in PhMe	τ (ns) in CH_2Cl_2	τ (ns) in CH_3CN
1	279	598	%[a]
2	304	643	330
3	318	732	412
4	374	827	605
5	288	611	314
6	317	742	380
7	448	873	581
8	877	1103	1068

[a]not soluble in this solvent.

strongly nonpolar fullerene unit and the external dendritic wedges; in this case a 45% lifetime prolongation is found in passing from **2** to **4** (23% and 28% only for $PhCH_3$ and CH_2Cl_2, respectively). It must be emphasized that the triplet lifetimes of **4** in the three solvents are rather different from each other, likely reflecting specific solvent-fullerene interactions that affect excited state deactivation rates. This suggests that, albeit a dendritic effect is evidenced, even the largest wedge is not able to provide a complete shielding of the central fulleropyrrolidine core in **4**. The latter hypothesis was confirmed by computational studies. As shown in Figure 3.2, the calculated structure of **4** reveals that the dendritic shell is unable to completely cover the fullerene core. In contrast, the triplet lifetimes of **8** [26] in the three solvents lead towards a similar value suggesting that the fullerene core is in a similar environment whatever the nature of the solvent is. In other words the C_{60} unit is, to a large extent, not surrounded by solvent molecules but substantially buried in the middle of the dendritic structure which is capable of creating a specific site-isolated microenvironment around the fullerene moiety. The latter hypothesis is quite reasonable based on the

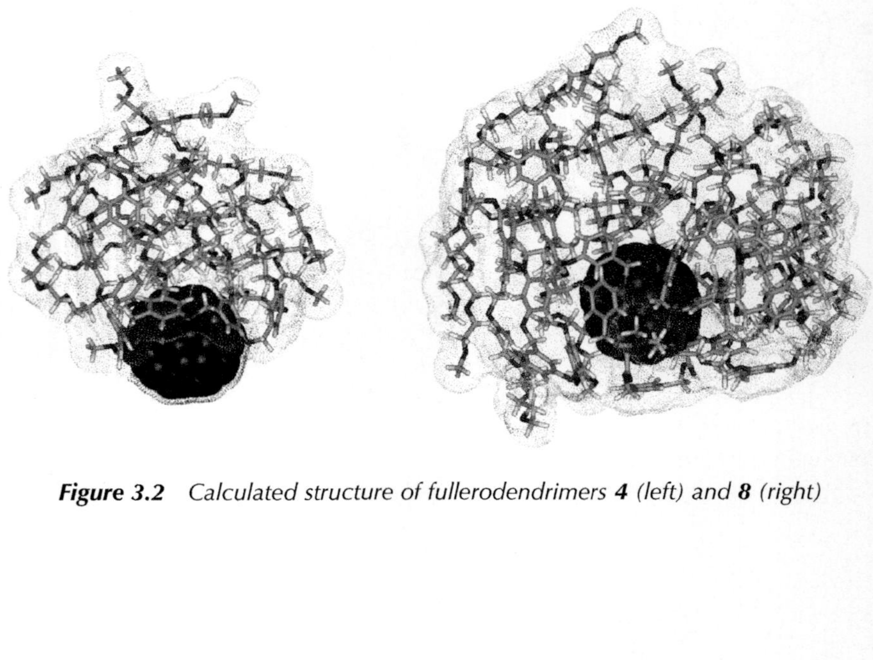

Figure 3.2 *Calculated structure of fullerodendrimers **4** (left) and **8** (right)*

calculated structure of **8** (Figure 3.2) showing that the dendritic branches are able to fully cover the central fullerene core.

The dendritic effect evidenced for **1–8** was found to be useful to optimize the optical limiting properties characteristic of fullerene derivatives. Effectively, the intensity dependant absorption of fullerenes originates from larger absorption cross sections of excited states compared to that of the ground state [30], therefore the increased triplet lifetime observed for the largest fullerodendrimers may allow for an effective limitation on a longer time scale. For practical applications, the use of solid devices is largely preferred to solutions and inclusion of fullerene derivatives in sol-gel glasses has shown interesting perspectives [31]. However, faster de-excitation dynamics and reduced triplet yields are typically observed for fullerene-doped sol-gel glasses when compared to solutions [31]. The latter observations are mainly explained by two factors: (i) perturbation of the molecular energy levels due to the interactions with the sol-gel matrix and (ii) interactions between neighboring fullerene spheres due to aggregation [31]. Therefore, the encapsulation of the C_{60} core evidenced by the photophysical studies for both series of fullerodendrimers might also be useful to prevent such undesirable effects. The incorporation of fullerodendrimers **1–4** in sol-gel glasses has been easily achieved by soaking mesoporous silica glasses with a solution of **1–4** [24, 32]. For the largest compounds, the resulting samples only contain well-dispersed fullerodendrimer molecules. Measurements on the resulting doped samples have revealed efficient optical limiting properties [32]. The transmission as a function of the fluence of the laser pulses remains nearly constant for fluences lower than 5 mJ/cm^2. When the intensity increases above this threshold, the effect of induced absorption appears, and the transmission diminishes rapidly, thus showing the potential of these materials for optical limiting applications.

Fullerodendrimers allow also an evaluation of the accessibility of the C_{60} core unit by studying bimolecular deactivation of its excited states by external quenchers. Recently Ito, Komatsu and co-workers have used this approach to investigate a series of fullerodendrimers (**9–11**) in which Fréchet-type dendrons have been connected to a fullerene moiety via an acetylene linker (Figure 3.3) [33, 34]. Both energy and electron transfer quenchers have been employed to show that the quenching rates of the fullerene triplet state are decreased as a function of the size of the dendrimer shell. These results further demonstrate that fullerene is an excellent functional group to probe the accessibility of a dendrimer core by external molecules.

3.2.2 Light Harvesting Dendrimers with a Fullerene Core

The fullerene C_{60} is also an attractive functional core for the preparation of light harvesting dendrimers. Effectively, its first singlet and triplet excited-states are relatively low in energy and photoinduced energy transfer events have been evidenced in some fullerene-based dyads [17]. In particular, photophysical investigations of some fulleropyrrolidine derivatives substituted with oligophenylenevinylene (OPV) moieties revealed a very efficient singlet-singlet OPV $\rightarrow$ C_{60} photoinduced energy-transfer [35–36]. Based on this observation, dendrimers **12–14** with a fullerene core and peripheral OPV subunits (Figure 3.4) have been prepared [37, 38].

The photophysical properties of fullerodendrimers **12–14** have been first investigated in CH_2Cl_2 solutions. Upon excitation at the OPV band maximum, dramatic quenching of OPV

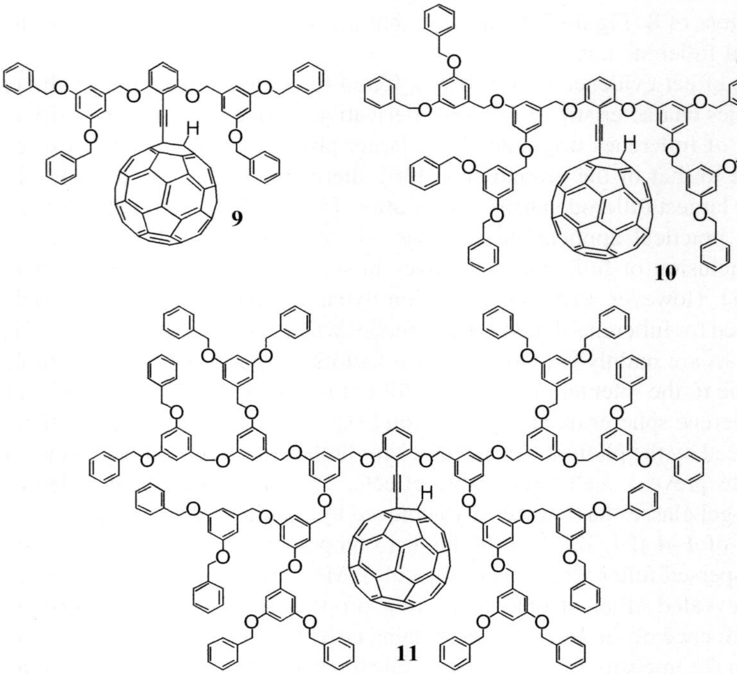

Figure 3.3 Fullerodendrimers **9–11**

fluorescence is observed for all fullerodendrimers. At 394 nm (corresponding to OPV band maxima) the molar absorptivities (ε) of these fullerodendrimers are 134,800 for **12**, 255,100 for **13** and 730,400 M^{-1} cm^{-1} for **14**. Since the ε of the ubiquitous *N*-methyl-fulleropyrrolidine at 394 nm is only 7600 a remarkable light harvesting capability of the peripheral units relative to the central core is evidenced along the series. UV-VIS-NIR luminescence and transient absorption spectroscopy have been used to elucidate in more details the photoinduced processes in fullerodendrimers **12–14** as a function of the dendritic generation and of the solvent polarity (toluene, CH_2Cl_2, benzonitrile), taking into account that the free energy change for electron transfer is the same along the series due to invariability of the donor-acceptor couple. In any solvents, all of the fullerodendrimers exhibit ultrafast OPV $\rightarrow$ C_{60} singlet energy transfer (k_{EnT} ca. 10^{10}–10^{12} s^{-1}). In CH_2Cl_2, a slightly exergonic OPV $\rightarrow$ C_{60} electron transfer from the lowest fullerene singlet level ($^1C_{60*}$) is made possible (ΔG_{CS} $\approx$ 0.07 eV), but it is observed, to an increasing extent, only in the largest systems **13–14** characterized by a lower activation barrier for electron transfer. This effect has been related to a decrease of the reorganization energy upon enlargement of the molecular architecture. Structural factors are also at the origin of an unprecedented OPV $\rightarrow$ C_{60} electron transfer observed for **13** and **14** in apolar toluene, whereas in benzonitrile electron transfer occurs in all cases.

Related compounds have been reported by Martin, Guldi and co-workers [39, 40]. The end-capping of the dendritic spacer with dibutylaniline units yielded the multicomponent photoactive system **15** in which the dendritic wedge plays at the same time the role of an

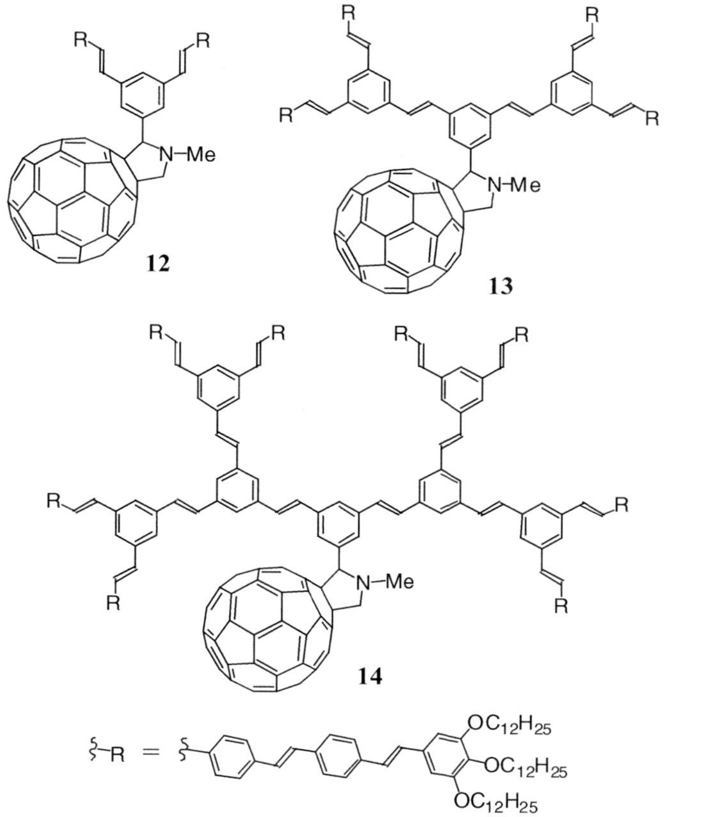

Figure 3.4 *Fullerodendrimers with peripheral OPV units*

antenna capable of channeling the absorbed energy to the fullerene core and of an electron donating unit (Figure 3.5). Photophysical investigations in benzonitrile solutions have shown that, upon photoexcitation, efficient and fast energy transfer takes place from the initially excited antenna moiety to the fullerene core. This process populates the lowest fullerene singlet excited state which is able to promote electron transfer from the dendritic unit to the fullerene core. Langa and co-workers [41] have prepared fullerodendrimer **16** in which the phenylenevinylene dendritic wedge is terminated with ferrocene subunits. Nanosecond transient absorption spectral studies have shown that efficient charge separation occurs in this system, even in apolar solvents.

3.3 Fullerene-Rich Dendrimers

Whereas the main part of the fullerene-containing dendrimers reported so far have been prepared with a C_{60} core, dendritic structures with fullerene units at their surface or with C_{60} spheres in the dendritic branches have been much scarcely considered. This is mainly

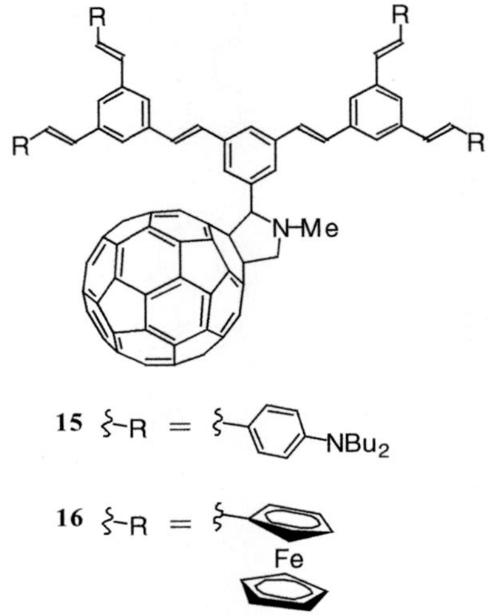

Figure 3.5 *Fullerodendrimers* ***15–16***

associated with the difficulties related to the synthesis of fullerene-rich molecules [14], the two major problems for the preparation of such dendrimers being the low solubility of C_{60} and its chemical reactivity limiting the range of reactions that can be used for the synthesis of branched structures bearing multiple C_{60} units. Over the past years, efficient synthesis of dendrons substituted with fullerene moieties have been reported [18]. These fullerodendrons are interesting building blocks for the preparation of monodisperse fullerene-rich macromolecules with intriguing properties. For example, fullerene-containing dendritic branches have been attached to an OPV core bearing two alcohol functions to yield dendrimers **17**, **18** and **19** with two, four or eight peripheral C_{60} groups, respectively (Figure 3.6) [42].

The photophysical properties of **17–19** have been systematically investigated in solvents of increasing polarity *i.e.* toluene, dichloromethane, and benzonitrile. Ultrafast OPV $\rightarrow$ C_{60} singlet energy transfer takes place upon photoexcitation of the OPV core for the whole series of dendrimers, whatever the solvent is. Electron transfer from the fullerene singlet is thermodynamically allowed in CH_2Cl_2 and benzonitrile, but not in apolar toluene. For a given solvent, the extent of electron transfer, signaled by the quenching of the fullerene fluorescence, is not the same along the series, despite the fact that identical electron transfer partners are present. By increasing the dendrimer size, electron transfer is progressively more difficult. Practically no electron transfer from the fullerene singlet occurs for **19** in CH_2Cl_2, whereas some of it is still detected in the more polar PhCN. These trends can be rationalized by considering increasingly compact dendrimer structures in more polar solvents [42]. This implies that the actual polarity experienced by the involved electron transfer partners, particularly the central OPV, is no longer that of the bulk solvent. This strongly affects electron transfer thermodynamics which, being reasonably located in the normal region of the Marcus parabola, becomes less exergonic and thus slower and less competitive towards intrinsic deactivation of the fullerene singlet state. This dendritic effect

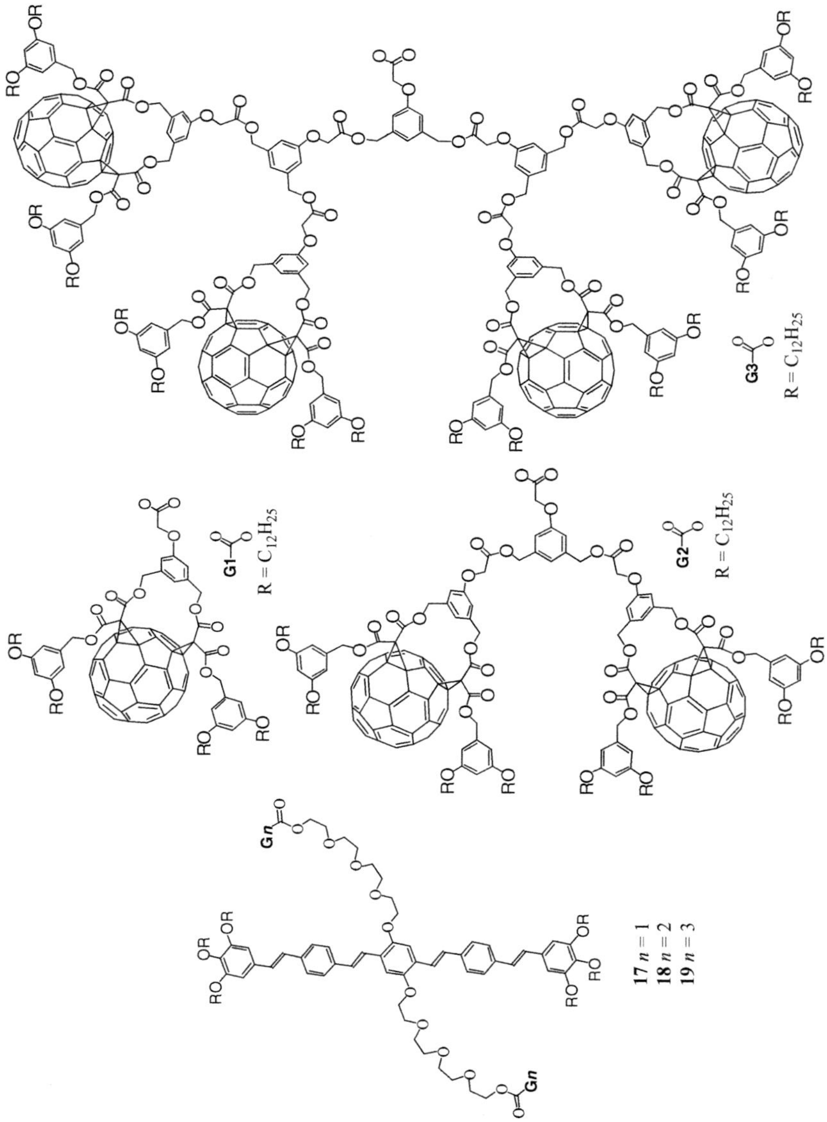

Figure 3.6 *Dendrimers with an OPV core and peripheral fullerene subunits*

is in line with the molecular dynamics studies which suggest that the central OPV unit is more and more protected by the dendritic branches when the generation number is increased. Indeed, the calculated structure of **19** shows that the two dendrons of third generation are able to fully cover the central OPV core.

Another interesting photoactive fullerene-rich dendrimers have been reported by Ito and co-workers [43]. A series of silicon-phthalocyanine (SiPc)-cored fullerodendrimers bearing up to eight axial fullerene subunits have been prepared from silicon phthalocyanine dichloride and the corresponding fullerodendrons bearing a phenol group at the focal point (Figure 3.7). The electrochemical properties of the $(C_{60})_n$-SiPc derivatives **20**, **21** and **22** have been investigated by cyclic voltammetry in benzonitrile. Whereas the fist reduction of all $(C_{60})_n$-SiPc conjugates is centered on the C_{60} subunit, the oxidation is centered on the SiPc core. The determination of the redox potentials of **20–22** was indeed important for the evaluation of the energetics of possible photoinduced electron-transfer processes in these systems. Actually, the driving force for the charge separation calculated from the Rehm–Weller equations suggests an exothermic charge-separation process from the first singlet excited state of the SiPc core as well as from the first singlet excited state of the peripheral fullerenes.

The photophysical properties of **20–22** have been investigated in details. Photoinduced electron transfer has been effectively evidenced in PhCN upon photoexcitation of either the SiPc core or the peripheral fullerene moieties for the whole series of dendrimers. Importantly, the nanosecond transient absorption studies revealed that the lifetimes of the formed radical ion pairs are prolonged on the order of **22** > **21** > **20**. The latter observation has been ascribed to the electron migration among the peripheral C_{60} subunits in **21** and **22**. Thus, fullerodendrons are not only interesting for their light harvesting capabilities but they are also capable of stabilizing charge separated states.

The preparation of covalent fullerene-rich dendrimers is rather difficult and involves a high number of synthetic steps thus limiting their accessibility and therefore their applications. The recent results on the self-assembly of fullerene-containing components by using supramolecular interactions rather than covalent bonds is an attractive alternative for their preparation. Indeed, fullerene-rich derivatives are thus easier to produce and the range of systems that can be prepared is not severely limited by the synthetic route. Indeed, the synthesis itself is restricted to the preparation of dendrons and self-aggregation leads to the dendritic superstructure thus avoiding tedious final synthetic steps with precursors incorporating potentially reactive functional groups such as C_{60}. For example, Nierengarten and co-workers have investigated the self-assembly of fullerene-functionalized dendritic branches **G(1-3)NH$_3$$^+$** bearing an ammonium function at the focal point on the fluorescent ditopic crown ether receptor **23** (Figure 3.8) [44, 45]. The resulting 2:1 supramolecular complexes are multicomponent photoactive devices in which the emission of the central ditopic receptor is dramatically quenched by the peripheral fullerene units (Figure 3.9) [44, 45]. This new property resulting from the association of the different molecular subunits allowed us to investigate in details the self-assembly process.

The complexation between the fullerodendrons **G(1-3)NH$_3$$^+$** and **23** has been investigated in CH_2Cl_2 by UV/vis and fluorescence binding studies. For comparison purposes, binding studies have been also performed with a reference unsubstituted benzylammonium guest (**G0NH$_3$$^+$**). The processing of the titration data led to the determination of two binding

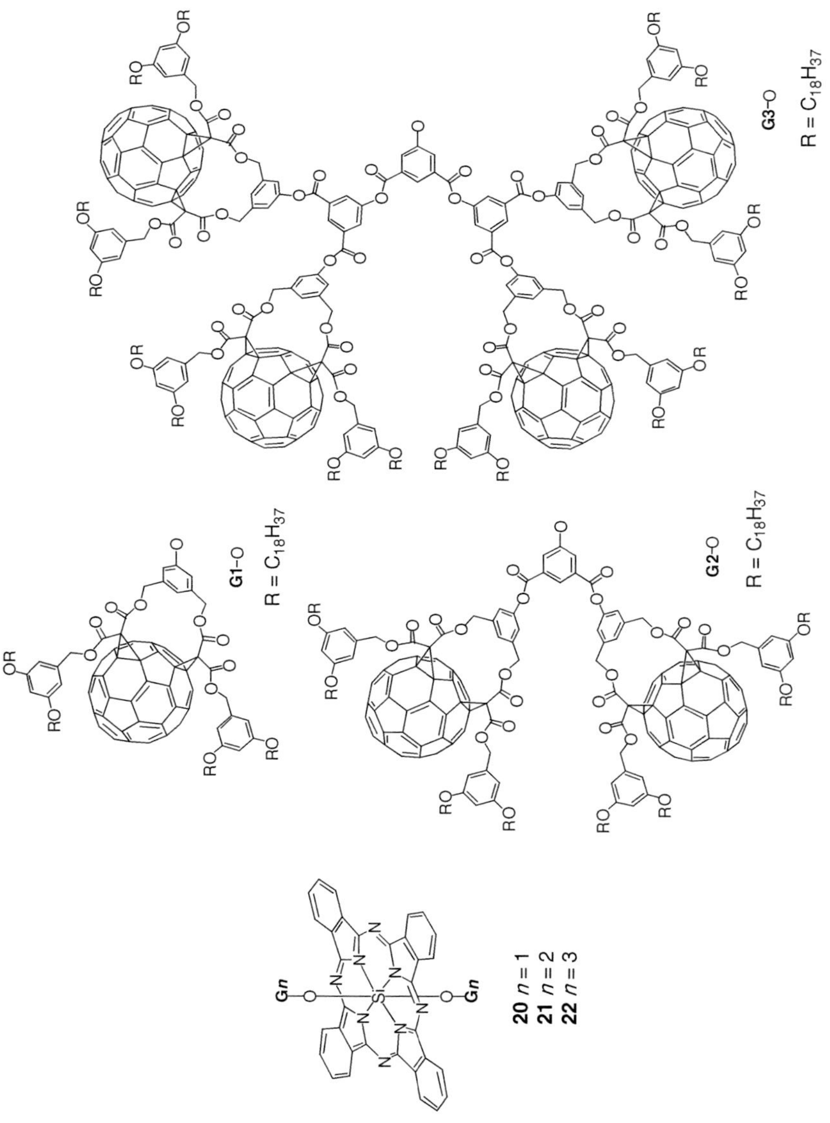

Figure 3.7 Dendrimers with a SiPc core and peripheral fullerene subunits

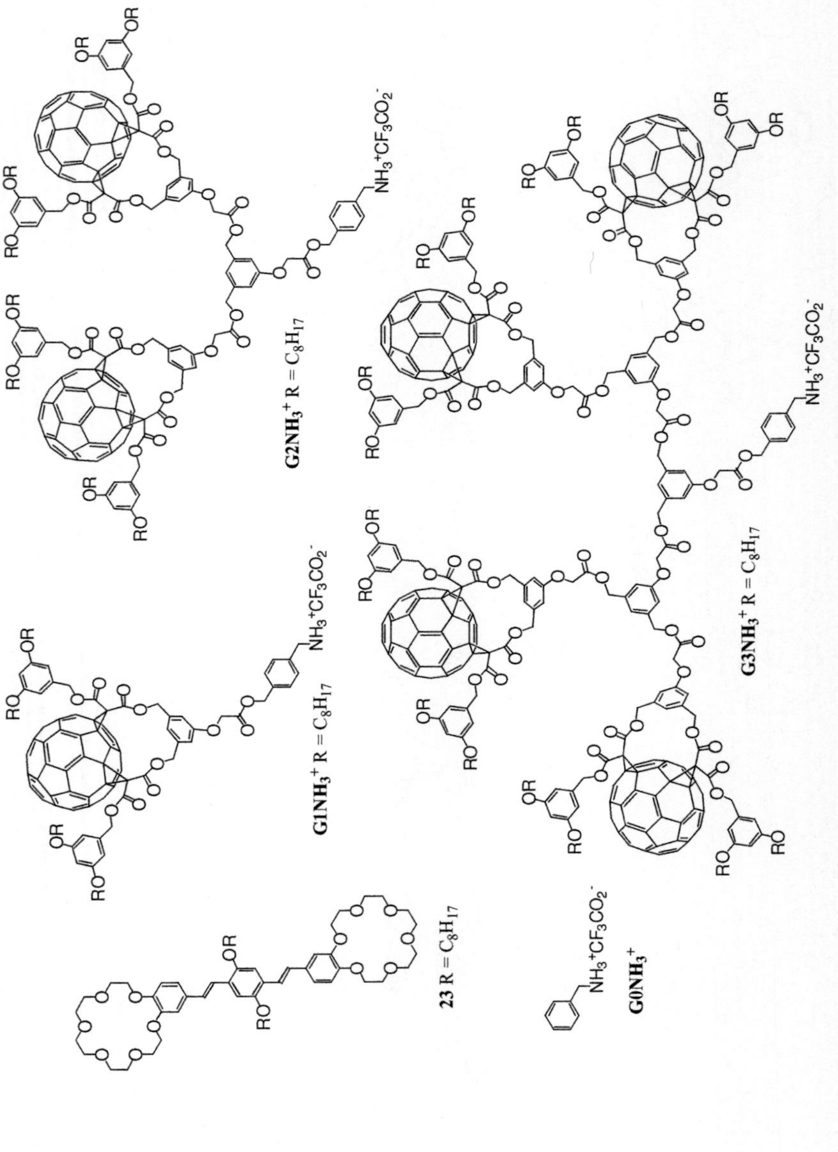

Figure 3.8 *Compounds **G(0-3)NH$_3^+$** and bis-crown ether **23***

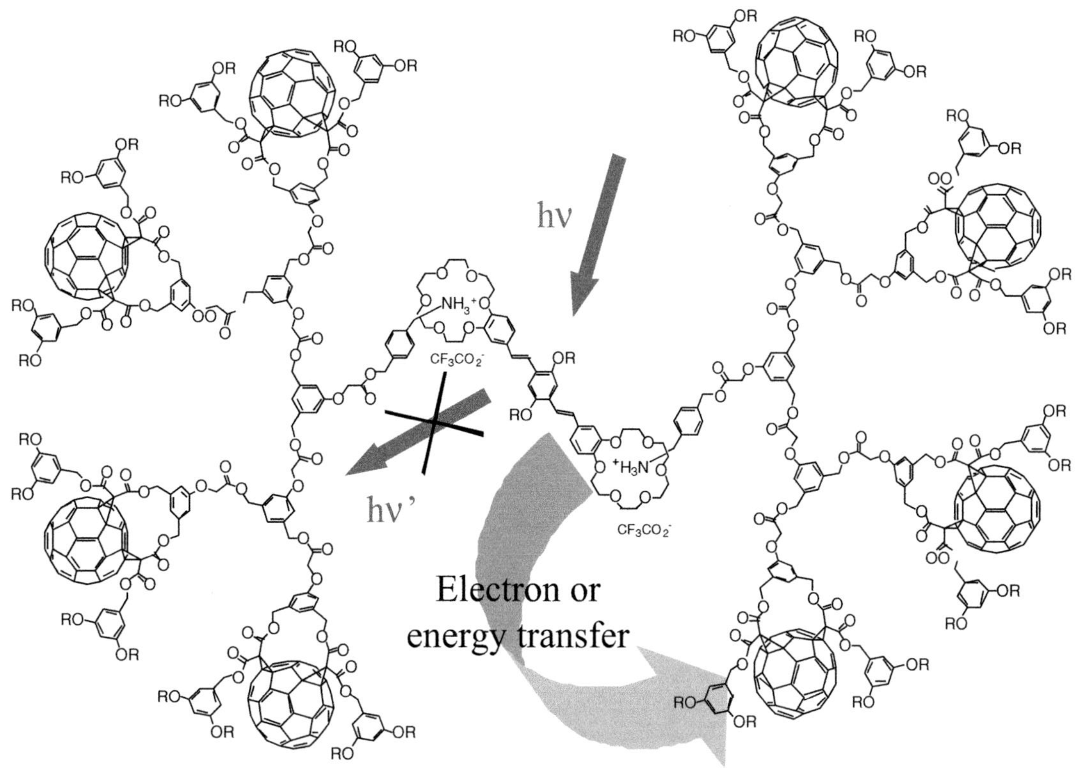

Figure 3.9 *Schematic representation of supramolecular complex [(**G3NH$_3^+$**)$_2$·(**23**)]. Upon selective excitation of the ditopic receptor, its emission is dramatically quenched by the peripheral fullerene units*

Table 3.2 *Stability constants determined by UV-vis and luminescence binding studies*[a]

	23		
	$\log K_1$	$\log K_2$	K_2/K_1
G0NH$_3^+$	4.5 (9)[b]	3.4 (1.8)[b]	0.08 (0.12)
	nd	*nd*	
G1NH$_3^+$	5.6 (8)[b]	6.5 (2)[b]	
	5.0 (1)[c]	5.6 (1)[c]	4.0 (1.2)
G2NH$_3^+$	5.8 (6)[b]	6.7 (8)[b]	
	5.33 (1)[c]	6.3 (1)[c]	9 (3)
G3NH$_3^+$	*nd*	$\log \beta_2 = 12.6$ (9)[b]	
	5.28 (7)[c]	6.48 (7)[c]	16 (4)

[a]All the measurements have been carried out in CH_2Cl_2 at 25 ± 0.2 C. The errors correspond to standard deviations given as 3σ. *nd* (not determined).
[b]Determined from the UV-visible absorption titration.
[c]Determined from the indirect luminescence titration.

constants defined by Equations (3.1) and (3.2); the binding constants deduced from the experimental data are summarized in Table 3.2.

$$\mathbf{23} + \mathbf{G}n\mathbf{NH_3}^+ \xrightleftharpoons{K_1} [(\mathbf{23}) \cdot (\mathbf{G}n\mathbf{NH_3}^+)] \qquad K_1 = \frac{[(\mathbf{23}) \cdot (\mathbf{G}n\mathbf{NH_3}^+)]}{[(\mathbf{23}) \times [(\mathbf{G}n\mathbf{NH_3}^+)]} \tag{3.1}$$

$$[(\mathbf{23}) \cdot (\mathbf{G}n\mathbf{NH_3}^+)] + \mathbf{G}n\mathbf{NH_3}^+ \xrightleftharpoons{K_2} [(\mathbf{23}) \cdot (\mathbf{G}n\mathbf{NH_3}^+)]$$

$$K_2 = \frac{[(\mathbf{23}) \cdot (\mathbf{G}n\mathbf{NH_3}^+)]}{[(\mathbf{23})] \times [(\mathbf{23}) \cdot (\mathbf{G}n\mathbf{NH_3}^+)]} \tag{3.2}$$

Several key points can be deduced from the results reported in Table 3.2. Interestingly, a strong stabilization of about one to two orders of magnitude is observed, when the $\log K_1$ values are compared to those generally reported in the literature for complexes formed between crown ether derivatives and various ammonium cations [46]. The $\log K_1$ values for the binding of fullerodendrons **G(1-3)NH$_3^+$** to **23** are also about one order of magnitude higher than that of the simple benzylammonium guest **G0NH$_3^+$**. Moreover, it is noteworthy that $\log K_1$ values slightly increase with the size of the branches. A sum of secondary weak intramolecular interactions such as π–π stacking and hydrophobic interactions within the supramolecular structures resulting from the association of **23** with **G(1-3)NH$_3^+$** must be at the origin of this stronger coordination. As far as the 2:1 non covalent arrays are concerned, the K_2/K_1 ratio provide a criterion to quantify the interactions between the two identical and independent binding sites [47]. For the binding of **G(1-3)NH$_3^+$** to **23**, the K_2/K_1 values summarized in Table 3.2 are significantly larger than 0.25 which is the value expected for a statistical model [47] and clearly indicates positive intramolecular interactions in the 2:1 associates [(**G(1-3)NH$_3^+$**)$_2$.(**23**)]. It can be added here that the averaged number of occupied sites of **23** calculated from Scatchard [48] or Hill [49] plots is close to 1 for **G0NH$_3^+$** and significantly higher than 1 for **G(1-**

3)NH_3^+ (**G1**: 1.59, **G2**: 1.72, and **G3**: 1.75), thus providing further evidence for the marked positive cooperative effect deduced from the binding constant analysis. The observed cooperativity may be ascribed to strong intramolecular fullerene–fullerene interactions between the two **G(1-3)**NH_3^+ guests within [(**G(1-3)**NH_3^+)$_2$·(**23**)]. This hypothesis is also supported by the absence of any positive interactions for the 2:1 complex obtained from **23** and the ammonium derivative **G0**NH_3^+ lacking the fullerene subunits for which the K_2/K_1 ratio ≈0.08(0.12). Finally, it is also important to highlight that the K_2/K_1 ratio is significantly increased when the size of the dendritic branches is increased. In other words, the cooperative effect is more and more effective when the number of C_{60} units is increased. This positive dendritic effect further confirms that intramolecular fullerene-fullerene interactions must be at the origin of the observed cooperative effect. These results show that the size of dendritic building blocks does not constitute a severe limitation for the self-assembly of large dendritic architectures. Furthermore, it appears that the stability of the highest generation supramolecular ensemble is increased due to the increased number of possible secondary interactions within the self-assembled structure.

Aida and co-workers have reported the preparation of fullerene-rich dendritic structures resulting from the apical coordination of C_{60} derivatives bearing pyridyl moieties to dendritic molecules appended with multiple Zn(II) porphyrin units [50]. For example, compound **24** bound **25** strongly to form stable [(**24**)·(**25**)$_{12}$] (Figure 3.10). Upon titration with **25** in $CHCl_3$ at 25 °C, **24** displayed a large spectral change in the Soret and Q-bands, characteristic of the axial coordination of zinc porphyrins, with a clear saturation profile at a molar ratio **25/24** exceeding 12. The average binding affinity (K), as estimated by simply assuming a one-to-one coordination between the individual zinc porphyrin and pyridine units, is $1.2 \times 10^6\,M^{-1}$. This value is more than 2 orders of magnitude greater than association constants reported for monodentate coordination between zinc porphyrins and pyridine derivatives [51]. The sizeable increase of stability can be ascribed to the simultaneous coordination of two Zn centers of **24** by the two pyridine moieties of **25**. Similar increases in the association constants have been reported for supramolecular systems resulting from the axial coordination of a bis-Zn(II)-porphyrinic receptor to substrates bearing two pyridine subunits [52].

Supramolecular assembly [(**24**)·(**25**)$_{12}$] combining C_{60} units and porphyrin moieties [50] is also a photochemical molecular device. Indeed, the photophysical properties of this system have been studied in detail and an almost quantitative intramolecular photoinduced electron transfer from the photoexcited porphyrins to the C_{60} units evidenced by means of steady-state emission spectroscopy and nanosecond flash photolysis measurements. Excited-state dynamic studies have been carried out to investigate both charge-separation and charge-recombination events in [(**24**)·(**25**)$_{12}$]. The charge-separation rate constants (k_{CS}) and the charge-recombination rate constants (k_{CR}) have been thus deduced. Importantly, the k_{CS}/k_{CR} ratio for [(**24**)·(**25**)$_{12}$] is more than an order of magnitude greater than those reported for precedent porphyrin-fullerene supramolecular dyads and triads [50]. It is obvious that a larger number of the fullerene units in [(**24**)·(**25**)$_{12}$] can enhance the probability of the electron transfer from the zinc porphyrin units. However, in addition to this, one can also presume that an efficient energy migration along the densely packed Zn(II) porphyrin array [50] may enhance the opportunity for this electron transfer.

25

24

Figure 3.10 Compounds 24–25

3.4 Conclusions

Owing to their special photophysical properties, fullerene derivatives are good candidates for evidencing dendritic effects. In particular, the triplet lifetimes of a C_{60} core can be used to evaluate its degree of isolation from external contacts. In addition, the protective effect observed for fullerodendrimers **4** and **8** might be useful for optical limiting applications. On the other hand, the fullerene core can act as a terminal energy receptor in dendrimer-based light-harvesting systems. When the fullerodendrimer is further functionalized with a suitable electron donor, it may exhibit the essential features of an artificial photosynthetic system where an initial photoinduced energy transfer from the antenna to the C_{60} core can be followed by electron transfer. C_{60} is not only an interesting functional core for dendrimer chemistry, it can also be incorporated either at the periphery or within the dendritic structure to produce nanostructures with original electronic properties. Despite some remarkable recent achievements, it is clear that the examples discussed herein represent only the first steps towards the design of fullerene-based molecular assemblies which can display functionality at the macroscopic level. More research in this area is clearly needed to fully explore the possibilities offered by these dendritic materials, for example, in nanotechnology or in photovoltaics.

Acknowledgements

This research was supported by the CNRS. We warmly thank all our co-workers and collaborators for their outstanding contributions, their names are cited in the references.

References

[1] G. R. Newkome, C. N. Moorefield, F. Vögtle, *Dendrimers and Dendrons: Concepts, Syntheses, Applications*, VCH, Weinheim, **2001**.
[2] J. M. J. Fréchet and D. A. Tomalia (eds), *Dendrimers and other Dendritic Polymers*, John Wiley & Sons, Ltd, Chichester, **2001**.
[3] S. Hecht, J. M. J. Fréchet, Dendritic encapsulation of function: applying nature's site isolation principle from biomimetics to materials science, *Angew. Chem.* **2001**, *113*, 76, *Int. Ed.* **2001**, *40*, 74–91.
[4] C. B. Gorman, J. C. Smith, Structure-property relationships in dendritic encapsulation, *Acc. Chem. Res.* **2001**, *34*, 60–71.
[5] J.-F. Nierengarten, Dendritic encapsulation of active core molecules, *Comptes Rendus Chimie* **2003**, *6*, 725–733.
[6] R. Roy, D. Zanini, S. J. Meunier, A. Romanowska, Solid-phase synthesis of dendritic sialoside inhibitors of influenza A virus haemagglutinin, *Chem. Commun.* **1993**, 1869–1872.
[7] M. Fisher, F. Vögtle, Dendrimers: from design to application – a progress report, *Angew. Chem. Int. Ed.* **1999**, *38*, 884–905
[8] A. Adronov, J. M. J. Fréchet, Light-harvesting dendrimers, *Chem. Commun.* **2000**, 1701–1710.
[9] V. Balzani, P. Ceroni, A. Juris, M. Venturi, S. Campagna, F. Puntoriero, S. Serroni, Dendrimers based on photoactive metal complexes – recent advances, *Coord. Chem. Rev.* **2001**, *219*, 545–572.
[10] S. E. Stiriba, H. Frey, R. Haag, Dendritic polymers in biomedical applications: from potential to clinical use in diagnostics and therapy, *Angew. Chem. Int. Ed.* **2002**, *41*, 1329–1334.
[11] V. Vicinelli, P. Ceroni, M. Maestri, V. Balzani, M. Gorka, F. Vögtle, Luminescent lanthanide ions hosted in a fluorescent polylysin dendrimer – Antenna-like sensitization of visible and near-infrared emission, *J. Am. Chem. Soc.* **2002**, *124*, 6461–6468.

[12] J.-F. Nierengarten, Fullerodendrimers: a new class of compounds for supramolecular chemistry and materials science applications, *Chem. Eur. J.* **2000**, *6*, 3667–3670.

[13] A. Hirsch, O. Vostrowsky, Dendrimers with carbon-rich cores, *Top. Curr. Chem.* **2001**, *217*, 51–93

[14] J.-F. Nierengarten, Fullerodendrimers: fullerene-containing macromolecules with intriguing properties, *Top. Curr. Chem.* **2003**, *228*, 87–110.

[15] J.-F. Nierengarten, Chemical modification of C_{60} for materials science applications, *New J. Chem.* **2004**, *28*, 1177–1191.

[16] U. Hahn, F. Cardinali, J.-F. Nierengarten, Supramolecular chemistry for the self-assembly of fullerene-rich dendrimers, *New J. Chem.* **2007**, *31*, 1128–1138.

[17] T. M. Figueira-Duarte, A. Gégout, J.-F. Nierengarten, Molecular and supramolecular C_{60}-oligophenylenevinylene conjugates, *Chem. Commun.* **2007**, 109–119.

[18] M. Holler, J.-F. Nierengarten, Synthesis and properties of fullerene-rich dendrimers, *Aust. J. Chem.* **2009**, *62*, 605–623.

[19] X. Camps, H. Schönberger, A. Hirsch, The C_{60} core: a versatile tecton for dendrimer chemistry, *Chem. Eur. J.* **1997**, *3*, 561–567.

[20] J. Iehl, J.-F. Nierengarten, A click-click approach for the preparation of functionalized [5:1]-hexaadducts of C_{60}, *Chem. Eur. J.* **2009**, *15*, 7306–7309.

[21] F. Cardullo, F. Diederich, L. Echegoyen, T. Habicher, N. Jayaraman, R. M. Leblanc, J. F. Stoddart, S. Wang, Stable Langmuir and Langmuir–Blodgett films of fullerene-glycodendron conjugates, *Langmuir* **1998**, *14*, 1955–1959.

[22] S. Zhang, Y. Rio, F. Cardinali, C. Bourgogne, J.-L. Gallani, J.-F. Nierengarten, Amphiphilic diblock dendrimers with a fullerene core, *J. Org. Chem.* **2003**, *68*, 9787–9797.

[23] T. Chuard, R. Deschenaux, Design, mesomorphic properties, and supramolecular organization of [60]fullerene-containing thermotropic liquid crystals, *J. Mater. Chem.* **2002**, *12*, 1944–1951 and references therein.

[24] Y. Rio, G. Accorsi, H. Nierengarten, J.-L. Rehspringer, B. Hönerlage, G. Kopitkovas, A. Chugreev, A. Van Dorsselaer, N. Armaroli, J.-F. Nierengarten, Fullerodendrimers with peripheral triethyleneglycol chains: synthesis, mass spectrometric characterisation, and photophysical properties, *New J. Chem.* **2002**, *26*, 1146–1154.

[25] Y. Rio, J.-F. Nicoud, J.-L. Rehspringer, J.-F. Nierengarten, Fullerodendrimers with peripheral triethyleneglycol chains, *Tetrahedron Lett.* **2000**, *41*, 10207–10210.

[26] Y. Rio, G. Accorsi, H. Nierengarten, C. Bourgogne, J.-M. Strub, A. Van Dorsselaer, N. Armaroli, J.-F. Nierengarten, A fullerene core to probe dendritic shielding effects, *Tetrahedron* **2003**, *59*, 3833–3844.

[27] M. Prato, M. Maggini, Fulleropyrrolidines: a family of full-fledged fullerene derivatives, *Acc. Chem. Res.* **1998**, *31*, 519–526.

[28] J.-F. Nierengarten, V. Gramlich, F. Cardullo, F. Diederich, Regio- and diastereoselective bisfunctionalization of C_{60} and enantioselective synthesis of a C_{60} derivative with a chiral addition pattern, *Angew. Chem. Int. Ed. Engl.* **1996**, *35*, 2101–2103.

[29] C. Bingel, Cyclopropanierung von fullerenen, *Chem. Ber.* **1993**, *126*, 1957–1959.

[30] L. W. Tutt, A. Krost, Optical limiting performance of C_{60} and C_{70} solutions, *Nature* **1992**, *356*, 225–226.

[31] J. Schell, D. Felder, J.-F. Nierengarten, J.-L. Rehspringer, R. Lévy, B. Hönerlage, Induced absorption of C_{60} and a water-soluble C_{60} derivative in SiO_2 sol-gel matrices, *J. Sol-gel Science and Technology* **2001**, *22*, 225–236.

[32] J.-F. Nierengarten, N. Armaroli, G. Accorsi, Y. Rio, J.-F. Eckert, [60]Fullerene: a versatile photoactive core for dendrimer chemistry, *Chem. Eur. J.* **2003**, *9*, 36–41

[33] R. Kunieda, M. Fujitsuka, O. Ito, M. Ito, Y. Murata, K. Komatsu, Photochemical and photophysical properties of C_{60} dendrimers studied by laser flash photolysis, *J. Phys. Chem. B* **2002**, *106*, 7193–7199.

[34] Y. Murata, M. Ito, K. Komatsu, Synthesis and properties of novel fullerene derivatives having dendrimer units and the fullerenyl anions generated therefrom, *J. Mater. Chem.* **2002**, *12*, 2009–2020.

[35] N. Armaroli, F. Barigelletti, P. Ceroni, J.-F. Eckert, J.-F. Nicoud, J.-F. Nierengarten, Photoinduced energy transfer in a fullerene-oligophenylenevinylene conjugate, *Chem. Commun.* **2000**, 599–600.

[36] J.-F. Eckert, J.-F. Nicoud, J.-F. Nierengarten, S.-G. Liu, L. Echegoyen, F. Barigelletti, N. Armaroli, L. Ouali, V. Krasnikov, G. Hadziioannou, Fullerene-oligophenylenevinylene hybrids: synthesis, electronic properties, and incorporation in photovoltaic devices, *J. Am. Chem. Soc.* **2000**, *122*, 7467–7469.

[37] G. Accorsi, N. Armaroli, J.-F. Eckert, J.-F. Nierengarten, Functionalization of [60]fullerene with new light-collecting oligophenylenevinylene-terminated dendritic branches, *Tetrahedron Lett.* **2002**, *43*, 65–68.

[38] N. Armaroli, G. Accorsi, J. N. Clifford, J.-F. Eckert, J.-F. Nierengarten, Structure-dependent photoinduced electron transfer in fullerodendrimers with light harvesting oligophenylenevinylene terminals, *Chem. Asian J.* **2006**, *1*, 564–574.

[39] J. L. Segura, R. Gomez, N. Martin, C. P. Luo, A. Swartz, D. M. Guldi, Variation of the energy gap in fullerene-based dendrons: competitive versus sequential energy and electron transfer events, *Chem. Commun.* **2001**, 707–708.

[40] D. M. Guldi, A. Swartz, C. Luo, R. Gomez, J. L. Segura, N. Martin, Rigid dendritic donor-acceptor ensembles: control over energy and electron transduction, *J. Am. Chem. Soc.* **2002**, *124*, 10875–10886.

[41] L. Pérez, J. C. Garcia-Martinez, E. Diez-Barra, P. Atienzar, H. Garcia, J. Rodriguez-Lopez, F. Langa, Electron transfer in nonpolar solvents in fullerodendrimers with peripheral ferrocene units, *Chem. Eur. J.* **2006**, *12*, 5149–5157.

[42] M. Gutierrez-Nava, G. Accorsi, P. Masson, N. Armaroli, J.-F. Nierengarten, Polarity effects on the photophysics of dendrimers with an oligophenylenevinylene core and peripheral fullerene units, *Chem. Eur. J.* **2004**, *10*, 5076–5086.

[43] M. E. El-Khouly, E. S. Kang, K.-Y. Kay, C. S. Choi, Y. Aaraki, O. Ito, Silicon-phthalocyanine-cored fullerene dendrimers: synthesis and prolonged charge-separated states with dendrimer generations, *Chem. Eur. J.* **2007**, *13*, 2854–2863.

[44] M. Elhabiri, A. Trabolsi, F. Cardinali, U. Hahn, A.-M. Albrecht-Gary, J.-F. Nierengarten, Cooperative recognition of C_{60}-ammonium substrates by a ditopic oligophenylenevinylene-crown ether host, *Chem. Eur. J.* **2005**, *11*, 4793–4798.

[45] J.-F. Nierengarten, U. Hahn, A. Trabolsi, H. Herschbach, F. Cardinali, M. Elhabiri, E. Leize, A. Van Dorsselaer, A.-M. Albrecht-Gary, Synthesis of fullerodendrons with an ammonium unit at the focal point and their cooperative self-assembly on a fluorescent ditopic crown ether receptor, *Chem. Eur. J.* **2006**, *12*, 3365–3373.

[46] M. Gutiérrez-Nava, H. Nierengarten, P. Masson, A. Van Dorsselaer, J.-F. Nierengarten, A supramolecular oligophenylenevinylene-C_{60} conjugate, *Tetrahedron Lett.* **2003**, *44*, 3043–3046.

[47] B. Perlmutter-Hayman, Cooperative binding to macromolecules – a formal approach, *Acc. Chem. Res.* **1986**, *19*, 90–96.

[48] G. Scatchard, The attraction of proteins for small molecules and ions, *Ann. N. Y. Acad. Sci.* **1949**, *51*, 660–672.

[49] C. A. Hunter, H. L. Anderson, What is cooperativity?, *Angew. Chem. Int. Ed.* **2009**, *48*, 7488–7499 and references therein

[50] W.-S. Li, K. S. Kim, D.-L. Jiang, H. Tanaka, T. Kawai, J. H. Kwon, D. Kim, T. Aida, Construction of segregated arrays of multiple donor and acceptor units using a dendritic scaffold: remarkable dendrimer effects on photoinduced charge separation, *J. Am. Chem. Soc.* **2006**, *128*, 10527–10532.

[51] N. Armaroli, F. Diederich, L. Echegoyen, T. Habicher, L. Flamigni, G. Marconi, J.-F. Nierengarten, A new pyridyl-substituted methanofullerene derivative: photophysics, electrochemistry and self-assembly with Zinc(II)-*meso*-tetraphenylporphyrin (ZnTPP), *New J. Chem.* **1999**, *23*, 77–83

[52] A. Trabolsi, M. Elhabiri, M. Urbani, J. L. Delgado de la Cruz, F. Ajamaa, N. Solladié, A.-M. Albrecht-Gary, J.-F. Nierengarten, Supramolecular click chemistry for the self-assembly of a stable Zn(II)-porphyrin-C_{60} conjugate, *Chem. Commun.* **2005**, 5736–5738.

4

Novel Electron Donor Acceptor Nanocomposites

Hiroshi Imahori[a], *Dirk M. Guldi*[b] *and Shunichi Fukuzumi*[c]

[a]*Institute for Integrated Cell-Material Sciences (iCeMS), Kyoto University, Nishikyo-ku, Kyoto, Japan and Department of Molecular Engineering, Graduate School of Engineering, Kyoto University, Nishikyo-ku, Kyoto, Japan*

[b]*Department of Chemistry and Pharmacy & Interdisciplinary Center for Molecular Materials (ICMM), Friedrich-Alexander-Universität Erlangen-Nürnberg, Erlangen, Germany*

[c]*Department of Material and Life Science, Graduate School of Engineering, Osaka University, SORST, Japan Science and Technology Agency, Osaka, Japan*

4.1 Introduction

The rapid consumption of fossil fuel is expected to cause unacceptable environmental problems such as the greenhouse effect, which could lead to disastrous climatic consequences [1, 2]. Thus, renewable and clean energy resources are definitely required in order to stop global warming [1, 2]. Among renewable energy resources, solar energy is by far the largest exploitable resource. Nature harnesses solar energy for its production by photosynthesis and fossil fuel is the product of photosynthesis [1, 2]. Thus, extensive efforts have been devoted to develop artificial systems for the efficient and economical conversion of solar energy into stored chemical fuels [3–7]. Obviously a solution to all global environment and energy resource problems is not easy to achieve [8, 9]. However, the importance and complexity of energy transfer and electron transfer processes in photosynthesis have inspired design and synthesis of a large number of donor-acceptor ensembles including nanocomposites that can mimic the energy transfer or electron transfer processes in photosynthesis. The specific objective of this chapter is to describe recent development

Chemistry of Nanocarbons Edited by Takeshi Akasaka, Fred Wudl and Shigeru Nagase

of electron donor acceptor nanocomposites and their applications aiming at efficient and economical conversion of solar energy into stored chemical fuels.

4.2 Electron Donor-Fullerene Composites

4.2.1 General

Photochemical and photophysical properties of donor-fullerene composites have been extensively studied in connection with solar energy conversion [10–28]. In particular, fullerenes have small reorganization energies of electron transfer, which results in remarkable acceleration of photoinduced charge separation and of charge shift as well as deceleration of charge recombination [29–34]. Thus, they have been frequently employed as an electron acceptor in donor-fullerene composites to yield a long-lived charge-separated state with a high quantum yield [10–28].

4.2.2 Donor-Fullerene Dyads for Photoinduced Electron Transfer

A number of donor-fullerene linked dyads with a different bridge have been prepared to disclose the photophysical properties. As donors a wide variety of chromophores including porphyrins [35–45], phthalocyanines [46, 47], amines [48], polycondensed aromatics [49], transition complexes [50–52], carotenoids [53], ferrocenes [54, 55], tetrathiafulvalenes (TTF) [56], and others [57] have been employed. Here the photodynamics of zinc porphyrin-C_{60} linked dyad **1** is presented to understand the typical relaxation processes of the excited states in donor-fullerene linked dyads (Figure 4.1) [37]. For instance, photoexcitation of **1** in polar solvents results in the occurrence of a photoinduced electron transfer from the zinc porphyrin excited singlet state ($^1ZnP^*$) ($k_{ET(CS1)} = 9.5 \times 10^9\ s^{-1}$) and the zinc porphyrin excited triplet state ($^3ZnP^*$) ($k_{ET(CS4)} > 1.5 \times 10^7\ s^{-1}$) to C_{60} as well as from ZnP to the C_{60} excited singlet state ($^1C_{60^*}$) ($k_{ET(CS2)} = 5.5 \times 10^8\ s^{-1}$) and the C_{60} excited triplet state ($^3C_{60^*}$) ($k_{ET(CS3)} = 1.5 \times 10^7\ s^{-1}$), yielding the same charge-separated state ($ZnP^{\cdot+}$-$C_{60}^{\cdot-}$). The energy levels in benzonitrile (PhCN) are shown in Scheme 4.1 to illustrate the different relaxation pathways of photoexcited **1**. The charge separation efficiencies initiated by $^1ZnP^*$ ($\Phi_{CS1}(^1ZnP^*)$) and $^1C_{60^*}$ ($\Phi_{CS1}(^1C_{60^*})$) were determined as 95% and 23%, respectively. The unquenched $^1ZnP^*$ and $^1C_{60^*}$ undergo an intersystem crossing to yield $^3ZnP^*$ and $^3C_{60^*}$, respectively, which then generate the charge-separated state quantitatively [37].

The total efficiency of $ZnP^{\cdot+}$-$C_{60}^{\cdot-}$ formation from the initial excited states in PhCN was estimated to be 99% based on Scheme 4.1. The resulting charge-separated state recombines to regenerate the ground state with a lifetime of 0.77 μs ($k_{ET(CR1)} = 1.3 \times 10^6\ s^{-1}$). This rate constant is nearly four orders of magnitude smaller than that of the charge separation from $^1ZnP^*$ [37]. Such fast charge separation and slow charge recombination for **1** in polar solvents are in marked contrast to conventional donor-acceptor linked dyads, in which the charge recombination rates are even larger than the charge separation rates in polar solvents [37].

It should be noted here that photoinduced electron transfer from the donor singlet excited state to the C_{60} moiety competes with the corresponding photoinduced energy transfer [10–28]. Thus, photoinduced events (i.e. electron transfer versus energy transfer) in

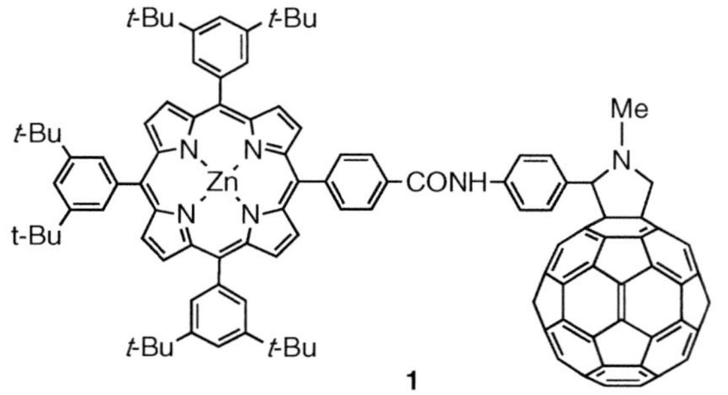

1

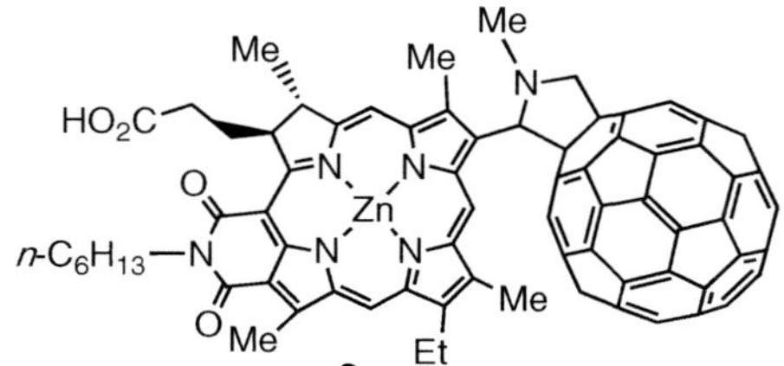

2

Figure 4.1 *Leading examples of donor-fullerene linked dyads exhibiting the formation of a long-lived charge separated state*

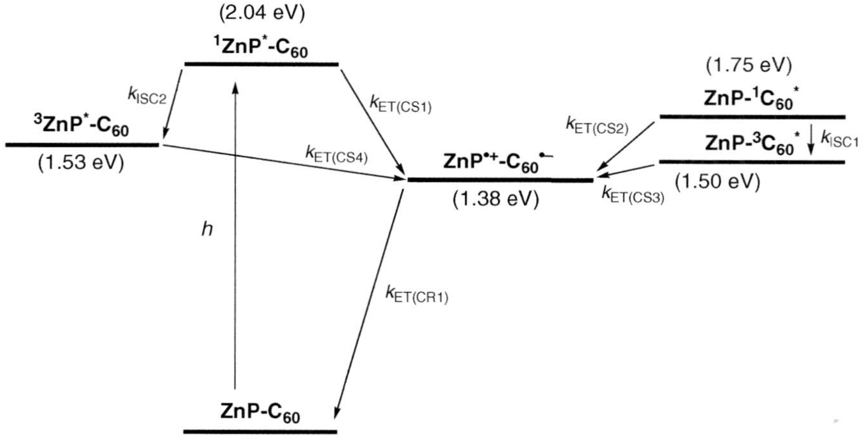

Scheme 4.1 *Reaction scheme and energy diagram for **1** in PhCN*

donor-fullerene linked dyads are influenced by the combination of donor and fullerene as well as the environments including solvent and bridge between donor and fullerene. Moreover, the resulting charge-separated state decays to different energy states rather than the ground state depending on the energy level of the charge-separated state relative to those of the singlet and triplet excited states of the donor and fullerene owing to the small reorganization energies of donor-fullerene linked dyads [10–28].

Photoexcitation of chlorin-C_{60} dyad **2** (Figure 4.1) resulted in formation of a charge-separated state with a lifetime of 120 s ($8.3 \times 10^{-3}\ s^{-1}$) in frozen PhCN at 123 K [45], which is the longest value of charge separation ever reported for donor-fullerene linked systems [10–28]. Unfortunately, however, the quantum yield of formation of the charge-separated state was as low as 12%, which is much smaller than the efficiency of the photoinduced charge separation (100%) estimated from the fluorescence lifetime of the porphyrin moiety [45]. Porphyrin-fullerene dyads and their analogs with a short spacer are known to form a short-lived exciplex from the excited states [42, 43]. Some part of the exciplex is converted into a long-lived charge-separated state, whereas the other part of the exciplex decays rapidly to the ground state [42, 43]. The low quantum yield may arise from the predominant decay of the exciplex state to the ground state over the charge-separated state.

4.2.3 Donor-Fullerene Linked Multicomponent Systems

The remarkable effect of fullerenes in electron transfer is quite similar to the situation in photosynthetic multistep electron transfer in nature. Therefore, utilization of fullerenes in multistep electron transfer systems is attractive in mimicking multistep electron transfer in photosynthetic reaction center. There have been a number of examples of C_{60}-based multichromophore linked systems including triad [37, 58, 59], tetrad [60, 61], pentad [62, 63], and hexad [64, 65]. Figure 4.2 illustrates leading examples of donor-fullerene linked multichromophore systems **3–6**.

The photophysical properties of carotenoid (C)-porphyrin (H_2P)-C_{60} triad **3** are described as a typical example [58]. The energy gradients of each state were designed in the order of $C{-}^1HP^*{-}C_{60} > C{-}H_2P^{\cdot+}{-}C_{60}^{\cdot-} > C^{\cdot+}{-}H_2P{-}C_{60}^{\cdot-}$. In 2-methyltetrahydrofuran, there exists a combination of two pathways to yield $C\text{-}H_2P^{\cdot+}\text{-}C_{60}^{\cdot-}$; i.e. electron transfer from the $^1HP^*$ to the C_{60} and from the H_2P to the $^1C_{60}^*$. A subsequent charge-shift from the carotenoid to the $H_2P^{\cdot+}$ produces $C^{\cdot+}{-}H_2P{-}C_{60}^{\cdot-}$ with a total quantum yield of 88%. The final charge-separated state decays by charge recombination (0.34 μs) to yield the $^3C^*$ rather than the ground state because of the small reorganization energy [58]. Similar multistep electron transfer was obtained for C_{60}-based tetrad, pentad, and hexad. The long lifetime of the final charge-separated state was attained for **4** (1.6 s at 163 K) and **5** (0.53 s at 163 K) with the total charge separation efficiencies of 34% and 83%, respectively [61, 62]. In contrast, ferrocene-butadiyne-linked zinc porphyrin tetramer-fullerene hexad **6** exhibited a very short lifetime of the final charge-separated state (7.8 ns) probably due to superexchange-mediated charge recombination [65].

4.2.4 Supramolecular Donor-Fullerene Systems

Self-assembly of donor and acceptor molecules has become a central theme of supramolecular chemistry in recent years. In this regard, a variety of noncovalently bonded donor-fullerene systems have been prepared to examine photoinduced energy and electron transfer processes in solutions toward the realization of efficient solar energy conversion [66–75]. Figure 4.3 illustrates leading examples of supramolecular donor-fullerene systems **7–9**. TTF as donor was assembled with C_{60} through a complementary guanidium-carboxylate ion pair to yield supramolecular dyad **7** [68]. The lifetime measured for the charge-separated state was 1.0 μs, which is one order of magnitude slower than those reported for covalently

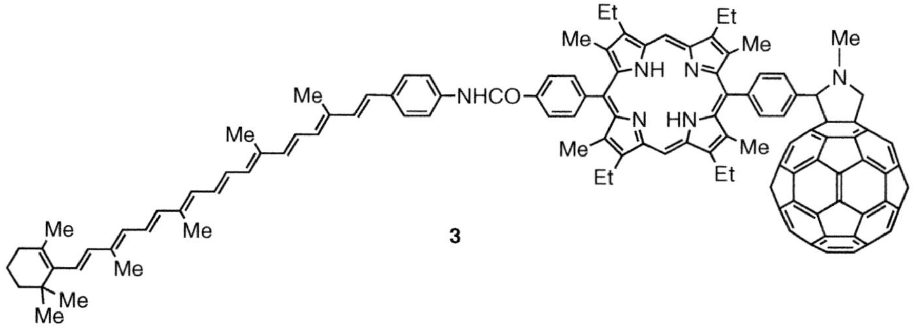

3

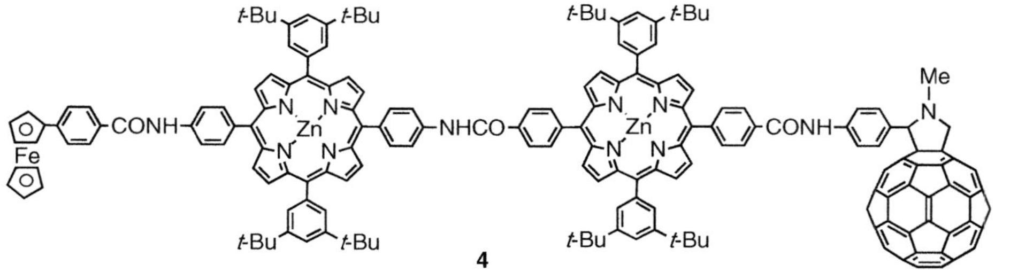

4

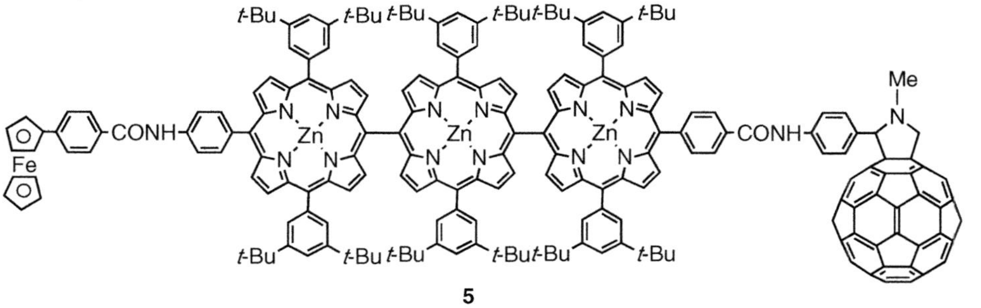

5

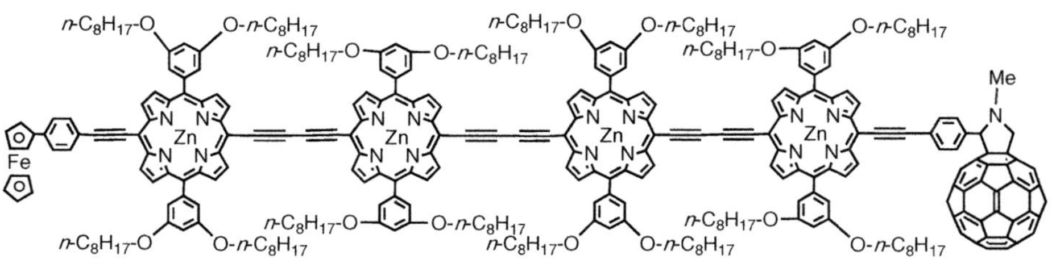

6

Figure 4.2 *Leading examples of donor-fullerene linked multichromophore systems* ***3–6***

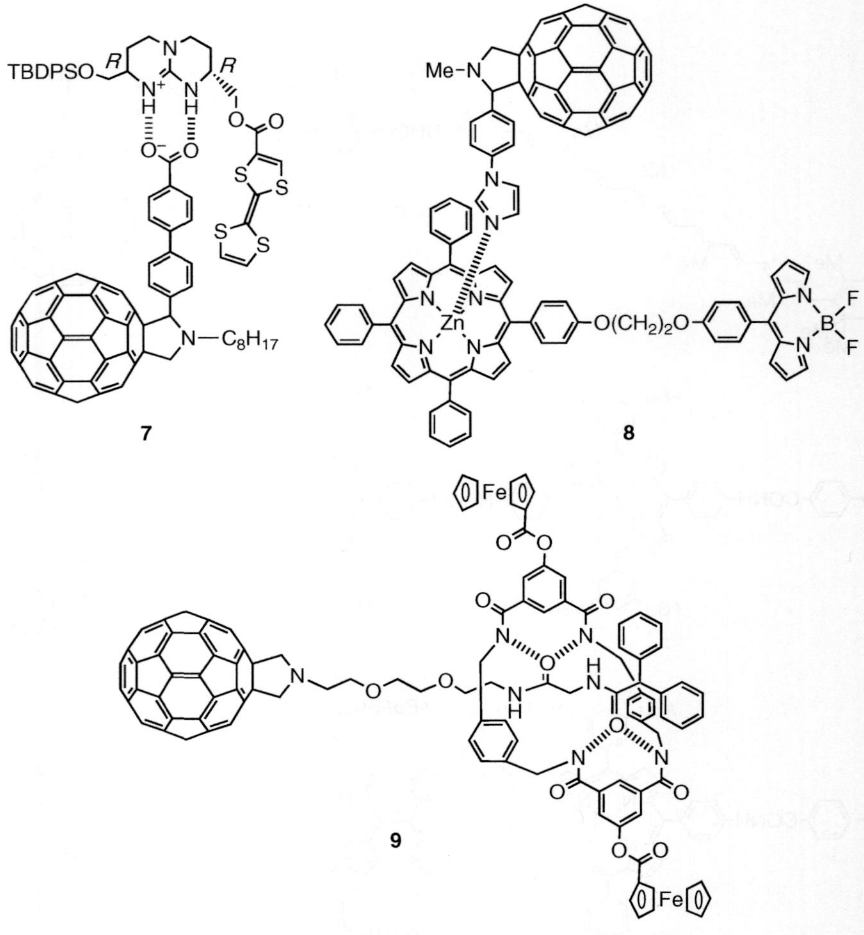

7 8 9

Figure 4.3 *Leading examples of supramolecular donor-fullerene systems* **7–9**

linked C_{60}-TTF dyads [68]. A supramolecular triad **8** was organized by axially coordinating imidazole-appended fulleropyrrolidine to the zinc atom of a covalently linked zinc porphyrin-boron dipyrrin dyad [69]. Selective excitation of the boron dipyrrin moiety in the dyad resulted in efficient energy transfer ($9.2 \times 10^9\ s^{-1}$) with a quantum yield of 83% creating the $^1ZnP^*$. Upon forming the supramolecular triad, subsequent electron transfer took place from the $^1ZnP^*$ to the C_{60}, generating a charge-separated state ($4.7 \times 10^9\ s^{-1}$) with a quantum yield of 90% [69]. A molecular rotaxane shuttle **9** consisting of two ferrocene-appended macrocycle and C_{60}-tethered axle was prepared to modulate the kinetics of charge separation and charge recombination [74]. In a nonpolar solvent hydrogen bonding between the macrocycle and the fumaramide template of the axle is strengthened. As a result the macrocycle shuttles to the opposite end of the thread far from the C_{60} moiety, leading to slow charge separation and charge recombination. On the other hand, in a polar solvent the hydrogen bonding interaction is weakened. Therefore, the macrocycle shuttles to the opposite end of the thread close to the C_{60} moiety, resulting fast charge separation and charge recombination [74].

4.2.5 Photoelectrochemical Devices and Solar Cells

Extensive efforts have also been made in recent years to explore the photovoltaic and photoelectrochemical properties of donor-fullerene composites [76–107]. To optimize the device performance, it is of importance to (i) collect the visible light extensively, (ii) produce a charge-separated state efficiently, and (iii) transport resultant electrons and holes into respective electrodes, minimizing undesirable charge recombination. Accordingly, the fabrication of donor-fullerene composites onto electrodes is a vital step for controlling the morphology of the composite assemblies on the electrode surface in molecular scale. Versatile methods such as Langmuir-Blodgett (LB) films [76], self-assembled monolayers (SAM) [77–85], layer-by-layer deposition [86, 87], vacuum deposition [88, 89], electrophoretic deposition [90–99], chemical adsorption, and spin coating [100–107], have been adopted to fabricate photoelectrochemical devices and solar cells.

4.2.5.1 Langmuir Blodgett Films

LB technique has been proven to be a powerful, convenient method to construct organic thin films controlled in the order of the molecular level. It has been frequently used to organize donor-fullerene composites on electrode surfaces [76].

Photoinduced electron transfer and photocurrent generation in LB films of phytochlorin-fullerene dyad and porphyrin-fullerene dyads were studied [76]. For instance, a mixed monolayer of octadecylamine and chlorin-fullerene dyad possessing a hydrophilic propionic acid residue can be transferred onto a solid substrate, which is characterized by uniform orientation of the dyad. From time-resolved Maxwell displacement charge measurements, the LB film was found to exhibit vectorial photoinduced electron transfer with a poor quantum yield of 0.2%. The lifetime of the charge-separated state in the LB film was ca. 30 ns, being almost independent of the concentration of the dyad (2–50 mol%) [76].

4.2.5.2 Self-assembled Monolayers

SAMs have recently attracted much attention as a new methodology for molecular assembly. They enable the molecules of interest to be bound covalently on the surface such as metals, semiconductors, and insulators in a highly organized manner. The well-ordered structure in SAMs is in striking contrast with conventional LB films and lipid bilayer membranes in terms of stability, uniformity, and manipulation. Therefore, they make it possible to arrange functional molecules unidirectionally at the molecular level on electrodes in the case that substituents, which would self-assemble covalently on the substrates, are attached to a terminal of the molecules. A wide variety of examples have been reported to date involving donor-fullerene composites on gold [77–81], indium tin oxide (ITO) [82–85], and other surfaces [86]. Leading examples of donor-fullerene linked molecules **10–12** that can self-assemble on a gold electrode are depicted in Figure 4.4.

A tripodal rigid anchor was employed to organized oligothiophene-fullerene dyad **10** vertically on a gold electrode [77]. The internal quantum yield of photocurrent generation (35%) is remarkable considering the rather simple molecular structure. An alkanethiol-

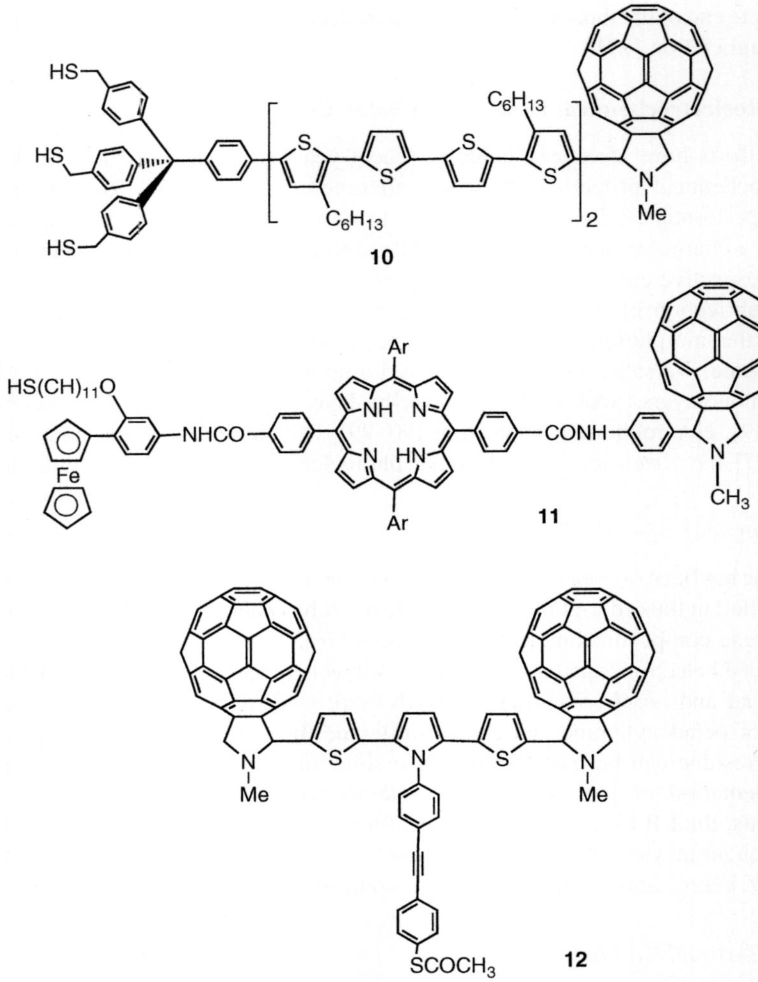

Figure 4.4 *Leading examples of donor-fullerene linked systems **10–12** that can self-assemble on a gold electrode*

attached triad **11** was designed with a linear array of ferrocene (Fc), porphyrin (H_2P), and C_{60} [78]. The triad molecules were densely packed with an almost perpendicular orientation on the gold surface. The internal quantum yield of photocurrent generation in the **11**-modified gold electrode in the presence of methylviologen was 25% [78]. Utilization of C_{60} with the small reorganization energy allows the device to accelerate the forward electron transfer and decelerate the undesirable back electron transfer, thus leading to the high quantum yields. The triad **11** was further incorporated into the boron dipyrrin-alkanethiol to mimic light harvesting in antenna complexes and multistep electron transfer in reaction centers [79]. The internal quantum yield ($50 \pm 8\%$) at 510 nm is one of the highest values ever reported for photocurrent generation at monolayer-modified metal electrodes using donor-acceptor linked molecules [77–81]. ATTF-porphyrin-fullerene triad was also developed as a nanoscale power supply for a supramolecular machine [80]. The rigid nanostructured triad **12** was presented in which the fixation of the anchoring site onto the nitrogen atom of the pyrrole moiety allows the conjugated chain to adopt an orientation parallel to the surface [81]. The internal quantum yield of photocurrent generation using **12** reached 51%, although the value may suffer from large uncertainty because of the poor absorbance on the gold surface [81].

Some of SAM devices on gold electrodes have exhibited efficient photocurrent generation [77–81]. However, strong quenching of the excited singlet state of the adsorbed dyes by gold surfaces has precluded achievement of a high internal quantum yield for charge separation on the surfaces as attained in natural photosynthesis (~100%). To surmount such a quenching problem, ITO which has high optical transparency and electrical conductivity seems to be attractive as an electrode, since the ITO electrode with a conduction band of which the level is higher than that of the excited states of adsorbed dye on the surface can suppress the quenching of the dye excited states on the surface.

An ITO electrode was also covalently modified with a C_{60}-metal cluster moiety that was further tethered with zinc porphyrin (Figure 4.5) [84]. The internal quantum yield of the anodic photocurrent generation was estimated to be 10.4%. Surprisingly, an addition of DABCO into the device resulted in the improvement of the quantum yield (19.5%), which is one of the highest quantum yields ever reported for molecular photoelectrochemical devices based on the covalently linked donor-acceptor molecules on ITO [82–85]. From the fluorescence lifetime measurements together with femtosecond transient absorption studies, the authors draw the conclusion that the complexation of DABCO between the two zinc porphyrins precludes aggregation with adjacent porphyrins and increase of the donor-acceptor separation, leading to the high performance in photocurrent generation [84].

An ITO electrode was fabricated with sodium 3-mercaptoethanesulfonate (first layer), hexacationic homooxacalix[3]arene-C_{60} 2:1 complex (second layer), and anionic zinc porphyrin polymers (third layer) sequentially [85]. The internal quantum yield of anodic photocurrent generation in the supramolecular device was estimated as 21% ($\lambda_{ex} = 430$ nm) [85].

The examples given demonstrate that SAMs of donor-fullerene composites on gold and ITO electrodes are excellent systems for the realization of efficient photocurrent generation

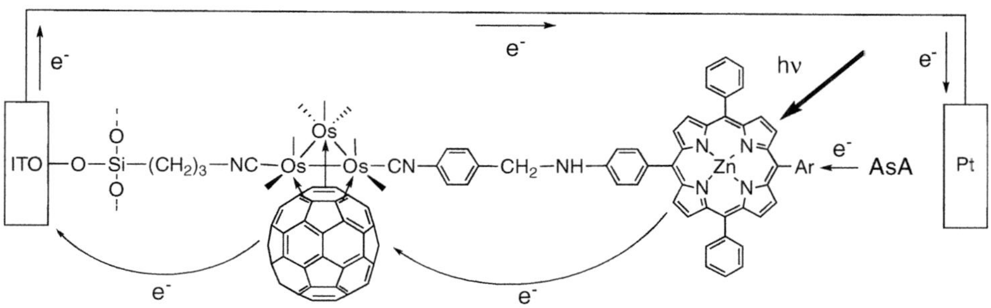

Figure 4.5 *Schematic diagram for photocurrent generation by ITO electrode modified with porphyrin-fullerene linked dyad*

on electrode surfaces. SAM will open a door for the development of photoactive molecular devices in which the highly ordered, well-designed architecture acts as efficient photocatalysts, photosensors, and photodiodes.

4.2.5.3 Layer-by-Layer Deposition

Weak interactions (i.e. coordination bonding, hydrogen bonding, electrostatic and *van der Waals* interactions) were employed to fabricate photoactive semiconducting electrodes deposited with donor-fullerene composites [86, 87]. C_{60} molecules with negatively charged moieties are adsorbed at the PDDA-modified, positively charged ITO surface [87], as illustrated in Figure 4.6. Strong *van der Waals* interactions between C_{60} cores facilitate the device layer formation, leaving the anionic dendrimer branches on the surface. In the next step octacationic porphyrins are deposited via electrostatic interactions with the anionic dendrimer branches in a monolayer fashion. Subsequent octaanionic porphyrin layers are built up analogously utilizing cationic/anionic contacts. The IPCE value of the anodic photocurrent generation was 1.6% at 440 nm with a bias potential of 0 V vs Ag/AgCl [87].

4.2.5.4 Vacuum Deposition

Vacuum deposition technique has been frequently employed to fabricate fullerenes on electrode surfaces [88, 89]. Typically, bulk heterojunction solar cells have been prepared using vacuum codeposition of donor (i.e. phthalocyanines) and pristine fullerenes onto electrode surfaces. The representative device structure of small molecule-based bulk heterojunction solar cells is shown in Figure 4.7. So far the power conversion efficiency has reached up to 5.0% for single cell [89] and 5.7% for tandem cell [88].

4.2.5.5 Electrochemical Deposition

A novel approach to enhance the light-harvesting efficiency was introduced into C_{60}-based photoelectrochemical devices by electrophoretically depositing donor-C_{60} composite clusters in acetonitrile/toluene (3/1, v/v) onto a semiconducting electrode [90–92]. Specifically, a toluene solution of donor-C_{60} molecules is rapidly injected into acetonitrile to form donor-C_{60} clusters due to the lyophobic nature in the mixed solvent. The clusters of donor-C_{60} are deposited as thin films on nanostructured SnO_2 electrodes under the influence of an electric field.

Porphyrin-fullerene composites were electrophoretically deposited onto a SnO_2 electrode to construct a novel organic solar cell (dye-sensitized bulk heterojuntion solar cell), possessing both the dye-sensitized and bulk heterojunction characters [93]. The photocurrent generation is initiated by electron transfer from the $^1H_2P^*$ ($^1H_2P^*/H_2P^{\cdot+}=-0.7$ V vs NHE) to C_{60} ($C_{60}/C_{60}^{\cdot-}=-0.2$ V vs NHE). The resulting $C_{60}^{\cdot-}$ transfers electrons to the conduction band of the SnO_2 ($E_{CB}=0$ V vs NHE), to generate the current in the circuit. The regeneration of H_2P clusters ($H_2P/H_2P^{\cdot+}=1.2$ V vs NHE) is achieved by the iodide/triiodide couple ($I^-/I_3^-=0.5$ V vs NHE) present in the electrolyte system (Figure 4.8) [93].

The present organic solar cell (dye-sensitized bulk heterojuntion solar cell) is unique in that it possesses both the dye-sensitized and bulk heterojunction characteristics. Moreover, the blend films exhibit the multilayer structure on the SnO_2, which presents a striking contrast to monolayer structure of adsorbed dyes on TiO_2 electrodes of dye-sensitized solar

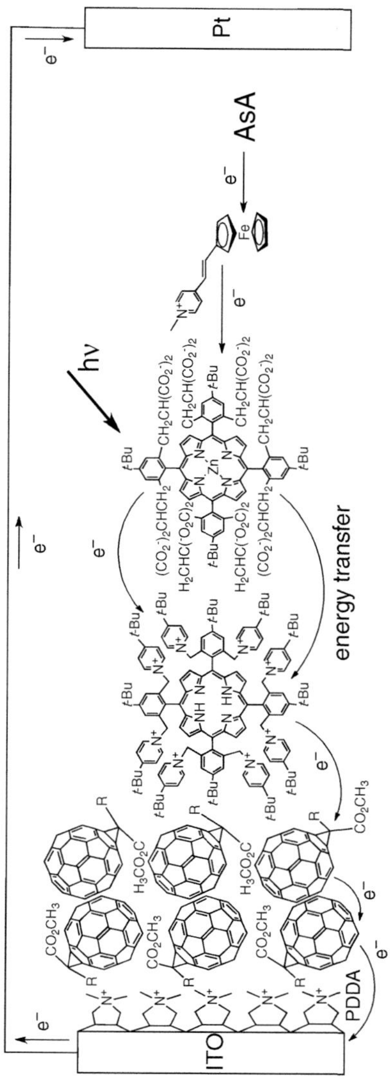

Figure 4.6 *Schematic diagram for photocurrent generation in fullerene-porphyrin-ferrocene system by layer-by-layer assembly*

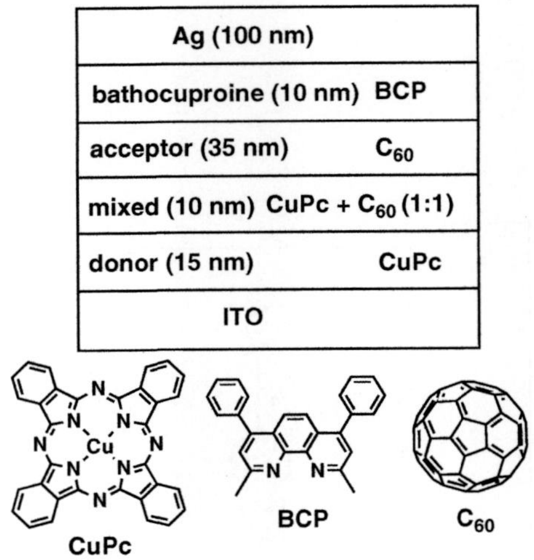

Figure 4.7 *Structure of single photovoltaic cell consisting of a mixed CuPc : C_{60} layer sandwiched between homogeneous CuPc and C_{60} layers as the photoactive layer with BCP layer serving electron blocking layer*

cells. Therefore, we can expect improvement of the photovoltaic properties by modulating both the structures of electrode surfaces and donor-acceptor multilayers [92].

Supramolecular assembly of donor-acceptor molecules is a potential approach to create a desirable phase-separated, interpenetrating network involving molecular-based nanostructured electron and hole highways. However, different, complex hierarchies of self-organization going from simple molecules to devices have limited improvement of the device performance. To construct such complex hierarchies comprising of donor and acceptor molecules on electrode surfaces pre-organized molecular systems are excellent candidates for achieving the molecular nanoarchitectures. In particular, porphyrins have been three-dimensionally organized using dendrimers [94], oligomers [95], and nanoparticles [96–98] to combine with fullerenes for organic solar cells.

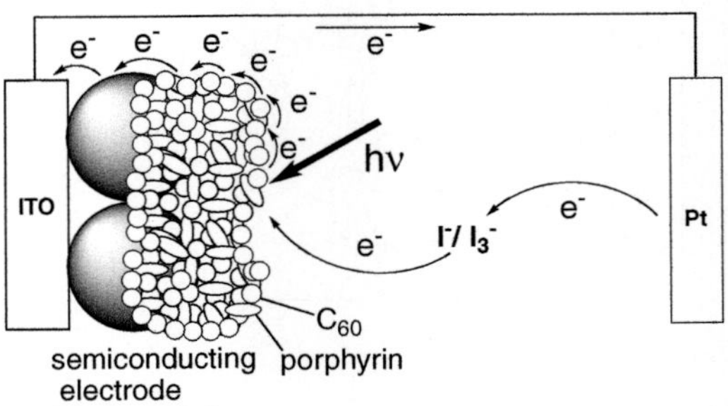

Figure 4.8 *Schematic diagram for photocurrent generation in dye-sensitized bulk heterojunction solar cell consisting of porphyrin-fullerene composites*

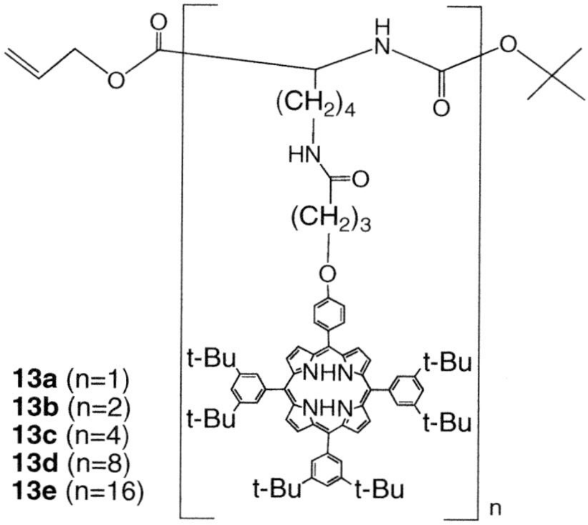

13a (n=1)
13b (n=2)
13c (n=4)
13d (n=8)
13e (n=16)

Figure 4.9 *Molecular structures of porphyrin oligomers* ***13a*** *(n = 1),* ***13b*** *(n = 2),* ***13c*** *(n = 4),* ***13d*** *(n = 8), and* ***13a*** *(n = 16) for dye-sensitized bulk heterojunction solar cells*

For instance, dye-sensitized bulk heterojunction solar cells using supramolecular complexes of porphyrin-peptide oligomers **13a-e** (n = 1,2,4,8,16) with C_{60} was fabricated (Figure 4.9) [95]. The SnO_2 electrodes modified with the composite clusters of **13a-e** with C_{60} were prepared by the electrophoretic deposition method. The η value of 1.6% (J_{SC} = 0.36 mA cm^{-2}, V_{OC} = 0.32 V, ff = 0.47, W_{IN} = 3.4 mW cm^{-2}) and IPCE of 48% at 600 nm were attained for the **13e**-C_{60} composite device. The η value (1.6%) is ca. 40 times as large as that (0.043%) of the reference device using porphyrin monomer [95]. These results explicitly exemplify that the formation of a molecular assembly between C_{60} and multi-porphyrin arrays with an oligopeptide backbone controls the charge separation efficiency in the supramolecular complex, which is essential for the efficient light-energy conversion.

Unique molecular arrangement of 5,10,15,20-(tetrakis(3,5-dimethoxyphenyl)porphyrinato zinc(II) and C_{60} on a SnO_2 electrode resulted in one of largest IPCE value (ca. 60%) among this type of photoelectrochemical devices [99]. Rapid formation of the composite clusters (~100 nm) and the micro-cocrystal (~2 μm) from the combination in the mixed solvent is notable. The unique association is accelerated by the hydrogen bonding interactions between the methoxy groups of the porphyrins, the CH-π interactions between the methoxy groups of the porphyrin and C_{60}, and the π–π interactions between the porphyrin-C_{60} as well as C_{60} molecules. The SnO_2 electrode modified with the composite clusters and the micro-cocrystal yields remarkably efficient photocurrent generation by the bicontinuous donor-acceptor network at molecular level [99].

4.2.5.6 Spin Coating Deposition

For typical bulk heterojunction solar cells involving blend films of conjugated polymers-fullerene derivatives or donor-fullerene linked dyads, spin-coating method has been used for the fabrication on electrode surfaces [100–104]. Bicontinuous, interpenetrating network composed of conjugate polymers and fullerene derivatives at nanometer scale in

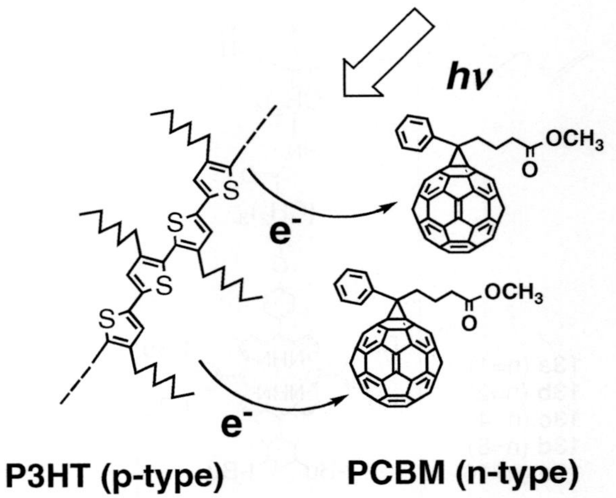

Figure 4.10 Schematic illustration of charge separation in large molecule-based bulk heterojunction solar cells consisting of P3HT [poly(3-hexylthiophene)] as a donor and PCBM ([6,6]-phenyl-C61-butyric acid methyl ester) as an acceptor

a photoactive layer is considered to be essential for the efficient charge separation and subsequent efficient hole- and electron-transportation (Figure 4.10). So far the power conversion efficiency has reached up to 6.1% for single cell [103] and 6.5% for tandem cell [104] with the photoactive layer of poly(3-hexylthiophene) (P3HT) as a donor and [6]-phenyl-C61-butyric acid methyl ester (PCBM) as an acceptor.

Hydrogen-bonded network of donor and fullerene was also formed on electrodes [105–107]. Specifically, porphyrin carboxylic acid and C_{60} carboxylic acid were spin-coated on TiO_2 and SnO_2 electrodes for dye-sensitized bulk heterojunction solar cells [105]. The TiO_2 cell yielded $\eta = 2.1\%$ with $J_{SC} = 5.1$ mA cm^{-2}, $V_{OC} = 0.58$ V, and $ff = 0.70$, whereas the SnO_2 cell exhibited $\eta = 0.31\%$ with $J_{SC} = 2.3$ mA cm^{-2}, $V_{OC} = 0.36$ V, and $ff = 0.39$. The large V_{OC} value of the TiO_2 cell relative to that of the SnO_2 cell is consistent with the higher conduction band of the TiO_2 than that of the SnO_2 [105].

4.3 Carbon Nanotubes

4.3.1 General

Conceptionally, single wall carbon nanotubes (SWNT) are considered as small strips of graphene sheets that have been rolled up to form perfect seamless single walled nanocylinders. SWNT are usually described using the chiral vector, which connects two crystallographically equivalent sites on a graphene sheet. The way the graphene sheets are wrapped varies largely and is represented by a pair of indices (*n*,*m*). These integers relate the structure of each SWNT to both its diameter and chirality. The diameter of most SWNT is about 1 nm, while their length reaches into the order of centimeters [108–117]. A multi wall carbon nanotube (MWNT) is similarly considered to be a coaxial assembly of cylinders of SWNT. The simplest representative of a MWNT is a double wall carbon nanotube (DWNT).

Depending on their helicity, SWNT are either electrically conductive or semiconductive [118, 119]. The electronic properties of a SWNT vary in periodic ways between metallic and semiconductor and follows a general rule. If $(n-m)$ is a multiple of 3, then the tube exhibits a metallic behavior. If $(n-m)$ is, on the other hand, *not* a multiple of 3, then the tube exhibits a semiconducting behavior. The electrical transport in good quality metallic SWNT is ballistic, that is, electrons do not suffer from any scattering over a length scale of several micrometers and/or from any electromigration, even at room temperature.

Electrostatic electron/hole interaction energies in form of exciton binding energies are significant in SWNT with values on the order of 0.3 to 0.5 eV at band gap energies of approximately 1 eV [120–123]. An immediate consequence of this strong electron/hole attraction is that photoexcited states of carbon nanotubes (CNT) are regarded as excitonic (i.e., electron/hole pairs) rather than as uncoupled electrons. In SWNT, excitons are characterized by electron/hole separations (i.e. Bohr radius) of approximately 2.5 nm [122, 124].

Fluorescence in SWNT bundles is hardly ever observed. Metallic SWNT that are statistically present in bundles are the inception for photoexcited carriers in semiconducting SWNT to relax along efficient nonradiative pathways [125]. On the contrary, fluorescence measurements with individual SWNT have revealed distinct fluorescing pattern for more than 30 different SWNT [126]. Interestingly, the fluorescence quantum yields are very low (i.e., 10^{-4}) – a finding which has been rationalized on the basis of multiple dark excitonic states that are situated below the lowest lying bright excitonic states [127]. In complementary work, which focused on the fluorescence of SWNT the excited states have been shown to decay on the time scale of 10 ps.

The outstanding photostability of SWNT, namely resistance to photobleaching, together with their versatile wavelength tunability – absorbing light between 800 and 1100 nm and emitting light between 1300 and 1500 nm – render them ideal single molecule fluorophores [128, 129].

Ultrafast transient absorption spectroscopic measurements complement the fluorescence studies and have been used to clarify the time scales and nature of ground state recoveries in SWNT and to extract information about excitonic lifetimes. For relaxation from the excited state, an omnipresent fast decay component (i.e. 300 to 500 fs) is likely due to the presence of bundled SWNT and/or metallic SWNT. A much slower decay component (i.e. 100 to 130 ps) only appears when semiconducting SWNT are probed and likely corresponds to the intrinsic excited state lifetime of photoexcited excitons.

The electron-accepting properties of CNT and the factors that control these, need careful considerations as they impact the reactivity of the reduced state, their photochemistry and photophysics. Reduction of SWNT bundles is achieved by exposure to molecules of different redox potentials, in form of intercalating with alkali metals and/or anion radicals [130, 131]. The course of these reactions – filling the density of states and thus modifying the conducting nature of individual SWNT – is typically monitored spectroscopically by visible/near-infrared absorption measurements [132] and by in situ Raman [133]. In the visible/near-infrared absorption spectrum, the most profound changes are seen as bleaching of the optical transitions that are associated with the filling of the corresponding electronic states. Owing to their special electronic structure, SWNT show characteristic Raman spectra, which are understood in terms of resonance enhancement in one-dimensional conductors with van Hove singularities in the electronic density of states.

At first glance, when SWNT are reduced, the radial breathing mode (RBM) intensities go down due to an inherent loss of resonant conditions (i.e. bleaching of the S_{11}, S_{22} and M_{11} optical transition) [134, 135]. A similar intensity loss characterizes the tangential vibrational modes (high energy mode (HEM)).

In-situ spectroelectrochemistry, where bulk spectroscopy and phase boundary electrochemistry are complementary combined, is, nevertheless, better suited to gather details about the reductive chemistry of SWNT. For example, individual SWNT were tested by in situ micro-Raman spectroelectrochemistry [136, 137]. Overall, the spectroelectrochemical results confirm the charging induced bleaching of transitions between van Hove singularities. The bleaching of optical transitions is mirrored by the quenching of resonance Raman scattering in the region of RBM and HEM.

Oxidation has been among the first reactions that were ever tested with CNT [138] and is still a key step in their purification [139, 140]. The oxidation processes are commonly accompanied by gradually disappearing S_{11}, S_{22} and M_{11} bands and concomitantly growing bands in the range between S_{11} and S_{22} [141]. Raman spectroelectrochemistry, on the other hand, inflicts a reversible drop of the RBM and HEM intensities [134, 141]. In addition, strong blue shifts are seen for the RBM in the potential range between 0.5 to 1 V. The HEM blue shifts too – stiffening is connected to the introduction of holes in the π-band [142].

4.3.2 Carbon Nanotube – Electron Donor Acceptor Conjugates

Via the cycloaddition of azomethine ylides several electron donors were attached to the sidewalls of CNT. In this context, ferrocene constitutes one of the first examples that was brought forward [143]. Spectroscopic and kinetic analyses of the photophysical properties of SWNT-Fc were interpreted in terms of intramolecular charge separation that evolves from photoexcited SWNT. The charge separation dynamics is very fast ($3.6 \times 10^9\,s^{-1}$), whereas the charge recombination kinetics is very slow ($9.0 \times 10^5\,s^{-1}$). Since the correct identification of the product was deemed to be critical, additional time resolved pulse radiolytic and steady state electrolytic experiments were carried out to establish the characteristic fingerprints of reduced SWNT. All techniques gave similar broad absorptions in the visible range and confirmed the formation of reduced SWNT.

In a modified strategy, namely purifying, shortening and endowing SWNT with carboxylic groups electron donors like TTF and extended tetrathiafulvalene (exTTF) were tested [144]. Photophysical investigation supported the occurrence of photoinduced charge transfer processes in SWNT-TTF and SWNT-exTTF and helped to identify the reduced SWNT and oxidized TTF and/or exTTF as metastable states. Overall, remarkable lifetimes – in the range of hundreds of nanoseconds – are noted. Most important is that we succeeded in the control over the rate of charge recombination by either systematically altering the relative donor acceptor separations ($6.3 \times 10^6\,s^{-1}$–$2.9 \times 10^6\,s^{-1}$) or integrating different electron donors ($5.2 \times 10^6\,s^{-1}$ versus $3.6 \times 10^6\,s^{-1}$) [145].

As a complement to the covalent approach, π-π interactions were pursued to anchor TTF and exTTF to the surface of SWNT by using a pyrene tether [146]. Nevertheless, π-π interactions between, for example, the concave hydrocarbon skeleton of exTTF and the convex surface of SWNT add further strength and stability to SWNT/pyrene-exTTF. In this context, for the first time a complete and concise characterization of the radical ion pair state has been achieved, especially in light of injecting electrons into the conduction band of SWNT. The close proximity between exTTF and the electron-accepting SWNT leads to very rapid charge transfer ($1.1 \times 10^{12}\,s^{-1}$) that affords, in turn, a short-lived radical ion pair state ($3.3 \times 10^{6}\,s^{-1}$). Significantly weaker are the interactions in SWNT/pyrene-TTF, which led to lifetimes exceeding that noted for SWNT/pyrene-exTTF [147].

More powerful is the concept of integrating porphyrins [148, 149] or phthalocyanines [150, 151] that serve as visible light harvesting chromophores/electron donors (Figure 4.11). A first example involved the efficient covalent tethering of SWNT with porphyrins through the esterification of SWNT bound carboxylic acid groups. The work started with two porphyrin derivatives containing terminal hydroxyl groups, that is, [5-(4-hydroxyhexyloxyphenyl)-10,15,20-tris(4-hexadecyloxyphenyl)porphyrin] and [5-(4-hydroxymethylphenyl)-10,15,20-tris(4-hexadecyloxyphenyl)porphyrin]. In the corresponding SWNT-H_2P conjugates the photoexcited porphyrins deactivate, through a transduction of excited state energy. Interestingly, the rates and efficiencies of the excited state transfer depend on the length of the tether that links the porphyrins with the SWNT: The shorter tethers gave rise to weaker fluorescence quenching [152, 153].

SWNT functionalization is also the basis when applying Suzuki coupling reactions [154]. Recent work documents that this type of coupling reactions represents an efficient method for introducing ZnP onto the SWNT sidewalls. Despite all perceptions, covalently functionalized SWNT were found to serve as efficient quenchers even after their perfectly conjugated sidewall structure has been disrupted [155].

Similarly, unsymmetrically substituted aminophthalocyanines ZnPc were linked to SWNT through a reaction with the terminal carboxylic acid groups of shortened SWNT. However, the resulting materials were found to be nearly insoluble in organic solvents [156, 157]. Reasonable alternatives involve a straightforward cycloaddition reaction with N-octylglycine and a formyl-containing ZnPc, a stepwise approach that involves cycloaddition of azomethine ylides to the double bonds of SWNT using *p*-formyl benzoic acid followed by esterification with an appropriate ZnPc [158, 159] or functionalization of SWNT with 4-(2-trimethylsilyl)ethynylaniline and the subsequent ZnPc attachment using the *Huisgen* 1,3-dipolar cycloaddition [160]. The occurrence of charge transfer from photoexcited ZnPc to SWNT is observed in transient absorption experiments, which reflect the absorption of the $ZnPc^{\bullet+}$ and the concomitant bleaching of the van Hove singularities of SWNT. Charge separation ($2.0 \times 10^{10}\,s^{-1}$) and charge recombination ($7 \times 10^{-5}\,s^{-1}$) dynamics reveal a notable stabilization of the radical ion pair product.

Quite different is the approach, which involves placing pyridyl isoxazolino functionalities along the sidewalls of short SWNT [161]. The synthesis is based on the cycloaddition of a nitrile oxide onto SWNT. The resulting SWNT-pyridine forms complexes with ZnP and/or RuP [161, 162]. Formation of the SWNT-pyridine/ZnP complex was firmly established by a detailed electrochemical study. However, photochemical excitation of SWNT-pyridine/ZnP does not lead to generation of the radical ion pair states. Instead, fluorescence and transient absorption studies indicate that the main process is energy transfer from the singlet excited state of ZnP to SWNT-pyridine.

An important consideration when associating SWNT with electron donors is to preserve the unique electronic structures of SWNT. A versatile approach involves grafting SWNT with polymers such as poly(sodium 4-styrenesulfonate) (PSS^{n-}) [163], poly(4-vinylpyridine) (PVP) [164], and poly((vinylbenzyl)trimethylammonium chloride) ($PVBTA^{n+}$) [165] to form highly dispersable SWNT-PSS^{n-}, SWNT-PVP, and SWNT-$PVBTA^{n+}$, respectively (Figure 4.12). In the next step, coulomb complex formation was achieved with SWNT-PSS^{n-} and cationic porphyrin (H_2P^{8+}). Likewise a donor acceptor

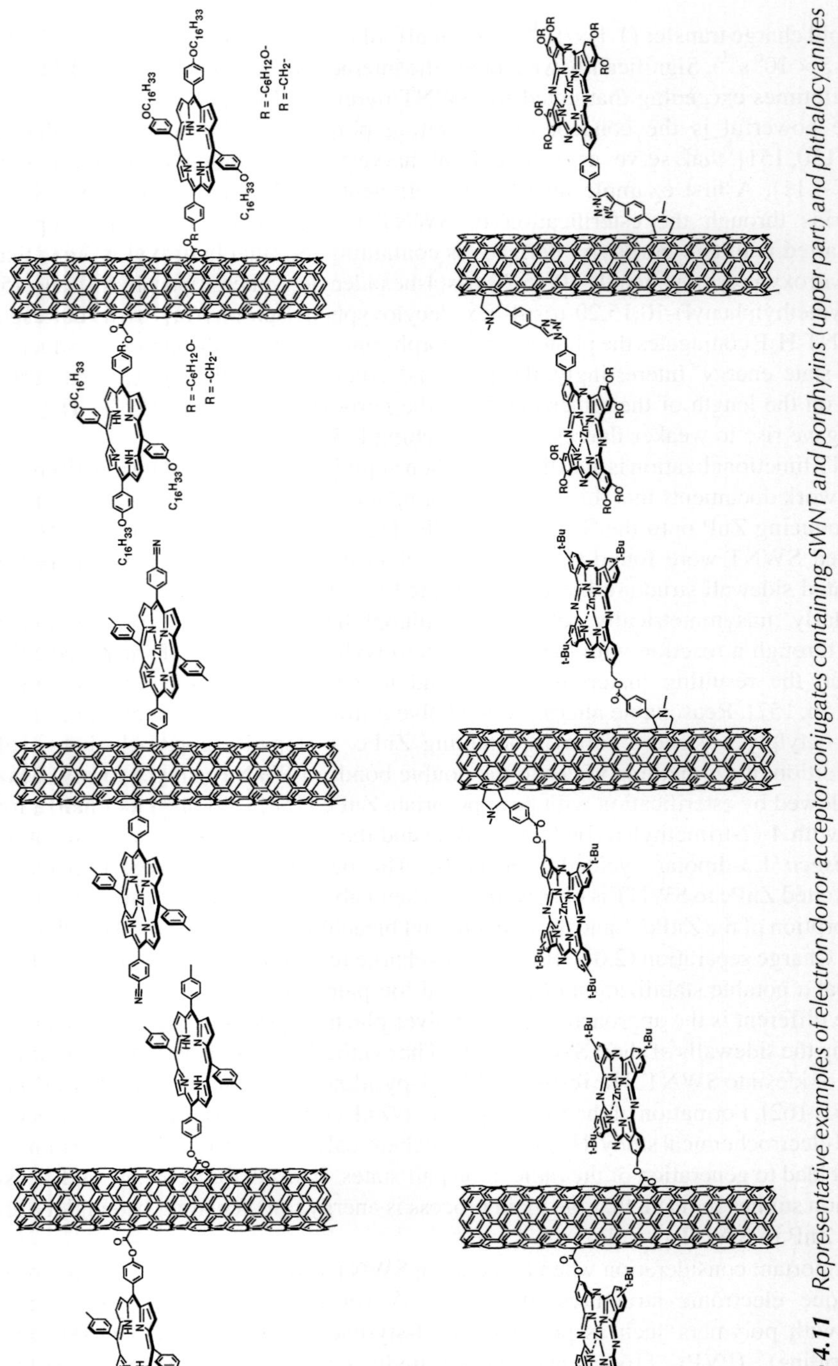

Figure 4.11 *Representative examples of electron donor acceptor conjugates containing SWNT and porphyrins (upper part) and phthalocyanines (lower part)*

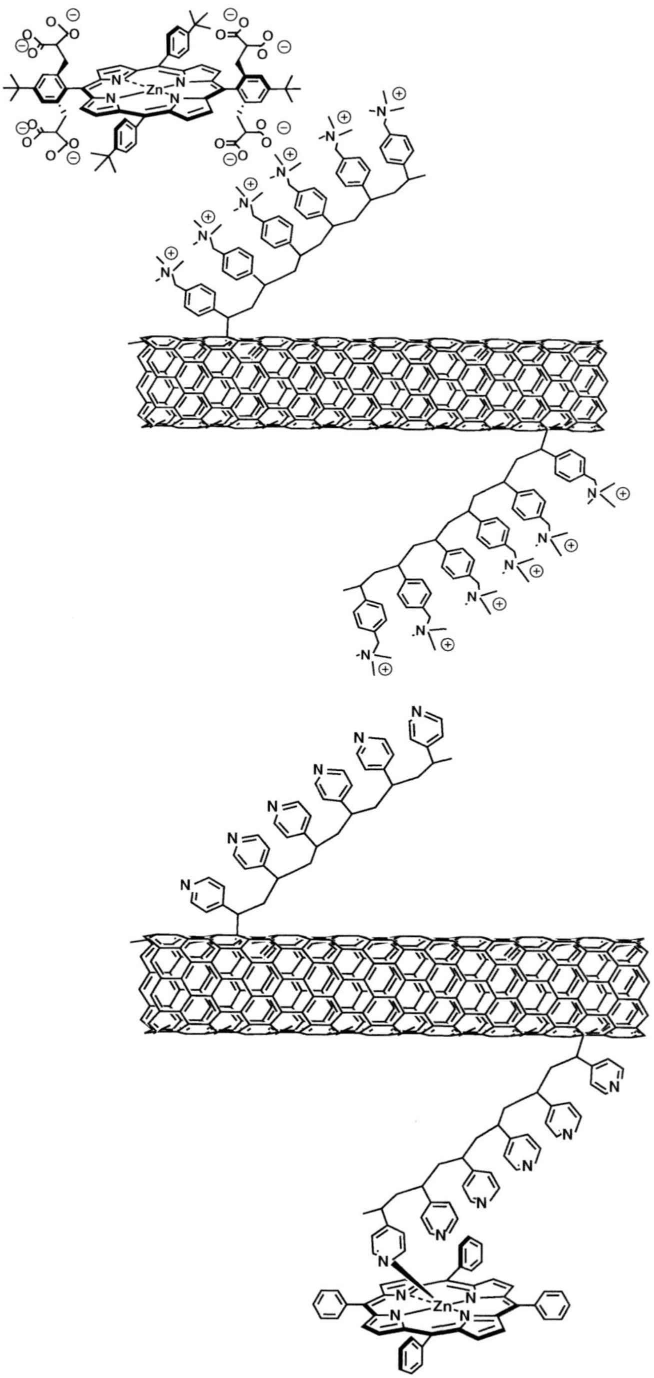

Figure 4.12 *Representative examples of electron donor acceptor hybrids containing SWNT and porphyrins*

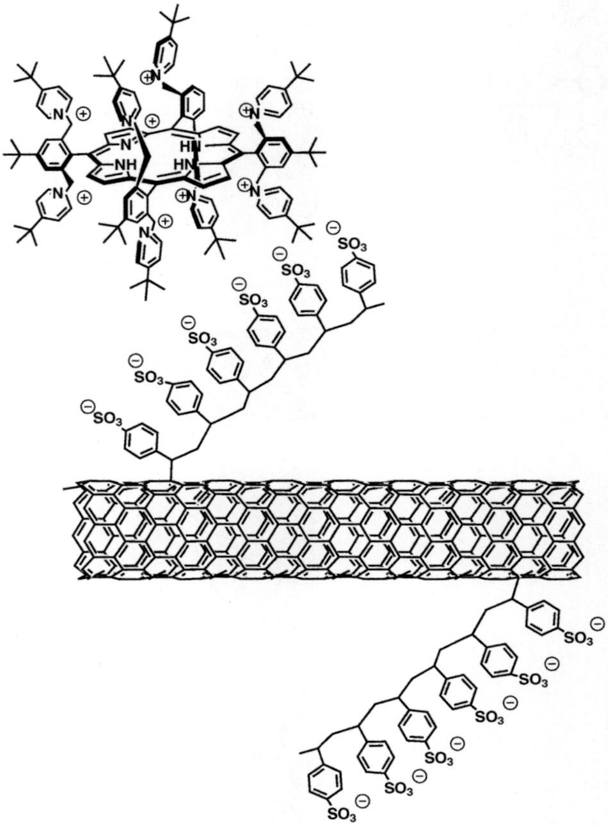

Figure 4.12 *(Continued)*

nanohybrid has been prepared using electrostatic/van der Waals interactions between SWNT-PVBTA^{n+} and 5,15-bis[2′,6′-bis{2″,2″-bis(carboxy)ethyl}-methyl-4′-*tert*-butyl-pheny]-10,20-bis(4′-*tert*-butylphenyl) porphyrin (H_2P^{8-}). Several spectroscopic techniques like absorption, fluorescence, and TEM were used to monitor the complex formations. Importantly, photoexcitation of H_2P^{8+} or H_2P^{8-} in the newly formed nanohybrid structure results in efficient charge separation (3.3×10^9 s^{-1}), which lead, subsequently, to the radical ion pair formation. In SWNT-PSS^{n-}/H_2P^{8+}, the newly formed radical ion pair exhibits a remarkably long lifetime (7.1×10^4 s^{-1}), which constitutes one of longest values reported for any CNT ensemble found so far. In SWNT-PVBTA^{n+}/H_2P^{8-} the charge separation tends to be slightly faster (4.5×10^5 s^{-1}). Differently, SWNT-PVP were assayed in coordination tests with ZnP. Kinetic and spectroscopic evidence corroborates the successful formation of SWNT-PVP/ZnP nanohybrids in solutions. Within this SWNT-PVP/ZnP nanohybrid, static charge transfer quenching (2.0×10^9 s^{-1}) converts the photoexcited ZnP chromophore into a microsecond-lived radical ion pair state, that is, one electron oxidized ZnP and reduced SWNT.

A series of SWNT that are functionalized with poly(amido amine) (PAMAM) dendrimers were recently described. Since the dendrimers are linked directly to the SWNT surface via a divergent methodology, it allows increasing the number of functional groups without implementing significant damages to the conjugated π-system of SWNT [166]. Photophysical investigations reveal that in SWNT-$(H_2P)_n$ some porphyrin units interact with SWNT while others do not. Those that react ($2.5 \times 10^{10}\,s^{-1}$) form a radical ion pair state that decays to ground state with a time constant of $2.9 \times 10^5\,s^{-1}$.

4.3.3 Carbon Nanotube – Electron Donor Acceptor Hybrids

Nevertheless, covalently modified CNT may not be suitable for applications, which are based on the high conductivity and/or mechanical strength of pristine SWNT. Noncovalent approaches might offer a solution to preserve the electronic and structural integrity of SWNT, permitting the use of both their conductivity and strength properties in future applications. Triggered by these incentives the noncovalent integration of a wide range of functional groups onto CNT emerged as viable alternatives [167–170].

As a leading example, the facile supramolecular association of SWNT with linearly polymerized porphyrin polymers should be considered. The target SWNT nanohybrids, which are dispersable in organic media, were realized through the use of soluble and redox-inert poly(methylmethacrylate) (PMMA) bearing surface immobilized porphyrins (i.e. H_2P-polymer). Conclusive evidence for H_2P-polymer/SWNT interactions came from absorption spectroscopy – a conclusion that was further corroborated by TEM and AFM. In fact, the latter illustrates the debundling of individual SWNT. An additional feature of H_2P-polymer/SWNT is charge separation, which has been shown to be long lived ($4.7 \times 10^5\,s^{-1}$) [171]. Using a flexible porphyrinic polypeptide $P(H_2P)_{16}$ ensures via supramolecular wrapping of the peptid backbone around SWNT extracting large-diameter SWNT (ca. 1.3 nm). Like in the aforementioned case, photoexcitation of $P(H_2P)_{16}$ affords a slowly decaying radical ion pair state ($2.7 \times 10^6\,s^{-1}$) [170]. Following basically a similar strategy, SWNT were found to strongly interact with a highly soluble, conjugated ZnP-polymer, a triply fused ZnP-trimer [172], and just ZnP (Figure 4.13). Unambiguous evidence for interactions between ZnP-polymer, ZnP-trimer and ZnP and SWNT came from fluorescence spectra: Quenching of the ZnP centered fluorescence was ascribed to energy transfer between the photoexcited porphyrin and SWNT.

Generally, SWNT interactions with H_2P are appreciably stronger than what is typically seen when employing ZnP. AFM images of, for example, SWNT/H_2P revealed smaller bundle sizes than those recorded for SWNT/ZnP with even some individual specimen (i.e. 1.5 ± 0.2 nm). The trend of stability was further corroborated with a water-soluble H_2P derivative (5,10,15,20-tetrakis(4-sulfonatophenyl)porphyrin dihydrochloride). In these SWNT suspensions, which were shown to be stable for several weeks, eminent interactions protect H_2P against protonations to the corresponding diacid. Moreover, these strongly interacting hybrids have been successfully aligned onto hydrophilic polydimethylsiloxane surfaces by flowing SWNT solutions along a desired direction and then transferred to silicon substrates by stamping [173].

Instead of immobilizing ZnP, H_2P, etc. directly onto the SWNT, ionic pyrene derivatives (i.e. 1-(trimethylammonium acetyl)pyrene (pyrene$^+$)) were used to solubilize

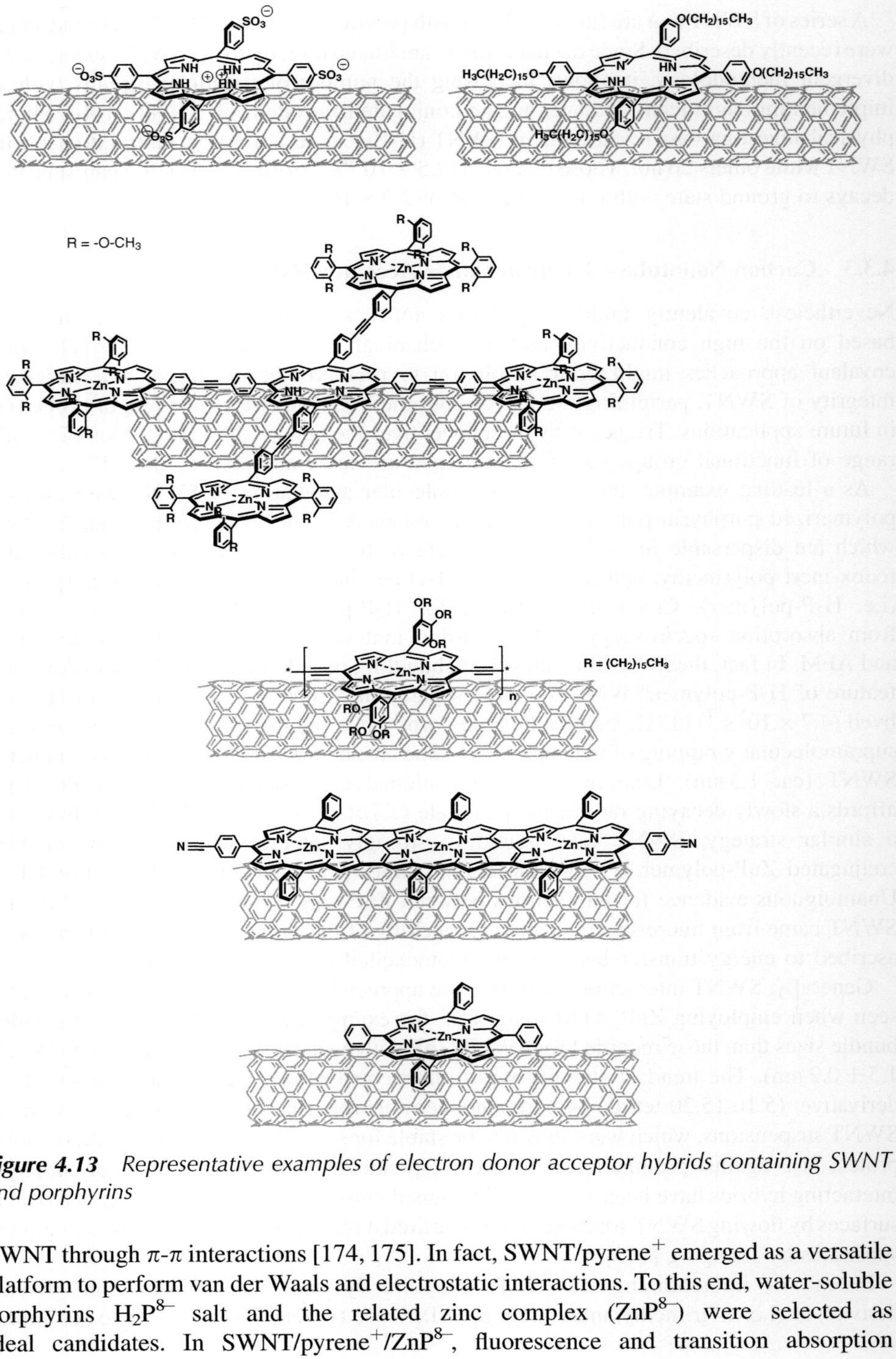

Figure 4.13 Representative examples of electron donor acceptor hybrids containing SWNT and porphyrins

SWNT through π-π interactions [174, 175]. In fact, SWNT/pyrene$^+$ emerged as a versatile platform to perform van der Waals and electrostatic interactions. To this end, water-soluble porphyrins H_2P^{8-} salt and the related zinc complex (ZnP^{8-}) were selected as ideal candidates. In SWNT/pyrene$^+$/ZnP^{8-}, fluorescence and transition absorption studies provided support for a rapid charge transfer ($5.0 \times 10^9\ s^{-1}$). MWNT interact similarly to SWNT with pyrene$^+$ and produce stable MWNT/pyrene$^+$. Interesting is the fact that a better delocalization of electrons in MWNT/pyrene$^+$/ZnP^{8-} helps to significantly enhance the stability of the radical ion pair state ($1.7 \times 10^5\ s^{-1}$) relative to SWNT/pyrene$^+$/ZnP^{8-} ($2.5 \times 10^6\ s^{-1}$). Percolation of the charge inside the concentric wires in MWNT decelerates the decay dynamics that are associated with the charge recombination [176].

Viable alternatives to porphyrins or phthalocyanines emerged around excited state electron donors such as copolymers of unsubstituted thiophene and 7-(thien-3-ylsulfanyl)heptanoic acid or size-quantized thioglycolic acid stabilized CdTe^{n-} nanoparticles [177, 178]. They share in common that they are water soluble and that they were successfully combined with SWNT/pyrene$^+$ via electrostatic forces. In the resulting nanohybrids strong electronic interactions were noted. In particular, polythiophene and/or CdTe tend to donate excited state electrons to SWNT in the ground state, which slowly recombine – $3.5 \times 10^7\ s^{-1}$ in SWNT/pyrene$^+$/CdTe^{n-}.

Applying similar π-π interactions, SWNT were integrated with a series of negatively charged pyrene derivatives (pyrene$^-$) – 1-pyreneacetic acid, 1-pyrenecarboxylic acid, 1-pyrenebutyric acid, 8-hydroxy-1,3,6-pyrenetrisulfonic acid. But none of the resulting SWNT/pyrene$^-$ showed the tremendous stability that was seen for SWNT/pyrene$^+$. Still, a series of water soluble, positively charged ZnP^{8+} were shown to form photoactive SWNT/pyrene$^-$/ZnP^{8+}, etc [179].

Considering the broad adaptability and the stability of the SWNT/pyrene motif, π-π stacking with pyrene-imidazole, pyrene- *pyro*-pheophorbide *a*, and pyrene-CdSe was demonstrated to solubilize SWNT [180–182]. Special mentioning deserves the case of pyrene-imidazole – through the use of the imidazole ligand naphthalocyanine (ZnNc) and ZnP were axially coordinated. Steady-state and time-resolved emission studies revealed efficient fluorescence quenching of the donor entities. Nanosecond transient absorption spectra revealed that the photoexcitation of ZnNc or ZnP resulted in the one electron oxidation of the donor unit with a simultaneous one electron reduction of SWNT. Both radical ion pair states decay on the nanosecond time scale (ZnPc: $1.7 \times 10^7\ s^{-1}$; ZnP: $1.1 \times 10^7\ s^{-1}$) to repopulate the ground state [180].

One of the few examples to this day, in which SWNT act as electron donor is based on integrating 5,15-bis(4-pyridyl)-2,8,12,18-tetraethyl-3,7,13,17-tetramethyl zinc porphyrin onto SWNT [183]. Electron transfer proceeds from SWNT to the photoexcited porphyrin that is self-assembled on the surface. In the second case, pyrene-NH_3^+ was first immobilized onto SWNT [184]. However, pyrene-NH_3^+ not only π-π stacks onto SWNT, but also complexes benzo-18-crown-6. Such ammonium/crown ether interactions were then used to yield stable SWNT/pyrene-NH_3^+/crown-C_{60}. Steady state and time resolved absorption spectroscopy prompted to a photoinduced charge transfer, during which SWNT and C_{60} are oxidized and reduced, respectively. The rates of charge separation and charge recombination were found to be 3.5×10^9 and $1.0 \times 10^7\ s^{-1}$, respectively [184]. In a final example, SWNT were found to make complexes with sapphyrins. The resulting SWNT/sapphyrins undergo photoexcited intramolecular charge transfer from SWNT to the sapphyrin moiety upon photoexcitation, for which proof came from a combination of steady state investigations, femtosecond transient absorption spectroscopies and pulse radiolysis experiments [185].

4.4 Other Nanocarbon Composites

The usefulness of carbon nanotubes in optical, electronic and catalytic applications has prompted researchers to synthesize various carbon nanostructures [186–194]. Of particular interest are the cup-stacked carbon nanotubes consisting of truncated conical graphene layers [187]. Although conventional carbon nanotubes are made up of seamless cylinders of hexagonal carbon network, the cup-stacked structure provides a hollow tubular morphology. This stacked-cup morphology provides a large portion of exposed and reactive edges in the outer and inner surfaces of the hollow tubes. The availability of inner and outer edges of these stacked-cups to chemical functionalization or surface modification opens up new avenues to utilize them in electronic and catalytic applications [188]. Application of carbon stacked cups in fuel cell has also been explored [189].

The optical properties of cup-stacked carbon nanotubes and the effectiveness of the semiconductor properties to generate photocurrent in a photoelectrochemical cell were investigated extensively (Figure 4.14) [190]. Stacked-cup carbon nanotubes (SCCNT) of short length (0.2–0.3 μm), referred as carbon nanobarrels, have been electrophoretically deposited as thin films on conducting glass electrodes from a THF suspension with the application of a dc field (200 V cm^{-1}) [190]. These SCCNT films undergo charge separation and deliver photocurrent under visible light irradiation [190]. Photocurrents up to 1 mA cm^{-2} under anodic bias show the importance of these nanostructures in direct conversion of light energy into electricity [190]. The maximum IPCE value of 17% observed with SCCNT

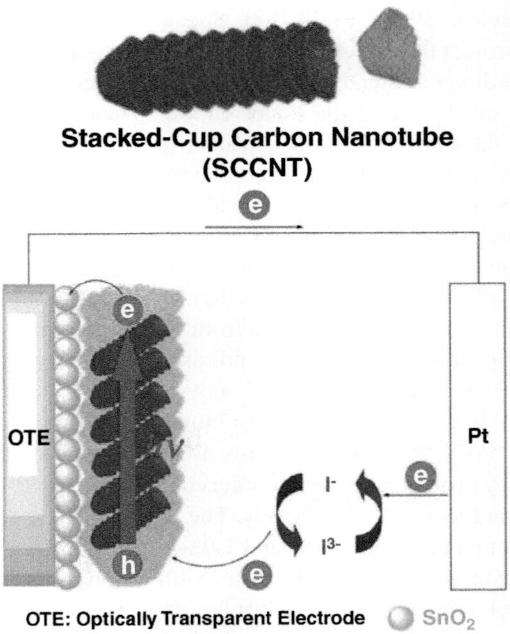

Figure 4.14 *Illustration of Stacked-Cup Carbon Nanotubes (SCCNT) and the application to the photoelectrochemical cell*

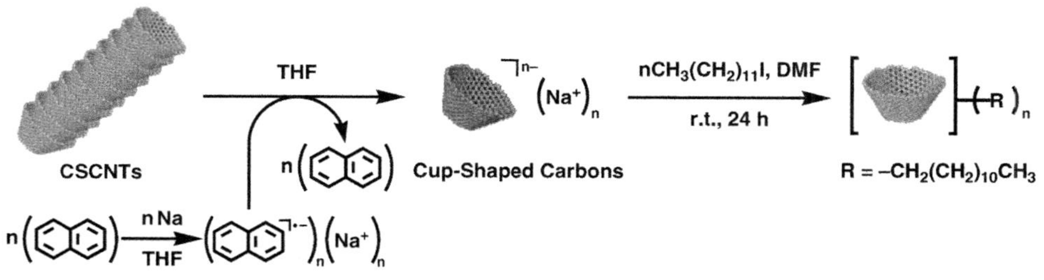

Figure 4.15 *Destacking of CSCNTs by the electron-transfer reduction and the dodecylation of the cup-shaped carbons*

system is two orders of magnitude greater than those obtained with single wall carbon nanotubes (SWCNT) [190]. The power conversion efficiencies (η) of SCCNT-modified electrodes increase with increasing the tube length [191]. The η value of SCCNT-modified electrodes with the longest tube length is determined to be 0.11%, which is about 6 times greater than that with the smallest tube length (0.018%) [191].

The electron-transfer reduction of CSCNTs with sodium naphthalenide and the subsequent treatment with 1-iodododecane results in electrostatically destacking CSCNTs to afford individual cup-shaped carbons with the controlled diameter and size as shown in Figure 4.15 [192]. The photoinduced electron-transfer from an NAD (nicotinamide adenine dinucleotide) dimer analogue, 1-benzyl-1,4-dihydronicotinamide dimer [$(BNA)_2$] [193] to CSCNTs also yields electrostatically destacked CSCNTs to afford CSCNTs with the controlled diameter and size [194].

The possibility of modifying these cup-shaped carbons with light harvesting molecules (e.g. porphyrin) with covalent or noncovalent interactions opens up new avenues to design hybrid assemblies for photoelectrochemical solar cells.

References

[1] N. S. Lewis and D. G. Nocera, Powering the planet: chemical challenges in solar energy utilization, *Proc. Natl. Acad. Sci., USA*, **103**, 15729–15735 (2006).
[2] D. G. Nocera, Living healthy on a dying planet, *Chem. Soc. Rev.*, **38**, 13–15 (2009).
[3] S. Fukuzumi, Bioinspired energy conversion systems for hydrogen production and storage, *Eur. J. Inorg. Chem.*, 1351–1362 (2008).
[4] M. Hambourger, G. F. Moore, D. M. Kramer, D. Gust, A. L. Moore and T. A. Moore, Biology and technology for photochemical fuel production, *Chem. Soc. Rev.*, **38**, 25–35 (2009).
[5] S. Fukuzumi, T. Honda, K. Ohkubo and T. Kojima, Charge separation in metallomacrocycle complexes linked with electron acceptors by axial coordination, *Dalton Trans.*, 3880–3889 (2009).
[6] S. Fukuzumi and T. Kojima, Photofunctional nanomaterials composed of multiporphyrins and carbon-based π-electron acceptors, *J. Mater. Chem.*, **18**, 1427–1439 (2008).
[7] D. Gust, D. Kramer, A. Moore, T. A. Moore, A. Thomas and W. Vermaas, Engineered and artificial photosynthesis: human ingenuity enters the game, *MRS Bull.*, **33**, 383–387 (2008).
[8] J. Turner, The other half of the equation, *Nature Mater.*, **7**, 770–771 (2008).
[9] D. M. Guldi, Let there be light – but not too much, *Nautre Nanotech.*, **3**, 257–258 (2008).
[10] H. Imahori and Y. Sakata, Donor-linked fullerenes: photoinduced electron transfer and its potential application, *Adv. Mater.*, **9**, 537–546 (1997).

[11] M. Prato, [60]Fullerene chemistry for materials science applications, *J. Mater. Chem.*, **7**, 1097–1109 (1997).
[12] N. Martín, L. Sánchez, B. Illescas and I. Pérez, C_{60}-based electroactive organofullerene, *Chem. Rev.*, **98**, 2527–2547 (1998).
[13] H. Imahori and Y. Sakata, Fullerenes as novel acceptors in photosynthetic electron transfer, *Eur. J. Org. Chem.*, 2445–2457 (1999).
[14] F. Diederich and M. Gómez-López, Supramolecular fullerene chemistry, *Chem. Soc. Rev.*, **28**, 263–277 (1999).
[15] D. M. Guldi, Fullerene: three dimensional electron acceptor materials, *Chem. Commun.*, 321–327 (2000).
[16] D. M. Guldi and M. Prato, Excited state properties of C_{60} fullerene derivatives, *Acc. Chem. Res.*, **33**, 695–703 (2000).
[17] D. Gust, T. A. Moore and A. L. Moore, Mimicking photosynthetic solar energy transduction, *Acc. Chem. Res.*, **34**, 40–48 (2001).
[18] N. Armaroli, Photoactive mono- and polynuclear Cu(I)-phenanthrolines. A viable alterative to Ru(II)-polypyridines?, *Chem. Soc. Rev.*, **30**, 13–124 (2001).
[19] D. M. Guldi, Fullerene-porphyrin architectures; photosynthetic antenna and reaction center models, *Chem. Soc. Rev.*, **31**, 22–36 (2002).
[20] H. Imahori, Y. Mori and Y. Matano, Nanostructured artificial photosynthesis, *J. Photochem. Photobiol., C*, **4**, 51–83 (2003).
[21] S. Fukuzumi, New perspective of electron transfer chemistry, *Org. Biomol. Chem.*, **1**, 609–620 (2004).
[22] H. Imahori, Porphyrin-fullerene linked systems as artificial photosynthetic mimics, *Org. Biomol. Chem.*, **2**, 1425–1433 (2004).
[23] N.-F. Nierengarten, Chemical modification of C_{60} for materials science applications, *New J. Chem.*, **28**, 1177–1191 (2004).
[24] M. E. El-Khouly, O. Ito, P. M. Smith and F. D'Souza, Intermolecular and supramolecular photoinduced electron transfer processes of fullerene-porphyrin/phthalocyanine systems, *J. Photochem. Photobiol. C*, **5**, 79–104 (2004).
[25] L. Sánchez, N. Martín and D. M. Guldi, Materials for organic solar cells: the C_{60}/π-conjugated oligomer approach, *Chem. Soc. Rev.*, **34**, 31–47 (2005).
[26] H. Imahori, Creation of fullerene-based artificial photosynthetic systems, *Bull. Chem. Soc. Jpn.*, **80**, 621–636 (2007).
[27] S. Fukuzumi, Development of bioinspired artificial photosynthetic systems, *Phys. Chem. Chem. Phys.*, **10**, 2283–2297 (2008).
[28] Y. Araki and O. Ito, Factors controlling lifetimes of photoinduced charge-separated states of fullerene-donor molecular systems, *J. Photochem. Photobiol. C*, **9**, 93–110 (2008).
[29] H. Imahori, K. Hagiwara, T. Akiyama, M. Aoki, S. Taniguchi, T. Okada, M. Shirakawa and Y. Sakata, The small reorganization energy of C_{60} in electron transfer, *Chem. Phys. Lett.*, **263**, 545–550 (1996).
[30] D. M. Guldi and K.-D. Asmus, Electron transfer from C_{76} ($C_{2v}{}'$) and C_{78} (D_2) to radical cations of various arenas: evidence for the Marcus inverted region, *J. Am. Chem. Soc.*, **119**, 5744–6745 (1997).
[31] N. V. Tkachenko, C. Guenther, H. Imahori, K. Tamaki, Y. Sakata, S. Fukuzumi and H. Lemmetyinen, Near infra-red emission of charge-transfer complexes of porphyrin-fullerene films, *Chem. Phys. Lett.*, **326**, 344–350 (2000).
[32] H. Imahori, N. V. Tkachenko, V. Vehmanen, K. Tamaki, H. Lemmetyinen, Y. Sakata and S. Fukuzumi, An Extremely small reorganization energy of electron transfer in porphyrin-fullerene dyad, *J. Phys. Chem. A*, **105**, 1750–1756 (2001).
[33] H. Imahori, H. Yamada, D. M. Guldi, Y. Endo, A. Shimomura, S. Kundu, K. Yamada, T. Okada, Y. Sakata and S. Fukuzumi, Comparison of reorganization energies for intra- and intermolecular electron transfer, *Angew. Chem. Int. Ed.*, **41**, 2344–2347 (2002).
[34] S. Fukuzumi, K. Ohkubo, H. Imahori and D. M. Guldi, Driving force dependence of intermolecular electron-transfer reactions of fullerene, *Chem. Eur. J.*, **9**, 1585–1593 (2003).

[35] D. Kuciauskas, S. Lin, G. R. Seely, A. L. Moore, T. A. Moore and D. Gust, Energy and electron transfer in porphyrin-fullerene dyads, *J. Phys. Chem.*, **100**, 15926–15932 (1996).
[36] H. Imahori, K. Hagiwara, M. Aoki, T. Akiyama, S. Taniguchi, T. Okada, M. Shirakawa and Y. Sakata, Linkage and solvent dependence of photoinduced electron transfer in zincporphyrin-C_{60} dyads, *J. Am. Chem. Soc.*, **118**, 11771–11782 (1996).
[37] Imahori, K. Tamaki, D. M. Guldi, C. Luo, M. Fujitsuka, O. Ito, Y. Sakata and S. Fukuzumi, Modulating charge separation and charge recombination dynamics in porphyrin-fullerene linked dyads and triads: Marcus-normal versus inverted region, *J. Am. Chem. Soc.*, **123**, 2607–2617 (2001).
[38] D. I. Schuster, P. Cheng, P. D. Jarowski, D. M. Guldi, L. Echegoyen, S. Pyo, A. R. Holzwarth, S. E. Braslavsky, R. M. Williams and G. Klihm, Design, synthesis, and photophysical studies of a porphyrin-fullerene dyad with parachute topology; charge recombination in the Marcus inverted region, *J. Am. Chem. Soc.*, **126**, 7257–7270 (2004).
[39] D. M. Guldi, A. Hirsch, M. Scheloske, E. Dietel, A. Troisi, F. Zerbetto and M. Prato, Modulating charge-transfer interactions in topologically different porphyrin-C_{60} dyads, *Chem. Eur. J.*, **9**, 4968–4979 (2003).
[40] L. R. Sutton, M. Scheloske, K. S. Pirner, A. Hirsch, D. M. Guldi and J. P. Gisselbrecht, Unexpected change in charge transfer behavior in a cobalt(II) porphyrin-fullerene conjugate that stabilize radical ion pair states, *J. Am. Chem. Soc.*, **126**, 10370–10381 (2004).
[41] G. de la Torres, F. Giacalone, J. L. Segura, N. Martín and D. M. Guldi, Electronic communication through pi-conjugated wires in covalently linked porphyrin/C_{60} ensembles, *Chem. Eur. J.*, **11**, 1267–1280 (2005).
[42] N. V. Tkachenko, L. Rantala, A. Y. Tauber, J. Helaja, P. H. Hynninen and H. Lemmetyinen, Photoinduced electron transfer in phtochlorin-[60]fullerene dyads, *J. Am. Chem. Soc.*, **121**, 9378–9387 (1999).
[43] N. V. Tkachenko, H. Lemmetyinen, J. Sonoda, K. Ohkubo, T. Sato, H. Imahori and S. Fukuzumi, Ultrafast photodynamics of exciplex formation and photoinduced electron transfer in porphyrin-fullerene dyads linked at close proximity, *J. Phys. Chem. A*, **107**, 8834–8844 (2003).
[44] N. Armaroli, G. Marconi, L. Echegoyen, J. P. Bourgeois and F. Diederlich, Charge-transfer interactions in face-to-face porphyrin-fullerene systems: solvent-dependent luminescence in the infrared spectral region, *Chem. Eur. J.*, **6**, 1629–1645 (2000).
[45] K. Ohkubo, H. Kotani, J. Shao, Z. Ou, K. M. Kadish, G. Li, R. K. Pandey, M. Fujitsuka, O. Ito, H. Imahori and S. Fukuzumi, Production of an ultra-long-lived charge-separated state in a zinc chlorin–C_{60} dyad by one-step photoinduced electron transfer, *Angew. Chem. Int. Ed.*, **43**, 853–856 (2004).
[46] D. Gonzalez-Rodriguez, T. Torres, D. M. Guldi, J. Rivera, M. A. Herranz and L. Echegoyen, Subphthalocyanines: tunable molecular scaffolds for intramolecular electron and energy transfer, *J. Am. Chem. Soc.*, **126**, 6301–6313 (2004).
[47] A. de la Escosura, M. V. Martinez-Diaz, D. M. Guldi and T. Torres, Stabilization of charge-separated states in phthalocyanine-fullerene ensembles through supramolecular donor-acceptor interactions, *J. Am. Chem. Soc.*, **128**, 4112–4118 (2006).
[48] R. M. Williams, M. Koeberg, J. M. Lawson, Y.-Z. An, Y. Rubin, M. N. Paddon-Row and J. W. Verhoeven, Photoinduced electron transfer to C_{60} across extended 3- and 11-bond hydrocarbon bridges: creation of a long-lived charge-separated state, *J. Org. Chem.*, **61**, 5055–5062 (1996).
[49] J. M. Lawson, A. M. Oliver, D. F. Rothenfluh, Y.-Z. An, G. A. Ellis, M. G. Ranasinghe, S. I. Khan, A. G. Franz, P. S. Ganapathi, M. J. Shephard, M. N. Paddon-Row and Y. Rubin, Synthesis of a variety of bichromophoric 'ball-and-chain' systems based on buckminsterfullerene (C_{60}) for the study of intramolecular electron and energy transfer processes, *J. Org. Chem.*, **61**, 5032–5054 (1996).
[50] N. Armaroli, F. Diederlich, C. O. Dietrich-Buchecker, L. Flamigni, G. Marconi, J.-F. Nierengarten and J.-P. Sauvage, A copper(I)-complexed rotaxane with two fullerene stoppers: synthesis, electrochemistry, and photoinduced processes, *Chem. Eur. J.*, **4**, 406–416 (1998).

[51] D. Armspach, E. C. Constable, F. Diederlich, C. E. Housecroft and J.-F. Nierengarten, Bucky ligands: synthesis, ruthenium(II) complexes, and electrochemical properties, *Chem. Eur. J.*, **4**, 723–733 (1998).
[52] M. Maggini, D. M. Guldi, S. Mondini, G. Scorrano, F. Paolucci, P. Ceroni and S. Roffia, Photoinduced electron transfer in a tris(2,2′-bipyridine)-C_{60}-ruthenium(II) dyad: evidence of charge recombination to a fullerene excited state, *Chem. Eur. J.*, **4**, 1992–2000 (1998).
[53] H. Imahori, S. Cardoso, D. Tatman, S. Lin, L. Noss, G. R. Seely, L. Sereno, J. C. de Silber, T. A. Moore, A. L. Moore and D. Gust, Photoinduced electron transfer in a carotenobuckminsterfullerene dyad, *Photochem. Photobiol.*, **62**, 1009–1014 (1995).
[54] D. M. Guldi, M. Maggini, G. Scorrano and M. Prato, Intramolecular electron transfer in fullerene/ferrocene based donor-bridge-acceptor dyads, *J. Am. Chem. Soc.*, **119**, 974–980 (1997).
[55] R. Marczak, M. Wielopolski, S. S. Gayathri, D. M. Guldi, Y. Matsuo, K. Matsuo, K. Tahara and E. Nakamura, Uniquely shaped double-decker buckyferrocenes-distinct electron donor-acceptor interactions, *J. Am. Chem. Soc.*, **130**, 16207–16215 (2008).
[56] N. Martín, L. Sánchez, M. A. Herranz and D. M. Guldi, Evidence for two separate one-electron transfer events in excited fulleropyrrolidine dyads containing tetrathiafulvalene (TTF), *J. Phys. Chem. A*, **104**, 4648–4657 (2000).
[57] M. Fujitsuka, O. Ito, T. Yamashiro, Y. Aso and T. Otsubo, Solvent polarity dependence of photoinduced charge separation in a tetrathiophene-C_{60} dyad studied by pico- and nanosecond laser flash photolysis in the near-IR region, *J. Phys. Chem. A*, **104**, 4876–4881 (2000).
[58] D. Kuciaukas, P. A. Liddell, S. Lin, S. G. Stone, A. L. Moore, T. A. Moore and D. Gust, Photoinduced electron transfer in carotenoporphyrin-fullerene triads: temperature and solvent effects, *J. Phys. Chem. B*, **104**, 4307–4321 (2000).
[59] D. Curiel, K. Ohkubo, J. R. Reimers, S. Fukuzumi and M. J. Crossley, Photoinduced electron transfer in a β,β–pyrrolic fused ferrocene-(zinc porphyrin)-fullerene, *Phys. Chem. Chem. Phys.*, **9**, 5260–5266 (2007).
[60] H. Imahori, D. M. Guldi, K. Tamaki, Y. Yoshida, C. Luo, Y. Sakata and S. Fukuzumi, Charge separation in a novel artificial photosynthetic reaction center lives 380 milliseconds, *J. Am. Chem. Soc.*, **123**, 6617–6618 (2001).
[61] D. M. Guldi, H. Imahori, K. Tamaki, Y. Kashiwagi, H. Yamada, Y. Sakata and S. Fukuzumi, A molecular tetrad allowing efficient energy storage for 1.6 s at 163 K, *J. Phys. Chem. A*, **108**, 541–548 (2004).
[62] H. Imahori, Y. Sekiguchi, Y. Kashiwagi, T. Sato, Y. Araki, O. Ito, H. Yamada and S. Fukuzumi, Long-lived charge-separated state generated in ferrocene-meso,meso-linked porphyrin trimer-fullerene pentad with a high quantum yield, *Chem. Eur. J.*, **10**, 3184–3196 (2004).
[63] S. D. Straight, G. Kodis, Y. Terazono, M. Hambourger, T. A. Moore, A. L. Moore and D. Gust, Self-regulation of photoinduced electron transfer by a molecular nonlinear transducer, *Nature Nanotech.*, **3**, 280–283 (2008).
[64] D. Kuciaukas, P. A. Liddell, S. Lin, T. E. Johnson, S. J. Weghorn, J. S. Lindsey, A. L. Moore, T. A. Moore and D. Gust, An artificial photosynthetic antenna-reaction center complex, *J. Am. Chem. Soc.*, **121**, 8604–8614 (1999).
[65] M. U. Winters, E. Dahlstedt, H. E. Blades, C. J. Wilson, M. J. Frampton, H. L. Anderson and B. Albinsson, Probing the efficiency of electron transfer through porphyrin-based molecular wires, *J. Am. Chem. Soc.*, **129**, 4291–4297 (2007).
[66] D. M. Guldi and N. Martín, Fullerene architectures made to order; biomimetic motifs – design and features, *J. Mater. Chem.*, **12**, 1978–1992 (2002).
[67] L. Sánchez, N. Martín and D. M. Guldi, Hydrogen-bonding motifs in fullerene chemistry, *Angew. Chem. Ind. Ed.*, **44**, 5374–5382 (2005).
[68] M. Segura, L. Sánchez, J. de Mendoza, N. Martín and D. M. Guldi, Hydrogen Bonding Interfaces in fullerene-TTF ensembles, *J. Am. Chem. Soc.*, **125**, 15093–15100 (2003).
[69] F. D'Souza, P. M. Smith, M. E. Zandler, A. L. McCarty, M. Itou, Y. Araki and O. Ito, Energy transfer followed by electron transfer in a supramolecular triad composed of boron dipyrrin, zinc porphyrin and fullerene: A model for the photosynthetic antenna-reaction center complex, *J. Am. Chem. Soc.*, **126**, 7898–7907 (2004).

[70] K. Li, D. I. Schuster, D. M. Guldi, M. A. Herranz and L. Echegoyen, Convergent synthesis and photophysics of [60]fullerene/porphyrin-based rotaxanes, *J. Am. Chem. Soc.*, **126**, 3388–3389 (2004).
[71] K. Li, P. J. Bracher, D. M. Guldi, M. A. Herranz, L. Echegoyen and D. I. Schuster, [60] Fullerene-stoppered porphyrinorotaxanes: pronounced elongation of charge-separated-state lifetimes, *J. Am. Chem. Soc.*, **126**, 9156–9157 (2004).
[72] L. Sánchez, M. Sierra, N. Martín, A. J. Myles, T. J. Dale, J. Rebek, W. Seitz and D. M. Guldi, Exceptionally strong electronic communication through hydrogen bonds in porphyrin-C_{60} pairs, *Angew. Chem. Int. Ed.*, **45**, 4637–4641 (2006).
[73] F. Wessendorf, J.-F. Gnichwitz, G. H. Sarova, K. Hager, U. Hartnagel, D. M. Guldi and A. Hirsch, Implementation of a Hamilton-receptor-based hydrogen-bonding motif toward a new electron donor-acceptor prototype: electron versus energy transfer, *J. Am. Chem. Soc.*, **129**, 16057–16071 (2007).
[74] A. Mateo-Alonso, C. Ehli, G. M. A. Rahman, D. M. Guldi, G. Fioravanti, M. Marcaccio, F. Paolucci and M. Prato, Tunning electron transfer through translational motion in molecular shuttles, *Angew. Chem. Int. Ed.*, **46**, 3521–3525 (2007).
[75] T. Oike, T. Kurata, K. Yakimiya, T. Otsubo, Y. Aso, H. Zhang, Y. Araki and O. Ito, Polyether-bridged sexithiophene as a complexation-gated molecular wire for intramolecular photoinduced electron transfer, *J. Am. Chem. Soc.*, **127**, 15372–15373 (2005).
[76] N. V. Tkachenko, E. Vuorimaa, T. Kesti, A. S. Alekseev, A. Y. Tauber, P. H. Hynninen and H. Lemmetyinen, Vectorial photoinduced electron transfer in phytochlorin-[60]fullerene Langmuir-Blodgett films, *J. Phys. Chem. B*, **104**, 6371–6379 (2000).
[77] D. Hirayama, K. Takimiya, Y. Aso, T. Otsubo, T. Hasobe, H. Yamada, H. Imahori, S. Fukuzumi and Y. Sakata, Large photocurrent generation of gold electrodes modified with [60]fullerene-linked oligothiophenes bearing a tripodal rigid anchor, *J. Am. Chem. Soc.*, **124**, 532–533 (2002).
[78] H. Imahori, H. Yamada, Y. Nishimura, I. Yamazaki and Y. Sakata, Vectorial multistep electron transfer at the gold electrodes modified with self-assembled monolayers of ferrocene-porphyrin-fullerene triads, *J. Phys. Chem. B*, **104**, 2099–2108 (2000).
[79] H. Imahori, H. Norieda, H. Yamada, Y. Nishimura, I. Yamazaki, Y. Sakata and S. Fukuzumi, Light-harvesting and photocurrent generation by gold electrodes modified with mixed self-assembled monolayers of boron-dipyrrin and ferrocene-porphyrin-fullerene triad, *J. Am. Chem. Soc.*, **123**, 100–110 (2001).
[80] S. Saha, E. Johansson, A. H. Flood, H.-R. Tseng, J. I. Zink and J. F. Stoddart, A photoactive molecular triad as a nanoscale power supply for a supramolecular machine, *Chem. Eur. J.*, **11**, 6846–6858 (2005).
[81] K.-S. Kim, M.-S. Kang, H. Ma and A. K. -Y. Jen, Highly efficient photocurrent generation from a self-assembled monolayer film of a novel C_{60}-tethered 2,5-dithienylpyrrole triad, *Chem. Mater.*, **16**, 5058–5062 (2004).
[82] H. Imahori, M. Kimura, K. Hosomizu, T. Sato, T. K. Ahn, S. K. Kim, D. Kim, Y. Nishimura, I. Yamazaki, Y. Araki, O. Ito and S. Fukuzumi, Vectorial electron relay at ITO electrodes modified with self-assembled monolayers of ferrocene-porphyrin-fullerene triads and porphyrin-fullerene dyads for molecular photovoltaic devices, *Chem. Eur. J.*, **10**, 5111–5122 (2004).
[83] M. Isosomppi, N. V. Tkachenko, A. Efimov, K. Kaunisto, K. Hosomizu, H. Imahori and H. Lemmetyinen, Photoinduced electron transfer in multilayer self-assembled structures of porphyrins and porphyrin-fullerene dyads on ITO, *J. Mater. Chem.*, **15**, 4546–4554 (2005).
[84] Y.-J. Cho, T. K. Ahn, H. Song, K. S. Kim, C. Y. Lee, W. S. Seo, K. Lee, S. K. Kim, D. Kim and J. T. Park, Unusually high performance photovoltaic cell based on a [60]fullerene metal cluster-porphyrin dyad SAM on an ITO electrode, *J. Am. Chem. Soc.*, **127**, 2380–2381 (2005).
[85] T. Konishi, A. Ikeda and S. Shinkai, Supramolecular design of photocurrent-generating devices using fullerenes aimed at modeling artificial photosynthesis, *Tetrahedron*, **61**, 4881–4889 (2005).

[86] A. Kira, T. Umeyama, Y. Matano, Y. Kaname, S. Isoda, J. K. Park, D. Kim and H. Imahori, Supramolecular donor-acceptor heterojunctions by vectorial stepwise assembly of porphyrins and coordination-bonded fullerene arrays for photocurrent generation, *J. Am. Chem. Soc.*, **131**, 3198–3200 (2009).
[87] D. M. Guldi, Biomimetic assemblies of carbon nanostructures for photochemical energy conversion, *J. Phys. Chem. B*, **109**, 11432–11441 (2005).
[88] J. Xue, S. Uchida, B. P. Rand and S. R. Forrest, Asymmetric tandem organic photovoltaic cells with hybrid planar-mixed molecular heterojunctions, *Appl. Phys. Lett.*, **85**, 5757–5759 (2004).
[89] J. Xue, B. P. Rand, S. Uchida and S. R. Forrest, A hybrid planar-mixed molecular heterojunction photovoltaic cell, *Adv. Mater.*, **17**, 66–71 (2005).
[90] P. V. Kamat, Meeting the clean energy demand: nanostructure architectures for solar energy conversion, *J. Phys. Chem. C*, **111**, 2834–2860 (2007).
[91] H. Imahori and S. Fukuzumi, Porphyrin- and fullerene-based molecular photovoltaic devices, *Adv. Funct. Mater.*, **14**, 525–536 (2004).
[92] H. Imahori and T. Umeyama, Donor-acceptor nanoarchitecture on semiconducting electrodes for solar energy conversion, *J. Phys. Chem. C*, **113**, 9029–9039 (2009).
[93] T. Hasobe, H. Imahori, S. Fukuzumi and P. V. Kamat, Light energy conversion using mixed molecular nanoclusters. Porphyrin and C_{60} cluster films for efficient photocurrent generation, *J. Phys. Chem. B*, **107**, 12105–12112 (2003).
[94] T. Hasobe, P. V. Kamat, M. A. Absalom, Y. Kashiwagi, J. Sly, M. J. Crossley, K. Hosomizu, H. Imahori and S. Fukuzumi, Supramolecular photovoltaic cells based on composite molecular nanoclusters: dendritic porphyrin and C_{60}, porphyrin dimer and C_{60}, and porphyrin-fullerene-C_{60} dyad, *J. Phys. Chem. B*, **118**, 12865–12872 (2004).
[95] T. Hasobe, K. Saito, P. V. Kamat, V. Troiani, H. Qiu, N. Solladié, K. S. Kim, J. K. Park, D. Kim, F. D'Souza and S. Fukuzumi, Organic solar cells. Supramolecular composites of porphyrins and fullerenes organized by polypeptide structures as light harvesters, *J. Mater. Chem.*, **17**, 4160–4170 (2007).
[96] T. Hasobe, H. Imahori, P. V. Kamat, T. K. Ahn, S. K. Kim, D. Kim, A. Fujimoto, T. Hirakawa and S. Fukuzumi, Photovoltaic cells using composite nanoclusters of porphyrins and fullerenes with gold nanoparticles, *J. Am. Chem. Soc.*, **127**, 1216–1228 (2005).
[97] H. Imahori, A. Fujimoto, S. Kang, H. Hotta, K. Yoshida, T. Umeyama, Y. Matano, S. Isoda, M. Isosomppi, N. V. Tkachenko and H. Lemmetyinen, Host-guest interactions in the supramolecular incorporation of fullerenes into tailored holes on porphyrin-modified gold nanoparticles in molecular photovoltaics, *Chem. Eur. J.*, **11**, 7265–7275 (2005).
[98] T. Hasobe, S. Fukuzumi, S. Hattori and P. V. Kamat, Shape- and functionality-controlled organization of TiO_2-porphyrin-C_{60} assemblies for improved performance of photochemical solar cells, *Chem. Asian J.*, **2**, 265–272 (2007).
[99] H. Imahori, M. Ueda, S. Kang, H. Hayashi, S. Hayashi, H. Kaji, S. Seki, A. Saeki, S. Tagawa, T. Umeyama, Y. Matano, K. Yoshida, S. Isoda, M. Shiro, N. V. Tkachenko and H. Lemmetyinen, Effects of porphyrin substituents on film structure and photoelectrochemical properties of porphyrin/fullerene composite clusters electrophoretically deposited on nanostructured SnO_2 electrodes, *Chem. Eur. J.*, **13**, 10182–10193 (2007).
[100] J.-F. Nierengarten, Fullerene-(π-conjugated oligomer) dyads as active photovoltaic materials, *Solar Energy Mater. Solar Cells*, **83**, 187–199 (2004).
[101] S. Gunes, H. Neugebauer and N. S. Sariciftci, Conjugated polymer-based organic solar cells, *Chem. Rev.*, **107**, 1324–1338 (2007).
[102] B. C. Thompson and J. M. Fréchet, Polymer-fullerene composite solar cells, *Angew. Chem. Int. Ed.*, **47**, 58–77 (2008).
[103] K. Kim, J. Liu, M. A. G. Namboothiry and D. L. Carroll, *Appl. Phys. Lett.*, **90**, 163511, (2007).
[104] J. Y. Kim, K. Lee, N. E. Coates, D. Moses, T.-Q. Nguyen, M. Dante and A. J. Heeger, Efficient tandem polymer solar cells fabricated by all-solution processing, *Science*, **317**, 222–225 (2007).
[105] A. Kira, M. Tanaka, T. Umeyama, Y. Matano, N. Yoshimoto, Y. Zhang, S. Ye, H. Lehtivuori, Isosomppi, N. V. Tkachenko H. Lemmetyinen and H. Imahori, Hydrogen-bonding effects on

film structure and photoelectrochemical properties of porphyrin and fullerene composites on nanostuctured TiO_2 electrodes, *J. Phys. Chem. B.*, **109**, 13618–13626 (2007).

[106] C.-H. Huang, N. D. McCenaghan, A. Kuhn, J. W. Hofstraat and D. M. Bassani, Enhanced photovoltaic response in hydrogen-bonded all-organic devices, *Org. Lett.*, **7**, 3409–3412 (2005).

[107] Y. Liu, Y. Li, L. Jiang, H. Gan, H. Liu, Y. Li, J. Zhuang, F. Lu and D. Zhu, Assembly and characterization of novel hydrogen-bond-induced nanoscale rods, *J. Org. Chem.*, **69**, 9049–9054 (2004).

[108] S. Reich, C. Thomsen and J. Maultzsch, *Carbon Nanotubes: Basic Concepts and Physical Properties*, Wiley-VCH, Weinheim, 2004.

[109] M. S. Dresselhaus, G. Dresselhaus and P. Avouris, *Carbon Nanotubes: Synthesis, Structure, Properties and Application*, Springer, Berlin, 2001.

[110] S. Roth and D. Carroll, *One-Dimensional Metals: Conjugated Polymers, Organic Crystals, Carbon Nanotube*, Wiley-VCH, Weinheim, 2004.

[111] M. Meyyappan, *Carbon Nanotubes, Science and Application*, Wiley-VCH, Weinheim, 2006.

[112] R. C. Haddon, Carbon nanotubes, *Acc. Chem. Res.*, **35**, 997–1113 (2002).

[113] V. Sgobba, G. M. A. Rahman, C. Ehli and D. M. Guldi,in *Fullerenes – Principles and Applications*, F. Langa and J. F. Nierengarten (eds), RSC, Cambridge, UK, 2007, Chapter 11.

[114] V. Sgobba, G. M. A. Rahman and D. M. Guldi,in *Chemistry of Carbon Nanotubes*, V. A. Basiuk and E. V. Basiuk (eds), American Scientific Publishers, North Lewis Way, California, USA, 2006, Chapter 6.

[115] L. J. Carlson and T. D. Krauss, Photophysics of individual single-walled carbon nanotubes, *Acc. Chem. Res.*, **41**, 235–243 (2008).

[116] S. V. Rotkin and S. Subramoney, *Applied Physics of Carbon Nanotubes*, Springer-Verlag, Berlin, 2005.

[117] S. J. Tans, M. H. Devoret, H. Dal, A. Thess, R. E. Smalley, L. J. Geerligs and C. Dekker, Individual single-wall carbon nanotubes as quantum wires, *Nature*, **386**, 474–477 (1997).

[118] P. M. Ajayan and O. Z. Zhou, Applications of carbon nanotubes, *Carbon Nanotubes*, **80**, 391–425 (2001).

[119] R. Krupke, F. Hennrich, H. v. Löhneysen and M. M. Kappes, Separation of metallic from semiconducting single-walled carbon nanotubes, *Science*, **301**, 344–347 (2003).

[120] V. Perebeinos, J. Tersoff and P. Avouris, Scaling of excitons in carbon nanotubes, *Phys. Rev. Lett.*, **92**, 257402/1–257402/4 (2004).

[121] H. Zhao and S. Mazumdar, Electron-electron interaction effects on the optical excitations of semiconducting single-walled carbon nanotubes, *Phys. Rev. Lett.*, **93**, 157402/1–157402/4 (2004).

[122] C. D. Spataru, S. Ismail-Beigi, L. X. Benedict and S. G. Louie, Excitonic effects and optical spectra of single-walled carbon nanotubes, *Phys. Rev. Lett.*, **92**, 077402/1–077402/4 (2004).

[123] Z. Wang, H. Zhao and S. Mazumdar, Quantitative calculations of the excitonic energy spectra of semiconducting single-walled carbon nanotubes within a π -electron model, *Phys. Rev. B: Condens. Matter*, **74**, 195406/1–195406/6 (2006).

[124] F. Wang, G. Dukovic, L. E. Brus and T. F. Heinz, The optical resonances in carbon nanotubes arise from excitons, *Science*, **308**, 838–841 (2005).

[125] Hartschuh, H. N. Pedrosa L. Novotny and T. D. Krauss, Simultaneous fluorescence and Raman scattering from single carbon nanotubes, *Science*, **301**, 1354–1357 (2003).

[126] S. M. Bachilo, M. S. Strano, C. Kittrell, R. H. Hauge, R. E. Smalley and R. B. Weisman, Structure-assigned optical spectra of single-walled carbon nanotubes, *Science*, **298**, 2361–2366 (2002).

[127] S. M. Dresselhaus, G. Dresselhaus, R. Saito and A. Jorio, Exciton photophysics of carbon nanotubes, *Ann. Rev. Phys. Chem.*, **58**, 719–747 (2007).

[128] P. W. Barone, R. S. Parker and M. S. Strano, In vivo fluorescence detection of glucose using a single-walled carbon nanotube optical sensor: design, fluorophore properties, advantages, and disadvantages, *Anal. Chem.*, **77**, 7556–7562 (2005).

[129] P. W. Barone, S. Baik, D. A. Heller and M. S. Strano, Near-infrared optical sensors based on single-walled carbon nanotubes, *Nature Mat.*, **4**, 86–92 (2005).
[130] P. Petit, C. Mathis, C. Journet and P. Bernier, Tuning and monitoring the electronic structure of carbon nanotubes, *Chem. Phys. Lett.*, **305**, 370–374 (1999).
[131] E. Jouguelet, C. Mathis and P. Petit, Controlling the electronic properties of single-wall carbon nanotubes by chemical doping, *Chem. Phys. Lett.*, **318**, 561–564 (2000).
[132] L. Kavan and L. Dunsch, Spectroelectrochemistry of carbon nanostructures, *ChemPhysChem*, **8**, 974–998 (2007).
[133] M. Rao, P. C. Eklund, S. Bandow, A. Thess and R. E. Smalley, Evidence for charge transfer in doped carbon nanotube bundles from Raman scattering, *Nature*, **388**, 257–259 (1997).
[134] L. Kavan, P. Rapta, L. Dunsch, M. J. Bronikowski, P. Willis and R. E. Smalley, Electrochemical tuning of electronic structure of single-walled carbon nanotubes: in-situ Raman and Vis-NIR study, *J. Phys. Chem. B*, **105**, 10764–10771 (2001).
[135] S. Gupta, M. Hughes, A. H. Windle and J. Robertson, Charge transfer in carbon nanotube actuators investigated using in situ Raman spectroscopy, *J. Appl. Phys.*, **95**, 2038–2048 (2004).
[136] K. Murakoshi and K. Okazaki, Electrochemical potential control of isolated single-walled carbon nanotubes on gold electrode, *Electrochim. Acta*, **50**, 3069–3075 (2005).
[137] K. Okazaki, Y. Nakato and K. Murakoshi, Absolute potential of the Fermi level of isolated single-walled carbon nanotubes, *Phys. Rev. B*, **68**, 035434/1–035434/5 (2003).
[138] P. M. Ajayan, T. W. Ebbesen, T. Ichihashi, S. Iijima, K. Tanigaki and H. Hiura, Opening carbon nanotubes with oxygen and implications for filling, *Nature*, **362**, 522–525 (1993).
[139] R. Sen, S. M. Rickard, M. E. Itkis and R. C. Haddon, Controlled purification of single-walled carbon nanotube films by use of selective oxidation and Near-IR spectroscopy, *Chem. Mater.*, **15**, 4273–4279 (2003).
[140] S. Huang and L. Dai, Plasma etching for purification and controlled opening of aligned carbon nanotubes, *J. Phys. Chem. B*, **106**, 3543–3545 (2002).
[141] L. Kavan and L. Dunsch, Ionic liquid for in situ Vis/NIR and Raman spectroelectrochemistry: doping of carbon nanostructures, *ChemPhysChem*, **4**, 944–950 (2003).
[142] L. Pietronero and S. Strassler, Bond-length change as tool to determine charge transfer and electron-phonon coupling in graphite intercalation compounds, *Phys. Rev. Lett.*, **47**, 593–596 (1981).
[143] D. M. Guldi, M. Marcaccio, D. Paolucci, F. Paolucci, N. Tagmatarchis, D. Tasis, E. Vázquez and M. Prato, Single-wall carbon nanotube-ferrocene nanohybrids: observing intramolecular electron transfer in functionalized SWNTs, *Angew. Chem. Int. Ed.*, **42**, 4206–4209 (2003).
[144] J. L. Segura and N. Martín, New concepts in tetrathiafulvalene chemistry, *Angew. Chem. Int. Ed.*, **40**, 1372–1409 (2001).
[145] M. Á. Herranz, N. Martín, S. Campidelli, M. Prato, G. Brehm and D. M. Guldi, Control over electron transfer in tetrathiafulvalene-modified single-walled carbon nanotubes, *Angew. Chem. Int. Ed.*, **45**, 4478–4482 (2006).
[146] M. A. Herranz, C. Ehli, S. Campidelli, M. Gutierrez, G. L. Hug, K. Ohkubo, S. Fukuzumi, M. Prato, N. Martín and D. M. Guldi, Spectroscopic characterization of photolytically generated radical ion pairs in single-wall carbon nanotubes bearing surface-immobilized tetrathiafulvalenes, *J. Am. Chem. Soc.*, **130**, 66–73 (2008).
[147] C. Ehli, D. M. Guldi, M. Angeles Herranz, N. Martín, S. Campidelli and M. Prato, Pyrene-tetrathiafulvalene supramolecular assembly with different types of carbon nanotubes, *J. Mat. Chem.*, **18**, 1498–1503 (2008).
[148] K. Kalyanasundaram, *Photochemistry of Polypyridine and Porphyrin Complexes*, Academic Press, London, 1997.
[149] K. M. Kadish, K. M. Smith and R. Guilard, *The Porphyrin Handbook*, Academic Press, New York, 2003.
[150] C. C. Leznoff and A. B. P. Lever, *Phthalocyanines: Properties and Applications*, VCH, Weinheim, 1996, vols. 1–4.

[151] G. de la Torre, P. Vázquez, F. Agulló-López and T. Torres, Role of structural factors in the nonlinear optical properties of phthalocyanines and related compounds, *Chem. Rev.*, **104**, 3723–3750 (2004).
[152] D. Baskaran, J. W. Mays, X. P. Zhang and M. S. Bratcher, Carbon nanotubes with covalently linked porphyrin antennae: photoinduced electron transfer, *J. Am. Chem. Soc.*, **127**, 6916–6917 (2005).
[153] H. Li, R. B. Martin, B. A. Harruff, R. A. Carino, L. F. Allard and Y.-P. Sun, Single-walled carbon nanotubes tethered with porphyrins: Synthesis and photophysical properties, *Adv. Mat.*, **16**, 896–900 (2004).
[154] F. Cheng and A. Adronov, Suzuki coupling reactions for the surface functionalization of single-walled carbon nanotubes, *Chem. Mater.*, **18**, 5389–5391 (2006).
[155] T. Umeyama and H. Imahori, Carbon nanotube-modified electrodes for solar energy conversion, *Energy Environ. Sci.*, **1**, 120–133 (2008).
[156] G. de la Torre, W. Blau and T. Torres, A survey on the functionalization of single-walled nanotubes. The chemical attachment of phthalocyanine moieties, *Nanotechnology*, **14**, 765–771 (2003).
[157] H.-B. Xu, H.-Z. Chen, M.-M. Shi, R. Bai and M. Wang, A novel donor-acceptor heterojunction from single-walled carbon nanotubes functionalized by erbium bisphthalocyanine, *Mat. Chem. Phys.*, **94**, 342–346 (2005).
[158] B. Ballesteros, S. Campidelli, G. de la Torre, C. Ehli, D. M. Guldi, M. Prato and T. Torres, Synthesis, characterization and photophysical properties of a SWNT-phthalocyanine hybrid, *Chem. Comm.*, 2950–2952 (2007).
[159] B. Ballesteros, G. De la Torre, C. Ehli, G. M. A. Rahman, F. Agullo-Rueda, D. M. Guldi and T. Torres, Single-wall carbon nanotubes bearing covalently linked phthalocyanines – photoinduced electron transfer, *J. Am. Chem. Soc.*, **129**, 5061–5068 (2007).
[160] S. Campidelli, B. Ballesteros, A. Filoramo, D. Díaz Díaz, G. De la Torre, T. Torres, G. M. A. Rahman, C. Ehli, D. Kiessling, F. Werner, V. Sgobba, D. M. Guldi, C. Cioffi, M. Prato and J.-P. Bourgoin, Facile decoration of functionalized single-wall carbon nanotubes with phthalocyanines via 'click chemistry', *J. Am. Chem. Soc.*, **130**, 11503–11509 (2008).
[161] M. Alvaro, P. Atienzar, P. de la Cruz, J. L. Delgardo, V. Troiani, H. Garcia, F. Langa, A. Palkar and L. Echegoyen, Synthesis, photochemistry, and electrochemistry of single-wall carbon nanotubes with pendent pyridyl groups and of their metal complexes with zinc porphyrin. Comparison with pyridyl-bearing fullerenes, *J. Am. Chem. Soc.*, **128**, 6626–6635 (2006).
[162] J. Yu, S. Mathew, B. S. Flavel, M. R. Johnston and J. G. Shapter, Ruthenium porphyrin functionalized single-walled carbon nanotube arrays: A step toward light harvesting antenna and multibit information storage, *J. Am. Chem. Soc.*, **130**, 8788–8796 (2008).
[163] D. M. Guldi, G. M. A. Rahman, J. Ramey, M. Marcaccio, D. Paolucci, F. Paolucci, S. Qin, W. T. Ford, D. Balbinot, N. Jux, N. Tagmatarchis and M. Prato, Donor-acceptor nanoensembles of soluble carbon nanotubes, *Chem. Commun.*, 2034–2035 (2004).
[164] D. M. Guldi, G. M. A. Rahman, S. Qin, M. Tchoul, W. T. Ford, M. Marcaccio, D. Paolucci, F. Paolucci, S. Campidelli and M. Prato, Versatile coordination chemistry towards multifunctional carbon nanotube nanohybrids, *Chem. Eur. J.*, **12**, 2152–2161 (2006).
[165] G. M. A. Rahman, A. Troeger, V. Sgobba, D. M. Guldi, N. Jux, M. N. Tchoul, W. T. Ford, A. Mateo-Alonso and M. Prato, Improving photocurrent generation: supramolecularly and covalently functionalized single-wall carbon nanotubes-polymer/porphyrin donor-acceptor nanohybrids, *Chem. Eur. J.*, **14**, 8837–8846 (2008).
[166] S. Campidelli, C. Sooambar, E. L. Diz, C. Ehli, D. M. Guldi and M. Prato, Dendrimer-functionalized single-wall carbon nanotubes: synthesis, characterization, and photoinduced electron transfer, *J. Am. Chem. Soc.*, **128**, 12544–12552 (2006).
[167] A. Hirsch, Functionalization of single-walled carbon nanotubes, *Angew. Chem. Int. Ed.*, **41**, 1853–1859 (2002).
[168] Star, J. F. Stoddart, D. Steuerman, M. Diehl, A. Boukai, E. W. Wong, X. Yang, S. W. Chung, H. Choi and J. R. Heath, Preparation and properties of polymer-wrapped single-walled carbon nanotubes, *Angew. Chem. Int. Ed.*, **40**, 1721–1725 (2001).

[169] D. M. Guldi, G. M. A. Rahman, N. Jux, N. Tagmatarchis and M. Prato, Integrating single-wall carbon nanotubes into donor-acceptor nanohybrids, *Angew. Chem. Int. Ed.*, **43**, 5526–5530 (2004).

[170] K. Saito, H. Qiu, V. Troiani, N. Solladié, T. Sakata, H. Mori, M. Ohama and S. Fukuzumi, Nondestructive formation of supramolecular nanohybrids of single-walled carbon nanotubes with flexible porphyrinic polypeptides, *J. Phys. Chem. C*, **111**, 1194–1199 (2007).

[171] D. M. Guldi, H. Taieb, G. M. A. Rahman, N. Tagmatarchis and M. Prato, Novel photoactive single-walled carbon nanotube – porphyrin polymer wraps: Efficient and long-lived intracomplex charge separation, *Adv. Mat.*, **17**, 871–875 (2005).

[172] F. Cheng, S. Zhang, A. Adronov, L. Echegoyen and F. Diederich, 'Triply fused ZnII-porphyrin oligomers: synthesis, properties, and supramolecular interactions with single-walled carbon nanotubes (SWNTs), *Chem. Eur. J.*, **12**, 6062–6070 (2006).

[173] J. Chen and C. P. Collier, Noncovalent functionalization of single-walled carbon nanotubes with water-soluble porphyrins, *J. Phys. Chem. B*, **109**, 7605–7609 (2005).

[174] N. Nakashima, Y. Tomonari and H. Murakami, Water-soluble single-walled carbon nanotubes via noncovalent sidewall-functionalization with a pyrene-carrying ammonium ion, *Chem. Lett.*, **6**, 638–639 (2002).

[175] Ehli, G. M. A. Rahman N. Jux, D. Balbinot, D. M. Guldi, F. Paolucci, M. Marcaccio, D. Paolucci, M. Melle-Franco, F. Zerbetto, S. Campidelli and M. Prato, Interactions in single wall carbon nanotubes/pyrene/porphyrin nanohybrids, *J. Am. Chem. Soc.*, **128**, 11222–11231 (2006).

[176] M. Guldi, G. M. A. Rahman, N. Jux, D. Balbinot, N. Tagmatarchis and M. Prato, Multiwalled carbon nanotubes in donor-acceptor nanohybrids-towards long-lived electron transfer products, *Chem. Commun.*, 2038–2040 (2005).

[177] G. M. A. Rahman, D. M. Guldi, R. Cagnoli, A. Mucci, L. Schenetti, L. Vaccari and M. Prato, Combining single wall carbon nanotubes and photoactive polymers for photoconversion, *J. Am. Chem. Soc.*, **127**, 10051–10057 (2005).

[178] M. Guldi, G. M. A. Rahman, V. Sgobba, N. A. Kotov, D. Bonifazi and M. Prato, CNT-CdTe Versatile donor-acceptor nanohybrids, *J. Am. Chem. Soc.*, **128**, 2315–2323 (2006).

[179] M. Guldi, G. M. A. Rahman, N. Jux, D. Balbinot, U. Hartnagel, N. Tagmatarchis and M. Prato, Functional single-wall carbon nanotube nanohybrids-associating SWNTs with water-soluble enzyme model systems, *J. Am. Chem. Soc.*, **127**, 9830–9838 (2005).

[180] R. Chitta, A. S. D. Sandanayaka, A. L. Schumacher, L. D'Souza, Y. Araki, O. Ito and F. D'Souza, Donor-acceptor nanohybrids of zinc naphthalocyanine or zinc porphyrin noncovalently linked to single-wall carbon nanotubes for photoinduced electron transfer, *J. Phys. Chem. C*, **111**, 6947–6955 (2007).

[181] J. S. Kavakka, S. Heikkinen, I. Kilpelaeinen, M. Mattila, H. Lipsanen and J. Helaja, Noncovalent attachment of pyro-pheophorbide a to a carbon nanotube, *Chem. Commun.*, 519–521 (2007).

[182] L. Hu, Y. L. Zhao, K. Ryu, C. Zhou, J. F. Stoddart and G. Gruner, Light-induced charge transfer in pyrene/CdSe-SWNT hybrids, *Adv. Mat.*, **20**, 939–946 (2008).

[183] D. S. Hecht, R. J. A. Ramirez, M. Briman, E. Artukovic, K. S. Chichak, J. F. Stoddart and G. Gruener, Bioinspired detection of light using a porphyrin-sensitized single-wall nanotube field effect transistor, *Nano Lett.*, **6**, 2031–2036 (2006).

[184] D'Souza, R. Chitta A. S. D. Sandanayaka, N. K. Subbaiyan, L. D'Souza, Y. Araki and O. Ito, Supramolecular carbon nanotube-fullerene donor-acceptor hybrids for photoinduced electron transfer, *J. Am. Chem. Soc.*, **129**, 15865–15871 (2007).

[185] P. J. Boul, D. G. Cho, G. M. A. Rahman, M. Marquez, Z. Ou, K. M. Kadish, D. M. Guldi and J. L. Sessler, Sapphyrin-nanotube assemblies, *J. Am. Chem. Soc.*, **129**, 5683–5687 (2007).

[186] C. Cioffi, S. Campidelli, C. Sooambar, M. Marcaccio, G. Marcolongo, M. Meneghetti, D. Paolucci, F. Paolucci, C. Ehli, G. M. A. Rahman, V. Sgobba, D. M. Guldi and M. Parto, Synthesis, characterization, and photoinduced electron transfer in functionalized single wall carbon nanohorns, *J. Am. Chem. Soc.*, **129**, 3938–3945 (2007).

[187] M. Endo, Y. A. Kim, T. Hayashi, Y. Fukai, K. Oshida, M. Terrones, T. Yanagisawa, S. Higaki and M. S. Dresselhaus, Structural characterization of cup-stacked-type nanofibers with an entirely hollow core, *Appl. Phys. Lett.*, **80**, 1267–1269 (2002).

[188] C. Kim, Y. J. Kim, Y. A. Kim, T. Yanagisawa, K. C. Park, M. Endo and M. S. Dresselhaus, High performance of cup-stacked-type carbon nanotubes as a Pt-Ru catalyst support for fuel cell applications, *J. Appl. Phys.*, **96**, 5903–5905 (2004).
[189] M. Endo, Y. A. Kim, M. Ezaka, K. Osada, T. Yanagisawa, T. Hayashi, M. Terrones and M. S. Dresselhaus, Selective and efficient impregnation of metal nanoparticles on cup-stacked-type carbon nanofibers, *Nano Lett.*, **3**, 723–726 (2003).
[190] T. Hasobe, S. Fukuzumi and P. V. Kamat, Stacked-cup carbon nanotubes for photoelectrochemical solar cells, *Angew. Chem. Int. Ed.*, **45**, 755–759 (2006).
[191] T. Hasobe, H. Murata and P. V. Kamat, Photoelectrochemistry of stacked-cup carbon nanotube films. Tube-length dependence and charge transfer with excited porphyrin, *J. Phys. Chem. C*, **111**, 16626–16634 (2007).
[192] K. Saito, M. Ohtani and S. Fukuzumi, Electron-transfer reduction of cup-stacked carbon nanotubes affording cup-shaped carbons with controlled diameter and size, *J. Am. Chem. Soc.*, **128**, 14216–14217 (2006).
[193] S. Fukuzumi, T. Suenobu, M. Patz, T. Hirasaka, S. Itoh, M. Fujitsuka and O. Ito, Selective one-electron and two-electron reduction of C_{60} with NADH and NAD dimer analogues via photoinduced electron transfer, *J. Am. Chem. Soc.*, **120**, 8060–8068 (1998).
[194] K. Saito, M. Ohtani and S. Fukuzumi, Nanostructural control of cup-stacked carbon nanotubes with 1-benzyl-1,4-dihydronicotinamide dimer via photoinduced electron transfer, *Chem. Commun.*, 55–57 (2007).

5

Higher Fullerenes: Chirality and Covalent Adducts

Agnieszka Kraszewska, François Diederich and Carlo Thilgen

Laboratory of Organic Chemistry, ETH Zurich, Zurich, Switzerland

5.1 Introduction

Fullerenes are attractive spheroidal molecules and constitute versatile scaffolds for the formation of three-dimensional structures. The highly symmetric buckminsterfullerene, C_{60}-I_h, is the archetype and major representative of this class of carbon molecules [1].

The higher fullerenes (C_{70} and beyond) [2–5] show a great diversity of larger and more complex – in part chiral [6–8] – structures which give rise to interesting differences in their properties as compared to C_{60}, e.g. their electrochemical redox potentials, thermodynamic stability, optical properties, or aromatic character [3, 9, 10]. The main difficulty in their investigation is the low abundance in fullerene soot [2, 3] and the increasing number of possible isomers for the homologues with increasing size [11]. Whereas C_{60}-I_h and C_{70}-D_{5h} exist as a single isomer each, four (out of five possible structures that are in accord with the isolated pentagon rule (IPR) [1, 12, 13]) cage isomers of C_{78} [14–19] and ten (out of 24) of C_{84} [4, 15, 20–24] were identified so far. The tedious separation of higher fullerenes makes commercial pure samples very expensive, and an unambiguous structural assignment is often impeded by the fact that several isomers have the same symmetry. The following empty fullerene cages were unambiguously proven to exist by single crystal X-ray diffraction studies of the pure allotropes or derivatives thereof: C_{60} and C_{70} [25], $C_{74}(1)$-D_{3h} [26, 27], $C_{76}(1)$-D_2 [28], $C_{78}(1)$-D_3 [29], $C_{78}(2)$-C_{2v} [29], $C_{78}(3)$-C_{2v} [30], $C_{78}(5)$-D_{3h} [26], $C_{84}(11)$-C_2 [26], $C_{84}(14)$-C_s [31], and $C_{84}(23)$-D_{2d} [32]. Further allotropes such as $C_{76}(2)$-T_d [17],

Chemistry of Nanocarbons Edited by Takeshi Akasaka, Fred Wudl and Shigeru Nagase

$C_{80}(1)$-D_{5d} [33], $C_{80}(2)$-D_2 [34], $C_{80}(5)$-C_{2v} [17], $C_{82}(3)$-C_2 [17], $C_{82}(5)$-C_2 [17], $C_{82}(9)$-C_{2v} [15], $C_{84}(4)$-D_{2d} [23], $C_{84}(5)$-D_2 [23], $C_{84}(16)$-C_s [23, 31], $C_{84}(19)$-D_{3d} [22], $C_{84}(22)$-D_2 [21], $C_{84}(24)$-D_{6h} [22], $C_{86}(16)$-C_s [35], $C_{86}(17)$-C_2 [35], $C_{88}(7)$-C_2 [36], $C_{88}(17)$-C_s [36], $C_{88}(33)$-C_2 [36] and $C_{90}(32)$-C_1 [29] have been identified using NMR spectra of the bare cages or derivatives thereof.

In addition to the empty carbon spheroids, endohedral inclusion compounds of fullerenes with new cage isomers have been isolated and characterized [37, 38], but their existence does not prove the independent incidence of the corresponding void cages. Due to the great variety of occurring parent fullerene structures and endohedrally included mono- and oligoatomic species, endohedral fullerene derivatives exist in an immense diversity and their discussion would exceed the scope of the present review. Their omission in this article is further justified by their properties strongly differing, in general, from those of empty fullerenes.

5.1.1 Fullerene Chirality – Classification and the Stereodescriptor System

Until now, carbon is the only element of which chiral molecular allotropes are known, and inherently chiral fullerenes show a chiral arrangement of formally sp^2-hybridized C-atoms constituting the fullerene scaffold. Well investigated are the isolated enantiomers of $C_{76}(1)$-D_2 [39–41] or $C_{84}(22)$-D_2 [42], as well as some derivatives thereof. With increasing cage size, the number of possible chiral fullerenes increases significantly [3, 11]: One out of five IPR-conforming structures [43] of C_{78} is chiral, as are two out of seven C_{80} isomers and three out of nine C_{82} isomers. The C_{84}, C_{86}, and C_{88} families include 10 (out of total 24), 14 (out of 19), and 21 (out of 35) chiral isomers, respectively.

The derivatization of achiral parent fullerene scaffolds can also generate chirality. Depending on the stereoelement(s) responsible for chirality, such derivatives can be divided into three classes [44, 45]:

- Fullerene derivatives with an inherently chiral functionalization pattern, in which the geometrical arrangement of the addends generates a chiral pattern, regardless of the addends being identical or not.
- Fullerene derivatives with a non-inherently chiral functionalization pattern, where the non-identity of at least some of the addends is a *conditio sine qua non* for chirality.
- Fullerene derivatives with the stereogenic element(s) located exclusively in the addend(s). Their functionalization pattern is achiral.

Derivatives of inherently chiral parent fullerenes have an inherently chiral functionalization pattern *per se*. A chiral fullerene functionalization pattern is either inherently or non-inherently chiral; both types exclude each other mutually. On the other hand, stereogenic elements in the addends can be superposed to a chiral functionalization pattern. A helpful tool for the classification of fullerene chirality is the flow chart shown in Figure 5.1.

The complexity of chiral fullerene structures requires a practicable stereodescriptor system. The CIP (Cahn, Ingold, Prelog) rules [46, 47] cannot be applied to inherently chiral parent fullerenes because no stereogenic center can be identified among the formally sp^2-hybridized carbon atoms of the spheroids. For fullerene derivatives with a chiral functionalization pattern, the configurational specification, according to the CIP rules, of stereogenic centers belonging to the fullerene cage would be complicated and the resulting

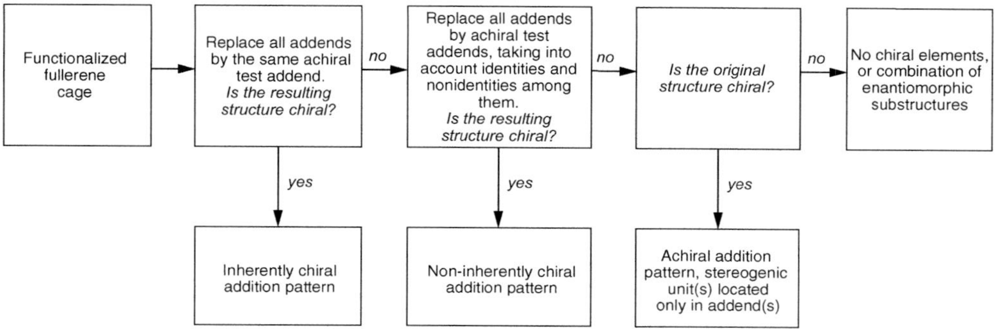

Figure 5.1 *Flow chart for the identification of different types of fullerene chirality by a stepwise, formal substitution test*

set of descriptors difficult to interpret. A feasible system, recommended by IUPAC, is based on the fact that the sequence of numbered fullerene carbon atoms, as used for nomenclature purposes, is helical and uses the 'numbering helices' as a reference for the stereodescription of chiral carbon spheroids or fullerene functionalization patterns [48, 49]. To assign a descriptor to a chiral fullerene unit, the viewer, looking from the outside of the cage at the polygon in which the numbering starts, traces a path from atom C(1) to C(2) to C(3) which are never aligned in a fullerene structure. If this path describes a clockwise direction, the configuration is specified by the descriptor $^{f,x}C$, where the superscript 'f' indicates that the descriptor refers to a fullerene and the superscript 'x' is either 's' for the systematic numbering [48, 49] ($^{f,s}C$) or 't' for the trivial numbering [50] ($^{f,t}C$) (the latter was originally denoted simply as ^{f}C [44, 51]) [48, 49]. If the path from C(1) to C(2) to C(3) describes an anticlockwise direction, the descriptor is accordingly $^{f,x}A$. The enantiomeric numbering schemes for a C_{70}-derivative with an inherently chiral functionalization pattern are shown in Figure 5.2.

5.1.2 Reactivity and Regioselectivity

Fullerenes in a stricter sense are made up exclusively of five- and six-membered rings, the pentagons being completely surrounded by hexagons (IPR) [12, 13]. Fullerene bonds correspond, therefore, to edges shared either by two hexagons (6–6 bond) or by a pentagon and a hexagon (5–6 bond). With decreasing symmetry of the fullerene, the number of distinct bonds increases and so does the number of theoretically possible reaction products. However, an often pronounced regioselectivity reduces the number of actually formed adducts, thereby facilitating their isolation and characterization. Nucleophilic or carbene additions, formation of transition metal complexes, cycloadditions, and lower levels of hydrogenation yield mostly 1,2-adducts across 6–6 bonds [10, 53], whereas radical reactions such as halogenation or trifluoromethylation give often rise to 1,4-type addition patterns.

In the well-investigated cycloaddition reactions, bond reactivities generally parallel local spheroid curvature, i.e. the pyramidalization angles of the involved carbon atoms. This is nicely seen in *Bingel* type cyclopropanations of C_{70}: although the molecule contains four different types of 6–6 bonds, the major product arises from addition across the C(8)–C(25)

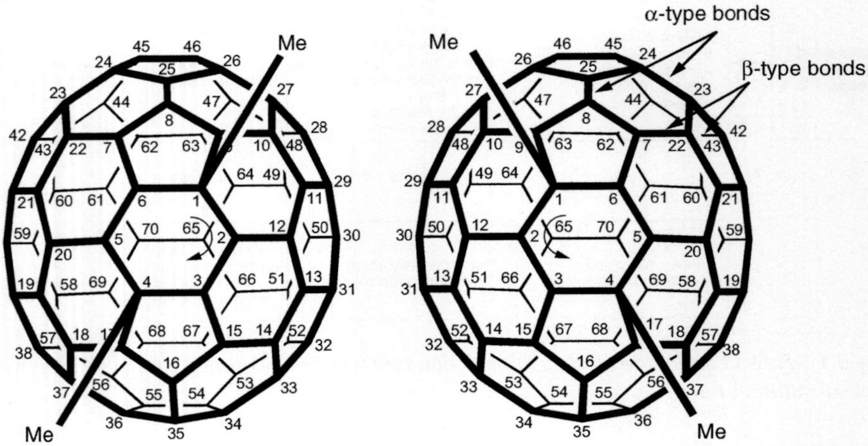

f,s*C* f,s*A*

Figure 5.2 Three-dimensional diagrams of the enantiomers of 1,4-dimethyl-1,4-dihydro (C_{70}-D_{5h})[5,6]fullerene [53]. In achiral parent C_{70}, both numberings are equivalent, but this equivalence vanishes in derivatives with a chiral functionalization pattern. The direction of the numbering (clockwise or anticlockwise) comes out of the requirement to afford the lowest locants for the addends [49, 50]. Two different bond types (α and β) are shown: α-type ≡ C(8)–C(25) bond and nine equivalent bonds (radiating from the polar pentagons); β-type ≡ C(7)–C(22) bond and nine equivalent non-equatorial bonds perpendicular to the main rotation axis

bond (for the atom numbering of C_{70}, see Figure 5.2) radiating from the polar pentagon, followed by minor amounts of C(7)–C(22) adduct [51, 54, 55]. Products resulting from addition across the remaining two 6–6 bonds in the flatter equatorial region have not been observed in this reaction. Similarly, 1,2-additions in higher fullerenes such as C_{76} tend to take place in the polar region displaying the highest local curvature [42, 56, 57].

For radical reactions, on the other hand, it was shown that the preferential addend arrangement depends on the size of the radical. Radicals such as Hal• or F_3C• lead to ribbon type addition patterns, preferentially in the equatorial region, with the addends attached to less pyramidalized C atoms on the fullerene core [58–62].

5.2 The Chemistry of C_{70}

5.2.1 C_{70}-Derivatives with an Inherently Chiral Functionalization Pattern

5.2.1.1 Bingel Cyclopropanation of C_{70}

5.2.1.1.1 ADDITION OF UNTETHERED MALONATES

The first *Bingel* cyclopropanation of C_{70} was carried out with diethyl 2-bromomalonate in the presence of 1,8-diazabicyclo[5.4.0]undec-7-ene (DBU) and afforded a single mono-adduct (**1**, Figure 5.3), which was identified as a C(8)–C(25) adduct [54]. This reaction shows a remarkable selectivity compared to the cycloaddition of nitrile oxides [63, 64],

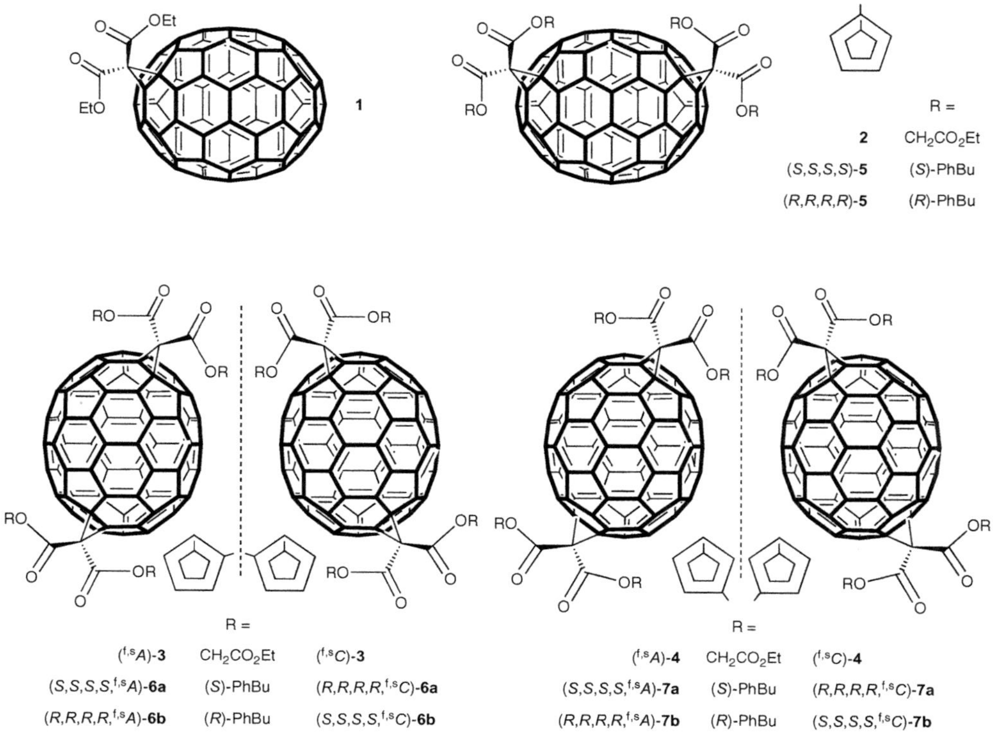

Figure 5.3 *Products of single and double Bingel addition of chiral and achiral malonates to C_{70}. PhBu = 2-phenylbutyl. The relative arrangement of the addends can easily be seen in a Newman-type projection along the C_5 rotation axis of the parent fullerene, showing the polar pentagons in a concentric fashion together with the functionalized bonds radiating from them*

(formation of C(8)–C(25)- and C(7)–C(22)-adducts) and that of azomethine ylides [65, 66], *o*-quinodimethanes [56], or benzyne [67, 68] (formation of C(8)–C(25)-, C(7)–C(22)-, and C(1)–C(2)-adducts, see also Section 5.2.1.2).

In a bis-functionalization, the second *Bingel* addition [51, 55] occurs in the unfunctionalized hemisphere of C_{70}, again at one of the bonds radiating from the polar pentagon. Three bis-adduct regioisomers resulted from double addition of bis[(ethoxycarbonyl) methyl] 2-bromomalonate to C_{70} (Figure 5.3) [51]. The relative position of the addends can be described with a simplified *Newman* type projection (Figure 5.3) as twelve-, two-, or five o'clock addition pattern. While the achiral twelve o'clock isomer **2** has C_{2v}-symmetry, (±)-**3** and (±)-**4** – the two- and five o'clock isomers, respectively – have C_2-symmetrical structures with an inherently chiral functionalization pattern and were obtained as racemic mixtures [51].

The reaction mixture contained bis-adducts **2**, (±)-**3**, and (±)-**4** in the ratio 1.4 : 5.3 : 1 which contrasts the statistically expected 1 : 2 : 2 relationship [51]. This deviation demonstrates the influence of the first addend on bond reactivities in the unfunctionalized hemisphere of the fullerene.

The resolution of racemates (±)-**3** and (±)-**4** was not possible. To circumvent the separation of enantiomeric adducts, the synthesis was repeated with a chiral, enantiopure malonate. Double addition of bis[(*S*)-1-phenylbutyl] 2-bromomalonate to C_{70} yielded five C_2-symmetric bis-adducts, the stereodescription of which must take into account both the fullerene functionalization pattern and the stereogenic centers in the addends (Figure 5.3). Whereas the twelve o'clock isomer (*S*,*S*,*S*,*S*)-**5** has its stereogenic elements located exclusively in the addends, the other two regioisomers occur as pairs of diastereoisomers: (*S*,*S*,*S*,*S*,$^{\mathrm{f,s}}$*A*)-**6a** and (*S*,*S*,*S*,*S*,$^{\mathrm{f,s}}$*C*)-**6b** (two o'clock isomers), as well as (*S*,*S*,*S*,*S*,$^{\mathrm{f,s}}$*A*)-**7a** and (*S*,*S*,*S*,*S*,$^{\mathrm{f,s}}$*C*)-**7b** (five o'clock isomers) [51]. All five isomers could be isolated by HPLC on unmodified silica gel, because no enantiomeric relationship exists among them.

The ^{13}C-NMR spectra of bis-adducts (*S*,*S*,*S*,*S*)-**5**, (*S*,*S*,*S*,*S*,$^{\mathrm{f,s}}$*A*)-**6a**, (*S*,*S*,*S*,*S*,$^{\mathrm{f,s}}$*C*)-**6b**, (*S*,*S*,*S*,*S*,$^{\mathrm{f,s}}$*A*)-**7a**, and (*S*,*S*,*S*,*S*,$^{\mathrm{f,s}}$*C*)-**7b** show similarities between isomers that have the same constitution ((*S*,*S*,*S*,*S*,$^{\mathrm{f,s}}$*A*)-**6a**/(*S*,*S*,*S*,*S*,$^{\mathrm{f,s}}$*C*)-**6b** on one hand and (*S*,*S*,*S*,*S*,$^{\mathrm{f,s}}$*A*)-**7a**/(*S*,*S*,*S*,*S*,$^{\mathrm{f,s}}$*C*)-**7b** on the other), and the CD spectra of such a pair show near-mirror image behavior. Only weak *Cotton* effects were observed for (*S*,*S*,*S*,*S*)-**5**, because there is no inherent contribution from the achiral residual fullerene chromophore, and the stereogenic centers located in the addends do not perturb the fullerene π-system significantly. It was concluded that the chiral functionalization pattern determines the shape of the CD spectra and that addends with stereogenic centers usually have only a minor influence. It ensues that although the isomers in the pairs (*S*,*S*,*S*,*S*,$^{\mathrm{f,s}}$*A*)-**6a**/(*S*,*S*,*S*,*S*,$^{\mathrm{f,s}}$*C*)-**6b** and (*S*,*S*,*S*,*S*,$^{\mathrm{f,s}}$*A*)-**7a**/(*S*,*S*,*S*,*S*,$^{\mathrm{f,s}}$*C*)-**7b** have a diastereoisomeric relationship, their CD spectra are nearly perfect mirror images, their fullerene functionalization patterns being enantiomeric [45, 51].

In a different experiment, an enantiomeric series to the above-mentioned bis-adducts (*S*,*S*,*S*,*S*)-**5**, (*S*,*S*,*S*,*S*,$^{\mathrm{f,s}}$*A*)-**6a**/(*S*,*S*,*S*,*S*,$^{\mathrm{f,s}}$*C*)-**6b**, and (*S*,*S*,*S*,*S*,$^{\mathrm{f,s}}$*A*)-**7a**/(*S*,*S*,*S*,*S*,$^{\mathrm{f,s}}$*C*)-**7b**), i.e. isomers (*R*,*R*,*R*,*R*)-**5**, (*R*,*R*,*R*,*R*,$^{\mathrm{f,s}}$*C*)-**6a**/(*R*,*R*,*R*,*R*,$^{\mathrm{f,s}}$*A*)-**6b**, and (*R*,*R*,*R*,*R*,$^{\mathrm{f,s}}$*C*)-**7a**/(*R*,*R*,*R*,*R*,$^{\mathrm{f,s}}$*A*)-**7b**, was synthesized by addition of enantiomerically pure bis[(*R*)-1-phenylbutyl] 2-bromomalonate to C_{70} [69]. The absolute configuration of the chiral bis-adducts was later assigned by comparison of the experimental CD data with calculated spectra [70]. As the chiroptical contributions from the optically active malonate addends are negligible, the CD calculations were performed for isomers incorporating the simpler ethyl instead of (*S*)-1-phenylbutyl ester residues.

In an extension of this work, higher adducts were prepared by further two *Bingel* additions to bis-adducts (*S*,*S*,*S*,*S*,$^{\mathrm{f,s}}$*A*)-**6a**, (*R*,*R*,*R*,*R*,$^{\mathrm{f,s}}$*C*)-**6a**, (*S*,*S*,*S*,*S*,$^{\mathrm{f,s}}$*C*)-**6b**, and (*R*,*R*,*R*,*R*,$^{\mathrm{f,s}}$*A*)-**6b**. It was found that if each hemisphere of C_{70} bears already an addend at an α-type bond, further addition occurs at β-type bonds (for the definition of bond types α and β, see Figure 5.2). Accordingly, four tetrakis-adducts, i.e. (*S*,*S*,*S*,*S*,$^{\mathrm{f,s}}$*C*)-**8a**, (*R*,*R*,*R*,*R*,$^{\mathrm{f,s}}$*A*)-**8a**, (*S*,*S*,*S*,*S*,$^{\mathrm{f,s}}$*A*)-**8b**, and (*R*,*R*,*R*,*R*,$^{\mathrm{f,s}}$*C*)-**8b** (Figure 5.4) with an inherently chiral functionalization pattern were obtained from addition of diethyl 2-bromomalonate to two o'clock isomers (*S*,*S*,*S*,*S*,$^{\mathrm{f,s}}$*C*)-**6b**, (*R*,*R*,*R*,*R*,$^{\mathrm{f,s}}$*A*)-**6b**, (*S*,*S*,*S*,*S*,$^{\mathrm{f,s}}$*A*)-**6a**, and (*R*,*R*,*R*,*R*,$^{\mathrm{f,s}}$*C*)-**6a** (Figure 5.3), respectively [51, 69]. Again, the CD spectra of the tetrakis-adducts demonstrate the enantiomeric relationships among fullerene chromophores and the negligible contribution of the addends to the *Cotton* effects.

5.2.1.1.2 ADDITION OF BIS-MALONATES TETHERED BY CROWN ETHERS

In a first application of the tether-directed remote functionalization [71, 72] to C_{70}, crown ether-derived bis-malonates were used to selectively generate bis-adducts with a defined

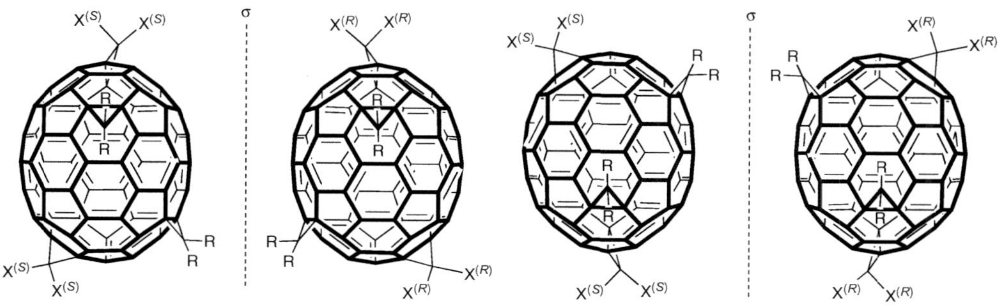

(S,S,S,S,f,sC)-**8a** (R,R,R,R,f,sA)-**8a** (S,S,S,S,f,sA)-**8b** (R,R,R,R,f,sC)-**8b**

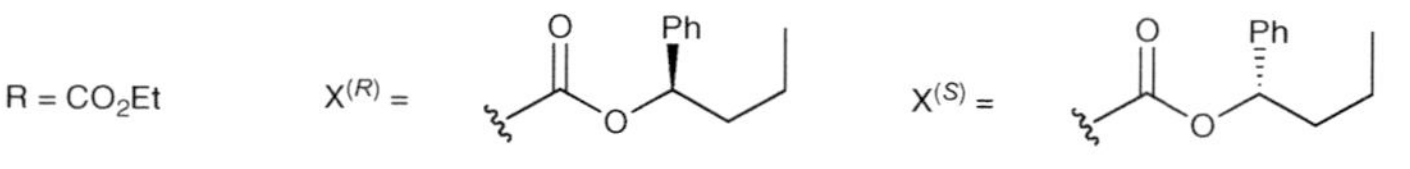

Figure 5.4 *Tetrakis-adducts resulting from twofold addition of diethyl 2-bromomalonate to enantiopure bis-adducts (S,S,S,S,^{f,s}C)-**6b**, (R,R,R,R,^{f,s}A)-**6b**, (S,S,S,S,^{f,s}A)-**6a**, and (R,R,R,R,^{f,s}C)-**6a** (Figure 5.3) with the two o'clock functionalization pattern*

functionalization pattern [73, 74]. With the *anti*-disubstituted dibenzo[18]crown-6 (DB18C6) derivative **9**, it was possible to obtain the kinetically disfavored (see Section 5.2.1.1.1) five o'clock bis-adducts (±)-**10a** and (±)-**10b** with complete regioselectivity (Scheme 5.1, top). The diastereoisomeric relationship between racemates (±)-**10a** and (±)-**10b** results from the different orientation of the conformationally locked *anti*-disubstituted crown ether which constitutes an extra stereogenic element (element of 'planar chirality' [46]). With the *syn*-disubstituted DB18C6 tether **11**, on the other hand, the regioisomeric bis-adducts (±)-**12** and **13** with the two o'clock- and twelve o'clock functionalization pattern, respectively, were obtained (Scheme 5.1, bottom). Transesterification of (±)-**10a**, (±)-**10b** (±)-**12**, and **13** with ethanol afforded the ethyl ester analogs of bis-adducts (±)-**4**, (±)-**3**, and **2** (Figure 5.3), which could not be selectively prepared by sequential addition of two independent malonate molecules.

The addition of a second crown ether-malonate conjugate **11** to isomers (±)-**10a** and (±)-**10b** provided tetrakis-adducts (±)-**14a** and (±)-**14b** (Scheme 5.2), respectively, which have the same fullerene functionalization pattern as tetrakis-adducts (*S,S,S,S*,^{f,s}*C*)-**8a**, (*R,R,R,R*,^{f,s}*A*)-**8a**, (*S,S,S,S*,^{f,s}*A*)-**8b**, and (*R,R,R,R*,^{f,s}*C*)-**8b** (Figure 5.4). The constitution of (±)-**14a** and (±)-**14b** was determined by recognition of the near-identity of their UV/Vis spectra and those of (±)-**8a** and (±)-**8b** [75].

5.2.1.1.3 ADDITION OF BIS-MALONATES TETHERED BY THE TRÖGER BASE

Provided a high asymmetric induction is operative, the addition to fullerenes of bis-malonates based on enantiopure tethers provides a means of diastereoselectively generating chiral fullerene functionalization patterns with a given configuration [71]. Enantiopure derivatives of the *Tröger* base appeared to be appropriate spacers for such a tether-directed remote functionalization because of their rigidity and folded geometry [76, 77].

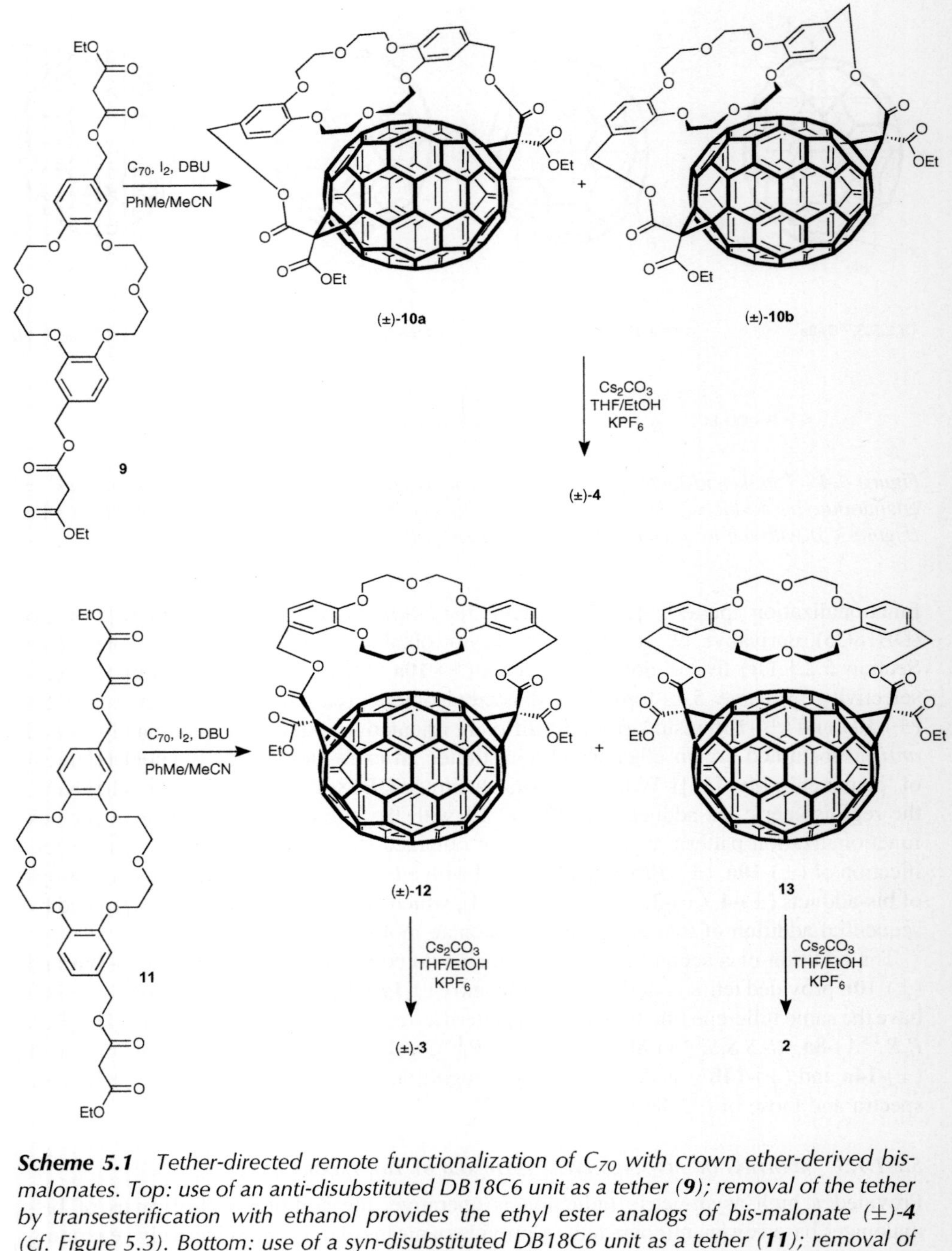

Scheme 5.1 *Tether-directed remote functionalization of C_{70} with crown ether-derived bis-malonates. Top: use of an anti-disubstituted DB18C6 unit as a tether (**9**); removal of the tether by transesterification with ethanol provides the ethyl ester analogs of bis-malonate (±)-**4** (cf. Figure 5.3). Bottom: use of a syn-disubstituted DB18C6 unit as a tether (**11**); removal of the tether by transesterification with ethanol provides the ethyl ester analogs of bis-malonates **2** and (±)-**3** (cf. Figure 5.3)*

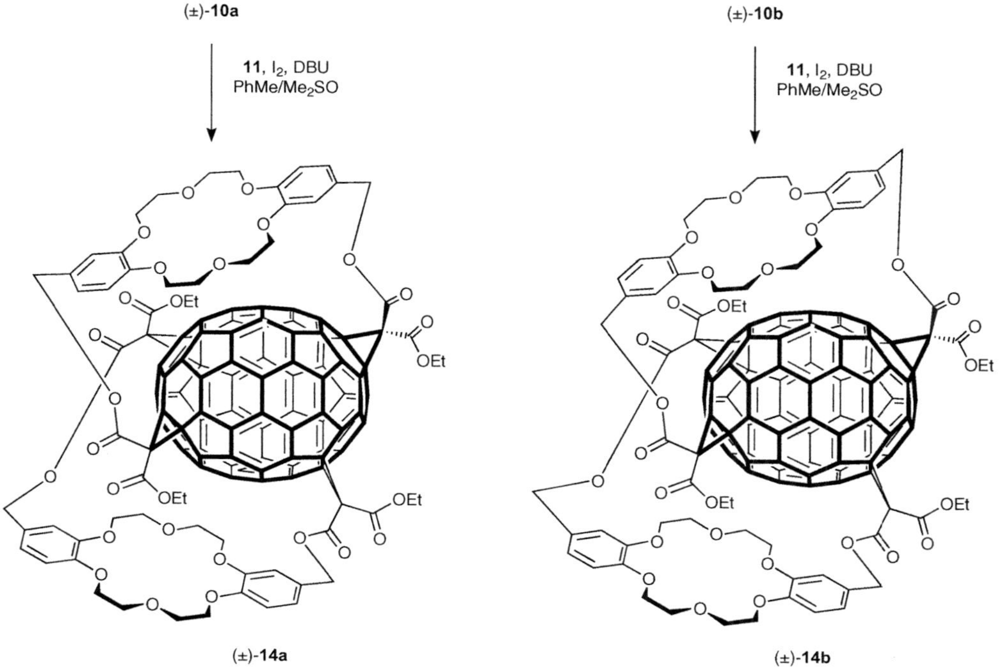

Scheme 5.2 *Tetrakis-adducts of C_{70} resulting from sequential addition, to the fullerene, of two constitutionally different (anti- (**9**) and syn-disubstituted (**11**)) crown ether-derived bis-malonates*

Addition of bis-malonates (*S*,*S*)-**15** and (*R*,*R*)-**15** to C_{70} in fact yielded enantiomeric bis-adducts (*S*,*S*,^{f,s}*A*)-**16** and (*R*,*R*,^{f,s}*C*)-**16**, respectively, with perfect regio- and diastereo-selectivity (Scheme 5.3, top) [78]. Constitution and configuration of their inherently chiral five o'clock functionalization pattern was determined by comparison of their UV/Vis and CD spectra with those obtained for (*S*,*S*,*S*,*S*,^{f,s}*A*)-**7a**/(*R*,*R*,*R*,*R*,^{f,s}*C*)-**7a** or (*S*,*S*,*S*,*S*,^{f,s}*C*)-**7b**/(*R*,*R*,*R*,*R*,^{f,s}*A*)-**7b** (Figure 5.3).

The addition to C_{70} of bis-malonates (*R*,*R*)-**17** and (*S*,*S*)-**17** derived from a shorter *Tröger* base proceeded again with complete regio- and diastereoselectivity (Scheme 5.3, center). The functionalization pattern of the obtained bis-adducts (*R*,*R*,^{f,s}*A*)-**18** and (*S*,*S*,^{f,s}*C*)-**18** turned out to be unknown at that point. They were preliminarily assigned as derivatives in which the cyclopropane rings are fused to two different types of 6–6 bonds in opposite hemispheres of C_{70}, an α-type and a β-type bond.

The addition of a second enantiopure bismalonate to (*R*,*R*,^{f,s}*A*)-**18** and (*S*,*S*,^{f,s}*C*)-**18** yielded tetrakis-adducts (*R*,*R*,*R*,*R*,^{f,s}*C*)-**19** and (*S*,*S*,*S*,*S*,^{f,s}*A*)-**19**, respectively, again with perfect regio- and stereoselectivity (Scheme 5.3, bottom) [78]. Tetrakis-adducts (*S*,*S*,*S*,*S*,^{f,s}*A*)-**19** and (*R*,*R*,*R*,*R*,^{f,s}*C*)-**19** were shown to have the same functionalization pattern as (*S*,*S*,*S*,*S*,^{f,s}*A*)-**7a** and (*R*,*R*,*R*,*R*,^{f,s}*C*)-**7a** (Figure 5.3) [69] which provided also a proof for the structure of the precursors (*S*,*S*,^{f,s}*C*)-**18** and (*R*,*R*,^{f,s}*A*)-**18** as enantiomers with a new inherently chiral functionalization pattern.

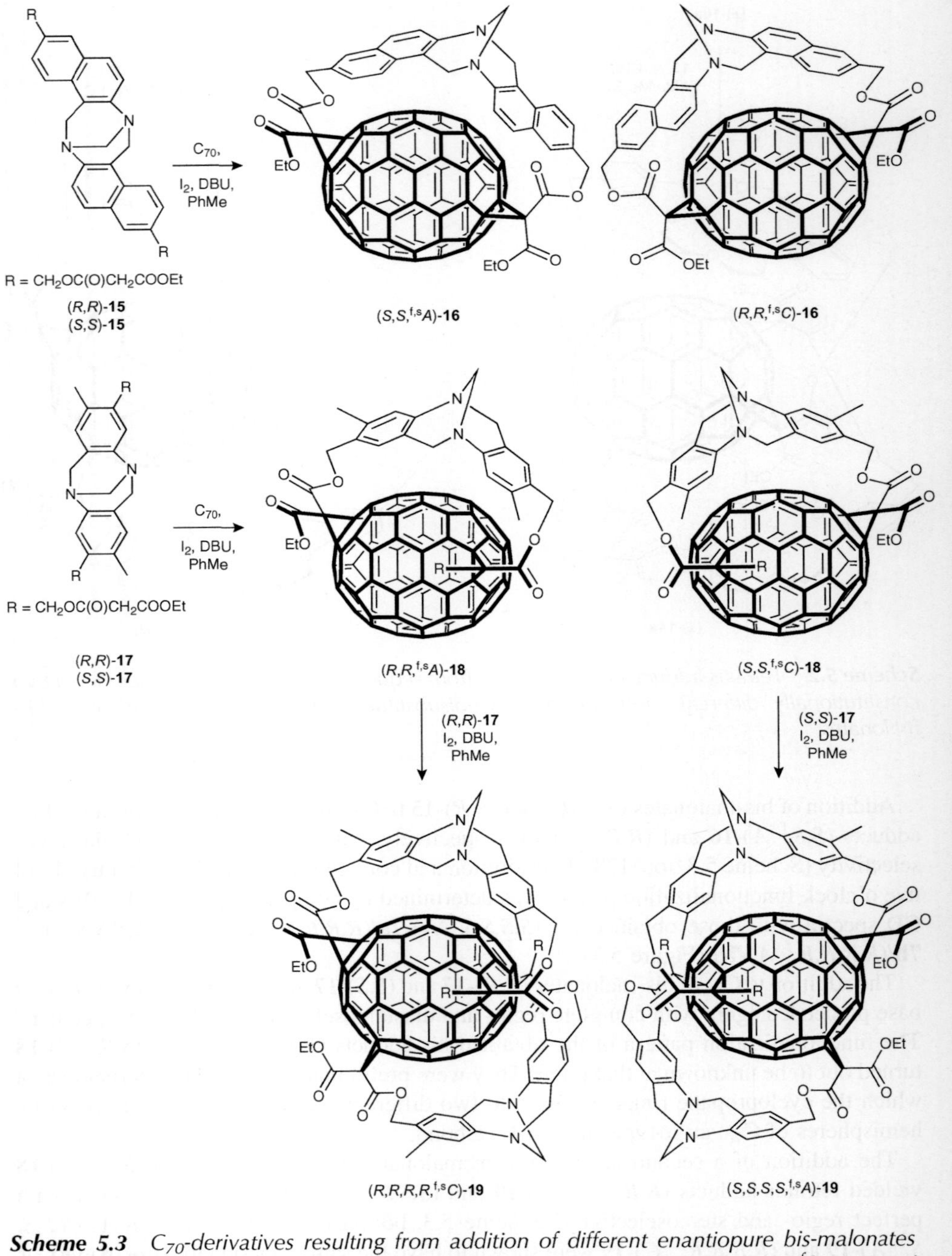

Scheme 5.3 *C_{70}-derivatives resulting from addition of different enantiopure bis-malonates tethered by Tröger base derivatives. Top: α,α-type five o'clock bis-adducts $(S,S,^{f,s}A)$-**16** and $(R,R,^{f,s}C)$-**16**. Center: Novel α,β-type adducts $(R,R,^{f,s}A)$-**18** and $(S,S,^{f,s}C)$-**18**, and their conversion to tetrakis-adducts $(R,R,R,R,^{f,s}C)$-**19** and $(S,S,S,S,^{f,s}A)$-**19**, respectively (bottom)*

These examples show that the refinement of the tether-directed remote functionalization [71, 72] makes C_{70} derivatives with a well-defined functionalization pattern available with a high regioselectivity and, in a number of cases, also a remarkable stereoselectivity. A remaining challenge in this field is the selective bis-functionalization of one hemisphere of [70]fullerene while the other one remains intact.

5.2.1.2 Further Chiral Addition Products of C_{70}

Addition across the C(1)–C(2) bond provides access to the only 6–6 mono-adducts of C_{70} with an inherently chiral functionalization pattern. Derivatives of this type were obtained as minor products during the *Diels-Alder* reaction of [70]fullerene with 4,5-dimethoxy-*o*-quinodimethane ((±)-**20**) [56], [3 + 2] cycloaddition of *N*-methylazomethine ylide ((±)-**21**) [65, 79], and [2 + 2] cycloaddition of benzyne ((±)-**22**) (Figure 5.5, top) [67, 68]. The formation of these products as minor isomers contrasts the statistical expectations and demonstrates the higher reactivity of bonds with higher curvature or higher double bond character [50] in cycloaddition reactions.

An inherently chiral functionalization pattern is also found in a number of 1,4-adducts of C_{70}. Depending on whether the two monovalent addends involved are identical or different, these C(1),C(4)-adducts (Figure 5.5, bottom) are either C_2- or C_1-symmetric. Dibenzylated [70]fullerene (±)-**23** was obtained by deprotonation of 8,25-dihydro (C_{70}-D_{5h})[5,6]fullerene in the presence of benzyl bromide [80]. A dimethylated compound with the same functionalization pattern ((±)-**24**) was suggested to be a product of fullerene reduction with Al–Ni alloy in a NaOH–dioxane–THF mixture, followed by reaction with MeI [52].

Reaction of C_{70} with chloroform in the presence of $AlCl_3$ afforded the C_1-symmetric mono-adduct (±)-**25** [81]. In contrast to nucleophilic additions, the associated electrophilic attack under *Friedel-Crafts* conditions is selective for C(1) of [70]fullerene. The corresponding alcohol (±)-**26** was produced during column chromatography of the initial

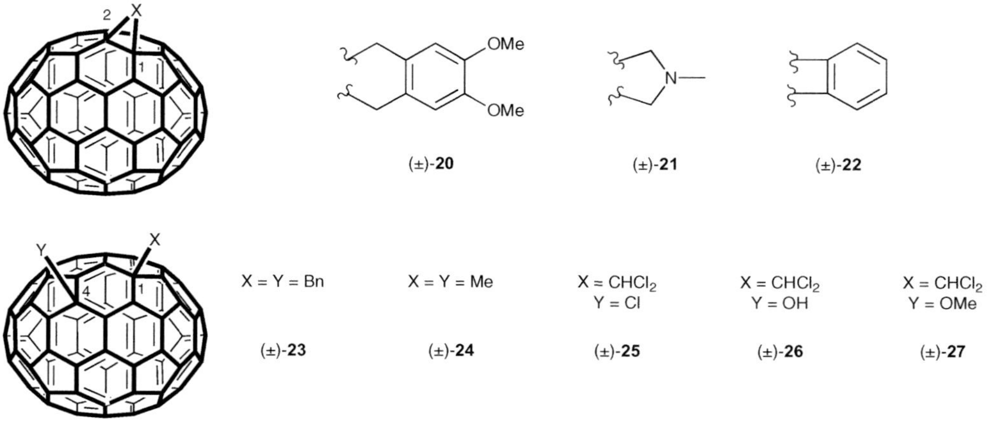

Figure 5.5 *C_{70}-adducts with an inherently chiral functionalization pattern resulting from addition across C(1)–C(2) (top) or functionalization of C(1) and C(4) (bottom)*

product (±)-**25**. Under acidic conditions, the hydroxy group was removed, and methyl ether (±)-**27** was obtained by quenching the acidic solution with methanol [81]. The latter transformation proceeds *via* a stable cation which can be stored in solution for several days.

5.2.1.3 Hydrogenation of C_{70}

Hydrogenation of fullerenes can be performed under different conditions, yielding a variety of $C_{70}H_{2n}$ hydrocarbons with $2n$ up to 44 [82–85]. Reported methods are the hydrogen transfer from 9,10-dihydroanthracene in the presence of [7*H*]benzanthrene at 250 °C [86], the hydroboration [82], the reduction with hydrazine hydrate [83], with Zn–conc. HCl in refluxing toluene [85, 87], or with Zn(Cu) in toluene, followed by aqueous work-up [84, 88, 89]. Some of the characterized derivatives have an inherently chiral functionalization pattern.

In independent studies, two C_s-symmetric isomers of $C_{70}H_2$ were obtained, namely 8,25-dihydro(C_{70}-D_{5h})[5,6]fullerene (α-type adduct) and 7,22-dihydro(C_{70}-D_{5h})[5,6]fullerene (β-type adduct) (for the numbering of C_{70} and bond assignment, see Figure 5.2). Their ratio depends on the reaction conditions [82–84].

The reduction of C_{70} with hydrazine hydrate afforded two $C_{70}H_4$ isomers: the C_s-symmetric 8,23,24,25-tetrahydro(C_{70}-D_{5h})[5,6]fullerene (with two adjacent α-bonds functionalized) and the chiral C_1-symmetric (±)-7,8,22,25-tetrahydro(C_{70}-D_{5h})[5,6]fullerene ((±)-**28**) (with an α-bond and an adjacent β-bond functionalized, Figure 5.6) [83]. Such bis-addition patterns with all addends located in one hemisphere of C_{70} are not known from the *Bingel* cyclopropanation [51, 55, 69] but they were found in Ir-complexes of the epoxyfullerene $C_{70}O$ [90].

The addition of a second pair of hydrogen atoms to 8,25-dihydro(C_{70}-D_{5h})[5,6]fullerene by reduction with Zn(Cu) in toluene, followed by aqueous work-up, was expected to provide three different isomers of $C_{70}H_4$ [84] that correspond to the double addition modes known from the *Bingel* cyclopropanation [51, 55] or the formation of transition metal complexes [91]. Six $C_{70}H_4$ isomers were detected by mass spectrometry, two of which were isolated and characterized [84]. The major isomer was assigned as (±)-8,22,33,34-tetrahydro(C_{70}-D_{5h})[5,6]fullerene ((±)-**29**) with the two o'clock functionalization pattern, followed by the five o'clock isomer (±)-8,25,53,54-tetrahydro(C_{70}-D_{5h})[5,6]fullerene ((±)-**30**) (Figure 5.6).

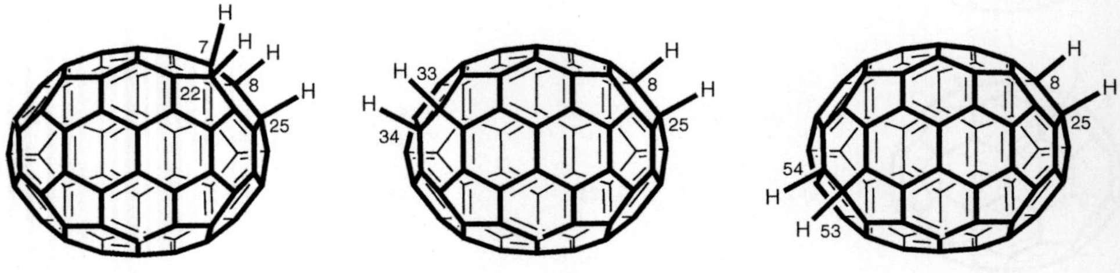

(±)-**28** (±)-**29** (±)-**30**

Figure 5.6 *$C_{70}H_4$ isomers with an inherently chiral functionalization pattern*

All four corresponding regioisomers were also detected in ^{3}He NMR studies of ^{3}He@$C_{70}H_4$ [88, 92].

In the higher adducts $C_{70}H_8$ [84] and $C_{70}H_{10}$ [89], a completely different addition pattern was observed. The resulting structures are C_s-symmetric, and the hydrogen atoms are arranged in a zigzag belt around the equator of C_{70} with no addends found in the polar regions of the fullerene. Therefore, none of the discussed $C_{70}H_4$ isomers can be the precursor of these higher adducts. Equivalent isomers also were found by ^{3}He NMR spectroscopy of the corresponding ^{3}He labeled derivatives [88, 92]. $C_{70}H_8$ is isomorphous with $C_{70}Me_8$ [52], $C_{70}Ph_8$ [93], $C_{70}(CF_3)_8$ (**31**, see Figure 5.10, Section 5.2.1.5) [94, 95] and $C_{70}(OO^tBu)_8$ (**32**, see Figure 5.13, Section 5.2.1.6) [96, 97], while $C_{70}H_{10}$ is isomorphous with $C_{70}Me_{10}$ [52], $C_{70}Ph_{10}$ [93], $C_{70}Cl_{10}$ [60], and $C_{70}Br_{10}$ [61] but was not found among the corresponding CF_3- or OO^tBu-derivatives.

With an increasing level of hydrogenation, the structure elucidation of hydro[70]fullerenes becomes complicated, even if pure isomers are available. Their rather low stability and decreasing solubility in common organic solvents hamper NMR spectroscopy, which is the most important tool for symmetry determination. In addition, several isomers can have the same symmetry, and their calculated differences in energy are often too small for a straightforward assignment [85–87, 98, 99].

Recently, $C_{70}H_{38}$ ((±)-**33**, Figure 5.7) was obtained by hydrogenation of C_{70} at 100–120 bar and 673 K for 72 h [100]. When handled carefully, the product was sufficiently stable for structure determination by extensive 2D NMR spectroscopy, carried out in part with ^{13}C-enriched samples. The unambiguously assigned C_2-symmetric isomer includes five benzenoid rings and two H-atoms attached directly to the equator.

Aromatic substructures seem to play an important role for the stability of highly substituted fullerenes. Not only $C_{70}H_{38}$ but also a great deal of other highly functionalized fullerenes with known structures such as the related fluorine derivatives $C_{70}F_{38}$ ((±)-**34** and (±)-**35**, see Figure 5.8, Section 5.2.1.4), contain isolated benzenoid rings. AM1 (Austin Model 1) calculations on the stabilities of $C_{70}X_{36/38}$ (X = H, F) carried out by *Clare* and *Kepert* predicted the most stable structures to contain aromatic substructures [101, 102]. In contrast, MNDO (Modified Neglect of Differential Overlap) calculations indicated structures without aromatic units to be more stable [103, 104]. It was assumed that isomers

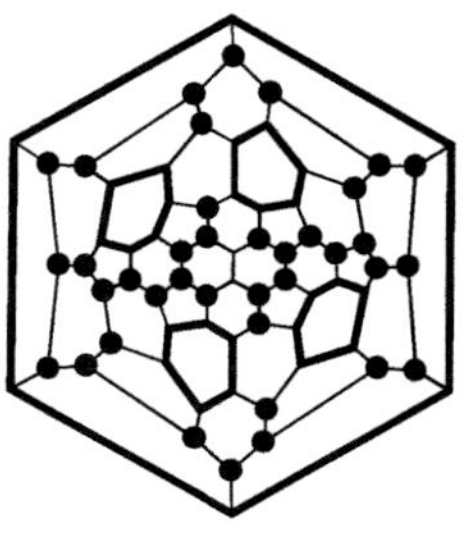

(±)-**33**

Figure 5.7 *Schlegel diagram of $C_{70}H_{38}$ ((±)-**33**). Black dots represent C-atoms with attached hydrogen. Isolated benzenoid rings are highlighted in bold*

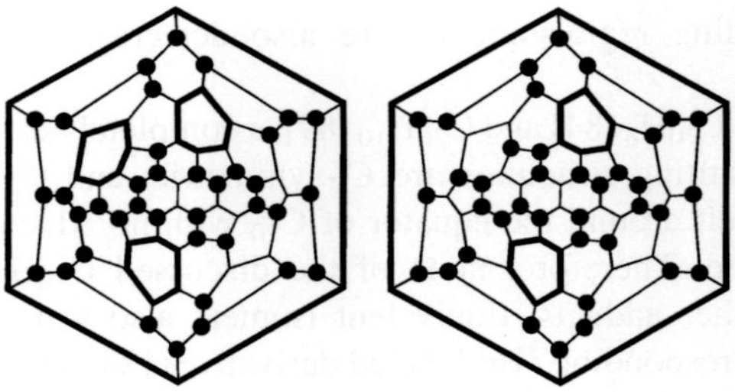

(±)-**34** (±)-**35**

Figure 5.8 *Schlegel diagrams of two chiral isomers of $C_{70}F_{38}$. Black dots represent carbon atoms with attached F-atoms. Isolated benzenoid rings are highlighted in bold*

with isolated double bonds, which cause relatively small amount of strain, are favored in comparison to structures with planar aromatic subunits, which introduce a higher steric strain in the fullerene cage.

5.2.1.4 Halogenation of C_{70}

The fluorination of C_{70} with MnF_3 at 450 °C led to the formation of 49 products, part of which was characterized by mass spectrometry and ^{19}F NMR spectroscopy [105]. In addition to 21 $C_{70}F_n$ derivatives ($n = 34, 36, 38, 40, 42, 44$), various oxides $C_{70}F_nO_x$ ($n = 34$, 36, 38, 40; $x = 1, 2, 3$; not all combinations of n and x were actually found) and hydroxides $C_{70}F_nO{\bullet}OH$ ($n = 35$, 37) were obtained. The latter are not formed during the fluorination reaction but at some point of the subsequent tedious HPLC separation. For many of the isolated compounds, the symmetry was shown to be low by 1D- and, in some cases, also by 2D ^{19}F NMR spectroscopy, but a conclusive structure determination was not possible. The structures of two out of eight isomers of $C_{70}F_{38}$ (Figure 5.8) were confirmed by X-ray crystallography. The C_1-symmetric isomer (±)-**34** contains four planar benzenoid rings and four isolated double bonds [58], while the C_2-symmetric isomer (±)-**35** includes three benzenoid rings and seven isolated double bonds (Figure 5.8) [59]. Both structures can formally be transformed into each other by only three 1,3-fluorine shifts. A surprising structural feature is the presence of two equatorial fluorines in both isomers. Addends in such a position were not observed before and they were assumed to be very unlikely, since their addition requires the rehybridization (sp^2 to sp^3) of a fullerene carbon atom in the flattest region of the fullerene, therefore introducing a considerable amount of strain. It was concluded that the destabilization resulting from equatorial addends is counterbalanced by the related increase of aromatic substructures.

In contrast to the manifold fluoro[70]fullerenes, only a few chlorine derivatives and a single bromine adduct are known.

Treatment of C_{70} with ICl in benzene gives $C_{70}Cl_{10}$ [60]. The ten chlorine atoms are arranged around the equatorial belt, showing nine 1,4- and one 1,2-relationships. This achiral functionalization pattern is identical with those of $C_{70}H_{10}$ [89], $C_{70}Ph_{10}$ [93], $C_{70}Me_{10}$ [52], and $C_{70}Br_{10}$ [61] which was characterized by X-ray crystallography and is the only known bromo[70]fullerene [106], obtained by treatment of the fullerene with neat bromine or with bromine in toluene.

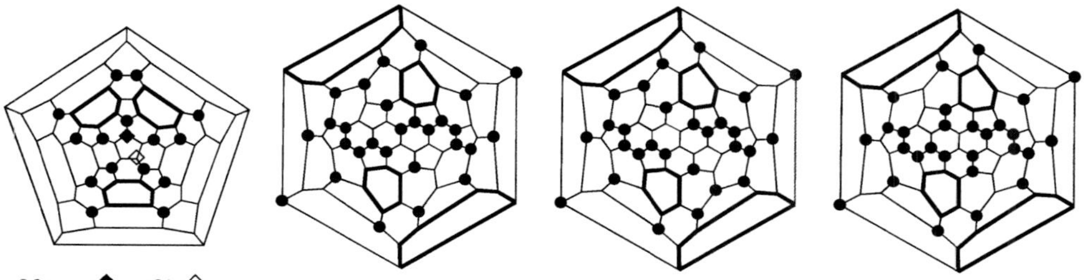

36: ◆ = Cl, ◇ = --
(±)-**37:** ◆ = --, ◇ = Cl

(±)-**38** (±)-**39** (±)-**40**

Figure 5.9 *Schlegel diagrams of polychlorinated C_{70}-derivatives. Far left (Schlegel diagram viewed along the C_5 symmetry axis): representation of* ***36*** *and (±)-**37**, two isomers of $C_{70}Cl_{16}$. Remainder (Schlegel diagrams viewed along a C_2 axis of the parent fullerene): C_1-symmetric (±)-**38**, C_2-symmetric (±)-**39**, and C_2-symmetric (±)-**40**, three isomers of $C_{70}Cl_{28}$. Black dots represent Cl-bearing C-atoms. Isolated benzenoid rings are marked in bold*

The next higher polychloro-[70]fullerene, $C_{70}Cl_{16}$, can be obtained by reaction of C_{70} with $Br_2/TiCl_4$, conditions designed originally for the synthesis of highly brominated fullerenes [107]. Its structure can be considered as an overlay of ten chlorine atoms arranged in the same type of belt as in $C_{70}Cl_{10}$ [60], and a cap of six chlorine atoms which corresponds to the chlorinated substructure of $C_{60}Cl_6$ [108, 109]. Two isomers of $C_{70}Cl_{16}$ were found that differ in the relative orientation of belt and cap: the C_s-symmetric **36** and C_1-symmetric (±)-**37**, which shows an inherently chiral functionalization pattern (Figure 5.9). Both structures are related by a single 1,3-chlorine shift, as shown in the *Schlegel* representation.

The highest degree of chlorination was achieved by reaction of C_{70} with $SbCl_5$, VCl_4, or PCl_5, or alternatively by treatment of $C_{70}Br_{10}$ with $SbCl_5$, yielding $C_{70}Cl_{28}$ which is composed of three isomers ((±)-**38**, (±)-**39**, and (±)-**40**) shown in Figure 5.9 [110]. Similar to the highly fluorinated $C_{70}F_{38}$ compounds ((±)-**34** and (±)-**35**, Figure 5.8), all isomers contain four isolated benzenoid rings, which compensate for the negative steric effects of the numerous 1,2-contacts among the Cl addends. These structures follow principles derived from DFT calculations, stating that each addend shows a maximum of two 1,2-contacts to adjacent groups and that no addend is located directly at an equatorial position. The only exception is (±)-**38** with an equatorial Cl-atom.

5.2.1.5 Perfluoralkylated Multi-Adducts of C_{70}

From 2001, the trifluormethylation of carbon cages opened up a new field of fullerene chemistry [111]. After initial work with C_{60} [112–115] afforded a multitude of compounds, the investigation of C_{70}-derivatives became even more challenging. The adducts were obtained by reaction of C_{70} with CF_3 radicals generated from thermal decomposition of $Ag(CF_3COO)$ at temperatures between 300 and 390 °C. The crude reaction mixture contained $C_{70}(CF_3)_n$ ($n = 2, 4, 6, 8, 10, 12$) [62, 94, 116], in ratios depending on reaction time, temperature, and the amount of CF_3 radical source. In a number of instances, partial separation was achieved by multi-stage HPLC, and several derivatives could be isolated in quantities sufficient for ^{19}F NMR spectroscopy, which was a key tool for symmetry determination and, in some cases, structure proposals [116]. In later experiments, fractional sublimation at 420–540 °C led to mixtures containing fewer and more stable derivatives, thereby simplifying the isolation of the main products [94] and minimizing the amounts of byproducts such as $C_{70}(CF_3)_nH_mO_x$. Higher $C_{70}(CF_3)_n$ adducts ($n = 14, 16, 18$) were obtained from the reaction of C_{70} with CF_3I at 390 °C [117, 118]. A combination of ^{19}F NMR spectroscopy, electronic absorption spectroscopy, AM1 and/or DFT calculations and, in several cases, X-ray crystallography [117–125], were used for structure determination. ^{19}F NMR spectroscopy, in particular, with its information on J_{FF} coupling constants, turned out to be an invaluable tool for structure elucidation.

All assigned derivatives (Figure 5.10) show arrangements of the CF_3 groups on the fullerene sphere obeying the following principles [62, 94]: The trifluoromethyl groups are

Figure 5.10 *Schlegel diagrams of trifluoromethyl derivatives of C_{70} (**31**, (±)-**41** – (±)-**59**). Black dots represent C-atoms with attached CF_3 groups. Ribbons are marked in gray*

located in ribbons or loops of edge-sharing *m*- and *p*-$C_6(CF_3)_2$ hexagons [126], the shared edges being $C(sp^2)$–$C(sp^3)$ bonds. In rare cases, an isolated *p*-$C_6(CF_3)_2$ hexagon is also present. The ribbons end always with a *p*-$C_6(CF_3)_2$ hexagon as a part of a p^3 (*para-para-para*) or *pmp* (*para-meta-para*) sequence; not known are *m*-$C_6(CF_3)_2$ hexagons at the terminus or an *mpp* arrangement as ending sequence. The CF_3 groups are not attached to triple hexagon junctions which are located in the flat equatorial part of [70]fullerene and would become much more strained through pyramidalization. There is only one $C_{70}(CF_3)_x$ isomer with an isolated (*i*) CF_3 group [62], and only in the two most highly functionalized $C_{70}(CF_3)_x$ isomers is there a pair of adjacent CF_3 groups [118]. The steric demand of the CF_3 group can often explain that functionalization patterns differ from those of the corresponding C_{70} derivatives with less bulky addends.

Two *p*-isomers of bis-trifluoromethylated C_{70} have been proposed. Whereas the structure of C_1-symmetric 7,24-bis(trifluoromethyl)-7,24-dihydro(C_{70}-D_{5h})[5,6]fullerene ((±)-**41**, Figure 5.10; for the numbering of C_{70}, see Figure 5.2) was deduced from ^{19}F NMR spectroscopy in combination with AM1 and DFT (Density Functional Theory) calculations [94], the structure of the second isomer (proposed as C_1-symmetric 2,23-bis(trifluoromethyl)-2,23-dihydro(C_{70}-D_{5h})[5,6]fullerene) could not be ascertained [62]. As to the next higher adduct, a combination of ^{19}F NMR spectroscopy and calculations strongly suggested the structure of C_1-*pmp*-$C_{70}(CF_3)_4$ ((±)-**42**, Figure 5.10) [94].

Three $C_{70}(CF_3)_6$ isomers are known. C_2-p^5-$C_{70}(CF_3)_6$ ((±)-**43**, Figure 5.10) is the most abundant one and contains a single ribbon of exclusively 1,4-functionalized hexagons [94]. The structure of C_1-p^3*mp*-$C_{70}(CF_3)_6$ ((±)-**44**) was first suggested by spectroscopic data [94] and later confirmed by X-ray crystallography [119]. The addition pattern of a third isomer (C_1-p^3,*p*-$C_{70}(CF_3)_6$) was proposed on the basis of spectroscopic data in combination with DFT calculations and remains to be confirmed [62].

Both isomers of $C_{70}(CF_3)_8$ have the CF_3 groups arranged in a p^7 ribbon and differ only in the position of a terminal addend. The achiral functionalization pattern of C_s-p^7-$C_{70}(CF_3)_8$ (**31**, Figure 5.10), known from $C_{70}H_8$ [83], was confirmed by X-ray crystallography [95]. The structure of chiral C_2-p^7-$C_{70}(CF_3)_8$ ((±)-**45**, Figure 5.10) was deduced from 1D and 2D ^{19}F NMR spectra and very recently confirmed by X-ray crystallography [127]. This pattern is unique among all known $C_{70}X_8$ derivatives [94, 127].

As shown by X-ray crystallography [121], the addends in the first isolated $C_{70}(CF_3)_{10}$ isomer ((±)-**46**, Figure 5.10) are arranged in a C_1-symmetric p^7*mp* ribbon [62, 94]. This pattern is new, in contrast to the known C_s-symmetric pattern of $C_{70}X_{10}$ with X = H [89], Cl [60], Br [61], where the addends form a p^9*o*-loop (p^9-*ortho*) around the equator of the fullerene with one 1,2- and nine 1,4-relationships. Although the C_1-p^7*mp* isomer (±)-**46** has two isolated double bonds in pentagons, it is more stable than the one with the p^9*o*-loop because the bulky CF_3 groups avoid 1,2-contacts. The structures of four other $C_{70}(CF_3)_{10}$ isomers were deduced by a combination of ^{19}F NMR and electronic absorption spectroscopy as well as by DFT calculations [62]. C_2-p^9-$C_{70}(CF_3)_{10}$ ((±)-**47**, Figure 5.10) contains only a single ribbon without a loop to avoid the steric hindrance of two adjacent CF_3 groups. The ^{19}F NMR spectra of each C_1-p^7,*p*-$C_{70}(CF_3)_{10}$ ((±)-**48**) and C_1-p^2*mpmp*,p^2-$C_{70}(CF_3)_{10}$ (structure not certain) reveal four terminal CF_3 groups, indicating a structure containing more than a single ribbon. The complete structures of (±)-**47** and (±)-**48** are based on an evaluation of the $^xJ_{FF}$ coupling constants. Isomer C_1-p^8,*i*-$C_{70}(CF_3)_{10}$ [62] (structure not certain) shows a singlet in the ^{19}F NMR spectrum, which demonstrates that one CF_3-group does not share the surrounding pentagon and hexagons with any of the other CF_3-addends. This is the first example of an isolated CF_3-group in a trifluoromethylated fullerene compound. For the sixth isomer, an X-ray crystal structure was obtained [120], showing that C_2-pmp^5mp-$C_{70}(CF_3)_{10}$ ((±)-**49**, Figure 5.10) has the addends arranged in a single ribbon.

Four C_1-symmetric $C_{70}(CF_3)_{12}$ isomers with a p^7mp-ribbon and an isolated 1,4-functionalized hexagon have been isolated ((±)-**50** – (±)-**53**, Figure 5.10) [62, 122]. Two of them – (±)-**50** [122, 123] and (±)-**51** [122, 124] – have been characterized by X-ray crystallography.

X-ray crystallographic characterization was also possible for four out of five isomers of $C_{70}(CF_3)_{14}$ [117, 125]. All four include the substructure of C_1-p^7mp-$C_{70}(CF_3)_{10}$ ((±)-**46**). Isomers (±)-**54** – (±)-**56** (Figure 5.10) have extended p^7mpmp ribbons in addition to an isolated hexagon each. Adduct (±)-**57** shows a complicated structure, the intricate functionalization pattern of which was assigned as pmp^9mp.

$C_{70}(CF_3)_{16}$ ((±)-**58**) and $C_{70}(CF_3)_{18}$ ((±)-**59**) are the first trifluoromethylated fullerenes with a pair of adjacent CF_3-groups (Figure 5.10). Their structures were determined by X-ray crystallography [118]. Both include the ribbon substructure of C_1-$C_{70}(CF_3)_{10}$ ((±)-**46**), but they are not related to any of the isolated isomers of $C_{70}(CF_3)_{12}$ or $C_{70}(CF_3)_{14}$.

An interesting issue is the control of product formation in the trifluormethylation of C_{70} and of fullerenes in general. It is not manifest whether the addition of CF_3 groups at higher temperatures occurs under kinetic or thermodynamic control. While products with a lower degree of trifluoromethylation are found to occur in the structures calculated to be clearly the most stable, this is not necessarily the case for the more highly functionalized $C_{70}(CF_3)_n$ with $n \geq 10$ [116, 121], where kinetic control has to be taken into account as well [117, 118, 122].

Remarkable is the effect of the functionalization pattern of a given $C_{70}X_n$ composition on the half-wave redox potential $E_{1/2}$. The largest reported $\Delta E_{1/2}$ (CV, dichloromethane, 0.1 M $N(nBu)_4BF_4$, $Fe(Cp^*)_2{}^{+/0}$ as internal standard) for the first reduction of several isomers of a given composition $C_{70}(X)_n$ is +0.15 V and was found among the three isomers of $C_{70}(Bn)_2$ [62, 128]. With the CF_3-functionalized C_{70} derivatives, studies were possible with up to five isomers of a given composition $C_{70}(CF_3)_n$. The $\Delta E_{1/2}$ values range between 0.16 V (first reduction of two isomers of $C_{70}(CF_3)_2$), and 0.45 V (second reduction of five isomers $C_{70}(CF_3)_{10}$) (for CV conditions, see above) [62]. The latter value was found for two isomers of decakis-trifluoromethyl-[70]fullerenes ((±)-**46** (Figure 5.10) and C_1-p^8, i-$C_{70}(CF_3)_{10}$ (not shown) [62]) with their structures differing only in the position of a single CF_3-group. Such an influence of the functionalization pattern on $\Delta E_{1/2}$ was also observed for $C_{60}(CF_3)_n$-isomers [115]. For a given type of functionalization, the functionalization pattern appears to be at least as important as the number of addends in determining $E_{1/2}$ values of fullerene derivatives.

From the reaction of C_{70} with C_2F_5I in a glass ampoule (350 °C, 40–60 h), seven $C_{70}(C_2F_5)_{10}$ isomers were isolated and characterized by X-ray diffraction methods (Figure 5.11) [129]. Isomer (±)-**60** shows the same addition pattern as the $C_{70}(CF_3)_{10}$ isomer (±)-**46** (Figure 5.10). The other six isomers ((±)-**61** – (±)-**66**, Figure 5.11) have unprecedented functionalization patterns with the addends forming three or four isolated domains on the fullerene surface. Five of the isomers – (±)-**61**, (±)-**62**, (±)-**63**, (±)-**65**, and (±)-**66** – include ribbons with an odd number of C_2F_5 groups, and isomer (±)-**62** contains

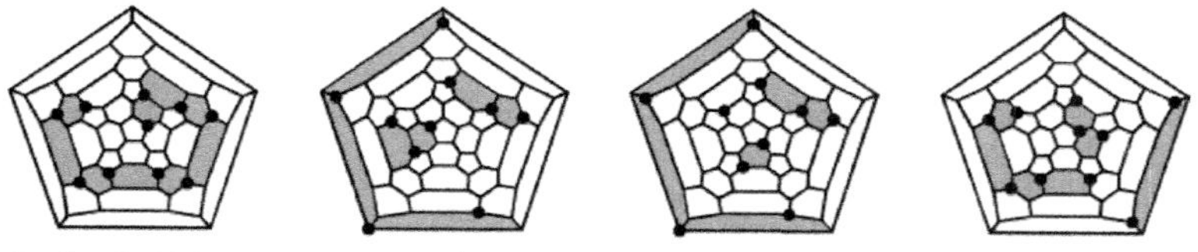

C_1-*p*^{7}*mp*-$C_{70}(C_2F_5)_{10}$ (±)-**50**

C_1-*pmp,p*^{2}*,mp*-1-$C_{70}(C_2F_5)_{10}$ (±)-**61**

C_1-*pmp,p*^{2}*,m,i*-$C_{70}(C_2F_5)_{10}$ (±)-**62**

C_1-*p*^{2}*,pm,p*-$C_{70}(C_2F_5)_{10}$ (±)-**63**

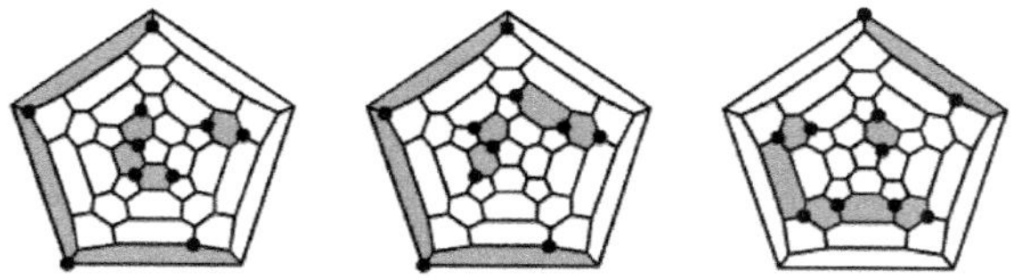

C_1-*pmp,pmp,p*-$C_{70}(C_2F_5)_{10}$ (±)-**64**

C_1-*pmp,p*^{2}*,pm*-2 $C_{70}(C_2F_5)_{10}$ (±)-**65**

C_1-*p*^{2}*,p,p*-$C_{70}(C_2F_5)_{10}$ (±)-**66**

Figure 5.11 *Seven isomers of $C_{70}(C_2F_5)_{10}$ ((±)-**60** – (±)-**66**). Only the addition pattern of (±)-**60** is known from the corresponding CF_3 derivative. Black dots represent C-atoms with attached C_2F_5 groups. Ribbons are marked in gray*

an isolated C_2F_5 group. Three of the isomers ((±)-**61**, (±)-**63**, and (±)-**65**) possess terminal C_2F_5 groups with a *m*-relationship to their nearest neighbor, an arrangement that is unknown in the CF_3 series [62]. The formation of the new functionalization patterns was explained by the bulkiness of the pentafluoroethyl as compared to the trifluoromethyl addend [129].

Addition of the even larger *n*-C_3F_7 groups to C_{70} provided again other functionalization patterns. Four out of at least sixteen $C_{70}(C_3F_7)_8$ isomers were isolated and studied by single-crystal X-ray crystallography [130]. Their functionalization patterns do not coincide with any of those known from trifluoromethylation. Isomers (±)-**67** and **68** (Figure 5.12) contain two *pmp* ribbons each, the relative addend positions leading to C_2- and C_s-symmetric molecules, respectively. The other two isomers ((±)-**69** and (±)-**70**) show C_1-symmetry and have their addends arranged in a *pmp,p,p* type (Figure 5.12).

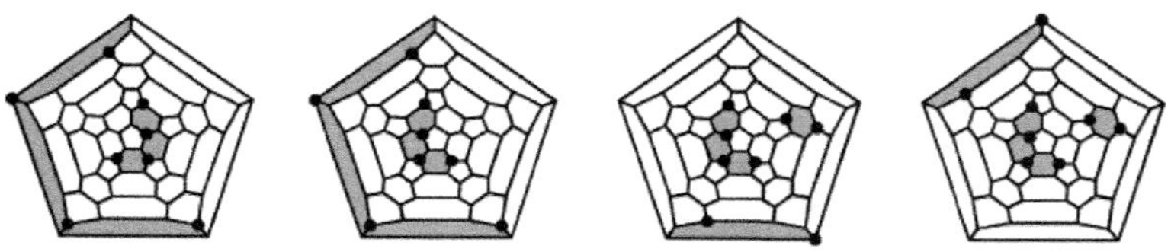

C_2-*pmp,pmp*-$C_{70}(C_3F_7)_8$ (±)-**67**

C_s-*pmp,pmp*-$C_{70}(C_3F_7)_8$ (±)-**68**

C_1-*pmp,p,p*-1-$C_{70}(C_3F_7)_8$ (±)-**69**

C_1-*pmp,p,p*-2-$C_{70}(C_3F_7)_8$ (±)-**70**

Figure 5.12 *Four isomers of $C_{70}(C_3F_7)_8$ ((±)-**67** – (±)-**70**). Black dots represent C-atoms with attached C_3F_7 groups. Ribbons are marked in gray*

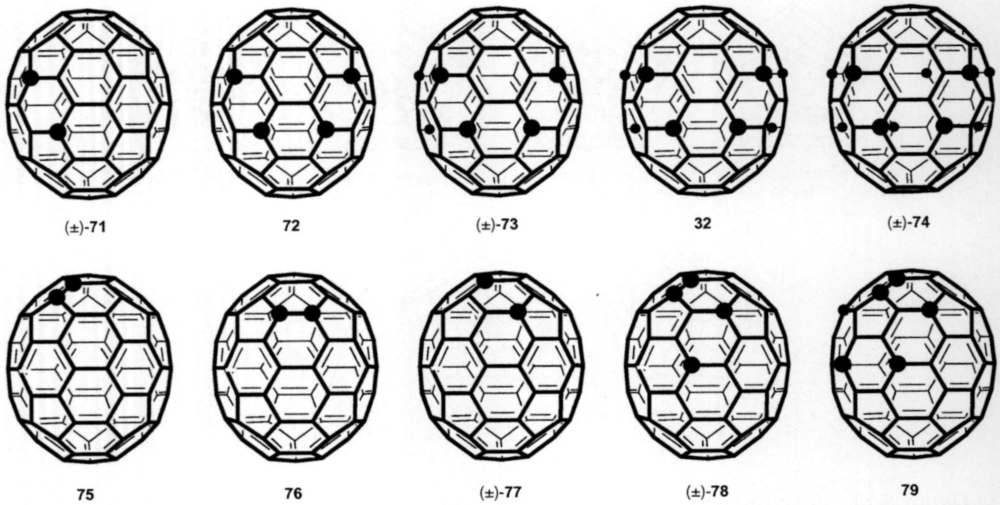

Figure 5.13 *Two series of five characterized* $(^{t}BuOO)_nC_{70}$ *derivatives. Top: equatorial addition mode, bottom: 'cyclopentadiene addition mode'*

5.2.1.6 Addition of tert-Butylperoxy Radicals to C_{70}

Ten compounds were isolated from the reaction of C_{70} with *tert*-butylhydroperoxide in the presence of ammonium cerium(IV) nitrate (CAN) [96, 97]. Their structures can be divided in two series (Figure 5.13).

Isomers **32** (see also Section 5.2.1.3), (±)-**71**, **72**, (±)-**73**, and (±)-**75** (Figure 5.13) have their addends arranged around the equatorial belt of the fullerene core. In this set, the lower adducts can be considered as precursors of the higher ones which are formed by successive addition of further ^{t}BuOO-groups. The achiral structure **32** is isomorphous with $C_{70}H_8$ [84] and also with $C_{70}(CF_3)_8$ isomer (±)-**31** (Figure 5.10). The C_2-symmetric structure of $C_{70}(OO^{t}Bu)_{10}$ ((±)-**74**), on the other hand, was not observed for hydrides but it is isomorphous with $C_{70}(CF_3)_{10}$ ((±)-**47**, Figure 5.10). The second series contains five derivatives (**75**, **76**, (±)-**77**, (±)-**78**, and **79**) that are based on the 'cyclopentadiene addition mode' which is well known from the chlorination of [60]fullerene [108, 109]. It should be noted that in the present case, this arrangement is not formed around the C_{60}-like pole but on the side of the ovoid [70]fullerene instead.

Further chiral C_{70}-derivatives with a modified fullerene core can be found among homo [70]fullerenes, aza- and oxahomo[70]fullerenes, as well as open-cage derivatives of the carbon spheroid [8].

5.2.2 C_{70}-Derivatives with a Non-Inherently Chiral Functionalization Pattern

The simplest non-inherently chiral addition pattern of C_{70} results from mono-addition of a C_s-symmetric, divalent addend across a nonequatorial bond that is perpendicular to the C_5 rotation axis. Bearing in mind the different bond reactivities of C_{70}, it is not surprising that only the functionalization of C(7)–C(22), which is known to be the second most reactive bond in many additions, will be discussed.

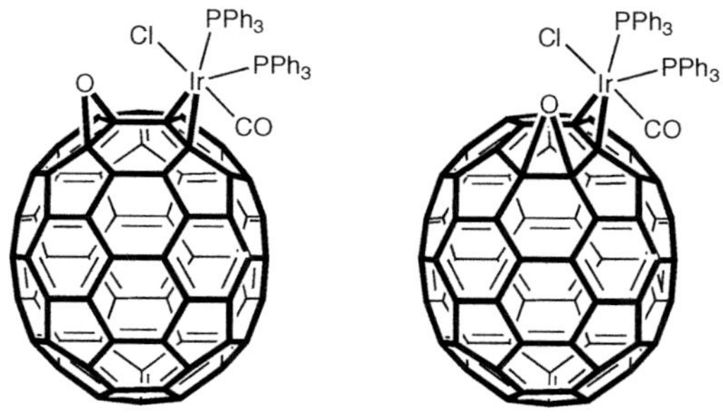

(±)-**80** (±)-**81**

Figure 5.14 *Structures of two Ir-complexes of $C_{70}O$, one with an inherently chiral ((±)-**81**), the other with a non-inherently chiral functionalization pattern ((±)-**80**)*

Until very recently, all of the C_{70}-derivatives with a non-inherently chiral addition pattern were isolated as racemic mixtures; their resolution was not attempted or not achieved.

The relative positions of the epoxy oxygen and the metal center in Ir-complexes of $C_{70}O$ provide chiral addition patterns (Figure 5.14) [90]. If the two non-identical addends are attached to an α-type bond of the same hemisphere, the resulting pattern is non-inherently chiral ((±)-**80**). The pattern in isomer (±)-**81**, on the other hand, is inherently chiral, regardless of the nature of the addends. A structurally related twofold addition of hydrogen within one hexagon was observed for the $C_{70}H_4$ derivatives 8,23,24,25-tetrahydro(C_{70}-D_{5h})[5,6]fullerene (achiral, not shown; for the atom numbering of C_{70}, see Figure 5.2) and 7,8,22,25-tetrahydro(C_{70}-D_{5h})[5,6]fullerene ((±)-**28**, Figure 5.6).

Several addition reactions (mostly cycloadditions) of C_s-symmetric addends provide mixtures of α- and β-type mono-adducts (Figure 5.15). Pyrazolo[70]fullerenes [131, 132] as well as triazolo[70]fullerenes [133–135] occur as C_s-symmetric α-adducts (**82/83**, in which two different orientations of the heterocycle lead to two different constitutional isomers, and **84/85**) in mixtures with racemic β-adducts ((±)-**86** and (±)-**87**).

Similarly, the [3 + 2] cycloaddition of a substituted trimethylenemethane yielded two C_s-symmetric α-adducts (**88/89**) and a pair of β-type enantiomers ((±)-**90**) [136–138]. Traces of water promote the formation of rearranged esters with a stereogenic center in the addend. As a consequence, the α-adducts (**88/89**) turn into a pair of C_1-symmetric enantiomers ((±)-**91**), while the transformed β-adducts (**92/93**), originating from (±)-**90**, now have a diastereoisomeric relationship.

Even more complex is the isomeric mixture obtained from the [2 + 2] cycloaddition between C_{70} and 3-methyl-2-cyclohexenone: the multitude of product isomers results from the different orientations of the cyclohexanone ring with respect to the fullerene and the superposition of possible configurations of fullerene addition pattern and the two stereogenic centers in the addend. As a result, a total of eight different racemic products ((±)-**94a/b**, (±)-**95a/b**, and (±)-**96a/b/c/d**) were obtained [139].

Hydroalkylation of [70]fullerene by alkyl halides under reductive conditions can also give rise to C_{70} derivatives with a non-inherently chiral functionalization pattern. Adduct (±)-**97** (β-type) was thus obtained through electron-transfer reaction between fullerene and zinc in the presence of methyl 2-bromoacetate [140], in addition to the achiral compound **98**

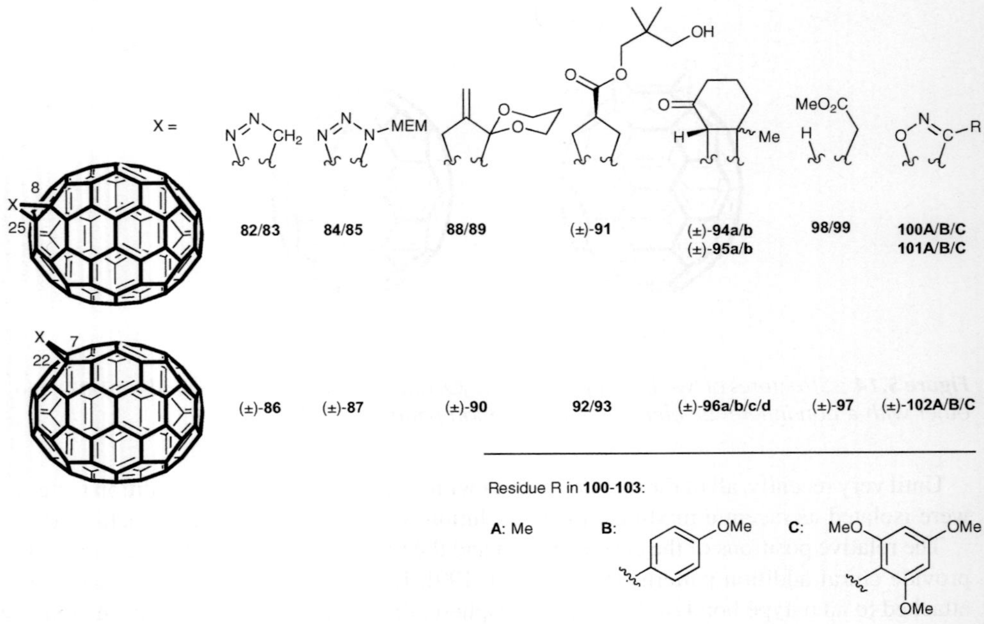

Figure 5.15 Addition of C_s-symmetric addends provides mixtures of α- and β-type mono-adducts, the latter having a non-inherently chiral addition pattern. MEM = [2-(methoxy)ethoxy] methyl

(α-type). The second α-type isomer, **99**, was not observed in this reaction, but it was found to be the only product when 8,25-dihydro(C_{70}-D_{5h})[5,6]fullerene was deprotonated with tetrabutylammonium hydroxide in PhCN, followed by alkylation with the methyl bromoacetate [141].

Similar to the other cycloadditions described in this section, the [3 + 2] cycloaddition of nitrile oxides to C_{70} provided mixtures of isoxazolofullerenes in the form of two constitutionally isomeric α- (**100A/B/C** and **101A/B/C**) and a pair of enantiomeric β-adducts ((±)-**102A/B/C**) [63, 142].

Very recently, isoxazolo[70]fullerenes of this type were reported as the first enantiopure C_{70}-derivatives with a non-inherently chiral addition pattern (Scheme 5.4) [64]. These results were achieved by i) addition to C_{70} of a chiral, enantiomerically pure nitrile oxide, generated *in situ* from hydroximoyl chloride (*S*)-**103**, ii) isolation of all adduct isomers, i.e. the α-adducts (*S*)-**104** and (*S*)-**105**, and in particular the two diastereoisomeric β-type C(7)–C(22) adducts ($S,{}^{f,s}A$)-**106** and ($S,{}^{f,s}C$)-**107**, and iii) separate removal, for each isomer, of the addend-based stereogenic element by deprotection of the alcohol function and subsequent oxidation, affording as pure compounds two C_s-symmetric α-adducts (**108** and **109**), as well as the enantiomers (${}^{f,s}A$)-**110** and (${}^{f,s}C$)-**110** (Scheme 5.4) [64].

The applied strategy is related to the '*Bingel*/retro-*Bingel*' routine used for the separation of enantiomers and different cage isomers of higer fullerenes (Section 5.3.2.1) or for the isolation of the enantiomers of C_{70}-derivatives with the inherently chiral two- and five o'clock patterns of double addition (Section 5.2.1.1.3).

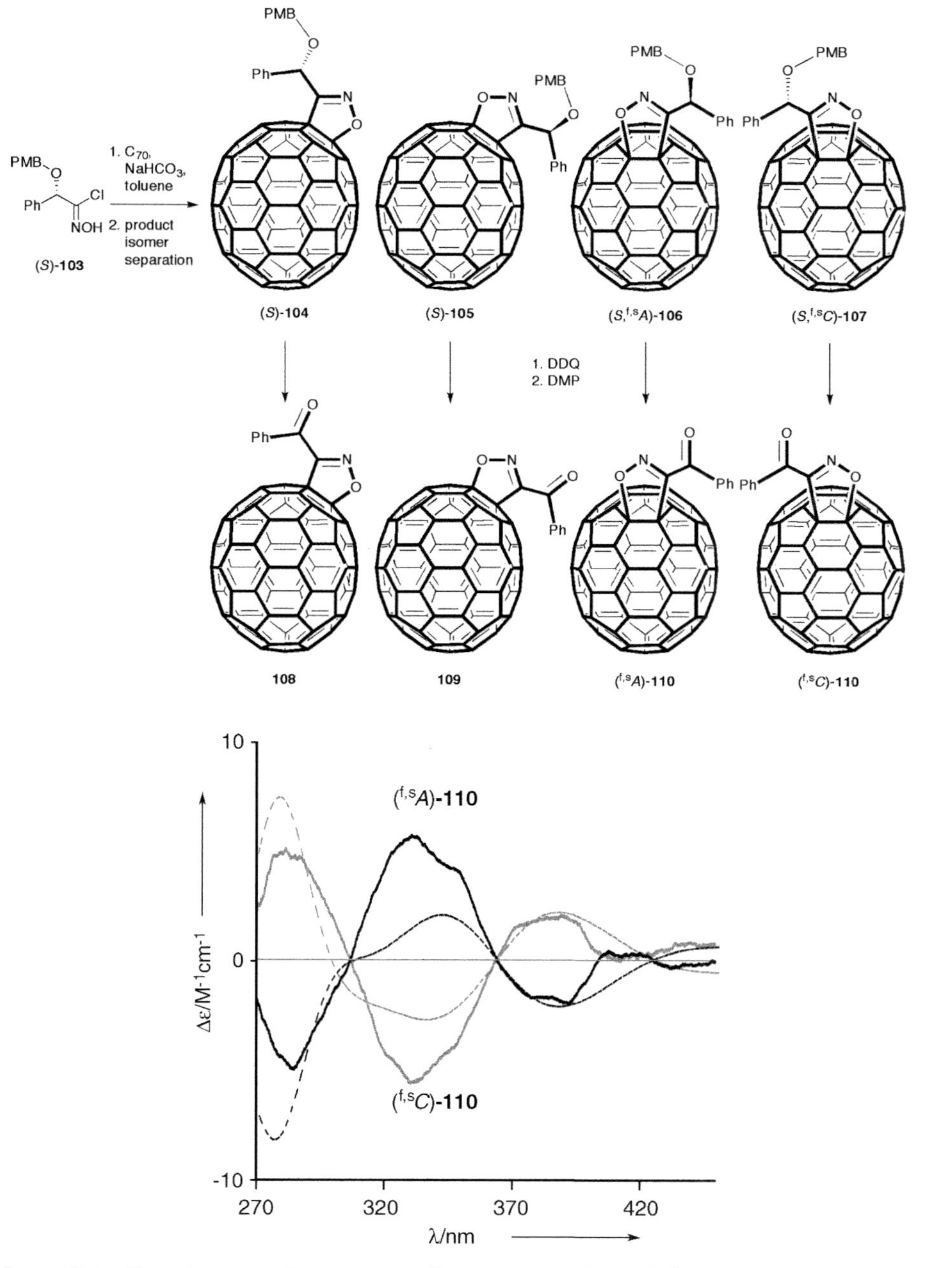

Scheme 5.4 *Top: Access to four isomerically pure isoxazolo[70]fullerenes, including the first enantiomerically pure derivatives of C_{70} with a non-inherently chiral functionalization pattern ((^{f,s}A)-**110** and (^{f,s}C)-**110**). PMB = p-methoxybenzyl; DDQ = 2,3-dichloro-5,6-dicyano-p-benzoquinone, DMP = Dess-Martin periodinane. Bottom: Experimental and calculated CD spectra of the enantiomers of benzoylisoxazolo[70]fullerene **110**. Experimental spectra (1,2-dichloroethane) (solid lines) of enantiomer (^{f,s}A)-**110** (black) and (^{f,s}C)-**110** (grey); calculated spectra (dashed lines) of the (^{f,s}A)- (black) and the (^{f,s}C)- (grey) enantiomers*

The absolute configuration of ($^{f,s}A$)-**110** and ($^{f,s}C$)-**110** was assigned by comparison of their CD-spectra with data obtained by calculations using the ZINDO [143] (*Zerner* Intermediate Neglect of Differential Overlap) method [64]. The magnitude of the observed *Cotton* effects ($\Delta\varepsilon$ up to $5\,\mathrm{L\,mol^{-1}\,cm^{-1}}$) is somewhat smaller than that measured for fullerene $C_{84}(22)$-D_2 ($\Delta\varepsilon$ up to $20\,\mathrm{L\,mol^{-1}\,cm^{-1}}$) [42], and significantly smaller than the largest *Cotton* effects recorded for C_{60}- and C_{70}-derivatives with an inherently chiral functionalization pattern ($\Delta\varepsilon$ up to $380\,\mathrm{L\,mol^{-1}\,cm^{-1}}$) [51, 78, 144–146] or for the chiral parent fullerene $C_{76}(1)$-D_2 ($\Delta\varepsilon$ up to $320\,\mathrm{L\,mol^{-1}\,cm^{-1}}$) [57].

Other C_{70}-derivatives with a non-inherently chiral functionalization pattern were identified [8] among the cyclobutadi[70]fullerene [147] and furanodi[70]fullerene 'cage dimers' [148].

5.2.3 Fullerene Derivatives with Stereogenic Centers in the Addends

In principle, any chiral residue can be attached to C_{70}, leading to a huge variety of possible products, especially if natural products such as sugars, amino acids, or steroids are taken into account. Some derivatives with stereogenic centers in the addends were already mentioned, e. g. appropriately substituted cyclopropa[70]fullerenes (Figure 5.3) [51, 69] or isoxazolo [70]fullerenes (Scheme 5.4) [64]. In both cases, the superposition of chiral functionalization pattern and stereogenic element(s) in the addend(s) was used to provide separable diastereoisomers with enantiomeric residual fullerene chromophores.

Beside this application, the presence of a stereogenic center in a side chain of a molecular scaffold is not a particularity of fullerene chemistry and, as seen above (Section 5.2.1.1), has only a minor influence on the chiroptical properties of the fullerene chromophore. Furthermore, the number of such compounds is countless and their discussion would exceed the scope of the present review.

5.3 The Higher Fullerenes Beyond C_{70}

5.3.1 Isolated and Structurally Assigned Higher Fullerenes

The most prominent fullerenes beyond C_{70} are C_{76}, C_{78}, and C_{84}. C_{76} was first isolated in 1991 and characterized by ^{13}C NMR spectroscopy [2, 19, 39, 149]. The structure was determined to be D_2-symmetric and, similar to the other chiral parent fullerenes, it is chiral without containing a stereogenic center [39]. The second IPR-structure of C_{76} was found recently in the form of the trifluoromethyl derivative C_s-p^9,p^2-($C_{76}(2)$-T_d)$(CF_3)_{12}$ (**111**, Figure 5.22, Section 5.3.2.4) [17, 29]. Five isomeric C_{78} fullerenes were obtained and structurally assigned, either as pure carbon allotropes ($C_{78}(1)$-D_3 [14, 15, 19, 150], $C_{78}(2)$-C_{2v} [14, 15, 19, 150], $C_{78}(3)$-C_{2v} [15, 19, 151]) or as exohedral adducts $C_{78}(5)$-D_{3h}) [16, 17]. C_{84} can theoretically occur as 24 IPR-conforming isomers [11]. Out of these, $C_{84}(22)$-D_2 and $C_{84}(23)$-D_{2d} [21, 32, 149] are generally the most abundant in fullerene soot [2–4, 15, 19] and were also calculated to be lowest in energy [152, 153]. In addition, a number of minor isomers have been isolated and characterized, either as pristine fullerenes, i.e. $C_{84}(4)$-D_{2d} [23], $C_{84}(5)$-D_2 [23], $C_{84}(14)$-C_s and $C_{84}(16)$-C_s [20, 31], $C_{84}(19)$-D_{3d} [20, 22], and $C_{84}(24)$-D_{6h} [20, 22], or as trifluoromethylated derivative (C_1-p^6,p^2,p-($C_{84}(11)$-C_2)$(CF_3)_{12}$, ($\pm$)-**112**, Figure 5.22, Section 5.3.2.4) [24].

Other higher fullerenes are less abundant in fullerene soot and some of them may not be very stable as empty, underivatized fullerenes. $C_{74}(1)$-D_{3h}, the only one IPR-conforming isomer of [74]fullerene, was isolated in the form of two derivatives, $C_{74}F_{38}$ [154] ((±)-**113**, Figure 5.18) and C_2-p^{11}-$C_{74}(CF_3)_{12}$ ((±)-**114**, Figure 5.22) [26]. Of the seven IPR-conforming isomers of C_{80}, only two were isolated as pure carbon cages. Whereas the first one has D_2-symmetry ($C_{80}(2)$-D_2) [34], the second isomer was identified as an ellipsoidal structure with D_{5d}-symmetry ($C_{80}(1)$-D_{5d}) [33]. A third isomer was found as polyfluoromethylated derivative of $C_{80}(5)$-C_{2v}, namely C_s-p^{10}-loop-,*p*-($C_{80}(5)$-C_{2v})$(CF_3)_{12}$ (**115**, Figure 5.22) [17]. The C_2-symmetry of an isolated isomer of C_{82} was determined by ^{13}C NMR spectroscopy [15]. The specific [82]fullerene cage isomer $C_{82}(3)$-C_2 was assigned with the help of ^{13}C NMR spectra calculated by density functional theory [155]. In addition, two trifluoromethyl derivatives of C_{82} were found in a mixture obtained by polytrifluormethylation of higher fullerenes (Section 5.3.2.4). One of them ((±)-**116**, Figure 5.22) is based on the mentioned cage isomer $C_{82}(3)$-C_2, the other ((±)-**117**, Figure 5.22) contains the hitherto unknown cage isomer $C_{82}(5)$-C_2 [17]. C_{86} is one of the largest fullerenes extracted from soot, and two isomers out of 19 possible IPR-conforming structures [11] were isolated by multi-stage HPLC [156]. Based on the good agreement of their calculated NMR chemical shifts with measured NMR data, they were assigned as $C_{86}(16)$-C_s and $C_{86}(17)$-C_2 [35]. With the same method, three isolated [149, 156] isomers of C_{88} were identified, namely $C_{88}(7)$-C_2, $C_{88}(17)$-C_s, and $C_{88}(33)$-C_2 [36]. A derivative of C_{90} was detected among the polytrifluormethylated higher fullerenes. ^{19}F NMR data in combination with DFT calculations allowed for structure determination of C_1-p^7,*p*,*p*-($C_{90}(32)$-C_1) ((±)-**118**, Figure 5.22) [29]. Until now, only one pure isomer of C_{92} was isolated. It was shown to possess C_2-symmetry but the identification of the precise cage isomer was impossible [157].

5.3.2 Inherently Chiral Fullerenes – Chiral Scaffolds

5.3.2.1 The 'Bingel/*retro*-Bingel' *Approach*

One option to obtain enantiopure compounds is the separation of racemic mixtures into the optical antipodes (resolution). The use of appropriate chiral stationary phases (CSP) proved effective for HPLC (high performance liquid chromatography) separation of a number of chiral fullerenes and fullerene derivatives. In cases where the difference in interactions between each enantiomer and the CSP is too small to allow for chromatographic resolution, separation may be achieved in the form of diastereoisomeric derivatives resulting from addition of further stereogenic elements of a given configuration (cf. Section 5.2.1.1.3). After the separation has been accomplished, the auxiliary chiral residue can be removed to obtain the individual enantiomers. For parent fullerenes, this strategy was first realized in a sequence of *Bingel* addition of enantiomerically pure malonates to the pristine carbon cage, separation of the resulting diastereoisomeric cyclopropafullerenes, and removal of the addends by electrochemical retro-*Bingel* reaction, leading to the denomination '*Bingel*/retro-*Bingel* approach' for the entire procedure (Section 5.3.2.1.1) [41, 158]. In the meantime, other reactions such as the 1,3-dipolar cycloaddition of enantiopure nitrile oxides [64] have been used in a similar way.

5.3.2.1.1 RESOLUTION OF THE ENANTIOMERS OF $C_{76}(1)$-D_2

The first (partial) separation of the enantiomers of C_{76} was achieved by asymmetric osmylation of the racemic fullerene using OsO_4 complexes with an enantiomerically pure ligand derived from a cinchona alkaloid [40]. This kinetic resolution provided enantiomerically enriched samples. Full separation was achieved by the so-called '*Bingel*/retro-*Bingel* approach' [41]: Mono-addition of enantiomerically pure bis[(*S*)-1-phenylbutyl] 2-bromomalonate to *rac*-C_{76} yielded seven adducts, all of which were isolated [57]. They included three constitutionally isomeric pairs of diastereoisomers which were easily identified by UV/Vis, NMR, and CD spectroscopy. Two of the pairs had C_1-symmetry, and the third pair was C_2-symmetrical; the latter must, therefore, have resulted from cyclopropanation of a bond bisected by a C_2-axis of the parent fullerene, but the exact addend position of the C_1-symmetric isomers could not be ascertained. After separation, the diastereoisomers ($S,S,^{f,s}A$)-**119** and ($S,S,^{f,s}C$)-**120** were independently submitted to constant potential electrolysis (CPE), yielding the pristine fullerene as separate enantiomers ($^{f,s}A$)-C_{76} and ($^{f,s}C$)-C_{76} (Scheme 5.5) [41]. In summary, the use of a chiral auxiliary afforded separable diastereoisomeric adducts of C_{76} which were individually retransformed to pristine [76]fullerene in the form of pure enantiomers.

The CD spectra of ($^{f,s}A$)-C_{76} and ($^{f,s}C$)-C_{76} show mirror image behavior (Scheme 5.5, right) and the band positions are in agreement with those reported for the enriched enantiomers obtained by kinetic resolution [40], but the magnitude of the *Cotton* effects

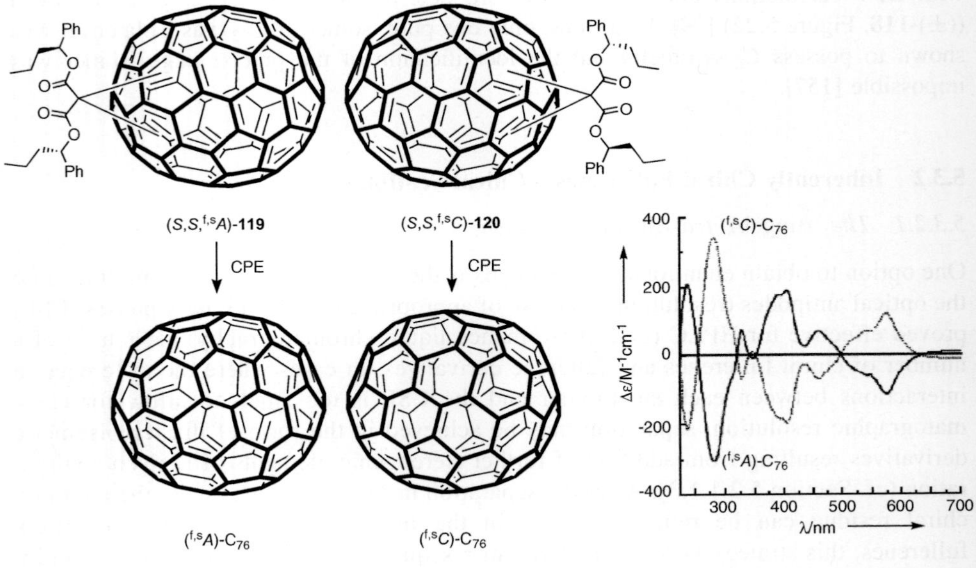

Scheme 5.5 *Left: Separate electrochemical retro-Bingel reactions of* ($S,S,^{f,s}A$)-***119*** *and* ($S,S,^{f,s}C$)-***120***, *two diastereoisomeric mono-adducts of* C_{76} *differing only in the configuration of the fullerene core. Even though the exact location of the malonate addend could not be ascertained, the diastereoisomeric relationship of the two compounds was proven spectroscopically. CPE = constant potential electrolysis. Right: Experimental CD spectra* (CH_2Cl_2) *of* [CD(−)282]-($^{f,s}A$)-C_{76} *and* [CD(+)281]-($^{f,s}C$)-C_{76}

differs by one order of magnitude ($\Delta\varepsilon \leq 32\,M^{-1}\,cm^{-1}$ [40] vs. $\Delta\varepsilon \leq 320\,M^{-1}\,cm^{-1}$ [41]). The absolute configuration of the C_{76} enantiomers was assigned as [CD(−)282]-($^{f,s}A$)-C_{76} and [CD(+)281]-($^{f,s}C$)-C_{76} by comparison of theoretically calculated CD data with the experimental spectra [70].

Judging from the inconspicuous differences in the shape of the twisted ovoid C_{76} enantiomers, their direct separation by HPLC on a CSP was expected to be difficult. However, amylose tris(3,5-dimethylphenyl carbamate) was found to be a suitable stationary phase for the chromatographic resolution of (±)-C_{76} with a hexane–$CHCl_3$ (80 : 20) mixture as the eluent [159].

5.3.2.1.2 SEPARATION OF DIFFERENT CAGE ISOMERS OF C_{78}

First separation and a structural assignment of a C_{2v}- and a D_3-symmetric isomer of C_{78} (IPR structures 2 and 1, respectively) [11] was achieved by multi-stage HPLC [14]. In later work, this separation was also accomplished by the '*Bingel*/retro-*Bingel* approach' [160]. The cage isomers C_{78}(1)-D_3 and C_{78}(2)-C_{2v} were separated as tris-malonates that had been obtained from *Bingel* reaction of a mixture of the two parent fullerenes with diethyl 2-bromomalonate in the presence of DBU (1,8-diazabicyclo[5.4.0]undec-7-ene) [160]. Three out of a total of eight obtained tris-adducts showed a symmetry higher than C_1, allowing the assignment of the respective cage isomers. The C_2-symmetric tris-adduct (±)-**121** has to be a derivative of C_{78}(2)-C_{2v} (Scheme 5.6), while the C_3-symmetric adduct (±)-**122a** or (±)-**122b** (no conclusive assignment was possible for these two plausible

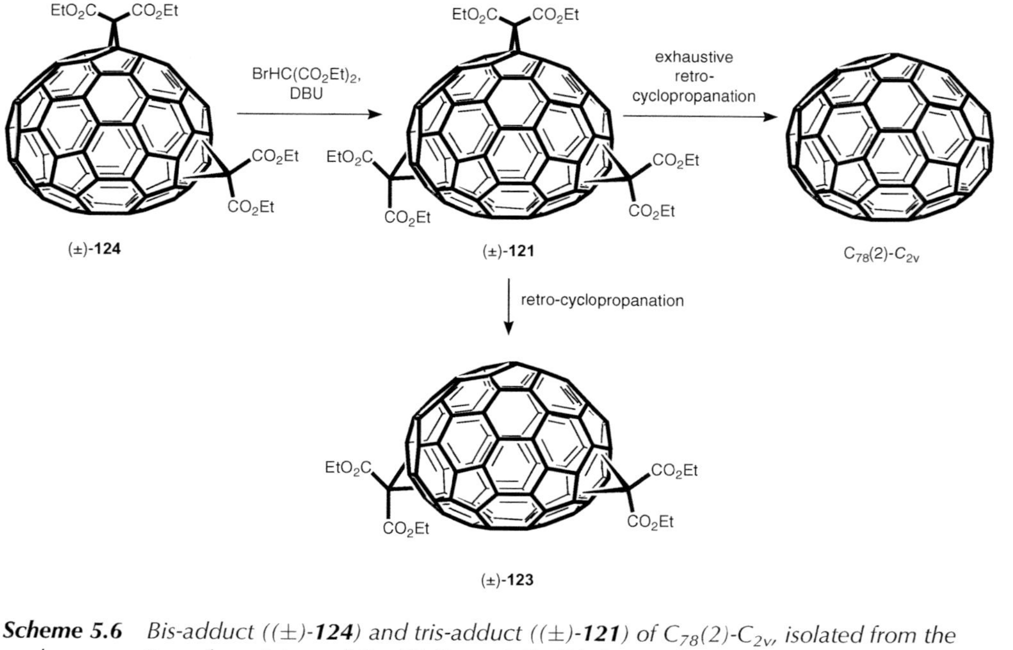

Scheme 5.6 *Bis-adduct ((±)-**124**) and tris-adduct ((±)-**121**) of $C_{78}(2)$-C_{2v}, isolated from the cyclopropanation of a mixture of $C_{78}(1)$-D_3 and $C_{78}(2)$-C_{2v}, and partial (→ (±)-**123**) and complete (→ $C_{78}(2)$-C_{2v}) electrochemical retro-cyclopropanation of **121***

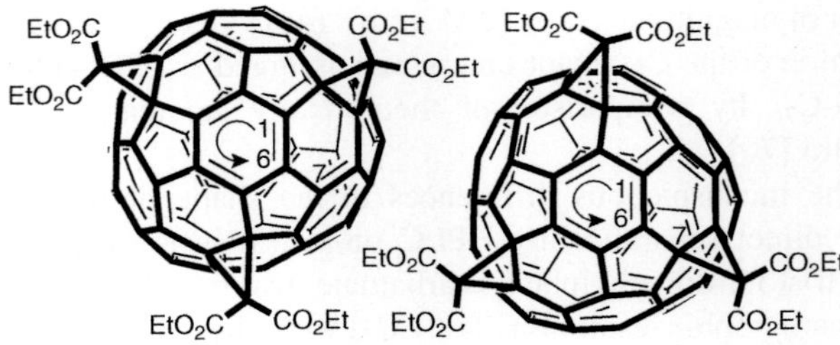

(±)-**122a** (±)-**122b**

*Figure 5.16 Two structures proposed for a tris-malonate of $C_{78}(1)$-D_3 ((±)-**122a/b**) isolated from cyclopropanation of a mixture of $C_{78}(1)$-D_3 and $C_{78}(2)$-C_{2v}*

structures) can only be based on $C_{78}(1)$-D_3 (Figure 5.16). Electrochemical retro-cyclopropanation of (±)-**121** afforded the pristine fullerene $C_{78}(2)$-C_{2v} as well as a new, C_2-symmetric bis-adduct ((±)-**123**) [158]. The latter was ruled out to be a precursor of (±)-**121** as it was not found in the initial reaction mixture from which bis-adduct (±)-**124** was isolated. As demonstrated by this case, the *Bingel*/retro-*Bingel* sequence can also be used for the separation of fullerene cage isomers (see also Section 5.3.2.1.3) and for the generation of new adducts, e.g. bis-adduct (±)-**123**, that are not accessible by direct functionalization.

5.3.2.1.3 RESOLUTION OF THE ENANTIOMERS OF $C_{84}(22)$-D_2

The separation of the two most abundant isomers of [84]fullerene, $C_{84}(22)$-D_2 and $C_{84}(23)$-D_{2d} was first achieved by multi-stage HPLC [21], and shortly afterwards by application of the *Bingel*/retro-*Bingel* strategy [42]. The latter method allowed also the isolation of the enantiomers of $C_{84}(22)$-D_2.

The addition of enantiomerically pure bis[(*S*)-1-phenylbutyl] 2-bromomalonate to C_{84} (main constituents: $C_{84}(22)$-D_2 and $C_{84}(23)$-D_{2d}) afforded several mono- and bis-adducts. Two of the latter were isolated and identified as ($S,S,S,S,^{f,s}A$)-**125** and ($S,S,S,S,^{f,s}C$)-**126** with a diastereoisomeric relationship (Scheme 5.7). As is typical for diastereoisomeric fullerene derivatives with enantiomorphic fullerene cores (see Section 5.2.1.1), their CD spectra showed mirror-image behavior because they are dominated by the inherently chiral cage chromophore and practically unaffected be the stereogenic elements in the malonate residues. Removal of the addends from each of the diastereoisomers by CPE afforded the individual enantiomers of $C_{84}(22)$-D_2 [42]. Their absolute configuration could be assigned by comparison of the experimental CD data with calculated spectra [161].

A further product isolated from the *Bingel* reaction of the [84]fullerene mixture with enantiomerically pure bis[(*S*)-1-phenylbutyl] 2-bromomalonate was a C_2-symetric mono-adduct derived from the parent fullerene $C_{84}(23)$-D_{2d} (not shown) which could be further transformed into a D_2-symmetric bis-adduct. After isolation, retro-*Bingel* reaction of the latter by CPE gave access to the pure achiral isomer $C_{84}(23)$-D_{2d} [42].

5.3.2.2 Covalent Adducts of the Higher Fullerenes Beyond C_{70}

Besides the higher fullerene derivatives discussed above in the context of the *Bingel*/retro-*Bingel* strategy, relatively few other derivatives had become available during the first dozen years of fullerene chemistry. It is only recently that their number is on a noticeable rise, thanks

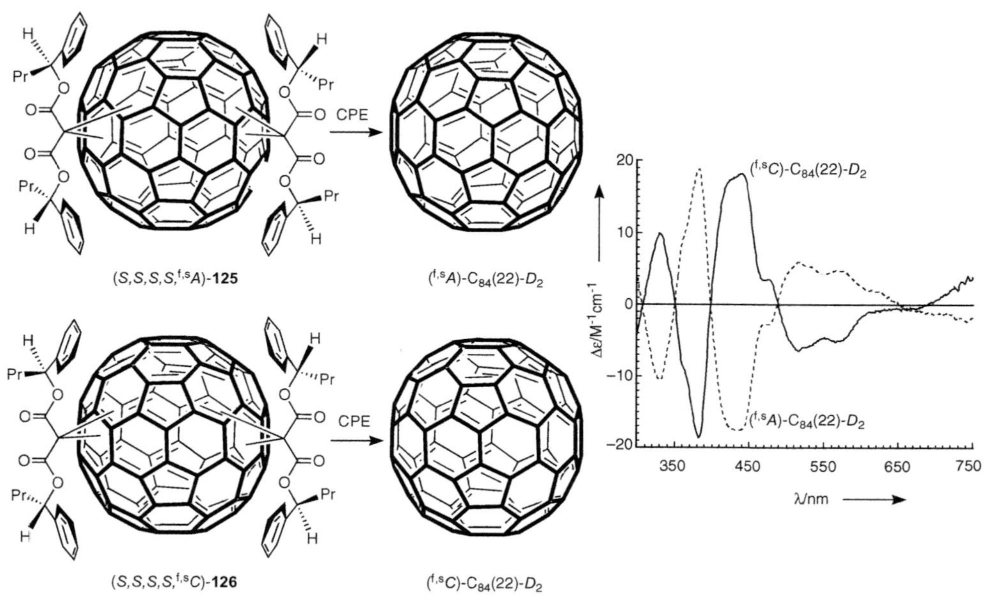

Scheme 5.7 *The Bingel/retro-Bingel approach applied to a mixture of C_{84} isomers afforded the two main constitutional isomers $C_{84}(22)$-D_2 and $C_{84}(23)$-D_{2d} (not shown) in pure form, in addition to a 3rd, achiral isomer (not identified). Electrochemical retro-Bingel reaction of the isolated diastereoisomers (S,S,S,S,^{f,s}A)-**125** and (S,S,S,S,^{f,s}C)-**126** allowed even the individual generation of the enantiomers of $C_{84}(22)$-D_2. CD spectra (CH_2Cl_2) of the enantiomers of [84] fullerene are displayed on the right side*

mainly to the isolation of halogenated and trifluoromethylated adducts. Among the chemically modified higher fullerenes, there is a large fraction of chiral compounds because all derivatives of a chiral parent automatically have an inherently chiral functionalization pattern.

At least six constitutionally isomeric mono-adducts were formed in the *Diels-Alder* reaction of $C_{76}(1)$-D_2 with 3,4-dimethoxy-*o*-quinodimethane [56] and two of them could be confidently assigned (Figure 5.17). The major product was isolated in pure form and identified by ^{1}H NMR spectroscopy as the C_1-symmetric C(25)–C(26) adduct (±)-**127**. A second isomer was characterized in a mixture with two C_1-symmetric derivatives. It was assigned as C_2-symmetric C(27)–C(50) adduct (±)-**128**. The addend positions of both derivatives correspond to a functionalization at the most pyramidalized positions of C_{76}.

5.3.2.3 Halogenation of Higher Fullerenes

5.3.2.3.1 FLUORINATION OF C_{74}-D_{3H}

Although C_{74} was detected by mass spectrometry in arc-generated fullerene soot [162], it could not be extracted with organic solvents although small amounts were obtained by sublimation [163]. The very small HOMO-LUMO gap (calculated to be 0.05 eV compared to 1.72 eV for C_{60}) is supposed to be responsible for this behavior, causing a kinetically unstable structure that has the tendency to polymerize [163]. After electrochemical reduction of the insoluble residue of fullerene soot, a stable di-anionic species of C_{74}

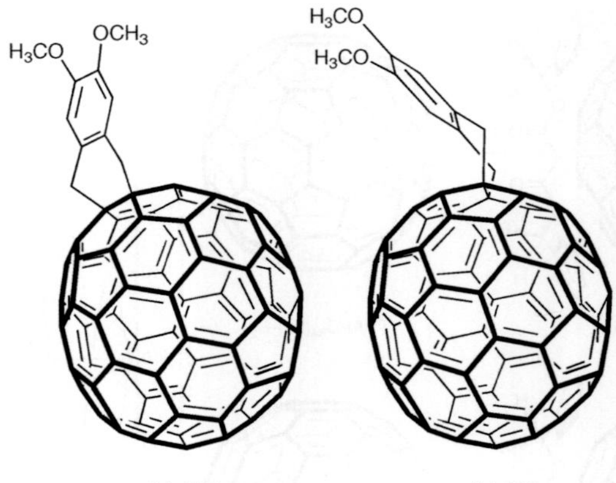

(±)-**127** (±)-**128**

Figure 5.17 *Two covalent derivatives of C_{76}(1)-D_2, obtained from Diels-Alder reaction of the fullerene with 3,4-dimethoxy-o-quinodimethane*

could be purified by HPLC under anaerobic conditions. Reoxidation of the anionic cage yielded pure but insoluble [74]fullerene [163]. Fluorination of the latter with K_2PtF_6 afforded D_3-symmetric $C_{74}F_{38}$ ((±)-**113**) as the first C_{74}-derivative isolated (Figure 5.18) [154]. This result provided evidence for the parent fullerene being $C_{74}(1)$-D_{3h}, the only IPR-conforming isomer of this carbon cage [11].

5.3.2.3.2 HALOGENATION OF C_{76}

Halogenofullerenes could have practical importance because of their potential to act as precursors for derivatives that may be made by subsequent replacement of halogen by other residues, e.g. aryl groups [93]. The greater solubility and higher reactivity of fluoro compounds compared to the parent fullerene assign them as suitable starting materials for further transformations. The lower steric hindrance towards fluorination generally provides different functionalization patterns as in chlorination. An access to fluorofullerenes with defined structures would therefore enable a route to so far unknown derivatives of higher fullerenes [164].

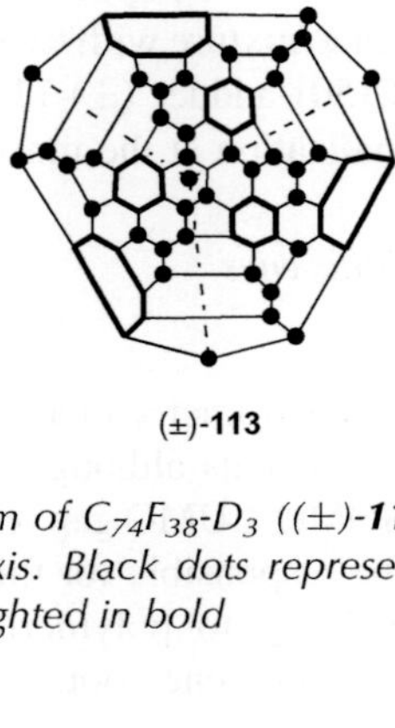

(±)-**113**

Figure 5.18 *Schlegel-type diagram of $C_{74}F_{38}$-D_3 ((±)-**113**), the first isolated C_{74}-derivative, viewed along the C_3 symmetry axis. Black dots represent C-atoms with attached F-atoms. Isolated benzenoid rings are highlighted in bold*

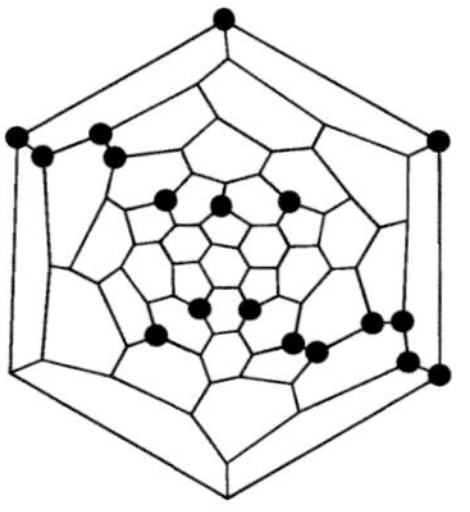

(±)-**129**

Figure 5.19 *Schlegel diagram of the C_2-symmetric ($^{f,s}C$)-enantiomer of $C_{76}Cl_{18}$. Black dots represent C-atoms with attached Cl-atoms*

The fluorination of $C_{76}(1)$-D_2 with MnF_3 at 450–500 °C afforded several species, among which $C_{76}F_{36}$, $C_{76}F_{38}$, $C_{76}F_{40}$ (five isomers), $C_{76}F_{42}$, and $C_{76}F_{44}$ were isolated and partially characterized [164]. Structure determination was not possible due to the small amounts of obtained samples.

$C_{76}Cl_{18}$ ((±)-**129**, Figure 5.19) was obtained by reaction of C_{76} with $Br_2/TiCl_4$ and it has C_2-symmetry. As shown by X-ray crystallography, the chlorine atoms are arranged in two ribbons around the $C_{76}(1)$-D_2 cage, forming clockwise helices in the case of the ($^{f,s}C$)-isomer and anticlockwise helices in the case of the ($^{f,s}A$)-isomer [28].

5.3.2.3.3 HALOGENATION OF C_{78}

Two fluorinated derivatives were detected by mass spectrometry in the mixtures obtained from fluorination of C_{78}-containig sample of C_{76}. Purification attempts provided enriched samples of $C_{78}F_{38}$ and $C_{78}F_{42}$ but their structures were not determined [164].

A well-characterized halogenofullerene is $C_{78}Br_{18}$, the only reported bromo-derivative of a higher fullerene [30]. X-ray crystal structure analysis indicates D_{3h}-symmetry which was interpreted as a result of statistical disordering of two $C_{78}Br_{18}$ isomers (**130** and **131**), both with C_{2v}-symmetry. The two parent cage isomers show the same symmetry and correspond to the IPR-structures $C_{78}(2)$-C_{2v} and $C_{78}(3)$-C_{2v} (Figure 5.20).

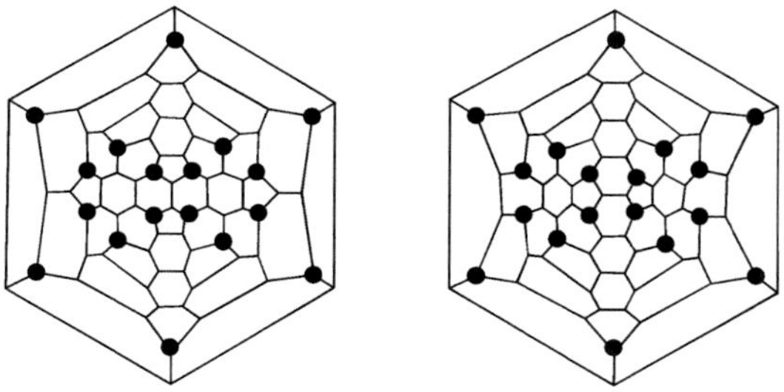

130 **131**

Figure 5.20 *Schlegel diagrams of two C_{2v}-symmetric isomers of $C_{78}Br_{18}$ (**130** and **131**). Black dots represent C-atoms with attached Br-atoms*

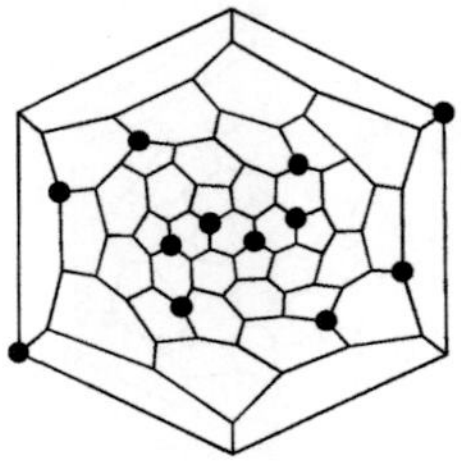

(±)-**132**

Figure 5.21 *Schlegel diagram of $C_{80}Cl_{12}$. Black dots represent C-atoms with attached Cl-atoms*

5.3.2.3.4 CHLORINATION OF C_{80}

Very recently, a D_2-symmetric chlorinated derivative of $C_{80}(2)$-D_2 was isolated and its structure determined by X-ray crystallography (Figure 5.21) [165]. It was obtained by reaction of the pristine fullerene with $Br_2/TiCl_4$ in a sealed ampoule heated for 5d to 150 °C. The chlorine atoms of $C_{80}Cl_{12}$ ((±)-**132**) are attached neither to the most pyramidalized carbon atoms nor, expectedly, to those at the junction of three hexagons, which are supposed to be the most inert.

5.3.2.4 *Multi-Adducts Resulting from Trifluoromethylation of Higher Fullerenes*

After extensive investigation of the trifluoromethylation of C_{60} and C_{70} (see Section 5.2.1.5), CF_3 derivatives were also prepared of the higher fullerenes by reaction of pure carbon cages with trifluoroiodomethane at 520–550 °C. So far, 28 compounds of composition $C_m(CF_3)_n$ ($m = 74, 76, 78, 80, 82, 84, 90$; $n = 6, 8, 10, 12, 14$; many but not all combinations were observed) have been reported [17, 24, 26, 29], part of which was unambiguously assigned by a combination of ^{19}F NMR spectroscopy and DFT calculations or by X-ray crystallography (Figure 5.22).

C_2-p^{11}-(C_{74}-D_{3h})$(CF_3)_{12}$ ((±)-**114**, Figure 5.22) was shown to contain a single C_2-p^{11} ribbon (for the description of trifluoromethylation patterns, cf. Section 5.2.1.5) [17], a structure that was verified by single crystal X-ray diffraction [26].

Ten isomers of composition $C_{76}(CF_3)_x$ ($x = 6, 8, 10, 12$) were found. Except for the achiral C_s-p^9,p^2-($C_{76}(2)$-T_d)$(CF_3)_{12}$ (**111**) [17], which has a p^9 loop (circular arrangement of nine '*para*'-functionalized hexagons) and a small p^2 ribbon with an odd number of CF_3-groups each, all derivatives are based on cage isomer $C_{76}(1)$-D_2, and the resulting symmetry of the derivatives is either C_1 or C_2. Four chiral octakis-, decakis-, or dodecakis-trifluoromethyl-[76]fullerene isomers were isolated and characterized very recently [29]. The two $C_{76}(CF_3)_8$ isomers ((±)-**133** and (±)-**134**) show C_1-symmetry and contain p^5,p arrangements of the addends [29]. The more highly functionalized derivatives occur as single isomers each, i.e. C_s-p^4,p^4-($C_{76}(1)$-D_2)$(CF_3)_{10}$ ((±)-**135**) and C_s-p^3mp,p^3mp-($C_{76}(1)$-D_2)$(CF_3)_{12}$ ((±)-**136**) [29].

The structures of four out of seven isomers of $C_{78}(CF_3)_x$ ($x = 8, 10, 12, 14$) were ascertained by X-ray crystallography: C_2-p^{11}-($C_{78}(5)$-D_{3h})$(CF_3)_{12}$ ((±)-**137**) is derived from the parent fullerene $C_{78}(5)$-D_{3h} [17, 26]. Compounds (±)-**138** and (±)-**139** are based

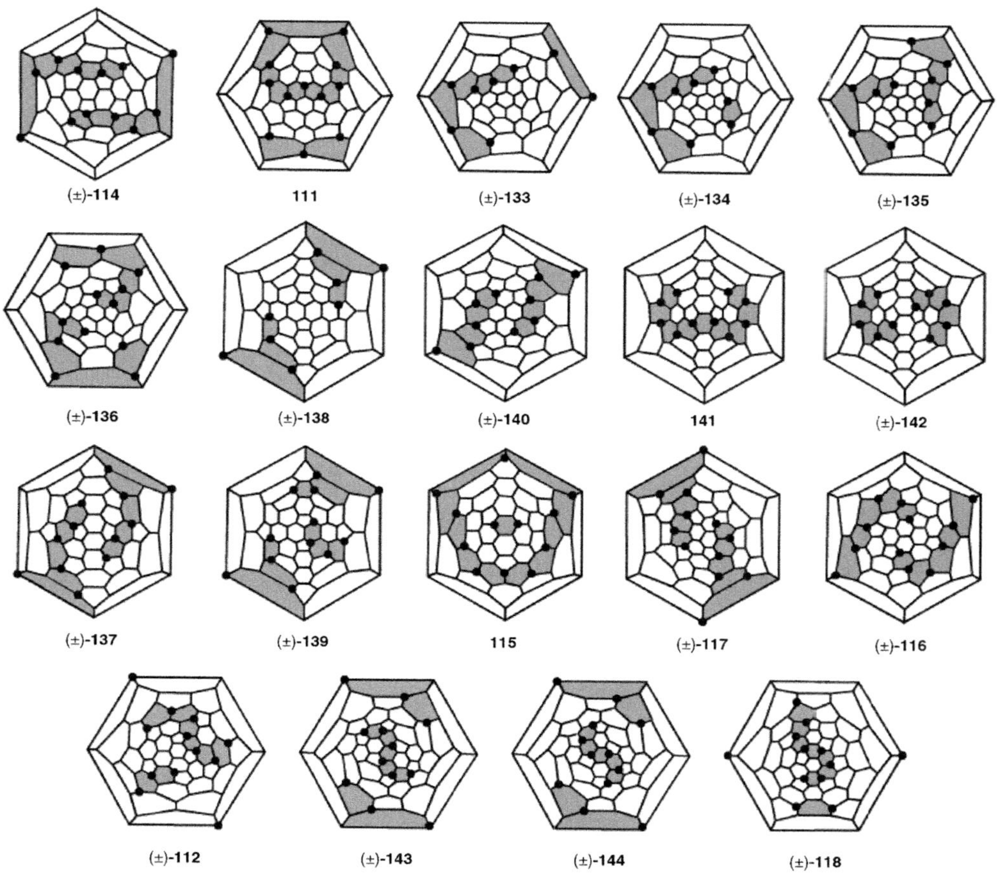

Figure 5.22 *Schlegel diagrams of trifluoromethylated higher fullerenes. Black dots represent C-atoms with attached CF_3-groups. Ribbons are marked in gray*

on cage isomer $C_{78}(3)$-C_{2v}, (±)-**140** is a derivative of $C_{78}(1)$-D_3, and **141** as well as (±)-**142** originate from $C_{78}(2)$-C_{2v} [29]. In the case of $C_{78}(CF_3)_{14}$, the cage isomer is unknown. C_s-p^{10}-loop-,p-($C_{80}(5)$-C_{2v})$(CF_3)_{12}$ (**115**) is the only trifluoromethylated derivative found for C_{80} [17]. It is the first isolated exohedral derivative of $C_{80}(5)$-C_{2v} and, at the same time, the first proof for the existence of the corresponding hollow carbon cage in fullerene soot.

The two known isomers of C_2-$C_{82}(CF_3)_{12}$ are based on two different C_2-symmetric cage isomers. They were assigned as C_2-p^{11}-($C_{82}(5)$-C_2)$(CF_3)_{12}$ ((±)-**117**) and C_2-p^5,p^5-($C_{82}(3)$-C_2)$(CF_3)_{12}$ ((±)-**116**) and contain one and two ribbons, respectively, of hexagons with a 1,4-relationship between CF_3-groups [17].

Five CF_3-isomers of C_{84} with 10, 12, or 14 addends were isolated [29]. A structure was determined only for the three $C_{84}(CF_3)_{12}$ isomers, the other two compounds remaining unclear in terms of their cage isomers. For C_1-p^6,p^2,p-($C_{84}(11)$-C_2)$(CF_3)_{12}$ ((±)-**112**), an X-ray crystal structure was obtained [24]. The other $C_{84}(CF_3)_{12}$-isomers ((±)-**143** and (±)-**144**) are based on cage isomer $C_{84}(22)$-D_2 and contain two p^5 ribbons each. Depending on the relative position of the CF_3-groups, adduct isomers with D_2-((±)-**143**) or C_2-symmetry ((±)-**144**)) result.

The largest trifluoromethylated fullerenes isolated so far are two isomers of $C_{90}(CF_3)_{12}$ [29]. While the cage isomer could not be determined in the first case, the second structure was assigned as C_1-p^7,p,p-(C_{90}(32)-C_1)(CF_3)$_{12}$ ((±)-**118**) by a combination of ^{19}F NMR spectroscopy and DFT calculations.

The advantage of poly-trifluoromethylated fullerene derivatives is their good solubility in organic solvents and the stability against air, light, and elevated temperatures, making their handling easy. Furthermore, they show a relatively high tendency to crystallize, which is an invaluable plus in the determination of the rather complex structures. In general, 1,4-addition of bulky residues does not take place at bonds involving the most pyramidalized fullerene cage C(sp^2)-atoms, although the corresponding C–C bonds are the most electron-rich and, therefore, the most reactive in many chemical transformations. Instead, the monovalent addends are attached to less pyramidalized cage atoms and form ribbons of edge-sharing *p*-hexagons, whereas bonds considered to be most reactive in other reactions such as cycloadditions remain intact [29].

The physicochemical properties of polyfluoroalkylated fullerenes are strongly influenced by the functionalization pattern, so it is important to understand which factors play a role for their formation [29]. The steric and electronic nature of addends determines their relative position on the fullerene surface. It is known that, in contrast to small addends, larger groups do not form structures with strings of contiguous cage C(sp^3)-atoms. For 4–12 larger addends like CF_3 groups or Br atoms, ribbons of edge-sharing *meta*- and/or *para*-hexagons are common. It was suggested, that terminal Br atoms, which are in a half boat conformation on the fullerene surface, might activate the next addition of bromine causing a ribbon formation. Considerably larger groups X, such as C_3F_7, show the tendency to form multiple isolated *p*-$C_6(X)_2$ hexagons.

Also, the structure of the fullerene affects the formation of the pattern on its surface. Addition on triple-hexagon junctions (THJs) is unfavored because it would provide undesirable strain due to rehybridization.

5.4 Concluding Remarks

During the last few years, several new cage isomers of higher fullerenes were discovered and structurally characterized, either in the form of pure carbon allotropes or of halogenated or trifluoromethylated cages. Structural elucidation of these fullerene derivatives greatly benefits from the fact that they have a relatively high tendency to provide single crystals suitable for X-ray analysis. In case of fluorine-containing fullerene derivatives, 1D and 2D ^{19}F NMR spectroscopy turned out to be a reliable tool for gaining a wealth of information which often allows for confident structure proposals.

From 2001 on, the investigation of poly-trifluoromethylated fullerenes opened up a fascinating field in fullerene chemistry, leading to the discovery and structural elucidation of several hitherto unknown cage isomers of higher fullerenes. Also, these derivatives generally exhibit functionalization patterns that differ considerably from those of adducts formed in cycloaddition or nucleophilic addition reactions. Arguments other than C-atom pyramidalization or double bond character must, therefore, be adopted to explain their formation, and these studies contribute significantly to the understanding of fullerene reactivity.

Regarding the chirality of higher fullerenes and of fullerene derivatives in general, the configurational assignment of resolved enantiomers has made considerable progress in recent years through the calculation of CD spectra at improved levels of theory and their comparison to experimental data. In this context, chiral parent fullerenes and many fullerene derivatives with an inherently or non-inherently chiral functionalization pattern benefit from the fact that the molecular scaffold is rather rigid in comparison to most other molecules, thereby smoothing the way for dependable calculations.

While C_{60} and C_{70} are available in multigram quantities, it is still difficult to obtain reasonable amounts of higher cage isomers in pure form to study their properties and chemical derivatization, in particular. Although there are several approaches for the selective synthesis of individual isomers, it still seems as promising to improve the efficiency of separation techniques by the development of new chromatographic stationary phases or supramolecular receptors for selective fullerene extraction. Finally, it is interesting to note that advances in other areas of fullerene chemistry, e.g. the preparation and purification of endohedral metallofullerenes or the derivatization of fullerenes directly from soot have unexpectedly led to the materialization of hitherto unknown cages.

Acknowledgement

We gratefully acknowledge continuing support of our fullerene research program by the Swiss National Science Foundation.

References

[1] H. W. Kroto, J. R. Heath, S. C. O'Brien, R. F. Curl and R. E. Smalley, C_{60}: Buckminsterfullerene, *Nature (London, U. K.)*, **318**, 162–163 (1985).

[2] F. Diederich, R. Ettl, Y. Rubin, R. L. Whetten, R. Beck, M. Alvarez, S. Anz, D. Sensharma, F. Wudl, K. C. Khemani and A. Koch, The higher fullerenes: isolation and characterization of C_{76}, C_{84}, C_{90}, and $C_{70}O$, an oxide of D_{5h}-C_{70}, *Science (Washington D.C.)*, **252**, 548–551 (1991).

[3] F. Diederich and R. L. Whetten, Beyond C_{60}: The higher fullerenes, *Acc. Chem. Res.*, **25**, 119–126 (1992).

[4] C. Thilgen, F. Diederich and R. L. Whetten, Buckminsterfullerenes, in *Buckminsterfullerenes*, W. E. Billups and M. A. Ciufolini (Eds.), VCH Publishers Inc., New York, 1993.

[5] C. Thilgen, A. Herrmann and F. Diederich, The covalent chemistry of higher fullerenes: C_{70} and beyond, *Angew. Chem. Int. Ed.*, **36**, 2268–2280 (1997).

[6] C. Thilgen and F. Diederich, The higher fullerenes: covalent chemistry and chirality, *Top. Curr. Chem.*, **199**, 135–171 (1999).

[7] F. Diederich, C. Thilgen and A. Herrmann, Chirality in fullerene chemistry, *Nachr. Chem. Tech. Lab.*, **44**, 9–16 (1996).

[8] C. Thilgen and F. Diederich, Structural aspects of fullerene chemistry – a journey through fullerene chirality, *Chem. Rev.*, **106**, 5049–5135 (2006).

[9] C. Boudon, J.-P. Gisselbrecht, M. Gross, A. Herrmann, M. Rüttimann, J. Crassous, F. Cardullo, L. Echegoyen and F. Diederich, Redox characteristics of covalent derivatives of the higher fullerenes C_{70}, C_{76}, and C_{78}, *J. Am. Chem. Soc.*, **120**, 7860–7868 (1998).

[10] A. Hirsch and M. Brettreich, *Fullerenes – Chemistry and Reactions*. Wiley-VCH, Weinheim, 2005.

[11] P. W. Fowler and D. E. Manolopoulos, *An Atlas of Fullerenes*. Dover Publications, Inc., Mineola, New York, USA, 2007.

[12] H. W. Kroto, The stability of the fullerenes C_{24}, C_{28}, C_{32}, C_{36}, C_{50}, C_{60} and C_{70}, *Nature (London, U. K.)*, **329**, 529–531 (1987).

[13] T. G. Schmalz, W. A. Seitz, D. J. Klein and G. E. Hite, Elemental carbon cages, *J. Am. Chem. Soc.*, **110**, 1113–1127 (1988).

[14] F. Diederich, R. L. Whetten, C. Thilgen, R. Ettl, I. Chao and M. M. Alvarez, Fullerene isomerism: isolation of C_{2v}-C_{78} and D_3-C_{78}, *Science (Washington D.C.)*, **254**, 1768–1770, (1991).

[15] K. Kikuchi, N. Nakahara, T. Wakabayashi, S. Suzuki, H. Shiromaru, Y. Miyake, K. Saito, I. Ikemoto, M. Kainosho and Y. Achiba, NMR characterization of isomers of C_{78}, C_{82} and C_{84} fullerenes, *Nature (London, U. K.)*, **357**, 142–145 (1992).

[16] K. S. Simeonov, K. Y. Amsharov, E. Krokos and M. Jansen, An epilogue on the C_{78}-fullerene family: the discovery and characterization of an elusive isomer, *Angew. Chem. Int. Ed.*, **47**, 9590 (2008).

[17] N. B. Shustova, I. V. Kuvychko, R. D. Bolskar, K. Seppelt, S. H. Strauss, A. A. Popov and O. V. Boltalina, Trifluoromethyl derivatives of insoluble small-HOMO–LUMO-gap hollow higher fullerenes. NMR and DFT structure elucidation of C_2-(C_{74}-D_{3h})(CF_3)$_{12}$, C_s-(C_{76}-T_d(2))(CF_3)$_{12}$, C_2-(C_{78}-D_{3h}(5))(CF_3)$_{12}$, C_s-(C_{80}-C_{2v}(5))(CF_3)$_{12}$, and C_2-(C_{82}-C_2(5))(CF_3)$_{12}$, *J. Am. Chem. Soc.*, **128**, 15793–15798 (2006).

[18] $C_{78}Cl_{18}$ was recently reported as a derivative of the elusive isomer C_{78}(4)-D_{3h} in K. S. Simeonov, K. Y. Amsharov, E. Krokos and M. Jansen, An epilogue on the C_{78}-fullerene family: the discovery and characterization of an elusive isomer, *Angew. Chem. Int. Ed.*, **47**, 6283–6285 (2008). The structure of the cage isomer was questioned in I. E. Kareev, A. A. Popov, I. V. Kuvychko, N. B. Shustova, S. F. Lebedkin, V. P. Bubnov, O. P. Anderson, K. Seppelt, S. H. Strauss and O. V. Boltalina, Synthesis and X-ray or NMR/DFT structure elucidation of 21 new trifluoromethyl derivatives of soluble cage isomers of C_{76}, C_{78}, C_{84}, and C_{90}, *J. Am. Chem. Soc.*, **130**, 13471–13489 (2008), where it was assigned as C_{78}(5)-D_{3h}.

[19] R. Taylor, G. J. Langley, A. G. Avent, T. J. S. Dennis, H. W. Kroto and D. R. M. Walton, ^{13}C NMR spectroscopy of C_{76}, C_{78}, C_{84} and mixtures of C_{86}-C_{102}; anomalous chromatographic behavior of C_{82}, and evidence for $C_{70}H_{12}$, *J. Chem. Soc., Perkin Trans.*, **2**, 1029–1036 (1993).

[20] A. G. Avent, D. Dubois, A. Penicaud and R. Taylor, The minor isomers and IR spectrum of [84] fullerene, *J. Chem. Soc., Perkin Trans.*, **2**, 1907–1910 (1997).

[21] T. J. S. Dennis, T. Kai, T. Tomiyama and H. Shinohara, Isolation and characterisation of the two major isomers of [84]fullerene (C_{84}), *Chem. Commun.*, 619–620 (1998).

[22] N. Tagmatarchis, A. G. Avent, K. Prassides, T. J. S. Dennis and H. Shinohara, Separation, isolation and characterisation of two minor isomers of the [84]fullerene C_{84}, *Chem. Commun.*, 1023–1024 (1999).

[23] T. J. S. Dennis, T. Kai, K. Asato, T. Tomiyama, H. Shinohara, T. Yoshida, Y. Kobayashi, H. Ishiwatari, Y. Miyake, K. Kikuchi and Y. Achiba, Isolation and characterization by ^{13}C NMR spectroscopy of [84]fullerene minor isomers, *J. Phys. Chem. A*, **103**, 8747–8752 (1999).

[24] I. E. Kareev, I. V. Kuvychko, N. B. Shustova, S. E. Lebedkin, V. P. Bubnov, O. P. Anderson, A. A. Popov, O. V. Boltalina and S. H. Strauss, C_1-(C_{84}-C_2(11))(CF_3)$_{12}$: trifluoromethylation yields structural proof of a minor C_{84} cage and reveals a principle of higher fullerene reactivity, *Angew. Chem. Int. Ed.*, **47**, 6204–6207 (2008).

[25] I. S. Neretin and Y. L. Slovokhotov, Chemical crystallography of fullerenes, *Russ. Chem. Rev.*, **73**, 455–486 (2004).

[26] N. B. Shustova, B. S. Newell, S. M. Miller, O. P. Anderson, R. D. Bolskar, K. Seppelt, A. A. Popov, O. V. Boltalina and S. H. Strauss, Discovering and verifying elusive fullerene cage isomers: structures of C_2-p^{11}-(C_{74}-D_{3h})(CF_3)$_{12}$ and C_2-p^{11}-(C_{78}-D_{3h}(5))(CF_3)$_{12}$, *Angew. Chem. Int. Ed.*, **46**, 4111–4114 (2007).

[27] Throughout this review, the numbers in parentheses following the fullerene formula refer to the numbering of the structures in P. W. Fowler and D. E. Manolopoulos, *An Atlas of Fullerenes*, Dover Publications, Inc., Mineola, New York, USA, 2007.

[28] K. S. Simeonov, K. Y. Amsharov and M. Jansen, Connectivity of the chiral D_2-symmetric isomer of C_{76} through a crystal-structure determination $C_{76}Cl_{18}$•$TiCl_4$, *Angew. Chem. Int. Ed.*, **46**, 8419–8421 (2007).
[29] I. E. Kareev, A. A. Popov, I. V. Kuvychko, N. B. Shustova, S. F. Lebedkin, V. P. Bubnov, O. P. Anderson, K. Seppelt, S. H. Strauss and O. V. Boltalina, Synthesis and X-ray or NMR/DFT structure elucidation of twenty-one new trifluoromethyl derivatives of soluble cage isomers of C_{76}, C_{78}, C_{84}, and C_{90}, *J. Am. Chem. Soc.*, **130**, 13471–13489 (2008).
[30] S. I. Troyanov and E. Kemnitz, The first crystal structure of a halogenated higher fullerene, $C_{78}Br_{18}$, obtained by bromination of a fullerene mixture, *Eur. J. Org. Chem.*, 3916–3919 (2003).
[31] L. Epple, K. Amsharov, K. Simeonov, I. Dix and M. Jansen, Crystallographic characterization and identification of a minor isomer of C_{84} fullerene, *Chem. Commun.*, 5610–5612 (2008).
[32] A. L. Balch, A. S. Ginwalla, J. W. Lee, B. C. Noll and M. M. Olmstead, Partial separation and structural characterization of C_{84} isomers by crystallisation of $(\eta^2\text{-}C_{84})Ir(CO)Cl(P(C_6H_5)_3)_2$, *J. Am. Chem. Soc.*, **116**, 2227–2228 (1994).
[33] C.-R. Wang, T. Sugai, T. Kai, T. Tomiyama and H. Shinohara, Production and isolation of an ellipsoidal C_{80} fullerene, *Chem. Commun.*, 557–558 (2000).
[34] F. H. Hennrich, R. H. Michel, A. Fischer, S. Richard-Schneider, S. Gilb, M. M. Kappes, D. Fuchs, M. Bürk, K. Kobayashi and S. Nagase, Isolation and characterization of C_{80}, *Angew. Chem. Int. Ed.*, **35**, 1732–1734 (1996).
[35] G. Sun and M. Kertesz, ^{13}C NMR spectra for IPR isomers of fullerene C_{86}, *Chem. Phys.*, **276**, 107–114 (2002).
[36] G. Sun, Assigning the major isomers of fullerene C_{88} by theoretical ^{13}C NMR spectra, *Chem. Phys. Lett.*, **367**, 26–33 (2003).
[37] T. Akasaka and S. Nagase, *Endofullerenes: A New Family of Carbon Clusters*, Kluwer Academic Publisher, Dordrecht, The Netherlands, 2002.
[38] L. Dunsch and S. F. Yang, Endohedral clusterfullerenes – playing with cluster and cage sizes, *Phys. Chem. Chem. Phys.*, **9**, 3067–3081 (2007).
[39] R. Ettl, I. Chao, F. Diederich and R. L. Whetten, Isolation of C_{76}, a chiral (D_2) allotrope of carbon, *Nature (London, U. K.)*, **353**, 149–153 (1991).
[40] J. M. Hawkins and A. Meyer, Optically active carbon: Kinetic resolution of C_{76} by asymmetric osmylation, *Science (Washington D.C.)*, **260**, 1918–1920 (1993).
[41] R. Kessinger, J. Crassous, A. Herrmann, M. Rüttimann, L. Echegoyen and F. Diederich, Preparation of enantiomerically pure C_{76} with a general electrochemical method for the removal of di(alkoxycarbonyl)methano bridges from methanofullerenes: the retro-*Bingel* reaction, *Angew. Chem. Int. Ed.*, **37**, 1919–1922 (1998).
[42] J. Crassous, J. Rivera, N. S. Fender, L. Shu, L. Echegoyen, C. Thilgen, A. Herrmann and F. Diederich, Chemistry of C_{84}: separation of three constitutional isomers and optical resolution of D_2-C_{84} by using the '*Bingel*-retro-*Bingel*' strategy, *Angew. Chem. Int. Ed.*, **38**, 1613–1617 (1999).
[43] Only IPR (isolated pentagon rule)-conforming fullerene structures are discussed in the present review.
[44] C. Thilgen, A. Herrmann and F. Diederich, Configurational description of chiral fullerenes and fullerene derivatives with a chiral functionalization pattern, *Helv. Chim. Acta*, **80**, 183–199 (1997).
[45] C. Thilgen, I. Gosse and F. Diederich, Chirality in fullerene chemistry, *Top. Stereochem.*, **23**, 1–124 (2003).
[46] R. S. Cahn, S. C. Ingold and V. Prelog, Specification of molecular chirality, *Angew. Chem. Int. Ed. Engl.*, **5**, 385–415 (1966).
[47] V. Prelog and G. Helmchen, Basic principles of the CIP-system and proposals for a revision, *Angew. Chem. Int. Ed. Engl.*, **21**, 567–583 (1982).
[48] W. H. Powell, F. Cozzi, G. P. Moss, C. Thilgen, R. J.-R. Hwu and A. Yerin, Nomenclature for the C_{60}-I_h and C_{70}-$D_{5h(6)}$ fullerenes – (IUPAC recommendations 2002), *Pure Appl. Chem.*, **74**, 629–695 (2002). www.chem.qmul.ac.uk/iupac/fullerene/Fu17.html.

[49] F. Cozzi, W. H. Powell and C. Thilgen, Numbering of fullerenes – (IUPAC recommendations 2005), *Pure Appl. Chem.*, **77**, 843–923 (2005).
[50] R. Taylor, C_{60}, C_{70}, C_{76}, C_{78} and C_{84}: Numbering, π-bond order calculations and addition pattern considerations, *J. Chem. Soc., Perkin Trans.*, **2**, 813–824 (1993).
[51] A. Herrmann, M. Rüttimann, C. Thilgen and F. Diederich, Multiple cyclopropanations of C_{70} – synthesis and characterization of bis-adducts, tris-adducts, and tetrakis-adducts and chiroptical properties of bis-adducts with chiral addends, including a recommendation for the configurational description of fullerene derivatives with a chiral addition pattern, *Helv. Chim. Acta*, **78**, 1673–1704 (1995).
[52] H. Al-Matar, A. K. Abdul-Sada, A. G. Avent, R. Taylor and X.-W. Wei, Methylation of [70] fullerene, *J. Chem. Soc., Perkin Trans.*, **2**, 1251–1256 (2002).
[53] F. Diederich and C. Thilgen, Covalent fullerene chemistry, *Science (Washington D.C.)*, **271**, 317–323 (1996).
[54] C. Bingel, Cyclopropanierung von Fullerenen, *Chem. Ber.*, **126**, 1957–1959 (1993).
[55] C. Bingel and H. Schiffer, Biscyclopropanation of C_{70}, *Liebigs Ann.*, 1551–1553 (1995).
[56] A. Herrmann, F. Diederich, C. Thilgen, H.-U. ter Meer and W. H. Müller, Chemistry of the higher fullerenes – preparative isolation of C_{76} by HPLC and synthesis, separation, and characterization of *Diels-Alder* monoadducts of C_{70} and C_{76}, *Helv. Chim. Acta*, **77**, 1689–1706 (1994).
[57] A. Herrmann and F. Diederich, Synthesis, separation, and characterization of optically pure C_{76} mono-adducts, *Helv. Chim. Acta*, **79**, 1741–1756 (1996).
[58] P. B. Hitchcock, A. G. Avent, N. Martsinovich, P. A. Troshin and R. Taylor, C_1 $C_{70}F_{38}$ contains four planar aromatic hexagons; the parallel between fluorination of [60]- and [70]fullerenes, *Org. Lett.*, **7**, 1975–1978 (2005).
[59] P. B. Hitchcock, A. G. Avent, N. Martsinovich, P. A. Troshin and R. Taylor, C_2 $C_{70}F_{38}$ is aromatic, contains three planar hexagons, and has equatorial addends, *Chem. Commun.*, 75–77 (2005).
[60] P. R. Birkett, A. G. Avent, A. D. Darwish, H. W. Kroto, R. Taylor and D. R. M. Walton, Formation and characterisation of $C_{70}Cl_{10}$, *J. Chem. Soc., Chem. Commun.*, 683–684 (1995).
[61] S. I. Troyanov, A. A. Popov, N. I. Denisenko, O. V. Boltalina, L. N. Sidorov and E. Kemnitz, The first X-ray crystal structures of halogenated [70]fullerene: $C_{70}Br_{10}$ and $C_{70}Br_{10}\cdot 3\ Br_2$, *Angew. Chem. Int. Ed.*, **42**, 2395–2398 (2003).
[62] A. A. Popov, I. E. Kareev, N. B. Shustova, S. F. Lebedkin, S. H. Strauss, O. V. Boltalina and L. Dunsch, Synthesis, spectroscopic and electrochemical characterization, and DFT study of seventeen $C_{70}(CF_3)_n$ derivatives ($n = 2, 4, 6, 8, 10, 12$), *Chem. Eur. J.*, **14**, 107–121 (2008).
[63] M. S. Meier, M. Poplawska, A. L. Compton, J. P. Shaw, J. P. Selegue and T. F. Guarr, Preparation and isolation of three isomeric C_{70} isoxazolines: strong deshielding in the polar region of C_{70}, *J. Am. Chem. Soc.*, **116**, 7044–7048 (1994).
[64] A. Kraszewska, P. Rivera-Fuentes, C. Thilgen and F. Diederich, First enantiomerically pure C_{70}-adducts with a non-inherently chiral addition pattern, *New J. Chem.*, **33**, 386–396 (2009).
[65] S. R. Wilson and Q. Lu, 1,3-dipolar cycloaddition of *N*-methylazomethine ylide to C_{70}, *J. Org. Chem.*, **60**, 6496–6498 (1995).
[66] F. Langa, P. de la Cruz, A. de la Hoz, E. Espĺdora, F. P. Cosśo and B. Lecea, Modification of regioselectivity in cycloadditions to C_{70} under microwave irradiation, *J. Org. Chem.*, **65**, 2499–2507 (2000).
[67] A. D. Darwish, A. G. Avent, R. Taylor and D. R. M. Walton, Reaction of benzyne with [70] fullerene gives four monoadducts: formation of a triptycene homologue by 1,4-cycloaddition of a fullerene, *J. Chem. Soc., Perkin Trans.*, **2**, 2079–2084 (1996).
[68] M. S. Meier, G. W. Wang, R. C. Haddon, C. Pratt Brock, M. A. Lloyd and J. P. Selegue, Benzyne adds across a closed 5–6 ring fusion in C_{70}: evidence for bond delocalization in fullerenes, *J. Am. Chem. Soc.*, **120**, 2337–2342 (1998).
[69] A. Herrmann, M. W. Rüttimann, T. Gibtner, C. Thilgen, F. Diederich, T. Mordasini and W. Thiel, Achiral and chiral higher adducts of C_{70} by *Bingel* cyclopropanation, *Helv. Chim. Acta*, **82**, 261–289 (1999).

[70] H. Goto, N. Harada, J. Crassous and F. Diederich, Absolute configuration of chiral fullerenes and covalent derivatives from their calculated circular dichroism spectra, *J. Chem. Soc., Perkin Trans.*, **2**, 1719–1723 (1998).
[71] C. Thilgen, S. Sergeyev and F. Diederich, Spacer-controlled multiple functionalization of fullerenes, *Top. Curr. Chem.*, **248**, 1–61 (2004).
[72] C. Thilgen and F. Diederich, Tether-directed remote functionalization of fullerenes C_{60} and C_{70}, *C. R. Chimie*, **9**, 868–880 (2006).
[73] M. J. van Eis, R. J. Alvarado, L. Echegoyen, P. Seiler and F. Diederich, First tether-directed regioselective bis-functionalisation of C_{70}: effects of cation complexation on the redox properties of diastereoisomeric fullerene crown ether conjugates, *Chem. Commun.*, 1859–1860 (2000).
[74] M. J. van Eis, P. Seiler, L. A. Muslinkina, M. Badertscher, E. Pretsch, F. Diederich, R. J. Alvarado, L. Echegoyen and I. P. Núñez, Supramolecular fullerene chemistry: a comprehensive study of cyclophane-type mono- and bis-crown ether conjugates of C_{70}, *Helv. Chim. Acta*, **85**, 2009–2055 (2002).
[75] M. J. van Eis, I. Pérez Núñez, L. A. Muslinkina, R. J. Alvarado, E. Pretsch, L. Echegoyen and F. Diederich, Sandwiching C_{70} between two crown ether-bound cations: regioselective synthesis, electrochemistry and cation binding properties of C_{70} bis-crown ether conjugates, *J. Chem. Soc., Perkin Trans.*, **2**, 1890–1892 (2001).
[76] S. Sergeyev, M. Schär, P. Seiler, O. Lukoyanova, L. Echegoyen and F. Diederich, Synthesis of *trans*-1, *trans*-2, *trans*-3, and *trans*-4 bisadducts of C_{60} by regio- and stereoselective tether-directed remote functionalization, *Chem. Eur. J.*, **11**, 2284–2294 (2005).
[77] S. Sergeyev and F. Diederich, Regio- and stereoselective tether-directed remote functionalization of C_{60} with derivatives of the *Tröger* base, *Angew. Chem. Int. Ed.*, **43**, 1738–1740 (2004).
[78] W. W. H. Wong and F. Diederich, Regio- and diastereoselective synthesis of bis- and tetrakisadducts of C_{70} by directed remote functionalization using *Tröger* base tethers, *Chem. Eur. J.*, **12**, 3463–3471 (2006).
[79] Q. Lu, D. I. Schuster and S. R. Wilson, Preparation and characterization of six bis(*N*-methylpyrrolidine)–C_{60} isomers: Magnetic deshielding in isomeric bisadducts of C_{60}, *J. Org. Chem.*, **61**, 4764–4768 (1996).
[80] M. S. Meier, R. G. Bergosh, M. E. Gallagher, H. P. Spielmann and Z. Wang, Alkylation of dihydrofullerenes, *J. Org. Chem.*, **67**, 5946–5952 (2002).
[81] T. Kitakawa, Y. Lee, N. Masaoka and K. Komatsu, Generation and properties of an alkylated C_{70} cation, *Angew. Chem. Int. Ed.*, **44**, 1398–1401 (2005).
[82] C. C. Henderson, C. M. Rohlfing, K. T. Gillen and P. A. Cahill, Synthesis, isolation, and equilibration of 1,9- and 7,8-$C_{70}H_2$, *Science (Washington D.C.)*, **264**, 397–399 (1994).
[83] A. G. Avent, A. D. Darwish, D. K. Heimbach, H. W. Kroto, M. F. Meidine, J. P. Parsons, C. Remars, R. Roers, O. Ohashi, R. Taylor and D. R. M. Walton, Formation of hydrides of fullerene C_{60} and fullerene C_{70}, *J. Chem. Soc., Perkin Trans.*, **2**, 15–22 (1994).
[84] H. P. Spielmann, G. W. Wang, M. S. Meier and B. R. Weedon, Preparation of $C_{70}H_2$, $C_{70}H_4$, and $C_{70}H_8$: three independent reduction manifolds in the Zn(Cu) reduction of C_{70}, *J. Org. Chem.*, **63**, 9865–9871 (1998).
[85] A. D. Darwish, A. K. Abdulsada, G. J. Langley, H. W. Kroto, R. Taylor and D. R. M. Walton, Polyhydrogenation of [60]- and [70]-fullerenes, *J. Chem. Soc., Perkin Trans.*, **2**, 2359–2365 (1995).
[86] M. Gerst, H. D. Beckhaus, C. Rüchardt, E. E. B. Campbell and R. Tellgmann, [7h]Benzanthrene, a catalyst for the transfer hydrogenation of C_{60} and C_{70} by 9,10-dihydroanthracene, *Tetrahedron Lett.*, **34**, 7729–7732 (1993).
[87] A. D. Darwish, A. K. AbdulSada, G. J. Langley, H. W. Kroto, R. Taylor and D. R. M. Walton, Polyhydrogenation of [60]- and [70]fullerenes with Zn/HCl and Zn/DCl, *Synth. Met.*, **77**, 303–307 (1996).
[88] G.-W. Wang, B. R. Weedon, M. S. Meier, M. Saunders and R. J. Cross, ^{3}He NMR study of $^3He@C_{60}H_6$ and $^3He@C_{70}H_{2-10}$, *Org. Lett.*, **2**, 2241–2243 (2000).

[89] H. P. Spielmann, B. R. Weedon and M. S. Meier, Preparation and NMR characterization of $C_{70}H_{10}$: cutting a fullerene π-system in half, *J. Org. Chem.*, **65**, 2755–2758 (2000).

[90] A. L. Balch, D. A. Costa and M. M. Olmstead, Structural characterization of $C_{70}O$. Remarkable disorder in crystalline $[(\eta^2\text{-}C_{70}O)Ir(CO)Cl(PPh_3)_2]\cdot 5C_6H_6$, *Chem. Commun.*, 2449–2450 (1996).

[91] A. L. Balch, J. W. Lee and M. M. Olmstead, Structure of a product of double addition to C_{70}: $[C_{70}\{Ir(CO)Cl(PPhMe_2)_2\}_2]\cdot 3\ C_6H_6$, *Angew. Chem. Int. Ed. Engl.*, **31**, 1356–1358 (1992).

[92] G.-W. Wang, X.-H. Zhang, H. Zhan, Q.-X. Guo and Y.-D. Wu, Accurate calculation, prediction, and assignment of ^{3}He NMR chemical shifts of helium-3-encapsulated fullerenes and fullerene derivatives, *J. Org. Chem.*, **68**, 6732–6738 (2003).

[93] A. G. Avent, P. R. Birkett, A. D. Darwish, H. W. Kroto, R. Taylor and D. R. M. Walton, Formation of $C_{70}Ph_{10}$ and $C_{70}Ph_8$ from the electrophile $C_{70}Cl_{10}$, *Tetrahedron*, **52**, 5235–5246 (1996).

[94] E. I. Dorozhkin, D. V. Ignat'eva, N. B. Tamm, A. A. Goryunkov, P. A. Khavrel, I. N. Ioffe, A. A. Popov, I. V. Kuvychko, A. V. Streletskiy, V. Y. Markov, J. Spandl, S. H. Strauss and O. V. Boltalina, Synthesis, characterization, and theoretical study of stable isomers of $C_{70}(CF_3)_n$ ($n = 2, 4, 6, 8, 10$), *Chem. Eur. J.*, **12**, 3876–3889 (2006).

[95] A. A. Goryunkov, E. I. Dorozhkin, D. V. Ignat'eva, L. N. Sidorov, E. Kemnitz, G. Sheldrick and S. I. Troyanov, Crystal and molecular structures of $C_{70}(CF_3)_8$•PhMe, *Mendeleev Commun.*, **15**, 225–227 (2005).

[96] L. Gan, S. Huang, X. Zhang, A. Zhang, B. Cheng, H. Cheng, X. Li and G. Shang, Fullerenes as a *tert*-butylperoxy radical trap, metal catalyzed reaction of *tert*-butyl hydroperoxide with fullerenes, and formation of the first fullerene mixed peroxides $C_{60}(O)(OO^tBu)_4$ and $C_{70}(OO^tBu)_{10}$, *J. Am. Chem. Soc.*, **124**, 13384–13385 (2002).

[97] Z. Xiao, F. Wang, S. Huang, L. Gan, J. Zhou, G. Yuan, M. Lu and J. Pan, Regiochemistry of [70]fullerene: preparation of $C_{70}(OO^tBu)_n$ ($n = 2, 4, 6, 8, 10$) through both equatorial and cyclopentadienyl addition mode, *J. Org. Chem.*, **70**, 2060–2066 (2005).

[98] R. Taylor, $C_{70}H_{36}$ is probably an aromatic compound, *J. Chem. Soc., Perkin Trans.*, **2**, 2497–2498 (1994).

[99] J. Nossal, R. K. Saini, L. B. Alemany, M. Meier and W. E. Billups, The synthesis and characterization of fullerene hydrides, *Eur. J. Org. Chem.*, 4167–4180 (2001).

[100] T. Wågberg, M. Hedenström, A. V. Talyzin, I. Sethson, Y. O. Tsybin, J. M. Purcell, A. G. Marshall, D. Noréus and D. Johnels, Synthesis and structural characterization of $C_{70}H_{38}$, *Angew. Chem. Int. Ed.*, **47**, (2008) 2796–2799.

[101] B. W. Clare and D. L. Kepert, Structures and isomerism in $C_{70}X_{36}$, X = H, F. A comparison with $C_{60}X_{36}$, *J. Mol. Struct. (Theochem)*, **583**, 45–62 (2002).

[102] B. W. Clare and D. L. Kepert, Structures of $C_{70}X_{38}$ and $C_{70}X_{40}$, X = H, F, *J. Mol. Struct. (Theochem)*, **583**, 19–30 (2002).

[103] P. W. Fowler, J. P. B. Sandall, S. J. Austin, D. E. Manolopoulos, P. D. M. Lawrenson and J. M. Smallwood, The isomer problem for fullerene derivatives: structural proposals for $C_{70}H_{36}$, *Synth. Met.*, **77**, 97–101 (1996).

[104] P. W. Fowler, J. P. B. Sandall and R. Taylor, Structural parallels in hydrogenated and fluorinated [60]- and [70]-fullerenes, *J. Chem. Soc., Perkin Trans.*, **2**, 419–423 (1997).

[105] R. Taylor, A. K. Abdul-Sada, O. V. Boltalina and J. M. Street, Isolation and characterisation of forty-nine fluorinated derivatives of [70]fullerene, *J. Chem. Soc., Perkin Trans.*, **2**, 1013–1021 (2000).

[106] The formation of $C_{70}Br_{14}$ was reported in G. Waidmann and M. Jansen, Darstellung und Charakterisierung von $C_{70}Br_{14}$, *Z. Anorg. Allg. Chem.*, **623**, 623–626 (1997). This compound was subsequently shown to be $C_{70}Br_{10}$, the only known bromide of [70]fullerene.

[107] S. I. Troyanov and A. A. Popov, A [70]fullerene chloride, $C_{70}Cl_{16}$, obtained by the attempted bromination of C_{70} in $TiCl_4$, *Angew. Chem. Int. Ed.*, **44**, 4215–4218 (2005).

[108] P. R. Birkett, A. G. Avent, A. D. Darwish, H. W. Kroto, R. Taylor and D. R. M. Walton, Preparation and ^{13}C NMR spectroscopic characterization of $C_{60}Cl_6$, *J. Chem. Soc., Chem. Commun.*, 1230–1232 (1993).

[109] R. Taylor, Surprises, serendipity, and symmetry in fullerene chemistry, *Synlett*, **6**, 776–793 (2000).

[110] S. I. Troyanov, N. B. Shustova, I. N. Ioffe, A. P. Turnbull and E. Kemnitz, Synthesis and structural characterization of highly chlorinated C_{70}, $C_{70}Cl_{28}$, *Chem. Commun.*, 72–74 (2005).

[111] L. S. Uzkikh, E. L. Dorozhkin, O. V. Boltalina and A. L. Boltalin, A new method for the synthesis of perfluoroalkyl derivatives of fullerenes, *Dokl. Chem.*, **379**, 204–207 (2001).

[112] A. A. Goryunkov, I. V. Kuvychko, I. N. Ioffe, D. L. Dick, L. N. Sidorov, S. H. Strauss and O. V. Boltalina, Isolation of $C_{60}(CF_3)_n$ ($n = 2, 4, 6, 8, 10$) with high compositional purity, *J. Fluorine Chem.*, **124**, 61–64 (2003).

[113] A. D. Darwish, A. K. Abdul-Sada, A. G. Avent, Y. Lyakhovetsky, E. A. Shilova and R. Taylor, Unusual addition patterns in trifluoromethylation of [60]fullerene, *Org. Biomol. Chem.*, **1**, 3102–3110 (2003).

[114] A. A. Goryunkov, I. N. Ioffe, I. V. Kuvychko, T. S. Yankova, V. Y. Markov, A. A. Streletskii, D. L. Dick, L. N. Sidorov, O. V. Boltalina and S. H. Strauss, Trifluoromethylated [60]fullerenes: synthesis and characterization, *Fullerenes, Nanotubes, Carbon Nanostruct.*, **12**, 181–185 (2004).

[115] A. A. Popov, I. E. Kareev, N. B. Shustova, E. B. Stukalin, S. F. Lebedkin, K. Seppelt, S. H. Strauss, O. V. Boltalina and L. Dunsch, Electrochemical, spectroscopic, and DFT study of $C_{60}(CF_3)_n$ frontier orbitals (n = 2–18): the link between double bonds in pentagons and reduction potentials, *J. Am. Chem. Soc.*, **129**, 11551–11568 (2007).

[116] A. D. Darwish, A. K. Abdul-Sada, A. G. Avent, N. Martsinovich, J. M. Street and R. Taylor, Novel addition in trifluoromethylation of [70]fullerene, *J. Fluorine Chem.*, **125**, 1383–1391 (2004).

[117] A. A. Goryunkov, D. V. Ignat'eva, N. B. Tamm, N. N. Moiseeva, I. N. Loffe, S. M. Avdoshenko, V. Y. Markov, L. N. Sidorov, E. Kemnitz and S. I. Troyanov, Preparation, crystallographic characterization, and theoretical study of $C_{70}(CF_3)_{14}$, *Eur. J. Org. Chem.*, 2508–2512 (2006).

[118] S. M. Avdoshenko, A. A. Goryunkov, I. N. Ioffe, D. V. Ignat'eva, L. N. Sidorov, P. Pattison, E. Kemnitz and S. I. Troyanov, Preparation, crystallographic characterization and theoretical study of $C_{70}(CF_3)_{16}$ and $C_{70}(CF_3)_{18}$, *Chem. Commun.*, 2463–2465 (2006).

[119] E. I. Dorozhkin, D. V. Ignat'eva, N. B. Tamm, N. V. Vasilyuk, A. A. Goryunkov, S. M. Avdoshenko, I. N. Ioffe, L. N. Sidorov, P. Pattison, E. Kemnitz and S. I. Troyanov, Structure of 1,4,10,19,25,41-$C_{70}(CF_3)_6$, isomer with unique arrangement of addends, *J. Fluorine Chem.*, **127**, 1344–1348 (2006).

[120] N. B. Shustova, D. V. Peryshkov, O. V. Boltalina and S. H. Strauss, 1,4,10,19,25,41,55,60, 67,69-Decakis(trifluoromethyl)-1,4,10,19,25,41,55,60,67,69-decahydro(C_{70}-D_{5h})[5,6]fuller-ene, *Acta Crystallogr., Sect. E: Struct. Rep. Online*, **63**, o4073–o4173 (2007).

[121] I. E. Kareev, I. V. Kuvychko, A. A. Popov, S. E. Lebedkin, S. M. Miller, O. P. Anderson, S. H. Strauss and O. V. Boltalina, High-temperature sythesis of the surprisingly stable C_1-$C_{70}(CF_3)_{10}$ isomer with a $para^7$-*meta-para* ribbon of nine $C_6(CF_3)_2$ edge-sharing hexagons, *Angew. Chem. Int. Ed.*, **44**, 7984–7987 (2005).

[122] D. V. Ignat'eva, A. A. Goryunkov, N. B. Tamm, I. N. Ioffe, S. M. Avdoshenko, L. N. Sidorov, A. Dimitrov, E. Kemnitz and S. I. Troyanov, Preparation, crystallographic characterization and theoretical study of two isomers of $C_{70}(CF_3)_{12}$, *Chem. Commun.*, 1778–1780 (2006) .

[123] I. E. Kareev, S. F. Lebedkin, S. M. Miller, O. P. Anderson, S. H. Strauss and O. V. Boltalina, 1,4,10,19,25,32,41,49,54,60,66,69-Dodecakis(trifluormethyl)-1,4,10,19,25,32,41,49,54,60, 66,69-dodecahydro(C_{70}-$D_{5h(6)}$)-[5,6]fullerene benzene 2.5-solvate, *Acta Crystallogr., Sect. E: Struct. Rep. Online*, **62**, o617–o619 (2006).

[124] I. E. Kareev, S. F. Lebedkin, S. M. Miller, O. P. Anderson, S. H. Strauss and O. V. Boltalina, 1,4,10,14,19,25,35,41,49,60,66,69-Dodecakis(trifluormethyl)-1,4,10,14,19,25,35,41,49,60, 66,69-dodecahydro(C_{70}-$D_{5h(6)}$)[5,6]-fullerene benzene disolvate, *Acta Crystallogr., Sect. E: Struct. Rep. Online*, **62**, o620–o622 (2006).

[125] N. B. Shustova, D. V. Peryshkov, I. E. Kareev, O. V. Boltalina and S. H. Strauss, 1,4,7,11,18,21,24,31,35,39,51,58,61,64-Tetradecakis(trifluoromethyl)-1,4,7,11,18,21,24,31,35, 39,51,58,61,64-tetradecahydro(C_{70}-D_{5h})[5,6]fullerene *p*-xylene trisolvate, *Acta Crystallogr., Sect. E: Struct. Rep. Online*, **63**, o3928–o3929 (2007).

[126] The designations *ortho* (*o*), *meta* (*m*) and *para* (*p*) are used by the authors of the original papers in analogy to the well-known substitution patterns of a benzene ring and designate the addition pattern of the functionalized hexagons on the fullerene spheroid.
[127] T. Mutig, I. N. Ioffe, E. Kemnitz and S. I. Troyanov, Crystal and molecular structures of C_2-$C_{70}(CF_3)_8$•1.5 PhMe, *Mendeleev Commun.*, **18**, 73–75 (2008).
[128] K. M. Kadish, X. Gao, O. Gorelik, E. Van Caemelbecke, T. Suenobu and S. Fukuzumi, Electrogeneration and characterization of $(C_6H_5CH_2)_2C_{70}$, *J. Phys. Chem. A*, **104**, 2902–2907 (2000).
[129] N. B. Tamm and S. I. Troyanov, Synthesis and molecular structure of seven isomers of $C_{70}(C_2F_5)_{10}$, *Mendeleev Commun.*, **17**, 172–174 (2007).
[130] T. Mutig, E. Kemnitz and S. I. Troyanov, Synthesis and structural characterization of four isomers of $C_{70}(n\text{-}C_3F_7)_8$, *Eur. J. Org. Chem.*, 3256–3259 (2008).
[131] A. B. Smith III, R. M. Strongin, L. Brard, G. T. Furst, W. J. Romanow, K. G. Owens, R. J. Goldschmidt and R. C. King, Synthesis of prototypical fullerene cyclopropanes and annulenes – isomer differentiation *via* NMR and UV spectroscopy, *J. Am. Chem. Soc.*, **117**, 5492–5502 (1995).
[132] A. B. Smith III, R. M. Strongin, L. Brard, G. T. Furst, W. J. Romanow, K. G. Owens and R. J. Goldschmidt, $C_{71}H_2$ cyclopropanes and annulenes: synthesis and characterization, *J. Chem. Soc., Chem. Commun.*, 2187–2188 (1994).
[133] C. Bellavia-Lund and F. Wudl, Synthesis of [70]azafulleroids: investigations of azide addition to C_{70}, *J. Am. Chem. Soc.*, **119**, 943–946 (1997).
[134] A. Hirsch and B. Nuber, Nitrogen heterofullerenes, *Acc. Chem. Res.*, **32**, 795–804 (1999).
[135] J. C. Hummelen, C. Bellavia-Lund and F. Wudl, Heterofullerenes, *Top. Curr. Chem.*, **199**, 93–134 (1999).
[136] E. Nakamura and S. Yamago, Cycloaddition of dipolar trimethylenemethane to C_{70} promoted by a trace amount of water, *Chem. Lett.*, 395–396 (1996).
[137] K. Irie, Y. Nakamura, H. Ohigashi, H. Tokuyama, S. Yamago and E. Nakamura, Photocytotoxicity of water-soluble fullerene derivatives, *Biosci., Biotechnol., Biochem.*, **60**, 1359–1361 (1996).
[138] E. Nakamura and S. Yamago, Thermal reactions of dipolar trimethylenemethane species, *Acc. Chem. Res.*, **35**, 867–877 (2002).
[139] J. Rosenthal, D. I. Schuster, R. J. Cross and A. M. Khong, ^{3}He NMR as a sensitive probe of fullerene reactivity: [2+2] photocycloaddition of 3-methyl-2-cyclohexenone to C_{70}, *J. Org. Chem.*, **71**, 1191–1199 (2006).
[140] Z. Wang and M. S. Meier, Monoalkylation of C_{60} and C_{70} with Zn and active alkyl bromide, *J. Org. Chem.*, **68**, 3043–3048 (2003).
[141] Z. W. Wang and M. S. Meier, Methodology for the preparation of C1-monoalkylated 1,2-dihydro[C_{70}] derivatives: Formation of the 'other' regioisomer, *J. Org. Chem.*, **69**, 2178–2180 (2004).
[142] H. Irngartinger, C.-M. Köhler, G. Baum and D. Fenske, [3+2] Cycloadditons of nitrile oxides to C_{70} affording 4,5-dihydroisoxazoles and their structure determination, *Liebigs Ann.*, 1609–1614 (1996).
[143] M. Zerner, *Semiempirical Molecular Orbital Methods.* VCH Publishers, Inc., New York, (1991).
[144] N. Chronakis and A. Hirsch, Regio- and stereoselective synthesis of enantiomerically pure [60]fullerene tris-adducts with an inherently chiral *e,e,e* addition pattern, *Chem. Commun.*, 3709–3711 (2005).
[145] F. Djojo and A. Hirsch, Synthesis and chiroptical properties of enantiomerically pure bis- and trisadducts of C_{60} with an inherent chiral addition pattern, *Chem. Eur. J.*, **4**, 344–356 (1998).
[146] F. Djojo, A. Hirsch and S. Grimme, The addition patterns of C_{60} trisadducts involving the positional relationships *e* and *trans-n* ($n = 2$–4): isolation, properties, and determination of the absolute configuration of tris(malonates) and tris[bis(oxazolines)], *Eur. J. Org. Chem.*, 3027–3039 (1999).
[147] G. S. Forman, N. Tagmatarchis and H. Shinohara, Novel synthesis and characterization of five isomers of $(C_{70})_2$ fullerene dimers, *J. Am. Chem. Soc.*, **124**, 178–179 (2002).

[148] T. Kudo, Y. Akimoto, K. Shinoda, B. Jeyadevan, K. Tohji, T. Nirasawa, M. Waelchli and W. Kratschmer, Characterization and structures of dimeric C_{70} oxides, $C_{140}O$, synthesized with hydrothermal treatment, *J. Phys. Chem. B*, **106**, 4383–4389 (2002).
[149] Y. Achiba, K. Kikuchi, Y. Aihara, T. Wakabayashi, Y. Miyake and M. Kainosho, Higher fullerenes: structure and properties, in *Materials Research Society Symposium Proceedings* P. Bernier, D. S. Bethune, L. Y. Chiang, T. W. Ebbesen, R. M. Metzger and J. W. Mintmire (Eds.), Vol. 359, p. 3. Materials Research Society, Pittsburgh, PA, (1995).
[150] R. Taylor, G. J. Langley, T. J. S. Dennis, H. W. Kroto and D. R. M. Walton, A Mass spectrometric-NMR study of fullerene-78 isomers, *J. Chem. Soc., Chem. Commun.*, 1043–1046 (1992).
[151] T. Wakabayashi, K. Kikuchi, S. Suzuki, H. Shiromaru and Y. Achiba, Pressure-controlled selective isomer formation of fullerene-C_{78}, *J. Phys. Chem.*, **98**, 3090–3091 (1994).
[152] B. L. Zhang, C. Z. Wang and K. M. Ho, Search for the ground-state structure of C_{84}, *J. Chem. Phys.*, **96**, 7183–7185 (1992).
[153] K. Raghavachari, Ground state of C_{84}: Two almost isoenergetic isomers, *Chem. Phys. Lett.*, **190**, 397–400 (1992).
[154] A. A. Goryunkov, V. Y. Markov, I. N. Ioffe, R. D. Bolskar, M. D. Diener, I. V. Kuvychko, S. H. Strauss and O. V. Boltalina, $C_{74}F_{38}$: an exohedral derivative of a small-bandgap fullerene with D_3 symmetry, *Angew. Chem. Int. Ed.*, **43**, 997–1000 (2004).
[155] G. Y. Sun and M. Kertesz, Identification for IPR isomers of fullerene C_{82} by theoretical ^{13}C NMR spectra calculated by density functional theory, *J. Phys. Chem. A*, **105**, 5468–5472 (2001).
[156] Y. Miyake, T. Minami, K. Kikuchi, M. Kainosho and Y. Achiba, Trends in structure and growth of higher fullerenes isomer structure of C_{86} and C_{88}, *Mol. Cryst. Liq. Cryst.*, **340**, 553–558 (2000).
[157] N. Tagmatarchis, D. Arcon, M. Prato and H. Shinohara, Production, isolation and structural characterization of [92]fullerene isomers, *Chem. Commun.*, 2992–.2993 (2002).
[158] M. Á. Herranz, F. Diederich and L. Echegoyen, Electrochemically induced retro-cyclopropanation reactions, *Eur. J. Org. Chem.*, 2299–2316 (2004).
[159] C. Yamamoto, T. Hayashi, Y. Okamoto, S. Ohkubo and T. Kato, Direct resolution of C_{76} enantiomers by HPLC using an amylose-based chiral stationary phase, *Chem. Commun.*, 925–926 (2001).
[160] A. Herrmann and F. Diederich, Synthesis, separation and characterisation of the first pure multiple adducts of C_{2v}- and D_3-C_{78}, *J. Chem. Soc., Perkin Trans.*, **2**, 1679–1684 (1997).
[161] F. Furche and R. Ahlrichs, Absolute configuration of D_2-symmetric fullerene C_{84}, *J. Am. Chem. Soc.*, **124**, 3804–3805 (2002).
[162] C. Yeretzian, W. J. B. K. Holczer, T. Su, S. Nguyen, R. B. Kaner and R. L. Whetten, Partial separation of fullerenes by gradiant sublimation, *J. Phys. Chem.*, **97**, 10097–10101 (1993).
[163] M. D. Diener and J. M. Alford, Isolation and properties of small-bandgap fullerenes, *Nature (London, U. K.)*, **393**, 668–671 (1998).
[164] A. K. Abdul-Sada, T. V. Avakyan, O. V. Boltalina, V. Y. Markov, J. M. Street and R. Taylor, Isolation and characterisation of fluorinated derivatives of [76]- and [78]fullerenes, *J. Chem. Soc., Perkin Trans.*, **2**, 2659–2666 (1999).
[165] K. S. Simeonov, K. Y. Amsharov and M. Jansen, $C_{80}Cl_{12}$: a chlorine derivative of the chiral D_2-C_{80} isomer – empirical rationale of halogen-atom addition pattern, *Chem. Eur. J.*, **15**, 1812–1815 (2009).

6

Application of Fullerenes to Nanodevices

Yutaka Matsuo and Eiichi Nakamura

Nakamura Functional Carbon Cluster Project, ERATO, Japan Science and Technology Agency and Department of Chemistry, University of Tokyo, Japan

6.1 Introduction

Fullerenes, nanometer-sized spherical carbon cluster molecules, are produced industrially by means of incomplete combustion of toluene. Thus, fullerenes have been extensively investigated as general industrial materials in recent years. Fullerenes have excellent electron-accepting properties, and they form stable anions upon reduction. Furthermore, fullerenes exhibit unique characteristics in response to light; specifically, they undergo photoexcitation, thereby generating a long-lived triplet excited state with almost 100% quantum yield. Because fullerenes have rich functions in terms of absorbing electrons and light, they have many useful applications with regard to photoelectronic functional materials [1].

Combining fullerenes with organic or inorganic functional units yields highly functionalized fullerene-based materials. For instance, combination of fullerenes with transition metal complexes can yield compounds with electrochemical and photophysical activity. Transition metal complexes have d-orbital electrons, and some late transition metal complexes, such as ferrocenes, have excellent electron-donating capability. Incorporating an electron-donating ferrocene with an electron-accepting fullerene produces a donor/acceptor-type metal-fullerene complex, which, under light irradiation, can undergo a photoinduced charge transfer to form a charge-separated state. By assembling such

Chemistry of Nanocarbons Edited by Takeshi Akasaka, Fred Wudl and Shigeru Nagase

photoexcited molecules on an electrode surface, a photocurrent can be generated. In this chapter, we describe synthesis, self-assembling capability, and photocurrent generation of metal-fullerene complexes [2].

6.2 Synthesis of Transition Metal Fullerene Complexes

Metal-fullerene complexes are a new class of organometallic compounds that have electrochemically active properties and sterically bulky ligands [3]. Most of these compounds are η^2-fullerene complexes that are based on π-coordination between an electron-rich transition metal and a fullerene double bond. This type of coordination bond is often kinetically and thermodynamically fragile. We have synthesized stable η^5-fullerene metal complexes by utilizing the cyclopentadiene part of penta(organo)[60]fullerenes, $C_{60}R_5H$ (R = alkyl, alkenyl, aryl groups, etc.; **1**: R = Me) (Figures 6.1 and 6.2) [4].

Considering their peculiar structure, the same method can not be applied in many cases for the synthesis of ordinary cyclopentadienyl metal complexes using C_5H_5 (= Cp) or a pentamethyl-cyclopentadienyl ligand, C_5Me_5. In such situations, organometallic complexes bearing labile ligands, such as $Mo(CO)_3(EtCN)_3$, $[RuCl_2(CO)_3]_2$, and $[RuCp(CH_3CN)_3][PF_6]$, are employed for the synthesis of metal fullerene complexes. For example, the reaction of $[K(thf)_n][C_{60}Me_5]$ (**2**) [5], which is derived from $C_{60}Me_5H$ and KH, with $[RuCl_2(CO)_3]_2$ in THF has produced a ruthenium penta(organo)[60]fullerene complex, $Ru(\eta^5\text{-}C_{60}Me_5)Cl(CO)_2$ (**3**) in 70–80% yield [6]. Similarly, the reaction of **2** with $Mo(CO)_3(EtCN)_3$, followed by treatment with Diazald has yielded an air-stable molybdenum nitrosyl complex, $Mo(\eta^5\text{-}C_{60}Me_5)(NO)(CO)_2$ (**4**). Synthesis of an iridium-$C_{60}Me_5$ complex **5** has been achieved by using a rather uncommon iridium dicarbonyl chloro dimer

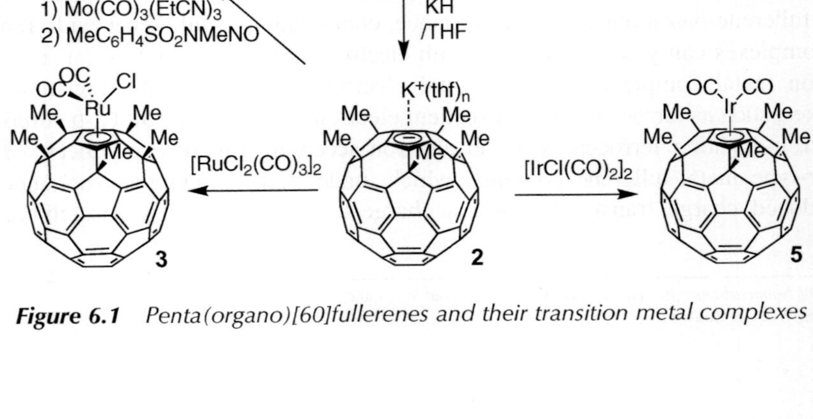

Figure 6.1 *Penta(organo)[60]fullerenes and their transition metal complexes*

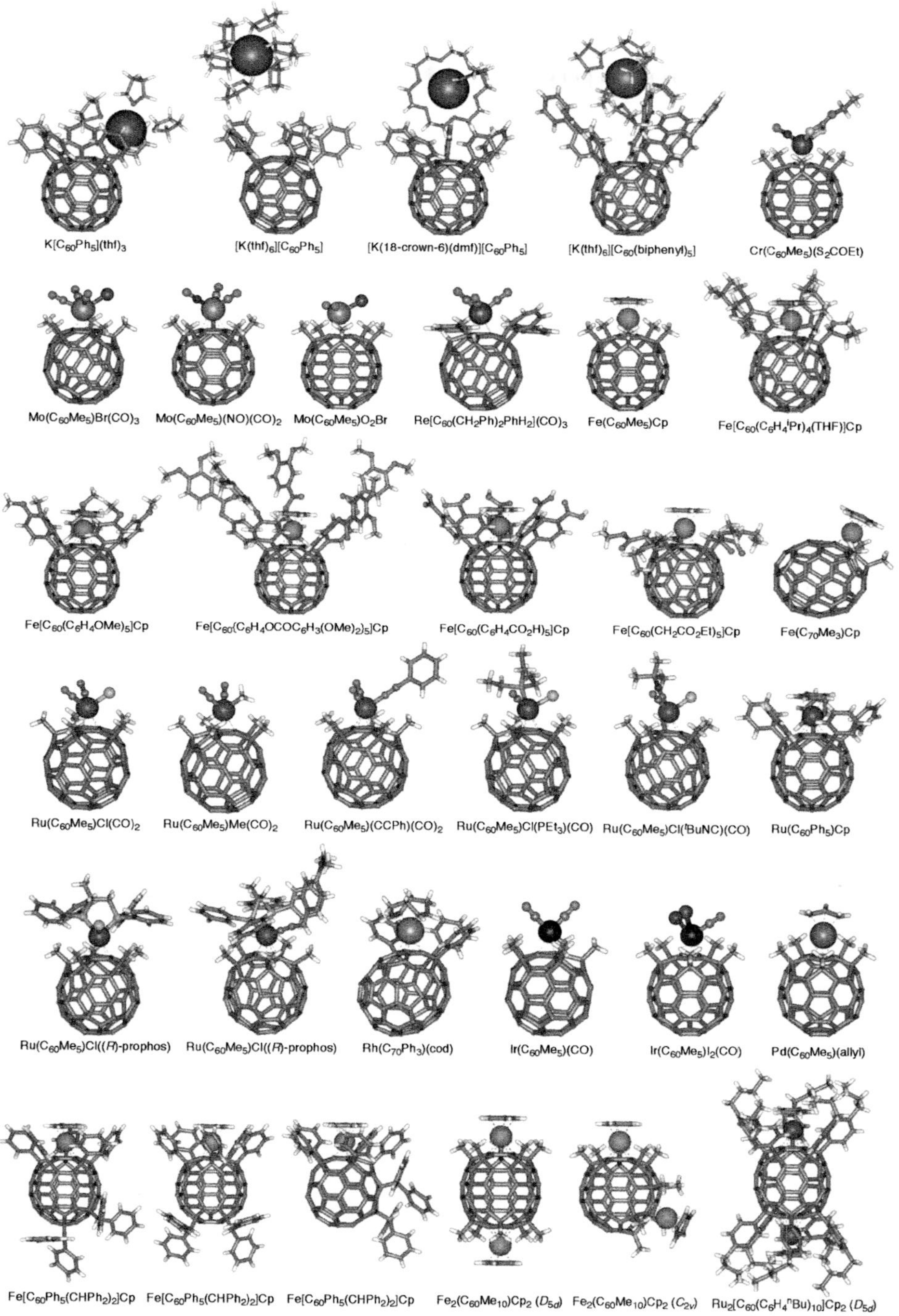

Figure 6.2 *X-ray crystallographic structures of metal-penta(organo)[60]fullerene complexes*

$[IrCl(CO)_2]_2$, which can be prepared by treatment of $[IrCl(coe)_2]_2$ (coe = cyclooctene) with carbon monoxide gas (1 atm) in acetonitrile.

A sandwich compound, buckyferrocene [7] (Buckminster fullerene + ferrocene), Fe($C_{60}R_5$)Cp (R = alkyl and aryl groups; **6**: R = Me) is a representative transition metal η^5-penta(organo)[60]fullerene complex, which exhibits a one reversible one-electron oxidation process at the ferrocene moiety and two reversible one-electron reduction processes at the fullerene moiety.

Although the mechanism for formation of buckyferrocenes is unclear, it is very convenient for obtaining the desired products. Buckyferrocenes can be obtained in 50–70% yield simply by heating a solution of a protio compound, $C_{60}R_5H$, and an iron(I) dimer complex, $[FeCp(CO)_2]_2$, in benzonitrile at 170–180 °C. We assume that electron transfer from the dimer complex to the fullerene part leads to the elimination of the hydrogen atom from the cyclopentadiene moiety to form an iron-fullerene bond; alternatively, decomposition of the dimer complex at high temperature could generate active species such as reducing iron(0) that remove the hydrogen atom of the cyclopentadiene moiety. The notable compound here is a hydrogen molecule-encapsulated buckyferrocene, Fe[H_2@$C_{60}Ph_5$]Cp [8]. The hydrogen molecule inside the structure influences the chemical and physical properties of the whole compound. Further research may reveal some interesting functions of this compound. A ruthenium congener, bucky ruthenocenes [9] Ru($C_{60}R_5$)Cp (R = Me, Ph, etc.), has also been synthesized in moderate to high yields by the reaction of $[K(thf)_n][C_{60}Me_5]$ with $[RuCp(CH_3CN)_3][PF_6]$ in THF at room temperature. Bucky metallocenes are very stable, even at 350 °C, while thermal treatment of these compounds in nitrogen at 500–700 °C yields metal nanoparticles embedded in carbonaceous substrates.

6.3 Organometallic Chemistry of Metal Fullerene Complexes

The stability of metal η^5-fullerene complexes facilitates the exploitation of the synthetic organometallic chemistry of these compounds. η^5-$C_{60}R_5$ complexes of group 6–10 transition metals – Cr [10], Mo [10], W [10], Re [11], Fe [12], Ru [6, 9], Co [12], Ir [13], Rh [14], Ni [15], Pd [15], and Pt [15] – have been prepared so far. In particular, half-sandwich-type complexes that have the halide or carbon monoxide ligands on the metal center can be derivatized widely into a variety of organometallic fullerene compounds. A ruthenium complex, Ru($C_{60}Me_5$)Cl(*R*)-prophos) (**7**), which has its central chirality on the ruthenium metal, has been obtained in a diastereoselective manner via reaction of the ruthenium chloro dicarbonyl complex, Ru(η^5-$C_{60}Me_5$)Cl$(CO)_2$, and the chiral diphosphine ligand, (*R*)-1,2-bis-diphenylphosphinopropane [(*R*)-prophos ligand] (Figure 6.3) [16]. The stereochemistry of the central metal has been determined by X-ray crystal structure analysis. It is believed that diastereoselectivity in the formation of the chiral ruthenium complex arises due to the steric bulkiness of the penta(organo)[60]fullerene ligands, which prevents the formation of a thermodynamically unstable isomer.

Various cationic complexes, $[Ru(C_{60}Me_5)((R)\text{-prophos})L]^+[SbF_6]^-$ (**8**: L = ligand), have also been obtained by the abstraction of the chloride ligands with a silver salt, $AgSbF_6$, in the presence of various coordinating ligands, such as acetonitrile, acetone, methacrolein, CO, and isonitriles. During this reaction, the stereochemistry on the metal center has been

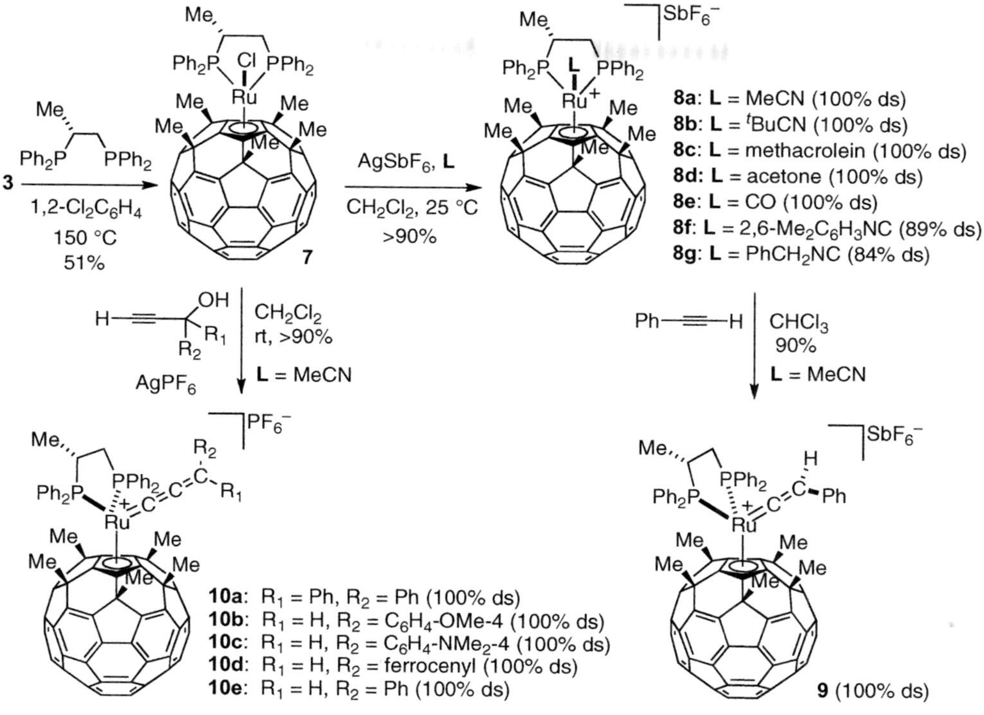

Figure 6.3 *Syntheses of chiral, cationic, and carbene ruthenium complexes*

maintained. Cationic complexes can be derivatized to vinylidene complexes, $[Ru(C_{60}Me_5)({=}CC{=}HPh)(R)\text{-prophos})]^+[SbF_6]^-$ (**9**), by reaction with terminal acetylenes.

In addition, allenylidene complexes, $[Ru(C_{60}Me_5)({=}C{=}C{=}CHR_2)(R)\text{-prophos})]^+[PF_6]^-$ (**10**), have been obtained by the reaction of the complex $Ru(C_{60}Me_5)Cl(R)$-prophos) with propargylalcohols and $AgPF_6$ [17]. Cationic complexes, as well as chiral carbene complexes, are of interest in transition metal-catalyzed organic synthesis.

6.4 Synthesis of Multimetal Fullerene Complexes

Dimetal fullerene complexes [18] have been obtained by using deca(organo)[60]fullerenes, which have two cyclopentadiene moieties in the Arctic and the Antarctic regions of the fullerene and a belt-shaped cyclic π-conjugated system called [10] cyclophenacene in the equatorial region (Figure 6.4) [19]. The reaction of the deca-(organo)[60]fullerene $C_{60}Me_5Ph_5H_2$ (**11**) with the iron complex $[FeCp(CO)_2]_2$ in benzonitrile at 180 °C produces the diiron complex $Fe_2(C_{60}Me_5Ph_5)Cp_2$ (**12**), known as double-decker buckyferrocene. Alternatively, the buckyferrocene $Fe(C_{60}Me_5)Cp$ is subjected to pentamethylation to obtain a decamethyl compound, $Fe(C_{60}Me_{10})CpH$, followed by metallation to produce the decamethyl[60]fullerene diiron complexes

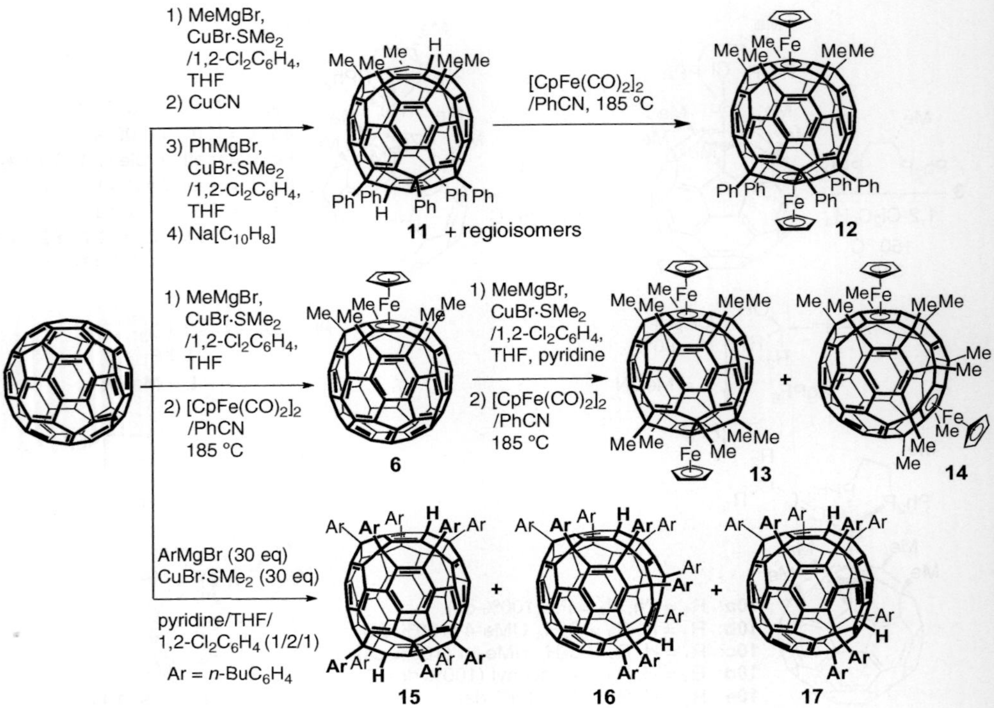

Figure 6.4 *Syntheses of deca(organo)[60]fullerenes and double-decker complexes*

$Fe_2(C_{60}Me_{10})Cp_2$ (**13** and **14**), which are D_{5d} [18] and C_{2v} [20] isomers. One-step synthesis of deca(organo)[60]fullerene is possible when one employs a pyridine-modified organocopper reagent with a mixed solvent composed of 1,2-dichlorobenzene, THF, and pyridine (volume ratio: 1:2:1), thereby yielding a cyclophenacene derivative **15** and its regioisomer **16**. When a large excess of pyridine is used (1,2-dichlorobenzene:THF:pyridine = 13:27:60), the reaction produces a mixture of **15** and the octa-adduct **17**. Decaaryl[60]fullerene **13** can be derivatized to a double-decker bis (ruthenocene) compound, $Ru_2(C_{60}Ar_{10})Cp_2$ (Ar = C_6H_4-nBu, etc.), as a nanometer scale motif.

Compound **13** has high symmetry (same D_{5d} symmetry as an armchair-type carbon nanotube), and the upper and lower iron atoms are in an equivalent environment. In this complex, electronic interaction of two metal atoms is observed through the fullerene π-system, i.e. electrochemical investigation has revealed separated the first and second oxidation processes for two symmetric ferrocene parts. The 110-mV separation of the two oxidation potentials (ΔE) is comparable to that of phenylene-linked diferrocenes, Fc-C_6H_4-Fc (ΔE = 131, 90, and 104 mV for *o*-, *m*-, and *p*-C_6H_4, respectively; Fc = ferrocenyl), but larger than that of a biphenylene-linked diferrocene, Fc-C_6H_4-C_6H_4-Fc (ΔE = 70 mV). This finding indicates that double-decker buckyferrocenes are promising candidates as molecular functional materials.

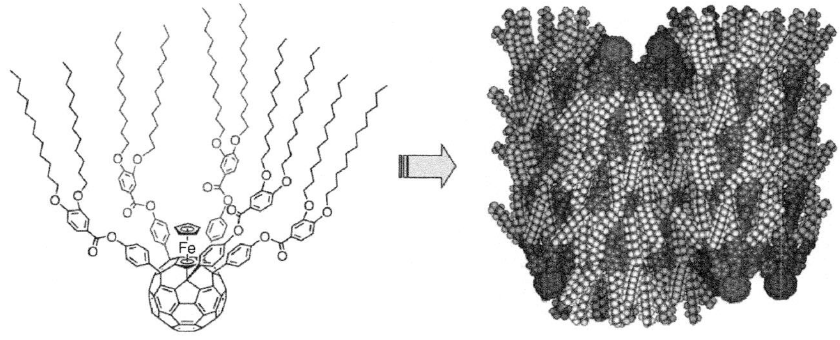

Figure 6.5 *Buckyferrocene-based metallomesogens and formation of columnar liquid crystals*

6.5 Supramolecular Structures of Penta(organo)[60]fullerene Derivatives

For practical use of functional materials in bulk form, such as in films, crystals, and liquid crystals, the construction of certain well-ordered molecular structures is essential. We found that badminton shuttlecock-shaped fullerene molecules with five organic feather-like addends can be stacked upon each other in a head-to-tail fashion to afford 1-dimensional columnar supramolecular structures, thereby forming liquid crystalline materials [21].

Liquid crystals that contain metal atoms are called metallomesogens. They are attracting considerable attention because they can enable the creation of liquid crystals that respond to stimuli via oxidization or reduction. For this purpose, we synthesized shuttlecock-shaped buckyferrocene **18** by attaching five 'feathers' onto the buckyferrocene structure to obtain liquid crystalline redox-active fullerene derivatives (Figure 6.5). These molecules form columnar liquid crystals in a temperature range of 10–100 °C, and they exhibit one reversible oxidation process and three reversible reduction processes. Thus, precise control of the donor/acceptor location can be achieved to build an 'electron highway' that effectively passes carriers.

Pentaaryl[60]fullerene anions $[K(thf)_n][C_{60}Ph_5]$ are soluble in water [22] and form spherical bilayer vesicles [23]. The fullerene bilayer is unusually watertight – over a thousand times more watertight than lipid vesicles. Water permeation is controlled by activation entropy [24]. X-ray crystallographic studies have been performed previously to investigate the potassium complexes of pentaaryl[60]fullerene anions, e.g. $K(C_{60}Ph_5)(thf)_3$, $[K(thf)_6][C_{60}Ph_5]$, and $[K(18\text{-crown-}6)(DMF)][C_{60}Ph_5]$ [5]. Potassium ions were found to be solvated by polar ligands such as THF.

6.6 Reduction of Penta(organo)[60]fullerenes to Generate Polyanions

An important characteristic of penta(organo)[60]fullerenes is their electron-accepting property. This property is essential for n-type materials, which are required in p-n junction-type organic thin-film devices, such as organic photovoltaic cells and organic

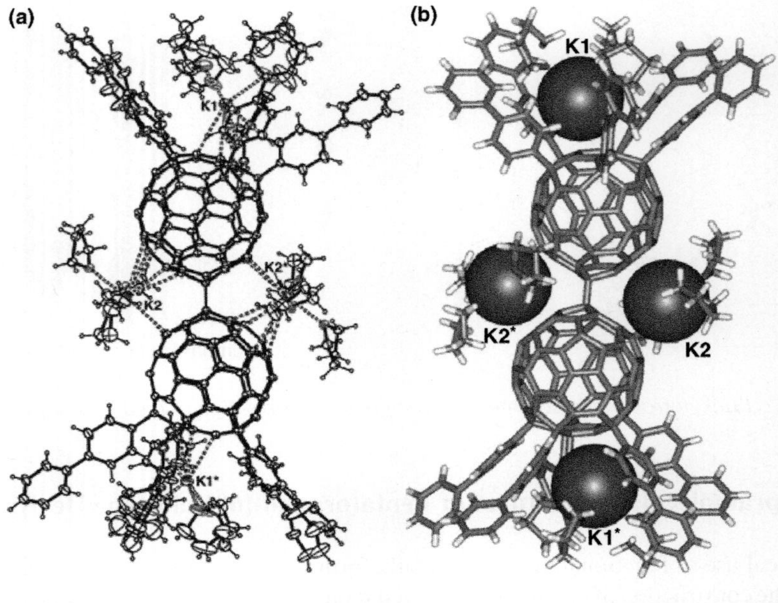

Figure 6.6 X-ray crystallographic structure of the dimeric compound of **19**. (a) ORTEP drawing. (b) Ball and stick model

light-emitting diode devices. Therefore, we have prepared and structurally characterized penta(organo)[60]fullerene-polyanions that are generated by the chemical reduction of penta(organo)[60]fullerenes [25].

Treatment of the potassium salt $[K(thf)_3][C_{60}(biphenyl)_5]$ with potassium/mercury amalgam in THF has produced a radical dianion $[K(thf)_n]_2[C_{60}(biphenyl)_5]$ (**19**). This compound forms a dimeric structure in the crystalline state, but equilibrium exists between the monomer and the dimer in solution. The single-bonded dimer $[K(thf)_n]_4[(biphenyl)_5C_{60}\text{-}C_{60}(biphenyl)_5]$ has been characterized by single crystal X-ray analysis (Figure 6.6), while its UV-vis spectrum in solution displays a broad absorption band in the near infrared region (ca. 1100 nm), owing to the monomeric open-shell compound. A trianion $[K(thf)_n]_3[C_{60}(biphenyl)_5]$ has also been obtained by the reduction of $[K(thf)_n]$ $[C_{60}(biphenyl)_5]$ with potassium metal. This compound has been utilized for synthesis of hepta(organo)[60]fullerenes.

6.7 Photoinduced Charge Separation

The complex of fullerene and ferrocene $Fe(C_{60}R_5)Cp$ (**6**) undergoes a photoinduced charge separation to generate charge-separated states under light irradiation, because an electron donor and an electron acceptor exist in the molecule. Because the ferrocene and the fullerene are directly connected, we observe very fast formation and deactivation of the charge-separated state [26]. More specifically, after irradiation by light, the charge-separated state

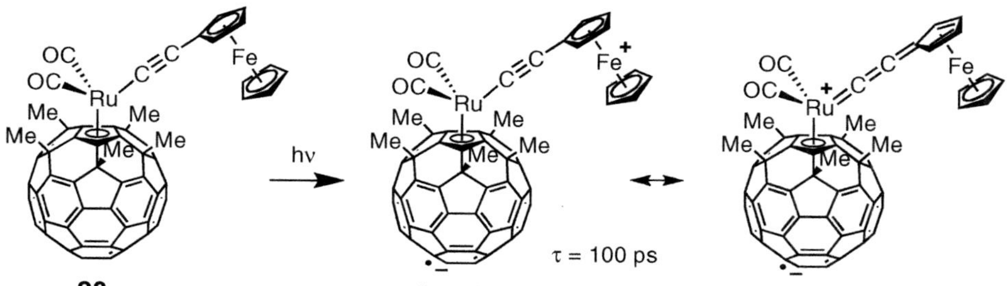

Figure 6.7 *Proposed charge separation state of* ***20***

generates in 0.8 ps via the singlet excitation state, and, thereafter, back electron transfer occurs in 35 ps. With regard to application of the fullerene complex to optoelectronic materials, the short lifetime of the charge-separated state is a disadvantage. Conversely, however, there is also a high possibility of taking advantage of the fact that the charge-separated state can be generated in a short period of time.

In the case of the dinuclear metal complex **13**, the lifetime of the charge-separated state is even shorter because of electronic interaction between the two metals [20]. In addition, the fullerene ruthenium complex, $Ru(C_{60}Me_5)(CCFc)(CO)_2$ (Fc = ferrocenyl) (**20**), which bears the ferrocene on the outside, shows a longer charge separation lifetime (100 ps) [27]. We assume that resonance stabilization between alkynyl and allenylidene forms contributes to the formation of a charge-separated state with a longer lifetime (Figure 6.7). These experimental facts indicate that photoelectrochemical properties can be controlled by changing the molecular design around the metal atom.

6.8 Photocurrent-Generating Organic and Organometallic Fullerene Derivatives

6.8.1 Attaching Legs to Fullerene Metal Complexes

To take out electric charge externally under the photoinduced charge separation state in a solvent, it is necessary to fix the molecule to the electrode. For this purpose, functional groups ('legs') that can be connected to the electrode surface must be introduced into the metal fullerene complex. Thus, five carboxylic acid groups [28] have been attached to the five phenyl groups to obtain nanometer-sized pentapod molecules that can stand upright on the electrode surface [29]. The molecules were synthesized by the following steps: First, a functionalized aryl Grignard reagent that has a carboxylate ester moiety was prepared by utilizing the iodine-magnesium exchange reaction at a low temperature. Next, the transmetallation of magnesium metal to copper was carried out, followed by the penta-addition addition reaction to [60]fullerene to obtain fullerene penta ester $C_{60}(C_6H_4C_6H_4CO_2Et)_5H$ (**21**) (Figure 6.8) [28]. Thereafter, the methyl group [30] or the ferrocene moiety was introduced, followed by hydrolysis of the ester groups to obtain pentapod molecules, a methylated compound $C_{60}(C_6H_4C_6H_4CO_2H)_5Me$ (**22**), and a buckyferrocene pentacarboxylic acid $Fe[C_{60}(C_6H_4C_6H_4CO_2H)_5]Cp$ (**23**). Furthermore,

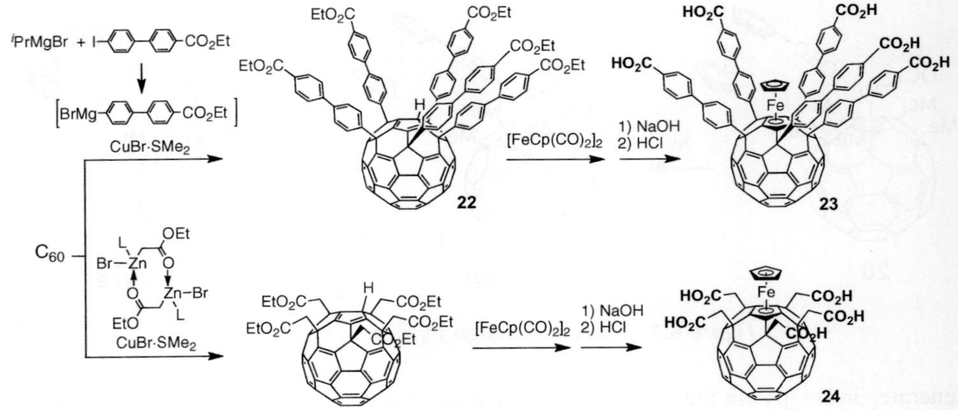

Figure 6.8 *Synthesis of buckyferrocene pentacarboxylic acids **23** and **24***

by using the Reformatsky reagent, a known functionalized organozinc reagent, pentacarboxylic acid derivatives $Fe[C_{60}(CH_2CO_2H)_5]Cp$ (**24**) and $C_{60}(CH_2CO_2H)_5Me$ were synthesized (Figure 6.8) [31].

Penta acid molecules with phenylene or biphenylene linkers are pentapod molecules that resemble the lunar lander. However, it appears that penta acid molecules with methylene linkers cannot stand upright because of their short legs. These penta acid molecules have a simpler and more rigid structure than those of the conventional self-assembling fullerene derivatives. This fact is advantageous for controlling molecular orientation. In addition, because of their unique shape, organic moieties or other transition metal complex moieties can be introduced in the pockets located among the five legs without changing the shape of the whole molecule.

6.8.2 Formation of Self-Assembled Monomolecular Films

Pentapod molecules are immobilized on the indium-tin oxide electrode (ITO electrode), thereby providing self-assembled monolayers (hereafter SAMs) of the molecules (Figure 6.9a). Preparation of SAMs is performed via the following steps: First, the ITO electrode is dipped in a 0.1 M THF solution of pentacarboxylic acids; next, the electrode is rinsed with a solvent to remove excess molecules; and, finally, the modified electrode is dried in argon.

Identification of SAMs of pentapod molecules is performed with electrochemical measurements using the modified ITO as a working electrode. In the scan at the oxidizing side, buckyferrocene pentacarboxylic acid **23** shows a reversible one-electron oxidation process, owing to the ferrocene moiety. The surface coverage (density) of the molecules on the electrode is calculated on the basis of the observed current and electrode area. In the case of buckyferrocene pentacarboxylic acid, the measurement revealed that three days are required to obtain saturation of coverage and that the surface coverage is approximately 10^{-10} mol/cm^2 (Figure 6.9b). This value agrees with the value obtained from a model where the pentapod molecules are filled almost evenly. Furthermore, formation of the monomolecular film was supported from the fact that the peak currents in the oxidization process

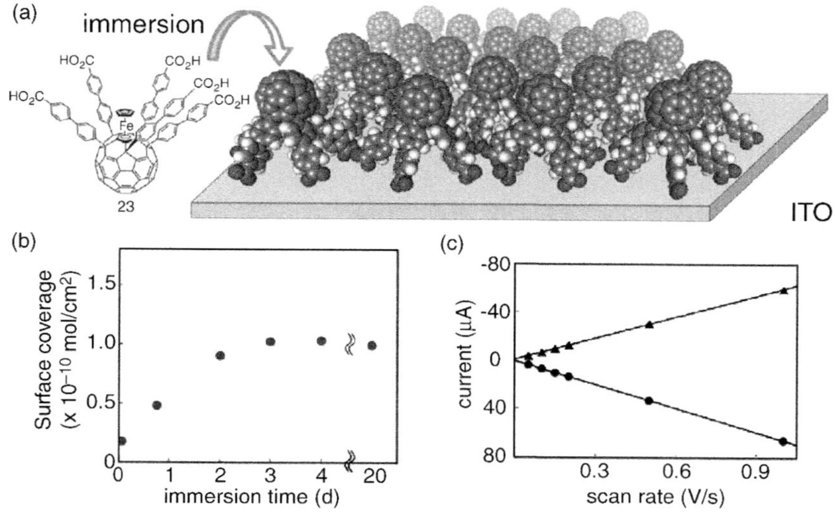

Figure 6.9 *Preparation of SAMs of **23** on the ITO electrode. (a) A model for the formation of SAMs by immersion of the ITO electrode into a solution of **23**. (b) A surface coverage vs. immersion time profile determined from oxidation of the ferrocene moiety of **23**. (c) A current vs. scan rate profile determined with anodic scan (oxidation of ferrocene) of **23***

from Fe(II) to Fe(III) and in the back reduction process from Fe(III) to Fe(II) are linearly proportional to the scan rate (Figure 6.9c), and that the potentials for both oxidation and reduction processes are almost identical.

6.8.3 Photoelectric Current Generation Function of Lunar Lander-Type Molecules

Photocurrent generation properties of SAMs of the pentapod molecules are characterized by using photoelectrochemical cells equipped with working, counter (platinum wire), and reference (Ag/Ag^+) electrodes. Measurements are performed using an aqueous medium that contains sodium sulfate as a supporting electrolyte and sacrificial reagents such as ascorbic acid (AsA) and methyl viologen (MV) for anodic current (reductive current from the molecules to the electrode) and cathodic current (oxidative current from the electrode to the molecules), respectively (Figure 6.10a–c).

The methyl-capped penta biphenylene carboxylic acid molecule $C_{60}(C_6H_4C_6H_4$-$CO_2H)_5Me$ (**22**) on ITO generated an anodic photocurrent to the ITO (Figure 6.10d) with a 7.2% quantum yield upon use of AsA as a sacrificial electron donor without bias voltage and with a 14.4% quantum yield upon application of a 0.1-V bias. It generated no cathodic current when O_2/MV was used as an electron acceptor. In contrast, the ferrocene molecule $Fe[C_{60}(C_6H_4C_6H_4CO_2H)_5]Cp$ (**23**) generated a cathodic current (Figure 6.10e) upon use of O_2/MV with up to 6.3% quantum yield but no anodic current upon use of AsA. The methylene ferrocene molecule $Fe[C_{60}(CH_2CO_2H)_5]Cp$ (**24**) generated only an anodic current (Figure 6.10f).

The mechanistic aspect of anodic and cathodic photocurrent generation have been discussed with regard to differences in excited states and molecular orientation.

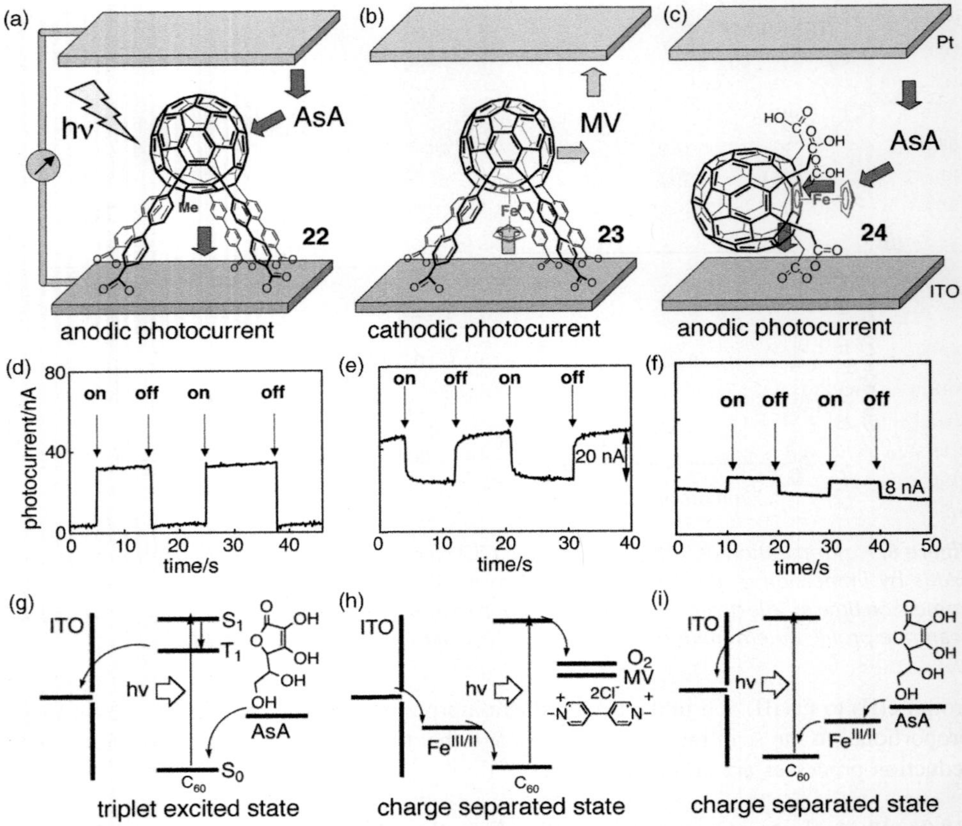

__Figure 6.10__ Photocurrent generation properties of fullerene pentacarboxylic acids __22–24__. (a–c) Molecular structures and orientation with direction of photocurrent. (d–f) On-off profiles of photocurrent generation (positive current: anodic; negative current: cathodic). (g–i) Difference in excited state and mechanism for photocurrent generation

Figure 6.10g illustrates a mechanism for anodic photocurrent generation by the methylated compound **22**, for which only a triplet photoexcited state with a microsecond-order lifetime is available [26]. Thus, a triplet photoexcited state of the methylated compound accepts an electron in its lower singly occupied orbital from the HOMO of the ambient AsA and donates an electron to ITO. Figure 6.10h illustrates a mechanism for cathodic current generation by buckyferrocene **23** in the presence of O_2/MV. Photoexcitation of the fullerene moiety generates a charge- separated state with a picosecond lifetime by rapid electron transfer from the ferrocene group [26]. Because the pentapod structure forces the cationic ferrocenium group to be sandwiched between the fullerene and the ITO surface, ITO supplies an electron to the ferrocenium group, and the overall result is the generation of a cathodic photocurrent. The anodic photocurrent observed for methylene ferrocene **24** has been ascribed to the inability of the molecule to stand upright on the ITO surface

(Figure 6.10i depicts a possible mechanism). Here, the ferrocene group is not sandwiched between fullerene and ITO but, rather, is exposed directly to the electrolyte. Therefore, Ferrocene accepts an electron from AsA rather than from ITO and, hence, generates only the anodic current. Thus, the direction of photocurrent has successfully been switched by changing the components (methylated compounds or iron complexes) and orientation (standing upright or lying down) of the molecules. Rigid and compact pentapod structures enable these modifications to control device functions.

6.9 Conclusion

The nanometer-sized, rigid molecular structure of [60]fullerene is useful for the creation of nanodevices such as double-decker metal complexes, liquid crystalline shuttlecock-shaped molecules, and photocurrent-generating pentapod molecules. Synthetic chemistry plays an important role in producing highly functionalized fullerene derivatives that have charge-separation and self-assembly capabilities. Given their varied photoelectrochemical functions and well-organized structure that can be controlled by molecular design, current and future fullerene derivatives will be of immense interest in the field of photoelectric conversion, particularly with regard to organic photovoltaics.

References

[1] D. M. Guldi and N. Martin (Eds.), *Fullerenes: From Synthesis to Optoelectronic Properties*, Kluwer Academic Publishers, Dordrecht, 2002.

[2] Y. Matsuo and E. Nakamura, Selective multi-addition of organocopper reagents to fullerenes, *Chem. Rev.*, **108**, 3016–3028 (2008).

[3] (a) A. H. H. Stephens and M. L. H. Green, Organometallic complexes of fullerenes, *Adv. Inorg. Chem.*, **44**, 1–43 (1997). (b) A. L. Balch and M. M. Olmstead, Reactions of transition metal complexes with fullerenes (C_{60}, C_{70}, etc.) and related materials, *Chem. Rev.*, **98**, 2123–2165 (1998).

[4] (a) M. Sawamura, H. Iikura, and E. Nakamura, The first pentahapto fullerene metal complexes, *J. Am. Chem. Soc.*, **118**, 12850–12851 (1996). (b) H. Iikura, S. Mori, M. Sawamura, and E. Nakamura, Endohedral homo-conjugation in cyclopentadiene embedded in C_{60}, *J. Org. Chem.*, **62**, 7912–7913 (1997). (c) Y. Matsuo, A. Muramatsu, K. Tahara, M. Koide, and E. Nakamura, Synthesis of 6,9,12,15,18-pentamethyl-1,6,9,12,15,18-hexahydro(C_{60}-I_h)[5,6]fullerene, *Org. Synth.*, **83**, 80–87 (2006).

[5] Y. Matsuo, K. Tahara, and E. Nakamura, X-ray crystallographic characterization of potassium pentaphenyl[60]fullerene, *Chem. Lett.*, **34**, 1078–1079 (2005).

[6] Y. Matsuo and E. Nakamura, Ruthenium(II) complexes of pentamethylated [60]fullerene. alkyl, alkynyl, chloro, isocyanide, phosphine complexes, *Organometallics*, **22**, 2554–2563 (2003).

[7] (a) M. Sawamura, Y. Kuninobu, M. Toganoh, Y. Matsuo, M. Yamanaka, and E. Nakamura, Hybrid of ferrocene and fullerene, *J. Am. Chem. Soc.*, **124**, 9354–9355 (2002). (b) M. Toganoh, Y. Matsuo, and E. Nakamura, Synthesis of ferrocene/hydrofullerene hybrid and functionalized bucky ferrocenes, *J. Am. Chem. Soc.*, **125**, 13974–13975 (2003).

[8] Y. Matsuo, H. Isobe, T. Tanaka, Y. Murata, M. Murata, K. Komatsu, and E. Nakamura, Organic and organometallic derivatives of dihydrogen-encapsulated [60]fullerene, *J. Am. Chem. Soc.*, **127**, 17148–17149 (2005).

[9] Y. Matsuo, Y. Kuninobu, S. Ito, and E. Nakamura, Synthesis and reactivity of bucky ruthenocene $Ru(\eta^5\text{-}C_{60}Me_5)(\eta^5\text{-}C_5H_5)$, *Chem. Lett.*, **33**, 68–69 (2004).

[10] Y. Matsuo, A. Iwashita, and E. Nakamura, Group 6 metal complexes of the η^5-pentamethyl[60] fullerene, *Organometallics*, **27**, 4611–4617 (2008).

[11] M. Toganoh, Y. Matsuo, and E. Nakamura, Rhenium-templated regioselective polyhydrogenation of [60]fullerene and derivatives. rhenium η^5-complexes of hydrofullerenes, *Angew. Chem. Int. Ed.*, **42**, 3530–3532 (2003).

[12] Y. Matsuo, Y. Kuninobu, A. Muramatsu, M. Sawamura, and E. Nakamura, Synthesis of metal fullerene complexes by the use of fullerene halides, *Organometallics*, **27**, 3403–3409 (2008).

[13] Y. Matsuo, A. Iwashita, and E. Nakamura, Synthesis and derivatization of Ir(I)- and Ir(III)-pentamethyl[60]fullerene complexes, *Organometallics*, **24**, 89–95 (2005).

[14] (a) M. Sawamura, Y. Kuninobu, and E. Nakamura, Half-sandwich metallocene embedded in spherically extended π-conjugate system. synthesis, structure, and electrochemistry of $Rh(\eta^5\text{-}C_{60}Me_5)(CO)_2$, *J. Am. Chem. Soc.*, **122**, 12407–12408 (2000). (b) Y. Matsuo and E. Nakamura, Synthesis of trialkyl[60]fullerene $C_{60}(CH_2SiMe_3)_3H$ and its potassium and rhodium(I) complexes, *Inorg. Chim. Acta*, **359**, 1979–1982 (2006).

[15] Y. Kuninobu, Y. Matsuo, M. Toganoh, M. Sawamura, and E. Nakamura, Nickel, palladium and platinum complexes of η^5-cyclopentadienide $C_{60}R_5$ ligands. kinetic and thermodynamic stabilization effects of $C_{60}Ph_5$ ligand, *Organometallics*, **23**, 3259–3266 (2004).

[16] Y. Matsuo, Y. Mitani, Y.-W. Zhong, and E. Nakamura, Remote chirality transfer within coordination sphere by the use of a ligand possessing a concave cavity, *Organometallics*, **25**, 2826–2832 (2006).

[17] Y.-W. Zhong, Y. Matsuo, and E. Nakamura, Chiral ruthenium allenylidene complexes bearing a fullerene cyclopentadienyl ligand: synthesis, characterization, and remote chirality transfer, *Chem. Asian J.*, **2**, 358–366 (2007).

[18] (a) Y. Matsuo, K. Tahara, and E. Nakamura, Synthesis and electrochemistry of double-decker buckyferrocenes, *J. Am. Chem. Soc.*, **128**, 7154–7155 (2006). (b) Y. Matsuo, K. Tahara, T. Fujita, and E. Nakamura, Di- and trinuclear [70]fullerene complexes: syntheses and metal–metal electronic interactions, *Angew. Chem. Int. Ed.*, **48**, 6239–6241 (2009).

[19] (a) E. Nakamura, K. Tahara, Y. Matsuo, and M. Sawamura, Synthesis, structure and aromaticity of a hoop-shaped cyclic benzenoid [10]cyclophenacene, *J. Am. Chem. Soc.*, **125**, 2834–2835 (2003). (b) Y. Matsuo, K. Tahara, M. Sawamura, and E. Nakamura, Creation of hoop- and bowl-shaped benzenoid systems by selective detraction of [60]fullerene conjugation. [10]cyclophenacene and fused corannulene derivatives, *J. Am. Chem. Soc.*, **126**, 8725–8734 (2004). (c) Y. Matsuo and E. Nakamura, Cyclophenacene cut out of fullerene in *Functional Organic Materials: Syntheses, Strategies and Applications*, T. J. J. Müller and U. H. F. Bunz (Eds.), Wiley-VCH Verlag, Weinheim, 2007. (d) Y. Matsuo, K. Tahara, K. Morita, K. Matsuo, and E. Nakamura, Regioselective octa- and deca-additions of pyridine-modified organocopper reagent to [60]fullerene, *Angew. Chem. Int. Ed.*, **46**, 2844–2847 (2007). (e) Y. Matsuo, Creation of cyclic π-electron conjugated systems through the functionalization of fullerenes and synthesis of their multinuclear metal complexes, *Bull Chem. Soc. Jpn.*, **81**, 320–330 (2008). (f) X. Zhang, Y. Matsuo, and E. Nakamura, Light emission of [10]cyclophenacene through energy transfer from neighboring carbazolylphenyl dendrons, *Org. Lett.*, **10**, 4145–4147 (2008).

[20] R. Marczak, M. Wielopolski, S. S. Gayathri, D. M. Guldi, Y. Matsuo, K. Matsuo, K. Tahara, and E. Nakamura, Uniquely shaped double-decker buckyferrocenes – distinct electron donor-acceptor interactions, *J. Am. Chem. Soc.*, **130**, 16207–16215 (2008).

[21] (a) M. Sawamura, K. Kawai, Y. Matsuo, K. Kanie, T. Kato, and E. Nakamura, Stacking of conical molecules with a fullerene apex into polar columns in crystals and liquid crystals, *Nature*, **419**, 702–705 (2002). (b) Y. Matsuo, A. Muramatsu, R. Hamasaki, N. Mizoshita, T. Kato, and E. Nakamura, Stacking of molecules possessing a fullerene apex and a cup-shaped cavity connected by silicon-connection, *J. Am. Chem. Soc.*, **126**, 432–433 (2004). (c) Y. Matsuo, A. Muramatsu, Y. Kamikawa, T. Kato, and E. Nakamura, Synthesis, structural, electrochemical and stacking properties of conical molecules possessing buckyferrocene on apex, *J. Am. Chem. Soc.*, **128**, 9586–9587 (2006). (d) Y.-W. Zhong, Y. Matsuo, and E. Nakamura, Lamellar assembly of conical molecules possessing a fullerene apex in crystals and liquid crystals, *J. Am. Chem. Soc.*, **129**, 3052–3053 (2007).

[22] M. Sawamura, N. Nagahama, M. Toganoh, U. E. Hackler, H. Isobe, E. Nakamura, S.-Q. Zhou, and B. Chu, Pentaorgano[60]fullerene $R_5C_{60}^-$. A water soluble hydrocarbon anion, *Chem. Lett.*, 1098–1099 (2000).
[23] S. Zhou, C. Burger, B. Chu, M. Sawamura, N. Nagahama, M. Toganoh, U. E. Hackler, H. Isobe, and E. Nakamura, Spherical bilayer vesicles of fullerene based surfactants in water: a laser light scattering study, *Science*, **291**, 1944–1947 (2001).
[24] H. Isobe, T. Homma, and E. Nakamura, Energetics of water permeation through fullerene membrane, *Proc. Natl. Acad. Sci. USA*, **104**, 14895–14899 (2007).
[25] Y. Matsuo and E. Nakamura, Syntheses, structure, and derivatization of potassium complexes of penta(organo)[60]fullerene-monoanion, -dianion, and -trianion into hepta- and octa(organo) fullerenes, *J. Am. Chem. Soc.*, **127**, 8457–8466 (2005).
[26] D. M. Guldi, G. M. A. Rahman, R. Marczak, Y. Matsuo, M. Yamanaka, and E. Nakamura, Sharing orbitals -ultrafast excited state deactivations with different outcome in bucky ferrocenes and ruthenocenes, *J. Am. Chem. Soc.*, **128**, 9420–9427 (2006).
[27] Y. Matsuo, K. Matsuo, T. Nanao, R. Marczak, S. S. Gayathri, D. M. Guldi, and E. Nakamura, Ruthenium connection in fullerene-ferrocene arrays. synthesis of $Ru(C_{60}Me_5)R(CO)_2$ (R = C_6H_4Fc and CCFc) and their charge transfer properties, *Chem. Asian J.*, **3**, 841–848 (2008).
[28] Y.-W. Zhong, Y. Matsuo, and E. Nakamura, Convergent synthesis of polyfunctionalized fullerene by regioselective five-fold addition of functionalized organocopper reagent to C_{60}, *Org. Lett.*, **8**, 1463–1466 (2006).
[29] Y. Matsuo, K. Kanaizuka, K. Matsuo, Y.-W. Zhong, T. Nakae, and E. Nakamura, Photocurrent-generating properties of organometallic fullerene molecules on an electrode, *J. Am. Chem. Soc.*, **130**, 5016–5017 (2008).
[30] R. Hamasaki, Y. Matsuo, and E. Nakamura, Synthesis of functionalized fullerene by mono-alkylation of fullerene cyclopentadienide, *Chem. Lett.*, **33**, 328–329 (2004).
[31] (a) T. Nakae, Y. Matsuo, and E. Nakamura, Synthesis of C_5-symmetric functionalized [60] fullerenes by copper-mediated five-fold addition of Reformatsky reagents, *Org. Lett.*, **10**, 621–623 (2008). (b) E. Nakamura, S. Mouri, Y. Nakamura, K. Harano, and H. Isobe, Mono- and penta-addition of enol silyl ethers to [60]fullerene, *Org. Lett.*, **10**, 4923–4926 (2008).

7

Supramolecular Chemistry of Fullerenes: Host Molecules for Fullerenes on the Basis of π-π Interaction

Takeshi Kawase

Graduate School of Engineering, University of Hyogo, Japan

7.1 Introduction

In 1990 fullerenes (C_{60} and C_{70}) were firstly extracted from carbon soot using benzene (Figure 7.1) [1, 2]. At the moment the supramolecular chemistry of fullerenes started off. The π-π interaction between the curved π surface of fullerenes and the flat π surface of aromatic compounds has attracted much attention, because the interaction should play an important role in the dissolution phenomenon. In the decade of 1990s the host-guest chemistry of fullerenes has been explored extensively for the sake of the separation and purification of fullerenes. Supramlecular chemists have found that host molecules with a bowl-shaped structure composed of electron rich aromatic units such as calix[n]arenes, calix[4]resorcinarenes and cyclotriveratrylenes can be employed for the purpose (Figure 7.2a). The crystallographic analyses of these host-guest complexes with C_{60} clearly indicate the importance of the π-π interactions. On the other hand, except a few examples, these traditional hosts bind fullerenes only in solid state. Interestingly, unmodified porphyrins and corannulene possess similar affinities for fullerene surfaces; they also show no evidence for complexation with fullerenes in solution state. In order to bind with

Chemistry of Nanocarbons Edited by Takeshi Akasaka, Fred Wudl and Shigeru Nagase

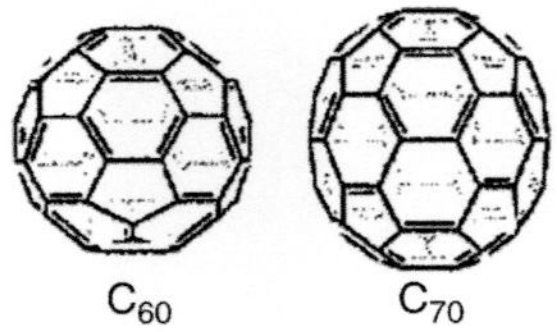

Figure 7.1 Fullerenes (C_{60} and C_{70})

Figure 7.2 Schematic representations of fullerene hosts (a) with a bowl shaped structure, (b) bearing appendants, and (c) bearing a dimeric structure

fullerenes more tightly, host molecules need appropriate modifications. The concept of the molecular designs is to construct the well preorganized π cavity for fullerenes. To achieve the purpose, electron rich aromatic units are introduced as appendants on the edge of the host (Figure 7.2b), and two host molecules are linked by appropriate tether(s) to form a dimeric structure with a creft-, ring- or ball-shaped cavity (Figure 7.2c). The binding abilities of these hosts fairly increased in comparison with those of simple traditional hosts. For example, the *K*a values of porphyrin-based host molecules are extremely large ($> 10^8$ dm^3 mol^{-1}). Moreover, brand-new host molecules bearing curved conjugated systems also form considerably stable complexes with fullerenes, where the novel concave-convex π-π interaction would be operative. The survey of these complexes should provide an insight into the supramolecular properties of convex π surface of fullerenes.

A number of excellent reviews have already been published [3–9]. This chapter provides the fullerene host molecules bearing a cavity surrounded by p-orbitals, so-called a 'π-cavity', and discusses the concept of molecular designs, the character of π-π interactions, and new development for construction of supramolecular architectures.

7.2 Fullerenes as a Electron Acceptor

Generally, nonbonded interaction between π conjugated systems can be considered on the basis of following three factors: the Van der Waals (VDW) interaction, the electrostatic (ES) interaction, and the charge transfer (CT) interaction. All the interactions should be operative between fullerene and the π electron systems of hosts. Fullerenes can be regarded as an electron acceptor, and C_{60} exhibits a CT absorption band in the 400–650 nm range in aromatic solvents. On the other hand, the CT bands show considerable solvatochromism [10, 11], because aromatic compounds form two types of complexes with C_{60}. First,

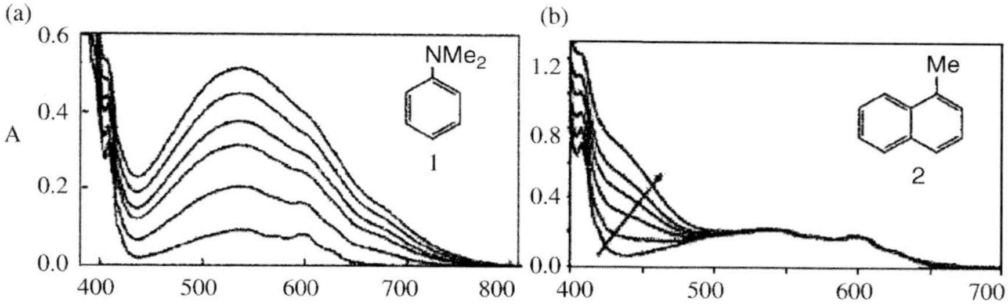

Figure 7.3 Absorption spectra of (a) C_{60} (4.9×10^{-4} M) as a function of N,N-dimethylaniline **1** concentration in toluene (Ref. 12), (b) C_{60} (2×10^{-4} M) as a function of 1-methylnaphthalene **2** concentration in toluene (Ref. 13)

strong electron donors such as aromatic amines provide exciplexes causing the hyperchromic shift in a 450–650 nm range of absorption. For example, the CT band with *N,N*-dimethylaniline **1** having a maximum located at 564 nm shows drastic change with increasing the aniline concentration (Figure 7.3a) [12]. Second, relatively poor electron donors such as 1-methylnaphthalene **2** provide simple 'contact complexes' causing the hyperchromic shift in a 400–500 nm range of absorption (Figure 7.3b) [13]. Phenol derivatives as building blocks for the traditional hosts are classified as poor electron donors. The affinity for fullerenes should depend on the electron donating property of the π-electron systems; however, the association constant (*K*a) of the complexes between C_{60} and **1** is relatively small ($0.07 \pm 0.01\ dm^3\ mol^{-1}$). Moreover, C_{60} forms only neutral molecular complexes ('contact complexes') with BEDT-TTF **3** (Figure 7.4) as a typical electron donor [14]. The low binding abilities suggest that the electron affinity of C_{60} is not so high. Thus, the CT interaction would not always play a decisive role in forming complexes with fullerene.

The first example of C_{60} involved in azacrown ethers **4** (Figure 7.5) with the lipophilic cavity in solution was reported by Ringsdorf and Diederich in 1992 [15]. Soon later, Wennerström's group reported a water-soluble complex between γ-cyclodextrin (γ-CD) **5** and C_{60} [16]. The discovery led to the extensive studies on various CD complexes related to water-soluble fullerene complexes [17, 18], tools for nano-composite [19], and a novel reducing reagent [20]. In these complexes, the attractive interactions between the heteroatoms (an n-donor) and the fullerene surface are operative, but the main driving force of these complexes is hydrophobic effect.

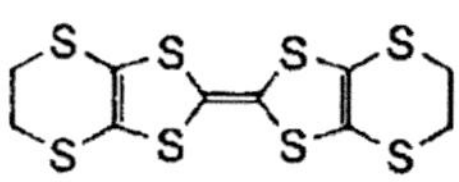

BEDT-TTF
3

Figure 7.4 C_{60} forms only neutral molecular complexes ('contact complexes') with BEDT-TTF **3** as a typical electron donor

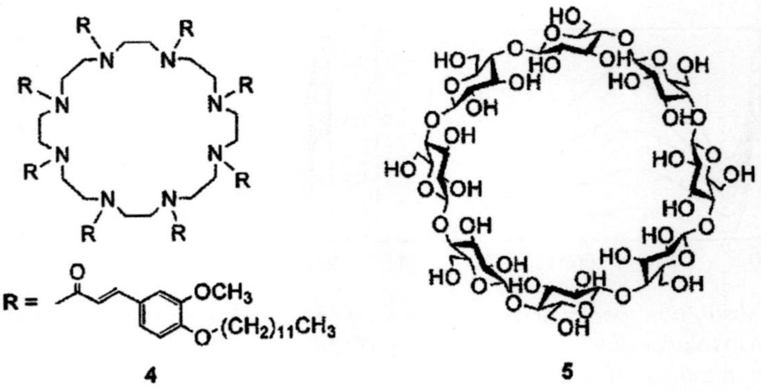

Figure 7.5 Compounds **4** and **5**

7.3 Host Molecules Composed of Aromatic π-systems

7.3.1 Hydrocarbon Hosts

It has been known that fullerenes are tightly solvated by aromatic solvents. The crystallographic analyses of the solvated C_{60} [21] and the (η^2-C_{60})Ir complex **6** [22] intuitively suggest that the face-to-face type interaction (π-π interaction) would be operative as a dominant force in the complex formation with fullerenes (Figure 7.6). The molecular structure of **6** also reveals that the phenyl rings lie above a 5:6 ring fusion in the chelated C_{60}. According to the theoretical calculation, the 5:6 ring fusion represents centers of positive charge on the C_{60} surface and the 6:6 fusion represents centers of negative charge (Figure 7.7). Thus, the electrostatic interaction between electropositive 5:6 fusion and

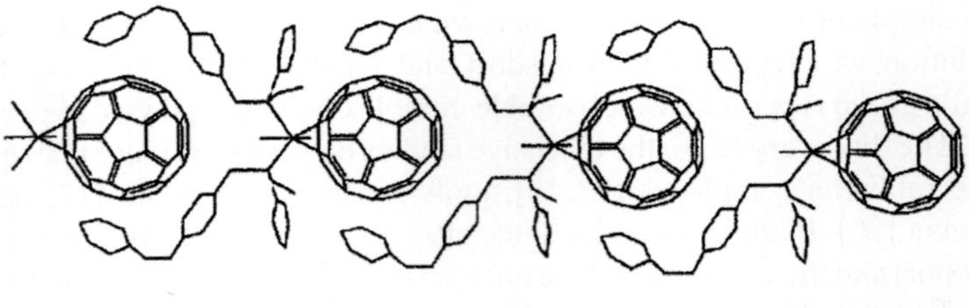

Figure 7.6 A drawing of molecular packing of (η^2-C_{60})Ir(CO)Cl(bobPPh$_2$)$_2$ **(6)**

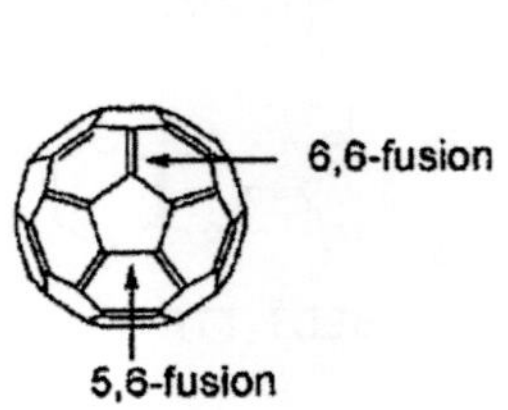

Figure 7.7 The 5:6 ring fusion represents centers of positive charge on the C_{60} surface and the 6:6 fusion represents center of negative charge

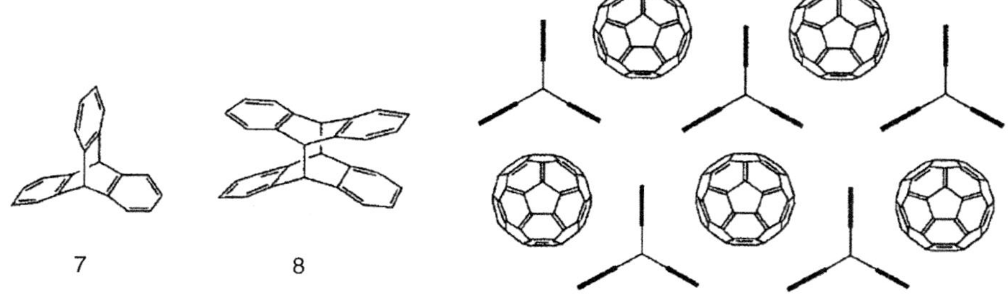

Figure 7.8 Molecular structures of *7* and ***8***, and packing arrangement for the complex of C_{60} and *7*

electronegative aromatic π-system has been thought to be important. A database study of crystal structure of fullerene compounds indicates that the hexagonal ring of C_{60} is more liable to interact with CH hydrogens though the contrast is not significant. As the hexagonal ring of C_{60} is more electron-rich than the pentagonal one, the results are compatible with the expected electrostatic potential of the fullerene surface [23].

Hydrocarbon molecules with a rigid framework such as triptycene **7** [24] and dianthracene **8** [25] can form inclusion complexes with fullerenes in the solid state. Each C_{60} molecule is sandwiched by two molecules of **7** in the crystals (Figure 7.8). However, these hosts do not form stable complexes in solution phase. In this context, host molecules based on a tribenzotriquinacene skeleton bearing dithia- or diaza-heteroaromatic rings **9** and **10** were prepared by Georghiou and Kuck's, and Volkmer's groups (Figure 7.9) [26, 27]. The concave shape of these molecules allows their efficient packing with C_{60} surface to gain wide van der Waals contact.These hosts exhibit moderately high association constants ($\sim 10^3$ dm^3 mol^{-1}) in solutions.

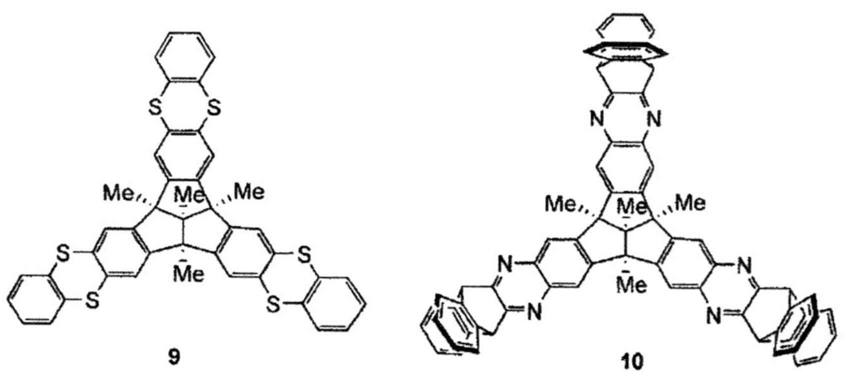

Figure 7.9 Host molecules based on a tribenzotriquinacene skeleton bearing dithia- or diaza-heteroaromatic rings **9** and **10** were prepared by Georghiou and Kuck's, and Volkmer's groups

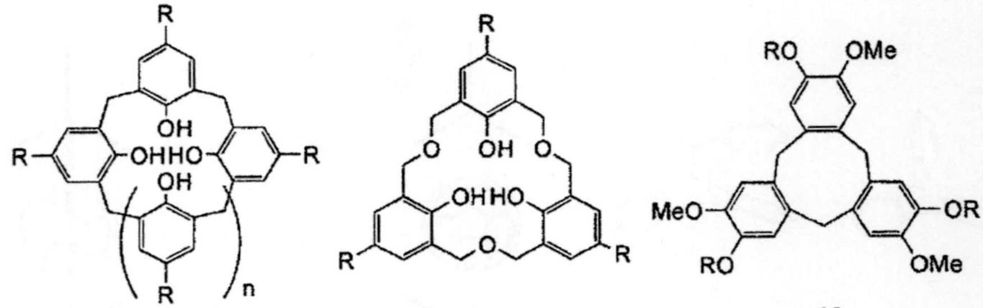

11: n = 5 13: n = 2
12: n = 3 16: n = 1

14

15

Figure 7.10 Cyclotriveratrilenes (CTV) [33] **12–15** form a variety of inclusion complexes with fullerenes

7.3.2 Hosts Composed of Electron Rich Aromatic π-Systems

In 1994 Atwood's [28] and Shinkai's [29] groups independently found that *p-tert*-butylcalix[8]arene **11** selectively precipitated with C_{60} from fullerite. The discovery has resulted in many studies concerning complexation behavior of various host molecules composed of electron rich aromatic systems; calix[6]- and -[5]arenes, oxacalix[3]arenes [30–32], and cyclotriveratrilenes (CTV) [33] **12–15** form a variety of inclusion complexes with fullerenes (Figure 7.10). The crystal structures of these complexes reveal that a C_{60} molecule is situated in the shallow cavity of the cone-shaped host to construct a 'ball and socket' nanostructure (Figure 7.11). The stabilities of complexes depend primarily on the size of the cavities. The theoretical and experimental studies indicate that calix[4]arenes **16** with relatively small cavities do not cover the fullerene molecule [34, 35].

Although alkyl- or halogen-substituted calix[4]-, -[6]- and -[8]arenes readily form inclusion complexes with fullerenes in solid state, the binding ability of these complexes in nonpolar organic solvents is generally poor. Contrary to the results, calix[5]arene derivatives **13** form rather stable complexes in solutions [36]. The association constants are in the range $2{\sim}3 \times 10^3\,\mathrm{dm^3\,mol^{-1}}$. Moreover, calix[4]naphthalene **17** [37] and trioxacalix[3]naphthylene **18** [38] derivatives afford more stable complexes than the corresponding calixarenes (Figure 7.12). These results indicate that the rigid and sizable cavity

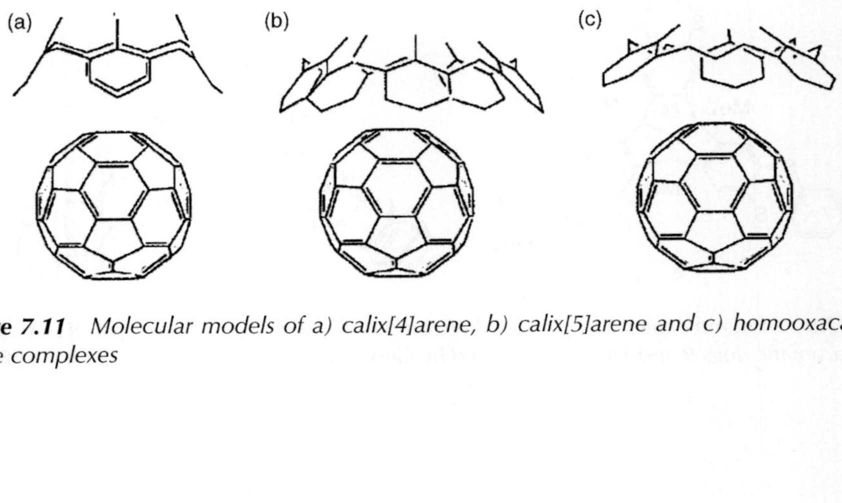

Figure 7.11 Molecular models of a) calix[4]arene, b) calix[5]arene and c) homooxacalix[3] arene complexes

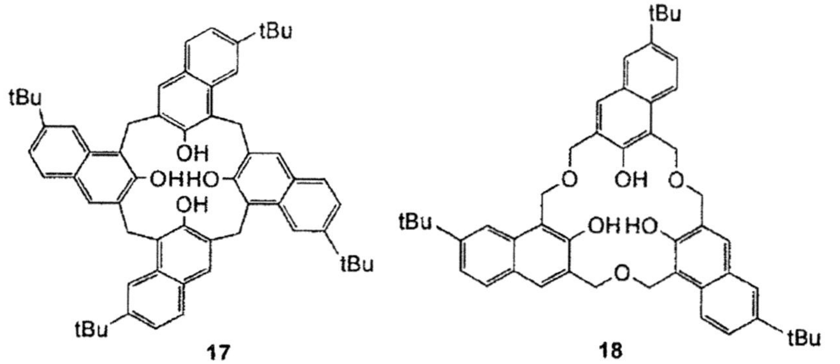

17 18

Figure 7.12 *Calix[4]naphthalene **17** [37] and trioxacalix[3]naphthylene **18** [38] derivatives afford more stable complexes than the corresponding calixarenes*

possessing electron negative potential would be important for the high affinity for fullerene surface.

7.3.3 Host Molecules Bearing Appendants

To prepare the host molecules with high affinity to fullerenes, Shinkai and co-workers have first designed calix[6]arene derivatives **19** bearing electron-rich aniline or 1,3-diaminobenzene units, expecting that the CT interaction would act as a driving force (Figure 7.13) [39]. The association constant (K_a) of **19** for C_{60} is up to 1.1×10^2 dm^3 mol^{-1}. The molecular design is applicable to the CTV derivatives. The CTV **20** having

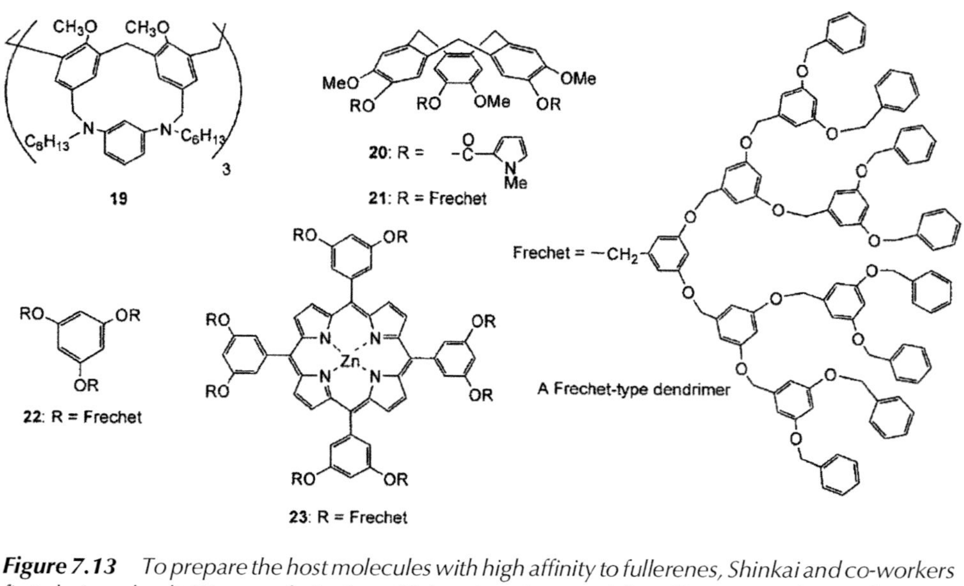

Figure 7.13 *To prepare the host molecules with high affinity to fullerenes, Shinkai and co-workers first designed calix[6]arene derivatives **19** bearing electron-rich aniline or 1,3-diaminobenzene units, expecting that the CT interaction would act as a driving force*

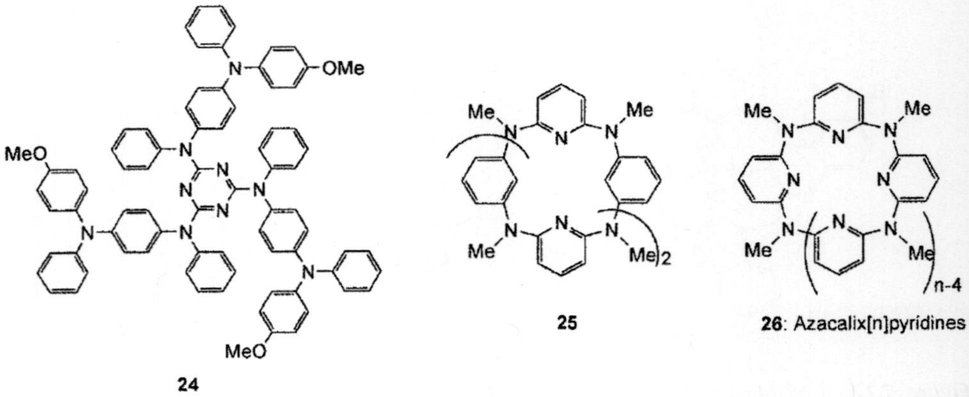

Figure 7.14 A dendrimer **24** [43] based on triarylamine units, azacalix[3]arene[3]pyridine **25** [44] and azacalix[n]pyridine **26** (n = 4~9) [45] can serve as efficient hosts for fullerenes

N-methylpyrrole appendants formed fairly stable complexes with fullerenes (K_a value for $C_{60} = 4.8 \times 10^4\,dm^3\,mol^{-1}$) [40]. Nierengarten's group synthesized CTV derivatives **21** functionalized by the Frechet type of dendrons [41]. The K_a values for C_{60} are significantly increased as the generation number of the dendric substituents is increased. The Frechet-type dendrimers bearing electron rich aromatic rings can provide the space size comparable with C_{60}. Shinkai's group independently found the host properties of the dendrimers with phloroglucinol **22** and tetraphenylporphyrin **23** cores for C_{60}. The K_a values of these dendrimers are not so high ($< 100\,dm^3\,mol^{-1}$) [42].

Contrary to these results, a dendrimer **24** [43] based on triarylamine units, azacalix[3]arene[3]pyridine **25** [44] and azacalix[n]pyridine **26** (n = 4~9) [45] can serve as efficient hosts for fullerenes (Figure 7.14). The stability of complexes is evaluated by the Stern-Volmer constants (K_{SV}) using emission spectra. Table 7.1 lists the K_{SV} values of the complexes **24** and **25**. As mentioned above, the stabilities of these fullerene complexes increase drastically due to the participation of additional attractive CT interaction. The K_{SV} values are informative to estimate the relative stability of these complexes; however, care should be taken to avoid several trivial artifacts [46].

Nakamura and co-workers have developed the reaction of C_{60} with organocupper reagents to afford molecules possessing the shape of a badminton shuttlecoch, $C_{60}R_5H$ [47]. Some derivatives **27** [48] and ferrocene analogues **28** [49] bearing aryl groups as appendants stack head-to-tail to form a one-dimensional array of fullerene molecules both in crystals and in liquid crystals; the concave surface constructed by the five aryl groups tightly binds the fullerene apex of adjacent molecule (Figure 7.15).

Table 7.1 *Stern-Volmer Constants (K_{SV}) of **22** and **23** and fullerenes at 300 K in toluene*

Compounds	fullerenes	K_{SV} ($dm^{-3}\,mol^{-1}$)	refs
22	C_{60}	180000 ± 8000	38
23	C_{60}	70680 ± 2060	39
23	C_{70}	136620 ± 3770	39

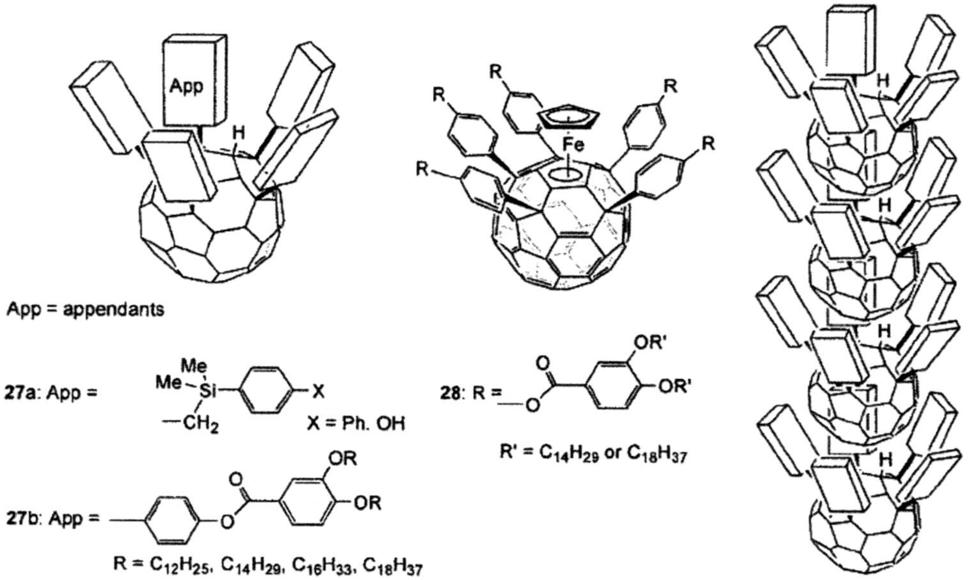

Figure 7.15 Fullerene derivatives with five appendants, **27** and **28**, and a stack of four molecules of **27b**

7.3.4 Host Molecules with Dimeric or Polymeric Structures

Fukazawa and Haino have found that the calix[5]arene **13** form a 1:1 complex in solution, but it does a 1:2 complex in the solid state [36]. One C_{60} molecule is sandwitched by two calix[5]arene cavities. A oxacalix[3]naphthalene **16** (R = ^{t}Bu) [38] and a CTV derivative [42] also form 1:2 complexes in the solid state. In this context, host molecules with a dimer structure were designed (Figure 7.16). In comparison with the corresponding calix[5] arene, the dimer **29** shows the high affinities for C_{60} and C_{70} [50]. Several calixarene dimers

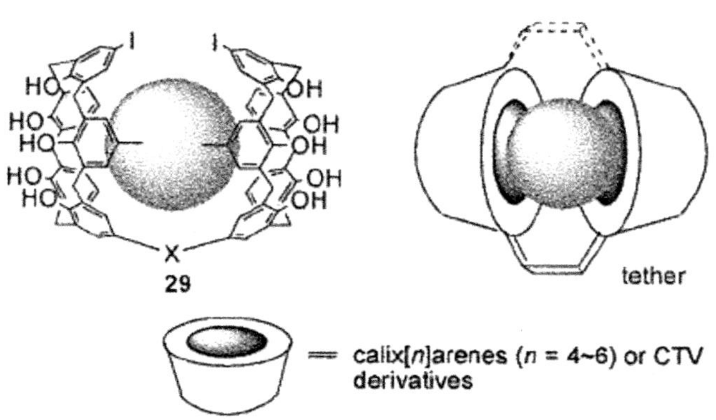

Figure 7.16 Structures of **29**, and schematic representations of dimeric hosts

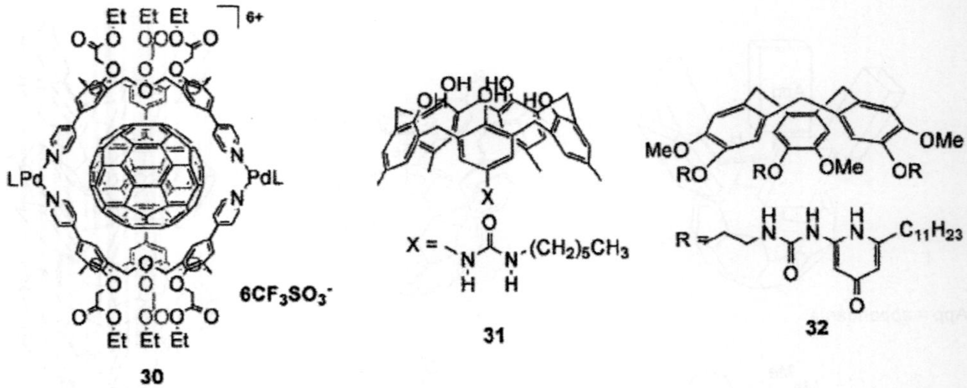

Figure 7.17 *Two calix[5]arenes with a urea functionality* ***31***

based on calix[4]-, -[5]- and -[6]arenes, and CTVs bearing an appropriate tether also afforded considerably stable complexes with fullerenes [51–53]. The association constants of these cage-shaped host molecules for fullerenes are up to 10^4–10^5 dm^3 mol^{-1}. The dimeric hosts bearing two diacetylenic tethers were also synthesized by Haino's and Yamamoto's groups. These cyclophane hosts show high binding abilities for higher fullerenes (C_{70}, C_{76}, C_{78}, etc.) corresponding to their expanded cavities [54, 55].

Cage-shaped host molecules constructed by coordination bonds of transition metal ions or by hydrogen bonds have been examined for the encapsulation of fullerene. Shinkai's group first reported a C_{60} complex with a self-assembled cage, the oxacalix[3]arene dimer **30** cross-linked by three Pd(II) complexes. (Figure 7.17a) [56]. Self-assembled cage-shaped molecules based on subphthalocyanine units [57], cavitand units [58], and calix[5]arene units [59] also constructed the dimeric structure in the presence of metal ions to catch a C_{60} molecule in the resulting cavity. These cage compounds involving metal cation(s) show relatively small association constants probably due to the high molecular mobility. Polar functional groups such as urea and amide play an important role in assembling subunits with hydrogen bonds. Two calix[5]arenes with a urea functionality **31** (Figure 7.17b) form a ternary complex with C_{60}; a C_{60} molecule is located in a self-assembling molecular capsule [60]. A CTV derivative with ureidopyrimidinone units **32** (Figure 7.17c) prepared by de Mendoza forms a dimeric hydrogen-bonded assembly that encapsulates a fullerene molecule within its large cavity. The system displays a remarkable selectivity for the encapsulation of C_{70} over C_{60} [61].

Naphthalene diimides (NDI) **33** functionalized with amino acids construct self-assembled helical nanotubes using hydrogen-bondings. The helical superstructures possess tubular cavities with a mean diameter of 12.4 Å, and capable of accommodating a string of C_{60} [62]. On the other hand, the NDIs **33** spontaneously form a new C_{70} receptor by changing the hydrogen bonding network (Figure 7.18) [63]. The counter-balance between van der Waals interaction and hydrogen-bonding arrangements play an important role in the formation of the different superstructures. Kawauchi, Kumaki and Yashima's group found that syndiotactic poly(methyl methacrylate) (st-PMMA) fold into a preferred-handed helical conformation assisted by encapsulated C_{60} within its

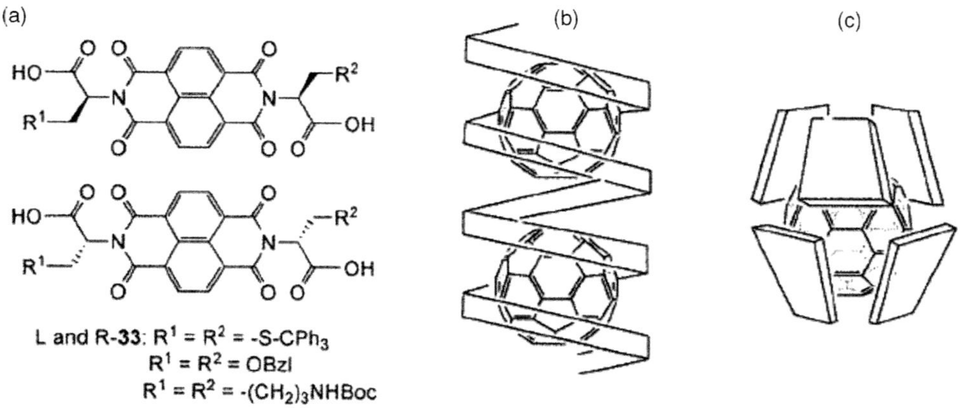

Figure 7.18 (a) L- and D-Naphthalene diimide (NDI) **33**, (b) helical superstructure of **33** and C_{60}, (c) a C_{70} receptor composed of **33**

helical cavity to form an optically active supramolecular peapod-like complex gel [64]. They also found that isotactic-PMMA replace the encapsulated C_{60} molecules within the st-PMMA cavity to form a stereocomplex [65]. Fullerene-containing polymers would have a great potential for practical purposes.

7.4 Complexes with Host Molecules Based on Porphyrin π Systems

7.4.1 Hosts with a Porphyrin π System

Fullerenes and porphyrins **34** are spontaneously attracted to each other. In 1997, Boyd, Reed and co-workers have firstly pointed out a close contact (2.75 Å) between porphyrin and C_{60} suggesting an attraction of C_{60} to the center of a porphyrin ring (Figure 7.19) [66]. They have also demonstrated that the graphitic and typical arene/arene distances are in the range 3.3–3.5 Å, fullerene/arene approaches lie in the range 3.0–3.5 Å and fullerene/fullerene separations are typically ca. 3.2 Å. The close approach is proposed to reflect an attractive π-π interaction. They concluded that, in the absence of steric effect, the hierarchy of interaction strengths is clearly porphyrin/porphyrin > porphyrin/fullerene > fullerene/fullerene. The fullerene-porphyrin cocrystals construct a variety of tape, sheet and 3D structural motifs [67]. However, the affinity between porphyrin and fullerene is not enough to form

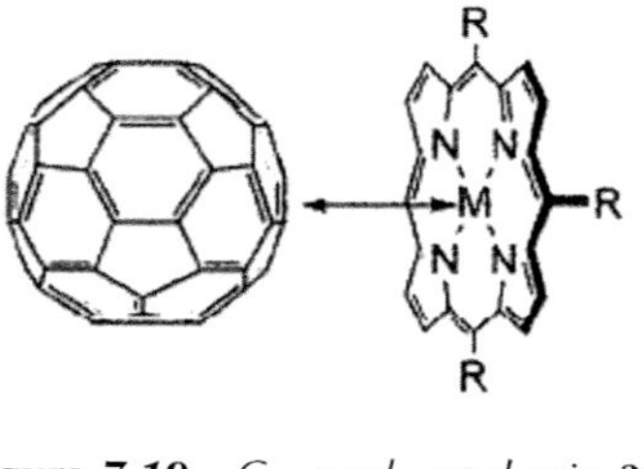

Figure 7.19 C_{60} and porphyrin **34**

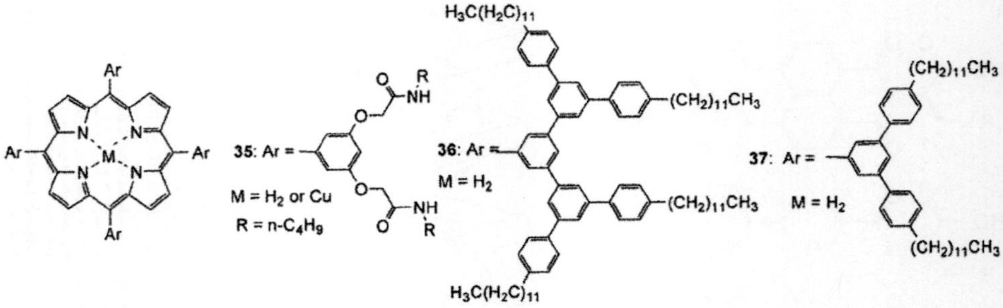

Figure 7.20 *Appropriately modified porphyrins provide various supramolecular structures with participation of fullerene*

complexes in nonpolar solvents. For example, a metalloporphyrin bearing a Rh-Me group **34** (M = Rh-Me) hardly interact with fullerenes in solution [68].

Appropriately modified porphyrins provide various supramolecular structures with participation of fullerene (Figure 7.20). Porphyrin-based gelators **35** bearing hydrogen-bonding sites were synthesized by Shinkai and co-workers. They tend to aggregate into a two-dimentional sheet-like structure utilizing the intermolecular hydrogen-bonding interaction. When C_{60} was added, the morphology was transformed to a one-dimentional fibrous structure [69]. Porphyrin **36** bearing phenylene-based rigid dendrimer with long alkyl chains forms a fullerene complex showing liquid crystalline properties [70]. The association constant in toluene solution was evaluated to be $2.7 \times 10^4\ dm^{-3}\ mol^{-1}$. However, a host **37** bearing the first generation dendrimers showed no evidence for complexation with C_{60}.

7.4.2 Hosts with Two Porphyrin π Systems

Reed, Boyd and co-workers have found that acyclic ('Jaws') bisporphyrins **38** and **39** (Figures 7.21 and 7.22) are employed as efficient hosts for fullerenes [71, 72]. The association constants in toluene solution span the range 490–5200 $dm^{-3}\ mol^{-1}$. The variable

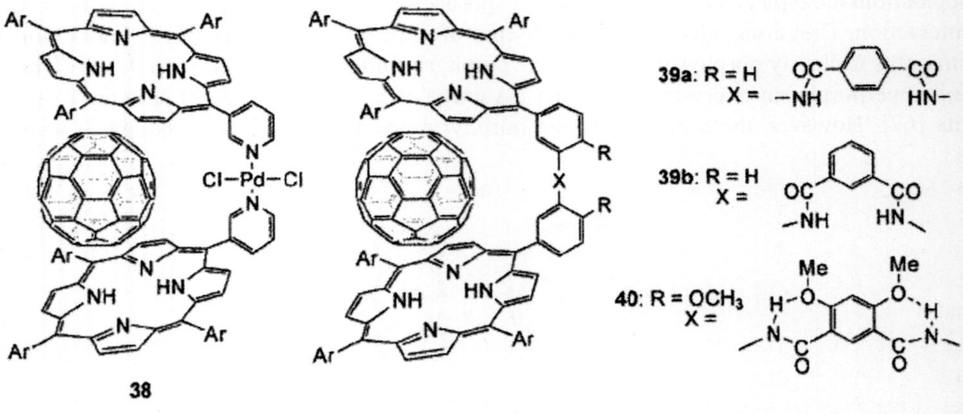

Figure 7.21 *'Jaws porphyrins' **38**, **39** and **40***

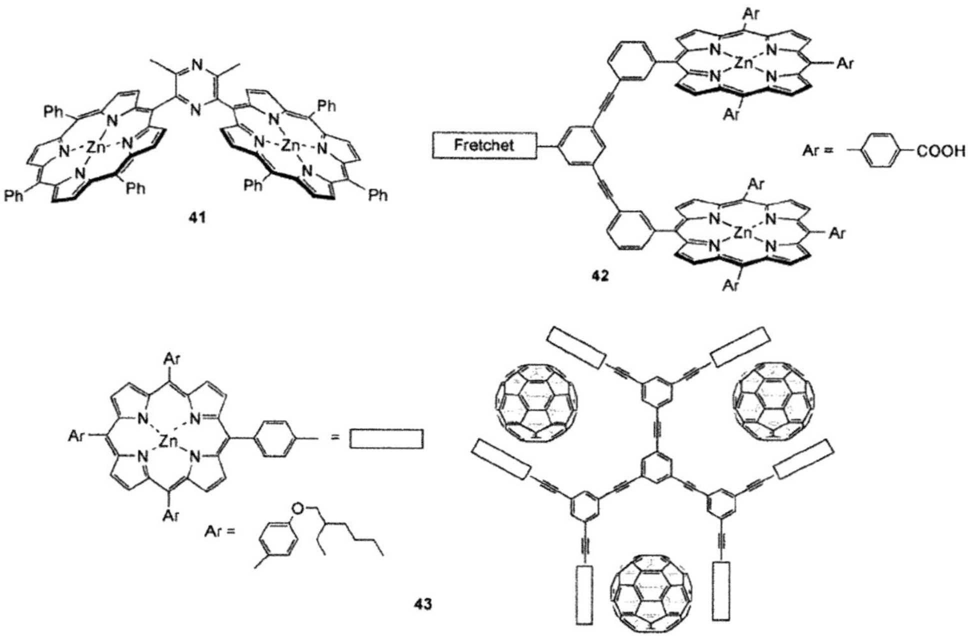

Figure 7.22 *V-shape Porphyrin **41**, and fullerene receptors **42**, and **43***

temperature ^{13}C NMR spectra reveal the coalescence of two fullerene signals corresponding to rapid exchange between the complexed and uncomplexed fullerenes at –55 °C for **39**·C_{60} [72]. The association constants and coalescence temperatures vary with the metal ion of host and the structure of tether [66]. The bisporphyrin receptors **40** with the preorganized U-shaped feature exhibit considerably strong binding abilities with fullerenes in solution ($\sim 3.4 \times 10^8$ dm^{-3} mol^{-1}) [73]. Contrary to these receptors, a V-shaped bisporphyrin receptor **41** exhibit relatively low binding ability (1500 ± 120 dm^{-3} mol^{-1}) [74].

A number of bisporphyrin and multiporphyrin receptors have been synthesized. Aida and Tashiro prepared an acyclic zinc porphyrin dimer **42** with a large [G4]-Frechet-type dendrimer (see Figure 7.12) and six carboxylic acid functionalities. In sharp contrast to the corresponding methyl ester, it forms one-dimensional supramolecular polymer, 'supramolecular peapods' with fullerene. Intermolecular hydrogen-bondings between carboxylic acid of **42** plays an important role in construction of the supramolecular architecture [75]. A rigid star-shaped *D*3-symmetric dendrimer **43** having three sets of bisporphyrin receptor units was prepared by Shinkai and co-workers (Figure 7.22). When two porphyrins sandwich one C_{60} molecule, the complexation site successively suppresses the rotational freedom of the remaining porphyrin tweezers. The domino effect would be effective for the binding of three equivalents of C_{60} in an allosteric manner [76].

Aida and co-workers found that a face-to-face cyclic dimer of zinc porphyrins **44**-Zn forms a highly stable 1:1 inclusion complex with C_{60} [68]. The K_a value (6.7×10^5 dm^{-3} mol^{-1}) exceeds those of the acyclic bisporphyrins [77]. The affinities to fullerenes are dependent upon central metal ions. The association constants of **44**-RhMe with C_{60} and C_{70}

are 10^7–10^8 dm^{-3} mol^{-1} in benzene. With respect to the high affinity, the coalescence of the NMR signals corresponding to in-and-out motion of the guest was not observed even at 100 °C [78]. Moreover, an iridium porphyrin dimer **44**-Ir exhibits the highest affinity toward fullerenes (> 10^9 dm^{-3} mol^{-1} in benzene) [79]. A crystallographic analysis reveals that each of iridium centers binds in an η^2 fashion to a 6:6 ring juncture. This bond formation causes an ellipsoidal deformation of fullerene molecules. They also synthesized a series of cyclic porphyrin dimers (Figure 7.23) with various cavity sizes to create the tailor made hosts for higher fullerenes and a fullerene dimer [78]. In particular, the host **45** bearing chiral *N*-methyldiarylporphyrin and methylrhodium diarylporphyrin units can spectroscopically discriminate enantiomers of C_{76} [80]. And, a host **46** composed of two fused zinc porphyrin dimers is capable of including C_{120} in its cavity with a considerably high association constant (> 10^8 dm^{-3} mol^{-1} in toluene). Considering the cavity size, it can accommodate two molecules of C_{60}; however, it includes only one molecule C_{60}. Thus, the host **46** displays strong negative homotropic cooperativity for the guest binding. The results indicate the presence of strong electronic interaction between the host and guest [81]. A porphyrin dimer

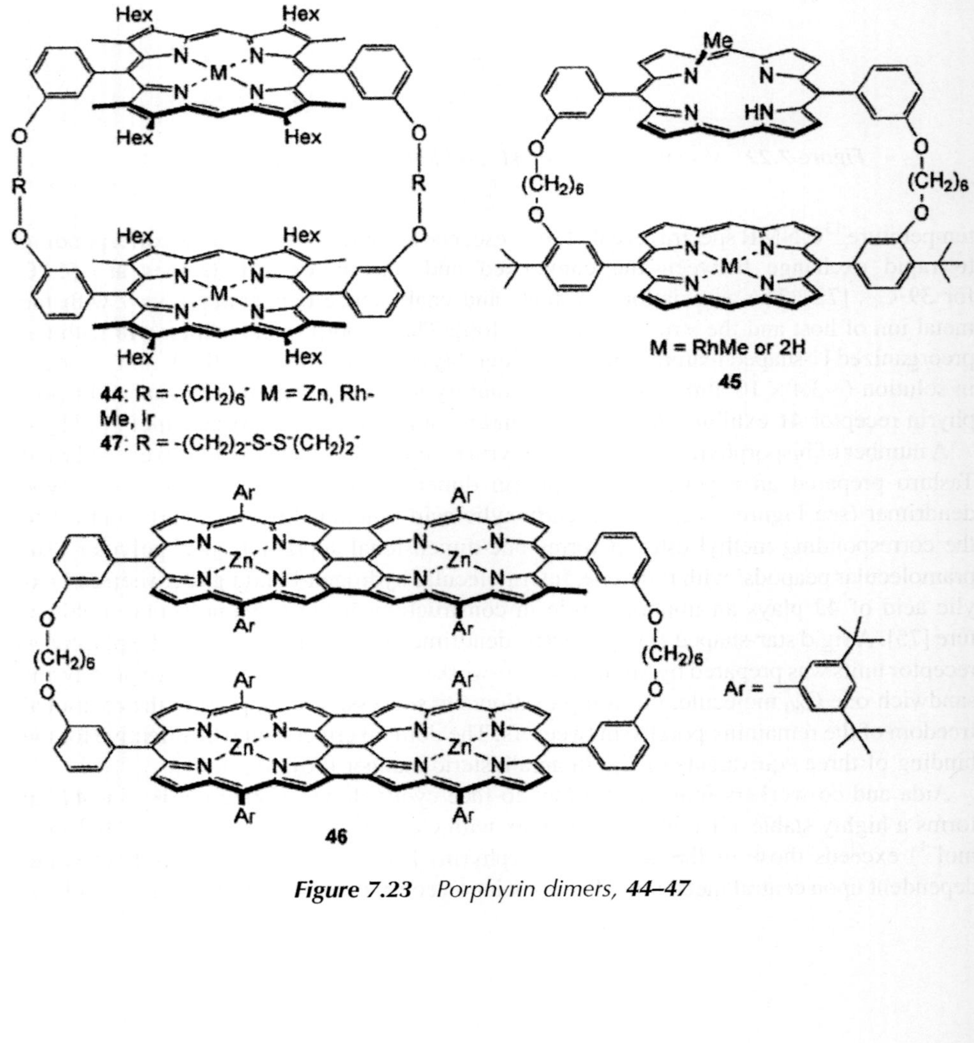

Figure 7.23 Porphyrin dimers, **44–47**

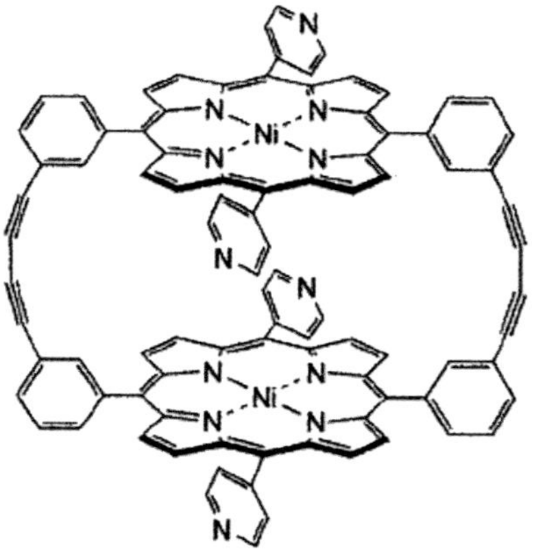

48

Figure 7.24 *A complex of **48** with C_{60} constructs organic nanotubes by nonclassical hydrogen bonds. A crystallographic analysis reveals C_{60} molecules located in the inner channel*

with adjustable linkers involving disulfide units **47** and diyne units **48** were prepared by Sanders' [82] and Tani's [83] groups, respectively. A complex of **48** with C_{60} constructs organic nanotubes by nonclassical hydrogen bonds. A crystallographic analysis reveals C_{60} molecules located in the inner channel (Figure 7.24).

There is an unexpectedly strong interaction between the curved π surface of C_{60} and the flat π surface of porphyrins. Recent theoretical [84] and experimental studies suggest that the considerably strong π-π interaction is largely the result of van der Waals dispersion force and is enhanced by weak electrostatic interaction. On the other hand, investigations of fullerene-porphyrin complexation suggest the presence of strong electronic interaction between the host and guest [73]. Thus, the stabilities of the fullerene complexes would be explained in terms of the participation of additional attractive CT interaction. It has been known that supramolecular properties of carbon nanotubes (CNs) are similar to those of fullerenes; porphyrin derivatives [85, 86] bind to the convex surface of carbon nanotubes, though CNs have little 5:6 fusions in the π system.

7.5 Complexes with Host Molecules Bearing a Cavity Consisting of Curved π System

7.5.1 Host with a Concave Structure

The curved π-surfaces of C_{60} and corannulene **49** are geometrically well matched; it forms a stable complex with $(C_{60})^+$ in the gas phase [87]. However, unsubstituted corannulene shows no evidence for complexation with fullerenes in solution as well as in the solid state. Corannulenes with expanded π-systems such as **50** [88] form inclusion complex with C_{60} (K_a = ca. 1400 dm^{-3} mol^{-1} in $CDCl_3$). A hexabenzocoronene derivative **51** with a double-concave structure forms an inclusion complex with C_{60}. The fullerene is positioned exactly on the central benzene rings of **51**, thus yielding a perfect columnar packing arrangement [89].

49

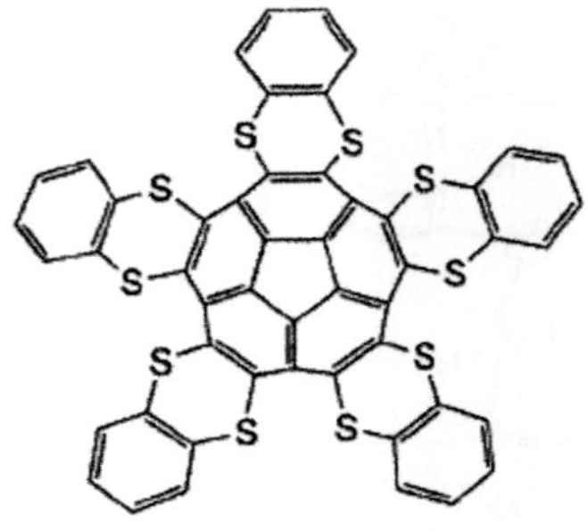

50

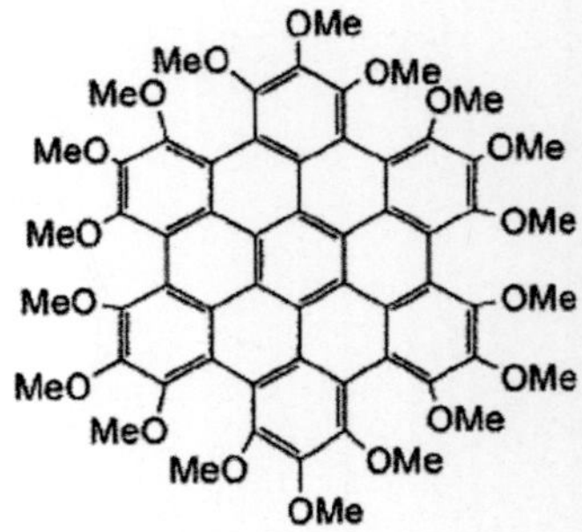

51

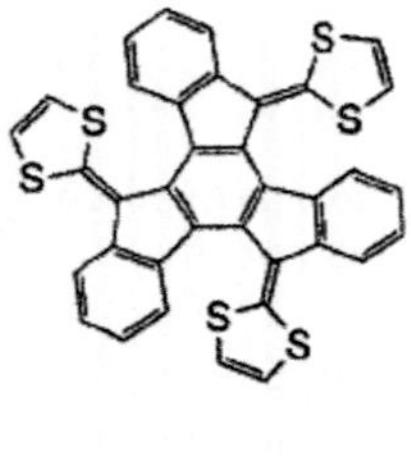

52

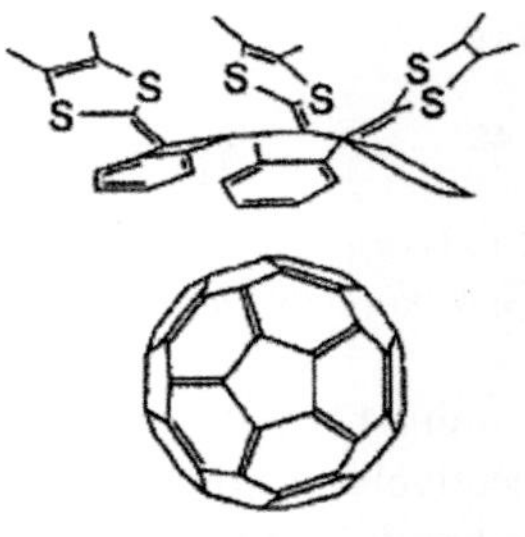

52 and C_{60} complex

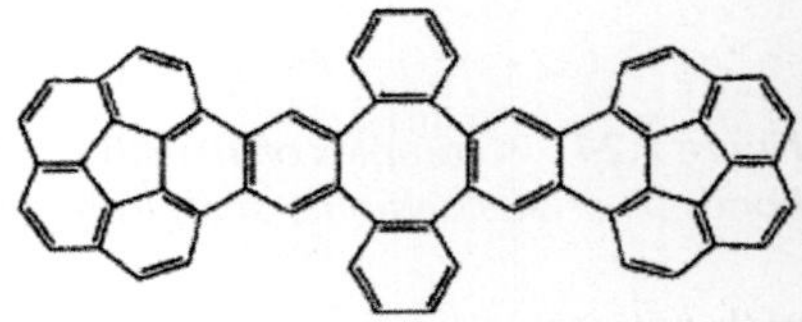

53

Figure 7.25 *49–53*

A concave tetrathiafulvalene-type donor **52** has two concave faces composed of three aromatic and dithiole rings. It also forms a 1:1 inclusion complex with fullerenes in solution ($K_a = (1.2 \pm 0.3) \times 10^3\ dm^{-3}\ mol^{-1}$ in $CDCl_3/CS_2$). A theoretical calculation suggests that the association preferentially occurs on the aromatic face of **52** [90].

The construction of a cage-shaped cavity by connecting two host molecules is also effective for high affinity toward fullerenes. A double concave hydrocarbon Buckycatcher **53** composed of two corannulene moieties prepared by Sygula includes a C_{60} molecule in the concave cavity in solid and solution ($K_a = 8600\ dm^{-3}\ mol^{-1}$) [91] (Figure 7.25). Martin's group found that a tweezers-shaped host **54** composed of two extended TTF derivatives exhibits high affinity for C_{60}. The stoicheometry varys a 1:1 complex to a 2:2 complex, depending upon the solvents [92]. They investigated the relative contributions of the concave-convex π-π interaction to bind C_{60} using the host **54** and the other tweezers-shaped hosts with various curvatures. The experimental and theoretical studies support the perceptible contribution of concave-convex complementarity to the stabilization of supramolecular associates [93] (Figure 7.26).

7.5.2 Complexes with Host Molecules Bearing a Cylindrical Cavity

Cyclic [n]paraphenyleneacetylenes ([n]CPPAs) possessing 1,4-phenylene and ethynylene units alternately adopt rigid and belt-shaped structures with well-defined cavities. Kawase and co-workers have synthesized [n]CPPAs ($n = 5{\sim}9$) **55–59** and the related compounds having naphthylene rings **60–62** (Figure 7.27) [94]. The carbon nanorings

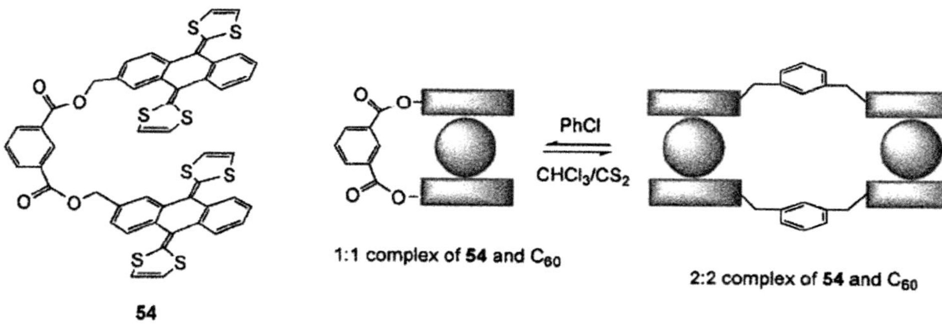

Figure 7.26 *Martin's group found that a tweezers-shaped host* ***54*** *composed of two extended TTF derivatives exhibits high affinity for* C_{60}*. The stoicheometry varys a 1:1 complex to a 2:2 complex, depending upon the solvents [92]. They investigated the relative contributions of the concave-convex* π-π *interaction to bind* C_{60} *using the host* ***54*** *and the other tweezers-shaped hosts with various curvatures*

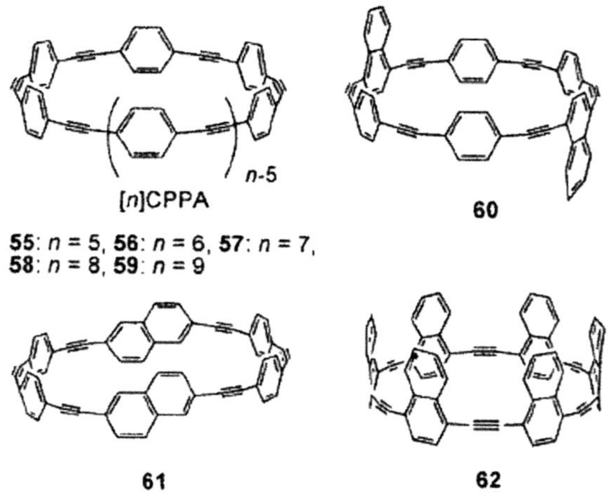

Figure 7.27 *CPPAs and the related compounds,* ***55–62***

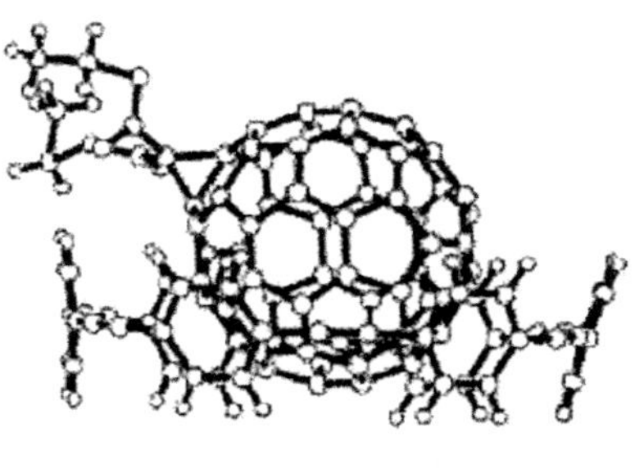

Figure 7.28 *Molecular structure of* ***56*** *and methanofullerene* ***63***

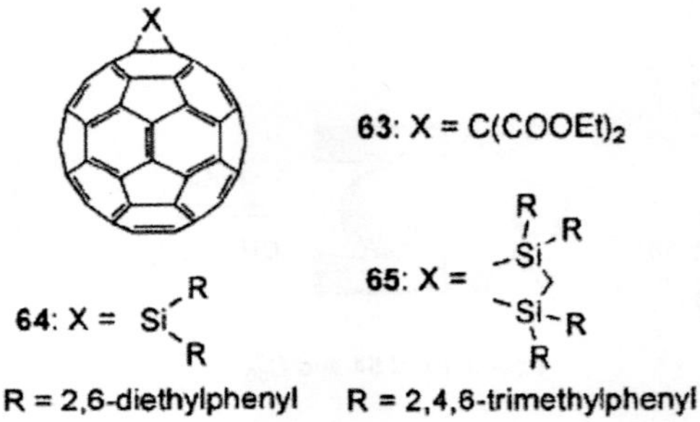

Figure 7.29

having an appropriate size of cavity form stable inclusion complexes with fullerenes in solution as well as in the solid state [95]. The molecular structure of the complex of [6] CPPA **56** and methanofullerene derivative **63** (Figure 7.28) clearly indicates that the concave–convex π-π interaction is operative between the host and guest as a major driving force [96]. The stability of complexes correlates well with the van der Waals contact between the host and guest (Table 7.2). The K_{SV} values of the complexes **62** for C_{60} are the largest in the known hosts for fullerenes, though the CPPAs are composed of only carbon and hydrogen atoms [97]. Moreover, a solid-to-liquid extraction experiment proved the considerably high selectivity of **61** for C_{70} against C_{60} (>10:1) [98].

It has been known that the electronic properties of C_{60} derivatives are correlated well with the electronegativity of the attached atoms. For example, the attachment of electron–positive silicon atom considerably increases the electron density of the π–systems of C_{60} derivatives (Figure 7.29). The values of Gibbs activation energies ($\Delta G^{\ddagger}_{dis}$) for dissociation of the complexes determined by the VT-NMR experiments can be evaluated the stability of these complexes. Thus, the $\Delta G^{\ddagger}_{dis}$ values of silylated fullerenes **64** (8.8 ± 0.2 kcal mol^{-1}) and **65** (8.5 ± 0.2 kcal mol^{-1}) are significantly smaller than those of C_{60} (9.9 ± 0.3 kcal

Table 7.2 *Diameters (Φ)[a] of the cavity of hosts, K_a, K_{SV} at 300 K and $\Delta G^{\ddagger}$ values[b] of the complexes*

Complex	Φ^a	K_a ($\times 10^{-4}$)^b	K_{SV} ($\times 10^{-4}$)^b	$\Delta G^{\ddagger}$ (CD_2Cl_2)
55·C_{60}	1.31	1.6 ± 0.2	7.0	9.9 ± 0.2
55·C_{70}		1.8 ± 0.2	14	9.6 ± 0.2
60·C_{60}	1.31	2.5~6.0	27	10.8 ± 0.3
60·C_{70}		—^{c}	26	10.1 ± 0.2
61·C_{60}	1.41	~10	26	<9
61·C_{70}		~100	430	11.9 ± 0.8
62·C_{60}	1.31	—^{c}	770	14.1 ± 0.3
62·C_{70}		—^{c}	1000	13.3 ± 0.3
56·C_{60}	1.53	—	5.6	<9
56·C_{70}		—	21	<9

^{a}nm, evaluated by AM1 calculations.
^{b}dm^3 mol^{-1}, in C_6H_6.
^{c}Undeterminable.

59•56 59•56•C_{60} 58•55

Figure 7.30 Ring-in-ring complexes of CPPAs

mol^{-1}) and methanofullerene derivative **63** ($9.4 \pm 0.2\,kcal\,mol^{-1}$). Therefore, an increase in electron density weakens the binding between the host and guest [99].

When the van der Waals distance (0.34 nm) is taken into account, [9]CPPA **59** is almost perfect complementarity to [6]CPPA **56**, and [8]- and [5]CPPAs, **58** and **55**, are under a similar condition. In fact, these CPPAs construct ring-in-ring complexes (Figure 7.30) [100, 101]. Moreover, **59** and **56** form an onion-type supramolecular structure with a C_{60} molecule [100]. The *K*a values and thermodynamic parameters of these complexes are summarized in Table 7.3. The similar K_a values of **59·56** in the presence and the absence of C_{60} suggest that a C_{60} molecule affords little electronic and structural perturbation to the host **56**. Together with the small ΔS value for the formation of **59·55**, the complex is characteristic of a van der Waals complex. Contrary to the results, the K_a value of **58·55** at 30 °C is about 200 times larger than that of **59·56**, although the contact area of **58·55** is apparently smaller than that of **59·56**. Moreover, the ΔS value for the formation of **58·55** is significantly large. The supramolecular property of **58·55** seems similar to that of fullerene complexes [93]. The results prove the substantial participation of the electrostatic interaction prior to the dispersion force. Planar phenylacetylene macrocycles without electron-withdrawing substituents on their aromatic rings do not aggregate in nonpolar solvent, because π-π stacking between planar aromatic hydrocarbons generally causes an electrostatically repulsive force [102]. On the basis of these results, the concave-convex π-π interaction should vary from repulsive to attractive with an increase in strain of the π electron system. The drastic increase of the association constants from **58·55** to **55·C_{60}** or **58·53** would be explained in terms of the participation of the additional electrostatic interaction corresponding to the increasing polarity of the π-systems.

Table 7.3 *The K_a values[a] and thermodynamic parameters of ring-in-ring complexes in $CDCl_3$*

	K_a	ΔH^b	ΔS^c
59·55	ca. 40	ca. –3.0	ca. –2.4
58·54	9200 ± 1400	–0.75	16

[a]at 303 K, $dm^3\,mol^{-1}$.
[b]$kcal\,mol^{-1}$.
[c]$kcal\,mol^{-1}\,K^{-1}$.

7.6 The Nature of Supramolecular Property of Fullerenes

The globular C_{60} is tightly solvated by aromatic solvents such as benzene, and contacts with the π-faces of the aromatics of solvents or hosts in crystal. Moreover, it favorably forms complexes with the hosts involving electron-rich aromatic systems. As mentioned above, the nonbonded interactions: the VDW, the ES, and the CT interactions are possibly operative between fullerene and π electron systems. Fullerenes construct considerably stable complexes with the host molecules based on aromatic amines or porphyrins. The high stabilities would be explained in terms of the participation of attractive CT interaction. Except these host-guest systems, however, the CT interaction would be negligible in the complexation. Because recent theoretical and experimental studies have revealed that the contributions associated with CT energy are much smaller than those arising from ES and VDW interactions. Moreover, it is difficult to rationalize the high affinity that exists between fullerene and well-designed fullerene host molecules based on VDW forces alone. Therefore, the ES interaction remains the most probable driving force for the complexation. On the other hand, C_{60} can be regarded as a special molecule with highly symmetric π-system; the electrostatic difference between 5:6 and 6:6 fusions is characteristic of the fullerene surface. However, the supramolecular properties of C_{60} are basically similar to those of CNs, although CNs have little 5:6 fusion in the π system. The ES interaction would be substantially operative between curved conjugated systems in addition to the dispersion force. The charge distribution owing to the difference between 5:6 and 6:6 ring fusions would control the mutual orientation in the crystal packing of complexes. The attractive interactions would also play an important role in the spontaneous formation of fullerene peapods and other new materials based on CNs. Further experimental and theoretical studies on these complexes and related substances will deepen understanding on the novel nature of fullerene and other curved π-electron systems.

References

[1] H. W. Kroto, J. R. Heath, S. C. O'Brien, R. F. Curl, and R. E. Smalley, C_{60}: Buckminsterfullerene, *Nature*, **318**, 162–163 (1985).

[2] W. Krätchmer, L. D. Lamb, K. Fostiropoulos, and D. R. Huffman, Solid C_{60}: a new form of carbon, *Nature*, **347**, 354–358 (1990).

[3] C. L. Raston,In *Comprehensive Supramolecular Chemistry*; J. L. Atwood, J. E. D. Davies, D. D. Macnicol and F. Vögtle, Eds.; Pergamon; Oxford, vol. 1, p. 777 (1996).

[4] F. Diederich and M. Gomez-Lopez, Supramolecular fullerene chemistry, *Chem. Soc. Rev.*, **28**, 263–277 (1999).

[5] M. J. Hardie and C. L. Raston, Confinement and recognition of icosahedral main group cage molecules: fullerene C60 and *o*-, *m*-, *p*-dicarbadodecacarborane(12), *Chem. Commun.*, 1153–1163 (1999).

[6] A. L. Balch and M. M. Olmstead, Reactions of transition metal complexes with fullerenes (C_{60}, C_{70}, etc.) and related materials, *Chem. Rev.*, **98**, 2123–2165 (1998).

[7] C. A. Reed and R. D. Bolsker, Discrete fulleride anions and fullerenium cations, *Chem. Rev.*, **100**, 1075–1120 (2000).

[8] K. E. Geckeler, *Advanced Macromolecular and Supramolecular Materials and Processes*; Ed.; Kluwer Academic/Plenim; New York, 2003.

[9] T. Da, Ros and M. Proto, Medicinal chemistry with fullerenes and fullerene derivatives, *Chem. Commun.*, 663–669 (1999).

[10] R. V. Bensasson, E. Bienvenue, M. Dellinger, S. Leach and P. Seta, C_{60} in model biological systems. A visible-UV absorption study of solvent-dependent parameters and solute aggregation, *J. Phys. Chem.*, **98**, 3492–3500 (1994).
[11] J. Catalan, J. L. Saiz, J. L. Laynez, N. Jagerovic and J. Elguero, The colors of C_{60} solutions, *Angew. Chem. Int. Ed. Engl.*, **34**, 105–107 (1995).
[12] Y.-P. Sun, C. E. Bunker and B. Ma, Quantitative studies of ground and excited state charge transfer complexes of fullerenes with *N,N*-dimethylaniline and *N,N*-diethylaniline, *J. Am. Chem. Soc.*, **116**, 9692–9699 (1994).
[13] G. Sarova and M. N. Berberan-Santos, Stable charge-transfer complexes versus contact complexes. Application to the interaction of fullerenes with aromatic hydrocarbons, *J. Phys. Chem. B*, **108**, 17261–17268 (2004).
[14] A. Izuoka, T. Tachikawa, T. Sugawara, Y. Suzuki, M. Konno, Y. Saito and H. Shinohara, An X-ray crystallographic analysis of a $(BEDT\text{-}TTF)_2C_{60}$ charge-transfer complex, *J. Chem. Soc., Chem. Commun.*, 1472–1473 (1992).
[15] F. Diederich, J. Effing, L. Jonas, L. Jullien, T. Plesnivy, H. Ringsdorf, C. Thilgen and D. Weinstein, *Angew. Chem. Int. Ed. Engl.*, **31**, 1599–1602 (1992).
[16] T. Andersson, K. Nilsson, M. Sundahl, G. Westman and O. Wennerström, C_{60} Embedded in γ-cyclodextrin: a water-soluble fullerene, *J. Chem. Soc., Chem. Commun.*, 604–606 (1992).
[17] Z. Yoshida, H. Takekuma, S. Takekuma and Y. Matsubara, Molecular recognition of C_{60} with γ-cyclodextrin, *Angew. Chem. Int. Ed. Engl.*, **33**, 1597–1599 (1994).
[18] A. Ikeda, T. Sue, M. Akiyama, K. Fujioka, T. Shigematsu, Y. Doi, J. Kikuchi, T. Konishi and R. Nakajima, Preparation of highly photosensitizing liposomes with fullerene-doped lipid bilayer using dispersion-controllable molecular exchange reactions, *Org. Lett.*, **10**, 4077–4080 (2008).
[19] Y. Liu, H. Wang, P. Liang and H.-Y. Zhang, Water-soluble supramolecular fullerene assembly mediated by metallobridged β-cyclodextrins, *Angew. Chem. Int. Ed.*, **43**, 2690–2694 (2004).
[20] Y. Nishibayashi, M. Saito, S. Uemura, S. Takekuma, H. Takekuma and Z. Yoshida, Buckminsterfullerenes: A non-metal system for nitrogen fixation, *Nature*, **428**, 279 (2004).
[21] M. F. Meidine, P. B. Hitchcock, H. W. Kroto, R. Taylor and D. R. M. Walton, Single crystal X-ray structure of benzene-solvated C_{60}, *J. Chem. Soc., Chem. Commun.*, 1534–1537 (1992).
[22] L. Balch, V. J. Catalano, J. W. Lee and M. M. Olmstead, Supramolecular Aggregation of an $(\eta^2\text{-}C_{60})$ Iridium complex involving phenyl chelation of the fullerene, *J. Am. Chem. Soc.*, **114**, 5455–5457 (1992).
[23] H. Suezawa, T. Yoshida, S. Ishihara, Y. Umezawa and M. Nishio, CH/π interactions as disclosed on the fullerene convex surface. A database study, *CrystEngComm.*, **5**, 514–518 (2003).
[24] E. M. Veen, P. M. Postma, H. T. Jonkman, A. L. Spek and B. L. Feringa, Solid state organization of C_{60} by inclusion crystallization with triptycenes, *Chem. Commun.*, 1709–1710 (1999).
[25] B. Kräutler, T. Müller, J. Maynollo, K. Gruber, C. Kratky, P. Ochsenbein, D. Schwarzenbach and H.-B. Bürgi, A topochemically controlled, regiospecific fullerene bisfunctionalization, *Angew. Chem. Int. Ed. Engl.*, **35**, 1204–1207 (1996).
[26] H. Bredenköuer, S. Henne and D. Volkmer, Nanosized ball joints constructed from C_{60} and tribenzotriquinacene sockets: synthesis, component self-assembly and structural investigations, *Chem.—Eur. J.*, **13**, 9931–9938 (2007).
[27] P. E. Georghiou, L. N. Dawe, H.-A. Tran, J. Strübe, B. Neumann, H.-G. Stammler and D. Kuck, C_{3v}-Symmetrical tribenzotriquinacenes as hosts for C_{60} and C_{70} in solution and in the solid state, *J. Org. Chem.*, **73**, 9040–9047 (2008).
[28] J. L. Atwood, G. A. Koutsantonis and C. L. Raston, Purification of C_{60} and C_{70} by selective complexation with calixarenes, *Nature*, **368**, 229–231 (1994); C. L. Raston, J. L. Atwood, P. J. Nichols and I. B. N. Sudria, Supramolecular encapsulation of aggregates of C_{60}, *Chem. Commun.*, 2615–2616 (1996).
[29] T. Suzuki, K. Nakashima and S. Shinkai, Very convenient and efficient purification method for fullerene (C_{60}) with 5,11,17,23,29,35,41,47-Octa-tert-butylcalix[8]arene-49,50,51,52,53,54,55,56-octol, *Chem. Lett.*, 699–702 (1994).
[30] J. L. Atwood, L. J. Barbour, C. L. Raston and I. B. N. Sudria, C_{60} and C_{70} Compounds in the pincerlike jaws of calix[6]arene, *Angew. Chem. Int. Ed.*, **37**, 981–983 (1998).

[31] J. L. Atwood, L. J. Barbour, M. W. Heaven and C. L. Raston, Controlling van der Waals contacts in complexes of fullerene C_{60}, *Angew. Chem. Int. Ed.*, **36**, 3254–3257 (2003).
[32] K. Tsubaki, K. Tanaka, T. Kinoshita and K. Fuji, Complexation of C_{60} with hexahomooxacalix [3]arenes and supramolecular structures of complexes in the solid state, *Chem. Commun.*, 895–896 (1998).
[33] J. W. Steed, P. C. Junk, J. L. Atwood, M. J. Barnes, C. L. Raston and R. S. Burkhalter, Ball and Socket Nanostructures: new supramolecular chemistry based on cyclotriveratrylene, *J. Am. Chem. Soc.*, **116**, 10346–10347 (1994).
[34] A. Ikeda, Y. Suzuki, M. Yoshimura and S. Shinkai, On the prerequisites for the formation of solution complexes from [60]fullerene and calix[n]arenes: a novel allosteric effect between [60]fullerene and metal cations in calix[n]aryl ester complexes, *Tetrahedron*, **54**, 2497–2508 (1998).
[35] L. J. Barbour, G. W. Orr and J. L. Atwood, Supramolecular intercalation of C_{60} into a calixarene bilayer—a well-ordered solid-state structure dominated by van der Waals contacts, *Chem. Commun.*, 1439–1450 (1997).
[36] T. Haino, M. Yanase and Y. Fukazawa, New supramolecular complex of C_{60} based on calix[5] arene–Its structure in the crystal and in solution, *Angew. Chem. Int. Ed. Engl.*, **36**, 259–260 (1997).
[37] P. E. Georghiou, A. H. Tran, S. S. Stroud and D. W. Thompson, Supramolecular complexation studies of [60]Fullerene with calix[4]naphthalenes–a reinvestigation, *Tetrahedron*, **62**, 2036–2044 (2006).
[38] S. Mizyed, M. Ashram, D. O. Miller and P. E. Georghiou, Supramolecular complexation of [60] fullerene with hexahomotrioxacalix[3]naphthalenes: a new class of naphthalene-based calixarenes, *J. Chem. Soc., Perkin Trans.*, **2**, 1916–1919 (2001).
[39] K. Araki, K. Akao, A. Ikeda, T. Suzuki and S. Shinkai, Molecular design of calixarene-based host molecules for inclusion of C_{60} in solution, *Tetrahedron Lett.*, **37**, 73–77 (1996); S. Shinkai and A. Ikeda, Novel interactions of calixarene p-systems with metal ions and fullerenes, *Pure Appl. Chem.*, **71**, 275–280 (1999).
[40] H. Matsubara, A. Hasegawa, K. Shiwaku, K. Asano, M. Uno, S. Takahashi and K. Yamamoto, Supramolecular inclusion complexes of fullerenes using cyclotriveratrylene derivatives with aromatic pendants, *Chem. Lett.*, 923–924 (1998).
[41] J.-F. Nierengarten, L. Oswald, J.-F. Eckert, J.-F. Nicoud and N. Armaroli, Complexation of fullerenes with dendritic cyclotriveratrylene derivatives, *Tetrahedron Lett.*, **40**, 5681–5684 (1999).
[42] M. Numata, A. Ikeda, C. Fukuhara and S. Shinkai, Dendrimers can act as a host for [60] fullerene, *Tetrahedron Lett.*, **40**, 6945–6948 (1999).
[43] D. I. Schuster, J. Rosenthal, S. MacMahon, P. D. Jarowski, C. A. Alabi and D. M. Guldi, Formation and photophysics of a stable concave-convex supramolecular complex of C_{60} and a substituted *s*-triazine derivative, *Chem. Commun.*, 2538–2539 (2002).
[44] M.-X. Wang, X.-H. Zhang and Q.-Y. Zheng, Synthesis, structure, and [60]fullerene complexation properties of azacalix[*m*]arene[*n*]pyridines, *Angew. Chem. Int. Ed.*, **43**, 838–843 (2004).
[45] E.-X. Zhang, D.-X. Wang, Q.-Y. Zheng and M.-X. Wang, Synthesis of large macrocyclic azacalix[*n*]arenes ($n = 6-9$) and their complexation with fullerenes C_{60} and C_{70}, *Org. Lett.*, **10**, 2565–2568 (2008).
[46] L. Stella, A. L. Capodilupo and M. Bietti, A reassessment of the association between azulene and [60]fullerene. Possible pitfalls in the determination of binding constants through fluorescence spectroscopy, *Chem. Commun.*, 4744–4746 (2008).
[47] Y. Matsuo and E. Nakamura, Selective multiaddition of organocopper reagents to fullerenes, *Chem. Rev.*, **108**, 3016–3028 (2008).
[48] M. Sawamura, K. Kawai, Y. Matsuo, K. Kanie and E. Nakamura, Stacking of conical molecules with a fullerene apex into polar columns in crystals and liquid crystals, *Nature*, **419**, 702–705 (2002).
[49] Y. Matsuo, A. Muramatsu, Y. Kamikawa, T. Kato and E. Nakamura, Synthesis and structural, electrochemical, and stacking properties of conical molecules possessing buckyferrocene on the apex, *J. Am. Chem. Soc.*, **128**, 9586–9587 (2006).

[50] T. Haino, M. Yanase and Y. Fukazawa, Fullerenes enclosed in bridged calix[5]arenes, *Angew. Chem. Int. Ed.*, **37**, 997–998 (1998).

[51] J. C. Iglesias-Sanchez, A. Fragoso, J. de Mendoza and P. Prados, Aryl-aryl linked Bi-5,5′-p-tert-butylcalix[4]arene tweezer for fullerene complexation, *Org. Lett.*, **8**, 2571 (2006).

[52] J. Wang and C. D. Gutsche, Complexation of fullerenes with bis-calix[n]arenes synthesized by tandem claisen rearangement, *J. Am. Chem. Soc.*, **120**, 12226–12231 (1998).

[53] H. Matsubara, T. Shimura, A. Hasegawa, M. Semba, K. Asano and K. Yamamoto, Syntheses of novel fullerene tweezers and their supramolecular inclusion complex of C_{60}, *Chem. Lett.*, 1099–1110 (1999).

[54] T. Haino, C. Fukunaga and Y. Fukazawa, A new calix[5]arene-based container: selective extraction of higher fullerenes, *Org. Lett.*, **8**, 3545–3548 (2006).

[55] H. Matsubara, S. Oguri, K. Asano and K. Yamamoto, Syntheses of novel cyclotriveratrylenophane capsules and their supramolecular complexes of fullerenes, *Chem. Lett.*, 431–432 (1999).

[56] A. Ikeda, M. Yoshimura, H. Uzdu, C. Fukuhara and S. Shinkai, Inclusion of [60]fullerene in a homooxacalix[3]arene-based dimeric capsules cross-linked by a Pd^{II}-pyridine interaction, *J. Am. Chem. Soc.*, **121**, 4296–4297 (1999).

[57] C. G. Claessens and T. Torres, Inclusion of C_{60} fullerene in a M_3L_2 subphthalocyanine cage, *Chem. Commun.*, 1298–1299 (2004).

[58] L. Pirondini, D. Bonifazi, B. Cantadori, P. Braiuca, M. Campagnolo, R. D. Zorzi, S. Geremia, F. Diederich and E. Dalcanale, Inclusion of methano[60]fullerene derivatives in cavitand-based coordination cages, *Tetrahedron*, **62**, 2008–2015 (2006).

[59] T. Haino, H. Araki, Y. Yamanaka and Y. Fukazawa, Fullerene receptor based on calyx[5]arene through metal-assisted self-assembly, *Tetrahedron Lett.*, **42**, 402 (2001).

[60] M. Yanase, T. Haino and Y. Fukazawa, A self-assembling molecular container for fullerenes, *Tetrahedron Lett.*, **40**, 2781–2784 (1999).

[61] E. Huerta, G. A. Metselaar, A. Fragoso, E. Santos, C. Bo and J. de Mendoza, Selective binding and easy separation of C_{70} by nanoscale self-assembled capsules, *Angew. Chem. Int. Ed.*, **46**, 202–205 (2007).

[62] G. Dan Pantos, J.-L. Wietor and J. K. M. Sanders, Filling helical nanotubes with C_{60}, *Angew. Chem. Int. Ed.*, **46**, 2238–2240 (2007).

[63] J.-L. Weitor, G. Dan Pantos and J. K. M. Sanders, Templated amplification of an unexpected receptor for C_{70}, *Angew. Chem. Int. Ed.*, **47**, 2689–2692 (2008).

[64] T. Kawauchi, J. Kumaki, A. Kitaura, K. Okoshi, H. Kusanagi, K. Kobayashi, T. Sugai, H. Shinohara and E. Yashima, Encapsulation of fullerenes in a helical PMMA cavity leading to a robust processable complex with a macromolecular helicity, *Angew. Chem. Int. Ed.*, **47**, 515–519 (2008).

[65] T. Kawauchi, A. Kitaura, J. Kumaki, H. Kusanagi and E. Yashima, Helix-sense-controlled synthesis of optically active poly(methyl methacrylate) stereocomplexes, *J. Am. Chem. Soc.*, **130**, 11889–11891 (2008).

[66] Y. Sun, T. Drovetskaya, R. D. Bolskar, R. Bau, P. D. W. Boyd and C. A. Reed, Fulleride of pyrrolidine-functionalized C_{60}, *J. Org. Chem.*, **62**, 3642–3649 (1997).

[67] P. D. W. Boyd and C. A. Reed, Fullerene – Porphyrin constructs, *Acc. Chem. Res.*, **38**, 235–242 (2005).

[68] K. Tashiro, T. Aida, J.-Y. Zheng, K. Kinbara, Saigo, S. Sakamoto and K. Yamaguchi, A cyclic dimer of metalloporphyrin forms a highly stable inclusion complex with C_{60}, *J. Am. Chem. Soc.*, **121**, 9477–9478 (1999).

[69] M. Shirakawa, N. Fujita, H. Shimakoshi, Y. Hisaeda and S. Shinkai, Molecular programming of organogelators which can accept [60]fullerene by encapsulation, *Tetrahedron*, **62**, 2016–2024 (2006).

[70] M. Kimura, Y. Saito, K. Ohta, K. Hanabusa, H. Shirai and N. Kobayashi, Self-organization of supramolecular complex composed of rigid dendritic porphyrin and fullerene, *J. Am. Chem. Soc.*, **124**, 5274–5275 (2002).

[71] D. Sun, F. S. Tham, C. A. Reed, L. Chaker, M. Burgess and P. D. W. Boyd, Porphyrin-fullerene host-guest chemistry, *J. Am. Chem. Soc.*, **122**, 10704–10705 (2000).

[72] D. Sun, F. S. Tham, C. A. Reed, L. Chaker and P. D. W. Boyd, Supramolecular fullerene-porphyrin chemistry. fullerene complexation by metalated 'jaws porphyrin' hosts, *J. Am. Chem. Soc.*, **124**, 6604–6612 (2002).
[73] Z.-Q. Wu, X.-B. Shao, C. Li, J.-L. Hou, K. Wang, X.-K. Jiang and Z.-T. Li, Hydrogen-bonding-driven preorganized zinc porphyrin receptors for efficient complexation of C_{60}, C_{70}, and C_{60} Derivatives, *J. Am. Chem. Soc.*, **127**, 17460–17468 (2005).
[74] Y. Eda, K. Itoh, Y. N. Ito and T. Kawata, 2,6-Bis(porphyrin)-substituted pyrazine: a new class of supramolecular synthon binding to a transition-metal ion and fullerene (C_{60}), *Tetrahedron*, **65**, 282–288 (2008).
[75] T. Yamaguchi, N. Ishii, K. Tashiro and T. Aida, Supramolecular peapods composed of a metalloporphyrin nanotube and fullerenes, *J. Am. Chem. Soc.*, **125**, 13934–13935 (2003).
[76] M. Ayabe, A. Ikeda, Y. Kubo, M. Takeuchi and S. Shinkai, A dendritic porphyrin receptor for C_{60} which features a profound positive allosteric effect, *Angew. Chem. Int. Ed.*, **41**, 2790–2792 (2002).
[77] J.-Y. Zheng, K. Tashiro, Y. Hirabayashi, K. Kinbara, K. Saigo, T. Aida, S. Sakamoto and K. Yamaguchi, Cyclic dimers of metalloporphyrins as tunable hosts for fullerenes: a remarkable effect of rhodium(III), *Angew. Chem. Int. Ed.*, **40**, 1857–1861 (2001).
[78] K. Tashiro and T. Aida, Metalloporphyrin hosts for supramolecular chemistry of fullerenes, *Chem. Soc. Rev.*, **36**, 189–197 (2007).
[79] M. Yanagisawa, K. Tashiro, M. Yamasaki and T. Aida, Hosting fullerenes by dynamic bond formation with an iridium porphyrin cyclic dimer: a 'chemical friction' for rotary guest motions, *J. Am. Chem. Soc.*, **129**, 11912–11913 (2007).
[80] Y. Shoji, K. Tashiro and T. Aida, Sensing of chiral fullerenes by a cyclic host with an asymmetrically distorted π-electronic component, *J. Am. Chem. Soc.*, **128**, 10690–10691 (2006).
[81] H. Sato, K. Tashiro, H. Shinmori, A. Osuka, Y. Murata, K. Komatsu and T. Aida, Positive heterotropic cooperativity for selective guest binding via electronic communications through a fused zinc porphyrin array, *J. Am. Chem. Soc.*, **127**, 13086–13087 (2005).
[82] A. L. Kieran, S. I. Pascu, T. Jarrosson and J. K. M. Sanders, Inclusion of C_{60} into an adjustable porphyrin dimmer generated by dynamic disulfide chemistry, *Chem. Commun.*, 1276–1278 (2005).
[83] H. Nobukuni, Y. Shimazaki, F. Tani and Y. Naruta, A nanotube of cyclic porphyrin dimers connected by nonclassical hydrogen bonds and its inclusion of C_{60} in a linear arrangement, *Angew. Chem. Int. Ed.*, **46**, 8975–8978 (2007).
[84] Y.-B. Wang and Z. Lin, Supramolecular interactions between fullerenes and porphyrins, *J. Am. Chem. Soc.*, **125**, 6072–6073 (2003).
[85] X. Peng, N. Komatsu, S. Bhattacharya, T. Shimawaki, S. Aonuma, T. Kimura and A. Osuka, Optically active single-walled carbon nanotubes, *Nature Nanotechnology*, **2**, 361–365 (2007); X. Peng, N. Komatsu, T. Kimura and A. Osuka, Improved optical enrichment of SWNTs through extraction with chiral nanotweezers of 2,6-pyridylene-bridged diporphyrins, *J. Am. Chem. Soc.*, **129**, 15947–15953 (2007).
[86] H. Li, B. Zhou, L. Gu, W. Wang, K. A. S. Fernando, S. Kumar, L. F. Allard and Y.-P. Sun, Selective interactions of porphyrins with semiconducting single-walled carbon nanotubes, *J. Am. Chem. Soc.*, **126**, 1014–1015 (2004).
[87] H. Becker, G. Javahery, S. Petrie, P. C. Cheng, H. Schwarz, L. T. Scott and D. K. Bohme, Gas-phase ion/molecular reactions of corannulene, a fullerene subunit, *J. Am. Chem. Soc.*, **115**, 11636–11637 (1993).
[88] P. E. Georghiou, A. H. Tran, S. Mizyed, M. Bancu and L. T. Scott, Concave polyarenes with sulfide-linked flaps and tentacles: new electron-rich hosts for fullerenes, *J. Org. Chem.*, **70**, 6158–6163 (2005).
[89] Z. Wang, F. Dötz, V. Enkelmann and K. Müllen, 'Double-concave' graphene: permethoxylated hexa-peri-hexabenzocoronene and its cocrystals with hexafluorobenzene and fullerene, *Angew. Chem. Int. Ed.*, **44**, 1247–1250 (2005).
[90] E. M. Perez, M. Sierra, L. Sanchez, M. Rosario, Torres, R. Viruela, P. M. Viruela, E. Orti and N. Martin, Concave tetrathiafulvalene-type donors as supramolecular partners for fullerenes, *Angew. Chem. Int. Ed.*, **46**, 1847–1851 (2007).

[91] A. Sygula, F. R. Fronczek, R. Sygula, P. W. Rabideau and M. M. Olmstead, A double concave hydrocarbon buckycatcher, *J. Am. Chem. Soc.*, **129**, 3842–3843 (2007).
[92] E. M. Perez, L. Sanchez, G. Fernadez and N. Martin, exTTF as a building block for fullerene receptors unexpected solvent-dependent positive homotropic cooperativity, *J. Am. Chem. Soc.*, **128**, 7172–7173 (2006).
[93] E. M. Ferez, A. L. Capodilupo, G. Fernandez, L. Sanchez, P. M. Viruela, R. Viruela, E. Orti, M. Bietti and N. Martin, Weighting non-covalent forces in the molecular recognition of C_{60}. Relevance of concave-convex complementarity, *Chem. Commun.*, 4567–4569 (2008).
[94] T. Kawase, H. R. Darabi and M. Oda, Cyclic [6]- and [8]Paraphenylacetylenes, *Angew. Chem. Int. Ed. Engl.*, **35**, 2662–2664 (1996).
[95] T. Kawase and H. Kurata, Carbon nanorings and their complexing abilities: exploration of the concave-convex π-π Interaction, *Chem. Rev.*, **106**, 5250–5273 (2006).
[96] T. Kawase, K. Tanaka, H. R. Darabi and M. Oda, Complexation of a carbon nanoring with fullerenes, *Angew. Chem. Int. Ed.*, **42**, 1624–1628 (2003).
[97] T. Kawase and M. Oda, Complexation of carbon nanorings with fullerenes, *Pure Appl. Chem.*, **78**, 831–839 (2006).
[98] T. Kawase, K. Tanaka, Y. Seirai, N. Shiono and M. Oda, Complexation of carbon nanorings with fullerenes. Novel supramolecular dynamics and structural tuning for a fullerene sensor, *Angew. Chem. Int. Ed.*, **42**, 5597–5600 (2003).
[99] T. Kawase, N. Fujiwara, M. Tsutsumi, M. Oda, Y. Maeda, T. Wakahara and T. Akasaka, Supramolecular dynamics of cyclic [6]paraphenyleneacetylene complexes with [60]- and [70] fullerene derivatives: electronic and structural effect to the complexation, *Angew. Chem. Int. Ed.*, **43**, 5060–5062 (2004).
[100] T. Kawase, K. Tanaka, N. Shiono, Y. Seirai and M. Oda, Onion-type complexation based on carbon nanorings and a buckminsterfullerene, *Angew. Chem. Int. Ed.*, **43**, 1722–1724 (2004).
[101] T. Kawase, Y. Nishiyama, T. Nakamura, T. Ebi, K. Matsumoto, H. Kurata and M. Oda, Cyclic [5]Paraphenyleneacetylene: synthesis, properties and formation of a ring-in-ring complex showing considerably large association constant and entropy effect, *Angew. Chem. Int. Ed.*, **46**, 1086–1088 (2007).
[102] D. Zhao and J. S. Moore, Shape-persistent aryl ethynylene macrocycles: syntheses and supramolecular chemistry, *Chem. Commun.*, 807–818 (2003).

8

Molecular Surgery toward Organic Synthesis of Endohedral Fullerenes

Michihisa Murata[a], Yasujiro Murata[a] and Koichi Komatsu[b]

[a]Institute for Chemical Research, Kyoto University, Japan
[b]Department of Environmental and Biotechnological Frontier Engineering, Fukui University of Technology, Japan

8.1 Introduction

Endohedral fullerenes, the cage-closed carbon molecules that incorporate atom(s) or molecule(s) inside the cage [1–6], are not only of scientific interest but are also expected to be important for their potential use in various fields such as electronics [7], magnetic resonance imaging as a contrast agent [8], and NMR analysis [9, 10]. However, development of their applications has been hampered by a severe limitation in their production, which has relied only on physical methods, such as co-vaporization of carbon and metal atoms [2, 3] and high-pressure/high-temperature treatment with noble gases [9–13], that are difficult to control and yield only milligram quantities of pure product after laborious isolation procedures.

Toward the solution of this issue Rubin proposed a concept to realize endohedral fullerenes by the use of organic reactions, that is, 'molecular surgery' [14–16]. This approach consists of a series of steps, which are 'incision' of the fullerene cage to form an opening on the surface, insertion of some small atom(s) or molecule(s) through the opening, and 'suture' of the opening to reproduce the fullerene cage while retaining the guest species inside. Toward this purpose, cage-opened C_{60} derivative **1** with a 14-membered-ring opening has been synthesized (Figure 8.1) [17], and the insertion of

Chemistry of Nanocarbons Edited by Takeshi Akasaka, Fred Wudl and Shigeru Nagase

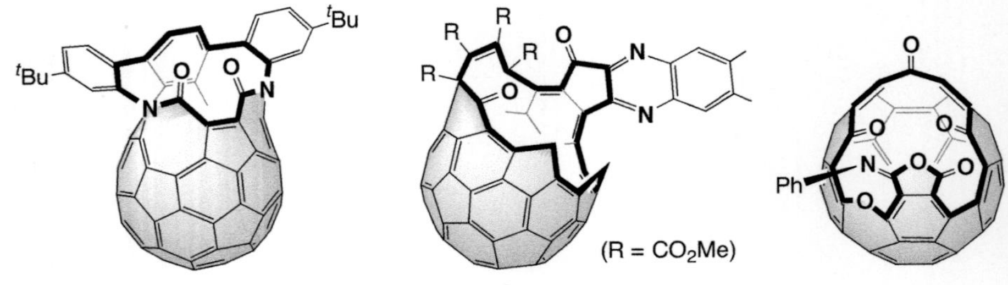

Figure 8.1 *Representative cage-opened C_{60} derivatives with a sufficiently large opening for insertion of an atom or a small molecule*

a H_2 molecule or a He atom into **1** has been achieved at occupation levels of 5% or 1.5%, respectively [18]. Thereafter, bowl-shaped compound **2** with a 20-membered-ring opening has been synthesized [19]. Although until then the species encapsulated through an opening on the C_{60} cage had been limited only to a He atom or a H_2 molecule, it has been demonstrated that **2** can encapsulate a larger molecule such as H_2O (occupation level, 75%) [19], CO (84%) [20], or even NH_3 (35–50%) [21]. In addition, compound **3** with a 19-membered-ring opening has been synthesized and the encapsulation of an H_2O molecule (88%) has been achieved [22]. These bowl-shaped compounds **2** and **3** derived from C_{60} are quite attractive as the novel host molecules. However, from the viewpoint of molecular surgery, the restoration of such severely ruptured π-systems to the original structure of C_{60} by means of organic synthetic procedures appears to be a highly difficult task.

In this chapter is described the first successful accomplishment of the molecular surgery operation to provide new endohedral fullerenes encapsulating molecular hydrogen. Fundamental properties and reactions of the endohedral fullerenes containing H_2 molecule(s) are also summarized.

8.2 Molecular-Surgery Synthesis of Endohedral C_{60} Encapsulating Molecular Hydrogen

8.2.1 Cage Opening

This section describes the first step of the molecular surgical operation toward endohedral C_{60} encapsulating molecular hydrogen, i.e., 'incision' of C_{60} cage.

In our previous work, reactions of C_{60} with diaza- [23], triaza- [24], and tetraaza-aromatic compounds [25, 26] had been conducted as shown in Scheme 8.1. Particularly, the thermal reactions with phthalazine (**4**) in refluxing 1-chloronaphthalene (1-Cl-Naph) or with 4,6-dimethyl-1,2,3-triazine (**6**) in refluxing *o*-dichlorobenzene (ODCB) resulted in the formation of cage-opened C_{60} derivatives **5** [23] and **7** [24], respectively, having an eight-membered-ring opening in one-pot. On the other hand, the solid-state reaction with 3,6-di (2-pyridyl)-1,2,4,5-tetrazine (**8**) using the high-speed vibration milling (HSVM) [27, 28] technique gave the monoadduct **9**, which was transformed upon contact with silica-gel to 1,2,3,4-tetrahydro-C_{60} derivative **10** [26]. The characteristic of the methods for the creation

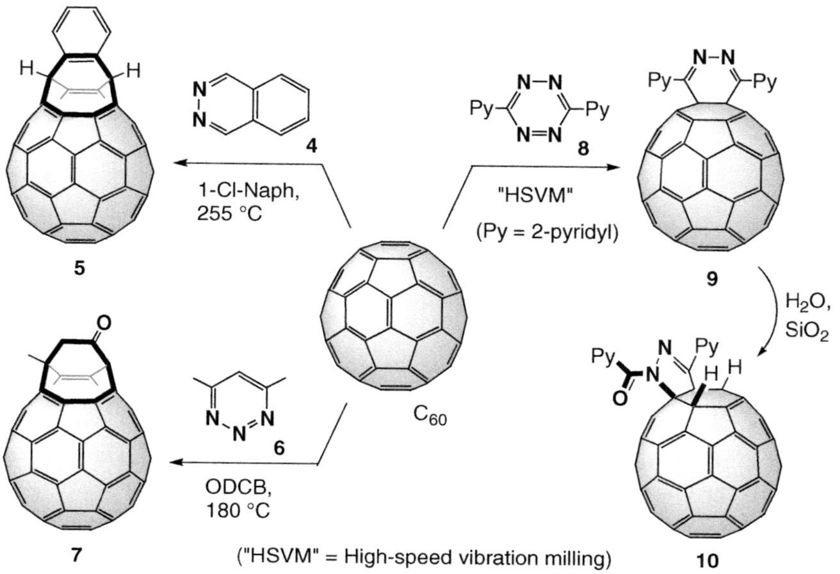

Scheme 8.1

of an opening resides in the simple and facile synthetic procedure. Hence, we searched for appropriate aza-aromatics and found that 5,6-diphenyl-3-(2-pyridyl)-1,2,4-triazine (**11**) is useful to make cage-opened fullerenes as will be described below.

A thermal reaction of C_{60} with **11** in refluxing ODCB for 17 hours proceeded to give cage-opened fullerene **15** in 85% yield based on consumed C_{60} (59% conversion) (Scheme 8.2) [29]. This reaction must have undergone through an initial [4+2] cycloaddition of **11** with C_{60} to give adduct **12** and the following extrusion of a N_2 molecule to give 2-aza-1,3-cyclohexadiene-fused C_{60} derivative **13**. Then, a 'formal' intramolecular [4+4] reaction at high temperature and subsequent retro[2+2+2] reaction provide **15** with an eight-membered-ring opening [29, 30]. The X-ray structure of **15** indicated that C=C double bonds on the rim of the opening are considerably distorted (twist angle, 39.0° and 38.8°) and the coefficients of the HOMO are relatively high on these carbons. Thus, singlet oxygen, generated by irradiation of visible light to the system, selectively reacts with one of these double bonds, resulting in enlargement of the opening to give diketones **16** and **17** with a 12-membered-ring opening in 60% and 31% yields, respectively, along with dioxa-

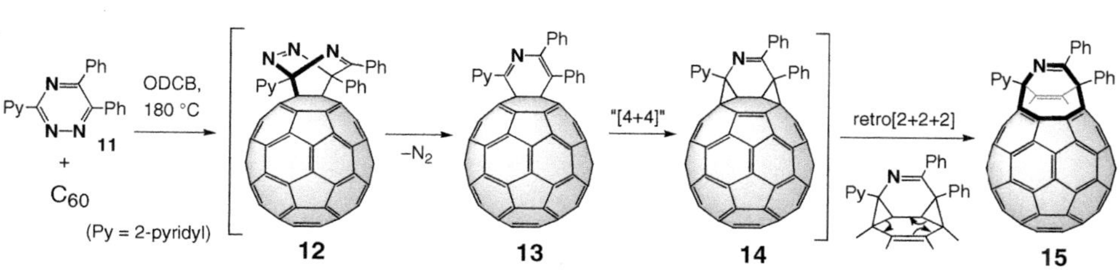

Scheme 8.2

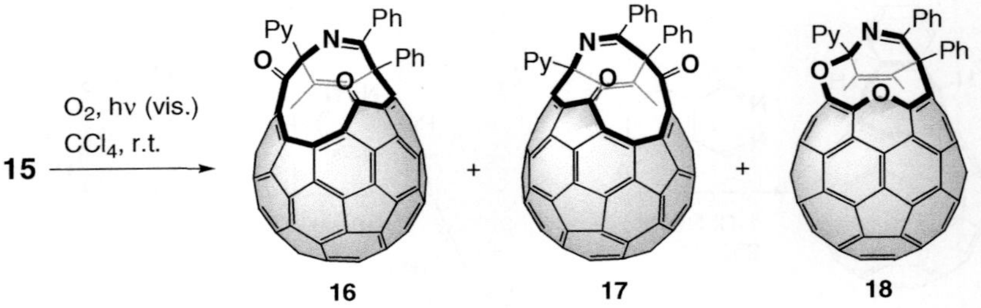

Scheme 8.3

compound **18** with a 10-membered-ring opening in 2% yield (Scheme 8.3) [29, 31, 32]. However, the 12-membered-ring opening in the major product **16** was not large enough even for a molecule of H_2 to pass through as indicated by a calculated high energy barrier (51.8 kcal mol^{-1}) (B3LYP/6-31G**//B3LYP/3-21G). Further enlargement of the opening was apparently necessary.

Compound **15** was expected to be activated by a typical electron donor such as tetrakis-(dimethylamino)ethylene (TDAE), since the cyclic voltammogram indicated that **16** is even a better π-electron acceptor than C_{60} as shown by the lower reduction potential by about 0.2 V possibly due to the presence of two carbonyl groups [29]. The LUMO coefficients of **16** were relatively high on the butadiene carbons in the rim of the opening [29]. Thus, in the presence of TDAE, **16** was found to react with elemental sulfur upon refluxing in ODCB for 0.5 hour to afford compound **19** having a 13-membered-ring opening as a single product in 77% yield (Scheme 8.4) [29]. The structure of **19** was unambiguously proved by X-ray crystallography. The size of the opening of **19** was 5.64 Å for a longer axis and 3.75 Å for a shorter axis. The calculated activation energy (B3LYP/6-31G**//B3LYP/3-21G) for the insertion of a H_2 molecule through the opening of **19** (30.1 kcal mol^{-1}) [33] was lower than that calculated for bislactam **1** (41.4 kcal mol^{-1}) [18] by 11.3 kcal mol^{-1} in support of the larger opening of **19**. Thus, the encapsulation of a H_2 molecule to **19** was expected to be easier.

The enlargement of the opening of **16** was also possible by the insertion of a selenium atom in place of a sulfur atom to the rim of the opening [34]. In this case, stronger activation by the use of sodium alkanethiolate [35] was needed. The reaction of **16** with elemental selenium in the presence of CH_3SNa in refluxing ODCB for 2 hours afforded cage-opened fullerene **20** in 46% yield (Scheme 8.4). The results of X-ray crystallography showed that

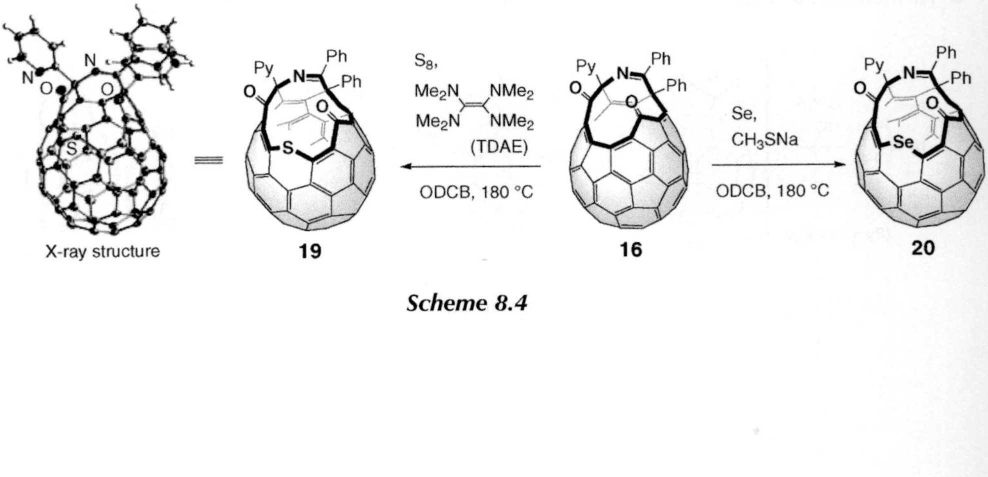

Scheme 8.4

the size of the opening in **20** was 5.72 Å for a longer axis and 3.88 Å for a shorter axis, which was slightly larger than that of sulfur analogue **19** because of a slightly longer C–Se bond than a C–S bond [34].

8.2.2 Encapsulation of a H_2 Molecule

As a guest molecule to be encapsulated in the cage-opened fullerene **19**, the smallest molecule, H_2, was selected. When a powder of **19** was treated with high-pressure H_2 gas (800 atm) at 200 °C for 8 hours in an autoclave, the 100% incorporation of a H_2 molecule inside the cage was realized [33]. The ^1H NMR of the resulting material in ODCB-d_4 showed an intense signal of the encapsulated H_2 molecule at δ –7.25 ppm. The occupation level was highly dependent on the pressure of H_2 gas: the yield of H_2@**19** was 90% at 560 atm and 51% at 180 atm, with all other conditions being the same. The integrated intensity of the H_2 signal relative to that of one of the pyridyl proton signals demonstrated that 100% encapsulation was achieved. This made it possible to directly observe a single H_2 molecule at the center of the fullerene cage with the synchrotron X-ray diffraction technique with MEM (maximum entropy method) analysis [36], and also with the solid-state NMR spectroscopy [37]. For a selenium analogue **20** with a slightly larger opening, the 100% insertion of a H_2 molecule was achieved under the slightly milder conditions, that is, at 190 °C under 760 atm of H_2 gas [34].

H_2@**19** is quite stable at room temperature, but it released H_2 slowly at the temperature above 160 °C [33]. The rate of ejection monitored at 160, 170, 180, and 190 °C followed first-order kinetics. The Arrhenius plot gave an excellent linear fit, with the pre-exponential factor (A) and the activation energy (E_a) being $10^{11.8}$ and 34.2 kcal mol^{-1}, respectively [33, 34]. It is to be noted that the E_a value is close to the calculated value (30.1 kcal mol^{-1}) for the insertion of a H_2 molecule through the opening of **19** as mentioned above. The release of a H_2 molecule from selenium analogue H_2@**20** was almost three times faster than that of H_2@**19** at 160, 170, and 180 °C and the activation energy was 32.4 kcal mol^{-1} [34], reflecting the slightly larger opening of H_2@**20**. Recent investigations demonstrated that the rate of release of the encapsulated H_2 molecule can be well correlated to the size of the opening on the fullerene cage [38].

Upon MALDI-TOF mass measurement on H_2@**19**, the molecular ion peak of H_2@**19** (m/z 1068) was clearly observed. When a laser power was increased, the peak height for the molecular ion decreased and, instead, the formation of C_{60} (m/z 720) was clearly observed. More remarkable is the appearance of a peak at m/z 722, corresponding to $H_2@C_{60}$. The intensity of this peak was approximately one-third of that for C_{60}, taking the isotope distribution of C_{60} into consideration. Thus, upon laser irradiation, generation of $H_2@C_{60}$ is possible in the gas phase by self-restoration of H_2@**19** having a large opening [33].

8.2.3 Encapsulation of a He Atom

Next, the insertion of the smallest noble-gas atom, ^3He, through the opening of **19** was investigated [39]. Upon heating an ODCB solution of **19** under ^3He gas (20 atm) at 80 °C for a few hours, ^3He@**19** at an occupation level of 0.1% was formed. The release of the ^3He atom from ^3He@**19** was found to occur at the temperature close to room temperature and the activation energy was determined to be 22.8 kcal mol^{-1}, which is much lower (by 11.4 kcal mol^{-1}) than that for the release of a H_2 molecule from H_2@**19** [33, 34]. Since the

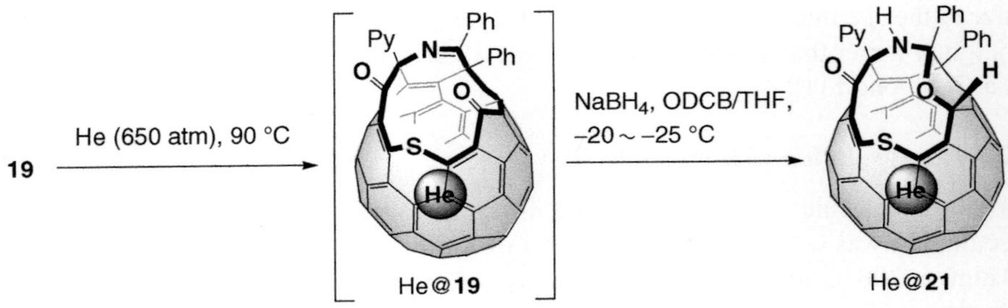

Scheme 8.5

half-life of ^{3}He release from ^{3}He@**19** was only 40.3 hours at 30 °C, it was required to reduce the opening-size of ^{3}He@**19** in order to keep the encapsulated ^{3}He atom within the cage.

Upon reduction of a carbonyl group in **19** with sodium borohydride, a transannular ether-bond formation readily took place at room temperature to give product **21** in 86% yield [40]. The structure of **21** was confirmed by the X-ray crystallography. The activation energy for release of a He atom from He@**21** was calculated to be 50.4 kcal mol^{-1} (B3LYP/6-31G**//B3LYP/3-21G), which is more than twice as large as that from He@**19** [33, 39], indicating the effective size-reduction of the opening of **19**. Thus, after treatment of **19** with He gas (650 atm) at 90 °C for 24 hours, the resulting material was immediately subjected to the sodium borohydride reduction at –20 to –25 °C (Scheme 8.5). He@**21** was successfully obtained as a stable complex in 90% yield and the occupation level of the He atom was determined to be 35% based on the mass spectroscopic analysis [40]. Although the noncovalent interaction between the encapsulated He atom and the fullerene cage of **21** was expected to be almost negligible, the NMR signal of a methine proton of He@**21** showed a slight downfield shift by 0.36 Hz as compared to that of empty **21**. The methine proton signal of hydrogen complex H_2@**21**, prepared separately, exhibited 1.9 Hz downfield shift relative to that of empty **21**. These results demonstrate that the noncovalent interaction of the encapsulated He atom with the fullerene cage of **21** is smaller than that of the encapsulated H_2 molecule. The NMR signal of the methine proton outside the cage is a good indicator of the electronic interaction inside the fullerene cage [40].

8.2.4 Closure of the Opening

As described above, C_{60} cage was incised for the creation of a 13-membered-ring opening and a H_2 molecule was completely inserted into the carbonaceous cage [29, 33]. In this section, a method to suture the opening while retaining the H_2 molecule inside the cage to complete the molecular surgery is described [41–43]. Prior to this study, there had been no report for the attempt at reducing an opening-size on the fullerene cage.

In order to restore the shape of C_{60} from cage-opened fullerene **19**, the first operation to do would be the removal of the sulfur atom out of the rim of the opening of **19** [41, 42]. An oxidation of the sulfide unit of H_2@**19** by *m*-chloroperbenzoic acid (MCPBA) was conducted to make the sulfur atom readily removable. The reaction proceeded at room temperature to give sulfoxide H_2@**22** in a quantitative yield (Scheme 8.6, step a). Then, irradiation of a benzene solution of H_2@**22** with visible light at room temperature caused elimination of the SO unit to give product H_2@**16** having a 12-membered-ring opening in

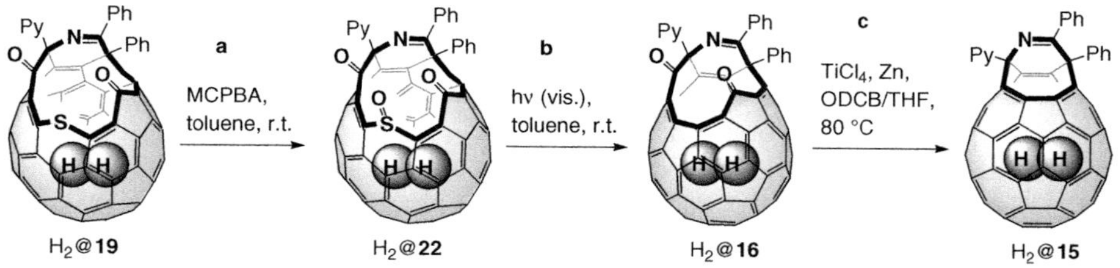

Scheme 8.6

42% yield (Scheme 8.6, step b). This removal of a sulfur atom makes the distance between two carbonyl carbons across the opening closer from 3.89 Å for H_2@**19** to 3.12 Å for H_2@**16** (B3LYP/6-31G*) [42]. Thus, the McMurry reaction [45] worked efficiently for reductive coupling of the two carbonyl groups at the opening of H_2@**16**, leading to the formation of H_2@**15** with an eight-membered-ring opening in 88% yield (Scheme 8.6, step c). Here, it is to be noted that the MALDI-TOF mass spectrum of H_2@**15** already exhibited an intense peak of $H_2@C_{60}$ together with a smaller molecular ion peak of H_2@**15** [42].

At each of these three steps, complete retention of the encapsulated H_2 molecule was confirmed by the integrated peak intensity (2.00 ± 0.05 H) of the characteristic upfield NMR signals of the encapsulated H_2 molecule (–6.33 ppm for H_2@**22**, –5.80 for H_2@**16**, and –2.95 for H_2@**15**) [41, 42].

The final step to remove all the remaining organic addends on the fullerene cage was performed by simply heating a powder of H_2@**15** (245 mg) in a glass tube under vacuum placed in an electric furnace at 340 °C for 2 hours. The crude product was dissolved in carbon disulfide and passed through a silica-gel column to give a purple solution containing desired $H_2@C_{60}$ (118 mg, contaminated by 9% empty C_{60}) in 67% yield [41, 42]. Similar results were obtained when H_2@**15** was heated at 300 °C for 24 hours, at 320 °C for 8 hours, or at 400 °C for 2 minutes. Thus, $H_2@C_{60}$ was synthesized in a total yield of 22% from H_2@**19**, which can be obtained in 40% yield from consumed C_{60} [29, 33].

The closure of the eight-membered-ring opening is considered to proceed according to the mechanism shown in Scheme 8.7 [41, 42], which is almost like a reversal of the reactions shown in Scheme 8.2. An initial [2+2+2] cyclization produces intermediate H_2@**14** having two cyclopropane rings, which undergo radical cleavage to give intermediate H_2@**13**. As to the following step, conceptually the most reasonable one is a retro[2+2+2] reaction to give

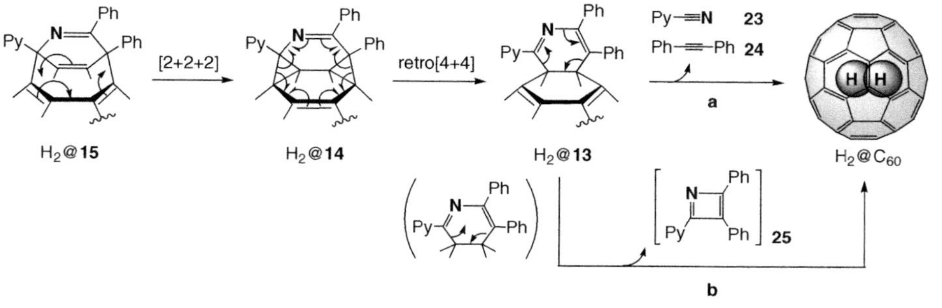

Scheme 8.7

$H_2@C_{60}$ together with 2-cyanopyridine (**23**) and diphenylacetylene (**24**) (Scheme 8.7a). Indeed, **23** and **24** were detected in the crude product. However, the reaction was not so clean, and, surprisingly, benzonitrile and 2-(phenylethynyl)pyridine were also detected together with an unknown compound having a molecular formula of 'Ph_2PyC_3N'. This latter fact indicates that occurrence of the reaction pathway shown in Scheme 8.7b cannot be ruled out, although this involves an extrusion of highly unstable species, such as azacyclobutadiene (azete) derivative **25** [42].

$H_2@C_{60}$ was completely separated from empty C_{60} by recycling HPLC on semipreparative Cosmosil Buckyprep columns (two directly connected columns; 250 mm length, 10 mm inner diameter; mobile phase, toluene; flow rate, 4 mL min^{-1}) to give $H_2@C_{60}$ as a pure material after 20 recycles (total retention time, 399 minutes; the retention time for empty C_{60}, 395 minutes). The adsorption mechanism of the Buckyprep column is largely based on the π–π interaction with pyrenyl groups in the stationary phase. A very weak van der Waals interaction between the encapsulated H_2 molecule and the C_{60} π-system must have contributed to this separation.

The ^1H NMR signal of the encapsulated H_2 molecule inside C_{60} appeared at δ –1.44 ppm in ODCB-d_4, which is 5.98 ppm upfield shifted from a signal of dissolved free H_2. This value is comparable to the 6.36 ppm upfield shift of a ^3He NMR signal for $^3He@C_{60}$ [9, 10], suggesting that this nearly 6 ppm upfield shift is a universal value corresponding to the interior magnetic shielding effect of C_{60}. The ^{13}C NMR spectrum of pure $H_2@C_{60}$ exhibited a signal at δ 142.844 ppm in ODCB-d_4, which is very slightly downfield shifted by 0.078 ppm relative to that of C_{60}. The IR spectrum of $H_2@C_{60}$ was almost the same as that of C_{60}, exhibiting four absorption bands at 1429.2, 1182.3, 576.7 and 526.5 cm^{-1} (to be compared with 1429.2, 1182.3, 575.7, and 526.5 cm^{-1} for empty C_{60} measured under the same conditions). Only the band at 576.7 cm^{-1} of $H_2@C_{60}$, corresponding to an out-of-plane vibration mode [46], is higher in energy than that of C_{60} by 1.0 cm^{-1}. This might be ascribed to a very slight repulsive interaction between C_{60} cage and inner H_2 molecule. The UV-Vis spectrum of $H_2@C_{60}$ was almost the same as that of C_{60}. The electrochemical behavior of $H_2@C_{60}$ was quite similar to that of C_{60} as far as three stepwise reduction waves (up to –2.0 V vs. Fc/Fc^+ in ODCB; –0.95, –1.37, –1.89 V) and one oxidation peak (at +1.62 V in 1,1,2,2-tetrachloroethane) are concerned [41]. When the cyclic voltammetry of $H_2@C_{60}$ was conducted in toluene–acetonitrile (5 : 1) at –10 °C under vacuum [47], the fourth (–2.39 V), fifth (–2.95 V), and sixth (–3.5 V) reduction waves appeared, which were shifted slightly and gradually to more negative potentials (by 0.04, 0.07, and 0.15 V, respectively) than C_{60}, indicating that a very weak repulsive interaction operates between the H_2 molecule and negatively multi-charged C_{60} cage [42].

$H_2@C_{60}$ is thermally stable. Upon heating $H_2@C_{60}$ at 500 °C for 10 minutes under vacuum, no decomposition or no release of the encapsulated H_2 molecule was observed at all [41, 42].

8.3 Chemical Functionalization of $H_2@C_{60}$

In order to examine the effect of encapsulated H_2 molecule on the chemical reactivity of the outer surface of C_{60} cage, several representative reactions that are well known for empty C_{60} were conducted for $H_2@C_{60}$ (Scheme 8.8).

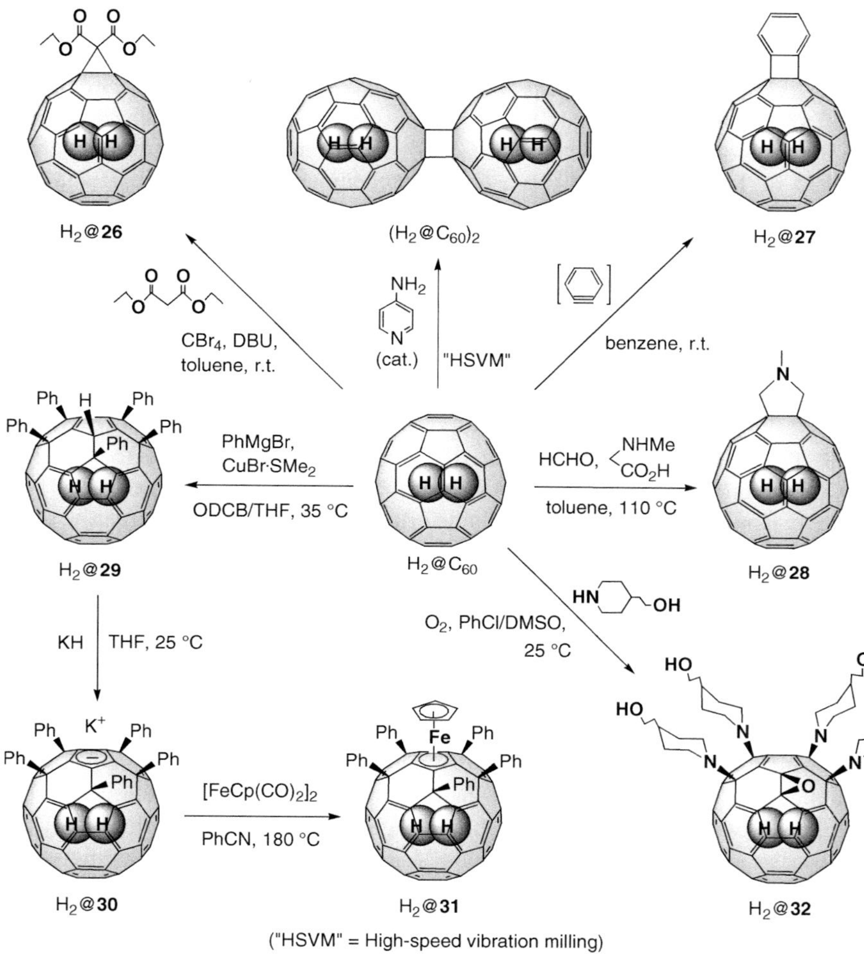

Scheme 8.8

The solid-state mechanochemical dimerization of $H_2@C_{60}$ (occupation level of 91%) using the HSVM (high-speed vibration milling) technique under the same conditions as reported previously [27, 28] afforded the dumbbell-shaped dimer $(H_2@C_{60})_2$ [41, 42] in 30% yield similarly to the reaction of empty C_{60}. The inside H_2 does not affect the reactivity of the C_{60} cage in this reaction. The NMR signal for the inside H_2 molecule of $(H_2@C_{60})_2$ was observed at δ –4.04 ppm, which is 8.58 ppm upfield shifted from that of free H_2 similarly to the case for $^3He@C_{120}$ [28] (8.81 ppm upfield shift from the signal of free ^3He). Three adducts $H_2@$**26**, $H_2@$**27** and $H_2@$**28** were also synthesized (Scheme 8.8) and their NMR signal for the encapsulated H_2 molecule in ODCB-d_4 appeared at δ –3.27, –4.30 and –4.64 ppm, respectively [42]. Since this chemical shift changes sensitively according to the difference in structures of the organic addends, the encapsulated H_2 molecule within C_{60} cage can also be used as a good probe to investigate the chemical reactions at the exterior of the cage, just as a ^3He atom inside C_{60} (occupation level of 0.1%) has been used for this purpose [48–52].

Furthermore, regioselective mulit-addition reactions of $H_2@C_{60}$ were conducted (Scheme 8.8) [53]. The NMR signals for the encapsulated H_2 molecule appeared at δ –10.39 ppm for compound $H_2@$**29** in $CDCl_3$–CS_2, δ –9.79 ppm for potassium cyclopentadienide $H_2@$**30** in THF-d_8, δ –10.44 ppm for pentaphenyl bucky ferrocene $H_2@$**31** in $CDCl_3$-CS_2, and δ –10.77 ppm for tetraaminofullerene epoxide $H_2@$**32** in $CDCl_3$. Although the chemical shifts for the encapsulated H_2 molecule of amphiphilic derivative $H_2@$**32** were measured in a variety of solvents, such as THF-d_8, DMSO-d_6–toluene-d_8 (1 : 1), DMSO-d_6, and D_2O–DMSO-d_6 (1 : 1), no specific solvent effect on the chemical shift was observed.

8.4 Utilization of the Encapsulated H_2 as an NMR Probe

The ^1H NMR chemical shift of the encapsulated H_2 inside C_{60} has been shown to be highly sensitive to the change in the chemical environment of the fullerene cage [42, 53] as described above. However, the magnetic shielding effect of ionic C_{60} and its derivatives has not been reported previously.

Fullerene C_{60} can accept one to six electrons in the threefold degenerate LUMOs. When C_{60} acquires six electrons, overall aromaticity of the fullerene π-system is known to increase drastically [54]. This was unequivocally disclosed by the fact that the ^3He NMR signal of $^3He@C_{60}$ (δ –6.36 ppm relative to the signal of dissolved free ^3He) [9, 10] shifted to dramatically higher field (δ –48.7 ppm) upon six-electron reduction, reflecting the strong shielding effect of $C_{60}{}^{6-}$ [54]. Theoretical as well as experimental studies indicated that all of the hexagons and pentagons of $C_{60}{}^{6-}$ show diamagnetic ring currents. Among the other possible anionic states of C_{60}, dianion $C_{60}{}^{2-}$ is particularly important in synthetic chemistry for introduction of two functional groups on C_{60} cage [55–57]. Although the '$2(N+1)^2$ rule' [58], describing the spherical aromaticity of I_h-symmetrical fullerenes, predicts that the 62-π-electron system should not have high aromaticity, little had been known about the aromaticity of $C_{60}{}^{2-}$.

To clarify this issue, the dianion of $H_2@C_{60}$ was generated by treating with an excessive amount of CH_3SNa [59, 60] in CD_3CN under vacuum [61]. The resulting dark red solution showed a Vis/NIR absorptions at $\lambda_{max} = 830$ and 944 nm [62, 63] and a broad ^{13}C NMR signal at around 183 ppm [64], indicating the generation of $H_2@C_{60}{}^{2-}$. The ^1H NMR signal of the encapsulated H_2 molecule of $H_2@C_{60}{}^{2-}$ was observed at surprisingly low field such as δ 26.36 ppm (Figure 8.2). This is downfield shifted by 27.8 ppm relative to that of neutral $H_2@C_{60}$ (δ –1.44 ppm in ODCB-d_4) [41, 42]. This result demonstrates that the overall aromaticity of C_{60} π-system decreases drastically upon two-electron reduction. The somewhat broadened NMR signal of $H_2@C_{60}{}^{2-}$ is most likely due to the thermal population of the triplet excited states [64], although the possibility of formation of the radical trianion cannot be rigorously ruled out.

The NICS (nucleus independent chemical shifts) [65] calculations (B3LYP/6-31G*) for all the hexagons and pentagons of $C_{60}{}^{2-}$ in the singlet state suggested that, upon two-electron reduction, the ring currents of all hexagons become paramagnetic while those of all pentagons become diamagnetic. Because there exist more hexagons than pentagons in a C_{60} cage, the antiaromatic character of hexagons overwhelms the aromatic character of pentagons,

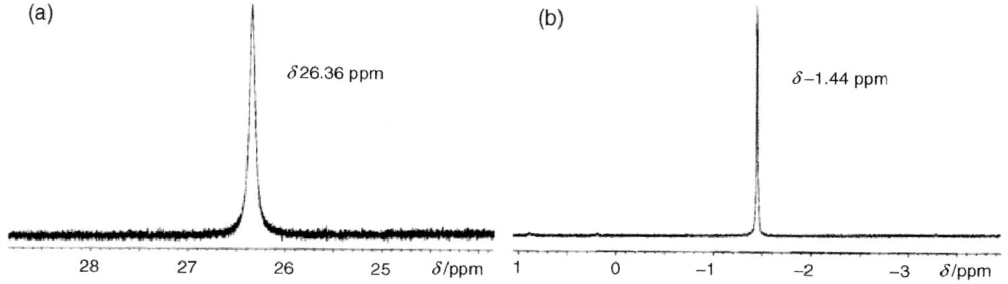

Figure 8.2 *^{1}H NMR spectra (300 MHz) of (a) $H_2@C_{60}^{2-}$ in CD_3CN and (b) $H_2@C_{60}$ in ODCB-d_4*

resulting in a strong deshielding effect inside the cage. This is the first time that the reversal of aromaticity/antiaromaticity for hexagons and pentagons of fullerenes was observed. The similar reversal was observed even for the dianion of heavily functionalized C_{60} such as cage-opened fullerene $H_2@$**19**. The NMR signal of $H_2@$**19**$^{2-}$ was observed at δ 8.10 ppm in CD_3CN, which is downfield shifted by 15.4 ppm relative to that of neutral $H_2@$**19** (δ −7.25 ppm in ODCB-d_4) [33]. The NICS calculations (B3LYP/6-31G*) again showed that the aromatic and antiaromatic characters of hexagons and pentagons are mostly reversed in the same way as those for $H_2@C_{60}^{2-}$, in spite of the highly ruptured π-system in $H_2@$**19**.

Next, generation of dichloromethyl-C_{60} cation [66] and (1-octynyl)-C_{60} anion [67] encapsulating a H_2 molecule ($H_2@$**35**$^+$ and $H_2@$**38**$^-$) was conducted (Scheme 8.9) [68] in order to examine the magnetic properties of the monofunctionalized C_{60} bearing a positive or negative charge.

Cation $H_2@$**35**$^+$ was generated in three steps from $H_2@C_{60}$. First, $H_2@C_{60}$ (occupation level of 9%) was treated with aluminum (III) chloride in $CHCl_3$ at room temperature for

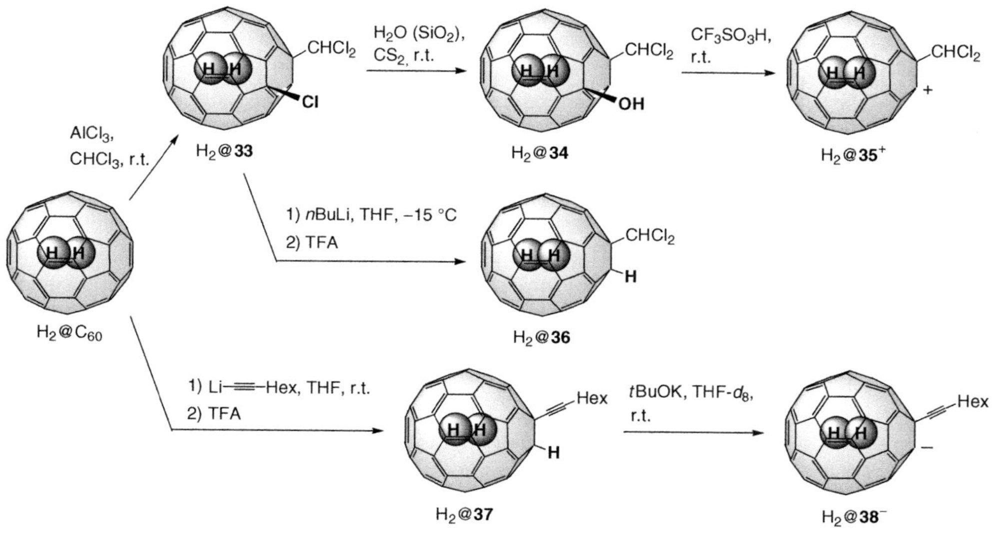

Scheme 8.9

2 hours to give 1,4-adduct H_2@**33** in 61% yield. H_2@**33** was readily converted into fullerenol H_2@**34** in 60% yield upon silica-gel column chromatography. Then, a brown powder of fullerenol H_2@**34** was added to triflic acid to give a reddish purple solution of cation H_2@**35**$^+$ (Scheme 8.9). The ^{1}H NMR signal of the encapsulated H_2 molecule of H_2@**35**$^+$ was observed at δ –2.89 ppm. This signal was downfield shifted only by 1.73 ppm from a H_2 signal of the corresponding neutral compound H_2@**36** (δ –4.62 ppm in CS_2-$CDCl_3$ (1 : 1)), prepared through halogen–lithium exchange reaction of H_2@**33**, indicating that the aromaticity of cation **35**$^+$ was slightly decreased as compared to that of **36**.

On the other hand, anion H_2@**38**$^-$ was generated as follows. The reaction of H_2@C_{60} (occupation level of 9%) with 1-octynyllithium at room temperature in THF and subsequent protonation with trifluoroacetic acid (TFA) afforded 1,2-adduct H_2@**37** in 47% yield. When a THF-d_8 solution of H_2@**37** was treated with *t*-BuOK, the brown solution immediately turned into a dark green solution, indicating the formation of desired anion H_2@**38**$^-$ (Scheme 8.9). The NMR signal for the encapsulated H_2 molecule of anion H_2@**38**$^-$ appeared at δ –0.60 ppm. This resonance is 4.15 ppm downfield shifted as compared with a H_2 molecule of neutral precursor H_2@**37** (δ –4.75 ppm in CS_2-$CDCl_3$ (1 : 1)), again indicating the slight decrease in aromaticity.

The relatively small difference in chemical shifts of the encapsulated H_2 molecules between cation H_2@**35**$^+$ and anion H_2@**38**$^-$ (absolute $\Delta\delta$ value, 2.29 ppm) demonstrates that the aromaticity of the fullerenes are affected to a comparative degree in these cationic and anionic π-systems. The effect of positive or negative charge on C_{60} cage was shown to be quite small upon the chemical shift of encapsulated H_2.

8.5 Physical Properties of an Encapsulated H_2 in C_{60}

The encapsulated H_2 molecule of H_2@**19** is considered to be isolated from the outside environment by the surrounding fullerene cage because the opening is so small that only a He atom or a H_2 molecule can go through. Actually, the nuclear spin–lattice relaxation time (T_1) of the encapsulated H_2 of H_2@**19** upon the ^{1}H NMR measurement was not affected by the presence of molecular oxygen as a paramagnetic species in the solution [33]. The T_1 values of the encapsulated H_2 molecule and one of the pyridyl proton of H_2@**19** in ODCB-d_4 are 0.2 and 3.9 seconds under vacuum and 0.2 and 0.9 seconds in an oxygen-saturated solution, respectively.

In H_2@C_{60}, synthesized by complete closure of the opening of H_2@**19**, the encapsulated H_2 is entirely isolated from the outside. As judged from the difference in chemical shift of ^{13}C NMR ($\Delta\delta = 0.078$ ppm; see above), the interaction of the encapsulated H_2 and the outer C_{60} cage in H_2@C_{60} appears to exist but should be very weak. To investigate the nature of such interaction, the T_1 values of H_2 molecule encapsulated in C_{60} cage as well as those of free H_2 molecule were measured for the first time [69]. The T_1 value of free H_2 at 300 K was found to depend significantly on the organic solvent, for example, from 1.44 seconds in benzene to 0.84 second in CCl_4. A somewhat larger variation of T_1 values was observed for H_2 in H_2@C_{60}; from 0.118 second in benzene to 0.046 second in CCl_4, which are 12–18 times smaller than those for free H_2. However, the value of T_1 for both H_2 and H_2@C_{60} does not significantly change between the solutions in benzene-h_6 and benzene-d_6. Therefore, the dominating interactions determining H_2 and H_2@C_{60} nuclear relaxation are concluded to be intramolecular. On the other hand, the T_1 value for both free H_2 and H_2 in $H_2@C_{60}$ was found to be temperature dependent with the maximum value observed at ~240 K. This kind of temperature dependence of T_1 is consistent with two different relaxation mechanisms dominantly operating at different temperature ranges, that is, below and above 240 K, for both H_2 and $H_2@C_{60}$. Qualitatively, the dipole–dipole interaction accounts for the observed increase in T_1 with temperature below 240 K, whereas the spin–rotation interaction accounts for the observed decrease in T_1 with temperature above 240 K. These facts and consideration derived therefrom imply that the H_2 in both environments rotates through large angles between collisions with the solvent shell or with the walls of the C_{60} cage.

Although the encapsulated H_2 molecule in C_{60} is completely isolated from the outside, it can communicate with the outside world [70]. First, no difference in the triplet lifetime was observed for C_{60}, $H_2@C_{60}$ and $D_2@C_{60}$ upon irradiation of laser pulse. Thus, the interaction of encapsulated H_2 and D_2 with the paramagnetic walls of the triplet fullerene is too weak to be determined by triplet lifetime measurements. However, clear difference in reactivity was observed for the quenching of singlet oxygen $^1O_2$ by C_{60}, $H_2@C_{60}$, and $D_2@C_{60}$. The absolute quenching rate constants k_q of $^1O_2$ by $H_2@C_{60}$, $D_2@C_{60}$, and C_{60} were determined using a time-resolved method in CS_2 to give the values of $k_q(H_2@C_{60}) = 1.5 \times 10^5\ M^{-1}\ s^{-1}$, $k_q(D_2@C_{60}) = 0.49 \times 10^5\ M^{-1}\ s^{-1}$, and $k_q(C_{60}) = 0.38 \times 10^5\ M^{-1}\ s^{-1}$, respectively. The results demonstrate that both $H_2@C_{60}$ and $D_2@C_{60}$ are better quenchers than empty C_{60}. Importantly, the $^1O_2$ can sense the difference between encapsulated H_2 and D_2. The rate constants for quenching of $^1O_2$ by free H_2 and D_2 in CCl_4 were also measured to afford the values of $k_q(H_2) = 0.81 \times 10^5\ M^{-1}\ s^{-1}$, $k_q(D_2) = 0.024 \times 10^5\ M^{-1}\ s^{-1}$, which are significantly smaller than the values by $H_2@C_{60}$ and $D_2@C_{60}$. This is a unique example of an encapsulated guest having a significantly larger rate constant for quenching than the free guest. Since $^1O_2$ might form an exciplex with the outer surface of fullerene, it is speculated that this unique behavior can be attributed to a significant lifetime to provide an opportunity for $^1O_2$ and the encapsulated H_2 to interact for a considerable period of time.

Interaction of the encapsulated H_2 molecule with another species outside the fullerene cage is also seen for the interaction with nitroxide radicals. In the presence of paramagnet nitroxide radicals, bimolecular contribution to the spin–lattice relaxation rate, $1/T_1$, for the protons of H_2 and $H_2@C_{60}$ dissolved in toluene-d_8 was investigated [71]. The measured relaxation rates depended on the concentration of the nitroxide, $[S]$, according to the relationship: $1/T_1 = 1/T_{1,0} + R_1[S]$, where $T_{1,0}$ is the relaxation time in the absence of paramagnetic relaxant and R_1 ($M^{-1}\ s^{-1}$) is the second-order relaxation coefficient, or relaxibity. It was found that the relaxation effect of the paramagnets is enhanced five-fold in $H_2@C_{60}$ compared to free H_2 under the same conditions.

8.6 Molecular-Surgery Synthesis of Endohedral C_{70} Encapsulating Molecular Hydrogen

8.6.1 Synthesis of $(H_2)_2@C_{70}$ and $H_2@C_{70}$

Taking the thickness of the π-electron cloud of fullerenes into consideration, the size of the inner cavity of C_{70} is estimated to be 4.6 Å along the long axis and 3.6 Å along the short axis, which is larger than that of C_{60} (3.6 Å in inner diameter). Therefore, it is expected to be possible to insert more than one small molecule [72, 73] through a chemically created opening on the surface of C_{70}. However, most studies on this line had previously been made only on C_{60} because of the wealth of knowledge about the chemical reactivity of C_{60} and also owing to its higher symmetry.

In this section we first describe the 'incision' of the C_{70} cage to create an opening in the way similar to that developed for the synthesis of cage-opened C_{60} **19** [29] and then the insertion of one and two molecules of H_2 inside the C_{70} cage [74]. Subsequently, 'suture' of the opening to the original C_{70} structure to realize novel endohedral C_{70} encapsulating H_2 molecule(s), $H_2@C_{70}$ and $(H_2)_2@C_{70}$, will be described [75].

As the first step of the 'incision', a thermal reaction of C_{70} with 3,6-di(2-pyridyl)-pyridazine (**39**) [76] was conducted in 1-chloronaphthalene (1-Cl-Naph) at 255 °C for 24 hours to give an isomeric mixture of **40** and **41** both having an eight-membered-ring opening, in 73% and 9% yields, respectively, based on consumed C_{70} (55% conversion) (Scheme 8.10). Then, the reaction of **40** with singlet oxygen was carried out in CS_2 under irradiation of visible light at room temperature for 5 hours to afford two isomeric products **42** and **43** having a 12-membered-ring opening in 60% and 22% yields, respectively, based on consumed **40** (81% conversion) (Scheme 8.10). The opening of **42** was enlarged by the insertion of a sulfur atom to the rim of the opening using TDAE as the π-electron donor to give **44** with a 13-membered-ring opening in 94% yield (Scheme 8.10) [74] in exactly the same manner as that developed for the synthesis of C_{60} analogue **19** [29]. The results of the X-ray crystallography of **44** showed that the opening size is almost the same as that of **19**.

The energies required for the insertion of one and two H_2 molecules into **44** were calculated to be 31.2 and 31.0 kcal mol^{-1} (B3LYP/6-31G**//B3LYP/3-21G) [74], which are comparable to the calculated value for **19** (31.1 kcal mol^{-1}) [33], reflecting that the opening-size of **44** is almost the same as that of **19**. The insertion of H_2 molecule(s) was carried out by applying 890 atm of H_2 gas at 230 °C for 8 hours. The successful encapsulation of H_2

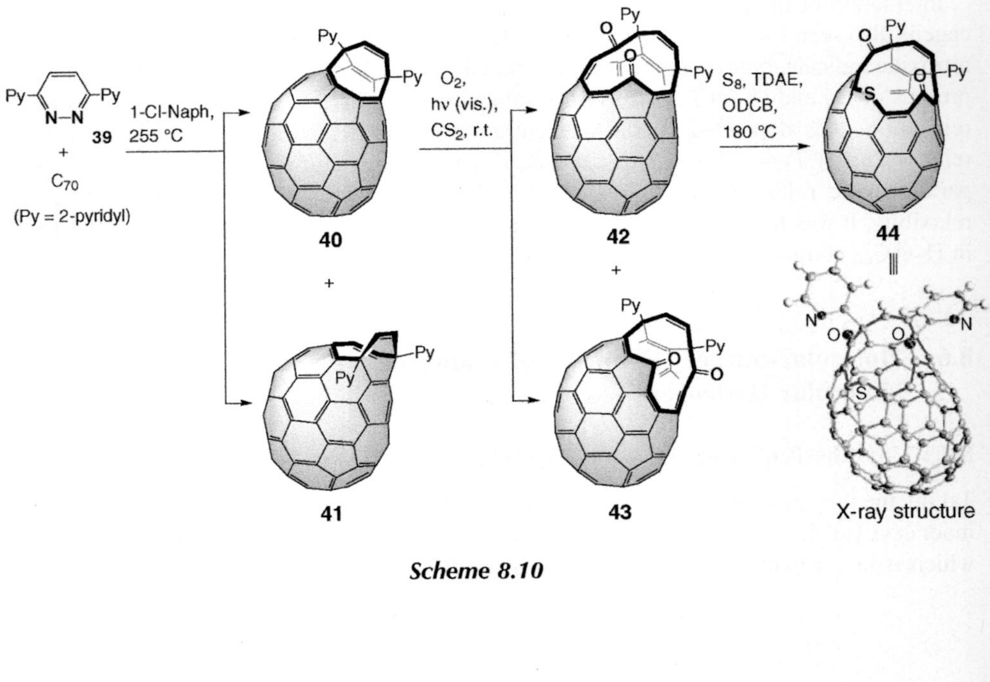

Scheme 8.10

molecule(s) was clearly demonstrated by appearance of a new intense signal in the ^{1}H NMR spectrum at unusually high field, δ –16.51 ppm in ODCB-d_4. This signal was assigned to the resonance of H_2@**44** based on the mass spectroscopic analysis. Noteworthy is that a small signal was also observed at δ –15.22 ppm. The integrated relative intensity of these two signals was determined to be 1.94 H and 0.13 H, respectively, by comparison with the intensity of a pyridyl proton signal at δ 8.68 ppm (1.00 H).

The GIAO (B3LYP/6-311G**//B3LYP/6-31G*) calculations of double hydrogen complex $(H_2)_2$@**44** predicted the chemical shifts to be δ –13.03 ppm for a H_2 molecule near the opening and δ –17.68 ppm for the other one. The rapid exchange of each position of these two molecules of H_2 must be occurring to give a time-averaged NMR signal at around δ –15.3 ppm, which is close to the experimentally observed value (δ –15.22 ppm in ODCB-d_4). Therefore, this signal for the two molecules of H_2 inside **44** was expected to split into two signals upon lowering the temperature.

Prior to the low-temperature NMR experiments, the concentration of the compound giving the H_2 signal at δ –15.22 ppm was enriched by the recycling HPLC on a Cosmosil Buckyprep column. Then, the low-temperature NMR experiments were conducted. As shown in Figure 8.3, the ^{1}H NMR spectrum at 20 °C in a solution of CS_2-CD_2Cl_2 (4 : 1)

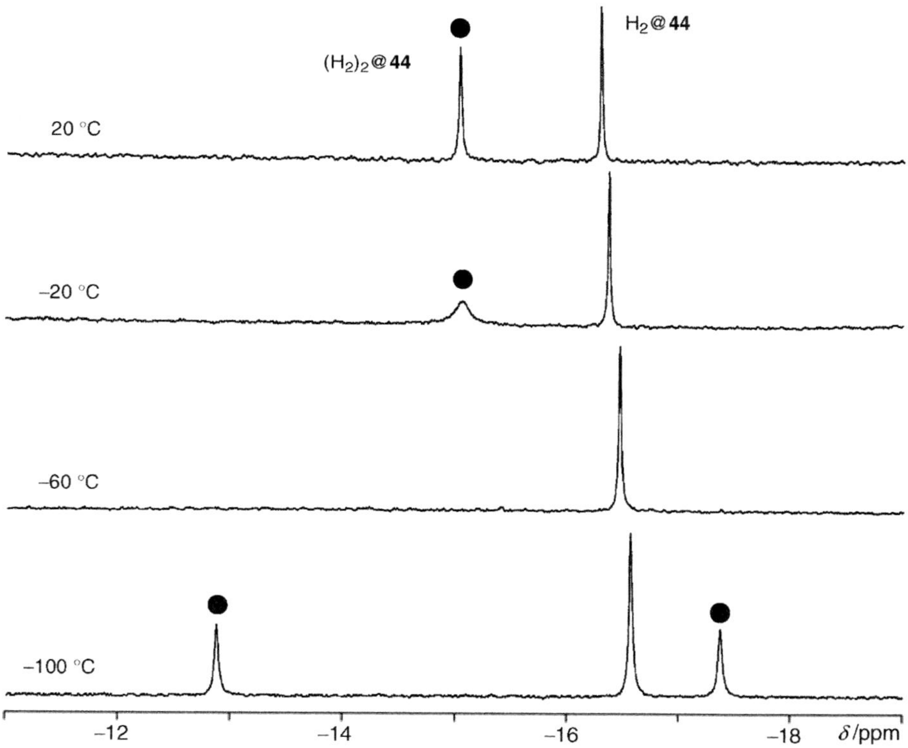

*Figure 8.3 Low temperature ^{1}H NMR spectra (400 MHz, CS_2-CD_2Cl_2 (4 : 1)) of the mixture of $(H_2)_2$@**44** and H_2@**44** (1 : 2)*

clearly exhibited a slightly broadened signal at δ –15.04 ppm along with the signal of H_2@**44** at δ –16.30 ppm. The difference in the line-shape indicates that the motion of the incorporated two molecules of H_2 is already slightly restricted. As expected, the signal of H_2@**44** became broader upon cooling the solution to –20 °C, and flattened at –60 °C. The existence of two molecules of H_2 inside **44** was indisputably proved by the appearance of new two signals at –100 °C. The observed chemical shifts (δ –12.87 and –17.38 ppm) are in good agreement with the predicted values by the DFT calculations as described above. The line-shape analysis indicated that the two molecules of H_2 inside C_{70} exchange their positions with each other at the rate of only 50 times per second at –100 °C, while the rate was increased up to 4.9×10^5 times per second at 20 °C. The Arrhenius plot gave an excellent linear fit with the pre-exponential factor (A) and the activation energy (E_a) being $10^{11.7}$ and 8.0 kcal mol^{-1}, respectively. The activation parameters were determined to be $\Delta G^{\ddagger}$ (25 °C) = 9.4 kcal mol^{-1}, $\Delta H^{\ddagger}$ = 7.4 kcal mol^{-1}, and $\Delta S^{\ddagger}$ = –7 cal K^{-1} mol^{-1}. It should be noted that the shape of the signal for H_2@**44** remained unchanged throughout these low-temperature NMR measurements, indicating that there is no restriction to the motion of a single hydrogen molecule inside the C_{70} cage.

The escaping rates of a H_2 molecule from H_2@**44** monitored at 160, 170, 180, and 190 °C followed the first order kinetics, and the activation parameters were determined to be E_a = 33.8 kcal mol^{-1} and $A = 10^{11.7}$.

The 'suture' of the opening of $(H_2)_2$@**44** and H_2@**44** was carried out by applying the same reactions as those employed for the synthesis of H_2@C_{60}. As shown in Scheme 8.11, the mixture of $(H_2)_2$@**44** and H_2@**44** (3 : 97) [74] was oxidized with MCPBA to give

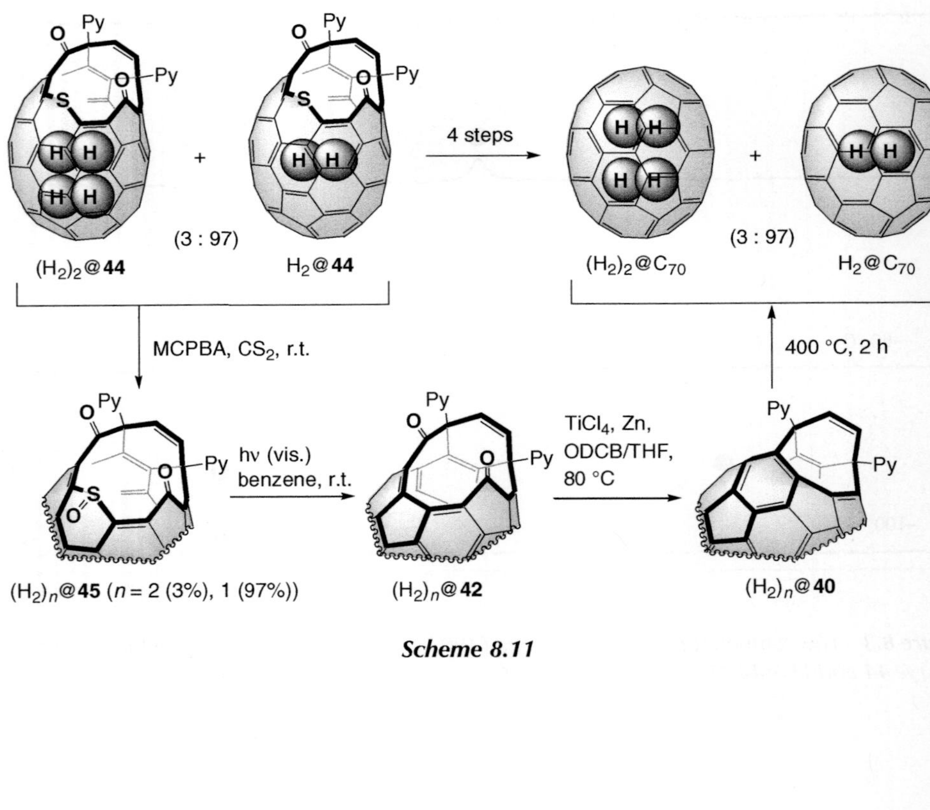

Scheme 8.11

sulfoxides $(H_2)_n@$**45** ($n=2$ (3%), 1 (97%)), which was subjected to the subsequent photoelimination of the resulting SO unit to afford 12-membered-ring compounds $(H_2)_n@$**42** ($n=2$ (3%), 1 (97%)) in 57% yield. Then, two carbonyl groups were coupled by McMurry reaction to give eight-membered-ring compounds $(H_2)_n@$**40** in 61% yield. Finally, thermolysis of $(H_2)_n@$**40** at 400 °C under vacuum for 2 hours provided endohedral fullerene $(H_2)_n@C_{70}$ ($n=2$ (3%), 1 (97%)) (contaminated by 10% empty C_{70}) in 56% yield as a dark brown powder.

The ^1H NMR spectrum of the crude product from the thermal reaction in ODCB-d_4 exhibited a small signal for $(H_2)_2@C_{70}$ at such a high field as δ –23.80 ppm, along with a signal for $H_2@C_{70}$ at δ –23.97 ppm with an integrated ratio being 6 : 97. This molar ratio is consistent with that of $(H_2)_2@$**44** and $H_2@$**44**. The difference in chemical shifts between $(H_2)_2@C_{70}$ and $H_2@C_{70}$ ($\Delta\delta$ 0.17 ppm) is apparently larger than that between $^3He@C_{70}$ and $(^3He)_2@C_{70}$ ($\Delta\delta$ 0.014 ppm) [72]. The two molecules of H_2 or two atoms of ^3He should be located along the longer axis of the oval cage with exchanging each position. Along this axis there exists a small gradient in the intensity of the magnetic field with the intensity being lower at the center of the C_{70} cage [72]. Therefore, the difference in $\Delta\delta$ values could be ascribed to the geometry of the two H_2 molecules of $(H_2)_2@C_{70}$, which should be more off-centered than that of the two ^3He atoms of $(^3He)_2@C_{70}$ due to the steric reason.

In the ^{13}C NMR spectrum of $H_2@C_{70}$ in ODCB-d_4, five signals appeared at δ 150.95, 148.39, 147.71, 145.72, and 131.24 ppm. All of the signals were slightly shifted to downfield as compared to those of C_{70} (δ 150.90, 148.36, 147.69, 145.67, and 131.17 ppm) in the range $\Delta\delta$ 0.02–0.07 ppm. It should be noted that the $\Delta\delta$ values are smaller than that between $H_2@C_{60}$ and C_{60} ($\Delta\delta$ 0.08 ppm) [41, 42]. This indicates that the van der Waals interaction between inner H_2 and outer C_{70} is quite minute, as compared to that of $H_2@C_{60}$, reflecting the larger space inside. In accord with this, the UV–Vis and IR spectra of $H_2@C_{70}$ were quite identical to those of C_{70}.

By applying the similar conditions for the separation of $H_2@C_{60}$ from C_{60} [41, 42], $H_2@C_{70}$ was separated from C_{70} by recycling HPLC on a semipreparative Cosmosil Buckyprep column (two directly connected columns; 250 mm length, 20 mm inner diameter; mobile phase, toluene; flow rate, 6 mL min^{-1}; 50 °C) after 15 recycles (total retention time, 1081 minutes; the retention time for empty C_{70}, 1073 minutes). Furthermore, a fraction eluted just after that of $H_2@C_{70}$ was found to contain $(H_2)_2@C_{70}$ with increased concentration (18%). By repeating this purification three times, $(H_2)_2@C_{70}$ was isolated as a pure material (< 1 mg), which exhibited the correct molecular-ion peak at m/z 844 ($C_{70}H_4$) upon MALDI-TOF mass spectrometry.

8.6.2 Diels-Alder Reaction of $(H_2)_2@C_{70}$ and $H_2@C_{70}$

Although the interaction between the encapsulated H_2 and C_{70} cage is quite minute for $H_2@C_{70}$, there is still a possibility that a difference in chemical reactivity of the outer cage becomes appreciable when the two molecules of H_2 exist inside the cage.

To clarify this, a Diels–Alder reaction of $(H_2)_2@C_{70}$ and $H_2@C_{70}$ with 9,10-dimethylanthracene (DMA) was investigated (Scheme 8.12). The addition of DMA to C_{60} is known to occur reversibly at room temperature [77]. Thus, a solution of a mixture of $(H_2)_2@C_{70}$, $H_2@C_{70}$, and C_{70} (molar ratio, 2 : 70 : 28; total concentration, 13.8 mM) and DMA

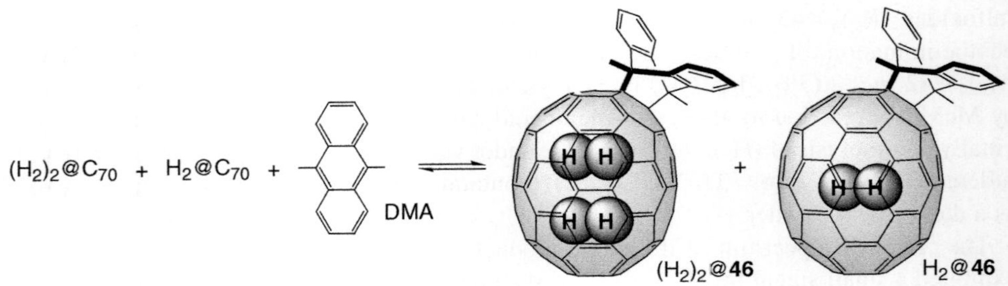

Scheme 8.12

(6.11 mM) in ODCB-d_4 was prepared [52]. The NMR spectrum exhibited new signals for the encapsulated H_2 of monoadducts $(H_2)_2$@**46** and H_2@**46** at δ –21.80 and –22.22 ppm, respectively, in addition to the signals of unreacted $(H_2)_2$@C_{70} and H_2@C_{70}. The equilibrium constants K_2 for the addition of DMA to $(H_2)_2$@C_{70} and K_1 for that to H_2@C_{70} were determined at 30, 40, and 50 °C and summarized in Table 8.1. As shown, the K_2 value is smaller than the K_1 value by more than 15% at each temperature, demonstrating the '*apparently*' decreased reactivity of $(H_2)_2$@C_{70} toward DMA. The van't Hoff plot of ln K_2 or ln K_1 vs. T^{-1} gave excellent linear fits and provided ΔH_2 and ΔH_1 as –13.4 and –13.8 kcal mol^{-1}, respectively, and ΔS_2 and ΔS_1 as –32.9 and –33.7 cal mol^{-1} K^{-1}, respectively. Thus, the two molecules of H_2 encapsulated inside C_{70} cage slightly affect both ΔH and ΔS in this addition reaction.

As a related study, ^{129}Xe@C_{60} was reported to exhibit smaller equilibrium constant in an addition reaction with DMA than that of ^{3}He@C_{60} [78]. The encapsulated ^{129}Xe atom was suggested to have substantial interaction with C_{60} and slightly change the electron distribution of C_{60} [78]. Theoretical calculations (MPWB1K/6-31G**) [79] of $(H_2)_2$@C_{70} and H_2@C_{70} showed that the longer axis of the C_{70} cage of $(H_2)_2$@C_{70} is elongated only by 0.02 Å, while the shorter axis is shortened by 0.02 Å, as compared to those of H_2@C_{70}. The energy levels of the frontier orbitals of $(H_2)_2$@C_{70} and H_2@C_{70} are also almost identical. However, the encapsulation of two molecules of H_2 into C_{70} is calculated to be exothermic by –9.3 kcal mol^{-1} after BSSE (Basis Set Superposition Error) correction. This stabilization energy of $(H_2)_2$@C_{70} is higher than that of H_2@C_{70} (–6.9 kcal mol^{-1}), indicating more interaction is present between two molecules of H_2 and the C_{70} cage. Hence, it would be also reasonable to ascribe the observed difference in the equilibrium constants K_2 and K_1 to the increased electron density on the exterior of the C_{70} cage of $(H_2)_2$@C_{70}.

Table 8.1 *Equilibrium constants K_2 and K_1 for addition of DMA to $(H_2)_2$@C_{70} and H_2@C_{70} in ODCB-d_4 at 30, 40, and 50 °C*

	T (°C)	30	40	50
$(H_2)_2$@C_{70}	K_2 (M^{-1})	296	143	74.7
H_2@C_{70}	K_1 (M^{-1})	364	177	88.4

8.7 Outlook

In this chapter 'molecular surgery' methods to realize novel endohedral fullerenes C_{60} and C_{70} encapsulating molecular hydrogen are outlined. This method can be applied to the preparation of endohedral fullerenes encapsulating atoms or molecules with sizes comparable or smaller than a H_2 molecule such as He, Ne, and D_2. To bring the molecular surgery method into the next level, it is important to make a larger opening that can be 'sutured' to the original form of fullerenes after the insertion of a larger molecule, such as O_2, H_2O, CO, NH_3, or CH_4. From the inside, these guest molecules should exert a great change in the electronic properties of the fullerene π-systems, which are intriguing from the view point of electronic and materials properties. While the recent synthesis of bowl-shaped compounds provided access to the encapsulation of molecules such as H_2O, CO, and NH_3, operation to suture the opening by organic synthetic procedure must be a highly difficult task. Of course the most challenging and important goal would be to develop a route toward the endohedral metallofullerenes. However, as far as the present methods are used, the insertion of metal ions such as Li^+ and Na^+ through the opening is hampered by their strong coordination to the carbonyl oxygen atoms at the opening of **19**. Further development of the technique for the modifications of fullerenes will be indispensable for this project to be accomplished.

References

[1] T. Akasaka and S. Nagase (eds), *Endofullerenes: A New Family of Carbon Clusters*, Kluwer Academic Publisher, Dordrecht, The Netherlands, 2002.
[2] K. M. Kadish and R. S. Ruoff (eds), *Fullerenes: Chemistry, Physics and Technology*, John Wiley & Sons, Inc., New York, 2000.
[3] H. Shinohara, Endohedral metallofullerenes, *Rep. Prog. Phys.*, **63**, 843–892 (2000).
[4] S. Liu and S. Sun, Recent progress in the studies of endohedral metallofullerenes, *J. Organomet. Chem.*, **599**, 74–86 (2000).
[5] S. Nagase, K. Kobayashi and T. Akasaka, Endohedral metallofullerenes: new spherical cage molecules with interesting properties, *Bull. Chem. Soc. Jpn.*, **69**, 2131–2142 (1996).
[6] D. S. Bethune, R. D. Johnson, J. R. Salem, M. S. de Vries and C. S. Yannoni, Atoms in carbon cages: the structure and properties of endohedral fullerenes, *Nature*, **366**, 123–128 (1993).
[7] T. Tsuchiya, R. Kumashiro, K. Tanigaki, Y. Matsunaga, M. O. Ishitsuka, T. Wakahara, Y. Maeda, Y. Takano, M. Aoyagi, T. Akasaka, M. T. H. Liu, T. Kato, K. Suenaga, J. S. Jeong, S. Iijima, F. Kimura, T. Kimura and S. Nagase, Nanorods of endohedral metallofullerene derivative, *J. Am. Chem. Soc.*, **130**, 450–451 (2008).
[8] C.-Y. Shu, C.-R. Wang, J.-F. Zhang, H. W. Gibson, H. C. Dorn, F. D. Corwin, P. P. Fatouros and T. J. S. Dennis, Organophosphonate functionalized Gd@C_{82} as a magnetic resonance imaging contrast agent, *Chem. Mater.*, **20**, 2106–2109 (2008).
[9] M. Saunders, R. J. Cross, H. A. Jiménez-Vázquez, R. Shimshi and A. Khong, Noble gas atoms inside fullerenes, *Science*, **271**, 1693–1697 (1996).
[10] M. Saunders, H. A. Jiménez-Vázquez, R. J. Cross, S. Mroczkowski, D. I. Freedberg and F. A. L. Anet, Probing the interior of fullerenes by ^{3}He NMR spectroscopy of endohedral ^{3}He@C_{60} and ^{3}He@C_{70}, *Nature*, **367**, 256–258 (1994).
[11] M. S. Syamala, R. J. Cross and M. Saunders, ^{129}Xe NMR spectrum of xenon inside C_{60}, *J. Am. Chem. Soc.*, **124**, 6216–6219 (2002).
[12] K. Yamamoto, M. Saunders, A. Khong, R. J. Cross, M. Grayson, M. L. Gross, A. F. Benedetto and R. B. Weisman, Isolation and spectral properties of Kr@C_{60}, a stable van der Waals molecule, *J. Am. Chem. Soc.*, **121**, 1591–1596 (1999).

[13] M. Saunders, H. A. Jiménez-Vázquez, R. J. Cross, S. Mroczkowski, M. L. Gross, D. E. Giblin and R. J. Poreda, Incorporation of helium, neon, argon, krypton, and xenon into fullerenes using high pressure, *J. Am. Chem. Soc.*, **116**, 2193–2194 (1994).

[14] Y. Rubin, Organic approaches to endohedral metallofullerenes: cracking open or zipping up carbon shells?, *Chem. –Eur. J.*, **3**, 1009–1016 (1997).

[15] Y. Rubin, Ring opening reactions of fullerenes: designed approached to endohedral metal complexes, *Top. Curr. Chem.*, **199**, 67–91 (1999).

[16] M. Murata, Y. Murata and K. Komatsu, Surgery of fullerenes, *Chem. Commun.*, 6083–6094 (2008).

[17] G. Schick, T. Jarrosson and Y. Rubin, Formation of an effective opening within the fullerene core of C_{60} by an unusual reaction sequence, *Angew. Chem. Int. Ed.*, **38**, 2360–2363 (1999).

[18] Y. Rubin, T. Jarrosson, G.-W. Wang, M. D. Bartberger, K. N. Houk, G. Schick, M. Saunders and R. J. Cross, Insertion of helium and molecular hydrogen through the orifice of an open fullerene, *Angew. Chem. Int. Ed.*, **40**, 1543–1546 (2001).

[19] S.-i. Iwamatsu, T. Uozaki, K. Kobayashi, S. Re, S. Nagase and S. Murata, A bowl-shaped fullerene encapsulates a water into the cage, *J. Am. Chem. Soc.*, **126**, 2668–2669 (2004).

[20] S.-i. Iwamatsu, C. M. Stanisky, R. J. Cross, M. Saunders, N. Mizorogi, S. Nagase and S. Murata, Carbon monoxide inside an open-cage fullerene, *Angew. Chem. Int. Ed.*, **45**, 5337–5340 (2006).

[21] K. E. Whitener Jr., M. Frunzi, S.-i. Iwamatsu, S. Murata, R. J. Cross and M. Saunders, Putting ammonia into a chemically opened fullerene, *J. Am. Chem. Soc.*, **130**, 13996–13999 (2008).

[22] Z. Xiao, J. Yao, D. Yang, F. Wang, S. Huang, L. Gan, Z. Jia, Z. Jiang, X. Yang, B. Zheng, G. Yuan, S. Zhang and Z. Wang, Synthesis of [59]fullerenones through peroxide-mediated stepwise cleavage of fullerene skeleton bonds and X-ray structures of their water-encapsulated open-cage complexes, *J. Am. Chem. Soc.*, **129**, 16149–16162 (2007).

[23] Y. Murata, N. Kato and K. Komatsu, The reaction of fullerene C_{60} with phthalazine: the mechanochemical solid-state reaction yielding a new C_{60} dimer versus the liquid-phase reaction affording an open-cage fullerene, *J. Org. Chem.*, **66**, 7235–7239 (2001).

[24] Y. Murata, M. Murata and K. Komatsu, The reaction of fullerene C_{60} with 4,6-dimethyl-1,2,3-triazine: formation of an open-cage fullerene derivative, *J. Org. Chem.*, **66**, 8187–8191 (2001).

[25] Y. Murata, M. Suzuki and K. Komatsu, Synthesis and electrochemical properties of novel dimeric fullerenes incorporated in a 2,3-diazabicyclo[2.2.2]oct-2-ene framework, *Chem. Commun.*, 2338–2339 (2001).

[26] Y. Murata, M. Suzuki, Y. Rubin and K. Komatsu, Structure of the hydration product of the C_{60}-di (2-pyridyl)-1,2,4,5- tetrazine adduct, *Bull. Chem. Soc. Jpn.*, **76**, 1669–1672 (2003).

[27] G.-W. Wang, K. Komatsu, Y. Murata and M. Shiro, Synthesis and X-ray structure of dumb-bell-shaped C_{120}, *Nature*, **387**, 583–586 (1997).

[28] K. Komatsu, G.-W. Wang, Y. Murata, T. Tanaka and K. Fujiwara, Mechanochemical synthesis and characterization of the fullerene dimer C_{120}, *J. Org. Chem.*, **63**, 9358–9366 (1998).

[29] Y. Murata, M. Murata and K. Komatsu, Synthesis, structure, and properties of novel open-cage fullerenes having heteroatom(s) on the rim of the orifice, *Chem. –Eur. J.*, **9**, 1600–1609 (2003).

[30] M.-J. Arce, A. L. Viado, Y.-Z. An, S. I. Khan and Y. Rubin, Triple scission of a six-membered ring on the surface of C_{60} *via* consecutive pericyclic reactions and oxidative cobalt insertion, *J. Am. Chem. Soc.*, **118**, 3775–3776 (1996).

[31] H. Inoue, H. Yamaguchi, S.-i. Iwamatsu, T. Uozaki, T. Suzuki, T. Akasaka, S. Nagase and S. Murata, Photooxygenative partial ring cleavage of bis(fulleroid): synthesis of a novel fullerene derivative with a 12-membered ring, *Tetrahedron Lett.*, **42**, 895–897 (2001).

[32] Y. Murata and K. Komatsu, Photochemical reaction of the open-cage fullerene derivative with singlet oxygen, *Chem. Lett.*, 896–897 (2001).

[33] Y. Murata, M. Murata and K. Komatsu, 100% Encapsulation of a hydrogen molecule into an open-cage fullerene derivative and gas-phase generation of $H_2@C_{60}$, *J. Am. Chem. Soc.*, **125**, 7152–7153 (2003).

[34] S.-C. Chuang, Y. Murata, M. Murata, S. Mori, S. Maeda, F. Tanabe and K. Komatsu, Fine tuning of the orifice size of an open-cage fullerene by placing selenium in the rim: insertion/release of molecular hydrogen, *Chem. Commun.*, 1278–1280 (2007).

[35] E. Allard, L. Riviére, J. Delaunay, D. Dubois and J. Cousseau, Chemical generation and reactivity of C_{60}^{2-}. High-yield and regioselective alkylation of [60, I_h] fullerene, *Tetrahedron Lett.*, **40**, 7223–7226 (1999).
[36] H. Sawa, Y. Wakabayashi, Y. Murata, M. Murata and K. Komatsu, Floating single hydrogen molecule in an open-cage fullerene, *Angew. Chem. Int. Ed.*, **44**, 1981–1983 (2005).
[37] M. Carravetta, Y. Murata, M. Murata, I. Heinmaa, R. Stern, A. Tontcheva, A. Samoson, Y. Rubin, K. Komatsu and M. H. Levitt, Solid-state NMR spectroscopy of molecular hydrogen trapped inside an open-cage fullerene, *J. Am. Chem. Soc.*, **126**, 4092–4093 (2004).
[38] S.-C. Chuang, Y. Murata, M. Murata and K. Komatsu, An orifice-size index for open-cage fullerenes, *J. Org. Chem.*, **72**, 6447–6453 (2007).
[39] C. M. Stanisky, R. J. Cross, M. Saunders, M. Murata, Y. Murata and K. Komatsu, Helium entry and escape through a chemically opened window in a fullerene, *J. Am. Chem. Soc.*, **127**, 299–302 (2005).
[40] S.-C. Chuang, Y. Murata, M. Murata and K. Komatsu, The outside knows the difference inside: trapping helium by immediate reduction of the orifice size of an open-cage fullerene and the effect of encapsulated helium and hydrogen upon the NMR of a proton directly attached to the outside, *Chem. Commun.*, 1751 (2007).
[41] K. Komatsu, M. Murata and Y. Murata, Encapsulation of molecular hydrogen in fullerene C_{60} by organic synthesis, *Science*, **307**, 238–240 (2005).
[42] M. Murata, Y. Murata and K. Komatsu, Synthesis and properties of endohedral C_{60} encapsulating molecular hydrogen, *J. Am. Chem. Soc.*, **128**, 8024–8033 (2006).
[43] K. Komatsu and Y. Murata, A new route to an endohedral fullerene by way of σ-Framework Transformations, *Chem. Lett.*, **34**, 886–891 (2005).
[44] K. Komatsu, Studies on two- and three-dimensional π-conjugated systems with unique structures and properties, *Bull. Chem. Soc. Jpn.*, **80**, 2285–2302 (2007).
[45] J. E. McMurry, Carbonyl-coupling reactions using low-valent titanium, *Chem. Rev.*, **89**, 1513–1524 (1989).
[46] D. Bakowies and W. Thiel, Theoretical infrared spectra of large carbon clusters, *Chem. Phys.*, **151**, 309–321 (1991).
[47] Q. Xie, E. Pérez-Cordero and L. Echegoyen, Electrochemical detection of C_{60}^{6-} and C_{70}^{6-}: Enhanced stability of fullerides in solution, *J. Am. Chem. Soc.*, **114**, 3978–3980 (1992).
[48] M. Saunders, H. A. Jiménez-Vázquez, B. W. Bangerter and R. J. Cross, ^{3}He NMR: a powerful new tool for following fullerene chemistry, *J. Am. Chem. Soc.*, **116**, 3621–3622 (1994).
[49] D. I. Schuster, J. Cao, N. Kaprinidis, Y. Wu, A. W. Jensen, Q. Lu, H. Wang and S. R. Wilson, [2 + 2] Photocycloaddition of cyclic enones to C_{60}, *J. Am. Chem. Soc.*, **118**, 5639–5647 (1996).
[50] R. J. Cross, H. A. Jiménez-Vázquez, Q. Lu, M. Saunders, D. I. Schuster, S. R. Wilson and H. Zhao, Differentiation of isomers resulting from bisaddition to C_{60} Using ^{3}He NMR Spectrometry, *J. Am. Chem. Soc.*, **118**, 11454 (1996).
[51] M. Rüttimann, R. F. Haldimann, L. Isaacs, F. Diederich, A. Khong, H. A. Jiménez-Vázquez, R. J. Cross and M. Saunders, π-Electron ring-current effects in multiple adducts of ^{3}He@C_{60} and ^{3}He@C_{60}: a ^{3}He NMR study, *Chem. –Eur. J.*, **3**, 1071–1076 (1997)
[52] G.-W. Wang, M. Saunders and R. J. Cross, Reversible diels-alder addition to fullerenes: a study of equilibria using ^{3}He NMR spectroscopy, *J. Am. Chem. Soc.*, **123**, 256 (2001).
[53] Y. Matsuo, H. Isobe, T. Tanaka, Y. Murata, M. Murata, K. Komatsu and E. Nakamura, Organic and organometallic derivatives of dihydrogen-encapsulated [60]fullerene, *J. Am. Chem. Soc.*, **127**, 17148–17149 (2005).
[54] E. Shabtai, A. Weitz, R. C. Haddon, R. E. Hoffman, M. Rabinovitz, A. Khong, R. J. Cross, M. Saunders, P.-C. Cheng and L. T. Scott, ^{3}He NMR of He@C_{60}^{6-} and He@C_{70}^{6-}. New records for the most shielded and the most deshielded ^{3}He inside a fullerene, *J. Am. Chem. Soc.*, **120**, 6389–6393 (1998).
[55] R. Subramanian, K. M. Kadish, M. N. Vijayashree, X. Gao and M. T. Jones, Chemical generation of C_{60}^{2-} and electron transfer mechanism for the reactions with alkyl bromides, *J. Phys. Chem.*, **100**, 16327–16335 (1996).

[56] S. Fukuzumi, T. Suenobu, T. Hirasaka, R. Arakawa and K. M. Kadish, Formation of C_{60} adducts with two different alkyl groups via combination of electron transfer and S_N2 reactions, *J. Am. Chem. Soc.*, **120**, 9220–9227 (1998).
[57] Y.-H. Zhu, L.-C. Song, Q.-M. Hu and C.-M. Li, Synthesis and isolation of σ-bonded fullerene metal derivatives from reactions of fullerene dianion C_{60}^{2-} with organometal halides, *Org. Lett.*, **1**, 1693–1695 (1999).
[58] A. Hirsch, Z. Chen and H. Jiao, Spherical aromaticity in I_h symmetrical fullerenes: the $2(N+1)^2$ Rule, *Angew. Chem. Int. Ed.*, **39**, 3915–3917 (2000).
[59] F. Cheng, Y. Murata and K. Komatsu, Synthesis, X-ray structure, and properties of the singly bonded C_{60} dimer having diethoxyphosphorylmethyl groups utilizing the chemistry of C_{60}^{2-}, *Org. Lett.*, **4**, 2541–2544 (2002).
[60] E. Allard, J. Delaunay, F. Cheng, J. Cousseau, J. Ordfflna and G. Javier, Novel C_{60}-based building blocks derived from C_{60}^{2-} anion, *Org. Lett.*, **3**, 3503–3506 (2001).
[61] M. Murata, Y. Ochi, F. Tanabe, K. Komatsu and Y. Murata, Internal magnetic fields of dianions of fullerene C_{60} and its cage-opened derivatives studied with encapsulated H_2 as an NMR probe, *Angew. Chem. Int. Ed.*, **47**, 2039–2041 (2008).
[62] P. C. Trulove, R. T. Carlin, G. R. Eaton and S. S. Eaton, Determination of the singlet-triplet energy separation for C_{60}^{2-} in DMSO by electron paramagnetic resonance, *J. Am. Chem. Soc.*, **117**, 6265–6272 (1995).
[63] R. Subramanian, P. Boulas, M. N. Vijayashree, F. D'Souza, M. T. Jones and K. M. Kadish, A facile and selective method for the solution-phase generation of C_{60}^{-} and C_{60}^{2-}, *J. Chem. Soc., Chem. Commun.*, 1847–1848 (1994).
[64] P. D.W. Boyd, P. Bhyrappa, P. Paul, J. Stinchcombe, R. D. Bolskar, Y. Sun and C. A. Reed, The C_{60}^{2-} fulleride ion, *J. Am. Chem. Soc.*, **117**, 2907–2914 (1995).
[65] P. v. R. Schleyer, C. Maerker, A. Dransfeld, H. Jiao and N. J. R. v. E. Hommes, Nucleus-independent chemical shifts: a simple and efficient aromaticity probe, *J. Am. Chem. Soc.*, **118**, 6317–6318 (1996).
[66] T. Kitagawa, H. Sakamoto and K. Takeuchi, Electrophilic addition of polychloroalkanes to C_{60}: direct observation of alkylfullerenyl cation intermediates, *J. Am. Chem. Soc.*, **121**, 4298–4299 (1999).
[67] Y. Murata, K. Motoyama, K. Komatsu and T. S. M. Wan, Synthesis, properties, and reactions of a stable carbanion derived from alkynyldihydrofullerene: 1-octynyl-C_{60} carbanion, *Tetrahedron*, **52**, 5077–5090 (1996).
[68] M. Murata, Y. Ochi, T. Kitagawa, K. Komatsu and Y. Murata, NMR studies on monofunctionalized fullerenyl cation and anion encapsulating a H_2 molecule, *Chem. –Asian J.*, **3**, 1336–1342 (2008).
[69] E. Sartori, M. Ruzzi, N. J. Turro, J. D. Decatur, D. C. Doetschman, R. G. Lawler, A. L. Buchachenko, Y. Murata and K. Komatsu, Nuclear relaxation of H_2 and $H_2@C_{60}$ in organic solvents, *J. Am. Chem. Soc.*, **128**, 14752–14753 (2006).
[70] J. López-Gejo, A. A. Martí, M. Ruzzi, S. Jockusch, K. Komatsu, F. Tanabe, Y. Murata and N. J. Turro, Can H_2 inside C_{60} communicate with the outside world?, *J. Am. Chem. Soc.*, **129**, 14554–14555 (2007).
[71] E. Sartori, M. Ruzzi, N. J. Turro, K. Komatsu, Y. Murata, R. G. Lawler and A. L. Buchachenko, Paramagnet enhanced nuclear relaxation of H_2 in organic solvents and in $H_2@C_{60}$, *J. Am. Chem. Soc.*, **130**, 2221–2225 (2008).
[72] A. Khong, H. A. Jiménez-Vázquez, M. Saunders, R. J. Cross, J. Laskin, T. Peres, C. Lifshitz, R. Strongin and A. B. SmithIII, An NMR study of He_2 inside C_{70}, *J. Am. Chem. Soc.*, **120**, 6380–6383 (1998).
[73] J. Laskin, T. Peres, C. Lifshitz, M. Saunders, R. J. Cross and A. Khong, An artificial molecule of Ne_2 inside C_{70}, *Chem. Phys. Lett.*, **285**, 7–9 (1998).
[74] Y. Murata, S. Maeda, M. Murata and K. Komatsu, Encapsulation and dynamic behavior of two H_2 molecules in an open-cage C_{70}, *J. Am. Chem. Soc.*, **130**, 6702–6703 (2008).
[75] M. Murata, S. Maeda, Y. Morinaka, Y. Murata and K. Komatsu, Synthesis and reaction of fullerene C_{70} encapsulating two molecules of H_2, *J. Am. Chem. Soc.*, **130**, 15800–15801 (2008).

[76] W. A. Butte and F. H. Case, The synthesis of some pyridylpyridazines and -pyrimidines, *J. Org. Chem.*, **26**, 4690–4692 (1961).
[77] I. Lamparth, C. Maichle-Mossmer and A. Hirsch, Reversible template-directed activation of equatorial double bonds of the fullerene framework: regioselective direct synthesis, crystal structure, and aromatic properties of T_h-$C_{66}(COOEt)_{12}$, *Angew. Chem. Int. Ed. Engl.*, **34**, 1607–1609 (1995).
[78] M. Frunzi, R. J. Cross and M. Saunders, Effect of xenon on fullerene reactions, *J. Am. Chem. Soc.*, **129**, 13343–13346 (2007).
[79] Z. Slanina, P. Pulay and S. Nagase, H_2, Ne, and N_2 energies of encapsulation into C_{60} evaluated with the MPWB1K functional, *J. Chem. Theory Comput.*, **2**, 782–785 (2006).

9

New Endohedral Metallofullerenes: Trimetallic Nitride Endohedral Fullerenes

Marilyn M. Olmstead[a], Alan L. Balch[a], Julio R. Pinzón[b], Luis Echegoyen[b], Harry W. Gibson[c] and Harry C. Dorn[c]

[a]*Department of Chemistry, University of California, Davis, CA USA*
[b]*Department of Chemistry, Clemson University, Clemson, SC, USA*
[c]*Department of Chemistry, Virginia Polytechnic Institute & State University, Blacksburg, VA, USA*

9.1 Discovery, Preparation, and Purification

In the period 1986–99, a number of classic endofullerenes were reported with both nonmetallic atoms as encapsulants (e.g. $He@C_{60}$) and simple metallic encapsulants (endohedral metallofullerenes, EMFs, $A_x@C_{2y}$, x = 1, 2, 3; A = metal, y = 30–50) were reported. Early reviews on endohedral metallofullerenes were published by Akasaka, Kobayashi, and Nagase and later by Shinohara [1, 2]. The discovery in 1999 of the trimetallic nitride templated endohedral metallofullerenes (TNT EMFs, $A_{3-x}B_xN@C_{2y}$, x = 0–3, A,B = metal, y = 34, 39–50) has opened new vistas in the field of endofullerenes [3, 4]. Initially, the TNT EMFs were prepared in a Krätschmer-Huffman electric-arc generator by vaporization of graphite rods containing metals and/or metal oxides in a He atmosphere by accidental inclusion of atmospheric air (N_2, O_2). This serendipitous discovery was improved by the introduction of pure N_2 (10–40 torr) in the electric-arc background gas of 100–400 torr He [4]. Dunsch and coworkers altered the process by using

Chemistry of Nanocarbons Edited by Takeshi Akasaka, Fred Wudl and Shigeru Nagase

graphite rods packed with calcium cyanamide, CaNCN, which provides a alternative nitrogen source in the plasma [5]. Later the Dunsch group made a significant improvement by employing the 'reactive gas', ammonia (NH_3), which provided the TNT EMFs as the dominant soot product(s) with concomitant diminished production of empty-cage fullerenes [5, 6]. More recently, Stevenson and coworkers have employed copper metal packed in the graphite rods to improve chemical yields of $Sc_3N@C_{80}$ by a factor of 3–5. The Stevenson group also reported a 'Chemically Adjusting Plasma Temperature, Energy, and Reactivity' (CAPTEAR) method that utilizes copper nitrate hydrate [$Cu(NO_3)_2 \cdot 2.5\ H_2O$] packed in the graphite rods. This approach reportedly provides an exothermic nitrate moiety in the plasma, which suppresses empty-cage fullerene formation. In addition, the latter group suggest a 'competitive link' between the formation of C_{60} and $Sc_3N@C_{80}$ based on an inverse yield relationship in the production of the two compounds [7]. An extensive review on the metal nitride cluster fullerenes was published in 2007 by Dunsch and Yang and featured a discussion of future applications [8].

Initially the difficulty in obtaining purified samples from the myriad of different products in the soot obtained from the electric-arc process was a daunting task involving extensive, time consuming chromatographic procedures that required large volumes of solvent. However, the purification of the TNT EMFs has been greatly simplified by a number of recent approaches. For example, Dorn and Gibson have developed a selective chemical reactivity approach for purification of carbon TNT EMFs in a single facile step [9]. This approach utilizes a cyclopentadiene modified Merrifield-type resin in which the more reactive empty-cage fullerenes C_{60}, C_{70}, etc. become attached to the resin (Diels-Alder reaction) and are not eluted from the column at room temperature. The more stable, less reactive TNT EMFs pass through unreacted. The empty-cage fullerenes can be readily recovered by retro-reactions at elevated temperatures [9]. Other related approaches have been reported using the reaction of amino-capped silica gel using a 'stir and filter' approach [10]. Using a variation of Kräutler's solvent-free method of forming anthracene-C_{60} adducts, the reaction of a large excess of a low melting aromatic diene, 9-methylanthracene, with Sc- and Lu-based soot extracts allows the separation of the reacted empty-cage fullerenes from the unreacted TNT EMFs [11]. All of these approaches rely on the inherent lower reactivity of the TNT EMFs relative to empty-cage fullerenes and more reactive classic metallofullerenes (EMFs, $A_x@C_{2y}$, x = 1,2, 3; A = metal, y = 30–50). For the $A_3N@C_{80}$ TNT EMF, it is well established that there are two isomers and the dominant $A_3N@I_h\text{-}C_{80}$ isomer is usually accompanied by minor amounts of the of $A_3N@D_{5h}\text{-}C_{80}$ isomer as described below. Echegoyen and coworkers have reported a chemical oxidation approach based on a 270 mV difference in the first oxidation potentials of the $A_3N@C_{80}$ isomers [12]. This electrochemical difference was exploited to separate the minor scandium isomer, $Sc_3N@D_{5h}\text{-}C_{80}$ from the dominant $Sc_3N@I_h\text{-}C_{80}$ isomer [12]. A similar electrochemical difference has also been reported for the corresponding dysprosium isomers [13].

9.2 Structural Studies

Experimental studies of the structures of endohedral fullerene have focused on examination of ^{13}C NMR spectra, infrared spectroscopy, and X-ray diffraction [14]. Computational studies add an important dimension to the understanding of the relative stabilities of various fullerene cage isomers and have made a number of important predictions in regard to cage structure [15–18]. While ^{13}C NMR spectra can give important information about the symmetry of the carbon cage, obtaining a sufficiently large sample for measurement can be challenging. Also, for the higher fullerenes, instances of the same point symmetry for different isomers are common. Moreover, the data can be difficult to interpret if the metal ion (s) involved are paramagnetic. Single crystal X-ray diffraction studies of cocrystals comprised of the fullerene and Ni^{II}(OEP) (OEP is the dianion of octaethylporphyrin) together with solvate molecules have proven to be effective, particularly when only sub-milligram quantities of sample are available. The nesting of the endohedral fullerene against the surface provided by the porphyrin provides a means of inducing sufficient order in the crystals to allow structural elucidation. Here, we will focus on the results obtained in this fashion. To date, eight different cage sizes ranging from C_{68} to C_{88} have been characterized by X-ray diffraction methods. Figures 9.1 and 9.2 show drawings of the eight different cages that have been investigated and their interactions with the metalloporphyrin, which always uses all eight ethyl groups to embrace the endohedral guest. The endohedral fullerenes shown in these figures include: $Sc_3N@D_3(6140)$-C_{68} [19], $Sc_3N@D_{3h}(5)$-C_{78} [20, 21], $Tb_3N@I_h$-C_{80} [22], $Tb_3N@D_{5h}$-C_{80} [22], $Gd_3N@C_s(39663)$-C_{82} [23], $Tb_3N@C_s(51365)$-

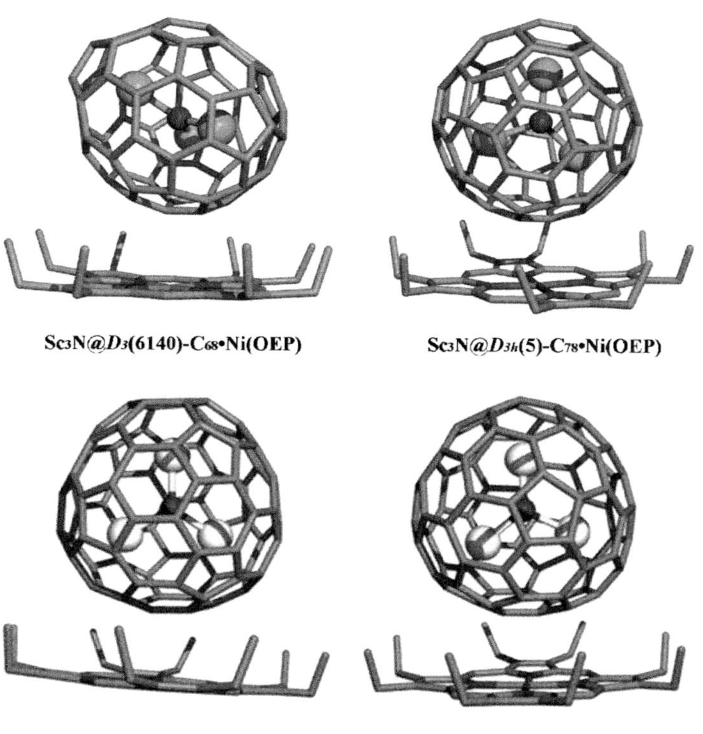

Figure 9.1 *The structures of $Sc_3N@D_3(6140)$-C_{68}•Ni(OEP), $Sc_3N@D_{3h}(5)$-C_{78}•Ni(OEP), $Tb_3N@I_h$-C_{80}•Ni(OEP), $Tb_3N@D_{5h}$-C_{80}•Ni(OEP)*

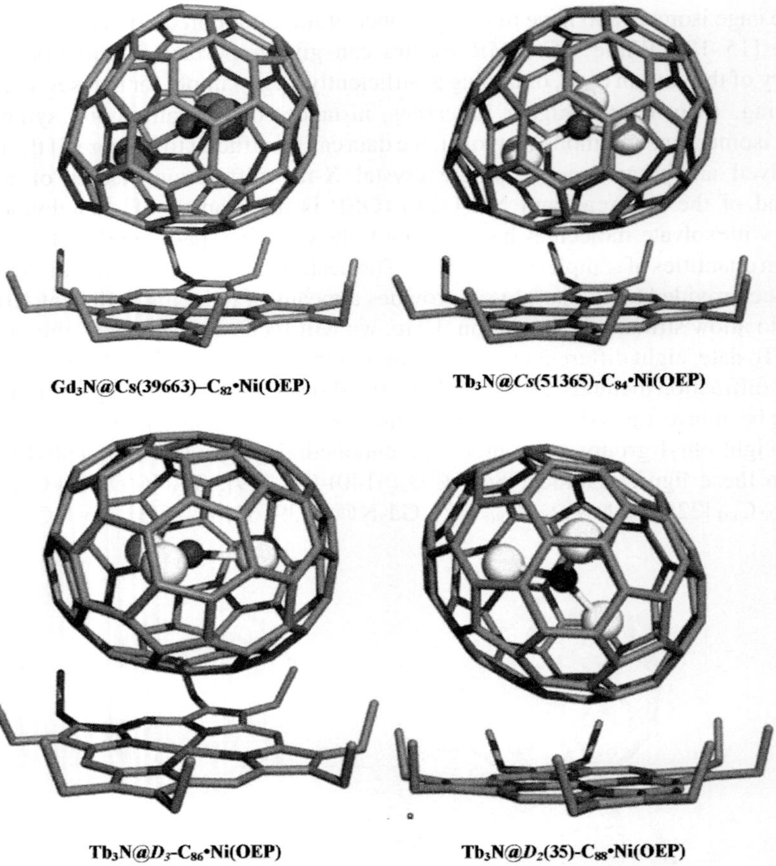

Figure 9.2 *The structures of* $Gd_3N@C_s(39663)–C_{82}$*•Ni(OEP),* $Tb_3N@C_s(51365)-C_{84}$*•Ni(OEP),* $Tb_3N@D_3-C_{86}$*•Ni(OEP), and* $Tb_3N@D_2(35)-C_{88}$*•Ni(OEP)*

C_{84} [24], $Tb_3N@D_3$-C_{86} [22], and $Tb_3N@D_2(35)$-C_{88} [22]. Table 9.1 lists all of the TNT endohedrals that have been structurally characterized to date and categorizes them by cage size.

Five of the fullerene cages in these endohedrals [$D_{3h}(5)$-C_{78}, I_h-C_{80}, D_{5h}-C_{80}, D_3-C_{86}, and $D_2(35)$-C_{88}] obey the isolated pentagon rule (IPR), which requires that each pentagon in a fullerene is surrounded by five hexagons. However, three of the cages [$D_3(6140)$-C_{68}, $C_s(39663)$-C_{82}, and $C_s(51365)$-C_{84}] do not follow the IPR. With this number of exceptions, it appears that the IPR is more of a suggestion than a rule for endohedral fullerenes. Of the three fullerene cages that defy the IPR, the cage in $Sc_3N@D_3(6140)$-C_{68} has three equivalent sites where pairs of pentagons abut. The scandium ions are each situated within the pentalene units formed by the abutting pentagons. In contrast, in $Gd_3N@C_s(39663)$-C_{82} and $Tb_3N@C_s(51365)$-C_{84}, there is only one place where two pentagons are joined. This situation gives these two cages distinct egg-like shapes. Notice how similar the shapes of $Gd_3N@C_s(39663)$-C_{82} and $Tb_3N@C_s(51365)$-C_{84} are. Both carbon cages also nestle into

Table 9.1 *Crystallographically characterized members of the $M_3N@C_{2n}$ family*

Cage Size	Examples
C_{68}	$Sc_3N@D_3(6140)$-C_{68} [19]
C_{78}	$Sc_3N@D_{3h}(5)$-C_{78} [20, 21]
C_{80}, I_h cage symmetry	$Sc_3N@I_h$-C_{80} [3, 25], $Lu_3N@I_h$-C_{80} [25], $Gd_3N@I_h$-C_{80} [26], $Tb_3N@I_h$-C_{80} [22], $Tm_3N@I_h$-C_{80} [27], $Dy_3N@I_h$-C_{80} [28], $ErSc_2N@I_h$-C_{80} [29], $CeSc_2N@I_h$-C_{80} [30], $GdSc_2N@I_h$-C_{80} [31], $Gd_2ScN@I_h$-C_{80} [31], $TbSc_2N@I_h$-C_{80} [31]
C_{80}, D_{5h} cage symmetry	$Sc_3N@D_{5h}$-C_{80} [32], $Tb_3N@D_{5h}$-C_{80} [22], $Tm_3N@D_{5h}$-C_{80} [27]
C_{82}	$Gd_3N@C_s(39663)$–C_{82} [23]
C_{84}	$Tb_3N@C_s(51365)$-C_{84} [24], $Tm_3N@C_s(51365)$-C_{84}, [33] $Gd_3N@C_s(51365)$-C_{84} [33]
C_{86}	$Tb_3N@D_3$-C_{86} [22]
C_{88}	$Tb_3N@D_2(35)$-C_{88} [22]

the Ni^{II}(OEP) molecule in nearly the same fashion. For each of these two non-IPR endohedrals, there is a metal ion that is positioned within the fold of the pair of adjacent pentagons.

As Table 9.1 shows, the largest number of structural results is available for C_{80} and both the I_h-C_{80} and D_{5h}-C_{80} cages have been studied. This is the only cage size in which more than a single cage isomer has been crystallographically characterized. Within the $M_3N@I_h$-C_{80} framework, the M-N distances increase gradually along the following series: $Sc_3N@I_h$-C_{80}, 1.9931(14)–2.0526(14) Å; $Dy_3N@I_h$-C_{80}, 2.004(8)–2.067(6) Å; $Lu_3N@I_h$-C_{80}, 2.001(3)–2.0819(8) Å; $Tm_3N@I_h$-C_{80}, 2.020(6)–2.058(6) Å; $Tb_3N@I_h$-C_{80}, 2.056(4)–2.089(4) Å; $Gd_3N@I_h$-C_{80}, 2.038(8)–2.117(5) Å. With the largest metal ions, Gd^{3+} and Tb^{3+}, the M_3N units are pyramidalized. For example, in $Gd_3N@I_h$-C_{80} the nitrogen atom is 0.522 (8) Å from the plane of the three Gd^{3+} ions, while in $Tb_3N@I_h$-C_{80} it is 0.453(4) Å from the plane of the Tb^{3+} ions. In the other cases the M_3N units are planar. For molecules of the $M_3N@D_{5h}$-C_{80} family, the M-N distances are similar to those noted above for the I_h-C_{80} cage: $Sc_3N@D_{5h}$-C_{80}, 2.014(2)–2.04(2) Å; $Tm_3N@D_{5h}$-C_{80}, 2.025(5)–2.062(2) Å; and $Tb_3N@D_{5h}$-C_{80}: 2.008(8)–2.130(6) Å. While the M_3N units are planar in $Sc_3N@D_{5h}$-C_{80} and $Tm_3N@D_{5h}$-C_{80}, in $Tb_3N@D_{5h}$-C_{80} it is pyramidal with the nitrogen atom 0.416 (13) Å from the Tb_3 plane.

It has also been possible to characterize the structures of mixed-metal endoheral fullerenes of the type, $M'M_2@I_h$-C_{80}. Five such species have been examined: $ErSc_2N@I_h$-C_{80} [29], $CeSc_2N@I_h$-C_{80} [30], $GdSc_2N@I_h$-C_{80} [31], $Gd_2ScN@I_h$-C_{80} [31], and $TbSc_2N@I_h$-C_{80} [31]. The M′M2 units within the I_h-C_{80} cage are remarkably well ordered and lie in a plane that is nearly perpendicular to the plane of the adjacent metalloporphyrin. The scandium ions preferentially lie closest to the porphyrin plane. For these mixed metal systems, the difference in sizes of the metal ions produce acentric arrangements inside the I_h-C_{80} cage, and the presence of a large metal ion generally results in shortening of the Sc-N distances. In the series $Sc_mGd_{3-m}@I_h$-C_{80}, the Sc-N distances shorten progressively in the series; $Sc_3N@I_h$-C_{80} (1.9931(14)–2.0526(14) Å), $GdSc_2N@I_h$-C_{80} (1.916(9)–1.919(8) Å), and $Gd_2ScN@I_h$-C_{80} (1.911(3) Å). Likewise the Gd-N distances lengthen progressively in the series: $Gd_3N@I_h$-C_{80} (2.038(8)–2.117(5) Å), $Gd_2ScN@I_h$-C_{80} (2.072(3)–2.102(3) Å), $GdSc_2N@I_h$-C_{80} (2.149(10) Å).

Examination of related $M_3N@C_{2n}$ families of endohedrals allows us to make comparisons of the effects of cage size on the M_3N group inside. The Sc_3N unit retains its planar geometry even as the cage size decreases in the series, $Sc_3N@I_h$-C_{80}, $Sc_3N@D_{5h}$-C_{80}, $Sc_3N@D_{3h}(5)$-C_{78}, and $Sc_3N@D_3(6140)$-C_{68}. Additionally, there is only a slight variation of the Sc-N distance as the cage size decreases: $Sc_3N@I_h$-C_{80}, 1.9931(14)–2.0526(14) Å; $Sc_3N@D_{5h}$-C_{80}, 2.014(2)–2.041(2) Å; $Sc_3N@D_{3h}$-C_{78}, 1.981(6)–2.127(4) Å; and $Sc_3N@D_3(6140)$-C_{68}, 1.961(4)–2.022(3) Å. The small cage in $Sc_3N@D_3(6140)$-C_{68} has a flattened aspect that provides added space for the M_3N unit in the region perpendicular to the direction of flattening with the scandium ions tucked into the corners provided by the fused pentagons.

The $Tb_3N@C_{2n}$ family of endohedrals provides a different look at the effects of cage size on the M_3N unit. For the three largest cages, $Tb_3N@C_s(51365)$-C_{84}, $Tb_3N@D_3$-C_{86}, and $Tb_3N@D_2(35)$-C_{88}, the Tb_3N units are planar, while for the two smallest, the I_h and D_{5h} isomers of $Tb_3N@C_{80}$, the Tb_3N units are significantly pyramidalized. The flattened shapes of the large cages, $Tb_3N@D_3$-C_{86}, and $Tb_3N@D_2(35)$-C_{88}, also assist in enlarging the interior space to accommodate planar Tb_3N groups.

As the size of the fullerene cage increases, the number of isomeric structures that can be constructed using hexagonal and pentagonal arrangements of carbon atoms also increases. For example, for C_{60} and C_{70}, the most abundant empty cage fullerenes, there is only one cage isomer that satisfies the IPR, but for a C_{84} cage, there are 24 IPR isomers. In this regard it is fortunate that for the largest endohedral fullerenes encountered to date only a few isomers of any one compound have been encountered and that it has been possible to separate these isomers by chromatography. The case of $M_3N@C_{84}$ is illustrative. For $Tb_3N@C_{84}$ and $Tm_3N@C_{84}$ only two isomers have been discovered, while for $Gd_3N@C_{84}$ three isomers have been detected and separated [23, 24]. Surprisingly, for each of these different metal ions, the most abundant isomers all have identical non-IPR fullerene cage structures: $Tb_3N@C_s(51365)$-C_{84}, $Tm_3N@C_s(51365)$-C_{84}, and $Gd_3N@C_s(51365)$-C_{84}. Thus, the same non-IPR cage was produced under different experimental conditions that employed three different metal oxide precursors, different packing materials inside the graphite rods and different arc-discharge generators in different laboratories. Moreover, since a non-IPR structure was involved, the isomeric possibilities had increased from the mere 24 IPR isomers for C_{84} to 51568 non-IPR isomers!

A number of exohedral adducts of $Sc_3N@D_{3h}(5)$-C_{78} and $M_3N@I_h$-C_{80} have been crystallographically characterized as shown in Table 9.2. Since the addends generally provide effective symmetry lowering, these modified endohedral fullerenes have been crystallized without the need for cocrystallization with Ni^{II}(OEP). While generally adduct formation proceeds without major alteration of the carbon cage, the Bingel-Hirsch and diazo adducts possess 'open structures' in which the cage sp^3-sp^3 cyclopropyl C–C bonds at a 6:6 ring junction have opened via a norcaradiene-type rearrangement, so as to preserve the aromaticity of the cage [40, 42].

As summarized by the ^{13}C NMR data in Table 9.3, there are several TNT EMF carbon cage motifs that are readily identifed by characteristic ^{13}C NMR chemical shift ranges. Of the seven isolated pentagon rule IPR isomers for the C_{80} cage, only the I_h isomer yields a ^{13}C NMR spectrum containing two lines with a 3 : 1 ratio. An initial observation for the

Table 9.2 *Crystallographically characterized exohedral mono adducts of the $M_3N@C_{2n}$ family*

Adduct	Comments
$Sc_3N@D_{3h}(5)$-C_{78}-6 : 6 $(CH_2)_2NCPh_3$ [34]	Prato adduct at a 6 : 6 ring junction offset from the horizontal mirror plane, Sc^{3+} ions reside in horizonal mirror plane
$Sc_3N@I_h$-C_{80}-5 : 6$C_{10}H_{12}O_2$ [35, 36]	Cyclo-addition of 4,5-dimethoxyquinodimethane at a 5 : 6 ring junction
$Sc_3N@I_h$-C_{80}-5 : 6$(CH_2)_2NCPh_3$ [37]	Prato adduct at a 5 : 6 ring junction, Sc^{3+} ions well ordered.
$Sc_3N@I_h$-C_{80}-6 : 6$(CH_2)_2NCPh_3$ [37]	Prato adduct at a 6 : 6 ring junction, Sc^{3+} ions disordered over multiple sites
$Sc_3N@I_h$-C_{80}-$(CH_2Ph)_2$ [38]	1,4-Addition to a six-membered ring
$Y_3N@I_h$-C_{80}-5,6$(CH_2)_2NEt$ [39]	Prato adduct at a 5 : 6 ring junction, Y^{3+} ions well ordered
$Y_3N@I_h$-C_{80}-6 : 6$C(CO_2CH_2Ph)$ [40]	Open Bingel-Hirsch adduct at a 6 : 6 ring junction
$Sc_3N@I_h$-C_{80}-6 : 6[C(Ph) $(CH_2)_3COOCH_3$] [41]	Open methanofullerene at a 6 : 6 ring junction
$Y_3N@I_h$-C_{80}-6 : 6[C(Ph) $(CH_2)_3COOCH_3$] [41]	Open methanofullerene at a 6 : 6 ring junction

$A_3N@I_h$-C_{80} family is the relatively small perturbation of the ^{13}C NMR shifts as a function of size and metal differences of the $(A_3N)^{+6}$ clusters. This is illustrated for the $Sc_3N@I_h$-C_{80} and $Lu_3N@I_h$-C_{80} cages as well as mixed clusters, $Lu_2YN@C_{80}$ and $LuY_2N@C_{80}$ that have been reported [3, 43, 44]. In these cases, the I_h-C_{80} cage with the corannulene type motif exhibits ^{13}C NMR shifts for the 6,6,5 and 6,6,6 carbon junctions in ranges from 142.8–144.7 and 135.9–138.2 ppm, respectively. These relatively small chemical shift ranges even include the weakly paramagnetic $CeSc_2N@C_{80}$ system [30]. The ^{13}C NMR spectrum for $Y_3N@I_h$-C_{80} supports an electronic distribution of $[Y_3N]^{+6}@[C_{80}]^{-6}$, a nearly spherical charge distribution over the fullerene cage, since the corannulene-type 6,6,6 carbon atoms (intersection of three hexagons, $\delta = 138.2$ ppm) and 6,6,5 carbon atoms (intersection of a

Table 9.3 *^{13}C NMR chemical shifts of TNT EMF cages*

TNT-EMF	^{13}C NMR Chemical Shifts (ppm)
$Sc_3N@D_3(6140)$-C_{68} [47]	158.49, 150.40, 149.51, 147.41, 145.52, 143.55, 142.92, 137.77, 137.62, 137.19, 137.12, 136.87(1/3)
$Sc_3N@D_{3h}(5)$-C_{78} [20]	155.60, 151.23 (1/2), 150.50, 143.16 (1/2), 142.48, 135.34, 133.84 (1/2), 132.97
$Sc_3N@I_h$-C_{80} [3]	144.57, 137.24
$Sc_3N@D_{5h}$-C_{80} [49]	149.8, 145.0, 143.9, 139.3, 138.5, 135.2
$LuSc_2N@I_h$-C_{80} [44]	143.99, 137.12
$Lu_2ScN@I_h$-C_{80} [44]	43.99, 137.12
$Lu_3N@I_h$-C_{80} [43]	144.0, 137.4
$Lu_3N@D_{5h}$-C_{80} [32]	149.0, 144.7, 143.2, 138.2, 138.1, 135.5
$Lu_2YN@C_{80}$ [44]	144.22, 137.66
$LuY_2N@I_h$-C_{80} [44]	144.41, 137.95
$Y_3N@I_h$-C_{80} [44]	144.44, 138.04
$CeSc_2N@I_h$-C_{80} [30]	142.85, 135.90

pentagon and two hexagons, $\delta = 144.6$ ppm) are so similar. These values are similar to those of $Sc_3N@I_h$-C_{80} (137.24 and 144.57 ppm) and $Lu_3N@I_h$-C_{80} (137.4 and 144.0 ppm), respectively [3, 43]. For the examples cited, the ^{13}C NMR spectra at ambient temperatures confirm isotropic motional averaging of the A_3N cluster inside the I_h cage. This description is supported by the computational studies, but requires DFT-NMR calculations on a large series of snapshots to reproduce the experimental two-line spectrum [45]. Even in the solid state, ^{13}C and ^{45}Sc NMR studies for $Sc_3N@I_h$-C_{80} suggest significant reorientation with a two-line ^{13}C spectrum similar to the values found for solution studies. Also, reorientational correlation times reported for the internal cluster (^{45}Sc NMR) and the carbon cage (^{13}C NMR) appear to be similar between 200–300 K [46]. Computational studies have indicated that the motion of the $[Sc_3N]^{+6}$ cluster is restricted for the lower symmetry $Sc_3N@D_3(5)$-C_{78} system [21]. The motion of the $[Sc_3N]^{+6}$ cluster is restricted in the the non-IPR $Sc_3N@D_3(6140)$-C_{68}, which exhibits 12 spectral lines consistent with its D_3 symmetry ($11 \times 6, 2 \times 1$ pattern) [47]. The observed and computationally predicted spectral lines and especially the signal at 158.49 ppm are consistent with the pentalene fused carbons in this motionally restricted non-IPR structure [47, 48].

As previously indicated, it has been established that there are two isomers of $Sc_3N@C_{80}$ (I_h and D_{5h} both obeying the IPR). These isomers have been isolated and characterized by ^{13}C NMR [3, 49]. The D_{5h} isomer has also been characterized by single crystal X-ray studies for other paramagnetic $A_3N@C_{80}$ systems, such as $Tm_3N@D_{5h}$-C_{80} and $Tb_3N@D_{5h}$-C_{80} [22, 27]. For $Y_3N@D_{5h}$-C_{80}, the data closely match the ^{13}C NMR data reported for other $A_3N@D_{5h}$-C_{80} isomers (intensity ratios 1 : 2 : 2 : 1 : 1 : 1) and do not significantly deviate as a function of the metal in the $[A_3N]^{+6}$ (M = Lu and Sc) cluster [44]. The detailed assignments of these carbon signals await future 2d-INADEQUATE carbon-carbon connectivity studies, but the current results clearly illustrate the importance of the position and motional process of the internal $[A_3N]^{+6}$ cluster in determining ^{13}C NMR chemical shifts.

The limited availability of some TNT-EMFs has hindered a complete study of their chemical reactivity. For some of these fullerenes, the development of improved synthetic methodologies and purification processes has allowed their preparation on multi-milligram scales. These developments have made it possible to make a general assessment of the chemical reactivity of these new fullerenes. So far the most common reactions observed for empty cage fullerenes such as cycloadditions, nucleophilic additions, free radical additions and redox reactions have been observed with TNT-EMFs. However, these reactions have been shown to be strongly dependent on the cage size, cage symmetry and nature of the encapsulated cluster. Scheme 9.1 shows the chemical reactions that have been explored with $Sc_3N@I_h$-C_{80}, which is the most abundant TNT-EMF [3].

9.2.1 Cycloaddition Reactions

The first fully characterized derivative of a TNT-EMF was a Diels-Alder adduct [35, 36], obtained by refluxing ^{13}C labeled 6,7-dimethoxyisochroman-3-one with $Sc_3N@I_h$-C_{80}. The icosahedral C_{80} fullerene cage has two different types of carbons: pyrene type carbon atoms located at the junction of three six-membered rings and corannulene type carbons at the intersection of two six-membered rings and a five membered ring. Reactions at these sites give rise to two different types of bonds: the [5,6]-bonds, which occur between a five- and a six-membered ring, and the [6,6]-bonds between two six-membered rings. ^{13}C-NMR

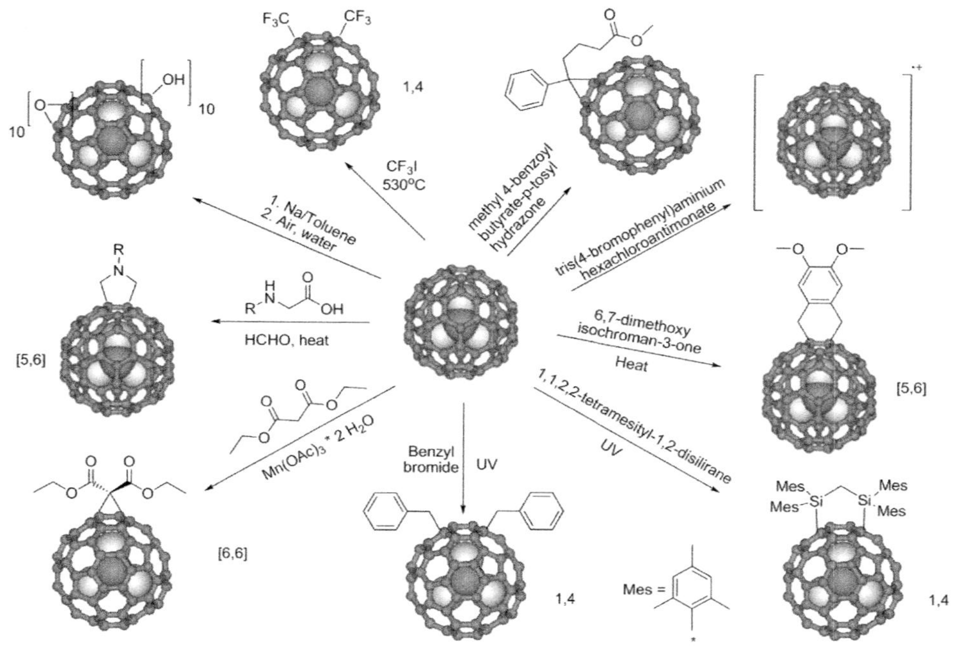

Scheme 9.1 *Representative chemical reactions studied for* $Sc_3N@I_h\text{-}C_{80}$

characterization of the Diels-Alder adduct in conjunction with a Hetero Multiple Bond Correlation (HMQC) experiment indicated the presence of a symmetry plane bisecting the molecule. Despite the fact that there are three possible distinct adducts that posses a symmetry plane, the known [4+2] mechanism of this reaction establishes that the compound is the 1,2-derivative on a [5,6]-bond. This assignment was later confirmed by single crystal X-ray analysis [36]. The same reaction was also used to prepare a mono and a bis-adduct of $Gd_3N@C_{80}$ [50]. However, the resulting compounds were not fully characterized; thus, the relative position of the addends is not known.

The second example of a cycloaddition reaction on a TNT-EMF was a 1,3-dipolar cycloaddition of an azomethine ylide (the Prato reaction) [51]. Echegoyen *et al.* reported the synthesis of the *N*-ethylpyrrolidine adduct of $Sc_3N@I_h\text{-}C_{80}$ and showed by ^{13}C-NMR and ^{1}H-^{13}C-HMQC that the cycloaddition reaction had occurred regioselectively on a [5,6]-bond as in the Diels Alder case. Most interestingly, the methylene diastereotopic protons on the pyrrolidine ring have a 1.2 ppm chemical shift difference, a consequence of the two different magnetic environments created by the $Sc_3N@I_h\text{-}C_{80}$ fullerene cage [52] and the effect of the nitrogen lone pair [53]. Independently, Dorn and coworkers reported the synthesis of the *N*-methylpyrrolidine derivative of $Sc_3N@I_h\text{-}C_{80}$ and $Er_3N@I_h\text{-}C_{80}$ [53]. For the $Sc_3N@I_h\text{-}C_{80}$ adduct, the addition occurred regioselectively on a [5,6]-bond as indicated by the NMR experiments [53]. See Figure 9.3a for the X-ray structure of this adduct. No NMR data was obtained for $Er_3N@I_h\text{-}C_{80}$ due to the paramagnetic nature of the encapsulated metal; thus, for $Er_3N@I_h\text{-}C_{80}$, it was not possible to establish whether the addition had occurred on a [5,6]- or a [6,6]-bond [53].

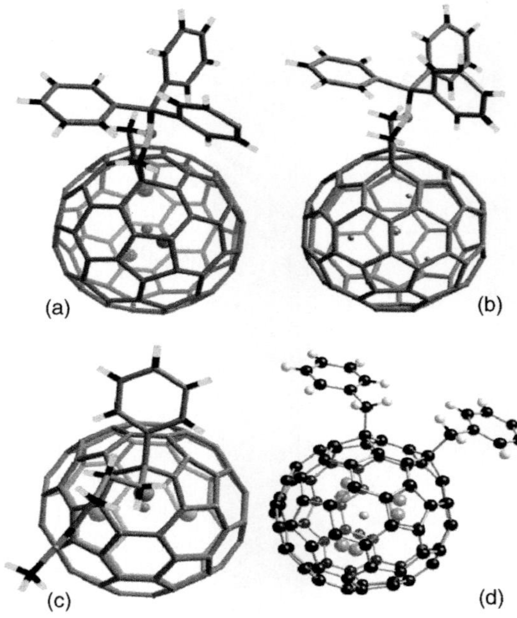

Figure 9.3 *X-ray structures of derivatives of C_{80} TNT EMFs: a) the 5,6-(N-tritylpyrrolidino) derivative of $Sc_3N@I_h$-C_{80}, b) the 6,6-(N-tritylpyrrolidino) derivative of $Sc_3N@I_h$-C_{80} c) the 6,6-'open' phenyl carbomethoxypropyl methano derivative of $Y_3N@I_h$-C_{80} and d) the 1,4-dibenzyl adduct of $Sc_3N@I_h$-C_{80}*

A year later, the same groups reported simultaneously the synthesis of [6,6]-pyrrolidine derivatives of $Sc_3N@I_h$-C_{80}, $Y_3N@I_h$-C_{80} and $Er_3N@I_h$-C_{80} [37, 54, 55]. The *N*-tritylpyrrolidino-$Sc_3N@I_h$-C_{80} derivative is so far the only example of a [6,6]-pyrrolidine derivative of $Sc_3N@I_h$-C_{80}. See Figure 9.3b for the X-ray structure of this adduct. Further studies are required to determine the factor(s) controlling the regioselectivity of addition reactions on $Sc_3N@I_h$-C_{80}. The electrochemical properties of TNT-EMF pyrrolidine derivatives were studied and showed that the [6,6]-regioisomers of $Y_3N@I_h$-C_{80} and $Er_3N@I_h$-C_{80} display irreversible reductive electrochemical behavior at normal scan rates (100 mV/s). Increasing the scan rate to 30 V/s did not change the appearance of the reduction waves [55]. All the [6,6]-$M_3N@I_h$-C_{80} (M = Sc, Y, Er) pyrrolidine derivatives isomerize upon heating to the corresponding [5,6]-$M_3N@I_h$-C_{80} (M = Sc, Y, Er) regioisomers, indicating that the [6,6]-regioisomers are the kinetic products, whereas the [5,6]-regioisomers are the thermodynamic products [37]. The electrochemical behavior of the [5,6]-regioisomers revealed reversible reduction waves for the $Sc_3N@I_h$-C_{80} and $Er_3N@I_h$-C_{80} derivatives at normal scan rates, while $Y_3N@I_h$-C_{80} shows irreversible electrochemical behavior at normal scan rates, but reversible reductive electrochemical behavior when the scan rate is increased to 30 V/s [55]. A comparative study of the reactivity of the series $Sc_{3\text{-}x}Y_xN@C_{80}$ showed a change of regioselectivity from the [5,6]-isomer to [6,6]-isomer induced by the encapsulated metals [56]. A similar study with the $Sc_xGd_{3\text{-}x}N@C_{80}$ series showed the same trend, but the [6,6]-$Gd_3N@I_h$-C_{80} does not isomerize upon heating [57].

The 1,3-dipolar cycloaddition reaction of an azomethine ylide was also conducted with $Sc_3N@D_{5h}$-C_{80} [32]. Due to the lowered symmetry of the fullerene cage, there are six types of nonequivalent carbons and nine types of bonds. Compared to $Sc_3N@I_h$-C_{80}, the reaction on $Sc_3N@D_{5h}$-C_{80} occurred faster under the same conditions, indicating a higher reactivity towards the 1,3-dipolar cycloadditions, probably as a result of the smaller HOMO-LUMO gap. Two mono-adduct fractions are dominant in the HPLC chromatogram, indicating a high degree of regioselectivity since there are nine possible positions available for functionalization. Neither of the two isomers was completely characterized; however, the ^{1}H-NMR spectrum of one of them shows a singlet around 3.2 ppm, which resembles a typical ^{1}H-NMR spectrum for a C_{60} pyrrolidine derivative. Only two of the nine possible regioisomers can give rise to that ^{1}H-NMR pattern; therefore, the pyrrolidine ring is connected to the bonds formed by either the carbons labeled **e** or **f** in Figure 9.4a [32]. Considering the bond lengths obtained from the X-ray crystal structure, the **e-e** bonds (pyracelene type) are shorter and more pyramidalized than the **f-f** bonds (see Figure 9.4a); hence, it is more likely that the obtained regioisomer is the one in which the pyrrolidine ring is attached to the pyracelene patch. The pyrrolidine NMR proton signals of the major regioisomer are observed as two sets of doublets, so they are in nonequivalent magnetic environments, and they could correspond to any of the remaining seven regioisomers. The same reaction on $Sc_3N@D_{3h}(5)$-C_{78}, which has eight different carbon types and 13 different sets of C–C bonds (see Figure 9.4b), yielded the two kinetically controlled monoadducts on the **c-f** and **b-d** bonds as confirmed by ^{1}H, COSY and HMQC NMR experiments and single crystal X-ray diffraction [34]. Contrary to the situation with $Sc_3N@I_h$-C_{80}, in which the Sc_3N cluster can rotate freely inside the cage [3, 45], in $Sc_3N@D_{3h}(5)$-C_{78} the scandium atoms are localized on the three pyracylene patches [20, 21]. In the pyrrolidine adduct, the cluster remains in the horizontal plane and the addend avoids the bonds close to the metal. Hence, this is the first example of regioselectivity controlled by the encapsulated cluster in the TNT-EMFs.

The 1,3-dipolar cycloaddition of azomethine ylides was used for the preparation of donor-acceptor conjugates for exploring the potential application of TNT-EMFs in the construction of organic solar cells [58, 59]. However, the pyrrolidine adducts formed by

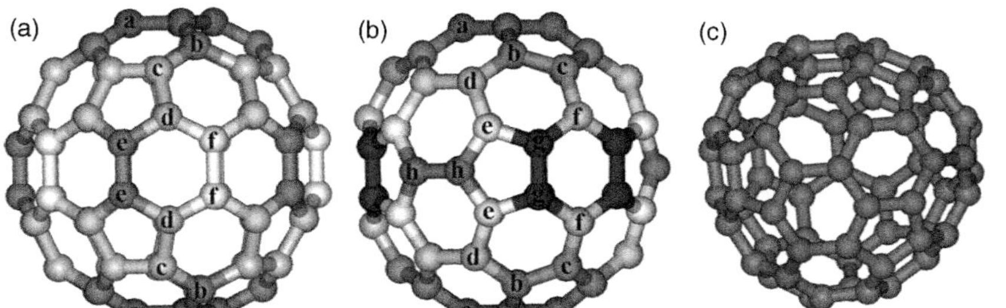

Figure 9.4 *(a) D_{5h}-C_{80} fullerene cage b) $D_{3h}(5)$-C_{78} fullerene cage and c) $D_3(6140)$-C_{68} fullerene cage. The bonds highlighted in red have been suggested as the preferred site for addition. The atoms highlighted in green correspond to the pentalene patches where the metal atoms of the cluster are bound*

TNT-EMFs are not thermally stable; the $CH_2N(R)CH_2$ moiety can be thermally removed [60] even in the absence of a catalyst or by electro oxidative conditions [61]. This might be useful for protecting/deprotecting strategies for functionalization of TNT-EMFs, but the low stability of this type of adduct may prevent their use in future photovoltaic and other applications.

A process that is believed to be a 1,3-dipolar cycloaddition involves reactions of the endohedral metallofullerenes with azoalkanes formed in situ by base treatment of *N*-tosylhydrazones. In recent work two different groups report preparation of $Lu_3N@C_{80}$- [62], $Sc_3N@C_{80}$- [41] and $Y_3N@C_{80}$-based [41] analogs of methanofullerene phenyl-C_{61}-butyric acid methyl ester (PCBM) [63] which has been widely used and an acceptor in bulk heterojunction solar cells [64]. The reactions take place at 6,6-bonds and yield open structures, as opposed to the closed or cyclopropyl structure of the dominant C_{60} compound. See Figure 9.3c for the X-ray structure of one of these adducts. The initially formed intermediate is believed to be a diazole, which undergoes extrusion of molecular nitrogen to yield the cyclopropyl compound, which in turn undergoes a norcaradiene-type rearrangement to the open homoaromatic structure [41]. These materials are believed to have great potential in solar cells, because their LUMO energy levels are closer to the LUMO levels of the conducting polymeric component of the cells, i. e. poly(3-hexylthiophene) [62].

9.2.2 Free Radical and Nucleophilic Addition Reactions

Water soluble derivatives of TNT-EMFs have potential applications in biological systems. In order to prepare water soluble fullerols, $Sc_3N@C_{80}$ was refluxed in toluene in the presence of sodium metal to produce poly(anionic radical) species that precipitate out of solution. The exposure of this material to air and water produced a golden colored aqueous solution of $Sc_3N@C_{80}(OH)_{\sim10}O_{\sim10}$ deduced from the X-ray photoelectron spectrum (XPS) [65].

The photochemical reaction between 1,1,2,2-tetramesityl-1,2-disilirane and $Sc_3N@I_h$-C_{80} yielded a mixture of two monoadducts with 1,2- and 1,4-addition patterns. The 1,2-addition occurred on a [5,6]-bond and the 1,4-addition occurred over two corannulene-type carbon atoms [66, 67]. This reaction does not occur thermally and the 1,2-isomer converts into the 1,4-isomer upon heating, indicating that the 1,4-isomer is thermodynamically more stable. The first reduction potential of the derivative was shifted cathodically by 230 mV due to the electron donating nature of the disilirane group. Nonetheless, the silane addend is removed under reductive conditions to yield the pristine fullerene.

The free radical trifluoromethylation of both I_h and D_{5h} isomers of $Sc_3N@C_{80}$ preferentially produced bis-adducts that display a single signal in the ^{19}F-NMR spectrum, indicating the presence of a symmetry plane. Based on DFT calculations it was suggested that 1,4-addition on $Sc_3N@I_h$-C_{80} took place on corannulene-type carbon atoms and over atoms marked as **d** on Figure 9.4a for the $Sc_3N@D_{5h}$-C_{80} isomer [68]. More recently, the reaction of $Sc_3N@I_h$-C_{80} and $Lu_3N@I_h$-C_{80} with benzyl radicals yielded the 1,4-dibenzyl adducts [38] as in the case of the disilirane additions [66, 67]. See Figure 9.3d for the X-ray structure of the $Sc_3N@I_h$-C_{80} adduct.

The cyclopropanation of fullerenes with malonates (Bingel-Hirsch reaction) [69, 70] is one of the most common methodologies used for the functionalization of fullerenes. This reaction was successfully applied to $Y_3N@I_h$-C_{80}, to yield an open cage fulleroid [40, 54], but failed on $Sc_3N@I_h$-C_{80} under the same experimental conditions. The obtained adduct does not undergo retro-cyclopropanation reaction under reductive conditions. The same reaction was applied to $Sc_3N@D_{3h}(5)$-C_{78} to prepare a monoadduct whose addend is attached to the **c-f** bond as in the 1,3-dipolar cycloaddition (Figure 9.3) [34]. Most importantly, a bis-adduct was also formed with high regioselectivity [71]; The second addition occurred on the *anti-* **1** position, which corresponds to the kinetically preferred site for nucleophilic attack [72], in total agreement with the theoretically predicted regiochemistry [73].

Water soluble derivatives of $Gd_3N@C_{80}$ were prepared via the Bingel-Hirsch reaction with poly(ethylene glycol) (PEG) malonates [74]. These were subsequently hydroxylated to afford materials that exhibitied very high magnetic resonance imaging (MRI) relaxivities and thus effective contrast agents [74]. More recently 'PEGylation' has been achieved on $Gd_3N@C_{80}$ by reaction with an amino poly(ethylene glycol) ether, butanone peroxide and 2,2,4-trimethyl-pentanediol diisobutyrate, yielding derivatives with up 22 ethyleneoxy moieties per cage with unknown structure; these materials also function as very good MRI contrast agents [75].

The Bingel reaction on $Sc_3N@D_3(6140)$-C_{68} yielded a monoadduct with remarkable regioselectivity [76] despite the fact that the $Sc_3N@D_3(6140)$-C_{68} contains 12 different types of carbons and 6 different types of C–C bonds. Based on LUMO electron density studies and ^{13}C-NMR data, it was proposed that the addition occurred at a bond exocyclic to the pentalene patch that contains a unique [5,5]-bond only found in non-IPR systems and where the scandium atoms are strongly coordinated in $Sc_3N@D_3(6140)$-C_{68} [19].

More recently differential reactivity towards the Bingel reaction was observed in the series $Gd_3N@I_h$-C_{80}, $Gd_3N@C_s(51365)$-C_{84} [33] and in an undetermined isomer of $Gd_3N@C_{88}$ [77], with the smallest cage being the most reactive. It was suggested that the difference in reactivity is a consequence of the decreased pyramidalization degree of the carbon atoms in the larger fullerene cages. Based on electrochemical data, it was determined that the [6,6]-bond in the I_h-C_{80} cage is the most reactive for addition. For $Gd_3N@C_s(51365)$-C_{84} a highly regioselective reaction was observed, yielding only one monoadduct. However, it was not possible to obtain NMR data due to the paramagnetic nature of the encapsulated metal; therefore, the position of the addition has not been established.

An analog of the Bingel reaction was conducted on $Sc_3N@I_h$-C_{80} by using free radicals generated from diethyl malonate and manganese(III) acetate [42]; polyadducts with up to 8 addends were detected by MALDI-TOF mass spectrometry. Two different monoadducts were obtained, $Sc_3N@I_h$-$C_{80}[^{13}C(COOC_2H_5)_2]$ and $Sc_3N@I_h$-$C_{80}[^{13}CHCOOC_2H_5]$. The addition proceeded regioselectively at the [6,6]-position and, based on NMR data, it was suggested that the adducts are open cage fulleroids as in the case of $Y_3N@I_h$-C_{80} [40]. The same reaction on $Lu_3N@I_h$-C_{80} yielded equivalent products and polyadducts with up to 10 addends.

Two other free radical-type reactions are noteworthy. In one a peroxide reaction was used to introduce several carboxyethyl groups onto $Gd_3N@I_h$-C_{80}; the resultant products are water-soluble and possess excellent magnetic resonance relaxivities [78]. In another study, the adhesive properties (measured by tack analysis) decreased in blends of $Sc_3N@I_h$-C_{80} and polystyrene-block-polyisoprene-block-polystyrene (SIS) copolymer (a pressure-sensitive adhesive) under white light irradiation in air; the authors conclude that the reduction of tack is attributable to the in situ generation of $^{1}O_2$ and subsequent photooxidative cross-linking of the adhesive film [79].

9.2.3 Electrochemistry Studies of TNT-EMFs

Electrochemical techniques are convenient to study the electronic properties of TNT-EMFs. After the successful separation of the I_h and D_{5h} $Sc_3N@C_{80}$ isomers [49, 80], CV experiments with $Sc_3N@I_h$-C_{80} in 0.10 M $TBABF_4$ in *o*-DCB at 10 mV/s scan rate revealed an electrochemical HOMO-LUMO gap of 1.86 V. This value is much smaller than that of C_{60} (2.35 V) [81], but considerably higher that the originally reported 0.8 V value [3]. The electrochemical measurement explains better the relatively high abundance, high thermal stability and lower reactivity of TNT-EMFs when compared to other endohedral metallofullerenes [49, 80]. The reported redox potentials for most of the $M_3N@I_h$-C_{80} and $M_3N@D_{5h}$-C_{80} TNT-EMFs isomers in *o*-DCB are listed in Table 9.4, These TNT-EMFs usually display irreversible reductive electrochemical behavior and reversible oxidative electrochemical behavior. There is a significant difference between the first oxidation potential of TNT-EMFs containing the same cluster but different fullerene cage. Therefore, the change of the fullerene cage symmetry seems to have a significant effect on the HOMO energy levels. A method for separating the I_h and D_{5h} isomers of $Sc_3N@C_{80}$ based on the oxidation potential difference between the two isomers was developed [12]. According to the oxidation potential values it seems that this methodology can be extended to the separation of I_h and D_{5h} isomers of $Lu_3N@C_{80}$, $Dy_3N@C_{80}$, and $Tm_3N@C_{80}$ as well. On the other hand, the reduction potentials are only slightly affected by the change of the symmetry of the fullerene cage or by the variation of the metal, because, except for $Sc_3N@C_{80}$, they all

Table 9.4 *Redox potentials of $M_3N@I_h$-C_{80} in volts Vs Fc/Fc^+ redox pair in o-DCB*

TNT-EMF	$E^{+/+2}$	$E^{0/+}$	$E^{0/-}$	$E^{-/2-}$	$E^{2-/3-}$	$\Delta(E^{0/+}-E^{0/-})$
$Sc_3N@I_h$-C_{80} [80]	–	+0.62	−1.24	−1.62	–	1.86
$Sc_3N@I_h$-C_{80} [12]	+1.09	+0.59	−1.26	−1.62	−2.37	1.85
$Sc_3N@I_h$-C_{80} [66]	–	+0.62	−1.22	−1.59	−1.90	1.84
$Sc_3N@I_h$-C_{80} [55]	–	+0.59	−1.29	−1.56	−2.32	1.88
$Sc_3N@I_h$-C_{80} [32]	–	+0.57	−1.27	–	–	1.84
$Y_3N@I_h$-C_{80} [55]	–	+0.64	−1.44	−1.83	−2.38	2.08
$Lu_3N@I_h$-C_{80} [32]	–	+0.64	−1.40	–	–	2.04
$Lu_3N@I_h$-C_{80} [62]	–	+0.64	−1.42	–	–	2.06
$Tm_3N@I_h$-C_{80} [82]	–	+0.68	−1.31	−1.76	–	1.99
$Tm_3N@I_h$-C_{80} [27]	–	+0.65	−1.43	–	–	2.08
$Er_3N@I_h$-C_{80} [55]	–	+0.63	−1.42	−1.80	–	2.05
$Dy_3N@I_h$-C_{80} [83]	–	+0.70	−1.37	−1.86	–	2.07
$Gd_3N@I_h$-C_{80} [84]	–	+0.58	−1.44	−1.86	−2.18	2.02
$Nd_3N@I_h$-C_{80} [85]	–	+0.63	−1.42	–	–	2.05
$Pr_3N@I_h$-C_{80} [85]	–	+0.59	−1.41	–	–	2.00
$Sc_3N@D_{5h}$-C_{80} [12]	+0.70	+0.35	–	–	–	–
$Sc_3N@D_{5h}$-C_{80} [32]	–	+0.34	−1.33	–	–	1.67
$Lu_3N@D_{5h}$-C_{80} [32]	–	+0.45	−1.41	–	–	1.86
$Dy_3N@D_{5h}$-C_{80} [81]	–	+0.41	−1.40	−1.85	–	1.81
$Tm_3N@D_{5h}$-C_{80} [27]	–	+0.39	−1.45	–	–	1.84

The oxidation values are half wave potentials whereas the reductions correspond to peak potentials.

have very similar reduction potentials. It is very interesting to note that, except for $Sc_3N@C_{80}$, the rest of the I_h and D_{5h} compounds also have very similar HOMO-LUMO gaps.

The EPR spectrum of the anion radical of $Sc_3N@I_h$-C_{80} has 22 lines with a hyperfine splitting constant of 55.6 Gauss, which is consistent with a high localization of the electronic spin on the Sc atoms (S = 7/2) [12, 86]. Hence, there is an important contribution of the Sc metal orbitals to the LUMO that must be responsible for the unique electronic properties observed when M = Sc.

The influence of the cage size for cages larger than C_{80} has been studied in isolated isomers of the series $Gd_3N@C_{2n}$ [84], $Nd_3N@C_{2n}$ [90], $Pr_3N@C_{2n}$ [85], $Ce_3N@C_{2n}$ [85], and $La_3N@C_{2n}$ [88]. The redox potentials in *o*-DCB vs Fc/Fc$^+$ are listed in Table 9.5.

There is a significant variation of the first oxidation potential throughout the series, confirming that the HOMO is a fullerene cage based orbital. The variation of the first reduction potentials among each of the series is important as well, but they correlate with the electronegativity of the metal [84]. A recent study concluded that, with the exception of $Sc_3N@C_{2n}$, neither the metal nor the cage size correlate with the observed HOMO-LUMO gap values and both the HOMO and the LUMO are cage based [17]. Most interesting is the fact that the $M_3N@C_{88}$ fullerenes exhibit both reversible and irreversible electrochemical behavior and very low HOMO-LUMO gaps, a consequence of their low oxidation potentials [84].

The EPR spectra of both the radical cation and radical anion of $Sc_3N@D_3(6140)$-C_{68} show hyperfine structure similar to that observed for the $Sc_3N@I_h$-C_{80} anion radical,

Table 9.5 *Redox potentials of $M_3N@C_{2n}$ in volts Vs Fc/Fc$^+$ redox pair in o-DCB*

TNT-EMF	$E^{+/+2}$	$E^{0/+}$	$E^{0/-}$	$E^{-/2-}$	$E^{2-/3-}$	$\Delta(E^{0/+}-E^{0/-})$
$Gd_3N@C_{80}$ [84]	–	+0.58	–1.44	–1.86	–2.18	2.02
$Gd_3N@C_{82}$ [87]	–	+0.37	–1.52	–1.86	–	1.89
$Gd_3N@C_{84}$ [84]	–	+0.32	–1.37	–1.76	–	1.69
$Gd_3N@C_{86}$ [87]	–	+0.35	–1.35	–1.70	–	1.70
$Gd_3N@C_{88}$ [84]	+0.49	+0.06	–1.43	–1.74	–	1.49
$Nd_3N@C_{80}$ [85]	–	+0.63	–1.42	–	–	2.05
$Nd_3N@C_{84}$ [85]	–	+0.31	–1.44	–	–	1.75
$Nd_3N@C_{86}$ [85]	–	+0.36	–1.46	–	–	1.82
$Nd_3N@C_{88}$ [85]	–	+0.07	–1.33	–	–	1.40
$Pr_3N@C_{80}$ [85]	–	+0.59	–1.41	–	–	2.00
$Pr_3N@C_{86}$ [85]	–	+0.31	–1.48	–	–	1.79
$Pr_3N@C_{88}$ [85]	–	+0.09	–1.31	–	–	1.40
$Pr_3N@C_{92}$ [17]	–	+0.35	–1.46	–	–	1.81
$Pr_3N@C_{96}$ [88]	+0.53	+0.14	–1.54	–1.77	–	1.68
$Ce_3N@C_{88}$ [85]	–	+0.08	–1.30	–	–	1.38
$Ce_3N@C_{92}$ [17]	–	+0.32	–1.48	–	–	1.80
$Ce_3N@C_{96}$ [88]	+0.67	+0.18	–1.50	–1.84	–	1.68
$La_3N@C_{88}$ [88]	+0.66	+0.21	–1.36	–1.67	–	1.57
$La_3N@C_{92}$ [88]	–	+0.36	–1.44	–1.64	–	1.80
$La_3N@C_{96}$ [88]	+0.53	+0.14	–1.54	–1.77	–	1.68
$Sc_3N@D_3(6140)$-C_{68} [89]	+0.85	+0.33	–1.38	–1.98	–	1.71
$Sc_3N@$ $D_{3h}(5)$-C_{78} [71]	–	+0.12	–1.54	–	–	1.66
$Dy_3N@C_{78}$ [83]	–	+0.47	–1.54	–1.93	–	2.01

indicating an important participation of the metal orbitals in both the HOMO and LUMO. However, much smaller hyperfine coupling constants (cation 1.28 Gauss, anion 1.75 Gauss) were observed, suggesting that most of the unpaired spin is delocalized on the fullerene cage [89, 91].

9.3 Summary and Conclusions

Co-crystallization of TNT-EMFs with Ni^{II}(OEP) (OEP is the dianion of octaethylporphyrin) has proved to be a powerful technique for unequivocally assigning their chemical structures even in the presence of paramagnetic metals or with small available quantities. So far, structures ranging from C_{68} to C_{88} have been characterized using this technique. The most typical reactions with empty cage fullerenes (Diels Alder, addition of dipolarophiles, nucleophilic additions, free radical additions and redox reactions) have also been observed with TNT-EMFs. However, the encapsulated metal and the cage symmetry play an important role, making the chemical reactivity different from empty cage fullerenes. Similar to what has been observed with empty cage fullerenes, the driving force controlling the reactivity of TNT-EMFs is the release of bond strain. However, the shorter bonds and the most pyramidalized bonds are not necessarily the most reactive. Hydrogenation, addition of other nucleophiles, preparation of open cage derivatives, and especially preparation of highly substituted adducts with specific regiochemistry and enhanced solubility need to be explored in order to yield derivatives for potential technological applications. Finally, the electronic properties of TNT-EMFs have been studied by using electrochemical techniques. Most of them, except for some $M_3N@C_{88}$ species, have irreversible reductive and reversible oxidative electrochemical behavior. It has also been demonstrated that the electrochemical properties can be tuned by selective functionalization. It has been found that, except for $Sc_3N@C_{2n}$, all TNT-EMF systems have both cage HOMO and LUMO, but the HOMO orbital energies show larger variations among different fullerene cages.

References

[1] (a) S. Nagase, K. Kobayashi, T. Akasaka, *The Chemistry and Physics of Fullerenes and Related Materials*, ed. by K. M. Kadish and R. S. Ruoff, The Electrochemical Society, Inc., Pennington, 1995, 747–776. (b) S. Nagase, K. Kobayashi, T. Akasaka, Endohedral Metallofullerenes: New Spherical Cage Molecules with Interesting Properties, *Bull. Chem. Soc. Jpn.*, **69**, 2131–2142 (1996).

[2] H. Shinohara, Endohedral Metallofullerenes: Production, separation, and structural properties, *Fullerenes: Chemistry, Physics and Technology*, 357–393 (2000). H. Shinohara, Endohedral metallofullerenes, *Rep. Prog. Phys.*, **63**, 843–892 (2000).

[3] S. Stevenson, G. Rice, T. Glass, K. Harich, F. Cromer, M. R. Jordan, J. Craft, E. Hadju, R. Bible, M. M. Olmstead, K. Maitra, A. J. Fisher, A. L. Balch, H. C. Dorn, Small-bandgap endohedral metallofullerenes in high yield and purity, *Nature*, **401**, 55–57 (1999).

[4] H. C. Dorn, E. B. Iezzi, S. Stevenson, A. L. Balch, J. C. Duchamp, Trimetallic nitride template (TNT) endohedral metallofullerenes, Ch. XX, in *Endohedral Metallofullerenes: A New Family of Carbon Clusters* T. Akasaka and S. Nagase, eds, Kluwer Academic Publishers *Developments in Fullerene Science*, vol 3, (Endofullerenes) 121–131 (2002).

[5] L. Dunsch, M. Krause, J. Noack, P. Georgi, Endohedral nitride cluster fullerenes. formation and spectroscopic analysis of $L_{3-x}MxN@C_{2n}$ ($0 < x < 3$; $n = 39, 40$), *J. Phys. Chem. Solids*, **65**, 309–315 (2004).

[6] L. Dunsch, S. Yang, The recent state of endohedral fullerene research, *Electrochem. Soc. Interface*, **15**, 34–39 (2006).

[7] S. Stevenson, M. Thompson, M. Corey, H. L. Coumbe, M. A. Mackey, C. E. Coumbe, J. P. Phillips, Chemically Adjusting Plasma Temperature, Energy, and Reactivity (CAPTEAR) method using NO_x and combustion for selective synthesis of $Sc_3N@C_{80}$ metallic nitride fullerenes, *J. Am. Chem. Soc.*, **129**, 16257–16262 (2007).

[8] L. Dunsch, S. Yang, Metal nitride cluster fullerenes: their current state and future prospects, *Small*, **3**, 1298–1320 (2007).

[9] Z. Ge, J. C. Duchamp, T. Cai, H. W. Gibson, H. C. Dorn, Purification of endohedral trimetallic nitride fullerenes in a single, facile step, *J. Amer. Chem. Soc.*, **127**, 16292–16298 (2005).

[10] S. Stevenson, K. Harich, H. Yu, R. R. Stephen, D. Heaps, C. Coumbe, J. P. Phillips, Non-chromatographic, 'Stir and Filter Approach' (SAFA) for isolating $Sc_3N@C_{80}$ metallofullerenes, *J. Am. Chem. Soc.*, **128**, 8829–8835 (2006).S. Stevenson, M. A. Mackey, C. E. Coumbe, J. P. Phillips, B. Elliott, L. Echegoyen, Rapid removal of D_{5h} isomer using the 'Stir and Filter Approach' and isolation of large quantities of isomerically pure $Sc_3N@C_{80}$ metallic nitride fullerenes, *J. Am. Chem. Soc.*, **129**, 6072–6073 (2007).

[11] C. D. Angelii, T. Cai, J. C. Duchamp, J. E. Reid, E. S. Singer, H. W. Gibson, H. C. Dorn, Purification of trimetallic nitride templated endohedral metallofullerenes by a chemical reaction of congeners with eutectic 9-methylanthracene, *Chem. Mater.*, **20**, 4993–4997 (2008).

[12] B. Elliott, L. Yu, L. Echegoyen, A simple isomeric separation of D_{5h} and I_h $Sc_3N@C_{80}$ by selective chemical oxidation, *J. Am. Chem. Soc.*, **127**, 10885–10888 (2005).

[13] S. Yang, M. Zalibera, P. Rapta, L. Dunsch, Charge-induced reversible rearrangement of endohedral fullerenes: electrochemistry of tridysprosium nitride clusterfullerenes $Dy_3N@C_{2n}$ (2n = 78, 80), *Chem. Eur. J.*, **12**.

[14] T. Akasaka, S. Nagase, *Endofullerenes: a New Family of Carbon Clusters*, Dordrecht, Kluwer Academic Publishers, c2002. (2002).

[15] J. M. Campanera, C. Bo, J. M. Poblet, General rule for the stabilization of fullerene cages encapsulating trimetallic nitride templates, *Angew. Chem. Int. Ed.*, **44**, 7230–7233 (2005). D. Liu, F. Hagelberg, S. S. Park, Charge transfer and electron backdonation in metallofullerenes encapsulating NSc_3, *Chem. Phys.*, **330**, 380–386 (2006)

[16] A. A. Popov, L. Dunsch, Structure, stability, and cluster-cage interactions in nitride clusterfullerenes $M_3N@C_{2n}$ (M = Sc, Y; 2n = 68–98): a density functional theory study, *J. Am. Chem. Soc.*, **129**, 11835–11849 (2007).

[17] M. N. Chaur, R. Valencia, A. Rodriguez-Fortea, J. M. Poblet, L. Echegoyen, Trimetallic nitride endohedral fullerenes: experimental and theoretical evidence for the $M_3N^{6+}@C_{2n}{}^{6-}$ model, *Angew. Chem. Int. Ed.*, **48**, 1425–1428 (2009).

[18] R. Valencia, A. Rodriguez-Fortea, J. M. Poblet, Large fullerenes stabilized by encapsulation of metallic clusters, *Chem. Commun.*, 4161–4163 (2007).

[19] M. M. Olmstead, H. M. Lee, J. C. Duchamp, S. Stevenson, D. Marciu, H. C. Dorn, A. L. Balch, $Sc_3N@C_{68}$: Folded pentalene coordination in an endohedral fullerene that does not obey the isolated pentagon rule, *Angew. Chem. Int. Ed.*, **42**, 900–903 (2003).

[20] M. M. Olmstead, A. De Bettencourt-Dias, J. C. Duchamp, S. Stevenson, D. Marciu, H. C. Dorn, A. L. Balch, Isolation and structural characterization of the endohedral fullerene $Sc_3N@C_{78}$, *Angew. Chem. Int. Ed.*, **40**, 1223–1225 (2001).

[21] J. M. Campanera, C. Bo, M. M. Olmstead, A. L. Balch, J. M. Poblet, Bonding within the endohedral fullerenes $Sc_3N@C_{78}$ and $Sc_3N@C_{80}$ as determined by density functional calculations and reexamination of the crystal structure of $\{Sc_3N@C_{78}\}.\{Co(OEP)\}.1.5(C_6H_6).0.3(CHCl_3)$, *J. Phys. Chem. A.*, **106**, 12356–12364 (2002).

[22] T. Zuo, C. M. Beavers, J. C. Duchamp, A. Campbell, H. C. Dorn, M. M. Olmstead, A. L. Balch, Isolation and Structural Characterization of a Family of Endohedral Fullerenes Including the Large, Chiral Cage Fullerenes $Tb_3N@C_{88}$ and $Tb_3N@C_{86}$ as well as the I_h and D_{5h} Isomers of $Tb_3N@C_{80}$, *J. Am. Chem. Soc.*, **129**, 2035–2043 (2007).

[23] B. Q. Mercado, C. M. Beavers, M. M. Olmstead, M. N. Chaur, K. Walker, B. C. Holloway, L. Echegoyen, A. L. Balch, Is the isolated pentagon rule merely a suggestion for endohedral

fullerenes? The structure of a second egg-shaped endohedral fullerene-$Gd_3N@Cs(39663)$-C_{82}, *J. Am. Chem. Soc.*, **130**, 7854–7855 (2008).

[24] C. M. Beavers, T. Zuo, J. C. Duchamp, K. Harich, H. C. Dorn, M. M. Olmstead, A. L. Balch, $Tb_3N@C_{84}$: An improbable, egg-shaped endohedral fullerene that violates the isolated pentagon rule, *J. Am. Chem. Soc.*, **128**, 11352–11353 (2006).

[25] S. Stevenson, H. M. Lee, M. M. Olmstead, C. Kozikowski, P. Stevenson, A. L. Balch, Preparation and crystallographic characterization of a new endohedral, $Lu_3N@C_{80}\cdot0.5$(o-xylene), and Comparison with $Sc_3N@C_{80}\cdot0.5$(o-xylene), *Chem. Eur. J.*, **8**, 4528–4535 (2002).

[26] S. Stevenson, J. P. Phillips, J. E. Reid, M. M. Olmstead, S. P. Rath, A. L. Balch, Pyramidalization of Gd_3N inside a C_{80} cage: the synthesis and structure of $Gd_3N@C_{80}$, *Chem. Commun.*, 2814–2815 (2004).

[27] T. Zuo, M. M. Olmstead, C. M. Beavers, A. L. Balch, G. Wang, G. T. Yee, C. Shu, L. Xu, B. Elliott, L. Echegoyen, J. C. Duchamp, H. C. Dorn, Preparation and structural characterization of the I_h and the D_{5h} isomers of the endohedral fullerenes $Tm_3N@C_{80}$: icosahedral C_{80} cage encapsulation of a trimetallic nitride magnetic cluster with three uncoupled Tm^{3+} Ions, *Inorg. Chem.*, **47**, 5234–5244 (2008).

[28] S. Yang, S. I. Troyanov, A. A. Popov, M. Krause, L. Dunsch, Deviation from planarity of a large Dy_3N cluster encapsulated in an I_h-C_{80} Cage: an X-ray crystallographic and vibrational spectroscopic study, *J. Am. Chem. Soc.*, **128**, 16733–16739 (2006).

[29] M. M. Olmstead, A. de Bettencourt-Dias, J. C. Duchamp, S. Stevenson, H. C. Dorn, A. L. Balch, Isolation and crystallographic characterization of $ErSc_2N@C_{80}$: an endohedral fullerene which crystallizes with remarkable internal order, *J. Am. Chem. Soc.*, **122**, 12220–12226 (2000).

[30] X. Wang, T. Zuo, M. M. Olmstead, J. C. Duchamp, T. E. Glass, F. Cromer, A. L. Balch, H. C. Dorn, Preparation and structure of $CeSc_2N@C_{80}$: an icosahedral carbon cage enclosing an acentric $CeSc_2N$ unit with buried f electron spin, *J. Am. Chem. Soc.*, **128**, 8884–8889 (2006).

[31] S. Stevenson, C. J. Chancellor, H. M. Lee, M. M. Olmstead, A. L. Balch, Internal and external factors in the structural organization in cocrystals of the mixed-metal endohedrals ($GdSc_2N@I_h$-C_{80}, $Gd_2ScN@I_h$-C_{80}, and $TbSc_2N@$ I_h-C_{80}) and Nickel(II) Octaethylporphyrin, *Inorg. Chem.*, **47**, 1420–1427 (2008).

[32] T. Cai, L. Xu, M. R. Anderson, Z. Ge, T. Zuo, X. Wang, M. M. Olmstead, A. L. Balch, H. W. Gibson, H. C. Dorn, Structure and enhanced reactivity rates of the D_{5h} $Sc_3N@C_{80}$ and $Lu_3N@C_{80}$ Metallofullerene isomers: the importance of the pyracylene motif, *J. Am. Chem. Soc.*, **128**, 8581–8589 (2006).

[33] T. Zuo, K. Walker, M. M. Olmstead, F. Melin, B. C. Holloway, L. Echegoyen, H. C. Dorn, M. N. Chaur, C. J. Chancellor, C. M. Beavers, A. L. Balch, A. J. Athans, New egg-shaped fullerenes: non-isolated pentagon structures of $Tm_3N@C(s)(51\ 365)$-C_{84} and $Gd_3N@C(s)(51\ 365)$-C_{84}, *Chem. Commun.*, 1067–1069 (2008).

[34] T. Cai, L. Xu, H. W. Gibson, H. C. Dorn, C. J. Chancellor, M. M. Olmstead, A. L. Balch, $Sc_3N@C_{78}$: Encapsulated cluster regiocontrol of adduct docking on an ellipsoidal metallofullerene sphere, *J. Am. Chem. Soc.*, **129**, 10795–10800 (2007).

[35] E. B. Iezzi, J. C. Duchamp, K. Harich, T. E. Glass, H. M. Lee, M. M. Olmstead, A. L. Balch, H. C. Dorn, a symmetric derivative of the trimetallic nitride endohedral metallofullerene, $Sc_3N@C_{80}$, *J. Am. Chem. Soc.*, **124**, 524–525 (2002).

[36] H. M. Lee, M. M. Olmstead, E. Iezzi, J. C. Duchamp, H. C. Dorn, A. L. Balch, Crystallographic characterization and structural analysis of the first organic functionalization product of the endohedral fullerene $Sc_3N@C_{80}$, *J. Am. Chem. Soc.*, **124**, 3494–3495 (2002).

[37] T. Cai, C. Slebodnick, L. Xu, K. Harich, T. E. Glass, C. Chancellor, J. C. Fettinger, M. M. Olmstead, A. L. Balch, H. W. Gibson, H. C. Dorn, A pirouette on a metallofullerene sphere: interconversion of isomers of *N*-tritylpyrrolidino I_h $Sc_3N@C_{80}$, *J. Am. Chem. Soc.*, **128**, 6486–6492 (2006).

[38] C. Shu, C. Slebodnick, L. Xu, H. Champion, T. Fuhrer, T. Cai, J. E. Reid, W. Fu, K. Harich, H. C. Dorn, H. W. Gibson, Highly regioselective derivatization of trimetallic nitride templated endohedral metallofullerenes via a facile photochemical reaction, *J. Am. Chem. Soc.*, **130**, 17755–17760 (2008).

[39] L. Echegoyen, C. J. Chancellor, C. M. Cardona, B. Elliott, J. Rivera, M. M. Olmstead, A. L. Balch, X-Ray Crystallographic and EPR spectroscopic characterization of a pyrrolidine adduct of $Y_3N@C_{80}$, *Chem. Commun.*, 2653–2655 (2006).
[40] O. Lukoyanova, C. M. Cardona, J. Rivera, L. Z. Lugo-Morales, C. J. Chancellor, M. M. Olmstead, A. Rodriguez-Fortea, J. M. Poblet, A. L. Balch, L. Echegoyen, 'Open rather than closed' malonate methano-fullerene derivatives. the formation of methanofulleroid adducts of $Y_3N@C_{80}$, *J. Am. Chem. Soc.*, **129**, 10423–10430 (2007).
[41] C. Shu, W. Xu, C. Slebodnick, H. Champion, W. Fu, T. Cai, J. Reid, H. Azurmendi, C.-R. Wang, K. Harich, H. W. Gibson, H. C. Dorn, Synthesis and structure of PCBM analogues from TNT metallofullerenes, *Org. Lett.*, **11**, ASAP (2009).
[42] C. Shu, T. Cai, L. Xu, T. Zuo, J. Reid, K. Harich, H. C. Dorn, H. W. Gibson, Manganese(III)-catalyzed free radical reactions on trimetallic nitride endohedral metallofullerenes, *J. Am. Chem. Soc.*, **129**, 15710–15717 (2007).
[43] E. B. Iezzi, J. C. Duchamp, K. R. Fletcher, T. E. Glass, H. C. Dorn, Lutetium-based trimetallic nitride endohedral metallofullerenes: new contrast agents, *Nano Lett.*, **2**, 1187–1190 (2002).
[44] S. Yang, A. A. Popov, L. Dunsch, Carbon Pyramidalization in fullerene cages induced by the endohedral cluster: non-scandium mixed metal nitride clusterfullerenes, *Angew. Chem., Int. Ed.*, **47**, 8196–8200 (2008).
[45] T. Heine, K. Vietze, G. Seifert, ^{13}C NMR fingerprint characterizes long time-scale structure of $Sc_3N@C_{80}$ endohedral fullerene, *Magn. Reson. Chem.*, **42**, S199–S201 (2004).
[46] K. R. Gorny, C. H. Pennington, J. A. Martindale, P. J. Phillips, S. Stevenson, I. Heinmaa, R. Stern, Molecular orientational dynamics of the endohedral fullerene $Sc_3N@C_{80}$ as probed by ^{13}C and ^{45}Sc NMR. *Los Alamos National Laboratory, Preprint Archive, Condensed Matter* 1-6, arXiv:cond-mat/0604365 (2006).
[47] S. Stevenson, P. W. Fowler, T. Heine, J. C. Duchamp, G. Rice, T. Glass, K. Harich, E. Hajdu, R. Bible, H. C. Dorn, A stable non-classical metallofullerene family, *Nature*, **408**, 427–428 (2000).
[48] J. U. Reveles, T. Heine, A. M. Koester, ^{13}C NMR Pattern of $Sc_3N@C_{68}$. Structural assignment of the first fullerene with adjacent pentagons, *J. Phys. Chem. A*, **109**, 7068–7072 (2005).
[49] J. C. Duchamp, A. Demortier, K. R. Fletcher, D. Dorn, E. B. Iezzi, T. Glass, H. C. Dorn, An isomer of the endohedral metallofullerene $Sc_3N@C_{80}$ with D_{5h} symmetry, *Chem. Phys. Lett.*, **375**, 655–659 (2003).
[50] S. Stevenson, R. R. Stephen, T. M. Amos, V. R. Cadorette, J. E. Reid, J. P. Phillips, Synthesis and purification of a metallic nitride fullerene bisadduct: exploring the reactivity of $Gd_3N@C_{80}$, *J. Am. Chem. Soc.*, **127**, 12776–12777 (2005).
[51] M. Maggini, G. Scorrano, M. Prato, Addition of azomethine ylides to C_{60}: synthesis, characterization, and functionalization of fullerene pyrrolidines, *J. Am. Chem. Soc.*, **115**, 9798–9799 (1993).
[52] C. M. Cardona, A. Kitaygorodskiy, A. Ortiz, M. A. Herranz, L. Echegoyen, The first fulleropyrrolidine derivative of $Sc_3N@C_{80}$: pronounced chemical shift differences of the geminal protons on the pyrrolidine ring, *J. Org. Chem.*, **70**, 5092–5097 (2005).
[53] T. Cai, Z. Ge, E. B. Iezzi, T. E. Glass, K. Harich, H. W. Gibson, H. C. Dorn, Synthesis and Characterization of the first trimetallic nitride templated pyrrolidino endohedral metallofullerenes, *Chem. Commun.*, 3594–3596 (2005).
[54] C. M. Cardona, A. Kitaygorodskiy, L. Echegoyen, Trimetallic nitride endohedral metallofullerenes: reactivity dictated by the encapsulated metal cluster, *J. Am. Chem. Soc.*, **127**, 10448–10453 (2005).
[55] C. M. Cardona, B. Elliott, L. Echegoyen, Unexpected chemical and electrochemical properties of $M_3N@C_{80}$ ($M = Sc, Y, Er$), *J. Am. Chem. Soc.*, **128**, 6480–6485 (2006).
[56] N. Chen, L. Fan, K. Tan, Y. Wu, C. Shu, X. Lu, C. Wang, Comparative spectroscopic and reactivity studies of $Sc_{3-x}Y_xN@C_{80}$ ($x = 0$–3), *J. Phys. Chem. C.*, **111**, 11823–11828 (2007).
[57] N. Chen, E. Zhang, K. Tan, C. Wang, X. Lu, Size effect of encaged clusters on the exohedral chemistry of endohedral fullerenes: a case study on the pyrrolidino reaction of $Sc_xGd_{3-x}N@C_{80}$ ($x = 0$–3), *Org. Lett.*, **9**, 2011–2013 (2007).

[58] J. R. Pinzón, M. E. Plonska-Brzezinska, C. M. Cardona, A. J. Athans, S. S. Gayathri, D. M. Guldi, M. A. Herranz, N. Martin, T. Torres, L. Echegoyen, $Sc_3N@C_{80}$-Ferrocene electron-donor/acceptor conjugates as promising materials for photovoltaic applications, *Angew. Chem. Int. Ed.*, **47**, 4173–4176 (2008).

[59] J. R. Pinzón, C. M. Cardona, M. A. Herranz, M. E. Plonska-Brzezinska, A. Palkar, A. J. Athans, N. Martin, A. Rodriguez-Fortea, J. M. Poblet, G. Bottari, T. Torres, S. S. Gayathri, D. M. Guldi, L. Echegoyen, Metal nitride cluster fullerene $M_3N@C_{80}$ (M = Y, Sc) based dyads: synthesis, and electrochemical, theoretical and photophysical studies, *Chem. Eur. J.*, **15**, 864–877 (2009).

[60] N. Martin, M. Altable, S. Filippone, A. Martin-Domenech, L. Echegoyen, C. M. Cardona, Retro-cycloaddition reaction of pyrrolidinofullerenes, *Angew. Chem. Int. Ed.*, **45**, 110–114 (2005).

[61] O. Lukoyanova, C. M. Cardona, M. Altable, S. Filippone, A. M. Domenech, N. Martin, L. Echegoyen, Selective electrochemical retro-cycloaddition reaction by pyrrolidinofullerenes, *Angew. Chem. Int. Ed.*, **45**, 7430–7433 (2006).

[62] R. B. Ross, C. M. Cardona, D. M. Guldi, S. G. Sankaranarayanan, M. O. Reese, N. Kopidakis, J. Peet, B. Walker, G. C. Bazan, E. Van Keuren, B. C. Holloway, M. Drees, Endohedral fullerenes for organic photovoltaic devices, *Nature Mater.*, **8**, 208–212 (2009).

[63] J. C. Hummelen, B. W. Knight, F. Lepeq, F. Wudl, J. Yao, C. L. Wilkins, Preparation and characterization of fulleroid and methanofullerene derivatives, *J. Org. Chem.*, **60**, 532–538 (1995).

[64] B. C. Thompson, J. M. J. Frechet, Polymer-fullerene composite solar cells, *Angew. Chem., Int. Ed.*, **47**, 58–77 (2008).

[65] E. B. Iezzi, F. Cromer, P. Stevenson, H. C. Dorn, Synthesis of the first water-soluble trimetallic nitride endohedral metallofullerols, *Synth. Met.*, **128**, 289–291 (2002).

[66] Y. Iiduka, O. Ikenaga, A. Sakuraba, T. Wakahara, T. Tsuchiya, Y. Maeda, T. Nakahodo, T. Akasaka, M. Kako, N. Mizorogi, S. Nagase, Chemical reactivity of $Sc_3N@C_{80}$ and $La_2@C_{80}$, *J. Am. Chem. Soc.*, **127**, 9956–9957 (2005).

[67] T. Wakahara, Y. Iiduka, O. Ikenaga, T. Nakahodo, A. Sakuraba, T. Tsuchiya, Y. Maeda, M. Kako, T. Akasaka, K. Yoza, E. Horn, N. Mizorogi, S. Nagase, Characterization of the bis-silylated endofullerene $Sc_3N@C_{80}$, *J. Am. Chem. Soc.*, **128**, 9919–9925 (2006).

[68] N. B. Shustova, A. A. Popov, M. A. Mackey, C. E. Coumbe, J. P. Phillips, S. Stevenson, S. H. Strauss, O. V. Boltalina, Radical trifluoromethylation of $Sc_3N@C_{80}$, *J. Am. Chem. Soc.*, **129**, 11676–11677 (2007).

[69] C. Bingel, Cyclopropanierung von fullerenes, *Chem. Ber.*, **126**, 1957–1959 (1993).

[70] A. Hirsch, I. Lamparth, H. R. Karfunkel, Fullerene chemistry in three dimensions: isolation of seven regioisomeric bisadducts and chiral trisadducts of C_{60} and Di(ethoxycarbonyl)methylene, *Angew. Chem. Int. Ed.*, **33**, 437–438 (1994).

[71] T. Cai, L. Xu, C. Shu, H. A. Champion, J. E. Reid, C. Anklin, M. R. Anderson, H. W. Gibson, H. C. Dorn, Selective formation of a symmetric $Sc_3N@C_{78}$ bisadduct: adduct docking controlled by an internal trimetallic nitride cluster, *J. Am. Chem. Soc.*, **130**, 2136–2137 (2008).

[72] A. Hirsch, I. Lamparth, T. Groesser, H. R. Karfunkel, Regiochemistry of multiple additions to the fullerene core: synthesis of a T_h-symmetric hexakis adduct of C_{60} with bis(ethoxycarbonyl) methylene, *J. Am. Chem. Soc.*, **116**, 9385–9386 (1994).

[73] S. Osuna, M. Swart, J. M. Campanera, J. M. Poblet, M. Sola, Chemical reactivity of D_{3h} C_{78} (metallo)fullerene: regioselectivity changes induced by Sc_3N encapsulation, *J. Am. Chem. Soc.*, **130**, 6206–6214 (2008).

[74] P. P. Fatouros, F. D. Corwin, Z.-J. Chen, W. C. Broaddus, J. L. Tatum, Z. Ge, H. W. Gibson, J. L. Kile, A. P. Leonard, J. C. Duchamp, H. C. Dorn, *In Vitro* and *In Vivo* imaging studies of a new gadolinium endohedral metallofullerene MRI contrast agent, *Radiology*, **240**, 756–764 (2006).

[75] D. K. MacFarland, K. L. Walker, R. P. Lenk, S. R. Wilson, K. Kuma, C. L. Kepley, J. R. Garbow, Hydrochalarones: a novel endohedral metallofullerene platform for enhancing magnetic resonance imaging contrast, *J. Med. Chem.*, **51**, 3681–3683 (2008).

[76] T. Cai, L. Xu, C. Shu, J. E. Reid, H. W. Gibson, H. C. Dorn, Synthesis and characterization of a non-IPR fullerene derivative: $Sc_3N@C_{68}[C(COOC_2H_5)_2]$, *J. Phys. Chem. C*, **112**, 19203–19208 (2008).

[77] M. N. Chaur, F. Melin, A. J. Athans, B. Elliott, K. Walker, B. C. Holloway, L. Echegoyen, The influence of cage size on the reactivity of trimetallic nitride metallofullerenes: a mono- and bis-methanoadduct of $Gd_3N@C_{80}$ and a monoadduct of $Gd_3N@C_{84}$, *Chem. Commun.*, 2665–2667 (2008).
[78] C. Shu, F. D. Corwin, J. Zhang, J. E. Reid, M. Sun, W. Xu, J. H. Sim, C.-R. Wang, P. P. Fatouros, A. R. Esker, H. W. Gibson, H. C. Dorn, Preparation of a new gadofullerene-based magnetic resonance imaging contrast agent with high ^1H relaxivity, *Bioconj. Chem.*, **20**, 1186–1193 (2009).
[79] D. M. McCluskey, T. N. Smith, P. K. Madasu, C. E. Coumbe, M. A. Mackey, P. A. Fulmer, J. H. Wynne, S. Stevenson, J. P. Phillips, Evidence for singlet-oxygen generation and biocidal activity in photoresponsive metallic nitride fullerene–polymer adhesive films, *ACS Appl. Mater. Interfaces*, **1**, ASAP (2009).
[80] M. Krause, L. Dunsch, Isolation and characterisation of two $Sc_3N@C_{80}$ Isomers, *ChemPhysChem*, **5**, 1445–1449 (2004).
[81] Q. Xie, E. Perez-Cordero, L. Echegoyen, Electrochemical Detection of $C_{60}{}^{6-}$ and $C_{70}{}^{6-}$: Enhanced stability of fullerides in solution, *J. Am. Chem. Soc.*, **114**, 3978–3980 (1992).
[82] M. Krause, X. Liu, J. Wong, T. Pichler, M. Knupfer, L. Dunsch, The electronic and vibrational structure of endohedral $Tm_3N@C_{80}$ (I) fullerene – proof of an encaged Tm^{3+}, *J. Phys. Chem. A.*, **109**, 7088–7093 (2005).
[83] S. Yang, M. Zalibera, P. Rapta, L. Dunsch, Charge-induced reversible rearrangement of endohedral fullerenes: electrochemistry of tridysprosium nitride clusterfullerenes $Dy_3N@C_{2n}$ ($2n = 78, 80$), *Chem. Eur. J.*, **12**, 7848–7855 (2006).
[84] M. N. Chaur, F. Melin, B. Elliott, A. J. Athans, K. Walker, B. C. Holloway, L. Echegoyen, $Gd_3N@C_{2n}$ ($n = 40$, 42, and 44): Remarkably low HOMO-LUMO gap and unusual electrochemical reversibility of $Gd_3N@C_{88}$, *J. Am. Chem. Soc.*, **129**, 14826–14829 (2007).
[85] M. N. Chaur, F. Melin, B. Elliott, A. Kumbhar, A. J. Athans, L. Echegoyen, New $M_3N@C_{2n}$ Endohedral metallofullerene families ($M =$ Nd, Pr, Ce; $n = 40$–53): expanding the preferential templating of the C_{88} cage and approaching the C_{96} Cage, *Chem. Eur. J.*, **14**, 4594–4599 (2008).
[86] P. Jakes, K. Dinse, Chemically induced spin transfer to an encased molecular cluster: an EPR study of $Sc_3N@C_{80}$ radical anions, *J. Am. Chem. Soc.*, **123**, 8854–8855 (2001).
[87] M. N. Chaur, A. J. Athans, L. Echegoyen, Metallic nitride endohedral fullerenes: synthesis and electrochemical properties, *Tetrahedron*, **64**, 11387–11393 (2008).
[88] M. N. Chaur, F. Melin, J. Ashby, B. Elliott, A. Kumbhar, A. M. Rao, L. Echegoyen, Lanthanum nitride endohedral fullerenes $La_3N@C_{2n}$ ($43 < n < 55$): preferential formation of $La_3N@C_{96}$, *Chem. Eur. J.*, **14**, 8213–8219 (2008).
[89] S. Yang, P. Rapta, L. Dunsch, The spin state of a charged non-IPR fullerene: the stable radical cation of $Sc_3N@C_{68}$, *Chem. Commun.*, 189–191 (2007).
[90] F. Melin, M. N. Chaur, S. Engmann, B. Elliott, A. Kumbhar, A. J. Athans, L. Echegoyen, the large $Nd_3N@C_{2n}$ ($40 < n < 49$) cluster fullerene family: preferential templating of a C_{88} cage by a trimetallic nitride cluster, *Angew. Chem. Int. Ed.*, **46**, 9032–9035 (2007).
[91] P. Rapta, A. A. Popov, S. Yang, L. Dunsch, Charged states of $Sc_3N@C_{68}$: an in situ spectroelectrochemical study of the radical cation and radical anion of a non-ipr fullerene, *J. Phys. Chem. A.*, **112**, 5858–5865 (2008).

10

Recent Progress in Chemistry of Endohedral Metallofullerenes

Takahiro Tsuchiya[a], Takeshi Akasaka[a] and Shigeru Nagase[b]

[a]Center for Tsukuba Advanced Research Alliance, University of Tsukuba, Japan
[b]Department of Theoretical and Computational Molecular Science, Institute for Molecular Science, Okazaki, Japan

10.1 Introduction

Since the first discovery of fullerenes by Smalley, Kroto, Curl, and co-workers in 1985 [1], the insertion of one or more atoms into the hollow fullerene cage has been attempted. Furthermore, synthesis and extraction of endohedral metallofullerene $La@C_{82}$ was reported by Smalley and co-workers in 1991 [2]. Among endohedral fullerenes, metal encapsulating fullerenes [3] especially attract broad attention because of their novel properties attributable to their intramolecular metal-fullerene cage interaction. In the following years, great efforts have been made for the synthesis of various endohedral metallofullerenes. The encapsulated species were found to cover Group-3 metals and most lanthanide metals, as well as their nitride clusters and carbide clusters. The encapsulated atoms or clusters were widely investigated using X-ray photoemission; photograph energy loss spectroscopy, and theoretical calculations [4, 5]. All these studies revealed an electron-transfer interaction between a fullerene cage and an entrapped cluster or metal atoms. Consequently, an onion-like model can be used to describe the electronic structure of metallofullerenes, in which the interior layer composed by metal atoms or a cluster is positively charged and the exterior layered fullerene cage is negatively charged. This electron transfer was regarded to stabilize not only the encapsulated species, but also the fullerene cage, which can be otherwise unstable in its

Chemistry of Nanocarbons Edited by Takeshi Akasaka, Fred Wudl and Shigeru Nagase

empty form. This structural feature contrasts remarkably with that of nonmetallic endohedral fullerenes, such as $N@C_{60}$ [6]. $P@C_{60}$ [7, 8], $He@C_{60}$ [9] and recently synthesized $H_2@C_{60}$ [10, 11], in which the nonmetal atoms have very weak interactions with the fullerene cage. In this sense, endohedral metallofullerenes more closely resemble chemically hybrid molecules, whereas nonmetallic endohedral fullerenes are physically hybrid molecules.

Because of their unique and complex three-dimensional structures, the structural determination of endohedral metallofullerenes, including the structure of fullerene cage and the position or motion of encapsulated metallic species, has become a great challenge in fullerene chemistry. Correct understanding of their structures is generally believed to provide important clues in disclosing the formation mechanism of endohedral metallofullerenes. It might therefore be helpful in improving their production yield. As exemplified by $La@C_{82}$, endohedral metallofullerenes of many kinds have a paramagnetic nature. Although the nature of endohedral metallofullerenes is attractive as a stable neutral radical [12–17], it prevents detailed experimental characterization such as structural determination by NMR measurement. In this context, the electrochemical reduction of paramagnetic $M@C_{82}$ (M = La [18, 19], Ce [20], Pr [21]) and determination of their structure for the obtained diamagnetic anions have succeeded. Furthermore, powder [22–33] and single crystal [34–53] X-ray diffraction method have been used and developed for structural determinations of metallofullerenes. Consequently, many unconventional features of metallofullerenes have been revealed and clearly demonstrated. For example, some metallofullerenes such as $Sc_2@C_{66}$ [28], $Tb_3N@C_{84}$ [42], $Sc_3N@C_{68}$ [48], and $La_2@C_{72}$ [54], were surprisingly found to violate the isolated pentagon rule (IPR).

On the other hand, the enrichment of endohedral metallofullerenes using methods of electrochemical reduction [55], sublimation followed by chemical oxidation [56, 57], chemical reduction [58], or dimethylformamide extraction of soot [59–61] has been reported and a selective redox-based procedure has been used to purify endohedral metallofullerenes from soot [62]. Moreover, the convenient isolation systems of pure endohedral metallofullerenes using selective reduction of endohedral metallofullerenes from extracts of soot have been developed [63, 64]. Consequently, since macroscopic quantities have become available in recent years, interest in their chemical properties has been rapidly aroused, inspired by endohedral metallofullerenes' potential applications in material science. Many recent studies specifically examine their different chemical reactivities induced by encapsulated metallic species. These findings are also regarded as an important aspect of metallofullerene science, and will be discussed in detail in this chapter.

10.2 Chemical Derivatization of Mono-Metallofullerenes

Since its first extraction in 1991 by Smalley and co-workers, $La@C_{2v}$-C_{82} has been recognized as a prototypical endohedral metallofullerene. An additional reaction to $La@C_{2v}$-C_{82} might take place at several sites to afford numerous possible mono-adduct isomers because 24 non-equivalent carbons and 35 non-equivalent bonds exist in $La@C_{2v}$-C_{82}. Indeed, the reaction of $La@C_{2v}$-C_{82} with disilirane [65] or diphenyldiazomethane [66]

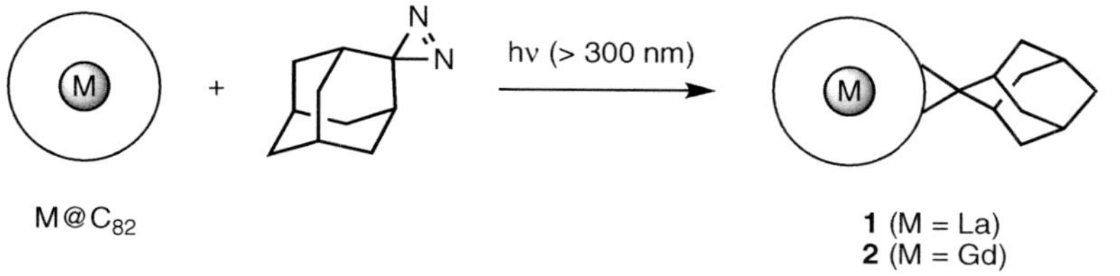

Scheme 10.1

gave several 1 : 1 adduct isomers. The ESR traces for the reactions reveal the formation of more than six or four regioisomers. We cannot isolate those isomers. Consequently, development of controlling the addition point for La@C_{2v}-C_{82} is the problem to be solved.

10.2.1 Carbene Reaction

The reaction of La@C_{2v}-C_{82} with adamantylidene carbene, which is formed by irradiation of 2-adamantane-2,3-[3H]-diazirine, achieved regiospecific addition (Scheme 10.1) [67]. The obtained adduct La@C_{82}Ad (**1**, Ad = adamantylidene) has been isolated. The local strain on each carbon atom of fullerenes plays an important role in determining their reactivity [68]. The pyramidalization angles from the p-orbital axis vector analysis POAV ($\theta_{\Delta\pi}$–90°) angles provide a useful index of the local strain [69]. The Mulliken charge densities and POAV angles in La@C_{2v}-C_{82} are presented in Figure 10.1. The negative charge and POAV angle are found to be large for the carbons A and B in the six-membered ring nearest to the La atom, which suggests that adamantyl carbene would selectively attack one of the six electron-rich strained carbons because it acts as an electrophile [70, 71]. In fact, the addition of adamantyl carbene to La@C_{2v}-C_{82} takes place between the carbon atoms C(1) and C(2), as indicated by the X-ray single crystal analysis results (Figure 10.2). The selective addition of adamantylidene carbene also proceeds in the reaction with Gd@C_{2v}-C_{82} [72]. The structure of Gd@C_{82}Ad (**2**) is determined using X-ray single crystal analysis. The structural aspects closely resemble those for La@C_{82}Ad.

10.2.2 Nucleophilic Reaction

The Bingel–Hirsh reaction is also a very efficient chemical modification method in fullerene chemistry [73–75]. Its mechanism involves the nucleophilic attack of a carbon anion that is produced in situ by deprotonation of α-halo esters or α-halo ketones. This method provides easy access to versatile fullerene derivatives as well as water-soluble fullerenes. The reaction is also performed on metallofullerenes with the goal of obtaining various adducts.

The theoretical calculations of the Mulliken charge densities of La@C_{2v}-C_{82} show that the carbon C in Figure 10.1 is the most positively charged. The POAV analysis shows that the carbon C also has large local strain. These make the carbon C in Figure 10.1 most reactive against the nucleophilic attack.

The nucleophilic reaction of diethyl bromomalonate with La@C_{2v}-C_{82} using diethyl bromomalonate in the presence of 1,8-diazabicyclo[5.4.0]undec-7-ene (DBU) yields a singly bonded monoadduct **3** as the major product (Scheme 10.2) [76]. Although pristine La@C_{2v}-C_{82} is paramagnetic, monoadduct **3** is diamagnetic; NMR measurements indicate

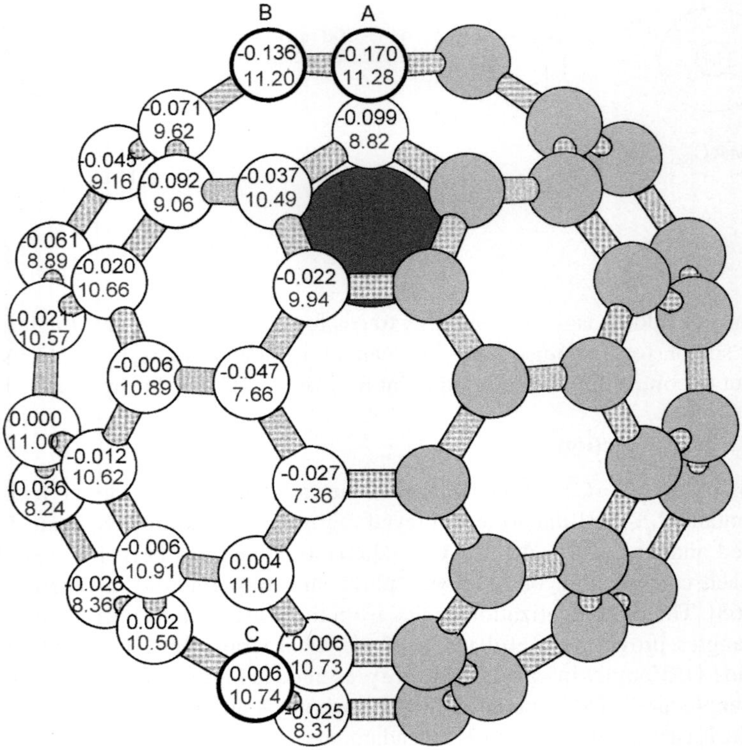

Figure 10.1 *Charge densities (upper) and POAV (q_{Dp}–90°) angles (lower) in La@C_{2v}-C_{82}*

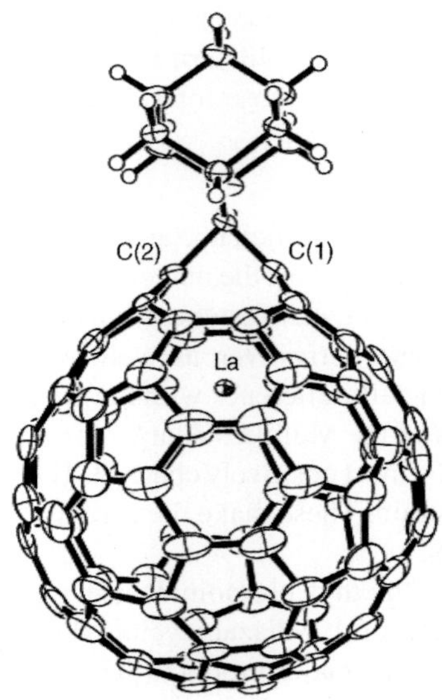

Figure 10.2 *ORTEP drawing of **2***

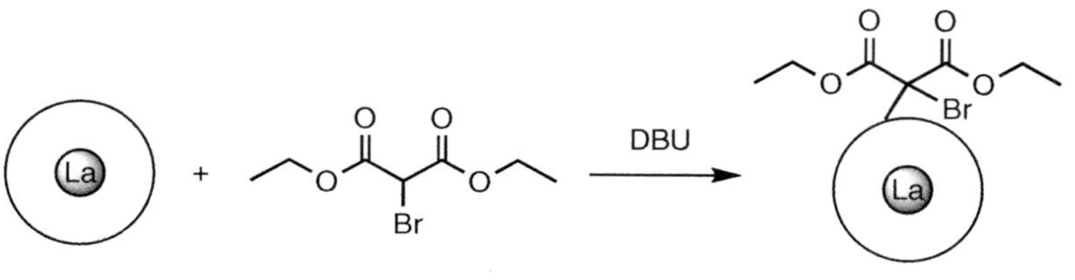

Scheme 10.2

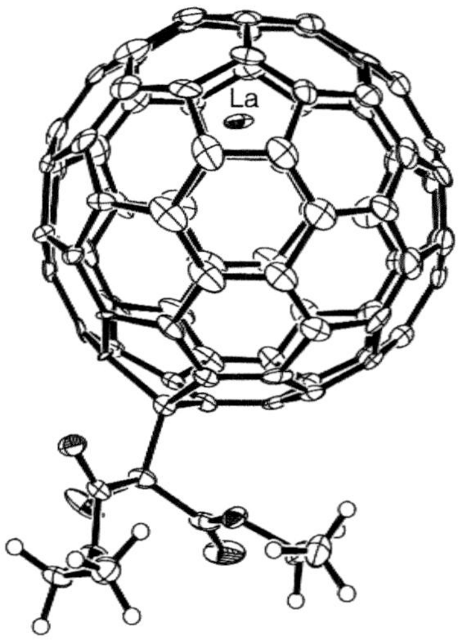

Figure 10.3 *ORTEP drawing of **3c***

that **3** has C_1 symmetry. X-ray single crystal analysis clearly reveals that a bromomalonate group is combined with the C_{82} cage at the carbon C in Figure 10.1, as presented in Figure 10.3. The nucleophilic reaction is followed by oxidation of an intermediate $[La@C_{82}Br(COOC_2H_5)_2]^-$ with oxidants (La@C_{2v}-C_{82} or trace oxygen in solvent) to afford the singly bonded final adduct. As described above, controlling the addition point for La@C_{2v}-C_{82} can be achieved using an electrophile or nucleophile.

10.3 Chemical Derivatization of Di-Metallofullerenes

Two metal atoms in $M_2@I_h$-C_{80} (M = La, Ce) are reportedly rotating three-dimensionally [77–80]. Controlling the motion of the 'untouchable' metal atoms inside the fullerene cage is expected to be an important stepping stone on the path to developing applications such as molecular switches with new electronic or magnetic properties [81, 82]. The motion of the

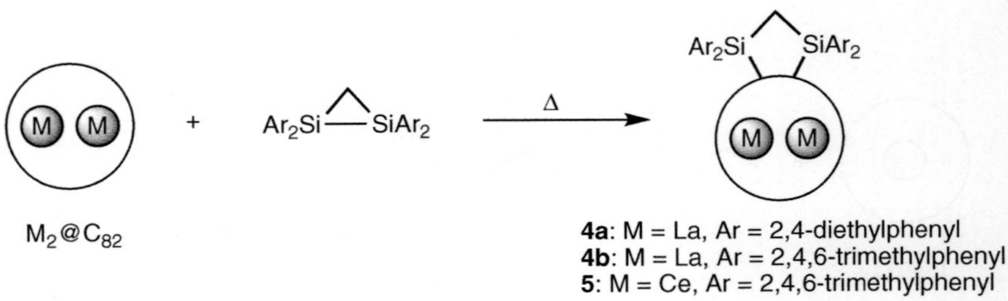

Scheme 10.3

encapsulated metal atoms depends on the electrostatic potential inside the fullerene cage. Therefore, it is possible to control the motion if the electrostatic potential was changed. Theoretical calculations show that attaching an electron-donating moiety such as a silyl group to the fullerene cage is effective [83].

10.3.1 Bis-silylation

The reaction of $La_2@I_h$-C_{80} with disilirane affords a 1:1 adduct **4** (Scheme 10.3) [84]. The molecular structure of the adduct is determined using NMR and X-ray crystallographic analyses. The crystal structure of the adduct indicates the 1,4-addition of disilirane to $La_2@I_h$-C_{80} and the two encapsulated La atoms are located at two positions directed toward the hexagonal ring at the equator, reflecting that these positions are energetically the most stable (Figure 10.4). The variable-temperature ^{139}La NMR spectra reveal the dynamic behavior of the La atoms inside the silylated C_{80} cage. For pristine $La_2@I_h$-C_{80}, a large broadening of the ^{139}La NMR linewidth with increasing temperature from 305 to 363 K is

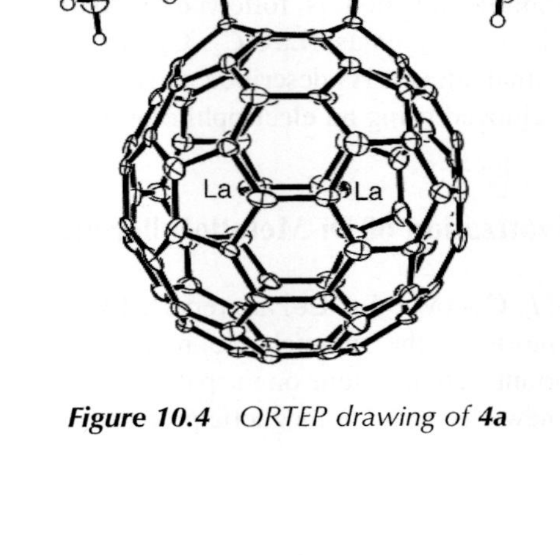

Figure 10.4 *ORTEP drawing of **4a***

observed because of the spin-rotation relaxation [78]. The VT-^{139}La NMR measurements of the adduct **4a** show a large broadening of the signal linewidth with increasing temperature from 183 to 308 K, indicating that two La atoms do not stand still but instead hop inside the silylated C_{80} cage in solution. It is also verified that the random motion of metal atoms of $Ce_2@I_h$-C_{80} is regulateed similarly by the disilirane addition [85]. The ^{1}H NMR measurement of adduct **5** reveals the paramagnetic shift derived from the *f*-electron of Ce atom: some signals are shifted considerably by changing the temperature. It is noteworthy that the effect of the *f*-electron extends to the disilirane moiety outside the fullerene cage. Consequently, the free random motion of two metal atoms in $M_2@I_h$-C_{80} is surely fixed at specific positions by exohedral chemical functionalization. Attachment of a silicon substituent can regulate the position of metal atoms under the equator inside the carbon cage.

10.3.2 Cycloaddition with Oxazolidinone

Meanwhile, the reaction of $La_2@I_h$-C_{80} with 3-triphenylmethyl-5-oxazolidinone affords pyrrolidinofullerene derivatives $La_2@C_{80}(CH_2)_2$NTrt (**6**, Trt = triphenylmethyl, Scheme 10.4) [86]. Both ^{1}H and ^{13}C NMR measurements show the formation of 6,6- and 5,6-pyrrolidinofullerene adducts (4 : 1). The ^{139}La NMR spectrum of the adduct measured at 278 K shows a broad signal at $\delta = -464$ ppm with a large linewidth of 570 Hz, indicative of overlapping of two nonequivalent La atoms. The dynamic behavior of La atoms is expected to be reflected in the ^{139}La NMR linewidth. Temperature-dependent signal broadening caused by the spin-rotation relaxation was not observed for the adduct at 278–313 K, suggesting that two La atoms do not circulate inside the cage, unlike the case of pristine $La_2@I_h$-C_{80}. X-ray crystallographic analysis of 6,6-adduct demonstrates that metal atoms are certainly stopped (Figure 10.5). The fixed position is supported by electrostatic calculations inside the cage of $[C_{80}(CH_2)_2NH]^{6-}$.

10.3.3 Carbene Reaction

Irradiation of a toluene solution of $M_2@I_h$-C_{80} (M = La, Ce) and an excess molar amount of 2-adamantane-2,3-[3H]-diazirine in a degassed sealed tube at room temperature using a high-pressure mercury-arc lamp (cutoff < 390 nm) caused the formation of the corresponding adduct, $M_2@C_{80}$Ad (**7**: M = La, **8**: M = Ce, Ad = adamantylidene) in 80% yield, which was purified by preparative HPLC (Scheme 10.5) [87]. The formation of **7** and **8** was confirmed using mass spectroscopic measurements. Subsequent NMR measurements revealed that the adducts have a 6,6-open structure. The single-crystal X-ray structure

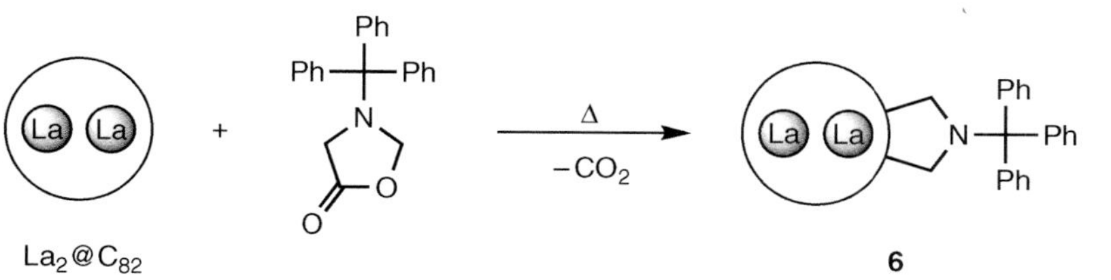

Scheme 10.4

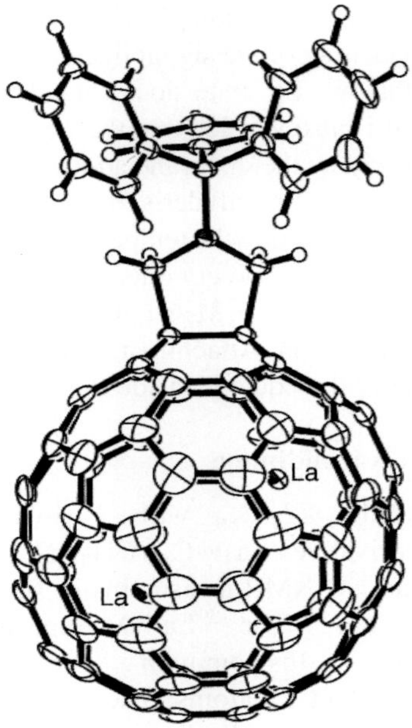

Figure 10.5 *ORTEP drawing of **6***

analysis confirms the 6,6-open structure of **7** (Figure 10.6). The opened C–C separation is 2.166 Å, whereas the 6,6-bond length on the addition site of 6,6-$La_2@C_{80}(CH_2)_2NTrt$ (**6**) is 1.635 Å. It is interesting that the two La atoms in **7** are collinear with the spiro carbon of the 6,6-open adduct. The metal positions differ greatly from those in $La_2@C_{80}(Ar_2Si)_2CH_2$ (**4a**), $Ce_2@C_{80}(Ar_2Si)_2CH_2$ (**5**) and 6,6-$La_2@C_{80}(CH_2)_2NTrt$ (**6**). The 6,6-bond cleavage results in the protrusion of the carbon atoms on the cage and the expansion of the cage's inner space, resulting in elongation of the La•••La distance.

Such control of the motion of metal atoms by chemical functionalization is expected to be of great help in designing novel molecular devices with new electronic or magnetic properties.

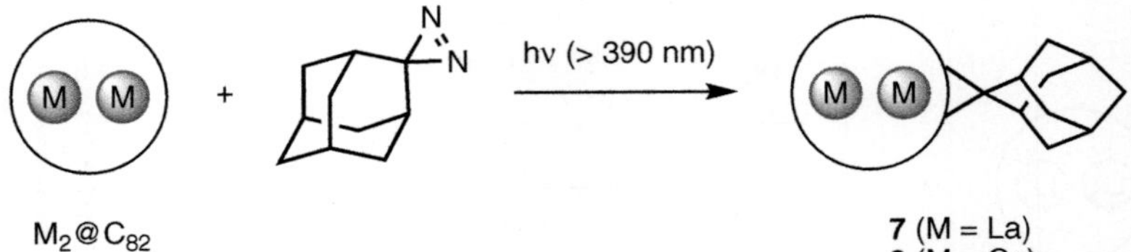

Scheme 10.5

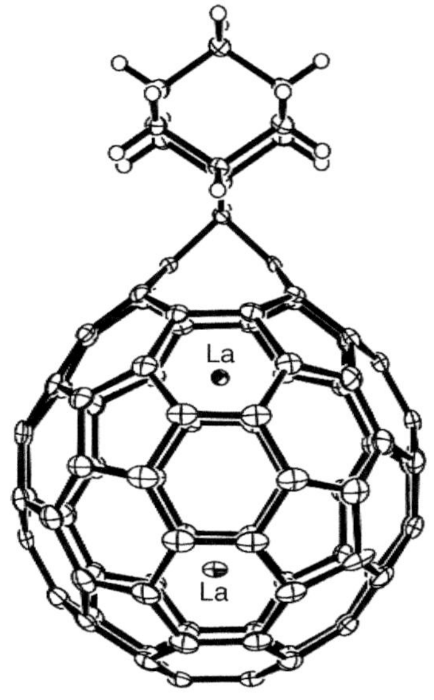

Figure 10.6 *ORTEP drawing of 7*

10.4 Chemical Derivatization of Trimetallic Nitride Template Fullerene

Dorn *et al.* developed a new synthetic method to afford a novel endohedral metallofullerene $Sc_3N@I_h\text{-}C_{80}$ [88]. In fact, $Sc_3N@I_h\text{-}C_{80}$ can be isolated in a remarkably high yield. Therefore, the design of $Sc_3N@C_{80}$ derivatives can be considerably beneficial for applications in material science and biochemistry. This has the same carbon cage (I_h) and electronic state (C_{80}^{6-}) as $La_2@I_h\text{-}C_{80}$ [77–80]. Therefore, it may be expected that $Sc_3N@I_h\text{-}C_{80}$ resembles $La_2@I_h\text{-}C_{80}$ in reactivity. Redox potential is important information related to the chemical reactivity of endohedral metallofullerenes as well as fullerenes [20, 65, 66, 89–91]. The oxidation potential of $Sc_3N@I_h\text{-}C_{80}$ is similar to that of $La_2@I_h\text{-}C_{80}$. However, the first reduction potential (–1.22 V) of $Sc_3N@I_h\text{-}C_{80}$ is much more negative than that of $La_2@I_h\text{-}C_{80}$ (–0.31 V vs. Fc/Fc^+), which suggests that $Sc_3N@I_h\text{-}C_{80}$ is much less reactive toward nucleophiles such as disilirane than $La_2@I_h\text{-}C_{80}$, in accordance with the fact that $Sc_3N@I_h\text{-}C_{80}$ does not react thermally with disilirane. Indeed, although the reaction of $La_2@I_h\text{-}C_{80}$ with disilirane affords the adduct both thermally and photochemically, the reaction of $Sc_3N@I_h\text{-}C_{80}$ with disilirane proceeds only photochemically. The MO diagrams calculated for $Sc_3N@I_h\text{-}C_{80}$ and $La_2@I_h\text{-}C_{80}$ are presented in Figure 10.7. Although $Sc_3N@I_h\text{-}C_{80}$ and $La_2@I_h\text{-}C_{80}$ have almost identical HOMO levels, $Sc_3N@I_h\text{-}C_{80}$ has a much higher LUMO level than $La_2@I_h\text{-}C_{80}$. These are consistent with the trends of the redox potentials, supporting the poor thermal reactivity of $Sc_3N@I_h\text{-}C_{80}$ toward disilirane. As Figure 10.8 shows, the LUMO of $Sc_3N@I_h\text{-}C_{80}$ is delocalized not only on the Sc_3N cation but also on the C_{80} cage. In contrast, the LUMO of $La_2@I_h\text{-}C_{80}$ is localized onto the two La^{3+} cations and is more suitable as an electron accommodation.

The photochemical reaction of disilirane with $Sc_3N@I_h\text{-}C_{80}$ proceeds via the 1,2- and 1,4-cycloadditions to form the mixture of 1,2(aa)-closed (**9a**) and 1,4(aa) (**9b**) adducts (Scheme 10.6). Adduct **9a** is thermodynamically less stable than **9b** but is more kinetically

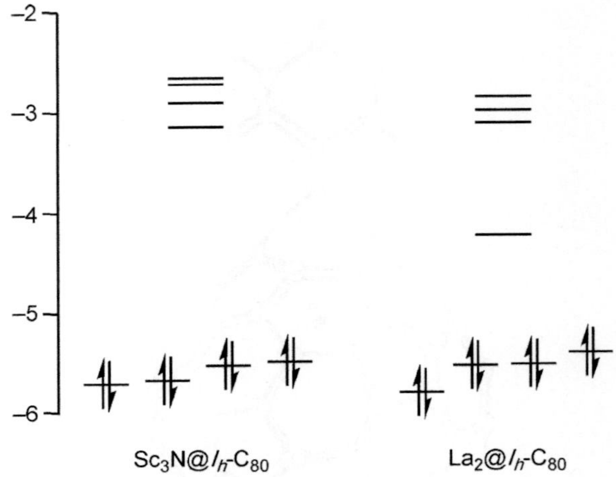

Figure 10.7 MO diagrams of $Sc_3N@I_h$-C_{80} and $La_2@I_h$-C_{80}

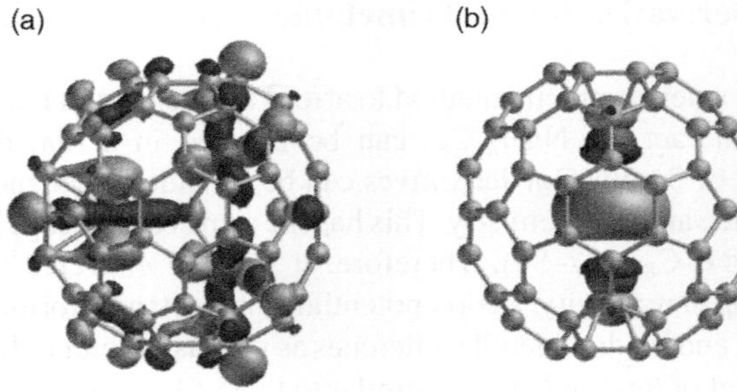

Figure 10.8 LUMOs of (a) $Sc_3N@I_h$-C_{80} and (b) $La_2@I_h$-C_{80}

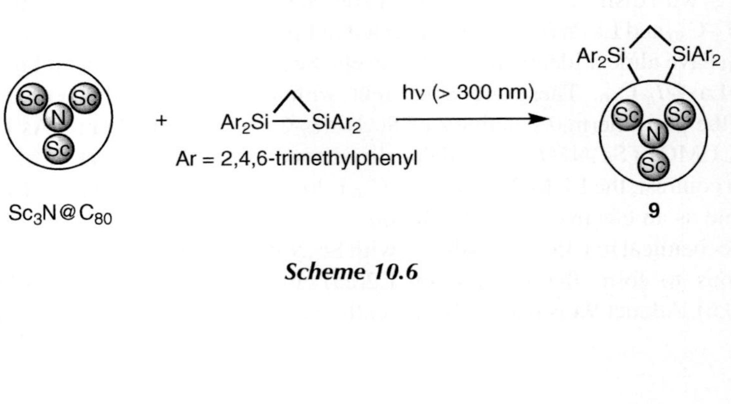

Scheme 10.6

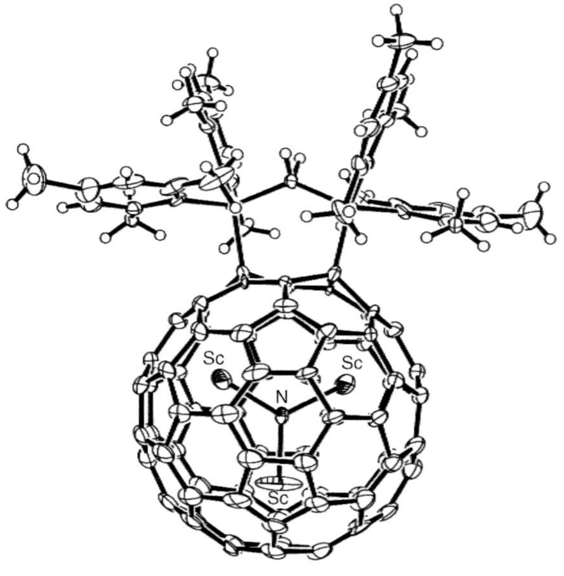

Figure 10.9 ORTEP drawings of **9b**

favorable. Figure 10.9 shows that the structure of **9b** is confirmed by single-crystal X-ray structure analysis.

10.5 Chemical Derivatization of Metallic Carbaide Fullerene

Endohedral metallofullerene Sc_3C_{82} continues to attract attention as a trimetallofullerene of which the atoms are equivalent [92, 93]; its structure has remained a controversial subject. Powder X-ray structure analysis shows than that three Sc atoms are encapsulated in the C_{3v}-C_{82} cage [30]; however, the structure does not correspond to the theoretically calculated energy minima or most stable structure [94]. In these circumstances, the Sc_3C_{82} structure was recently verified as $Sc_3C_2@C_{80}$ by ^{13}C NMR measurement of its anion and finally the single crystal X-ray structure analysis of its adamantylidene adduct **10** (Figure 10.10) [95].

10.6 Missing Metallofullerene

In 1991, Smalley and co-workers reported that $La@C_{72}$, $La@C_{74}$, and $La@C_{82}$ were produced especially abundantly in soot, but only $La@C_{82}$ was extracted with toluene [2]. To date, many soluble endohedral metallofullerenes have been separated and characterized [96]. However, insoluble endohedral metallofullerenes, such as $La@C_{72}$ and $La@C_{74}$, have not yet been isolated, although they are detected regularly in raw soot using mass spectrometry. Recently, $La@C_{72}$ and $La@C_{74}$, the so-called missing metallofullerenes, have been isolated and characterized as derivatives [97, 98].

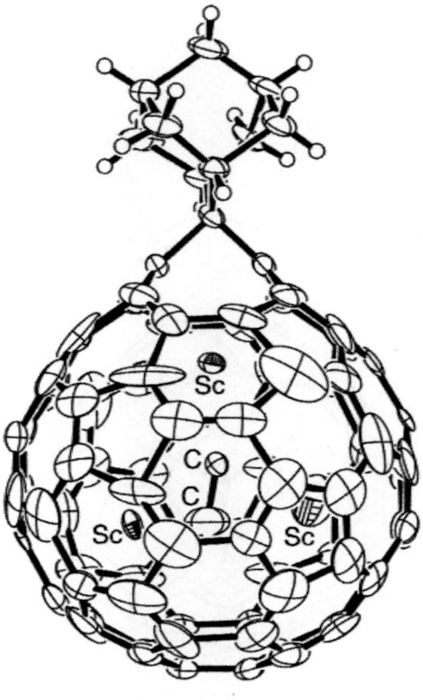

Figure 10.10 ORTEP drawings of **10**

Soot containing lanthanum metallofullerenes is produced using the DC arc discharge method, and La@C_{72} and La@C_{74} are observed in the raw soot using LD–TOF mass spectrometry. Endohedral metallofullerenes and empty fullerenes are extracted using 1,2,4-trichlorobenzene (TCB) under reflux. The soluble fraction is separated using a multistage high performance liquid chromatograph (HPLC), and fractions, which show a molecular ion peak at *m/z* 1148 or 1172 attributable to the dichlorophenyl group ($C_6H_3Cl_2$, mass *m/z* 145) adducts of La@C_{72} (*m/z* 1003), or La@C_{74} (*m/z* 1027), respectively, on MALDI–TOF mass measurements. The EPR measurement of the fractions presented no signals, indicative of a closed-shell electronic structure. These results suggest that La@C_{72} and La@C_{74} react with TCB in the process of the extraction to produce the adducts La@C_{72}($C_6H_3Cl_2$) (**11**) and La@C_{74}($C_6H_3Cl_2$) (**12**). Results of NMR studies show that they have C_1 symmetry. Their structures are finally confirmed using X-ray single crystal structure analyses. Figures 10.11 and 10.12 show that La@C_{72} and La@C_{74} respectively have C_2 and D_{3h} cage symmetries. It is noteworthy that the carbon cage of La@C_{72} has fused pentagons despite the fact that D_{6d}-C_{72} has a structure satisfying the isolated-pentagon rule (IPR).

Theoretical calculation of La@C_{2v}-C_{82} shows that the spin densities are distributed onto all the carbons of C_{82}, i.e. each carbon has a small spin density. In contrast to La@C_{2v}-C_{82}, La@D_{3h}-C_{74} is calculated as that about 50% of the total spin densities on C_{74} is localized on the three types of carbon, allowing these carbons to have high radical character. In fact, the dichlorophenyl radical, which may be produced by the reaction of TCB with reductant, such as lanthanum carbide in the raw soot, adds to one of these carbons to give the stable adduct. From these results, unconventionally high reactivity of La@D_{3h}-C_{74} is ascribed to the high radical character of the C_{74} cage. Meanwhile, La@C_2-C_{72} is calculated to have the smallest ionization potential (IP) among the reported lanthanum metallofullerenes [98]. Therefore,

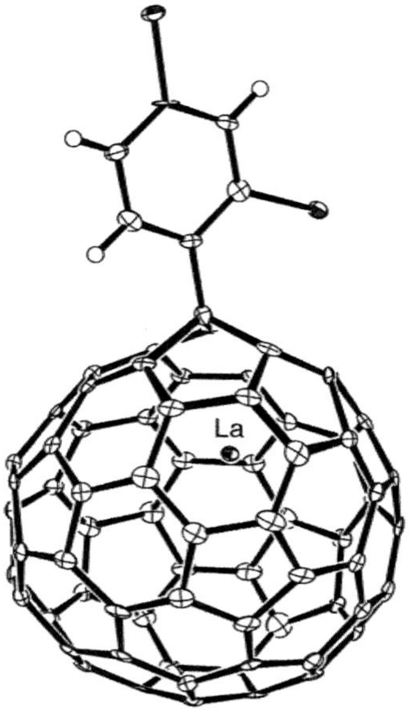

Figure 10.11 ORTEP drawings of **11**

La@C_2-C_{72} may interact strongly with amorphous carbon in soot and thereby become insoluble in common organic solvents. Then, the adduct La@$C_{72}(C_6H_3Cl_2)$ has a higher IP than that of La@C_2-C_{72}, this being supported by their redox potentials. These results suggest that the addition of a dichlorophenyl group to La@C_2-C_{72} engenders stable endohedral metallofullerene derivatives, which can be extracted in common organic solvents.

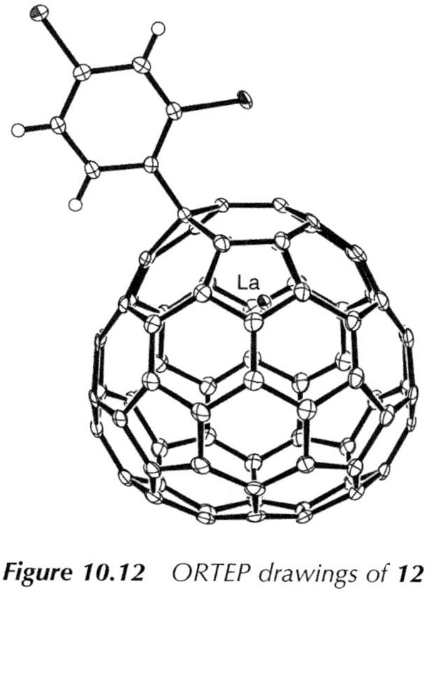

Figure 10.12 ORTEP drawings of **12**

The isolations of the La@C_2-C_{72} and La@D_{3h}-C_{74} as dichlorophenyl adducts suggest that many other insoluble and unknown endohedral metallofullerenes remain in raw soot, which will open up the new material science of metallofullerenes.

10.7 Supramolecular Chemistry

Supramolecular chemistry based on *empty* fullerenes such as C_{60} and C_{70} has been investigated extensively [99–114] with diverse objectives including purification [115, 116], enzyme mimicry [117, 118], and magnetic behavior [119]. Fullerenes are known to form host–guest complexes with crown ether [120, 121], calixarene [122–126], cyclodextrin [127, 128], porphyrin derivatives [130, 131], and so on. Sophisticated host molecules for empty fullerenes have been synthesized [131–138]. Furthermore, supramolecular systems that exhibit photoinduced electron and energy transfer have been studied actively using fullerene as an acceptor. To date, electron transfer reactions from organic donors to C_{60} have been widely reported [139–144]. In those systems, most electron transfer reactions proceed in a photoinduced, excited state. Electron transfers only slightly exist in the ground state. Correspondingly, a supramolecular donor–acceptor system based on endohedral metallofullerenes is expected to exhibit predominant electron transfer behavior.

10.7.1 Supramolecular System with Macrocycles

Mixing of La@C_{2v}-C_{82} with 1,4,7,10,13,16-hexaazacyclooctadecane (**13**, Figure 10.13) in toluene at ambient temperature yields precipitates of their complex, although no precipitates are formed in the case of C_{60} and **13** [145]. The precipitates are soluble in polar solvents, particularly in nitrobenzene. The vis-NIR spectra of their nitrobenzene solution are almost identical to that of the electrochemically produced [La@C_{82}]$^-$ [18], suggesting the formation of an electron-transfer complex of La@C_{2v}-C_{82} with **13**. Formation of the anion species of La@C_{2v}-C_{82} was also confirmed using ^{13}C NMR measurement. A 1 : 1 stoichiometry for complexation is established using Job's plot, and the binding ability of La@C_{2v}-C_{82} to **13** in nitrobenzene can be estimated as log $K = 5.7$ using a titration technique with vis-NIR spectroscopy. The value is larger than that of C_{60} with **13**.

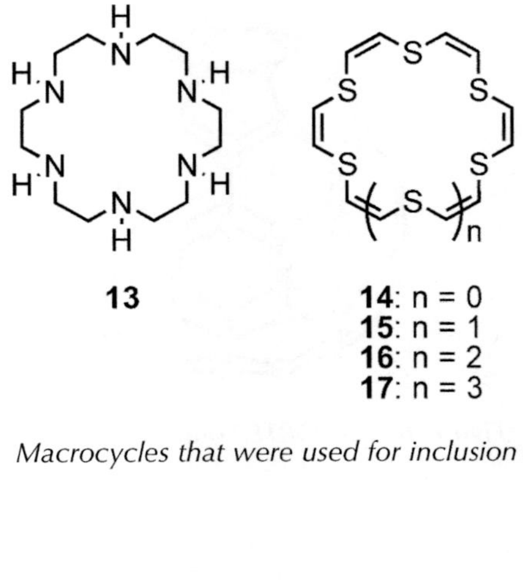

Figure 10.13 *Macrocycles that were used for inclusion of La@C_{2v}-C_{82}*

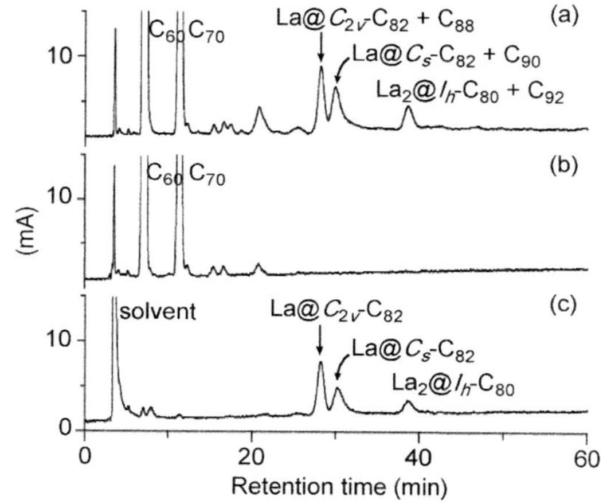

Figure 10.14 *HPLC profiles for (a) toluene extracts of carbon soot, (b) filtrate and (c) CS_2 extracts of complex*

As described previously, La@C_{2v}-C_{82} forms precipitates by complexing with **13** in toluene, although C_{60} does not. Based on this fact, the selective isolation of endohedral metallofullerene from soot extracts is also conducted using complexation. The addition of **13** to a toluene solution of the extracts containing lanthanum metallofullerenes predictably affords precipitates of the complex of endohedral metallofullerenes with **13**. Furthermore, the metallofullerene complexes and empty fullerene can be separated easily by filtration. HPLC profile and LD-TOF mass spectra of the filtrate (sample 1, Figures 10.14b and 10.15b) show that La@C_{2v}-C_{82}, La@C_s-C_{82}, and La_2@I_h-C_{80} are removed from extracts (Figures 10.14a and 10.15a). The free lanthanum metallofullerenes are extracted from precipitates using CS_2 with ultrasonication (Figures 10.14c and 10.15c).

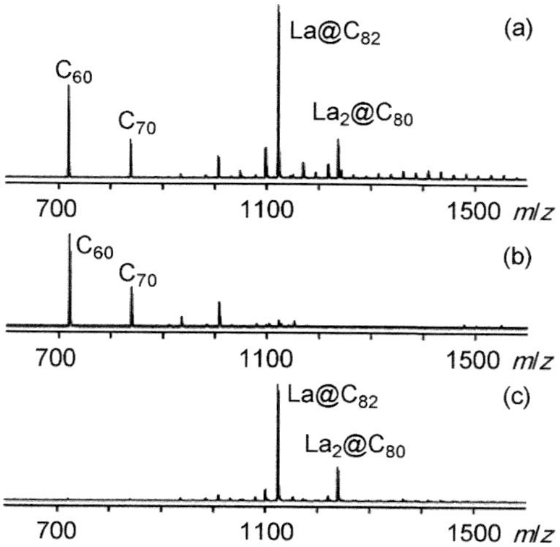

Figure 10.15 *Negative ion laser desorption mass spectra of (a) toluene extracts of carbon soot, (b) filtrate and (c) CS_2 extracts of complex*

The formation of 1 : 1 complex accompanying the electron transfer in nitrobenzene is also observed with 15-, 18-, 21-, and 24-membered unsaturated thiacrown ethers **14–17** (Figures 10.13) [146]. Among these, 21-membered **16** has the best ring-size for inclusion of La@C_{2v}-C_{82}. The ring-size effect indicates the formation of the inclusion complex.

10.7.2 Supramolecular System with Organic Donor

Through electron transfer, La@C_{2v}-C_{82} forms a complex with azacrown and unsaturated thiacrown ethers. Facile electron transfer is characteristic of endohedral metallofullerenes having low reduction potentials. In these systems, the formation of [La@C_{82}]$^-$ is confirmed by vis-NIR absorption, ESR, and NMR measurements. However, identification of the oxidized species of crown ethers is difficult because of their instability. In this context, complexation and electron transfer behaviors between La@C_{2v}-C_{82} and organic donor molecules such as *N,N,N′,N′*-tetramethyl-*p*-phenylenediamine (TMPD) [147], which forms a stable radical cation (Scheme 10.7), are examined [148]. The photoinduced electron transfer from TMPD to triplet C_{60} in nitrobenzene has been reported by Foote *et al.* [149]; in this report, no electron transfer occurs thermally.

A titration experiment using vis-NIR spectroscopy in nitrobenzene demonstrates the disappearance of the characteristic absorption maxima of La@C_{2v}-C_{82} accompanying the appearance of new absorption maxima corresponding to [TMPD]$^{\bullet+}$ radical cation and [La@C_{82}]$^-$ anion with increasing amounts of TMPD. The equilibrium constant of the La@C_{82}/TMPD system is evaluated as $\log K_{obs} = 5.4$. This value depends on the measurement solvent; the equilibrium constants in benzonitrile and *o*-dichlorobenzene are obtained respectively as $\log K_{obs} = 5.0$ and 3.1. In contrast, the equilibrium constant in toluene is too small to be detected. The values show good correspondence with the permittivity ε_r of the measurement solvents. Results show that the vis-NIR absorptions of the La@C_{82}/TMPD pair and the [La@C_{82}]$^-$/[TMPD]$^{\bullet+}$ pair differ, which reveals a solvatochromism.

Figure 10.16 shows that the spin-site exchange process between La@C_{2v}-C_{82} and TMPD is also confirmed by ESR measurement. Variable temperature ESR measurements from 320 to 240 K in *o*-dichlorobenzene/benzonitrile (=4 : 1) show that the equilibrium shifts to the formation of the ion pair at low temperatures (Figure 10.16b). Repeated temperature changes afforded the same spectra, indicating that La@C_{2v}-C_{82} and TMPD are in equilibrium with [La@C_{82}]$^-$/[TMPD]$^{\bullet+}$ in solution. The electron transfer can be controlled

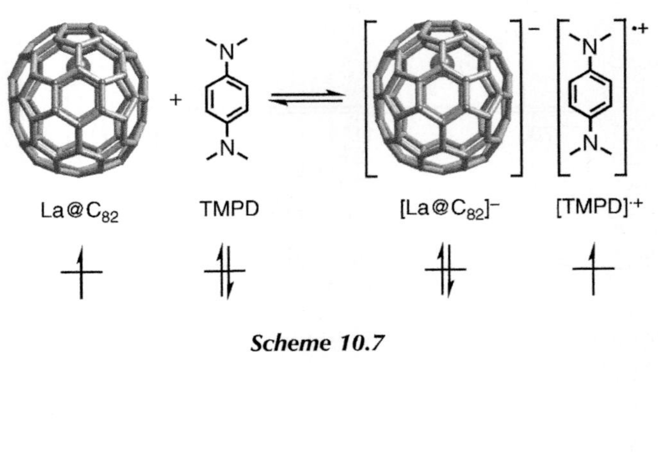

Scheme 10.7

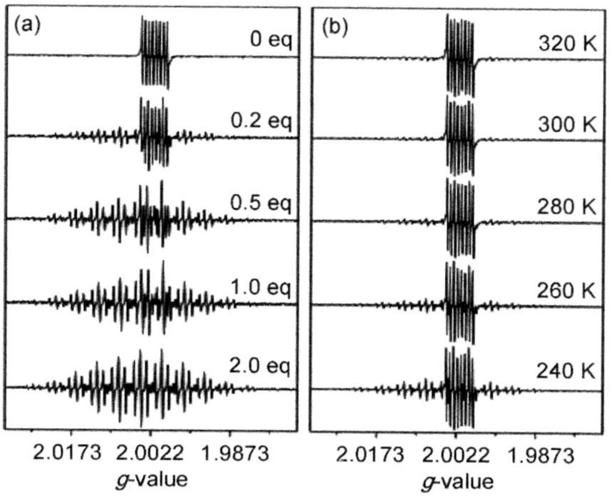

Figure 10.16 *EPR spectra of (a) La@C_{2v}-C_{82} in the presence of 0–2 equiv of TMPD in nitrobenzene at 296 K (b) La@C_{2v}-C_{82} with 1 equiv of TMPD in o-dichlorobenzene/benzonitrile (= 4 : 1) at 323–243 K*

reversibly by changing the temperature, which results in the occurrence of thermochromism.

Consequently, reversible intermolecular electron transfer systems at complete equilibrium in solution are first accomplished using La@C_{2v}-C_{82} with donor molecules, forming stable diamagnetic/paramagnetic anions and radical cations, respectively. These reversible electron transfer systems are stable even in air.

10.8 Conclusion

Regioselective chemical modification of monometallofullerene such as M@C_{2v}-C_{82} (M = La, Gd) is accomplished by using an electrophile or nucleophile. Controlling the motion of metal atoms in M_2@I_h-C_{80} (M = La, Ce) is also achieved by chemical functionalization. In addition, it is revealed that the reactivity of M_2@I_h-C_{80} with disilirane is higher than that of trimetallic nitride endohedral metallofullerene Sc_3N@I_h-C_{80}. The higher reactivity of M_2@I_h-C_{80} is caused by its much lower LUMO level. Furthermore, isolation of missing metallofullerenes is succeeded as their derivatives. The successful chemical functionalizations of endohedral metallofullerenes are of great help in designing future materials, as well as catalytic and biological applications using these materials.

Construction of supramolecular system based on endohedral metallofullerene is also examined. Consequently, La@C_{2v}-C_{82} is found to form an inclusion complex with azacrown and unsaturated thiacrown ethers by electron transfer. The electron transfer was revealed to proceed very easily even in the ground state. Reversible intermolecular spin-site exchange systems are accomplished using paramagnetic La@C_{2v}-C_{82} and organic donors. The realization of the stable and reversible electron transfer systems based on endohedral metallofullerene and organic donors would be an important stepping-stone toward developing materials for optical and magnetic applications.

References

[1] J. R. Heath, Q. Zhang, Y. Liu, R. F. Curl, F. K. Tittel, and R. E. Smalley, Lanthanum complexes of spheroidal carbon shells, *J. Am. Chem. Soc.*, **107**, 7779–7780 (1985).

[2] Y. Chai, T. Guo, C. Jin, R. E. Haufler, L. P. F. Chibante, J. Fure, L. Wang, J. M. Alford, and R. E. Smalley, Fullerenes with metals inside, *J. Phys. Chem.*, **95**, 7564–7568 (1991).

[3] F. T. Edelmann, Filled buckeyballs–recent developments from the endohedral metallofullerenes of lanthanides, *Angew. Chem. Int. Ed.*, **34**, 981–985 (1995).

[4] T. Akasaka and S. Nagase (eds), *Endofullerenes: A New Family of Carbon Clusters*, Kluwer Academic Publishers, Dordrecht, 2002.

[5] H. Shinohara, Endohedral metallofullerenes, *Rep. Prog. Phys.*, **63**, 843–892 (2000).

[6] B. Pietzak, M. Waiblinger, T. A. Murphy, A. Weidinger, M. Hohne, E. Dietel, and A. Hirsch, Properties of endohedral N@C_{60}, *Carbon*, 613–615 (1998).

[7] J. A. Larsson, J. C. Greer, W. Harneit, and A. Weidinger, Phosphorous trapped within buckminsterfullerene, *J. Chem. Phys.*, **116**, 7849–7854 (2002).

[8] M. Scheloske, B. Naydenov, C. Meyer, and W. Harneit, Synthesis and functionalization of fullerenes encapsulating atomic phosphorus, *Isra. J. Chem.*, **46**, 407–412 (2006).

[9] E. Shabtai, A. Weitz, R. C. Haddon, R. E. Hoffman, M. Rabinovitz, A. Khong, R. J. Cross, M. Saunders, P. C. Cheng, and L. T. Scott, ^{3}He NMR of He@$C_{60}{}^{6-}$ and He@$C_{70}{}^{6-}$. New records for the most shielded and the most deshielded ^{3}He inside a fullerene, *J. Am. Chem. Soc.*, **120**, 6389–6393 (1998).

[10] K. Komatsu, M. Murata, and Y. Murata, Encapsulation of molecular hydrogen in fullerene C_{60} by organic synthesis, *Science*, **307**, 238–240 (2005).

[11] M. Murata, Y. Murata, and K. Komatsu, Synthesis and properties of endohedral C_{60} encapsulating molecular hydrogen, *J. Am. Chem. Soc.*, **128**, 8024–8033 (2006).

[12] Y. Morita, S. Suzuki, K. Fukui, S. Nakazawa, H. Kitagawa, H. Kishida, H. Okamoto, A. Naito, A. Sekine, Y. Ohashi, M. Shiro, K. Sasaki, D. Shiomi, K. Sato, T. Takui, and K. Nakasuji, Thermochromism in an organic crystal based on the coexistence of sigma- and pi-dimers, *Nature Mater.*, **7**, 48–51 (2008).

[13] S. Suzuki, Y. Morita, K. Fukui, K. Sato, D. Shiomi, T. Takui, and K. Nakasuji, Aromaticity on the pancake-bonded dimer of neutral phenalenyl radical as studied by MS and NMR spectroscopies and NICS analysis, *J. Am. Chem. Soc.*, **128**, 2530–2531 (2006).

[14] S. Nishida, Y. Morita, K. Fukui, K. Sato, D. Shiomi, T. Takui, and K. Nakasuji, Spin transfer and solvato-/thermochromism induced by intramolecular electron transfer in a purely organic open-shell system, *Angew. Chem. Int. Ed.*, **44**, 7277–7280 (2005).

[15] Y. Apeloig, D. B. -Zhivotovskii, M. Bendikov, D. Danovich, M. Botoshansky, T. Vakul'skaya, M. Voronkov, R. Samoilova, M. Zdravkova, V. Igonin, V. Shklover, and Y. Struchkov, Synthesis and X-ray molecular structure of the first stable organic radical lacking resonance stabilization, *J. Am. Chem. Soc.*, **121**, 8118–8119 (1999).

[16] B. T. King, B. C. Noll, A. J. McKinley, and J. Michl, Dodecamethylcarba-closo-dodecaboranyl ($CB_{11}Me_{12}{}^{\bullet}$), a stable free radical, *J. Am. Chem. Soc.*, **118**, 10902–10903 (1996).

[17] K. Hatanaka, Y. Morita, T. Ohba, K. Yamaguchi, T. Takui, M. Kinoshita, and K. Nakasuji, Detection of new neutral radicals: 2-phenyl- and 2-p-methoxyphenyl-3-oxophenalenoxyl radicals, *Tetrahedron Lett.*, **37**, 873–876 (1996).

[18] T. Akasaka, T. Wakahara, S. Nagase, K. Kobayashi, M. Waelchli, K. Yamamoto, M. Kondo, S. Shirakura, S. Okubo, Y. Maeda, T. Kato, M. Kako, Y. Nakadaira, R. Nagahata, X. Gao, E. Van Caemelbecke, and K. M. Kadish, La@C_{82} anion. An unusually stable metallofullerene, *J. Am. Chem. Soc.*, **122**, 9316–9317 (2000).

[19] T. Akasaka, T. Wakahara, S. Nagase, K. Kobayashi, M. Waelchli, K. Yamamoto, M. Kondo, S. Shirakura, Y. Maeda, T. Kato, M. Kako, Y. Nakadaira, X. Gao, E. V. Caemelbecke, and K. M. Kadish, Structural determination of the La@C_{82} isomer, *J. Phys. Chem. B*, **105**, 2971–2974 (2001).

[20] T. Wakahara, J. Kobayashi, M. Yamada, Y. Maeda, T. Tsuchiya, M. Okamura, T. Akasaka, M. Waelchli, K. Kobayashi, S. Nagase, T. Kato, M. Kako, K. Yamamoto, and K. M. Kadish, Characterization of Ce@C_{82} and its anion, *J. Am. Chem. Soc.*, **126**, 4883–4887 (2004).

[21] T. Wakahara, S. Okubo, M. Kondo, Y. Maeda, T. Akasaka, M. Waelchli, M. Kako, K. Kobayashi, S. Nagase, T. Kato, K. Yamamoto, X. Gao, E. V. Caemelbecke, and K. M. Kadish, Ionization and structural determination of the major isomer of Pr@C_{82}, *Chem. Phys. Lett.*, **360**, 235–239 (2002).

[22] E. Nishibori, M. Ishihara, M. Takata, M. Sakata, Y. Ito, T. Inoue, and H. Shinohara, Bent $(metal)_2C_2$ clusters encapsulated in (Sc_2C_2)@C_{82}(III) and (Y_2C_2)@C_{82}(III) metallofullerenes, *Chem. Phys. Lett.*, **433**, 120–124 (2006).

[23] E. Nishibori, I. Terauchi, M. Sakata, M. Takata, Y. Ito, T. Sugai, and H. Shinohara, High-resolution analysis of (Sc_3C_2)@C_{80} metallofullerene by third generation synchrotron radiation X-ray powder diffraction, *J. Phys. Chem. B*, **110**, 19215–19219 (2006).

[24] E. Nishibori, S. Narioka, M. Takata, M. Sakata, T. Inoue, and H. Shinohara, A C_2 molecule entrapped in the pentagonal-dodecahedral Y_2 cage in Y_2C_2@C_{82}(III), *ChemPhysChem*, **7**, 345–348 (2006).

[25] M. Takata, E. Nishibori, M. Sakata, C. R. Wang, and H. Shinohara, Sc_2 dimer in IPR-violated $C6_6$ fullerene: a covalent bonded metallofullerene, *Chem. Phys. Lett.*, **372**, 512–518 (2003).

[26] E. Nishibori, M. Takata, M. Sakata, A. Taninaka, and H. Shinohara, Pentagonal-dodecahedral La_2 charge dens$_{\mathrm{it}}$y in [80-I_h]fullerene: La_2@C_{80}, *Angew. Chem. Int. Ed.*, **40**, 2998–2999 (2001).

[27] C. R. Wang, T. Kai, T. Tomiyama, T. Yoshida, Y. Kobayashi, E. Nishibori, M. Takata, M. Sakata, and H. Shinohara, A scandium carbide endohedral metallofullerene: (Sc_2C_2)@C_{84}, *Angew. Chem. Int. Ed.*, **40**, 397–399 (2001).

[28] C. R. Wang, T. Kai, T. Tomiyama, T. Yoshida, Y. Kobayashi, E. Nishibori, M. Takata, M. Sakata, and H. Shinohara, Materials science – C_{66} fullerene encaging a scandium dimmer, *Nature*, **408**, 426–427 (2000).

[29] E. Nishibori, M. Takata, M. Sakata, H. Tanaka, M. Hasegawa, and H. Shinohara, Giant motion of La atom inside C_{82} cage, *Chem. Phys. Lett.*, **330**, 497–502 (2000).

[30] M. Takata, E. Nishibori, M. Sakata, M. Inakuma, E. Yamamoto, and H. Shinohara, Triangle scandium cluster imprisoned in a fullerene cage, *Phys. Rev. Lett.*, **83**, 2214–2217 (1999).

[31] E. Nishibori, M. Takata, M. Sakata, M. Inakuma, and H. Shinohara, Determination of the cage structure of Sc@C_{82} by synchrotron powder diffraction, *Chem. Phys. Lett.*, **298**, 79–84 (1998).

[32] M. Takata, E. Nishibori, B. Umeda, M. Sakata, E. Yamamoto, and H. Shinohara, Structure of endohedral dimetallofullerene Sc_2@C_{84}, *Phys. Rev. Lett.*, **78**, 3330–3333 (1997).

[33] M. Takata, B. Umeda, E. Nishibori, M. Sakata, Y. Saito, M. Ohno, and H. Shinohara, Conformation by X-ray-diffraction of the4 endohedral nature of the metallofullerene Y@C_{82}, *Nature*, **377**, 46–49 (1995).

[34] H. Yang, C. Lu, Z. Liu, H. Jin, Y. Che, M. M. Olmstead, and A. L. Balch, Detection of a family of gadolinium-containing endohedral fullerenes and the isolation and crystallographic characterization of one member as a metal-carbide encapsulated inside a large fullerene cage, *J. Am. Chem. Soc.*, **130**, 17296–17300 (2008).

[35] S. Stevenson, M. A. Mackey, M. A. Stuart, J. P. Phillips, M. L. Easterling, C. J. Chancellor, M. M. Olmstead, and A. L. Balch, A distorted tetrahedral metal oxide cluster inside an icosahedral carbon cage. Synthesis, isolation, and structural characterization of $Sc_4(\mu_3\text{-}O)_2$@I_h-C_{80}, *J. Am. Chem. Soc.*, **130**, 11844–11845 (2008).

[36] B. Q. Mercado, C. M. Beavers, M. M. Olmstead, M. N. Chaur, K. Walker, B. C. Holloway, L. Echegoyen, and A. L. Balch, Is the isolated pentagon rule merely a suggestion for endohedral fullerenes? The structure of a second egg-shaped endohedral fullerene-Gd_3N@C_s(39663)-C_{82}, *J. Am. Chem. Soc.*, **130**, 7854–7855 (2008).

[37] T. Zuo, M. M. Olmstead, C. M. Beavers, A. L. Balch, G. Wang, G. T. Yee, C. Shu, L. Xu, B. Elliott, L. Echegoyen, J. C. Duchamp, and H. C. Dorn, Preparation and structural characterization of the I_h and the D_{5h} isomers of the endohedral fullerenes Tm_3N@C_{80}: Icosahedral C_{80} cage encapsulation of a trimetallic nitride magnetic cluster with three uncoupled Tm^{3+} ions, *Inorg. Chem.*, **47**, 5234–5244 (2008).

[38] T. Zuo, K. Walker, M. M. Olmstead, F. Melin, B. C. Holloway, L. Echegoyen, H. C. Dorn, M. N. Chaur, C. J. Chancellor, C. M. Beavers, A. L. Balch, and A. J. Athans, New egg-shaped fullerenes: non-isolated pentagon structures of Tm_3N@C_s(51365)-C_{84} and Gd_3N@C_s(51365)-C_{84}, *Chem. Commun.*, 1067–1069 (2008).

[39] S. Stevenson, C. J. Chancellor, H. M. Lee, M. M. Olmstead, and A. L. Balch, Internal and external factors in the structural organization in cocrystals of the mixed-metal endohedrals ($GdSc_2N@I_h$-C_{80}, $Gd_2ScN@I_h$-C_{80}, and $TbSc_2N@I_h$-C_{80}) and nickel(II) octaethylporphyrin, *Inorg. Chem.*, **47**, 1420–1427 (2008).
[40] T. Cai, L. Xu, H. W. Gibson, H. C. Dorn, C. J. Chancellor, M. M. Olmstead, and A. L. Balch, $Sc_3N@C_{78}$: Encapsulated cluster regiocontrol of adduct docking on an ellipsoidal metallofullerene sphere, *J. Am. Chem. Soc.*, **129**, 10795–10800 (2007).
[41] T. Zuo, C. M. Beavers, J. C. Duchamp, A. Campbell, H. C. Dorn, M. M. Olmstead, and A. L. Balch, Isolation and structural characterization of a family of endohedral fullerenes including the large, chiral cage fullerenes $Tb_3N@C_{88}$ and $Tb_3N@C_{86}$ as well as the I_h and D_{5h} isomers of $Tb_3N@C_{80}$, *J. Am. Chem. Soc.*, **129**, 2035–2043 (2007).
[42] C. M. Beavers, T. Zuo, J. C. Duchamp, K. Harich, H. C. Dorn, M. M. Olmstead, and A. L. Balch, $Tb_3N@C_{84}$: An improbable, egg-shaped endohedral fullerene that violates the isolated pentagon rule, *J. Am. Chem. Soc.*, **128**, 11352–11353 (2006).
[43] X. Wang, T. Zuo, M. M. Olmstead, J. C. Duchamp, T. E. Glass, F. Cromer, A. L. Balch, and H. C. Dorn, Preparation and structure of $CeSc_2N@C_{80}$: An icosahedral carbon cage enclosing an acentric $CeSc_2N$ unit with buried f electron spin, *J. Am. Chem. Soc.*, **128**, 8884–8889 (2006).
[44] T. Cai, L. Xu, M. R. Anderson, Z. Ge, T. Zuo, X. Wang, M. M. Olmstead, A. L. Balch, H. W. Gibson, and H. C. Dorn, Structure and enhanced reactivity rates of the D_{5h} $Sc_3N@C_{80}$ and $Lu_3N@C_{80}$ metallofullerene isomers: The importance of the pyracylene motif, *J. Am. Chem. Soc.*, **128**, 8581–8589 (2006).
[45] L. Echegoyen, C. J. Chancellor, C. M. E. B. Cardona, J. Rivera, M. M. Olmstead, and A. L. Balch, X-Ray crystallographic and EPR spectroscopic characterization of a pyrrolidine adduct of $Y_3N@C_{80}$, *Chem. Commun.*, 2653–2655 (2006).
[46] T. Cai, C. Slebodnick, L. Xu, K. Harich, T. E. Glass, C. Chancellor, J. C. Fettinger, M. M. Olmstead, A. L. Balch, H. W. Gibson, and H. C. Dorn, A pirouette on a metallofullerene sphere: Interconversion of isomers of *N*-tritylpyrrolidino I_h $Sc_3N@C_{80}$, *J. Am. Chem. Soc.*, **128**, 6486–6492 (2006).
[47] S. Stevenson, J. P. Phillips, J. E. Reid, M. M. Olmstead, S. P. Rath, and A. L. Balch, Pyramidalization of Gd_3N inside a C_{80} cage. The synthesis and structure of $Gd_3N@C_{80}$, *Chem. Commun.*, 2814–2815 (2004).
[48] M. M. Olmstead, H. M. Lee, J. C. Duchamp, S. Stevenson, D. Marciu, H. C. Dorn, and A. L. Balch, $Sc_3N@C_{68}$: Folded pentalene coordination in an endohedral fullerene that does not obey the isolated pentagon rule, *Angew. Chem. Int. Ed.*, **42**, 900–903 (2003).
[49] J. M. Campanera, C. Bo, M. M. Olmstead, A. L. Balch, and J. M. Poblet, Bonding within the endohedral fullerenes $Sc_3N@C_{78}$ and $Sc_3N@C_{80}$ as determined by density functional calculations and reexamination of the crystal structure of $\{Sc_3N@C_{78}\}\cdot Co(OEP)\}\cdot 1.5(C_6H_6)\cdot 0.3(CHCl_3)$, *J. Phys. Chem. A*, **106**, 12356–12364 (2002).
[50] M. M. Olmstead, H. M. Lee, S. Stevenson, H. C. Dorn, and A. L. Balch, Crystallographic characterization of isomer 2 of $Er_2@C_{82}$ and comparison with isomer 1 of $Er_2@C_{82}$, *Chem. Commun.*, 2688–2689 (2002).
[51] S. Stevenson, H. M. Lee, M. M. Olmstead, C. Kozikowski, P. Stevenson, and A. L. Balch, Preparation and crystallographic characterization of a new endohedral, $Lu_3N@C_{80}\cdot 5$(*o*-xylene), and comparison with $Sc_3N@C_{80}\cdot 5$(*o*-xylene), *Chemistry-Eur. J.*, **8**, 4528–4535 (2002).
[52] M. M. Olmstead, A. de, B. -Dias, S. Stevenson, H. C. Dorn, and A. L. Balch, Crystallographic characterization of the structure of the endohedral fullerene $\{Er_2@C_{82}$ Isomer I$\}$ with C_s cage symmetry and multiple sites for erbium along a band of ten contiguous hexagons, *J. Am. Chem. Soc.*, **124**, 4172–4173 (2002).
[53] M. M. Olmstead, A. de, B. -Dias, J. C. Duchamp, S. Stevenson, H. C. Dorn, and A. L. Balch, Isolation and crystallographic characterization of $ErSc_2N@C_{80}$: an endohedral fullerene which crystallizes with remarkable internal order, *J. Am. Chem. Soc.*, **122**, 12220–12226 (2000).
[54] H. Kato, A. Taninaka, T. Sugai, and H. Shinohara, Structure of a missing-caged metallofullerene: $La_2@C_{72}$, *J. Am. Chem. Soc.*, **125**, 7782–7783 (2003).

[55] M. D. Diener and J. M. Alford, Isolation and properties of small-bandgap fullerenes, *Nature*, **393**, 668–671 (1998).
[56] J. W. Raebiger and R. D. Bolskar, Improved production and separation processes for gadolinium metallofullerenes, *J. Phys. Chem. C*, **112**, 6605–6612 (2008).
[57] R. D. Bolskar and J. M. Alford, Chemical oxidation of endohedral metallofullerenes: identification and separation of distinct classes, *Chem. Commun.*, 1292–1293 (2003).
[58] B. Sun and Z. Gu, Solvent-dependent anion studies on enrichment of metallofullerene, *Chem. Lett.*, 1164–1165 (2002).
[59] B. Sun, L. Feng, Z. Shi, and Z. Gu, Improved extraction of metallofullerenes with DMF at high temperature, *Carbon*, **40**, 1591–1595 (2002).
[60] V. P. Bubnov, E. E. Laukhina, I. E. Kareev, V. K. Koltover, T. G. Prokhorova, E. B. Yagubskii, and Y. P. Kozmin, Endohedral metallofullerenes: A convenient gram-scale preparation, *Chem. Mater.*, **14**, 1004–1008 (2002).
[61] Y. F. Lian, S. F. Yang, and S. H. Yang, Revisiting the preparation of La@C_{82} (I and II) and La_2@C_{80}: Efficient production of the 'minor' isomer La@C_{82} (II), *J. Phys. Chem. B*, **106**, 3112–3117 (2002).
[62] M. D. Diener, R. D. Bolskar, and J. M. Alford, Redox properties and purification of endohedral metallofullerenes in *Endofullerenes: A New Family of Carbon Clusters*, T. Akasaka and S. Nagase (eds), Kluwer Academic Publishers: Dordrecht, (2002) pp. 133–151.
[63] T. Tsuchiya, T. Wakahara, S. Shirakura, Y. Maeda, T. Akasaka, K. Kobayashi, S. Nagase, T. Kato, and K. M. Kadish, Reduction of endohedral metallofullerenes: A convenient method for isolation, *Chem. Mater.*, **16**, 4343–4346 (2004).
[64] T. Tsuchiya, T. Wakahara, Y. Lian, Y. Maeda, T. Akasaka, T. Kato, N. Mizorogi, and S. Nagase, Selective extraction and purification of endohedral metallofullerene from carbon soot, *J. Phys. Chem. B*, **110**, 22517–22520 (2006).
[65] (a) T. Akasaka, T. Kato, K. Kobayashi, S. Nagase, K. Yamamoto, H. Funasaka, and T. Takahashi, Exohedral adducts of La@C_{82}, *Nature*, **374**, 600–601 (1995). (b) T. Akasaka, S. Nagase, K. Kobayashi, T. Suzuki, T. Kato, K. Kikuchi, Y. Achiba, K. Yamamoto, H. Funasaka, and T. Takahashi, Synthesis of the first adducts of the dimetallofullerenes La_2@C_{80} and Sc_2@C_{84} by addition of a disilirane, *Angew. Chem. Int. Ed.*, **34**, 2139–2141 (1995). (c) Y. Sasaki, M. Fujitsuka, O. Ito, Y. Maeda, T. Wakahara, T. Akasaka, K. Kobayashi, S. Nagase, M. Kako, and Y. Nakadaira, Photoinduced electron-transfer reactions between C_{60} and cyclic disiliranes (c-R_2Si-X-SiR_2; X = SiR_2, CH_2, O, NPh, S), *Heterocycles*, **54**, 777–787 (2001). (d) T. Akasaka, Y. Maeda, T. Wakahara, M. Okamura, M. Fujitsuka, O. Ito, K. Kobayashi, S. Nagase, M. Kako, Y. Nakadaira, and E. Horn, Novel metal-free bis-silylation: C_{60}-sensitized reaction of disilirane with benzonitrile, *Org. Lett.*, **1**, 1509–1512 (1999). (e) T. Akasaka, Y. Maeda, T. Wakahara, T. Mizushima, W. Ando, M. Waelchli, T. Suzuki, K. Kobayashi, S. Nagase, M. Kako, Y. Nakadaira, M. Fujitsuka, O. Ito, Y. Sasaki, K. Yamamoto, and T. Erata, A first photochemical bis-germylation of C_{60} with digermirane, *Org. Lett.*, **2**, 2671–2674 (2000). (f) T. Akasaka, W. Ando, K. Kobayashi, and S. Nagase, Organic chemical derivatization of fullerenes. 2. Photochemical [2 + 3] cycloaddition of C_{60} with disiliranes., *J. Am. Chem. Soc.*, **115**, 10366–10367 (1993). (g) T. Akasaka, E. Mitsuhida, W. Ando, K. Kobayashi, and S. Nagase, Adduct of C_{70} at the equatorial belt: Photochemical cycloaddition with disilirane, *J. Am. Chem. Soc.*, **116**, 2627–2628 (1994).
[66] (a) T. Akasaka, S. Okubo, M. Kondo, Y. Maeda, T. Wakahara, T. Kato, T. Suzuki, K. Yamamoto, K. Kobayashi, and S. Nagase, Isolation and characterization of two Pr@C_{82} isomers, *Chem. Phys. Lett.*, **319**, 153–156 (2000). (b) T. Suzuki, Y. Maruyama, T. Kato, T. Akasaka, K. Kobayashi, S. Nagase, K. Yamamoto, H. Funasaka, and T. Takahashi, Chemical reactivity of a metallofullerene: EPR study of diphenylmethano-La@C_{82} radicals, *J. Am. Chem. Soc.*, **117**, 9606–9607 (1995).
[67] Y. Maeda, Y. Matsunaga, T. Wakahara, S. Takahashi, T. Tsuchiya, M. O. Ishitsuka, T. Hasegawa, T. Akasaka, M. T. H. Liu, K. Kokura, E. Horn, K. Yoza, T. Kato, S. Okubo, K. Kobayashi, S. Nagase, and K. Yamamoto, Isolation and characterization of a carbene derivative of La@C_{82}, *J. Am. Chem. Soc.*, **126**, 6858–6859 (2004).

[68] R. C. Haddon, Chemistry of the fullerenes: The manifestation of strain in a class of continuous aromatic-molecules, *Science*, **261**, 1545–1550 (1993).
[69] T. Suzuki, Q. Li, K. C. Khemani, and F. Wudl, Synthesis of *m*-phenylene and *p*-phenylenebis (phenylfulleroids): Two-pearl sections of pearl necklace polymers, *J. Am. Chem. Soc.*, **114**, 7300–7301 (1992).
[70] R. A. Moss and M. J. Chang, Intermolecular chemistry of a dialkylcarbene: Adamantanylidene, *Tetrahedron Lett.*, **22**, 3749–3752 (1981).
[71] R. Bonneau, B. Hellrung, M. T. H. Liu, and J. Wirz, Adamantylidene revisited: flash photolysis of adamantanediazirine, *J. Photochem. Photobiol. A: Chemistry*, **116**, 9–19 (1998).
[72] T. Akasaka, T. Kono, Y. Takematsu, H. Nikawa, T. Nakahodo, T. Wakahara, M. O. Ishitsuka, T. Tsuchiya, Y. Maeda, M. T. H. Liu, K. Yoza, T. Kato, K. Yamamoto, N. Mizorogi, Z. Slanina, and S. Nagase, Does Gd@C_{82} have an anomalous endohedral structure? Synthesis and single crystal X-ray structure of the carbene adduct, *J. Am. Chem. Soc.*, **130**, 12840–12841 (2008).
[73] C. Bingel, Cyclopropylation of fullerenes, *Chem. Ber.*, **126**, 1957–1959 (1993).
[74] A. Hirsch, I. Lamparth, and H. R. Karfunkel, Fullerene chemistry in Three dimensions: Isolation of seven regioisomeric bisadducts and chiral trisadducts of C_{60} and di(ethoxycarbonyl)methylene, *Angew. Chem. Int. Ed.*, **33**, 437–438 (1994).
[75] A. Hirsch, I. Lamparth, T. Grösser, and H. R. Karfunkel, Regiochemistry of multiple additions to the fullerene core: Synthesis of a T_h-symmetric hexakis adduct of C_{60} with bis(ethoxycarbonyl)methylene, *J. Am. Chem. Soc.*, **116**, 9385–9386 (1994).
[76] (a) L. Feng, T. Nakahodo, T. Wakahara, T. Tsuchiya, Y. Maeda, T. Akasaka, T. Kato, E. Horn, K. Yoza, N. Mizorogi, and S. Nagase, A singly bonded derivative of endohedral metallofullerene: La@C_{82}CBr($COOC_2H_5$)$_2$, *J. Am. Chem. Soc.*, **127**, 17136–17137 (2005). (b) L. Feng, T. Tsuchiya, T. Wakahara, T. Nakahodo, Q. Piao, Y. Maeda, T. Akasaka, T. Kato, K. Yoza, E. Horn, N. Mizorogi, and S. Nagase, Synthesis and characterization of a bisadduct of La@C_{82}, *J. Am. Chem. Soc.*, **128**, 5990–5991 (2006). (c) L. Feng, T. Wakahara, T. Nakahodo, T. Tsuchiya, Q. Piao, Y. Maeda, Y. Lian, T. Akasaka, E. Horn, K. Yoza, T. Kato, N. Mizorogi, and S. Nagase, *Chem. Eur. J.*, **12**, 5578–5586 (2006).
[77] K. Kobayashi, S. Nagase, and T. Akasaka, Endohedral dimetallofullerenes Sc_2@C_{84} and La_2@C_{80}. Are the metal atoms still inside the fullerene cages? *Chem. Phys. Lett.*, **261**, 502–506 (1996).
[78] T. Akasaka, S. Nagase, K. Kobayashi, M. Waelchli, K. Yamamoto, H. Funasaka, M. Kako, T. Hoshino, and T. Erata, ^{13}C and ^{139}La NMR studies of La_2@C_{80}: First evidence for circular motion of metal atoms in endohedral dimetallofullerenes, *Angew. Chem. Int. Ed.*, **36**, 1643–1645 (1997).
[79] E. Nishibori, M. Takata, M. Sakata, A. Taninaka, and H. Shinohara, Pentagonal-dodecahedral La_2 charge density in [80-I_h]fullerene: La_2@C_{80}, *Angew. Chem. Int. Ed.*, **40**, 2998–2999 (2001).
[80] H. Shimotani, T. Ito, Y. Iwasa, A. Taninaka, H. Shinohara, E. Nishibori, M. Takata, and M. Sakata, Quantum chemical study on the configurations of encapsulated metal ions and the molecular vibration modes in endohedral dimetallofullerene La_2@C_{80}, *J. Am. Chem. Soc.*, **126**, 364–369 (2004).
[81] V. Balzani, A. Credi, F. M. Raymo, and J. F. Stoddard, Artificial molecular machines, *Angew. Chem. Int. Ed.*, **39**, 3348–3391 (2000).
[82] Molecular machines special issue, *Acc. Chem. Res.*, **34**, 409–522 (2001).
[83] K. Kobayashi, S. Nagase, Y. Maeda, T. Wakahara, and T. Akasaka, La_2@C_{80}: Is the circular motion of two La atoms controllable by exohedral addition? *Chem. Phys. Lett.*, **374**, 562–566 (2003).
[84] T. Wakahara, M. Yamada, S. Takahashi, T. Nakahodo, T. Tsuchiya, Y. Maeda, T. Akasaka, M. Kako, K. Yoza, E. Horn, N. Mizorogi, and S. Nagase, Two-dimensional hopping motion of encapsulated La atoms in silylated La_2@C_{80}, *Chem. Commun.*, 2680–2682 (2007).
[85] M. Yamada, T. Nakahodo, T. Wakahara, T. Tsuchiya, Y. Maeda, T. Akasaka, M. Kako, K. Yoza, E. Horn, N. Mizorogi, K. Kobayashi, and S. Nagase, Positional control of encapsulated atoms inside fullerene cage by exohedral addition, *J. Am. Chem. Soc.*, **127**, 14570–14571 (2005).

[86] M. Yamada, T. Wakahara, T. Nakahodo, T. Tsuchiya, Y. Maeda, T. Akasaka, K. Yoza, E. Horn, N. Mizorogi, and S. Nagase, Synthesis and structural characterization of endohedral pyrrolidinodimetallofullerene: $La_2@C_{80}(CH_2)_2NTrt$, *J. Am. Chem. Soc.*, **128**, 1402–1403 (2006).

[87] M. Yamada, C. Someya, T. Wakahara, T. Tsuchiya, Y. Maeda, T. Akasaka, K. Yoza, E. Horn, M. T. H. Liu, N. Mizorogi, and S. Nagase, Metal atoms collinear with the spiro carbon of 6,6-open adducts, $M_2@C_{80}(Ad)$ (M = La and Ce, Ad = Adamantylidene), *J. Am. Chem. Soc.*, **130**, 1171–1176 (2008).

[88] S. Stevenson, G. Rice, T. Glass, K. Harich, F. Cromer, M. R. Jordan, J. Craft, E. Hadju, R. Bible, M. M. Olmstead, K. Maitra, A. J. Fisher, A. L. Balch, and H. C. Dorn, Small-bandgap endohedral metallofullerenes in high yield and purity, *Nature*, **401**, 55–57 (1999).

[89] T. Akasaka, S. Nagase, K. Kobayashi, T. Suzuki, T. Kato, K. Yamamoto, H. Funasaka, and T. Takahashi, Exohedral derivatization of an endohedral metallofullerene $Gd@C_{82}$, *J. Chem. Soc., Chem. Commun.*, 1343–1344 (1995).

[90] T. Suzuki, K. Kikuchi, F. Oguri, Y. Nakao, S. Suzuki, Y. Achiba, K. Yamamoto, H. Funasaka, and T. Takahashi, Electrochemical properties of fullerenolanthanides, *Tetrahedron*, **52**, 4973–4982 (1996).

[91] M. Yamada, L. Feng, T. Wakahara, Y. Maeda, Y. Lian, M. Kako, T. Akasaka, T. Kato, K. Kobayashi, and S. Nagase, Synthesis and characterization of exohedrally silylated $M@C_{82}$ (M = Y and La), *J. Phys. Chem. B*, **109**, 6049–6501 (2005).

[92] H. Shinohara, H. Sato, M. Ohkohchi, Y. Ando, T. Kodama, T. Shida, T. Kato, and Y. Saito, Encapsulation of scandinum trimer in C_{82}, *Nature*, **357**, 52–54 (1992).

[93] C. S. Yannoni, M. Hoinkis, M S de Vries, D. S. Bethune, J. R. Salem, M. S. Crowder, and R. D. Johnson, Scandinum clusters in fullerene cages, *Science*, **256**, 1191–1192 (1992).

[94] K. Kobayashi and S. Nagase, Theoretical study of structures and dynamic properties of $Sc_3@C_{82}$, *Chem. Phys. Lett.*, **313**, 45–51 (1999).

[95] Y. Iiduka, T. Wakahara, T. Nakahodo, T. Tsuchiya, A. Sakuraba, Y. Maeda, T. Akasaka, K. Yoza, E. Horn, T. Kato, M. T. H. Liu, N. Mizorogi, K. Kobayashi, and S. Nagase, Structural determination of metallofullerene Sc_3C_{82} revisited: A surprising finding, *J. Am. Chem. Soc.*, **127**, 12500–12501 (2005).

[96] S. Okubo, T. Kato, M. Inakuma, and H. Shinohara, Separation and characterization of ESR-active lanthanum endohedral fullerenes, *New Diam. Front. Carbon Technol.*, **11**, 285–294 (2001).

[97] T. Wakahara, H. Nikawa, T. Kikuchi, T. Nakahodo, G MA. Rahman, T. Tsuchiya, Y. Maeda, T. Akasaka, K. Yoza, E. Horn, K. Yamamoto, N. Mizorogi, Z. Slanina, and S. Nagase, $La@C_{72}$ having a non-IPR carbon cage, *J. Am. Chem. Soc.*, **128**, 14228–14229 (2006).

[98] H. Nikawa, T. Kikuchi, T. Wakahara, T. Nakahodo, T. Tsuchiya, G MA. Rahman, T. Akasaka, Y. Maeda, K. Yoza, E. Horn, N. Mizorogi, and S. Nagase, Missing metallofullerene $La@C_{74}$, *J. Am. Chem. Soc.*, **127**, 9684–9685 (2005).

[99] F. Langa and J.-F. Nierengarten (Eds.), *Fullerenes: Principles and Applications*, RSC Nanoscience and Nanotechnology, Cambridge, 2007.

[100] U. Hahn, F. Cardinali, and J.-F. Nierengarten, Supramolecular chemistry for the self-assembly of fullerene-rich dendrimers, *New J. Chem.*, **31**, 1128–1138 (2007).

[101] K. Tashiro and T. Aida, Metalloporphyrin hosts for supramolecular chemistry of fullerenes, *Chem. Soc. Rev.*, **36**, 189–197 (2007).

[102] D. Bonifazi, O. Enger, and F. Diederich, Supramolecular [60]fullerene chemistry on surfaces, *Chem. Soc. Rev.*, **36**, 390–414 (2007).

[103] T. Kawase and H. Kurata, Ball-, bowl-, and belt-shaped conjugated systems and their complexing abilities: Exploration of the concave-convex π-π interaction, *Chem. Rev.*, **106**, 5250–5273 (2006).

[104] A. Hirsch, Functionalization of fullerenes and carbon nanotubes, *Phys. Stat. Sol. B*, **243**, 3209–3212 (2006).

[105] N. Tagmatarchis and M. Prato, Carbon-based materials: From fullerene nanostructures to functionalized carbon nanotubes, *Pure Appl. Chem.*, **77**, 1675–1706 (2005).

[106] L. Sánchez, N. Martín, and D. M. Guldi, Hydrogen-bonding motifs in fullerene chemistry, *Angew. Chem. Int. Ed.*, **44**, 5374–5382 (2005).
[107] F. D'Souza and O. Ito, Photoinduced electron transfer in supramolecular systems of fullerenes functionalized with ligands capable of binding to zinc porphyrins and zinc phthalocyanines, *Coord. Chem. Rev.*, **249**, 1410–1422 (2005).
[108] T. Konishi, A. Ikeda, and S. Shinkai, Supramolecular design of photocurrent-generating devices using fullerenes aimed at modelling artificial photosynthesis, *Tetrahedron*, **61**, 4881–4889 (2005).
[109] D. M. Guldi, F. Zerbetto, V. Georgakilas, and M. Prato, Ordering fullerene materials at nanometer dimensions, *Acc. Chem. Res.*, **38**, 38–43 (2005).
[110] E. Nakamura and H. Isobe, Functionalized fullerenes in water. The first 10 years of their chemistry, biology, and nanoscience, *Acc. Chem. Res.*, **36**, 807–815 (2003).
[111] D. M. Guldi and N. Martín, Fullerene architectures made to order; biomimetic motifs – design and features, *J. Mater. Chem.*, **12**, 1978–1992 (2002).
[112] F. Diederich and M. G. -Lopez, Supramolecular fullerene chemistry, *Chem. Soc. Rev.*, **28**, 263–277 (1999).
[113] A. L. Balch and M. M. Olmstead, Structural chemistry of supramolecular assemblies that place flat molecular surfaces around the curved exteriors of fullerenes, *Coord. Chem. Rev.*, **185-6**, 601–617 (1999).
[114] P. L. Boulas, M. G. -Kaifer, and L. Echegoyen, Electrochemistry of supramolecular systems, *Angew. Chem. Int. Ed.*, **37**, 216–247 (1998).
[115] J. L. Atwood, G. A. Koutsantonis, and C. L. Raston, Purification of C_{60} and C_{70} by selective complexation with calixarenes, *Nature*, **368**, 229–231 (1994).
[116] T. Suzuki, K. Nakashima, and S. Shinkai, Very convenient and efficient purification method for fullerene (C_{60}) with 5,11,17,23,29,35,41,47-octa-*tert*-butylcalix[8]arene-49,50,51,52,53,54, 55,56-octol, *Chem. Lett.*, 699–702 (1994).
[117] S. Bosi, T. D. Ros, G. Spalluto, and M. Prato, Fullerene derivatives: an attractive tool for biological applications, *Eur. J. Med. Chem.*, **38**, 913–923 (2003).
[118] S. D. Eastburn and B. Y. Tao, Applications of modified cyclodextrins, *Biotech. Adv.*, **12**, 325–339 (1994).
[119] R. E. Dinnebier, O. Gunnarsson, H. Brumm, E. Koch, P. W. Stephens, A. Huq, and M. Jansen, Structure of haloform intercalated C_{60} and its influence on superconductive properties, *Science*, **269**, 109–113 (2002).
[120] S. Bhattacharya, A. Sharma, S. K. Nayak, S. Chattopadhyay, and A. K. Mukherjee, NMR study of complexation of crown ethers with [60]- and [70]fullerenes, *J. Phys. Chem. B*, **107**, 4213–4217 (2003).
[121] M J van Eis, P. Seiler, L. A. Muslinkina, M. Badertscher, E. Pretsch, F. Diederich, R. J. Alvarado, L. Echegoyen, and I. P. Nunez, Supramolecular fullerene chemistry: A comprehensive study of cyclophane-type mono- and bis-crown ether conjugates of C_{70}, *Helv. Chim. Acta*, **85**, 2009–2055 (2002).
[122] S. Bhattacharya, S. K. Nayak, S. Chattopadhyay, M. Banerjee, and A. K. Mukherjee, Study of novel interactions of calixarene π-systems with [60]- and [70]fullerenes by the NMR spectrometric method, *J. Phys. Chem. B*, **107**, 11830–11834 (2003).
[123] B. Poór, L. Biczók, and M. Kubinyi, Interaction of triplet C_{60} with *p-tert*-butyl-calixarenes and their complexes with pyridine derivatives, *Phys. Chem. Chem. Phys.*, **5**, 2047–2052 (2003).
[124] A. Ikeda, Y. Suzuki, M. Yoshimura, and S. Shinkai, On the prerequisites for the formation of solution complexes from [60]fullerene and calix[n]arenes: A novel allosteric effect between [60]fullerene and metal cations in calix[n]aryl ester complexes, *Tetrahedron*, **54**, 2497–2508 (1998).
[125] T. Haino, M. Yanase, and Y. Fukazawa, New supramolecular complex of C_{60} based on calix[5] arene – Its structure in the crystal and in solution, *Angew. Chem. Int. Ed. Engl.*, **36**, 259–260 (1997).
[126] K. Araki, K. Akao, A. Ikeda, T. Suzuki, and S. Shinkai, Molecular design of calixarene-based host molecules for inclusion of C_{60} in solution, *Tetrahedron Lett.*, **37**, 73–76 (1996).

[127] C. N. Murthy and K. E. Geckeler, The water-soluble β-cyclodextrin-[60]fullerene complex, *Chem. Commun.*, 1194–1195 (2001).
[128] K. Komatsu, K. Fujiwara, Y. Murata, and T. Braun, Aqueous solubilization of crystalline fullerenes by supramolecular complexation with γ-cyclodextrin and sulfocalix[8]arene under mechanochemical high-speed vibration milling, *J. Chem. Soc., Perkin Trans.*, **1**, 2963–2966 (1999).
[129] Z. I. Yoshida, H. Takekuma, S. I. Takekuma, and Y. Matsubara, Molecular recognition of C_{60} with γ-cyclodextrin, *Angew. Chem. Int. Ed.*, **33**, 1597–1599 (1994).
[130] M. Yanagisawa, K. Tashiro, M. Yamasaki, and T. Aida, Hosting fullerenes by dynamic bond formation with an iridium porphyrin cyclic dimer: A 'chemical friction' for rotary guest motions, *J. Am. Chem. Soc.*, **129**, 11912–11913 (2007).
[131] Y. Kubo, A. Sugasaki, M. Ikeda, K. Sugiyasu, K. Sonoda, A. Ikeda, M. Takeuchi, and S. Shinkai, Cooperative C_{60} binding to a porphyrin tetramer arranged around a *p*-terphenyl axis in 1:2 host-guest stoichiometry, *Org. Lett.*, **4**, 925–928 (2002).
[132] O. D. Fox, J. Cookson, E JS. Wilkinson, M GB. Drew, E. J. MacLean, S. J. Teat, and P. D. Beer, Nanosized polymetallic resorcinarene-based host assemblies that strongly bind fullerenes, *J. Am. Chem. Soc.*, **128**, 6990–7002 (2006).
[133] M.-X. Wang, X.-H. Zhang, and Q.-Y. Zheng, Synthesis, structure, and [60]fullerene complexation properties of azacalix[m]arene[n]pyridines, *Angew. Chem. Int. Ed.*, **43**, 838–842 (2004).
[134] T. Kawase, K. Tanaka, N. Shiono, Y. Seirai, and M. Oda, Onion-type complexation based on carbon nanorings and a buckminsterfullerene, *Angew. Chem. Int. Ed.*, **43**, 1722–1724 (2004).
[135] D. Sun, F. S. Tham, C. A. Reed, L. Chaker, and P DW. Boyd, Supramolecular fullerene-porphyrin chemistry. Fullerene complexation by metalated 'jaws porphyrin' hosts, *J. Am. Chem. Soc.*, **124**, 6604–6612 (2002).
[136] T. Haino, Y. Yamanaka, H. Araki, and Y. Fukazawa, Metal-induced regulation of fullerene complexation with double-calix[5]arene, *Chem. Commun.*, 402–403 (2002).
[137] D. I. Schuster, J. Rosenthal, S. MacMahon, P. D. Jarowski, C. A. Alabi, and D. M. Guldi, Formation and photophysics of a stable concave-convex supramolecular complex of C_{60} and a substituted *s*-triazine derivative, *Chem. Commun.*, 2538–2539 (2002).
[138] A. Ikeda, H. Uzdu, M. Yoshimura, and S. Shinkai, Inclusion of [60]fullerene in a self-assembled homooxacalix[3]arene-based dimeric capsule constructed by a Pd^{II}–pyridine interaction. The Li^{+}-binding to the lower rims can improve the inclusion ability, *Tetrahedron*, **56**, 1825–1832 (2000).
[139] K. G. Thomas, M. V. George, and P. V. Kamat, Photoinduced electron-transfer processes in fullerene-based donor-acceptor systems, *Helv. Chim. Acta*, **88**, 1291–1308 (2005).
[140] H. Imahori, Porphyrin-fullerene linked systems as artificial photosynthetic mimics, *Org. Biomol. Chem.*, **2**, 1425–1433 (2004).
[141] M. E. E.-Khouly, O. Ito, P. M. Smith, and F. D'Souza, Intermolecular and supramolecular photoinduced electron transfer processes of fullerene-porphyrin/phthalocyanine systems, *J. Photochem. Photobiol. C*, **5**, 79–104 (2004).
[142] D. M. Guldi, Fullerene-porphyrin architectures; photosynthetic antenna and reaction center models, *Chem. Soc. Rev.*, **31**, 22–36 (2002).
[143] N. Martín, L. Sánchez, B. Illescas, and I. Perez, C_{60}-based electroactive organofullerenes, *Chem. Rev.*, **98**, 2527–2547 (1998).
[144] J. P. Mittal, Excited-states and electron-transfer reactions of fullerenes, *Pure Appl. Chem.*, **67**, 103–110 (1995).
[145] T. Tsuchiya, K. Sato, H. Kurihara, T. Wakahara, T. Nakahodo, Y. Maeda, T. Akasaka, K. Ohkubo, S. Fukuzumi, T. Kato, N. Mizorogi, K. Kobayashi, and S. Nagase, Host-guest complexation of endohedral metallofullerene with azacrown ether and its application, *J. Am. Chem. Soc.*, **128**, 6699–6703 (2006).
[146] T. Tsuchiya, H. Kurihara, K. Sato, T. Wakahara, T. Akasaka, T. Shimizu, N. Kamigata, N. Mizorogi, and S. Nagase, Supramolecular complex of $La@C_{82}$ with unsaturated thiacrown ethers, *Chem. Commun.*, 3585–3587 (2006).

[147] A. M. M. Rawashdeh, C. S.-Leventis, X. Gao, and N. Leventis, One-step synthesis and redox properties of dodecahydro-3a,9a-diazaperylene – The most easily oxidized *p*-phenylenediamine, *Chem. Commun.*, 1742–1743 (2001).

[148] T. Tsuchiya, K. Sato, H. Kurihara, T. Wakahara, Y. Maeda, T. Akasaka, K. Ohkubo, S. Fukuzumi, T. Kato, and S. Nagase, Spin-site exchange system constructed from endohedral metallofullerenes and organic donors, *J. Am. Chem. Soc.*, **128**, 14418–14419 (2006).

[149] J. W. Arbogast, C. S. Foote, and M. Kao, Electron-transfer to triplet C_{60}, *J. Am. Chem. Soc.*, **114**, 2277–2279 (1992).

11

Gadonanostructures as Magnetic Resonance Imaging Contrast Agents

Jeyarama S. Ananta and Lon J. Wilson

Department of Chemistry & Smalley Institute of Nanoscale Science and Technology Rice University, Houston, Texas, USA

11.1 Magnetic Resonance Imaging (MRI) and the Role of Contrast Agents (CAs)

Magnetic resonance imaging (MRI) has evolved into one of the most powerful, noninvasive imaging modalities used in diagnostic medicine and biomedical research [1]. The superior resolution and greater anatomical details provided by MRI are essential for early diagnosis of many diseases. MRI is an extension of nuclear magnetic resonance (NMR) spectroscopy, a characterization technique used extensively in the field of chemistry. Nuclear spin, an inherent property of water protons, is manipulated by an external magnetic field in MRI to obtain images. Each nuclear spin acts like a magnetic dipole, and in the absence of an external magnetic field, they are oriented in random directions. For a single spin system like ^1H, present in abundance in the human body, the application of an external field results in two different energy states: (1) a low-energy state corresponding to the alignment of nuclear spins parallel to the applied field and (2) a quantized high-energy state arising from anti-parallel alignment. The population distribution of nuclear spins of these two energy states is determined by Boltzmann distribution. Usually a greater number of spins is present in the low energy state, resulting in a net magnetization. These nuclear spins precess around the applied magnetic field (B_0) at a particular frequency known as the Larmor frequency given by the expression $\omega = \gamma B_0$ (γ is the gyromagnetic ratio specific for the NMR active isotope).

Chemistry of Nanocarbons Edited by Takeshi Akasaka, Fred Wudl and Shigeru Nagase

Application of a radiofrequency (RF) pulse at the Larmor frequency excites the proton from the low-energy ground state to the higher-energy excited state. The time taken for the excited state protons to return to equilibrium is termed the relaxation time. The relaxation process can occur via two mechanisms: (1) relaxation along the axis of the applied magnetic field (longitudinal relaxation time; T_1) and (2) relaxation along the axis perpendicular to the applied magnetic field (transverse relaxation time; T_2). The intensity of the MR signal is dependent on the above mentioned relaxation times, along with proton density. A variety of mathematical sequences have been employed to preferentially differentiate between these three parameters to obtain MR images, since the difference in the number of water protons in different tissues is subtle. A more detailed description of MRI can be found elsewhere [2].

In order to improve the sensitivity and diagnostic confidence of MRI, chemical contrast agents (CAs) have been widely used [2–4]. Annually, there are about 60 million MRI scans performed worldwide, and approximately 30% of them use chemical CAs. Chemical CAs improve the sensitivity of MRI by decreasing the proton relaxation time of water protons in and around their vicinity. All of the clinically used contrast agents are paramagnetic in nature. They exhibit very large lattice fields and decrease the T_1 relaxation time of water protons in their vicinity to produce contrast enhancement. The ability of a paramagnetic material to act as an MRI contrast agent is expressed in terms of its relaxivity (r_1 for T_1-relaxation). Relaxivity is the change in the relaxation rate ($1/T_1$; s^{-1}) of water protons per molar concentration of the paramagnetic CA and has the units of $mM^{-1}s^{-1}$.

The high-spin paramagnetic Gd^{3+} ion is the most effective and extensively used T_1 relaxation agent in MRI, since it has:

1. seven unpaired f-electrons, the maximum number of unpaired electrons observed for any atom or metal ion; the proton relaxation rate is directly proportional to the electron-spin quantum number;
2. a large magnetic moment ($63\ \mu_B^2$); the proton relaxation rate is directly proportional to the square of the magnetic moment for a paramagnetic material;
3. a slow-relaxing, high-symmetry ground sate (^8S); slow relaxation produces strong oscillations near the Larmor frequency and has a pronounced effect on the T_1 relaxation process.

Aqueous Gd^{3+} ion is toxic and has to be sequestered for biological use. Traditional sequestering methods use a variety of linear and macrocyclic chelates, and these Gd^{3+} chelate compounds have been extensively studied and characterized [2–4]. In spite of the enormous progress achieved in their synthesis and design, current clinical CAs have several limitations. Chelation of Gd^{3+} ion with multidentate ligands decreases the number of coordination sites available for water proton exchange (8 sites for free Gd^{3+} ion compared to 1–2 sites for Gd^{3+} chelate compounds) resulting in reduced relaxivity. Also, almost all of the clinically-used CAs are extracellular fluid space (ECF) agents. These ECF agents have very low blood retention times (ca. 60 s), and they distribute extracellularly and excrete via the kidneys. However, for applications such as magnetic resonance angiography (MRA), agents with longer blood retention times are preferred. In addition, the number of paramagnetic Gd^{3+} centers that can be attached or delivered to the cell surface is limited to the nM range due to biological constraints [5]. In order to image cells for advanced applications such as cell tracking, a very high relaxivity ($\geq 100\ mM^{-1}s^{-1}$) for the CA is required to compensate for the nM restriction in cell-surface receptor site concentration. Currently, clinical CAs suffer from very low relaxivities ($r_1 \sim 4\ mM^{-1}s^{-1}$ for $[Gd(DTPA)(H_2O)]^{2-}$ or Magnevist®),

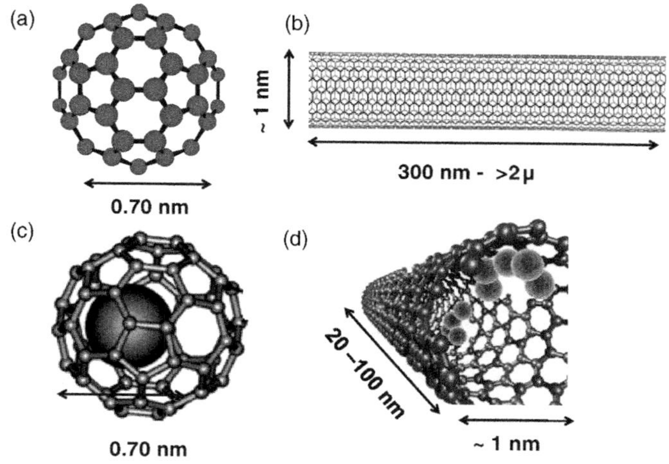

Figure 11.1 *Nanoscale building blocks of carbon nanotechnology: (a) C_{60} fullerene, (b) SWNT, (c) Gadofullerene, and (d) Gadonanotube*

and therefore cannot be used for such applications. Hence, there is an important need to develop new MRI CAs with superior efficacies and more desirable physiochemical properties.

The discovery of fullerenes, a new allotrope of carbon, by Smalley, Curl and Kroto in 1985 marked the beginning of the field of 'Carbon Nanotechnology' [6]. Ever since their discovery, carbon nanostructures have been one of most widely studied materials [7–9]. The two main building blocks of Carbon Nanotechnology are: (1) fullerenes (C_{60}, C_{70}, C_{74}, C_{76} etc.), which are hollow and spherical or nearly-spherical molecules (Figure 11.1a) and (2) carbon nanotubes, which are hollow all-carbon cylindrical materials (Figure 11.1b). Single-walled carbon nanotubes (SWNTs) can be visualized as being a single sheet of graphene rolled upon itself seamlessly to assume the shape of a drinking straw. All the carbon atoms of these nanostructures are surface atoms, which produces a hollow, internal space within the nanostructures which can be potentially filled with medically-interesting atoms, ions and even small molecules [10, 11]. When magnetically-active Gd^{3+} ions are encapsulated within the interior of these highly-ordered nanostructures (Gadonanostructures), a new class of MRI CA results. This chapter reviews and projects two of these new Gadonanostructures, Gadofullerenes (Figure 11.1c) and Gadonanotubes (Figure 11.1d), as new nanoscale paradigms in high-performance MRI CA probe design.

11.2 The Advantages of Gadonanostructures as MRI Contrast Agent Synthons

Gadonanostructures as MRI CAs offer several distinct advantages over today's clinically-used Gd^{3+}-ion-based CAs:

1. The gadolinium ions are trapped within a biologically-stable carbon cage (fullerenes and carbon nanotubes), preventing the release of toxic, free Gd^{3+} ion. In contrast, some

clinical CAs are susceptible to the release of Gd^{3+} ion [12], especially in patients with kidney disease. The release of free Gd^{3+} ions from Gd^{3+} chelate compounds has been linked to the development of nephrogenic systemic fibrosis (NSF) [13]. Hence, the thermodynamic and metabolic stability of the Gadonanostructures are important properties for their further development as CAs with longer blood retention times (blood pool agents).

2. The external carbon cage of the Gadonanostructures can be chemically derivatized to control toxicity, enhance biocompatibility and provide targeting ability using antibodies and/or peptides [14–16].
3. The nanoscale confinement of Gd^{3+} ions inside the carbon nanostructures provides them with an unusual metal-ion environment, resulting in remarkably high relaxivities ($r_1 > 100\,mM^{-1}s^{-1}$ per Gd^{3+} ion). In fact, the Gadonanotubes are the highest performing T_1-weighted MRI CA known, making them especially desirable candidates for future applications in molecular imaging.
4. The inherent lipophilicity of carbon nanostructures, including Gadonanotubes and Gadofullerenes, provides them the ability to efficiently translocate across cell membranes, without evidence of cytotoxicity [17–20]. Once target cells are internally labeled with Gadonanostructures in sufficient concentration, high resolution molecular imaging using MRI may well be achievable.

11.3 Gadofullerenes as MRI Contrast Agents

The first reported medical application of endohedral metallofullerenes was their use as MRI CAs [21–23]. Polyhydroxylated Gadofullerenes ($Gd@C_{82}(OH)_x$) were the first water-soluble metallofullerenes studied as MRI CAs [21, 22, 24]. There have been different values of relaxivity reported for $Gd@C_{82}(OH)_x$. Zhang *et al.* reported a value of $47\,mM^{-1}s^{-1}$ at 9.4 T and 27 °C [24], Wilson *et al.* reported $20\,mM^{-1}s^{-1}$ at 0.4 T and 40 °C [22], and Shinohara *et al.* reported a particularly high value of $81\,mM^{-1}s^{-1}$ at 1.0 T and 25 °C [21]. In spite of these initially observed variations in relaxivity, it should be noted that all of these Gadofullerene CA materials greatly outperformed (5–20 times greater relaxivity) clinical Gd^{3+}-based MRI CAs such as $[Gd(DTPA)(H_2O)]^{2-}]$ ($r_1 = {\sim}4\,mM^{-1}s^{-1}$). However, $Gd@C_{82}$ makes up only 10% of the metallofullerenes produced by electric-arc discharge. The remaining 90% consists mainly of $Gd@C_{60}$, $Gd@C_{70}$ and $Gd@C_{74}$. Initially, $Gd@C_{82}$ was the only metallofullerene component that could be extracted from the carbon-arc soot using laborious techniques including multistep, high-pressure liquid chromatography (HPLC). Even then, the process yielded only milligram quantities of material. The difficulty in procuring Gadofullerenes in sufficient quantities was initially viewed as an obstacle in measuring and understanding their ^1H relaxation properties and acquiring *in vivo* MR images using animal models. However, recent developments in the purification of endohedral metallofullerenes have yielded gram quantities of previously insoluble $Gd@C_{60}$ [25]. In addition, a new variant of endohedral metallofullerenes called trimetallic nitride template endohedral fullerenes (TNTs) have recently been produced in greater yields [26, 27]. Instead of a single Gd^{3+} ion encapsulated within the fullerene cage, TNTs have three Gd^{3+} ions bonded to a central nitrogen atom [28]. These advances in the production and purification of endohedral metallofullerenes have recently refuelled the interest in the

Table 11.1 *Relaxivity values for various Gadofullerenes*

Gadofullerene	r_1 ($mM^{-1}s^{-1}$) per Gd^{3+} ion	Field strength (T)	Reference
$Gd@C_{82}(OH)_x$	81	1.0	[21]
$Gd@C_{82}O_6(OH)_{16}$-$(NHCH_2CH_2COOH)_8$	9.1	1.5	[29]
$Gd@C_{60}(OH)_x$	83.2	1.5	[25]
$Gd@C_{60}[C(COOH)_2]_{10}$	24.0	1.5	[25]
$Gd_3N@C_{80}[DiPEG5000(OH)_x]$	48	2.4	[28]
$ScGd_2N@C_{80}O_{12}(OH)_{26}$	8.8	14.1	[30]

development of these nanoscale materials as high-performance MRI CAs. Representative relaxivity values for different variations of Gadofullerenes are presented in Table 11.1.

11.4 Understanding the Relaxation Mechanism of Gadofullerenes

Paramagnetic relaxation enhancement (PRE) is strongly influenced by the proton exchange properties of the CA. The proton exchange mechanism of any paramagnetic material can be classified into two types: (1) the inner-sphere mechanism and (2) outer-sphere mechanism. The inner-sphere mechanism is when the electron spins of the paramagnetic metal ion interact directly with the water protons in the first coordination sphere and the effect is transferred to the bulk by the chemical exchange of protons from the first coordination sphere. The outer-sphere contribution to proton exchange arises from the electron spin-proton relaxation of the bulk water protons assisted by ligand-mediated proton exchange. For Gadofullerenes, where a direct interaction between the Gd^{3+} ion and water molecules is prevented by the fullerene cage, only an outer-sphere mechanism is expected. In fact, ^{17}O studies of Gadofullerenes have supported an outer-sphere mechanism [23]. However, the overall effect of this outer-sphere mechanism for the Gadofullerenes is about 10 times greater than that observed for Gd^{3+} chelate compounds with an exclusively outer-sphere contribution [31].

The first detailed studies of ^1H relaxivity as a function of variable magnetic fields and temperature were performed on $Gd@C_{60}(OH)_x$ (Figure 11.2a) and $Gd@C_{60}[C(COOH)_2]_{10}$ (Figure 11.2b). Both of these Gadofullerene derivatives showed a characteristic broad peak in relaxivity between 30 MHz (0.7 T) and 60 MHz (1.4 T) (Figure 11.3) [23].

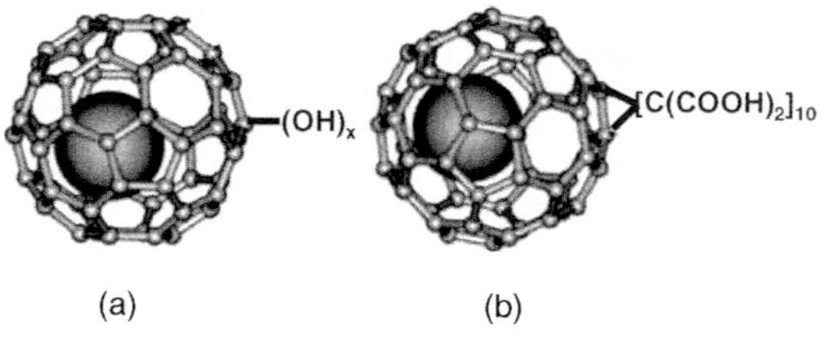

Figure 11.2 *Schematic representation of the $Gd@C_{60}$ derivatives (a) $Gd@C_{60}(OH)_x$ and (b) $Gd@C_{60}[C(COOH)_2]_{10}$*

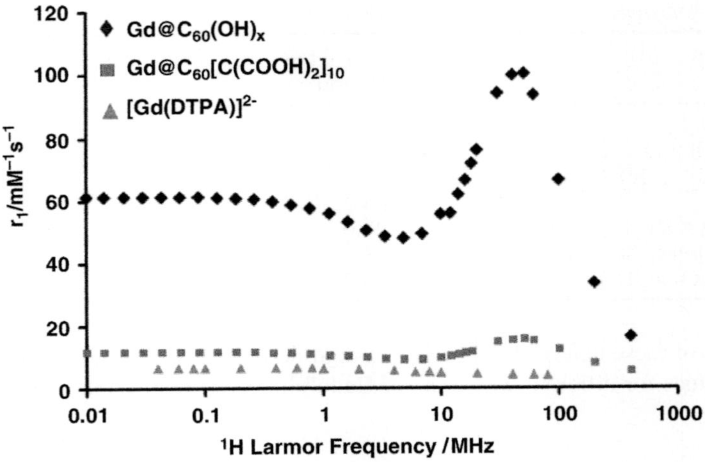

Figure 11.3 *^{1}H NMRD profiles of Gadofullerenes in aqueous solutions at 25 °C and pH = 7.4. For comparison, the profile for a clinical agent (Magnevist®) is also shown*

Such peaks were previously observed for slow-tumbling Gd^{3+} chelate compounds and nanoparticle-based systems [2–4]. Even when functionalized, Gadofullerenes tend to aggregate and form clusters in aqueous solution. Such aggregation has also been observed for various functionalized empty fullerenes [32, 33]. Aggregation of this nature could lead to slowly-tumbling Gadofullerene molecules within the aggregate, and as a consequence, to higher relaxivities.

Molecular rotation and proton exchange rate are the two important parameters that affect the relaxation properties of MRI CAs, and they have opposite temperature dependencies. Hence, observation of relaxivity as a function of temperature should reveal which factor is more important in determining relaxivity. $Gd@C_{60}(OH)_x$ did not show any perceptible temperature dependency. However, $Gd@C_{60}[C(COOH)_2]_{10}$ showed decreasing relaxivity with increasing temperature [23]. For proton exchange to be the reason for the observed increase in relaxivity, the relaxivity should increase with temperature, since the exchange rate is directly proportional to temperature. Hence, the observed negative temperature dependency for $Gd@C_{60}[C(COOH)_2]_{10}$ indicates that proton exchange is not the relaxivity-limiting step. The rotational correlation time (τ_R) of $Gd@C_{60}[C(COOH)_2]_{10}$ has been estimated to be 2.6 ns, which is longer than the τ_R observed for Gd^{3+} chelate compounds (ps range) [12], indicating that slow molecular rotation is, indeed, the determining factor for the increased relaxivity of $Gd@C_{60}[C(COOH)_2]_{10}$. Similar temperature-independent relaxation behavior was observed for $Gd@C_{82}(OH)_x$ at 200 MHz, and a slightly faster tumbling time of 0.8 ns was reported [34].

Tumor tissue has a slightly lower pH than normal tissue due to increased production of lactic acid [35]. Many CAs have been developed to exploit this difference in pH using MRI [36]. Gadofullerene CAs display remarkable pH-sensitive relaxivities (Figure 11.4a). With proton exchange for Gadofullerenes excluded as a major factor for their higher relaxivities, pH-induced aggregation might explain the observed pH dependency of the relaxivity [23]. Aggregation of Gadofullerenes is likely controlled by many factors such as

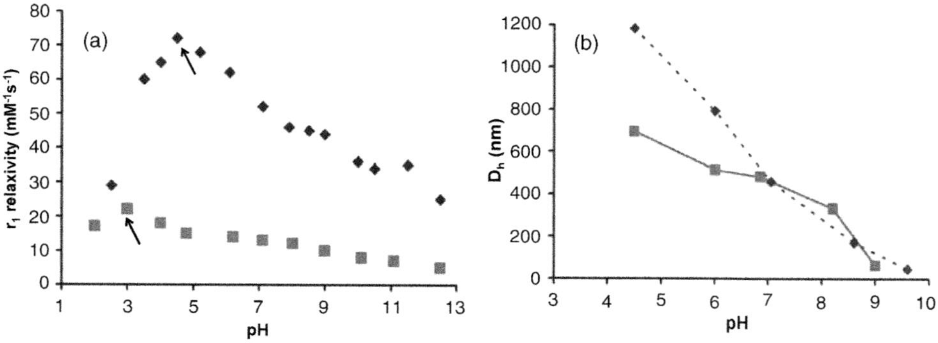

Figure 11.4 (a) pH-dependent relaxivity of $Gd@C_{60}(OH)_x$ [blue diamonds] and $Gd@C_{60}[C(COOH)_2]_{10}$ [purple squares] (arrows mark the pH threshold below which irreversible precipitation occurs). (b) effect of pH on the hydrodynamic radius (D_h) of the aggregates of $Gd@C_{60}(OH)_x$ [black diamonds] and $Gd@C_{60}[C(COOH)_2]_{10}$ [gray squares]

intermolecular hydrogen bonding and hydrophobic fullerene-fullerene interactions, with pH having a strong influence on the interactions. Dynamic light scattering (DLS) experiments conducted on the Gadofullerenes at different pH's have substantiated this aggregate size/relaxivity hypothesis (Figure 11.4b), since the aggregate size varies strongly as a function of pH. At acidic pH, larger aggregates are formed, resulting in slower tumbling aggregates and higher relaxivities. Similar pH-induced relaxivity changes have been observed for Gadofullerenes derivatized with other functionalities. For example, Shu *et al.* has reported similar pH-responsive relaxivities for $Gd@C_{82}O_6(OH)_{16}(NHCH_2CH_2COOH)_8$ where the relaxivity tripled upon decreasing pH ($r_1 = 2.52\,mM^{-1}s^{-1}$ at pH = 9 to $r_1 = 7.68\,mM^{-1}s^{-1}$ at pH = 2) [29]. However, the aggregation behavior of this Gadofullerene is more complicated, involving polymerization of clusters at acidic pH.

Similar dependency of relaxivity on aggregate size was observed when salts were added to aqueous solutions of Gadofullerenes to produce disaggregated species with lower relaxivities [37]. Interestingly, the effect of salt on the aggregation properties is not strictly ionic-strength dependent, since phosphate buffered saline (PBS) has a more pronounced effect than NaCl, even at lower concentrations (Table 11.2). The enhanced disaggregation induced by PBS could, for example, be due to the intercalation of phosphate anions between fullerene substituents through hydrogen bonding rather than resulting from ionic strength effects. Salt-induced disaggregation might also explain the observed variations in

Table 11.2 *Effect of salt concentration on the relaxivity of two Gadofullerenes*

	$Gd@C_{60}(OH)_x$		$Gd@C_{60}[C(COOH)_2]_{10}$	
	D_h (nm)	$r_1(mM^{-1}s^{-1})$	D_h (nm)	$r_1(mM^{-1}s^{-1})$
No added salt	811	83.2	721	24
10 mM PBS	91	14.1	32	6.8
50 mM NaCl	409	47.2	595	20.6
150 mM NaCl	121	31.6	107	16

relaxivity among the different Gadofullerene species reported in Table 11.1, since most of the different fullerene functionalization methods used different salts.

Initially, the effect of phosphate buffer on the relaxivity of Gadofullerenes was seen as an obstacle to their development as CAs; however, the PBS concentration used in the disaggregation studies was far greater than that present in a human body. In addition, the disaggregation reaction is not spontaneous, since it takes about 30–45 minutes to achieve complete disaggregation, and in fact, Gadofullerenes have been shown to be efficient CAs when used in animal model studies [21, 28].

The above studies suggest that the observed relaxivities of aggregated Gadofullerenes could be due to slow tumbling of the aggregates. However, such studies do not necessarily represent the true relaxation behavior of individual Gadofullerene molecules, since aggregate size varies from batch to batch. In order to determine the relaxation properties of individual Gadofullerene molecules, salt-induced disaggregation has been used. Complete disaggregation resulting in mostly individual Gadofullerene molecules (i.e. relaxivity values decrease and then plateau) has been induced by the addition of various salts. ^{17}O NMR studies performed on the disaggregated Gadofullerenes have revealed the unique presence of water molecules confined to the interstices inside the Gadofullerene aggregates where the diffusion rate of these confined water protons to the bulk solution is slower than the translational diffusion rate of bulk water molecules for a purely outer-sphere mechanism [38].

The rotational correlation time (τ_R) of the disaggregated Gadofullerenes has been estimated to be 1.2 ns [38]. This is not significantly different from the $\tau_R = 2.6$ ns value estimated for aggregated $Gd@C_{60}[C(COOH)_2]_{10}$ [23]. The lack of a significant difference in τ_R between Gadofullerene aggregates and individual Gadofullerene molecules suggests that the confined water molecules in Gadofullerene aggregates could play a vital role in the relaxation mechanism in addition to molecular rotation of the aggregate. In the case of disaggregated Gadofullerenes, the ^1H relaxation has two components: (1) an outer-sphere contribution due to the random translational motion of bulk water molecules and (2) an 'inner-sphere-like' mechanism arising from chemical exchange between protons of the functional groups on the fullerene surface and bulk water. It will be interesting to see if TNT Gadofullerene ($Gd_3N@C_{80}$; GadoTNT) will possess the same ^1H relaxation mechanisms as $Gd@C_{60}$. Unlike the present Gadofullerenes, $[Gd^{3+}@C_{60}{}^{3-}]$, where the Gd^{3+} ion donates three electrons to the fullerene cage inducing spin density on the cage, GadoTNT, $[(Gd_3N)^{6+}@C_{80}{}^{6-}]$, donates six electrons to the fullerene cage to produce a presumably diamagnetic cage structure. This might alter the ^1H relaxation mechanism for GadoTNTs.

11.5 Gadonanotubes as MRI Contrast Agents

Single-walled carbon nanotubes (SWNTs) are one of the most investigated nanomaterials for biomedical applications because their unique characteristics make them suitable for different applications [39, 40]. The ideal length of SWNTs for biomedical applications is uncertain; however, ultra-short single-walled carbon nanotubes (US-tubes) with a length of 20–100 nm may be the best suited for cellular uptake, biocompatibility, and elimination from the body [41]. The hollow interior of SWNTs and US-tubes can be used to encapsulate ions and small molecules [10, 11, 42], whereby their cytotoxicity is sequestered by the protective carbon sheath. Additionally, the exterior of SWNT materials can be chemically

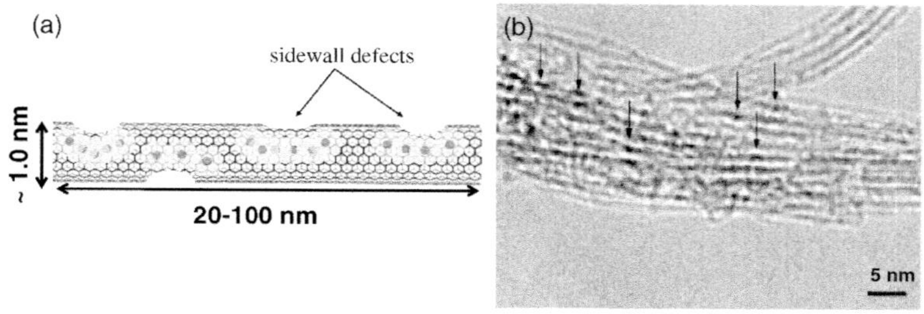

Figure 11.5 *(a) Schematic representation of a Gadonanotube (not to scale and Cl^- ions are not shown). (b) TEM image of bundled Gadonanotubes (arrows indicate the locations of the internal Gd^{3+}-ion clusters)*

modified with peptides, antibodies, and other small molecules to provide biocompatibility and cellular targeting [14–16].

Nanoscale loading and confinement of Gd^{3+} ions inside US-tubes resulted in a high-performance MRI CA called 'Gadonanotubes' (Figure 11.5a and 11.5b) [42]. The Gd^{3+} ions are present in the form of clusters located at the defect sides on the sidewalls of the US-tubes which are produced during the SWNT cutting process [43]. The Gadonanotubes resemble linear superparamagnetic molecular magnets with ^1H relaxivities 40–90 times greater than any current Gd^{3+}-based CA in clinical use. Because of the hydrophobic nature of the US-tube sheath, the Gadonanotubes do not disperse well in water. However, a variety of biocompatible surfactants have been used to impart biocompatibility to the Gadonanotubes. The relaxivities (per Gd^{3+} ion) of the Gadonanotubes dispersed in different surfactants, along with the aqueous $[Gd(H_2O)_8]^{3+}$ ion, are presented in Table 11.3 at a clinically-relevant field strength of 1.41 T. The r_1 relaxivity of the Gadonanotubes is about 20 times greater than that of the $[Gd(H_2O)_8]^{3+}$ ion at 1.5 T. When compared to clinically-used Gd^{3+}-based contrast agents, Gadonanotubes are about 40 times more efficacious at 1.5 T.

The nuclear magnetic relaxation dispersion (NMRD) profile of the Gadonanotubes at 37 °C is shown in Figure 11.6. For comparative purposes, the NMRD profile of the clinical agent, $[Gd\text{-}DTPA]^{2-}$ (Magnevist®), is also shown. At all field strengths, the Gadonanotubes far outperform the clinical Gd^{3+}-based CA. The enhancement is especially pronounced at very low-field strengths (~90 times greater than $[Gd\text{-}DTPA]^{2-}$ at 0.01 MHz). Such large ^1H relaxivities ($>600\ mM^{-1}s^{-1}$ @ 0.01 MHz!) are without precedent for MRI CAs. The current trend in molecular imaging is to use higher fields to counter the loss of resolution at low fields, but the outstanding performance in relaxivity of the Gadonanotubes, especially at low fields, could catalyze the development of low-cost, low-field molecular imaging by MRI.

Table 11.3 *Proton relaxivities of the Gadonanotubes and the free aqueous Gd^{3+} ion*

	$[Gd^{3+}]$ (mM)	Relaxation rate (s^{-1})	$r_1(mM^{-1}s^{-1})$ (per Gd^{3+} ion)
Gadonanotubes □	0.044	7.85	173
Gadonanotubes +	0.049	8.29	164
$[Gd(H_2O)_8]^{3+}$	1.99	16.95	8.4

Measurements at 1.41 T and 37 C. □: dispersed in sodium dodecyl benzene sulfonate (SDBS), +: dispersed in Pluronic.

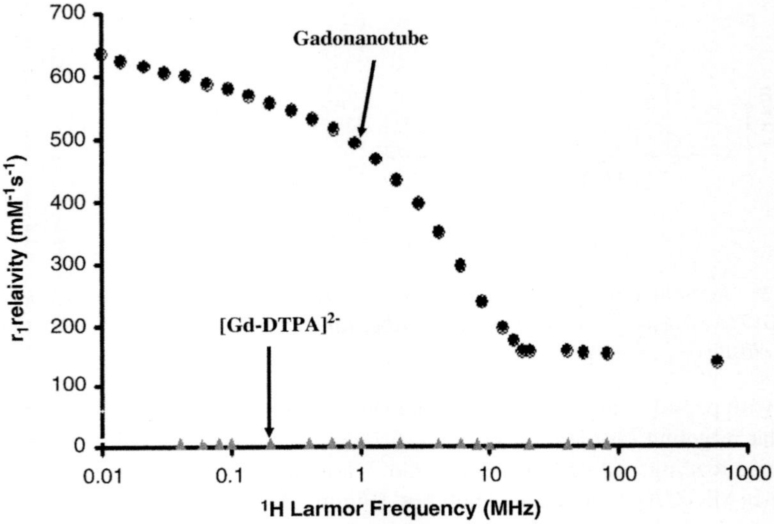

Figure 11.6 *NMRD profile of Gadonanotubes dispersed in SDBS surfactant. Measurements are at pH = 6.5 and at 37 °C. For comparison, the profile for a clinical agent (Magnevist®) is also shown*

The shape of the NMRD curve for the Gadonanotubes in Figure 11.6 is very different from those reported for other Gd^{3+}-based CAs. The classical Solomon-Bloembergen-Morgan (SBM) theory of relaxivity has not yet successfully interpreted the NMRD profile exhibited by the Gadonanotubes. The observed huge increase in relaxivity with decreasing magnetic fields below 1 MHz is especially different from other Gd^{3+}-based systems, where relaxivity values normally remain fairly constant at low fields. In addition, the Gadonanotubes show a fairly constant relaxivity at higher field strengths (> 10 MHz), again signaling that the Gadonanotubes present a new paradigm for CA development, since the performance of all other known MRI CAs fall off at higher fields.

Contrary to the Gadofullerenes, the observed relaxivities for the Gadonanotubes cannot be attributed to aggregation effects, since solution DLS measurements (D_h 20–100 nm) and cryo-TEM images reveal mostly individual nanotube structures. Unlike Gadofullerenes, for which the direct interaction of water molecules with Gd^{3+} ions is absent, for Gadonanotubes, water molecules have direct access to the Gd^{3+} centers of the internal clusters because of the 'ballistic' movement of water molecules through SWNTs [44]. Hence, the role of proton exchange on relaxivity for the Gadonanotubes cannot be neglected. In fact, the Gadonanotubes are the only class of MRI CA, where unchelated (naked) Gd^{3+} ions have direct access to many (up to 8) exchanging water molecules.

The Gadonanotubes also display relaxivities that are extremely sensitive to pH (Figure 11.7a) [45]. The relaxivity of the Gadonanotubes undergoes a threefold change in value over the range of pH = 8.3 ($r_1 = 40\ mM^{-1}s^{-1}$) to pH = 6.7 ($r_1 = 133\ mM^{-1}s^{-1}$). An even more dramatic change is observed between pH 7.0 and 7.4 with a slope of 98 $mM^{-1}s^{-1}$ per pH unit. Such dramatic response of relaxivity to pH has not been previously observed, and its presence for Gadonanotubes could pave the way for the development of ultra-

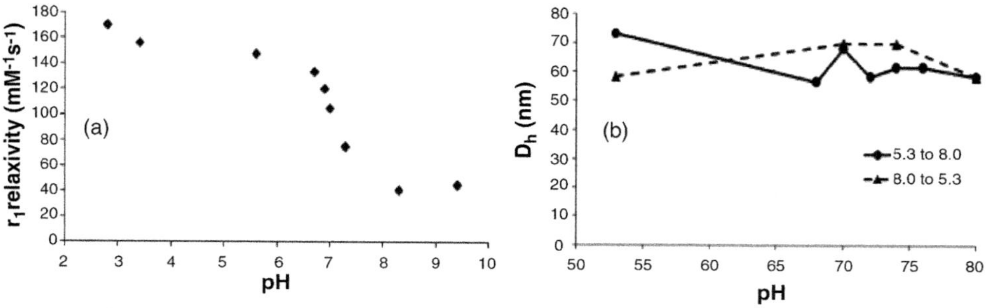

Figure 11.7 (a) pH-dependent relaxivities of the Gadonanotubes suspended with SDBS surfactant (b) Effect of pH on the aggregation state of the Gadonanotubes suspended with SDBS

sensitive, pH-dependent MR imaging. DLS studies on the Gadonanotubes at different pH's have eliminated the possibility of aggregation being responsible for the pH-dependent relaxivity (Figure 11.7b).

Temperature-dependent relaxivity studies of Gadonanotubes could give valuable information about the proton exchange rate. The Gadonanotubes show an intriguing temperature dependency at varying pHs. Under basic conditions, the relaxivity appears to be temperature-independent, while under acidic conditions, the relaxivity shows a marked temperature dependency, with values reaching as high as 500 $mM^{-1}s^{-1}$ at 5 °C [45]. A similar temperature dependency, but of a smaller magnitude (3.2–8.1 $mM^{-1}s^{-1}$), has been observed over the same temperature range for Gd^{3+} chelate compounds conjugated to a large protein [46]. The increased relaxivity at low temperature for the Gadonanotubes could be due to aggregation leading to a slower tumbling time or slowing down of the proton exchange rate. Clearly, further investigations are needed to better understand the extremely high relaxivities and the magnetic-field dependency of relaxivity for these remarkable materials. It is important to achieve this understanding in detail for the Gadonanotubes, since it could well lead to the design of even higher-performing MRI CA probes in the future.

Acknowledgement

We gratefully acknowledge the Robert A. Welch Foundation (Grant C-0627) and the Nanoscale Science and Engineering Initiative of the National Science Foundation under NSF Award Number EEC-0647452 at Rice University for the support that helped produce many of the results reported in this chapter.

References

[1] P. Mansfield, Snapshot magnetic resonance imaging. *Angew. Chem., Int. Ed.*, **43**, 5456–5464 (2004).

[2] A. E. Merbach, E. Toth, Editors, *The Chemistry of Contrast Agents in Medical Magnetic Resonance Imaging*. 471 pp. (2001).

[3] P. Caravan, J.J. Ellison, T.J. McMurry, R.B. Lauffer, Gadolinium(III) Chelates as MRI Contrast Agents: Structure, Dynamics, and Applications. *Chem. Rev.*, **99**, 2293–2352 (1999).

[4] R.B. Lauffer, Paramagnetic metal complexes as water proton relaxation agents for NMR imaging: theory and design. *Chem. Rev.*, **87**, 901–927 (1987).

[5] A.D. Nunn, K.E. Linder, M.F. Tweedle, Can receptors be imaged with MRI agents?. *Q J Nucl Med*, **41**, 155–162 (1997).

[6] H.W. Kroto, J.R. Heath, S.C. O'Brien, R.F. Curl, R.E. Smalley, C60: buckminsterfullerene. *Nature*, **318**, 162–163 (1985).

[7] P.M. Ajayan, Nanotubes from Carbon. *Chem. Rev.*, **99**, 1787–1799 (1999).

[8] K.B. Hartman, L.J. Wilson, M.G. Rosenblum, Detecting and treating cancer with nanotechnology. *Mol. Diagn. Ther.*, **12**, 1–14 (2008).

[9] C.R. Martin, P. Kohli, The emerging field of nanotube biotechnology. *Nat. Rev. Drug Discovery*, **2**, 29–37 (2003).

[10] J.M. Ashcroft, K.B. Hartman, K.R. Kissell, Y. Mackeyev, S. Pheasant, S. Young, P.A.W. Van der Heide, A.G. Mikos, L.J. Wilson, Single-molecule I2@US-tube nanocapsules: a new X-ray contrast-agent design. *Adv. Mater.*, **19**, 573–576 (2007).

[11] K.B. Hartman, D.K. Hamlin, S.D. Wilbur, L.J. Wilson, 211AtCl@US-tube nanocapsules: a new concept in radiotherapeutic-agent design. *Small*, **3**, 1496–1499 (2007).

[12] S. Laurent, L. Vander Elst, R.N. Muller, Comparative study of the physicochemical properties of six clinical low molecular weight gadolinium contrast agents. *Contrast Media Mol. Imaging*, **1**, 128–137 (2006).

[13] M.A. Perazella, Nephrogenic systemic fibrosis, kidney disease, and gadolinium: is there a link?. *Clin. J. Am. Soc. Nephrol.*, **2**, 200–202 (2007).

[14] Y. Mackeyev, K.B. Hartman, J.S. Ananta, A.V. Lee, L.J. Wilson, Catalytic synthesis of amino acid and peptide derivatized gadonanotubes. *J. Am. Chem. Soc.*, **131**, 8342–8343 (2009).

[15] Z. Ou, B. Wu, D. Xing, F. Zhou, H. Wang, Y. Tang, Functional single-walled carbon nanotubes based on an integrin alpha vbeta 3 monoclonal antibody for highly efficient cancer cell targeting. *Nanotechnology*, **20**, 105102/1–105102/7 (2009).

[16] A.A. Bhirde, V. Patel, J. Gavard, G. Zhang, A.A. Sousa, A. Masedunskas, R.D. Leapman, R. Weigert, J.S. Gutkind, J.F. Rusling, Targeted killing of cancer cells in vivo and in vitro with EGF-directed carbon nanotube-based drug delivery. *ACS Nano*, **3**, 307–316 (2009).

[17] D. Pantarotto, J.P. Briand, M. Prato, A. Bianco, Translocation of bioactive peptides across cell membranes by carbon nanotubes. *Chem. Commun.*, **1**, 16–17 (2004).

[18] B. Sitharaman, L.A. Tran, Q.P. Pham, R.D. Bolskar, R. Muthupillai, S.D. Flamm, A.G. Mikos, L.J. Wilson, Gadofullerenes as nanoscale magnetic labels for cellular MRI. *Contrast Media Mol. Imaging*, **2**, 139–146 (2007).

[19] N.W.S. Kam, T.C. Jessop, P.A. Wender, H. Dai, Nanotube molecular transporters: internalization of carbon nanotube-protein conjugates into mammalian cells. *J. Am. Chem. Soc.*, **126**, 6850–6851 (2004).

[20] L.L. Dugan, D.M. Turetsky, C. Du, D. Lobner, M. Wheeler, C.R. Almli, C.K.F. Shen, T.Y. Luh, D.W. Choi, T.S. Lin, Carboxyfullerenes as neuroprotective agents. *Proc. Natl. Acad. Sci. U. S. A.*, **94**, 9434–9439 (1997).

[21] M. Mikawa, H. Kato, M. Okumura, M. Narazaki, Y. Kanazawa, N. Miwa, H. Shinohara, Paramagnetic Water-Soluble Metallofullerenes Having the highest relaxivity for MRI contrast agents. *Bioconjugate Chem.*, **12**, 510–514 (2001).

[22] L.J. Wilson, Medical applications of fullerenes and metallofullerenes. *Electrochem. Soc. Interface*, **8**, 24–28 (1999).

[23] E. Toth, R.D. Bolskar, A. Borel, G. Gonzalez, L. Helm, A.E. Merbach, B. Sitharaman, L.J. Wilson, Water-soluble gadofullerenes: toward high-relaxivity, pH-responsive MRI contrast agents. *J. Am. Chem. Soc.*, **127**, 799–805 (2005).

[24] S. Zhang, D. Sun, X. Li, F. Pei, S. Liu, Synthesis and solvent enhanced relaxation property of water-soluble endohedral metallofullerenols. *Fullerene Sci. Technol.*, **5**, 1635–1643 (1997).

[25] R.D. Bolskar, A.F. Benedetto, L.O. Husebo, R.E. Price, E.F. Jackson, S. Wallace, L.J. Wilson, J.M. Alford, First soluble M@C60 derivatives provide enhanced access to metallofullerenes and

permit in vivo evaluation of Gd@C60[C(COOH)2]10 as a MRI contrast agent. *J. Am. Chem. Soc.*, **125**, 5471–5478 (2003).

[26] S. Stevenson, G. Rice, T. Glass, K. Harlch, F. Cromer, M.R. Jordan, J. Craft, E. Hadju, R. Bible, M.M. Olmstead, K. Maltra, A.J. Fisher, A.L. Balch, H.C. Dorn, Small-bandgap endohedral metallofullerenes in high yield and purity. *Nature*, **401**, 55–57 (1999).

[27] L. Dunsch, S. Yang, Metal nitride cluster fullerenes: their current state and future prospects. *Small*, **3**, 1298–1320 (2007).

[28] R.P. Fatouros, F.D. Corwin, Z.J. Chen, B.C. William, J.L. Tatum, B. Kettenmann, Z. Ge, H.W. Gibson, J.L. Russ, A.P. Leonard, J.C. Duchamp, H.C. Dorn, In vitro and in vivo imaging studies of a new endohedral metallofullerene nanoparticle. *Radiology*, **240**, 756–764 (2006).

[29] C.Y. Shu, E.Y. Zhang, J.F. Xiang, C.F. Zhu, C.R. Wang, X.L. Pei, H.B. Han, Aggregation studies of the water-soluble gadofullerene magnetic resonance imaging contrast agent: $[Gd@C_{82}O_6(OH)_{16}(NHCH_2CH_2COOH)_8]_x$. *J. Phys. Chem. B*, **110**, 15597–15601 (2006).

[30] E.Y. Zhang, C.Y. Shu, L. Feng, C.R. Wang, Preparation and characterization of two new water-soluble endohedral metallofullerenes as magnetic resonance imaging contrast agents. *J. Phys. Chem. B*, **111**, 14223–14226 (2007).

[31] K. Micskei, L. Helm, E. Brucher, A.E. Merbach, Oxygen-17 NMR study of water exchange on gadolinium polyaminopolyacetates [Gd(DTPA)(H2O)]2- and [Gd(DOTA)(H2O)]- related to NMR imaging. *Inorg. Chem.*, **32**, 3844–3850 (1993).

[32] D.M. Guldi, H. Hungerbuehler, K.D. Asmus, Unusual redox behavior of a water soluble malonic acid derivative of C60: evidence for possible cluster formation. *J. Phys. Chem.*, **99**, 13487–13493 (1995).

[33] B. Sitharaman, R.D. Bolskar, I. Rusakova, L.J. Wilson, $Gd@C_{60}[C(COOH)_2]_{10}$ and $Gd@C_{60}(OH)_x$: Nanoscale aggregation studies of two metallofullerene MRI contrast agents in aqueous solution. *Nano Lett.*, **4**, 2373–2378 (2004).

[34] H. Kato, Y. Kanazawa, M. Okumura, A. Taninaka, T. Yokawa, H. Shinohara, Lanthanoid endohedral metallofullerenols for MRI contrast agents. *J. Am. Chem. Soc.*, **125**, 4391–4397 (2003).

[35] I.F. Tannock, D. Rotin, Acid pH in tumors and its potential for therapeutic exploitation. *Cancer Res*, **49**, 4373–4384 (1989).

[36] S. Aime, S.G. Crich, M. Botta, G. Giovenzana, G. Palmisano, M. Sisti, A macromolecular Gd (III) complex as pH-responsive relaxometric probe for MRI applications. *Chem. Commun.*, **16**, 1577–1578 (1999).

[37] S. Laus, B. Sitharaman, E. Toth, R.D. Bolskar, L. Helm, S. Asokan, M.S. Wong, L.J. Wilson, A.E. Merbach, Destroying gadofullerene aggregates by salt addition in aqueous solution of Gd@C60 (OH)x and Gd@C60[C(COOH2)]10. *J. Am. Chem. Soc.*, **127**, 9368–9369 (2005).

[38] S. Laus, B. Sitharaman, E. Toth, R.D. Bolskar, L. Helm, L.J. Wilson, A.E. Merbach, Understanding paramagnetic relaxation phenomena for water-soluble gadofullerenes. *J. Phys. Chem. C*, **111**, 5633–5639 (2007).

[39] C.J. Gannon, P. Cherukuri, B.I. Yakobson, L. Cognet, J.S. Kanzius, C. Kittrell, R.B. Weisman, M. Pasquali, H.K. Schmidt, R.E. Smalley, S.A. Curley, Carbon nanotube-enhanced thermal destruction of cancer cells in a noninvasive radiofrequency field. *Cancer*, **110**, 2654–2655 (2007).

[40] C. Zavaleta, A. de la Zerda, Z. Liu, S. Keren, Z. Cheng, M. Schipper, X. Chen, H. Dai, S.S. Gambhir, Noninvasive Raman spectroscopy in living mice for evaluation of tumor targeting with carbon nanotubes. *Nano Lett.*, **8**, 2800–2805 (2008).

[41] B. Sitharaman, L.J. Wilson, Gadofullerenes and gadonanotubes: a new paradigm for high-performance magnetic resonance imaging contrast agent probes. *J. Biomed. Nanotechnol.*, **3**, 342–352 (2007).

[42] B. Sitharaman, K.R. Kissell, K.B. Hartman, L.A. Tran, A. Baikalov, I. Rusakova, Y. Sun, H.A. Khant, S.J. Ludtke, W. Chiu, S. Laus, E. Toth, L. Helm, A.E. Merbach, L.J. Wilson, Superparamagnetic gadonanotubes are high-performance MRI contrast agents. *Chem. Commun.*, **31**, 3915–3917 (2005).

[43] Z. Gu, H. Peng, R.H. Hauge, R.E. Smalley, J.L. Margrave, Cutting single-wall carbon nanotubes through fluorination. *Nano Lett.*, **2**, 1009–1013 (2002).
[44] A. Striolo, The Mechanism of Water Diffusion in Narrow Carbon Nanotubes. *Nano Lett.*, **6**, 633–639 (2006).
[45] K.B. Hartman, S. Laus, R.D. Bolskar, R. Muthupillai, L. Helm, E. Toth, A.E. Merbach, L.J. Wilson, Gadonanotubes as Ultrasensitive pH-Smart Probes for Magnetic Resonance Imaging. *Nano Lett.*, **8**, 415–419 (2008).
[46] T.H. Cheng, T.W. Lee, J.S. Jeng, C.M. Wu, G.C. Liu, M.Y.N. Chiang, Y.M. Wang, Synthesis and characterization of a novel paramagnetic macromolecular complex [Gd(TTDASQ-protamine)]. *Dalton Trans.*, **43**, 5149–5155 (2006).

12

Chemistry of Soluble Carbon Nanotubes: Fundamentals and Applications

Tsuyohiko Fujigaya[a] *and Naotoshi Nakashima*[a,b,†]

[a]*Department of Applied Chemistry, Graduate School of Engineering, Kyushu University, Fukuoka, Japan*

[b]*Japan Science and Technology Agency, CREST, Tokyo, Japan*

12.1 Introduction

Carbon nanotubes (CNTs) are made of rolled-up graphene sheets with one-dimensional extended π-conjugated structures, discovered in 1991 by Iijima [1]. They are classified into mainly three types of CNTs in terms of the number of graphene layers within a CNTs, that is, single-walled carbon nanotubes (SWNTs), double-walled carbon nanotubes (DWNTs) and multi-walled carbon nanotubes (MWNTs), which have one, two and more than three walls, respectively (Figure 12.1). CNTs have been central materials in the field of nanomaterials science and nanotechnology because of their remarkable electronic, mechanical and thermal properties that far exceed existing materials. Theoretical and experimental values of CNTs' physical properties are summarized in Table 12.1. It is noted that the experimental values vary paper by paper, which are mainly caused by the difference of CNT purity as well as measurement methods. The purity of CNTs has been improved as the production method sophisticated and further improvements of the values would be expected. One of the key issues in the utilization of such a seminal materials for basic researches together with the

[†]Corresponding author: N. Nakashima, email: nakashima-tcm@mail.cstm.kyushu-u.ac.jp

Chemistry of Nanocarbons Edited by Takeshi Akasaka, Fred Wudl and Shigeru Nagase

Table 12.1 *List of physical properties of SWNTs, MWNTs and metals as reference materials*

	SWNTs	MWNTs	Metals
Tensile strength	10–100 [GPa] [10]	11–63 [GPa] [11]	1.3 [GPa] (steel)
Young modulus	0.6–3.4 [TPa] [12, 13]	0.3–1.3 [TPa] [11, 14, 15]	0.2 [TPa] (steel)
Current density	10^9 [A/cm^2] [16]	10^9 [A/cm^2] [17]	10^6 [A/cm^2] (cupper)
Thermal conductivity	3500 [W/mK] [18]	3000 ~ [W/mK] [19]	420 [W/mK] (silver)

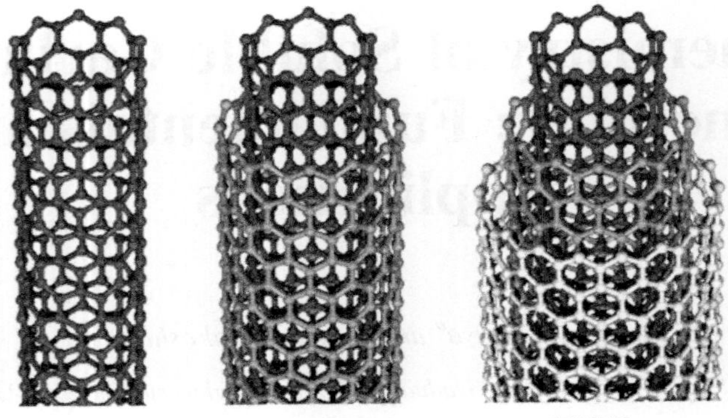

Figure 12.1 *Structures of SWNTs, DWNTs, and MWNTs*

material applications is to develop a methodology to solubilize/disperse them in solvents (Figure 12.2) [2–4] since as-synthesized CNTs form tight bundled structures [5] due to their strong van der Waals interaction (0.9 eV/nm) [6]. Solubilization/dispersion techniques can be categorized mainly into two methods, namely 'chemical' and 'physical' modification. Solubilization/dispersion of CNTs based on physical adsorption of dispersant molecules possesses several advantages such as the ease of preparation process and maintaining intrinsic CNT properties, which show sharp contrast with the chemical modification [7–9].

In this chapter, general strategies for CNT solubilization as well as the applications of solubilized CNTs are described.

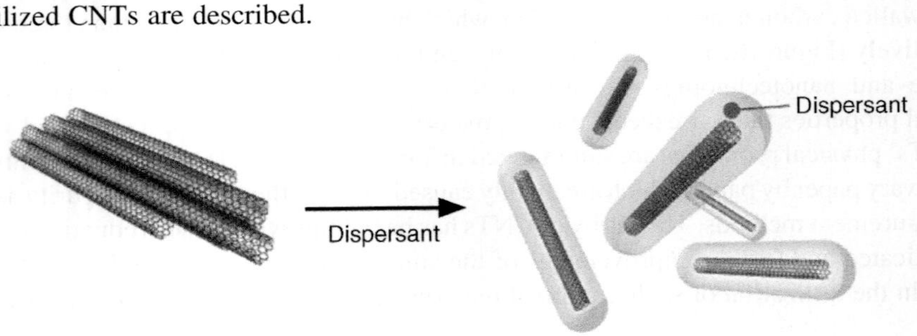

Figure 12.2 *Schematic illustration of the solubilization of CNTs through physical adsorption of the dispersant molecules on the surfaces on CNTs*

12.2 Characterizations of Dispersion States

The typical procedures for the preparation of individual solution of CNTs are ultrasonication of the CNTs in a dispersant solution, followed by centrifugation to give grey-transparent supernatant solutions. Individually solubilized/dispersed CNTs are often visualized by atomic force microscope (AFM) [20, 21] and transmission electron microscope (TEM) [22, 23] after casting on substrates. Near-IR (NIR) absorption and photoluminescence (PL) spectroscopy are strong tools for the direct observation of the dispersion nature of SWNTs associated with the allowed transition (van Hove transition) of SWNTs. Wiseman *et al.* [24, 25] have found that PL in the NIR region can be detected from surfactant-dissolved SWNTs and then have succeeded in the determination of the SWNT chirality indices (n,m) in the solution. Notably only individually dissolved semiconducting-SWNTs, and small bundled SWNTs in some case [26], exhibit PL because the bundled SWNTs end up with quenching their PL by the metallic-SWNTs in the bundle. Thus the PL observation can serve as a good indicator of the individual solubilization of SWNTs in solution as well as in films [27, 28]. Recent advance in the NIR detection technique allows to see PL from a single SWNT swimming in the solution and gel by combining an inversed microscope technique [29–31]. Small angle neutron scattering (SANS) technique is even more powerful method to figure out the degree of dispersion states together with the wrapping structures by dispersants [23, 32–37]. Even in the absence of PL signals due to the bundling, UV-visible NIR absorption spectroscopy is quite helpful for roughly evaluating the degree of bundling of SWNTs both in the solution and film states [38]. SWNT bundling renders the red-shifted and broadened features of the absorption spectra compared to that of the isolated SWNTs [39, 40]. In some cases, the electron conductivity measurement allows brief estimation of dispersion of the SWNTs in polymer films by evaluating the concentration of the SWNTs at the electron percolation threshold [41, 42]. A lower threshold concentration is a consequence of greater dispersion of the SWNTs in the matrices.

12.3 CNT Solubilization by Small Molecules

12.3.1 Surfactants

The most convenient and frequently-used dispersant for CNTs in aqueous media is surfactants such as sodium dodecyl sulfate (SDS) [43–45], sodium dodecylbenzene sulfonate (SDBS) [22, 46–49], cethyltrimethylammonium bromide (CTAB) [22, 50], Brij [22, 51], Tween [22, 51], and Triton X (Figure 12.3) [22, 46, 51, 52]. An early attempt in preparing CNT dispersions using a surfactant was explored by Bandow *et al.* for the purification of SWNTs from carbon soot material [53]. The suggested mechanism of the individual dispersion is the encapsulation of SWNTs in the hydrophobic interiors of the micelles, which results in the formation of a stable dispersion [39]. Among the conventional surfactants, SDBS is one of the most efficient SWNT solubilizer, that is, it has been reported that even in the concentration of 20 mg/mL of SWNTs in an SDBS micelle, no aggregation of the SWNTs occurs for more than 3 months [46].

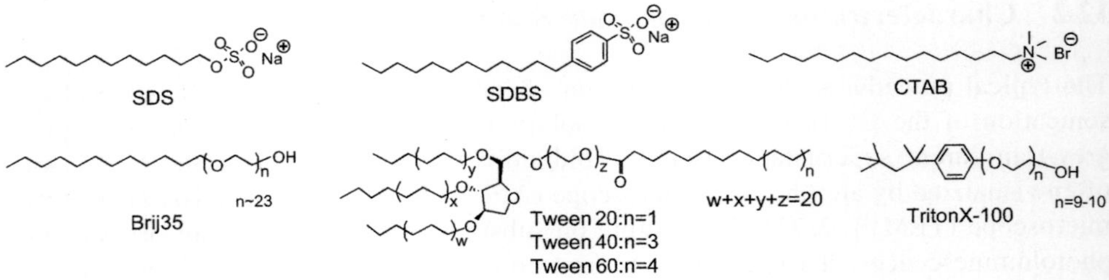

Figure 12.3 *Chemical structures of surfactants for CNTs solubilization*

Biological surfactants such as bile salts are act as SWNT solubilizers in water (Figure 12.4) [51, 54, 55]. Among the biological surfactants, micelles of anionic biosurfactants including sodium cholate (SC), sodium deoxycholate (SDC), sodium taurocholate (STC), sodium taurodeoxycholate (STDC), sodium glycocholate (SGC) *N,N*-bis(3-D-gluconamidopropyl) cholamide (BIGCHAP) and *N,N*-bis(3-D-gluconamidopropyl) deoxycholamide (deoxy-BIGCHAP) possess high solubilization ability, in which the PL in the NIR region guarantees the individual solubilization. The PL spectral analysis has revealed that the chiral indices of the SWNTs solubilized by the biosurfactants depend on the chemical structures of surfactants [55]. Simple dialysis for CNT solutions in surfactant aqueous solutions results in the flocculation of the CNTs [55]. This fact implies the surfactant molecules dynamically replace between the surface of CNTs and the bulk solution [56]. Assembled structures of surfactants on the CNT sidewalls are still under discussion and

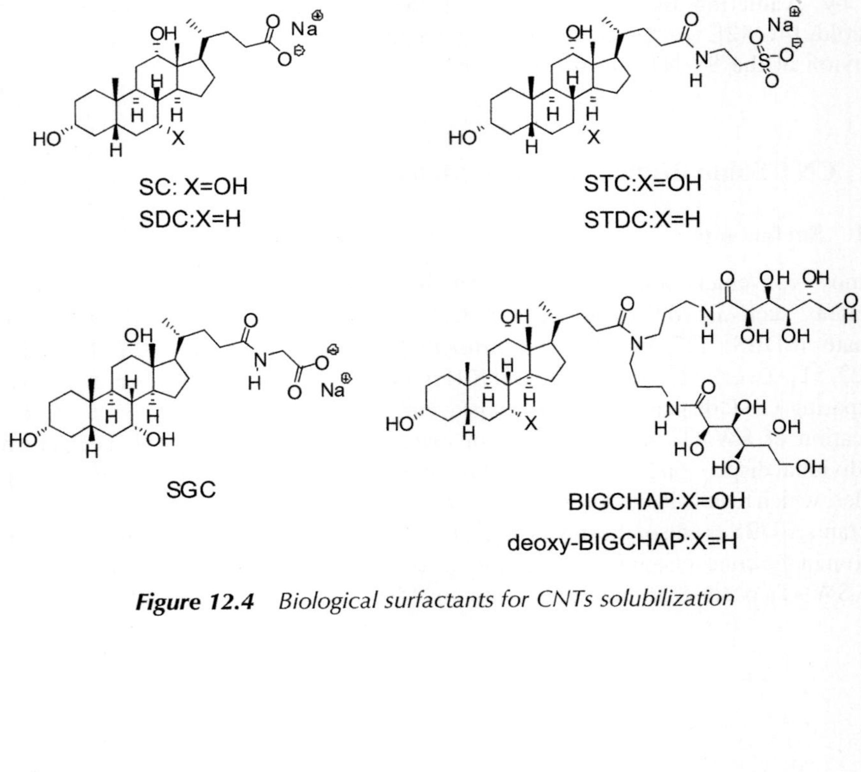

Figure 12.4 *Biological surfactants for CNTs solubilization*

several models such as cylinder, random, and hemimicelle type of structures have been proposed [36, 40, 43, 57, 58].

Chirality recognition of SWNTs with the aid of surfactant molecules is of interst [59, 60]. Niyogi *et al.* successfully separated larger diameter of SWNTs as a precipitate by the addition of salts into a SDS-dispersed SWNT solution [61]. They suggest tight assembling of SDS molecules onto the smaller diameter of the SWNTs leads such a difference in solubility. In 2006, almost perfect separation of metallic and semiconducting SWNTs has been achieved by using an aqueous SWNT solution of SC/SDS mixture in conjunction with the density-gradient ultracentrifugation technique [62]. The idea takes advantage of the small difference of the surfactant assembling density onto the different types of SWNTs, which enhance the difference in the gravity on each SWNTs [60]. Furthermore, the addition of SDC is found to realize the separation of high-pure metallic SWNTs [63]. It is very fascinating to recognize the SWNTs can 'feel' tiny structural differences in the surfactant molecules and their molecular assembly. As stated above, individual dispersion is a quite essential step for the solution-based separation and purification of SWNTs.

12.3.2 Aromatic Compounds

12.3.2.1 Polycyclic Aromatic Compounds

The surfaces of CNTs can be readily functionalized through π–π interactions with compounds having π-electron-rich structures due to the highly delocalized π-electrons of CNTs. The π–π interaction between polycyclic aromatic compounds and CNT sidewalls has been discussed based on both theoretical [64] and experimental [65] approaches. We reported that a pyrene-based ammonium salt (compound **1** in Table 12.2) is able to solubilize SWNTs [66] and fullerene-filled CNTs (so-called peapods) [67] in water. The pyrene-carrying compound acts as an efficient dispersant compared to naphthyl- and phenyl-based ammonium salts [68]. This is due to the strong binding affinity between the pyrene group and the CNT sidewalls. Now pyrene derivatives have been widely recognized as excellent solubilizers for CNTs as summarized in Table 12.2 [66–74]. By taking advantage of the efficient adsorbing capability on the CNT surfaces, pyrene derivatives have been used as decent interlinkers to anchor functional materials that can communicate with CNTs as summarized in Table 12.3 [75–94]. Pioneering work demonstrating that the pyrene derivative functioned as an interlinker was carried out by Dai *et al.* [75] They successfully attached a protein on the surface of the SWNTs with the aid of a pyrene-carrying succinidyl compound (compound **8** in Table 12.3). Pyrene-ammonium **1** was also used in many researches for anchoring anionic functional molecules on the surface of SWNTs and MWNTs [80, 81, 95]. Other polycyclic aromatic moieties such as anthracene [96, 97], terphenyl [97, 98], perylene [99], triphenylene [100], phenanthrene [101], and pentacene [102] also have affinity for the sidewalls of CNTs and various solubilizers bearing these molecules have been developed. Green tea solution also acts as excellent SWNT dispersant [103]. Our dissolution scenario is that catechin, a polycyclic aromatic compound, mainly contributes to the dispersion because epigallocatechin gallate also disperses the SWNTs in water.

The degree of the interactions between these polycyclic aromatic moieties and the CNT sidewalls has been accessed based on Raman spectroscopic analysis by monitoring the shift of radial breathing mode (RBM) [102]. HPLC technique using CNTs as a stationary phase is

Table 12.2 *Pyrene-based dispersants for CNTs*

	Pyrene	Research targets	Ref.
1	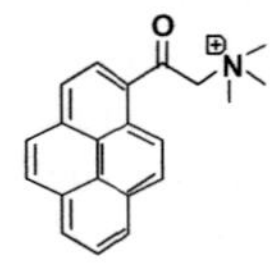	Long-lived charge separation between porphyrin and SWNTs Solubilization of C70@SWNTs into aqueous system	[66, 68] [67]
2	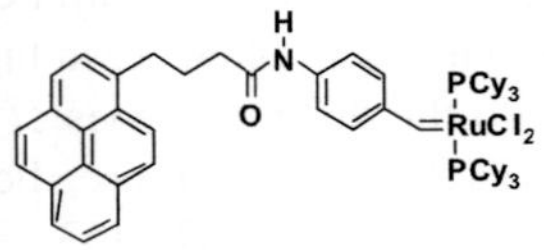	ROMP on SWNTs	[69]
3	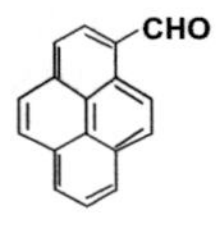	Functionalization of MWNTs in supercritical fluids	[70]
4	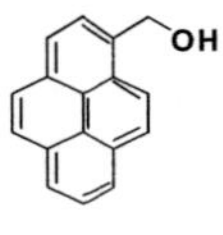	Functionalization of MWNTs in supercritical fluids	[70]
5	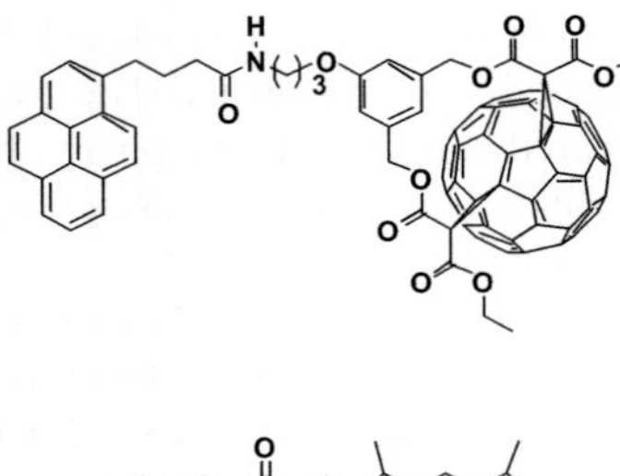	Electrochemical responce of fullerene/SWNTs hybrid	[71]
6	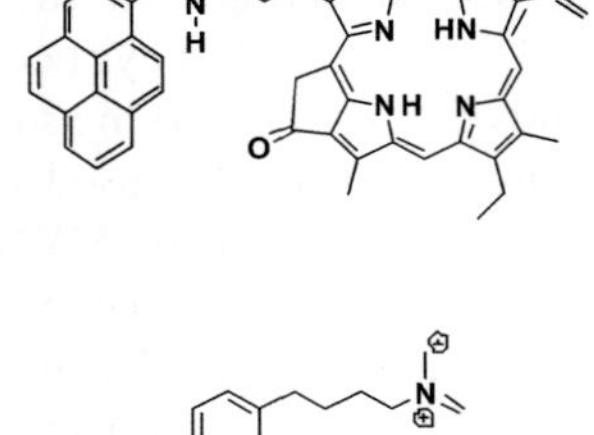	Attachment of pyrene-modified chlorophyll derivative	[72]
7	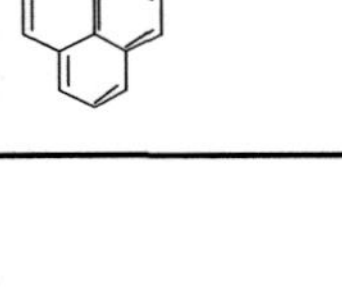	Selective fictionalization of SWNTs	[73, 74]

Table 12.3 *Pyrene-based interlinkers for immobilizing functional molecules on the CNTs surface*

	Pyrenes	Research target	Ref.
8	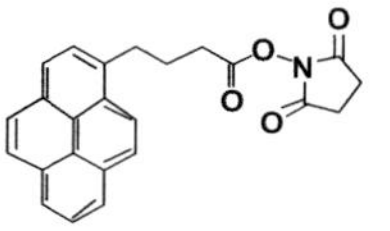	Immobilization of protein onto SWNTs surface	[75]
9	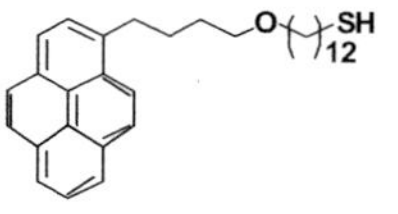	Au nanoparticle immobilization on MWNTs	[76]
10	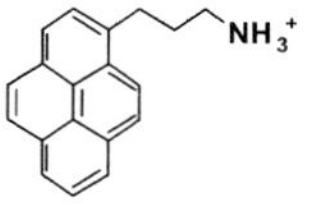	Layer-by-layer assembly of SWNTs with polyanion	[77]
11	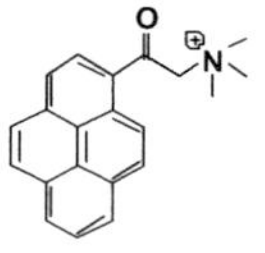	Photoinduced electron transfer from porphyrin to SWNTs	[78–81]
		Photoinduced electron transfer between CdTe and MWNTs	[82]
		Photoinduced electron transfer from polythiophene	[83]
12	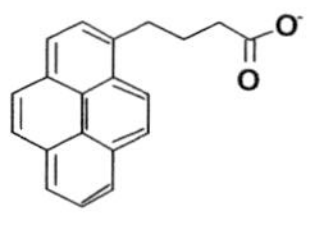	Photoinduced electron transfer from porphyrin to SWNTs	[84]
13	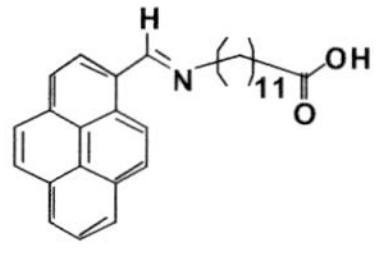	Immobilization of magnetic particle	[85]
14	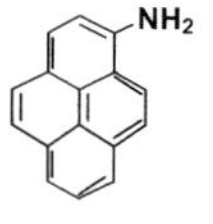	Immobilization of metal nanopartices onto SWNTs	[86]
15	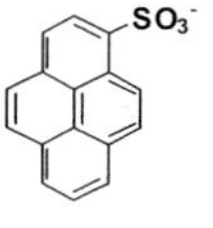	Enhancing of bioelectrocatalyzed oxidation	[87]

(continued)

Table 12.3 *(Continued)*

	Pyrenes	Research target	Ref.
16	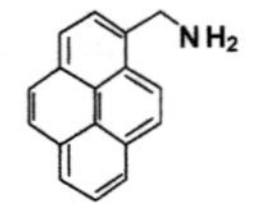	Immobilization of Au nanoparticle onto MWNTs surface	[88]
17	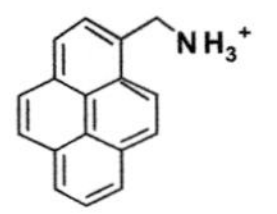	Layer-by-layer assembly of polyelectrolytes	[89]
		DNA adsorption and gene transcription	[289]
18	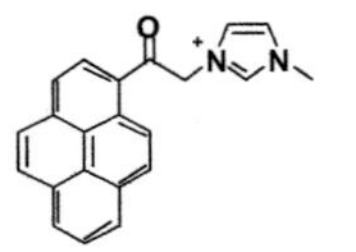	Layer-by-layer assembly of polyelectrolytes	[89]
19	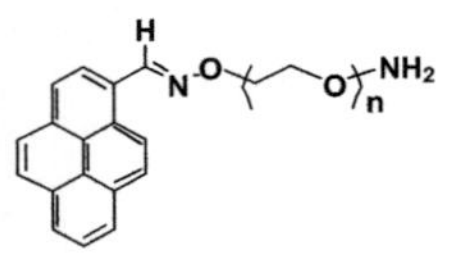	Immobilization of tabacco mosaic virus onto SWNTs	[90]
20	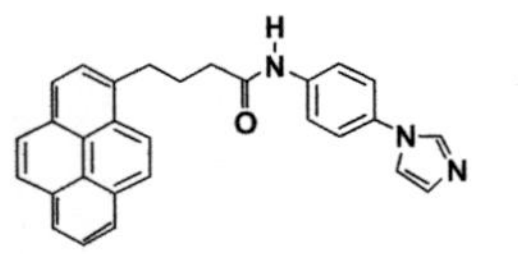	Photoinduced electron transfer from naphthalocyanine to SWNTs	[91]
21	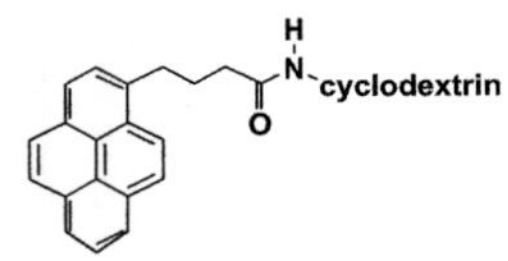	Modification of cyclodextrin on SWNTs surface	[92]
22	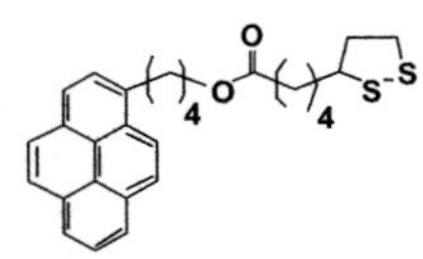	CdSe immobilization	[93]
23	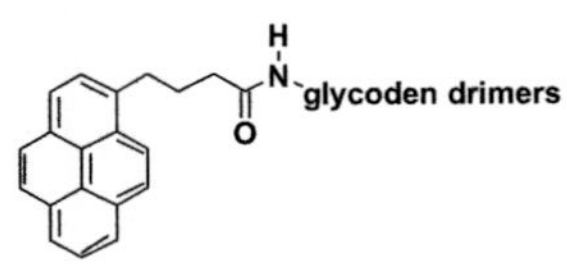	Binding to cell surface	[94]

a powerful tool to rank the affinity of the several dispersants on the surface of the CNTs at one time [104, 105]. Chang *et al.* evaluated the degree of affinity on the surface of SWNTs using a HPLC technique and found the polycyclic aromatic system shows better interaction than monocyclic compound [104] in accordance with several papers [68, 97]. The difference in affinity worked for CNTs is fascinating in the view of chirality recognition and enrichment of specific SWNTs [106].

12.3.2.2 Porphyrins

Porphyrin compounds are able to individually solubilize SWNTs [107]. Zinc protoporphyrin IX (compound **24** in Table 12.4) was used and found that a resulting **24**/SWNT solution is stable even after 6 months. Fluorescence quenching of the porphyrin in the **24**/SWNT evidenced the adsorption of the porphyrin onto the surfaces of the SWNTs. A series of porphyrin derivatives were tested and revealed that a wide range of porphyrin derivatives including **25** and **26** (Table 12.4) can also act as effective dispersants for SWNTs [108, 109]. The finding lead the theoretical as well as experimental attempts to understand the interaction between porphyrins and CNTs [110–112]. Importantly, not only the π–π interaction but also charge-transfer interaction have been pointed out to serve the adsorption of porphyrin derivatives on the surfaces of the CNTs [112]. The center metals in the porphyrin affect the degree of solubilization of the CNTs [112, 113]. Porphyrin derivatives [113–121] as well as their analog molecules such as phthalocyanines [122–124] and sapphyrin[125] have been reported to serve as dispersants for CNTs.

The combination of porphyrin and CNTs has attracted extensive interest due to their unique photophysical [78, 126, 127], electrochemical [128–130], electronic [118, 126, 131, 132], and optical [133, 134] properties of the composites. Extensive efforts have been carried out on the photoinduced electron transfer from porphyrins to CNTs including not only for physically connected but covalently bonded porphyrin/CNT hybrids [126]. Dye sensitize organic solar cell is emerged as a potential application for porphyrin/CNTs hybrids. On the other hand, similar to the other solubilizers, a porphyrin compound was used for a separation media for CNTs based on molecular recognition [113]. One of the striking results realized in these studies is the separation of optical active SWNTs reported by Osuka *et al.* [135]. They found that optically active porphyrin dimers (compound **41** in Table 12.4) can pick up SWNTs with right- or left-handed helicity structure from racemic SWNT mixtures depending on the chirality of compound 41. Closer look of the system additionally revealed compound 41 also recognized and enriched the specific diameter of the SWNTs [136].

12.4 Solubilization by Polymers

12.4.1 Vinyl Polymers

Commercially available poly(styrene sulfonate) [57], poly(vinyl alcohol) [137], poly (vinylpyrrolidone) [57] enable CNTs being dispersible in solution through polymer wrapping. In the case of polybutadiene, polyisoprene, polystyrene, poly(methyl methacrylate), and poly(ethylene oxide) [138], the importance of the CH-π interaction was pointed out for the dispersion mechanism. On the other hand, introduction of a π–π interaction is the

Table 12.4 Porphyrin-based solubilizers for CNTs

	Prophyrin	M	R	Topic	Ref
24	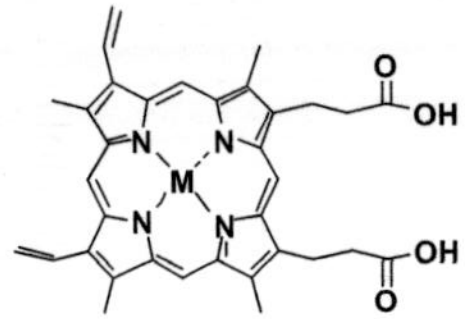	Zn		First porhyrin-based solubilizer Solubilization of SWNTs	[107]
25		2H			[108]
26		FeCl		Electrochemical response	[108]
27	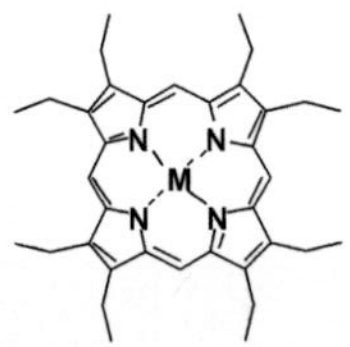	Co		EPR study	[109]
28		2H	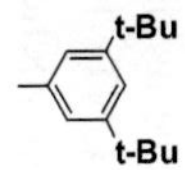	Photoinduced charge Injection	[126]
29	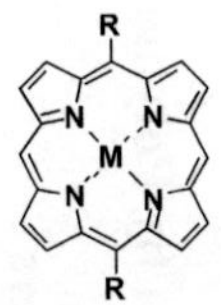	2H	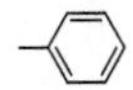	Porphyrin driven supramolecular assembly	[119]
30		Zn	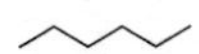	Direct observation of adsorved porphyrin	[131]
31		2H	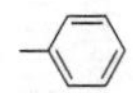	Photoinduced charge injection	[126]
32		2H, Zn	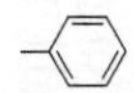	Nonlinear optical properties	[134]

33		2H, Zn	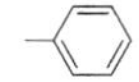	Interaction study	[112]
34		2H	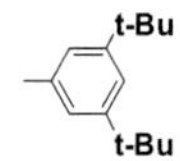	Photoinduced charge injection from excited porphyrin into SWNTs	[126]
35		$4H^{2+}$	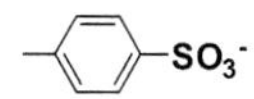	Solubilization of SWNTs in water	[114]
36		2H, Zn	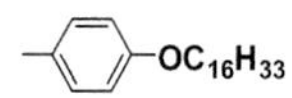	Separation of semiconducting SWNTs	[113]
37		2H, Zn	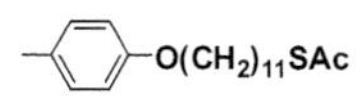	J-aggregation on the SWNTs surface	[290]
38	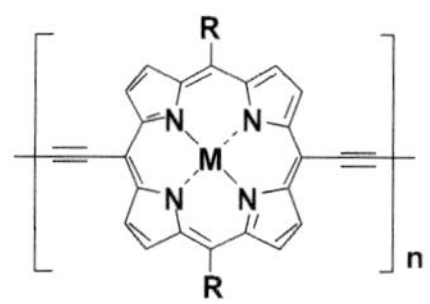	Zn	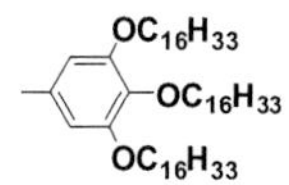	Solubilization by conjugated porphyrin polymer	[116]
39	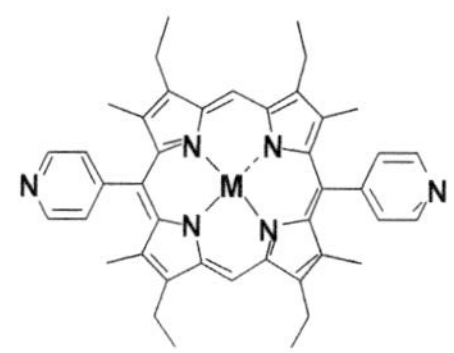	Zn		Supramolecular solubilization	[115]
40	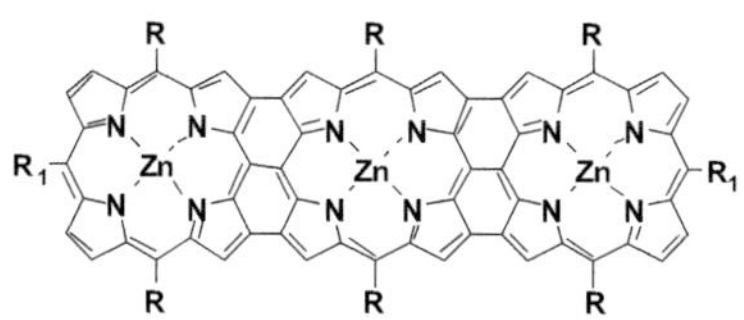		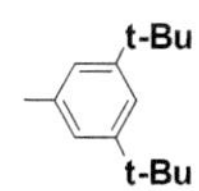 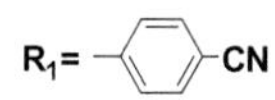	Solubilization by fully-fused porphyrin	[120]

(continued)

Table 12.4 *(Continued)*

	Prophyrin	M	R	Topic	Ref
41	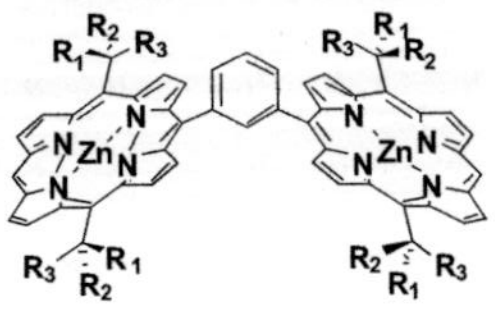	(R) : R_1=CH_2Ph, R_2=H, R3=$NHCO_2$t-Bu (S) : R_1=H, R_2=CH_2Ph, R3=$NHCO_2$t-Bu		Separation of optically active SWNTs	[135]
42	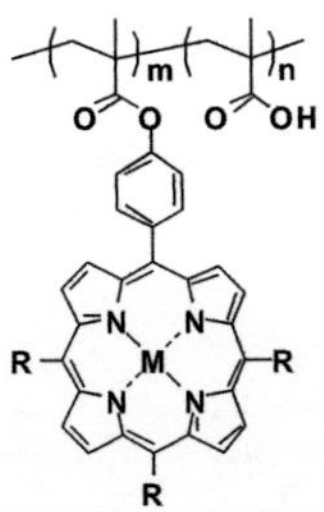	2H	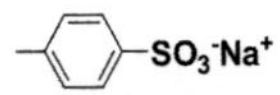	Long-lived charge separation	[121]

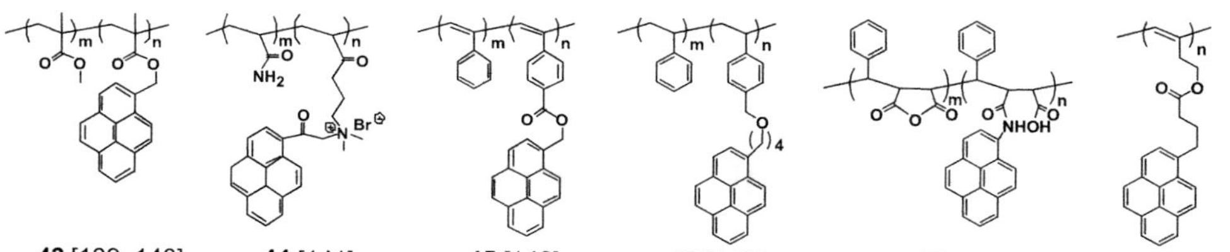

43 [139, 140] **44** [141] **45** [142] **46** [143] **47** [144] **48** [145]

Figure 12.5 *Pyrene-pendanted vinyl polymers for CNTs solubilization*

reliable strategy for CNT dissolution by vinyl polymers. For example, pyrene moieties are often employed in the polymer structures as a pendant group to offer better dispersion states of CNTs (Figure 12.5) [139–146]. Porphyrin and anthracene moieties are also employed as pendant groups as well [121, 147]. In our pyrene-carrying copolymer (compound **44** in Figure 12.5), no SWNT precipitate was observed on heating up to 95 °C in an aqueous solution, while dispersion in the corresponding monomer produced a precipitate around 50 °C [141]. This result clearly indicates the one of the general advantage on the thermodynamical stability of polymer-based solubilizers when it compared to the monomeric type.

12.4.2 Conducting Polymers

Pioneer works on CNT dissolution by polymer were realized with the *p*-phenylenevinylene derivatives (PPVs) [148–150]. Large π-conjugation on the PPVs may play a crucial role to interact with CNT surfaces. Unique opto-electronic properties of PPVs have been fascinating the PPV/CNT complex as a key component for the organic electronic application such as EL and solar cell [151–167]. In addition, the structural rigidity of the PPVs provide a unique opportunity to disperse the SWNTs with specific chiral indices by aligning their backbones along the SWNT surfaces in order to maximize the interaction between polymer and CNTs [168]. Coleman *et al.* used the conjugated polymer poly(*m*-phenylene-co-2,5-dioctoxy-*p*-phenylenevinylene) (PmPV) to preferentially disperse SWNTs with specific chiral indices leaving the others in the precipitate [169–172]. Recently, two different groups reported impressive works, in which they described that the poly(9,9-dioctylfluorenyl-2,7-diyl) (**PFO**) and poly[(9,9-dioctylfluorenyl-2,7-diyl)-alt-co-(1,4-benzo-2,1,3-thiadiazole)] (**PFO-BT**) (Figure 12.6) have strong selectivity to enrich the SWNTs with a single chirality index [173, 174]. In there, **PFO** selectively wraps only SWNTs with high chiral angles (close to an armchair configuration) and **PFO-BT** preferentially wraps the SWNTs with a diameter of around 1.05 nm. Further understanding of the interaction mechanism could

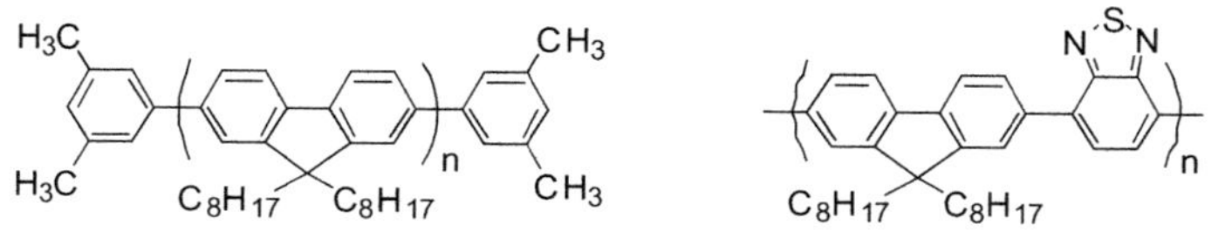

Figure 12.6 *Chemical structures of PFO (left) and PFO-BT (right)*

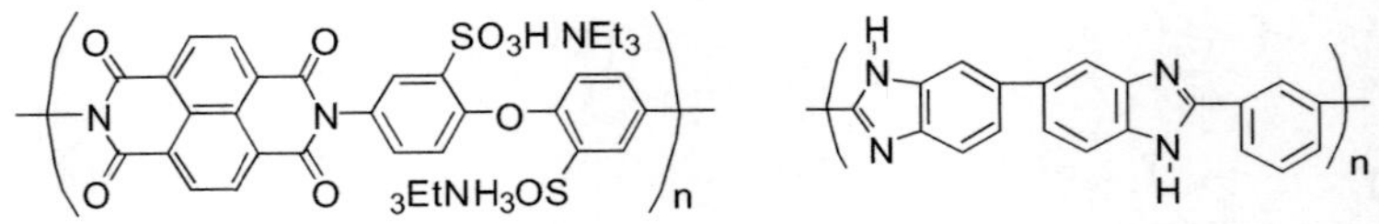

Figure 12.7 Chemical structure of PI-1 (left) and PBI (right)

lead to a strategic approach for extracting a single chirality index at one's request by the properly-designed polymers.

12.4.3 Condensation Polymers

Many papers describing the nanocomposite formation of CNTs with condensation polymers such as polyesters and polyamides have been published [175–179]. Most of them were prepared by melt mixing, polymer grafting and in situ polymerization methods by using oxidized CNTs due to the ease of sample preparation. We have reported an extremely efficient individual dissolution of SWNTs by a totally aromatic polyimide (PI-1 in Figure 12.7) [180]. As much as 2.0 mg/mL of the SWNTs is individually dissolved in the 1.0 mg/mL DMSO solution of **PI-1**. The major driving force for the solubilization of SWNTs is attributed to a π–π interaction between the condensed aromatic moieties on the polyimide and the surfaces of SWNTs. Generally speaking, the composite films consist from the individual dispersion of CNTs would maximize the performances of the materials such as mechanical properties with minimum addition of the CNTs. For this reason, the precise analysis of the degree of dispersion will become a strong focus of interest also for other polyimide/CNTs [181–191] since polyimides are widely known to possess an excellent mechanical strength and heat resistance [192].

Polybenzimidazole (**PBI** in Figure 12.7) is also recognized as a highly thermal stable polymer and widely used for firefighter's protective clothing, high-temperature gloves, and astronaut flight suits [193]. Different from the typical aromatic polyimides, **PBI** is soluble in common organic solvents such as DMAc, DMSO and DMF. We have reported that the **PBI** acts as a good dispersant for SWNTs due to the π–π interaction between the polymer and SWNT sidewalls. The vis-NIR absorption and PL spectra of a **PBI**/SWNT solution in DMAc clearly show the characteristic absorption peaks and strong PL spots, respectively, derived from the individual SWNTs (Figure 12.8). Effective dispersion of the SWNTs in the matrix **PBI** results in the dramatic reinforcement in the composite film. We have found that the addition of very small amounts of SWNTs (0.06 wt%) reinforces the mechanical properties of the original polymer by ca. 150 % without reducing their thermal stabilities.

12.4.4 Block Copolymers

The amphiphilicity of polymers is important for the dispersion of CNTs through a micelle-encapsulation mechanism. Taton and co-worker found that the micelle formation of polystyrene-*b*-poly(acrylic acid) in a DMF solution induced by water addition encapsulated the SWNTs to give the dissolution of SWNTs [194]. Up to date, wide range of block copolymers have been reported to disperse CNTs through the micelle encapsulation mechanism, especially polystyrene (PS) containing copolymers, such as polystyrene-*b*-poly(methacrylic acid) (PS-PMAA) [195], polystyrene-*b*-polybutadiene-*b*-polystyrene

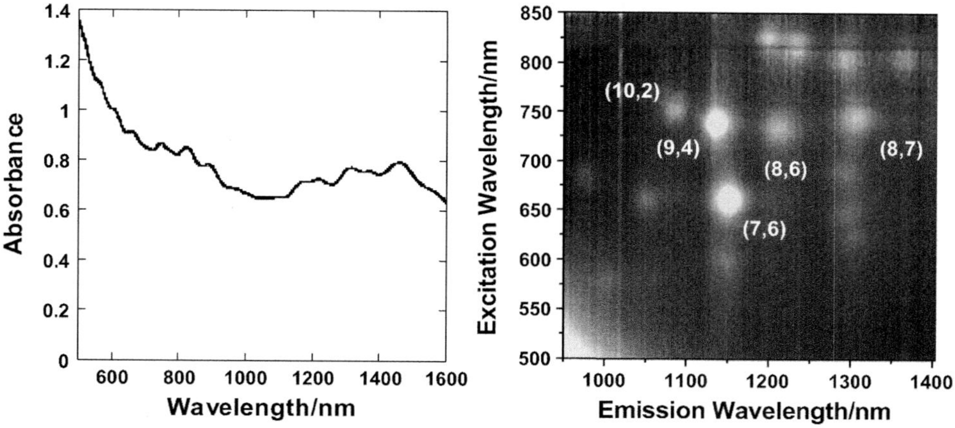

Figure 12.8 *Absorption and photoluminescence spectra of PBI/SWNT composite solution in DMAc*

(PS-PBD-PS) [194, 196, 197], polystyrene-*b*-poly(ethylene oxide) (PS-PEO) [198], polystyrene-*b*-poly(*tert*-butyl acrylate) (PS-PBA) [199], polystyrene-*b*-polyisoprene (PS-PI) [200], polystyrene-*b*-poly(4-vinylpyridine) (PS-P4VP) [201], polystyrene-*b*-poly-[sodium(2-sulfamate-3-carboxylate)isoprene] (PS-PSCI) [202, 203]. Polyethylene oxide blocks are also utilized as an effective segment for the CNTs dissolution in aqueous media, and numbers of block copolymers such as poly(ethylene oxide)-*b*-poly(propylene oxide) (PEO-PPO) [199], poly(methylmethacrylate)-*b*-poly(ethylene oxide) (PMMA-PEO) [198]. poly(ethyleneoxide)-*b*-poly(dimethylsiloxane)-*b*-poly(ethylene oxide) (PEO-PDMS-PEO) [195, 199], poly(ethylene oxide)-*b*-poly(propylene oxide)-*b*-poly(ethylene oxide) (PEO-PPO-PEO) [199, 204], poly(ethylene oxide)-*b*-poly[2-(*N*,*N*-dimethylamino)ethyl methacrylate] (PEO-PDEM) [205] have been reported as dispersants (Figure 12.9). An interesting example of the dispersion with the aid of block copolymers is the PS-P4VP dispersion reported by Shin *et al.*, in which the SWNTs are exfoliated both in polar and nonpolar solvents [201]. TEM observation revealed that the PS block contributes to the dissolution in toluene by exposing the segment outside, while the P4VP formed a shell in an ethanol solution, resulting in the stable dissolution of the SWNTs.

12.5 Nanotube/Polymer Hybrids and Composites

12.5.1 DNA/Nanotube Hybrids

CNTs have a high potential in the biological area since the outstanding findings of the DNA-assisted dissolution of SWNT in 2003 [206, 207]. Individual dissolution of SWNTs using double-strand DNA (dsDNA) and single-strand DNA (ssDNA) were reported from our group [206] and Zheng's group [207], respectively. Thereafter, ssRNA was also reported to dissolve SWNTs into aqueous media by the same procedure [208]. Together with the early attempts of direct observation of DNA adsorbed on the MWNTs [209, 210], unexpected combination of DNA and CNTs triggered vast number of researches to explore the novel

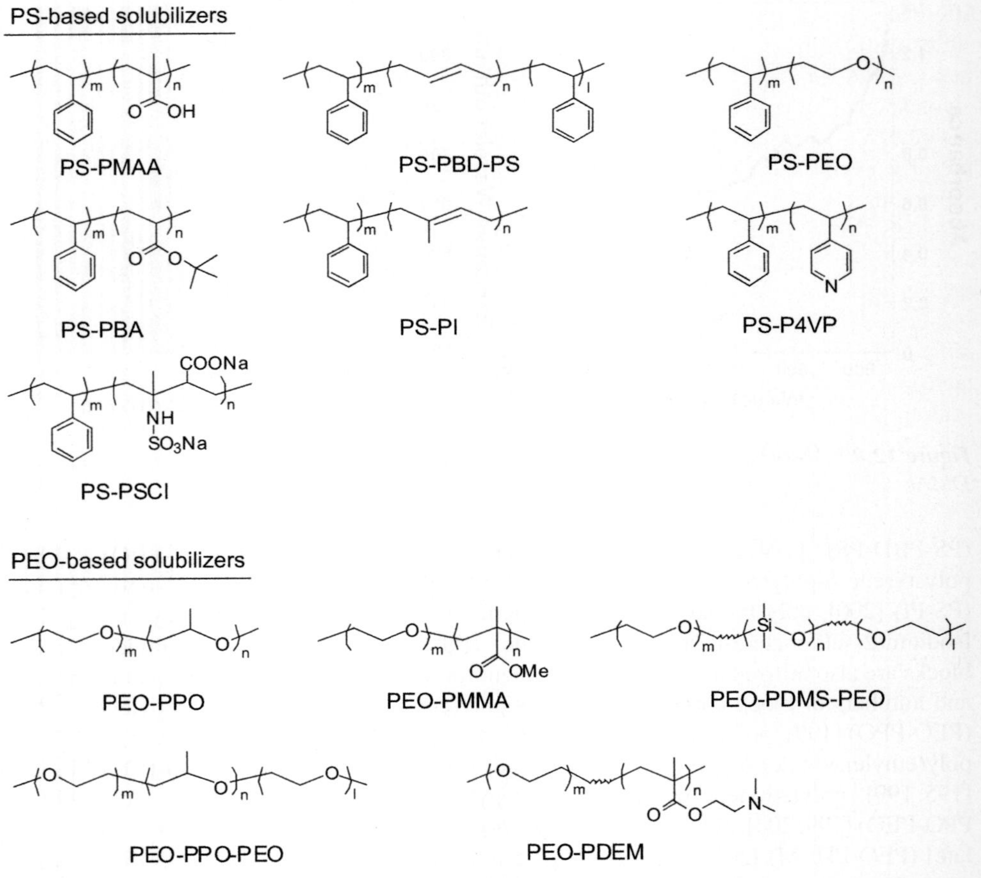

Figure 12.9 *List of block copolymer CNTs solubilizers*

applications of CNTs for biology. As for the mechanism of the individual dissolution of CNTs with ssDNA, DNA base stacking on the SWNT surfaces has been proposed in both experimentally [209–212] and theoretically [213–215]. SWNTs wrapped helically by the ssDNA have been observed by means of the AFM technique [216]. Direct interaction of DNA lead the strong dependency for the dissolution efficiency of SWNTs [207, 216–218]. Initial studies by Zheng *et al.* reported that ssDNA dissolution of CNTs are highly sequence dependant and poly-d(T) and $d(GT)_{10\text{-}45}$ provide a highest concentration of individual SWNTs aqueous solutions [207, 218].

On the other hand, detailed understanding of the dissolution mechanism of CNTs by dsDNA is still lacking. Wrapping mechanism by the denatured DNA generated on the surface of SWNTs is proposed [219], and this was supported by the HRTEM observation [220]. Dissolution efficiency of CNTs by DNA vary with the type of dsDNA and short dsDNA shows higher efficiency of dissolution than that of genomic long dsDNA [217]. Accurate understanding of the dissolution mechanism is necessary to explain this difference. It was noticed that SWNTs form relatively good dispersion and, without ultracentrifugation typically used to collect the individually isolated DNA/SWNT composites (>16000g) [207, 217, 221], the dispersion efficiency is so high to reach to exhibit a lyotropic LC phase in high concentration region [222]. The degree of debundling of dsDNA/SWNTs in solution prepared in such an ultracentrifugation-free procedure was carefully examined by utilizing AFM technique as a function of concentration. The dispersions are consisted from the mixture of small bundled and individually isolated dsDNA/SNWTs. Surprisingly, simple dilution of this dispersion gives the individually dispersed dsDNA/SWNT solutions [219]. This result provides a extremely simple and waste-minimized method to prepare individual dispersion of SWNTs.

Thermodynamical stability of the DNA/SWNT is another feature of the complex. By using a GPC technique, we have proved that the binding of dsDNA and SWNTs is highly stable, namely the detachment of the dsDNA from the surface of SWNTs is ignorable at least 1 month [223]. Thanks to the formation of a stable complex between the DNA and CNTs, wide range of researches have been achieved in view of the biological applications, such as the conformation transition monitoring of DNA [224], redox sensing of glucose and hydrogen peroxide [225], hybridization detection between ssDNA and their complimentary DNA [226], and uptake estimation of DNA/SWNTs into the cell [227]. As increased the possibility of DNA/CNTs as a gene delivery carrier, a strong demand to avoid the DNA damage during sonication arises. Modified DNA-wrapping protocol [228] using surfactant-dissolved SWNTs followed by the exchange of DNA in a dialysis membrane realized the sonication free process [225, 229].

As mentioned in the previous session (Section 12.3.), individual dispersion is an imperative step for the separation of SWNTs depending on the length and/or chirality of SWNTs. For this purpose, stable DNA/SWNT dispersions are quite suitable for the chromatography-based separation of SWNTs. Zheng *et al.* developed enrichment of SWNTs having specific chiral indices as well as the removal of free DNA by anion exchange chromatography for ssDNA/SWNTs [218, 230–232]. Furthermore, stable ssDNA/SWNT dispersions also enable the length sorting and removal of free DNA by the size-exclusion chromatography [233, 234]. Finally, their dedicated studies led the excellent result of chromatography-based enrichment of single chirality of SWNTs [230]. Especially, length separation of ssDNA/SWNTs composite is expected to provide a significant opportunity for the precise assessment of biological activity of the composites since the size effect is a general factor in nanomaterials for such cell uptake, retention, and distribution [235, 236]. Indeed, Becker *et al.* reported length dependent uptake of a ssDNA/SWNT composite in the cell [221].

12.5.2 Curable Monomers and Nanoimprinting

Heat- and photo-curable resins have been interesting as a promising matrix for the CNT composite owing to the several advantages.

1. Most of these monomers are viscose liquid and, principally, there are no need to add any solvents to obtain polymer/CNTs composites.
2. Mixing with the small monomers is expected to have lower entropic barriers to disperse compared to polymer melt mixing.
3. Quick solidification especially in photo-curable system can avoid the re-aggregation often occurring during solvent evaporation process.

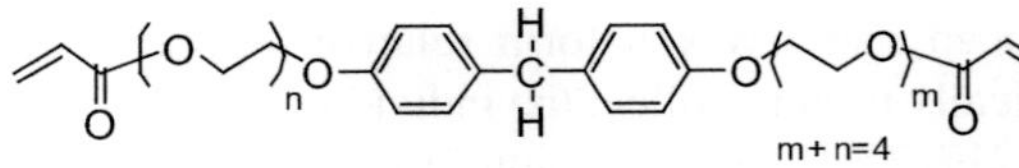

Figure 12.10 *Chemical structure of bisacrylate photocurable monomer (**UV-1**)*

Especially, epoxy/CNTs are the one of the most extensively-researched thermoset composites so far [237–244]. The combination of rheological study [245] and SANS measurements [35] are strong tools to understand the degree of dispersion in the composite [245]. A quick solidification without solvent removal process was utilized to keep the CNT alignment formed prior to the polymerization via magnetic or electronic induced orientation [246, 247]. The combination has been demonstrated to yield a good processability [248]. We have reported the mold-assisted photolithography of bis-acrylate/SWNT composites (**UV-1** in Figure 12.10) by using a PDMS stamp [249] and clear 2D patterns with a submicron scale were easily fabricated on a silicone wafer in few seconds (Figure 12.11). As expected, the degree of dispersion showed no change upon polymerization, which was proved by the monitoring of vis-NIR absorption spectroscopy. The composite present an extremely low electric percolation threshold (0.05–0.1 wt%) as well as low surface resistance accompanied by the nice dispersion compared to the other systems (in a order of 10^{-2} ohm/square), suggesting effective dispersion of the SWNTs in the matrix. These patterned polymer/SWNT composites with high conductivity may offer novel potential applications including an optical waveguide utilizing the nonlinear response of SWNTs [250], a scaffold for cell culture media [251, 252], a thin film transistor composed of an SWNTs network in the insulating resin, a separator for fuel cell, a chemical/biological sensor [253, 254], etc.

12.5.3 Nanotube/Polymer Gel-Near IR Responsive Materials

CNTs are characterized to their intense absorption in the NIR region and this absorption gives a potential use for the NIR functional materials. Mainly two of NIR-responsive

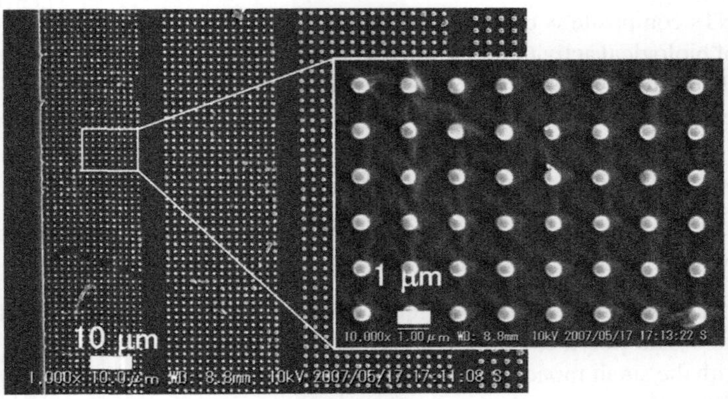

Figure 12.11 *SEM images of the nanoimprint patterns prepared from UV-**1**/SWNT using PDMS stamps*

materials have been explored. One is the NIR-saturable absorber necessary for solid state lasers based on the saturable absorption property of CNTs [255–257]. Sakakibara *et al.* demonstrated that the SWNTs composite dispersed in a polyimide matrix are well suited for the reproducible construction of mode locked fiber lasers and the generation of extremely short pulse durations [258–260]. Homogeneous dispersion of SWNTs in the polyimide matrix serve to minimize the loss of the light caused by the scattering and to realize such an excellent property. This application is quite unique and gives requisite optical devices such as laser and optical switches for NIR high-speed optical communication systems.

Another unique application of CNTs working in the NIR region is the photon-to-heat convertor utilizing an efficient photo-absorption and photothermal conversion of CNTs in the NIR region. Boldor *et al.* reported that MWNTs showed higher photothermal conversion efficiency than that of graphite [261]. Among the various light sources, NIR laser light is a fascinating stimulus especially from a biomedical point of view, because biomedical tissues have only a slight absorption in the NIR region, which enable remote stimulation of the NIR absorbent in the body from the outside. Dai *et al.* reported a NIR induced release of ssDNA from a ssDNA/SWNT composite dispersed in an aqueous media [262]. Photothermal conversion occurred due to the effective nonradiative process of excited-SWNTs generating intense heat in a very short period. As a result, the wrapped polymer is dissociated from the composites and the SWNTs start to aggregate through strong van der Waals interactions. They demonstrated that the photothermal conversion of CNTs irradiated by NIR light is effective to kill cancer cells stained with CNTs [262]. Clear unwrapping of the dispersant polymer induced by the NIR photothermal conversion was reported by our group [263]. We described that NIR light irradiation to the SWNTs solubilized with an anthracene-carring vinyl polymer (Anth-P in Figure 12.12) caused flocculation of the SWNTs. With increasing irradiation time, black flocculates are generated in the solution (Figure 12.13), indicating that the photothermal conversion of SWNTs provided intense heat just around them and, as a result, Anth-P was dissociated from the irradiated SWNTs. Furthermore, we have proposed the utilization of photothermal conversion of CNTs to thermoresponsive polymer materials. Poly(*N*-isopropylacrylamide) (PNIPAM) [264] and its derivatives are well-known thermoresponsive materials, which show a phase transition triggered by external stimuli such as the solvent composition [265], pH [266], ionic strength [266], electric field [267] and light [268]. Upon irradiation with the NIR light centered at 1064 nm, the PNIPAM/SWNT composite gel (200 μm in diameter) containing the SWNTs in the PNIAPM matrix immediately shrunk to a narrower gel (Figure 12.14) after 15 sec. After turning off the irradiation, the shrunken gel gradually

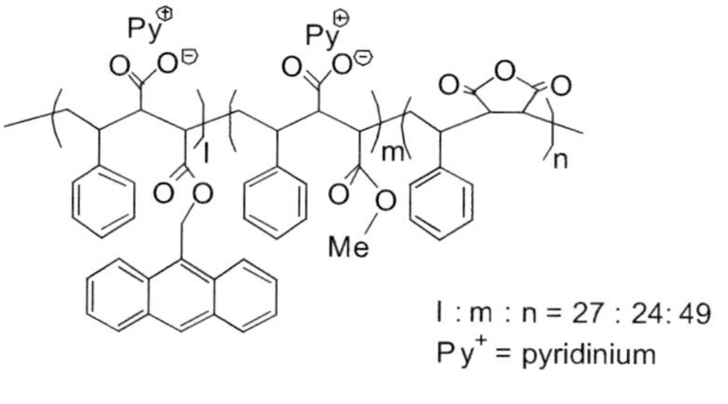

Figure 12.12 *Chemical structure of Anth-P*

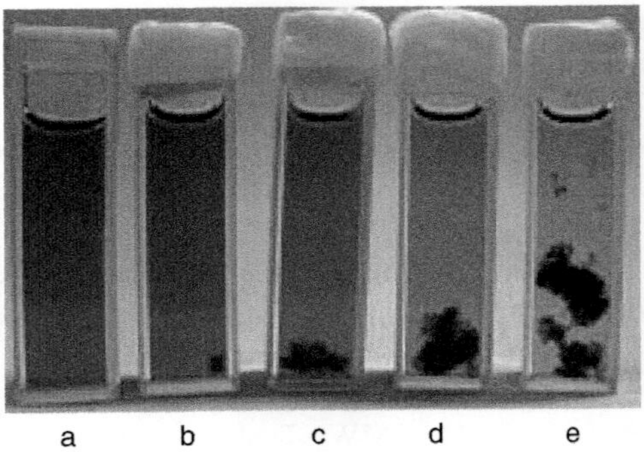

Figure 12.13 Photographs of the Anth-P/SWNT solutions in DMF before (a) and after laser irradiation for (b) 5 min, (c) 10 min, (d) 30 min, and (e) 60 min

swells and becomes around 200 μm in diameter after about 67 sec. The response time of the volume change is controllable by changing the concentration of the SWNTs as well as the power of the NIR laser light. Amazingly, no notable deterioration of the gel actuation is observed even after the 1200-cycle operation; namely the SWNT-composite gels are highly durable due to the toughness of the CNTs. In fact, the Raman spectra of the gels before and after the endurance test supports exhibit virtually identical G/D (Graphite/Defect) ratios, which guarantee that the SWNTs remain structurally intact. Very recently Miyako *et al.* [269] reported two different kinds of smart polymer gels (agarose and PNIPAM gels) containing SWNTs and single-walled nanohorns (SWNHs) that show marked phase transitions upon NIR irradiation; namely, they found that under NIR-laser irradiation (1064 nm), the nanocarbon–agarose gel hybrids exhibit a gel-to-sol transition, whereas control agarose gel (without the nanocarbons) does not show any phase transition. Such NIR actuation of the polymer/CNTs composites covers both a soft gel-type and solid film materials [270–276]. Wide range of absorption on CNTs provides an opportunity for a 'molecular heater' to work at the various wavelengths of the light source.

12.5.4 Conductive Nanotube Honeycomb Film

Honeycomb structures from organic (polymer) and organic/inorganic hybrid materials are of interest due to their unique structures and functions. Since the first report by François *et al.* [277] that self-organized honeycomb structures are formed from star-shaped

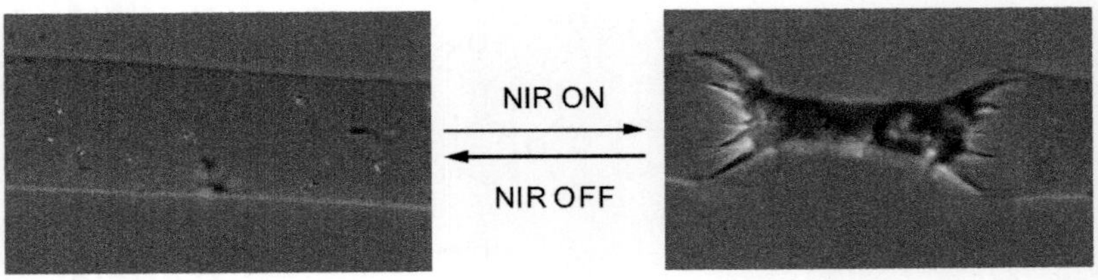

Figure 12.14 PNIPAM/SWNT gel that shows NIR laser-triggered volume phase transition. Left: before irradiation; right: after irradiation

polystyrene or poly(styrene)-poly(paraphenylene) block copolymers in carbon disulfide under flowing moist gas, many papers have been published describing the formation of similar honeycomb structures using different kinds of organic (polymer) materials including symmetric diblock copolymers [278], rod-coil diblock copolymers [279], a coil-like polymer [280], ion-complexed polymers [281], lipid-packaged Pt complexes [282], poly(D, L-lactic-co-glycolic acid) [283], polysulfone [284], amphiphilic poly(*p*-phenylenes) [285], and a poly(ε-caprolactone)/amphiphilic copolymer [286].

We have reported the discovery that the self-assembly of SWNTs with a honeycomb structure is spontaneously formed on glass substrates [287] and transparent plastic films like of poly(ethylene terephthalate) (PET), which is a widely used engineering plastic in the industrial field, by a simple solution casting method using a single-walled CNTs (SWNTs)/lipid conjugate (complex **1**, Figure 12.15) as the material, which is an ion-complex of shortened SWNTs and tridodecylmethylammonium chloride, a molecular-bilayer-forming amphiphile and available from our previous study. Complex **1** is soluble in several organic solvents including dichloromethane, chloroform, benzene and toluene. We recently

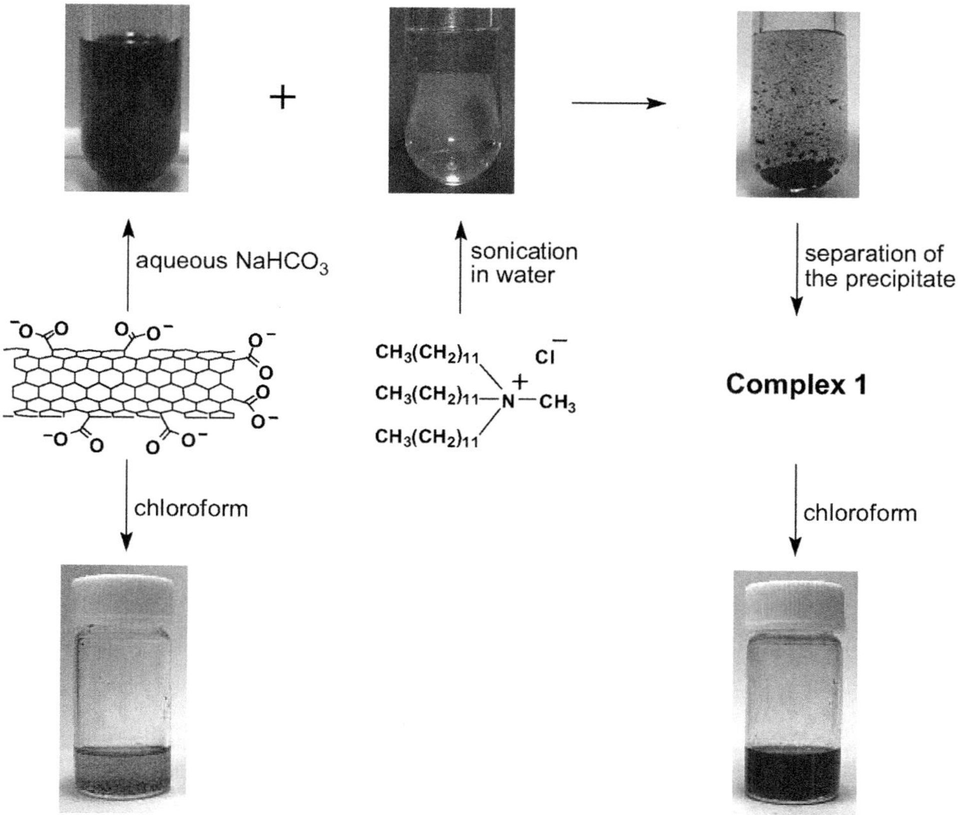

Figure 12.15 Preparation of Complex **1**

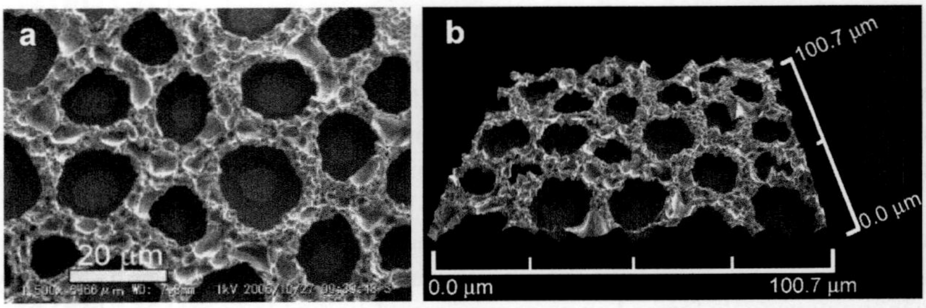

Figure 12.16 *SEM images of Complex **1** on a glass substrate before the ion-exchange*

reported the formation of (semi)conducting SWNT honeycomb structures on flexible transparent polymer films [288]. As the film, we have chosen the film of poly(ethylene terephthalate) (PET), which is a widely used engineering plastic in the industrial field. This study should be important from viewpoints of potential applications of conducting SWNTs with honeycomb structures for the fabrication of conducting plastic films with transparent flexible properties. Such films might be useful in many areas of application that require flexible conducting films as materials. The typical SEM images of honeycomb structure are shown in Figure 12.16. The sizes of the unit cells are controllable by changing the experimental conditions.

The surface resistivity (Rs) of the cast films of complex **1** with honeycomb structures is insulating (Rs $>10^8$ ohm/square due to the coating of the tube surfaces with the ammonium lipid. We developed a method to remove the lipid from the films by employing the 'ion-exchange method' as shown by Scheme 12.1. The experimental procedure is very simple, namely, each cast film is immersed overnight in a *p*-toluenesulfonic acid methanol solution, and then rinsed with methanol followed by air-drying. By this procedure, the methylene stretching vibrations in the FT-IR of the film almost disappear. The Raman spectra of complex **1** before and after the ion exchange are virtually identical. The SWNTs remain intact during all the processes. The SEM images of the cast films after the ion exchange are shown in Figure 12.17. After the ion exchange, the skeletons with the honeycomb-structures become thin due to the removal of the lipid. Higher magnification SEM measurements show oriented nanotubes along the honeycomb skeletons.

After ion-exchange, a dramatic change in the Rs values is observed. The Rs values decrease with increasing concentration of complex **1** due to the formation of network structures in larger areas on the films. When the film is prepared from complex **1**=3.0 mg mL^{-1} in chloroform, the Rs reached high conducting value, $3{\cdot}2 \times 10^2$ ohm/square. Similar behavior is observed when dichloromethane and benzene were used in place of chloroform. Interestingly, after the ion exchange, the Rs values of the films decrease more than 10^4–10^6 fold compared to the original values.

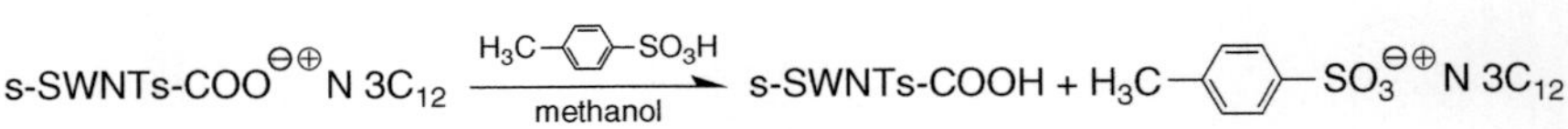

Scheme 12.1

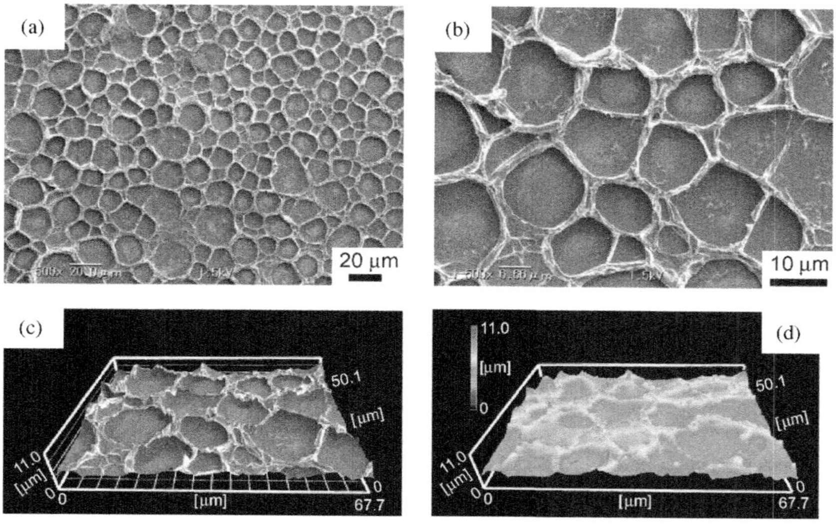

Figure 12.17 *SEM images of Complex **1** on a glass substrate after the ion-exchange*

The conductive SWNT honeycomb films on glass substrates and plastic films fabricated by the self-organization from nanotube solutions are useful in many areas of nanoscience and technology.

12.6 Summary

In this review article, we summarized recent progress on soluble CNTs based on non-covalent modification, in which π–π, CH–π and charge-transfer interactions play an important role. Individual solubilization of CNTs is necessary for a wide range of science and technology because the preparation of individually dissolved SWNTs is the first step to afford CNTs to practical use as well as fundamental studies. The individual solubilization based on physical modification maintains CNTs intact and is an attractive route for taking advantage of their intrinsic properties. Tremendous numbers of paper describing the applications of soluble CNTs have been reported and the some of them are unique for the CNT properties. Among wide range of applications, we highlighted: (i) DNA/SWNT hybrids not containing free DNA, (ii) NIR-responsive application based on unique saturable absorption and photothermal conversion properties of the CNTs in the NIR range, (iii) CNT/UV-curable resin composites having high conductivity and its application to nanoimprinting, and (iv) the formation of conducting CNT-honeycomb films on a transparent plastic film in order to demonstrate high potential applications of CNT/hybrids and CNT/composites in the areas of nanomaterials science and technology and bio-science.

References

[1] S. Iijima, *Nature* **354**, 56 (1991).
[2] N. Nakashima, *Int. J. Nanosci.* **4**, 119 (2005).

[3] H. Murakami, and N. Nakashima, *J. Nanosci. Nanotechnol.* **6**, 16 (2006).
[4] N. Nakashima, and T. Fujigaya, *Chem. Lett.* **36**, 692 (2007).
[5] C.A. Dyke, and J.M. Tour, *J. Phys. Chem. A* **108**, 11151 (2004).
[6] L.A. Girifalco, M. Hodak, and R.S. Lee, *Physical Review B* **62**, 13104 (2000).
[7] D. Tasis, N. Tagmatarchis, A. Bianco, and M. Prato, *Chem. Rev.* **106**, 1105 (2006).
[8] K. Balasubramanian, and M. Burghard, *Small* **1**, 180 (2005).
[9] D. Tasis, N. Tagmatarchis, V. Georgakilas, and M. Prato, *Chem. Eur. J.* **9**, 4000 (2003).
[10] S. Ogata, and Y. Shibutani, *Phys. Rev. B: Condens. Matter Mater. Phys.* **68**, 165409/1 (2003).
[11] M.-F. Yu, O. Lourie, M.J. Dyer, K. Moloni, T.F. Kelly, and R.S. Ruoff, *Science* **287**, 637 (2000).
[12] T. Uchida, and S. Kumar, *J. Appl. Polym. Sci.* **98**, 985 (2005).
[13] K. Enomoto, S. Kitakata, T. Yasuhara, N. Ohtake, T. Kuzumaki, and Y. Mitsuda, *Appl. Phys. Lett.* **88**, 153115 (2006).
[14] B.G. Demczyk, Y.M. Wang, J. Cumings, M. Hetman, W. Han, A. Zettl, and R.O. Ritchie, *Mater. Sci. Eng. A* **334**, 173 (2002).
[15] E.W. Wong, P.E. Sheehan, and C.M. Lieber, *Science* **277**, 1971 (1997).
[16] Z. Yao, C.L. Kane, and C. Dekker, *Phys. Rev. Lett.* **84**, 2941 (2000).
[17] B.Q. Wei, R. Vajtai, and P.M. Ajayan, *Appl. Phys. Lett.* **79**, 1172 (2001).
[18] E. Pop, D. Mann, Q. Wang, K. Goodson, and H. Dai, *Nano Lett.* **6**, 96 (2006).
[19] P. Kim, L. Shi, A. Majumdar, and P.L. McEuen, *Phys. Rev. Lett.* **87**, 215502 (2001).
[20] P.J. Boul, J. Liu, E.T. Mickelson, C.B. Huffman, L.M. Ericson, I.W. Chiang, K.A. Smith, D.T. Colbert, R.H. Hauge, J.L. Margrave, and R.E. Smalley, *Chem. Phys. Lett.* **310**, 367 (1999).
[21] C.A. Furtado, U.J. Kim, H.R. Gutierrez, L. Pan, E.C. Dickey, and P.C. Eklund, *J. Am. Chem. Soc.* **126**, 6095 (2004).
[22] V.C. Moore, M.S. Strano, E.H. Haroz, R.H. Hauge, R.E. Smalley, J. Schmidt, and Y. Talmon, *Nano Lett.* **3**, 1379 (2003).
[23] Y. Dror, W. Pyckhout-Hintzen, and Y. Cohen, *Macromolecules* **38**, 7828 (2005).
[24] S.M. Bachilo, M.S. Strano, C. Kittrell, R.H. Hauge, R.E. Smalley, and R.B. Weisman, *Science* **298**, 2361 (2002).
[25] M.J. O'Connell, S.M. Bachilo, C.B. Huffman, V.C. Moore, M.S. Strano, E.H. Haroz, K.L. Rialon, P.J. Boul, W.H. Noon, C. Kittrell, J. Ma, R.H. Hauge, R.B. Weisman, and R.E. Smalley, *Science* **297**, 593 (2002).
[26] O.N. Torrens, D.E. Milkie, M. Zheng, and J.M. Kikkawa, *Nano Lett.* **6**, 2864 (2006).
[27] Y. Kim, N. Minami, and S. Kazaoui, *Appl. Phys. Lett.* **86**, 073103/1 (2005).
[28] S. Kazaoui, N. Minami, B. Nalini, Y. Kim, and K. Hara, *J. Appl. Phys.* **98**, 084314/1 (2005).
[29] L. Cognet, D.A. Tsyboulski, and R.B. Weisman, *Nano Lett.* **8**, 749 (2008).
[30] D.A. Tsyboulski, J.-D.R. Rocha, S.M. Bachilo, L. Cognet, and R.B. Weisman, *Nano Lett.* **7**, 3080 (2007).
[31] L. Cognet, D.A. Tsyboulski, J.-D.R. Rocha, C.D. Doyle, J.M. Tour, and R.B. Weisman, *Science* **316**, 1465 (2007).
[32] W. Zhou, M.F. Islam, H. Wang, D.L. Ho, A.G. Yodh, K.I. Winey, and J.E. Fischer, *Chem. Phys. Lett.* **384**, 185 (2004).
[33] B.J. Bauer, E.K. Hobbie, and M.L. Becker, *Macromolecules* **39**, 2637 (2006).
[34] H. Wang, W. Zhou, D.L. Ho, K.I. Winey, J.E. Fischer, C.J. Glinka, and E.K. Hobbie, *Nano Lett.* **4**, 1789 (2004).
[35] T. Chatterjee, A. Jackson, and R. Krishnamoorti, *J. Am. Chem. Soc.* **130**, 6934 (2008).
[36] K. Yurekli, C.A. Mitchell, and R. Krishnamoorti, *J. Am. Chem. Soc.* **126**, 9902 (2004).
[37] J.A. Fagan, B.J. Landi, I. Mandelbaum, J.R. Simpson, V. Bajpai, B.J. Bauer, K. Migler, A.R.H. Walker, R. Raffaelle, and E.K. Hobbie, *J. Phys. Chem. B* **110**, 23801 (2006).
[38] Y. Tan, and D.E. Resasco, *J. Phys. Chem. B* **109**, 14454 (2005).
[39] M.J. O'Connell, S.M. Bachilo, C.B. Huffman, V.C. Moore, M.S. Strano, E.H. Haroz, K.L. Rialon, P.J. Boul, W.H. Noon, C. Kittrell, J. Ma, R.H. Hauge, R.B. Weisman, and R.E. Smalley, *Science* **297**, 593 (2002).
[40] M.S. Strano, V.C. Moore, M.K. Miller, M.J. Allen, E.H. Haroz, C. Kittrell, R.H. Hauge, and R. E. Smalley, *J. Nanosci. Nanotechnol.* **3**, 81 (2003).

[41] F. Du, C. Guthy, T. Kashiwagi, J.E. Fischer, and K.I. Winey, *J. Polym. Sci., Part B: Polym. Phys.* **44**, 1513 (2006).
[42] F. Du, J.E. Fischer, and K.I. Winey, *Phys. Rev. B: Condens. Matter Mater. Phys.* **72**, 121404/1 (2005).
[43] C. Richard, F. Balavoine, P. Schultz, T.W. Ebbesen, and C. Mioskowski, *Science* **300**, 775 (2003).
[44] G.S. Duesberg, M. Burghard, J. Muster, G. Philipp, and S. Roth, *Chem. Commun.* 435 (1998).
[45] M. Burghard, G. Duesberg, G. Philipp, J. Muster, and S. Roth, *Adv. Mater.* **10**, 584 (1998).
[46] M.F. Islam, E. Rojas, D.M. Bergey, A.T. Johnson, A.G. Yodh, and *Nano Lett.* **3**, 269 (2003).
[47] J.I. Paredes, and M. Burghard, *Langmuir* **20**, 5149 (2004).
[48] L.A. Hough, M.F. Islam, B. Hammouda, A.G. Yodh, and P.A. Heiney, *Nano Lett.* **6**, 313 (2006).
[49] M.F. Islam, M. Nobili, F. Ye, T.C. Lubensky, and A.G. Yodh, *Phys. Rev. Lett.* **95**, 148301/1 (2005).
[50] Y. Kim, S. Hong, S. Jung, M.S. Strano, J. Choi, and S. Baik, *J. Phys. Chem. B* **110**, 1541 (2006).
[51] W. Wenseleers, I.I. Vlasov, E. Goovaerts, E.D. Obraztsova, A.S. Lobach, and A. Bouwen, *Adv. Funct. Mater.* **14**, 1105 (2004).
[52] M.O. Lisunova, N.I. Lebovka, O.V. Melezhyk, and Y.P. Boiko, *J. Colloid Interface Sci.* **299**, 740 (2006).
[53] S. Bandow, A.M. Rao, K.A. Williams, A. Thess, R.E. Smalley, and P.C. Eklund, *J. Phys. Chem. B.* **101**, 8839 (1997).
[54] A. Ishibashi, and N. Nakashima, *Chem. Eur. J.* **12**, 7595 (2006).
[55] A. Ishibashi, and N. Nakashima, *Bull. Chem. Soc. Jpn.* **79**, 357 (2006).
[56] W. Barone Paul, S. Baik, A. Heller Daniel, and S. Strano Michael, *Nat. Mater.* **4**, 86 (2005).
[57] M.J. O'Connell, P. Boul, L.M. Ericson, C. Huffman, Y. Wang, E. Haroz, C. Kuper, J. Tour, K.D. Ausman, and R.E. Smalley, *Chem. Phys. Lett.* **342**, 265 (2001).
[58] P.C. Ke, *Phys. Chem. Chem. Phys.* **9**, 439 (2007).
[59] T.J. McDonald, C. Engtrakul, M. Jones, G. Rumbles, and M.J. Heben, *J. Phys. Chem. B.* **110**, 25339 (2006).
[60] N. Nair, W.J. Kim, R.D. Braatz, and M.S. Strano, *Langmuir* **24**, 1790 (2008).
[61] S. Niyogi, S. Boukhalfa, S.B. Chikkannanavar, T.J. McDonald, M.J. Heben, and S.K. Doorn, *J. Am. Chem. Soc.* **129**, 1898 (2007).
[62] M.S. Arnold, A.A. Green, J.F. Hulvat, S.I. Stupp, and M.C. Hersam, *Nature Nanotech.* **1**, 60 (2006).
[63] K. Yanagi, Y. Miyata, and H. Kataura, *Applied Physics Express* **1**, 034003/1 (2008).
[64] K.A. Park, S.M. Lee, S.H. Lee, and Y.H. Lee, *J. Phys. Chem. C* **111**, 1620 (2007).
[65] P. Leyton, J.S. Gomez-Jeria, S. Sanchez-Cortes, C. Domingo, and M. Campos-Vallette, *J. Phys. Chem. B* **110**, 6470 (2006).
[66] N. Nakashima, Y. Tomonari, and H. Murakami, *Chem. Lett.* 638 (2002).
[67] N. Nakashima, Y. Tanaka, Y. Tomonari, H. Murakami, H. Kataura, T. Sakaue, and K. Yoshikawa, *J. Phys. Chem. B* **109**, 13076 (2005).
[68] Y. Tomonari, H. Murakami, and N. Nakashima, *Chem. Eur. J.* **12**, 4027 (2006).
[69] F.J. Gomez, R.J. Chen, D. Wang, R.M. Waymouth, and H. Dai, *Chem. Commun.* 190 (2003).
[70] L.S. Fifield, L.R. Dalton, R.S. Addleman, R.A. Galhotra, M.H. Engelhard, G.E. Fryxell, and C. L. Aardahl, *J. Phys. Chem. B* **108**, 8737 (2004).
[71] D.M. Guldi, E. Menna, M. Maggini, M. Marcaccio, D. Paolucci, F. Paolucci, S. Campidelli, M. Prato, G.M.A. Rahman, and S. Schergna, *Chem. Eur. J.* **12**, 3975 (2006).
[72] J.S. Kavakka, S. Heikkinen, I. Kilpelaeinen, M. Mattila, H. Lipsanen, and J. Helaja, *Chem. Commun.* 519 (2007).
[73] C. Menard-Moyon, N. Izard, E. Doris, and C. Mioskowski, *J. Am. Chem. Soc.* **128**, 6552 (2006).
[74] S. Campidelli, M. Meneghetti, and M. Prato, *Small* **3**, 1672 (2007).
[75] R.J. Chen, Y. Zhang, D. Wang, and H. Dai, *J. Am. Chem. Soc.* **123**, 3838 (2001).
[76] L. Liu, T. Wang, J. Li, Z.-X. Guo, L. Dai, D. Zhang, and D. Zhu, *Chem. Phys. Lett.* **367**, 747 (2002).
[77] A.B. Artyukhin, O. Bakajin, P. Stroeve, and A. Noy, *Langmuir* **20**, 1442 (2004).

[78] V. Sgobba, G.M.A. Rahman, D.M. Guldi, N. Jux, S. Campidelli, and M. Prato, *Adv. Mater.* **18**, 2264 (2006).

[79] C. Ehli, G.M.A. Rahman, N. Jux, D. Balbinot, D.M. Guldi, F. Paolucci, M. Marcaccio, D. Paolucci, M. Melle-Franco, F. Zerbetto, S. Campidelli, and M. Prato, *J. Am. Chem. Soc.* **128**, 11222 (2006).

[80] D.M. Guldi, G.M.A. Rahman, M. Prato, N. Jux, S. Qin, and W. Ford, *Angew. Chem. Int. Ed.* **44**, 2015 (2005).

[81] D.M. Guldi, G.M.A. Rahman, N. Jux, D. Balbinot, N. Tagmatarchis, and M. Prato, *Chem. Commun.* 2038 (2005).

[82] D.M. Guldi, G.M.A. Rahman, V. Sgobba, N.A. Kotov, D. Bonifazi, and M. Prato, *J. Am. Chem. Soc.* **128**, 2315 (2006).

[83] G.M.A. Rahman, D.M. Guldi, R. Cagnoli, A. Mucci, L. Schenetti, L. Vaccari, and M. Prato, *J. Am. Chem. Soc.* **127**, 10051 (2005).

[84] D.M. Guldi, G.M.A. Rahman, N. Jux, D. Balbinot, U. Hartnagel, N. Tagmatarchis, and M. Prato, *J. Am. Chem. Soc.* **127**, 9830 (2005).

[85] V. Georgakilas, V. Tzitzios, D. Gournis, and D. Petridis, *Chem. Mater.* **17**, 1613 (2005).

[86] X. Li, Y. Liu, L. Fu, L. Cao, D. Wei, and Y. Wang, *Adv. Funct. Mater.* **16**, 2431 (2006).

[87] E. Granot, B. Basnar, Z. Cheglakov, E. Katz, and I. Willner, *Electroanalysis* **18**, 26 (2006).

[88] Y.-Y. Ou, and M.H. Huang, *J. Phys. Chem. B* **110**, 2031 (2006).

[89] H. Paloniemi, M. Lukkarinen, T. Aeaeritalo, S. Areva, J. Leiro, M. Heinonen, K. Haapakka, and J. Lukkari, *Langmuir* **22**, 74 (2006).

[90] P.G. Holder, and M.B. Francis, *Angew. Chem. Int. Ed.* **46**, 4370 (2007).

[91] R. Chitta, A.S.D. Sandanayaka, A.L. Schumacher, L. D'Souza, Y. Araki, O. Ito, and F. D'Souza, *J. Phys. Chem. C* **111**, 6947 (2007).

[92] T. Ogoshi, Y. Takashima, H. Yamaguchi, and A. Harada, *J. Am. Chem. Soc.* **129**, 4878 (2007).

[93] L. Hu, Y.-L. Zhao, K. Ryu, C. Zhou, J.F. Stoddart, and G. Gruner, *Adv. Mater.* **20**, 939 (2008).

[94] P. Wu, X. Chen, N. Hu, U.C. Tam, O. Blixt, A. Zettl, and C.R. Bertozzi, *Angew. Chem. Int. Ed.* **47**, 5022 (2008).

[95] D.M. Guldi, G.M.A. Rahman, N. Jux, N. Tagmatarchis, and M. Prato, *Angew. Chem. Int. Ed.* **43**, 5526 (2004).

[96] J. Zhang, J.K. Lee, Y. Wu, and R.W. Murray, *Nano Lett.* **3**, 403 (2003).

[97] E. Gregan, S.M. Keogh, A. Maguire, T.G. Hedderman, L.O. Neill, G. Chambers, and H.J. Byrne, *Carbon* **42**, 1031 (2004).

[98] T.G. Hedderman, S.M. Keogh, G. Chambers, and H.J. Byrne, *J. Phys. Chem. B* **108**, 18860 (2004).

[99] W. Feng, A. Fujii, M. Ozaki, and K. Yoshino, *Carbon* **43**, 2501 (2005).

[100] T. Yamamoto, Y. Miyauchi, J. Motoyanagi, T. Fukushima, T. Aida, M. Kato, and S. Maruyama, *Jpn. J. Appl. Phys.* **47**, 2000 (2008).

[101] S. Gotovac, Y. Hattori, D. Noguchi, J.-i. Miyamoto, M. Kanamaru, S. Utsumi, H. Kanoh, and K. Kaneko, *J. Phys. Chem. B* **110**, 16219 (2006).

[102] S. Gotovac, H. Honda, Y. Hattori, K. Takahashi, H. Kanoh, and K. Kaneko, *Nano Lett.* **7**, 583 (2007).

[103] G. Nakamura, K. Narimatsu, Y. Niidome, and N. Nakashima, *Chem. Lett.* **36**, 1140 (2007).

[104] Y.X. Chang, L.L. Zhou, G.X. Li, L. Li, and L.M. Yuan, *J. Liq. Chromatogr. Relat. Technol.* **30**, 2953 (2007).

[105] A. Fonverne, F. Ricoul, C. Demesmay, C. Delattre, A. Fournier, J. Dijon, and F. Vinet, *Sens. Actuators, B* **B129**, 510 (2008).

[106] R. Marquis, C. Greco, I. Sadokierska, S. Lebedkin, M.M. Kappes, T. Michel, L. Alvarez, J.-L. Sauvajol, S.h. Meunier, and C. Mioskowski, *Nano Lett.* **8**, 1830 (2008).

[107] H. Murakami, T. Nomura, and N. Nakashima, *Chem. Phys. Lett.* **378**, 481 (2003).

[108] H. Murakami, G. Nakamura, T. Nomura, T. Miyamoto, and N. Nakashima, *J. Porphyrins Phthalocyanines* **11**, 418 (2007).

[109] S. Cambre, W. Wenseleers, J. Culin, S. Van Doorslaer, A. Fonseca, J.B. Nagy, and E. Goovaerts, *ChemPhysChem* **9**, 1930 (2008).

[110] V.A. Basiuk, *J. Comput. Theor. Nanosci.* **3**, 767 (2006).
[111] Y. Yamaguchi, *J. Chem. Phys.* **120**, 7963 (2004).
[112] G.M.A. Rahman, D.M. Guldi, S. Campidelli, and M. Prato, *J. Mater. Chem.* **16**, 62 (2006).
[113] H. Li, B. Zhou, Y. Lin, L. Gu, W. Wang, K.A.S. Fernando, S. Kumar, L.F. Allard, and Y.-P. Sun, *J. Am. Chem. Soc.* **126**, 1014 (2004).
[114] J. Chen, and C.P. Collier, *J. Phys. Chem. B.* **109**, 7605 (2005).
[115] K.S. Chichak, A. Star, M.V.P. Altoe, and J.F. Stoddart, *Small* **1**, 452 (2005).
[116] F. Cheng, and A. Adronov, *Chem. Eur. J.* **12**, 5053 (2006).
[117] A. Satake, Y. Miyajima, and Y. Kobuke, *Chem. Mater.* **17**, 716 (2005).
[118] D.S. Hecht, R.J.A. Ramirez, M. Briman, E. Artukovic, K.S. Chichak, J.F. Stoddart, and G. Gruener, *Nano Lett.* **6**, 2031 (2006).
[119] T. Hasobe, S. Fukuzumi, and P.V. Kamat, *J. Am. Chem. Soc.* **127**, 11884 (2005).
[120] F. Cheng, S. Zhang, A. Adronov, L. Echegoyen, and F. Diederich, *Chem. Eur. J.* **12**, 6062 (2006).
[121] D.M. Guldi, H. Taieb, G.M.A. Rahman, N. Tagmatarchis, and M. Prato, *Adv. Mater.* **17**, 871 (2005).
[122] X. Wang, Y. Liu, W. Qiu, and D. Zhu, *J. Mater. Chem.* **12**, 1636 (2002).
[123] A. Ma, J. Lu, S. Yang, and K.M. Ng, *J. Cluster Sci.* **17**, 599 (2006).
[124] Y. Wang, H.-Z. Chen, H.-Y. Li, and M. Wang, *Mater. Sci. Eng., B.* **B117**, 296 (2005).
[125] P.J. Boul, D.-G. Cho, G.M.A. Rahman, M. Marquez, Z. Ou, K.M. Kadish, D.M. Guldi, and J.L. Sessler, *J. Am. Chem. Soc.* **129**, 5683 (2007).
[126] T. Hasobe, S. Fukuzumi, and P.V. Kamat, *J. Phys. Chem. B.* **110**, 25477 (2006).
[127] K. Saito, V. Troiani, H. Qiu, N. Solladie, T. Sakata, H. Mori, M. Ohama, and S. Fukuzumi, *J. Phys. Chem. C.* **111**, 1194 (2007).
[128] M. Alvaro, P. Atienzar, P. De la Cruz, J.L. Delgado, V. Troiani, H. Garcia, F. Langa, A. Palkar, and L. Echegoyen, *J. Am. Chem. Soc.* **128**, 6626 (2006).
[129] Q. Zhao, Z.-N. Gu, and Q.-K. Zhuang, *Electrochem. Commun.* **6**, 83 (2004).
[130] J. Qu, Y. Shen, X. Qu, and S. Dong, *Electroanalysis* **16**, 1444 (2004).
[131] H. Tanaka, T. Yajima, T. Matsumoto, Y. Otsuka, and T. Ogawa, *Adv. Mater.* **18**, 1411 (2006).
[132] H. Tanaka, T. Yajima, M. Kawao, and T. Ogawa, *J. Nanosci. Nanotechnol.* **6**, 1644 (2006).
[133] Z. Guo, F. Du, D. Ren, Y. Chen, J. Zheng, Z. Liu, and J. Tian, *J. Mater. Chem.* **16**, 3021 (2006).
[134] E.M.N. Mhuircheartaigh, S. Giordani, and W.J. Blau, *J. Phys. Chem. B.* **110**, 23136 (2006).
[135] X. Peng, N. Komatsu, S. Bhattacharya, T. Shimawaki, S. Aonuma, T. Kimura, and A. Osuka, *Nature Nanotech.* **2**, 361 (2007).
[136] X. Peng, N. Komatsu, T. Kimura, and A. Osuka, *ACS Nano* **2**, 2045 (2008).
[137] X. Zhang, T. Liu, T.V. Sreekumar, S. Kumar, V.C. Moore, R.H. Hauge, and R.E. Smalley, *Nano Lett.* **3**, 1285 (2003).
[138] D. Baskaran, J.W. Mays, and M.S. Bratcher, *Chem. Mater.* **17**, 3389 (2005).
[139] P. Petrov, F. Stassin, C. Pagnoulle, and R. Jerome, *Chem. Commun.* 2904 (2003).
[140] X. Lou, R. Daussin, S. Cuenot, A.-S. Duwez, C. Pagnoulle, C. Detrembleur, C. Bailly, and R. Jerome, *Chem. Mater.* **16**, 4005 (2004).
[141] N. Nakashima, S. Okuzono, Y. Tomonari, and H. Murakami, *Trans. Mater. Res. Soc. Jpn.* **29**, 525 (2004).
[142] W.Z. Yuan, J.Z. Sun, Y. Dong, M. Haeussler, F. Yang, H.P. Xu, A. Qin, J.W.Y. Lam, Q. Zheng, and B.Z. Tang, *Macromolecules* **39**, 8011 (2006).
[143] G.J. Bahun, C. Wang, and A. Adronov, *J. Polym. Sci., Part A: Polym. Chem.* **44**, 1941 (2006).
[144] D. Wang, W.-X. Ji, Z.-C. Li, and L. Chen, *J. Am. Chem. Soc.* **128**, 6556 (2006).
[145] W.Z. Yuan, Y. Mao, H. Zhao, J.Z. Sun, H.P. Xu, J.K. Jin, Q. Zheng, and B.Z. Tang, *Macromolecules* **41**, 701 (2008).
[146] Q. Yang, L. Shuai, and X. Pan, *Biomacromolecules* **9**, 3422 (2008).
[147] K. Narimatsu, J. Nishioka, H. Murakami, and N. Nakashima, *Chem. Lett.* **35**, 892 (2006).
[148] D.B. Romero, M. Carrard, W. De Heer, and L. Zuppiroli, *Adv. Mater.* **8**, 899 (1996).
[149] S.A. Curran, P.M. Ajayan, W.J. Blau, D.L. Carroll, J.N. Coleman, A.B. Dalton, A.P. Davey, A. Drury, B. McCarthy, S. Maier, and A. Strevens, *Adv. Mater.* **10**, 1091 (1998).

[150] J.N. Coleman, S. Curran, A.B. Dalton, A.P. Davey, B. McCarthy, W. Blau, and R.C. Barklie, *Phys. Rev. B: Condens. Matter Mater. Phys.* **58**, R7492 (1998).
[151] H. Ago, K. Petritsch, M.S.P. Shaffer, A.H. Windle, and R.H. Friend, *Adv. Mater.* **11**, 1281 (1999).
[152] J.N. Coleman, S. Curran, A.B. Dalton, A.P. Davey, B. Mc Carthy, W. Blau, and R.C. Barklie, *Synth. Met.* **102**, 1174 (1999).
[153] A.B. Dalton, H.J. Byrne, J.N. Coleman, S. Curran, A.P. Davey, B. McCarthy, and W. Blau, *Synth. Met.* **102**, 1176 (1999).
[154] S. Curran, A.P. Davey, J. Coleman, A. Dalton, B. McCarthy, S. Maier, A. Drury, D. Gray, M. Brennan, K. Ryder, M.L. de La Chapelle, C. Journet, P. Bernier, H.J. Byrne, D. Carroll, P.M. Ajayan, S. Lefrant, and W. Blau, *Synth. Met.* **103**, 2559 (1999).
[155] H. Ago, M.S.P. Shaffer, D.S. Ginger, A.H. Windle, and R.H. Friend, *Phys. Rev. B: Condens. Matter Mater. Phys.* **61**, 2286 (2000).
[156] B. McCarthy, A.B. Dalton, J.N. Coleman, H.J. Byrne, P. Bernier, and W.J. Blau, *Chem. Phys. Lett.* **350**, 27 (2001).
[157] B. McCarthy, J.N. Coleman, R. Czerw, A.B. Dalton, H.J. Byrne, D. Tekleab, P. Iyer, P.M. Ajayan, W.J. Blau, and D.L. Carroll, *Nanotechnology* **12**, 187 (2001).
[158] B. Mc Carthy, J.N. Coleman, R. Czerw, A.B. Dalton, D.L. Carroll, and W.J. Blau, *Synth. Met.* **121**, 1225 (2001).
[159] B. McCarthy, J.N. Coleman, R. Czerw, A.B. Dalton, M. in het Panhuis, A. Maiti, A. Drury, P. Bernier, J.B. Nagy, B. Lahr, H.J. Byrne, D.L. Carroll, and W.J. Blau, *J. Phys. Chem. B* **106**, 2210 (2002).
[160] D.W. Steuerman, A. Star, R. Narizzano, H. Choi, R.S. Ries, C. Nicolini, J.F. Stoddart, and J.R. Heath, *J. Phys. Chem. B* **106**, 3124 (2002).
[161] A. Star, and J.F. Stoddart, *Macromolecules* **35**, 7516 (2002).
[162] A. Star, Y. Liu, K. Grant, L. Ridvan, J.F. Stoddart, D.W. Steuerman, M.R. Diehl, A. Boukai, and J.R. Heath, *Macromolecules* **36**, 553 (2003).
[163] A. Drury, S. Maier, M. Ruether, and W.J. Blau, *J. Mater. Chem.* **13**, 485 (2003).
[164] C. Yang, M. Wohlgenannt, Z.V. Vardeny, W.J. Blau, A.B. Dalton, R. Baughman, and A.A. Zakhidov, *Physica B: Condensed Matter* **338**, 366 (2003).
[165] E. Mulazzi, R. Perego, H. Aarab, L. Mihut, S. Lefrant, E. Faulques, and J. Wery, *Phys. Rev. B: Condens. Matter Mater. Phys.* **70**, 155206/1 (2004).
[166] S.M. Keogh, T.G. Hedderman, M.G. Ruether, F.M. Lyng, E. Gregan, G.F. Farrell, G. Chambers, and H.J. Byrne, *J. Phys. Chem. B* **109**, 5600 (2005).
[167] S.M. Keogh, T.G. Hedderman, P. Lynch, G.F. Farrell, and H.J. Byrne, *J. Phys. Chem. B* **110**, 19369 (2006).
[168] J. Chen, H. Liu, W.A. Weimer, M.D. Halls, D.H. Waldeck, and G.C. Walker, *J. Am. Chem. Soc.* **124**, 9034 (2002).
[169] J.N. Coleman, D.F. O'Brien, B. McCarthy, B. Lahr, A. Drury, R.C. Barklie, W.J. Blau, and A.B. Dalton, *Chem. Commun.* 2001 (2000).
[170] J.N. Coleman, A.B. Dalton, S. Curran, A. Rubio, A.P. Davey, A. Drury, B. McCarthy, B. Lahr, P. M. Ajayan, S. Roth, R.C. Barklie, and W.J. Blau, *Adv. Mater.* **12**, 213 (2000).
[171] A.B. Dalton, C. Stephan, J.N. Coleman, B. McCarthy, P.M. Ajayan, S. Lefrant, P. Bernier, W.J. Blau, and H.J. Byrne, *J. Phys. Chem. B* **104**, 10012 (2000).
[172] M. in het Panhuis, A. Maiti, A.B. Dalton, A. van den Noort, J.N. Coleman, B. McCarthy, and W. J. Blau, *J. Phys. Chem. B.* **107**, 478 (2003).
[173] A. Nish, J.-Y. Hwang, J. Doig, and R.J. Nicholas, *Nature Nanotech.* **2**, 640 (2007).
[174] F. Chen, B. Wang, Y. Chen, and L.-J. Li, *Nano Lett.* **7**, 3013 (2007).
[175] Y. Lin, M.J. Meziani, and Y.-P. Sun, *J. Mater. Chem.* **17**, 1143 (2007).
[176] J.N. Coleman, U. Khan, and Y.K. Gun'ko, *Adv. Mater.* **18**, 689 (2006).
[177] J.N. Coleman, U. Khan, W.J. Blau, and Y.K. Gun'ko, *Carbon* **44**, 1624 (2006).
[178] R. Yerushalmi-Rozen, and I. Szleifer, *Soft Matter* **2**, 24 (2006).
[179] M. Moniruzzaman, and K.I. Winey, *Macromolecules* **39**, 5194 (2006).
[180] M. Shigeta, M. Komatsu, and N. Nakashima, *Chem. Phys. Lett.* **418**, 115 (2006).

[181] P.T. Lillehei, C. Park, J.H. Rouse, and E.J. Siochi, *Nano Lett.* **2**, 827 (2002).
[182] C. Park, R.E. Crooks, E.J. Siochi, J.S. Harrison, N. Evans, and E. Kenik, *Nanotechnology* **14**, L11 (2003).
[183] C. Park, Z. Ounaies, K.A. Watson, R.E. Crooks, J. Smith, S.E. Lowther, J.W. Connell, E.J. Siochi, J.S. Harrison, and T.L.St. Clair, *Chem. Phys. Lett.* **364**, 303 (2002).
[184] Z. Ounaies, C. Park, K.E. Wise, E.J. Siochi, and J.S. Harrison, *Compos. Sci. Technol.* **63**, 1637 (2003).
[185] L. Qu, Y. Lin, D.E. Hill, B. Zhou, W. Wang, X. Sun, A. Kitaygorodskiy, M. Suarez, J.W. Connell, L.F. Allard, and Y.-P. Sun, *Macromolecules* **37**, 6055 (2004).
[186] X. Jiang, Y. Bin, and M. Matsuo, *Polymer* **46**, 7418 (2005).
[187] Y. Lin, B. Zhou, R.B. Martin, K.B. Henbest, B.A. Harruff, J.E. Riggs, Z.-X. Guo, L.F. Allard, and Y.-P. Sun, *J. Phys. Chem. B* **109**, 14779 (2005).
[188] C. Park, J.H. Kang, J.S. Harrison, R.C. Costen, and S.E. Lowther, *Adv. Mater.* **20**, 2074 (2008).
[189] S.-M. Yuen, C.-C.M. Ma, and C.-L. Chiang, *Compos. Sci. Technol.* **68**, 2842 (2008).
[190] W.-J. Chou, C.-C. Wang, and C.-Y. Chen, *Compos. Sci. Technol.* **68**, 2208 (2008).
[191] K.J. Sun, R.A. Wincheski, and C. Park, *J. Appl. Phys.* **103**, 023908/1 (2008).
[192] K.L. Mittal, *Polyimides and Other High Temperature Polymers: Synthesis, Characterization and Applications*. Brill Academic Pub. (2003), Vol. 2.
[193] T.-S. Chung, *Plastics Engineering* **41**, 701 (1997).
[194] Y. Kang, and T.A. Taton, *J. Am. Chem. Soc.* **125**, 5650 (2003).
[195] E. Nativ-Roth, R. Shvartzman-Cohen, C. Bounioux, M. Florent, D. Zhang, I. Szleifer, and R. Yerushalmi-Rozen, *Macromolecules* **40**, 3676 (2007).
[196] L. Lu, Z. Zhou, Y. Zhang, S. Wang, and Y. Zhang, *Carbon.* **45**, 2621 (2007).
[197] K.J. Gilmore, S.E. Moulton, and G.G. Wallace, *Carbon* **45**, 402 (2007).
[198] I. Cotiuga, F. Picchioni, U.S. Agarwal, D. Wouters, J. Loos, and P.J. Lemstra, *Macromol. Rapid Commun.* **27**, 1073 (2006).
[199] R. Shvartzman-Cohen, Y. Levi-Kalisman, E. Nativ-Roth, and R. Yerushalmi-Rozen, *Langmuir* **20**, 6085 (2004).
[200] N.N. Slusarenko, B. Heurtefeu, M. Maugey, C. Zakri, P. Poulin, and S. Lecommandoux, *Carbon* **45**, 903 (2007).
[201] H.-i. Shin, B.G. Min, W. Jeong, and C. Park, *Macromol. Rapid Commun.* **26**, 1451 (2005).
[202] G. Mountrichas, S. Pispas, and N. Tagmatarchis, *Small* **3**, 404 (2007).
[203] G. Mountrichas, N. Tagmatarchis, and S. Pispas, *J. Phys. Chem. B* **111**, 8369 (2007).
[204] R. Shvartzman-Cohen, E. Nativ-Roth, E. Baskaran, Y. Levi-Kalisman, I. Szleifer, and R. Yerushalmi-Rozen, *J. Am. Chem. Soc.* **126**, 14850 (2004).
[205] Z. Wang, Q. Liu, H. Zhu, H. Liu, Y. Chen, and M. Yang, *Carbon* **45**, 285 (2007).
[206] N. Nakashima, S. Okuzono, H. Murakami, T. Nakai, and K. Yoshikawa, *Chem. Lett.* **32**, 456 (2003).
[207] M. Zheng, A. Jagota, E.D. Semke, B.A. Diner, R.S. McLean, S.R. Lustig, R.E. Richardson, and N.G. Tassi, *Nat. Mater.* **2**, 338 (2003).
[208] J.C.G. Jeynes, E. Mendoza, D.C.S. Chow, P.C.P. Watts, J. McFadden, and S.R.P. Silva, *Adv. Mater.* **18**, 1598 (2006).
[209] S.C. Tsang, Z. Guo, Y.K. Chen, M.L.H. Green, H.A.O. Hill, T.W. Hambley, and P.J. Sadler, *Angew. Chem. Int. Ed.* **36**, 2198 (1997).
[210] Z. Guo, P.J. Sadler, and S.C. Tsang, *Adv. Mater.* **10**, 701 (1998).
[211] E. Buzaneva, A. Karlash, K. Yakovkin, Y. Shtogun, S. Putselyk, D. Zherebetskiy, A. Gorchinskiy, G. Popova, S. Prilutska, O. Matyshevska, Y. Prilutskyy, P. Lytvyn, P. Scharff, and P. Eklund, *Mater. Sci. Eng., C* **C19**, 41 (2002).
[212] S. Meng, P. Maragakis, C. Papaloukas, and E. Kaxiras, *Nano Lett.* **7**, 45 (2007).
[213] Y. Wang, and Y. Bu, *J. Phys. Chem. B* **111**, 6520 (2007).
[214] Y. Wang, and Y. Bu, *J. Phys. Chem. B* **111**, 6520 (2007).
[215] R.R. Johnson, A.T.C. Johnson, and M.L. Klein, *Nano Lett.* **8**, 69 (2008).
[216] B. Gigliotti, B. Sakizzie, D.S. Bethune, R.M. Shelby, and J.N. Cha, *Nano Lett.* **6**, 159 (2006).
[217] S.R. Vogel, M.M. Kappes, F. Hennrich, and C. Richert, *Chem. Eur. J.* **13**, 1815 (2007).

[218] M. Zheng, A. Jagota, M.S. Strano, A.P. Santos, P. Barone, S.G. Chou, B.A. Diner, M.S. Dresselhaus, R.S. McLean, G.B. Onoa, G.G. Samsonidze, E.D. Semke, M. Usrey, and D.J. Walls, *Science* **302**, 1545 (2003).
[219] H. Cathcart, S. Quinn, V. Nicolosi, J.M. Kelly, W.J. Blau, and J.N. Coleman, *J. Phys. Chem. C* **111**, 66 (2007).
[220] H. Cathcart, V. Nicolosi, J.M. Hughes, W.J. Blau, J.M. Kelly, S.J. Quinn, and J.N. Coleman, *J. Am. Chem. Soc.* **130**, 12734 (2008).
[221] M.L. Becker, J.A. Fagan, N.D. Gallant, B.J. Bauer, V. Bajpai, E.K. Hobbie, S.H. Lacerda, K.B. Migler, and J.P. Jakupciak, *Adv. Mater.* **19**, 939 (2007).
[222] S. Badaire, C. Zakri, M. Maugey, A. Derre, J.N. Barisci, G. Wallace, and P. Poulin, *Adv. Mater.* **17**, 1673 (2005).
[223] Y. Noguchi, T. Fujigaya, Y. Niidome, and N. Nakashima, *Chem. Phys. Lett.* **455**, 249 (2008).
[224] D.A. Heller, E.S. Jeng, T.-K. Yeung, B.M. Martinez, A.E. Moll, J.B. Gastala, and M.S. Strano, *Science* **311**, 508 (2006).
[225] Y. Xu, P.E. Pehrsson, L. Chen, R. Zhang, and W. Zhao, *J. Phys. Chem. C* **111**, 8638 (2007).
[226] E.S. Jeng, A.E. Moll, A.C. Roy, J.B. Gastala, and M.S. Strano, *Nano Lett.* **6**, 371 (2006).
[227] N. Kam, S. Wong, Z. Liu, and H. Dai, *Angew. Chem. Int. Ed.* **45**, 577 (2006).
[228] E.S. Jeng, P.W. Barone, J.D. Nelson, and M.S. Strano, *Small* **3**, 1602 (2007).
[229] G.O. Gladchenko, M.V. Karachevtsev, V.S. Leontiev, V.A. Valeev, A.Y. Glamazda, A.M. Plokhotnichenko, and S.G. Stepanian, *Mol. Phys.* **104**, 3193 (2006).
[230] M. Zheng, and E.D. Semke, *J. Am. Chem. Soc.* **129**, 6084 (2007).
[231] M.S. Strano, M. Zheng, A. Jagota, G.B. Onoa, D.A. Heller, P.W. Barone, and M.L. Usrey, *Nano Lett.* **4**, 543 (2004).
[232] S.R. Lustig, A. Jagota, C. Khripin, and M. Zheng, *J. Phys. Chem. B* **109**, 2559 (2005).
[233] X. Huang, R.S. McLean, and M. Zheng, *Anal. Chem.* **77**, 6225 (2005).
[234] B.J. Bauer, M.L. Becker, V. Bajpai, J.A. Fagan, E.K. Hobbie, K. Migler, C.M. Guttman, and W. R. Blair, *J. Phys. Chem. C* **111**, 17914 (2007).
[235] X. Sun, R. Rossin, J.L. Turner, M.L. Becker, M.J. Joralemon, M.J. Welch, and K.L. Wooley, *Biomacromolecules* **6**, 2541 (2005).
[236] Y. Matsumura, and H. Maeda, *Cancer Res.* **46**, 6387 (1986).
[237] A.S. dos Santos, T.d.O.N. Leite, C.A. Furtado, C. Welter, L.C. Pardini, and G.G. Silva, *J. Appl. Polym. Sci.* **108**, 979 (2008).
[238] L.-Q. Liu, and H.D. Wagner, *Composite Interfaces* **14**, 285 (2007).
[239] E. Bekyarova, E.T. Thostenson, A. Yu, H. Kim, J. Gao, J. Tang, H.T. Hahn, T.W. Chou, M.E. Itkis, and R.C. Haddon, *Langmuir* **23**, 3970 (2007).
[240] A. Moisala, Q. Li, I.A. Kinloch, and A.H. Windle, *Compos. Sci. Technol.* **66**, 1285 (2006).
[241] R. Guzman de Villoria, A. Miravete, J. Cuartero, A. Chiminelli, and N. Tolosana, *Composites, Part B: Engineering* **37B**, 273 (2006).
[242] N. Li, Y. Huang, F. Du, X. He, X. Lin, H. Gao, Y. Ma, F. Li, Y. Chen, and P.C. Eklund, *Nano Lett.* **6**, 1141 (2006).
[243] J. Zhu, H. Peng, F. Rodriguez-Macias, J.L. Margrave, V.N. Khabashesku, A.M. Imam, K. Lozano, and E.V. Barrera, *Adv. Funct. Mater.* **14**, 643 (2004).
[244] J. Zhu, J. Kim, H. Peng, J.L. Margrave, V.N. Khabashesku, and E.V. Barrera, *Nano Lett.* **3**, 1107 (2003).
[245] Y.S. Song, and J.R. Youn, *Carbon* **43**, 1378 (2005).
[246] V.N. Bliznyuk, S. Singamaneni, R.L. Sanford, D. Chiappetta, B. Crooker, and P.V. Shibaev, *Polymer* **47**, 3915 (2006).
[247] C. Park, J. Wilkinson, S. Banda, Z. Ounaies, K.E. Wise, G. Sauti, P.T. Lillehei, and J.S. Harrison, *J. Polym. Sci., Part B: Polym. Phys.* **44**, 1751 (2006).
[248] J. Sandoval, K. Soto, L. Murr, and R. Wicker, *J. Mater. Sci.* **42**, 156 (2007).
[249] T. Fujigaya, S. Haraguchi, T. Fukumaru, and N. Nakashima, *Adv. Mater.* **20**, 2151 (2008).
[250] T.R. Schibli, K. Minoshima, H. Kataura, E. Itoga, N. Minami, S. Kazaoui, K. Miyashita, M. Tokumoto, and Y. Sakakibara, *Opt. Express* **13**, 8025 (2005).

[251] N. Aoki, A. Yokoyama, Y. Nodasaka, T. Akasaka, M. Uo, Y. Sato, K. Tohji, and F. Watari, *J. Biomed. Nanotechnol.* **1**, 402 (2005).
[252] N. Aoki, A. Yokoyama, Y. Nodasaka, T. Akasaka, M. Uo, Y. Sato, K. Tohji, and F. Watari, *Chem. Lett.* **35**, 508 (2006).
[253] B.L. Allen, P.D. Kichambare, and A. Star, *Adv. Mater.* **19**, 1439 (2007).
[254] K. Balasubramanian, and M. Burghard, *Anal. Bioanal. Chem.* **385**, 452 (2006).
[255] Y.C. Chen, N.R. Raravikar, L.S. Schadler, P.M. Ajayan, Y.P. Zhao, T.M. Lu, G.C. Wang, and X. C. Zhang, *Appl. Phys. Lett.* **81**, 975 (2002).
[256] S. Tatsuura, M. Furuki, Y. Sato, I. Iwasa, M. Tian, and H. Mitsu, *Adv. Mater.* **15**, 534 (2003).
[257] Y. Sakakibara, S. Tatsuura, H. Kataura, M. Tokumoto, and Y. Achiba, *Jpn. J. Appl. Phys., Part 2* **42**, L494 (2003).
[258] Y. Sakakibara, A.G. Rozhin, H. Kataura, Y. Achiba, and M. Tokumoto, *Jpn. J. Appl. Phys., Part 1* **44**, 1621 (2005).
[259] T.R. Schibli, K. Minoshima, H. Kataura, E. Itoga, N. Minami, S. Kazaoui, K. Miyashita, M. Tokumoto, and Y. Sakakibara, *Opt. Express* **13**, 8025 (2005).
[260] N. Nishizawa, Y. Seno, K. Sumimura, Y. Sakakibara, E. Itoga, H. Kataura, and K. Itoh, *Opt. Express* **16**, 9429 (2008).
[261] D. Boldor, N.M. Gerbo, W.T. Monroe, J.H. Palmer, Z. Li, and A.S. Biris, *Chem. Mater.* **20**, 4011 (2008).
[262] N. Kam, S. Wong, M. O'Connell, J. A. Wisdom, and H. Dai, *Proc. Natl. Acad. Sci. U.S.A.* **102**, 11600 (2005).
[263] K. Narimatsu, Y. Niidome, and N. Nakashima, *Chem. Phys. Lett.* **429**, 488 (2006).
[264] S. Hirotsu, Y. Hirokawa, and T. Tanaka, *J. Chem. Phys.* **87**, 1392 (1987).
[265] S. Katayama, Y. Hirokawa, and T. Tanaka, *Macromolecules* **17**, 2641 (1984).
[266] T. Tanaka, D. Fillmore, S.-T. Sun, I. Nishio, G. Swislow, and A. Shah, *Phys. Rev. Lett.* **45**, 1636 (1980).
[267] T. Tanaka, I. Nishio, S.T. Sun, and S. Ueno-Nishio, *Science* **218**, 467 (1982).
[268] A. Suzuki, and T. Tanaka, *Nature* **346**, 345 (1990).
[269] E. Miyako, H. Nagata, K. Kirano, and T. Hirotsu, *Small* **4**, 1711 (2008).
[270] S.V. Ahir, and E.M. Terentjev, *Nat. Mater.* **4**, 491 (2005).
[271] S.V. Ahir, and E.M. Terentjev, *Phys. Rev. Lett.* **96**, 133902/1 (2006).
[272] S.V. Ahir, A.M. Squires, A.R. Tajbakhsh, and E.M. Terentjev, *Phys. Rev. B: Condens. Matter Mater. Phys.* **73**, 085420/1 (2006).
[273] S. Lu, and B. Panchapakesan, *Nanotechnology* **16**, 2548 (2005).
[274] S. Lu, and B. Panchapakesan, *Nanotechnology* **18**, 305502 (2007).
[275] L. Yang, K.S.A. Li, and S.G.J. Chen, *Adv. Mater.* **20**, 2271 (2008).
[276] H. Koerner, G. Price, N.A. Pearce, M. Alexander, and R.A. Vaia, *Nat Mater.* **3**, 115 (2004).
[277] G. Widawski, M. Rawiso, and B. Francois, *Nature* **369**, 387 (1994).
[278] Z. Li, W. Zhao, Y. Liu, M.H. Rafailovich, J. Sokolov, K. Khougaz, A. Eisenberg, R.B. Lennox, and G. Krausch, *J. Am. Chem. Soc.* **118**, 10892 (1996).
[279] S.A. Jenekhe, and X.L. Chen, *Science* **283**, 372 (1999).
[280] M. Srinivasaro, D. Collings, A. Philips, and S. Patel, *Science* **292**, 79 (2001).
[281] H. Yabu, M. Tanaka, K. Ijiro, and M. Shimomura, *Langmuir* **19**, 6297 (2003).
[282] C.-S. Lee, and N. Kimizuka, *Proc. Natl. Acad. Sci. U.S.A.* **99**, 4922 (2002).
[283] X. Zhao, Q. Cai, G. Shi, Y. Shi, and G. Chen, *J. Appl. Polym. Sci.* **90**, 1846 (2003).
[284] Y. Xu, B. Zhu, and Y. Xu, *Polymer* **46**, 713 (2005).
[285] M. H. Nurmawati, R. Renu, P. K. Ajikumar, S. Sindhu, F. C. Cheong, C. H. Sow, and S. Valiyaveettil, *Adv. Funct. Mater.* **16**, 2340 (2006).
[286] K. Arai, M. Tanaka, S. Yamamoto, and M. Shimomura, *Colloids Surf. A* **313–314**, 530 (2008).
[287] H. Takamori, T. Fujigaya, Y. Yamaguchi, and N. Nakashima, *Adv. Mater.* **19**, 2535 (2007).
[288] N. Wakamatsu, H. Takamori, T. Fujigaya, and N. Nakashima, *Adv. Funct. Mater.* **19**, 311 (2009).
[289] K. Rege, G. Viswanathan, G. Zhu, A. Vijayaraghavan, P.M. Ajayan, and J.S. Dordick, *Small* **2**, 718 (2006).
[290] J. Geng, Y.K. Ko, S.C. Youn, Y.-H. Kim, S.A. Kim, D.-H. Jung, and H.-T. Jung, *J. Phys. Chem. C* **112**, 12264 (2008).

13

Functionalization of Carbon Nanotubes for Nanoelectronic and Photovoltaic Applications

Stéphane Campidelli[a] *and Maurizio Prato*[b]

[a]*CEA, IRAMIS, Laboratoire d'Electronique Moléculaire, Service de Physique de l'Etat Condensé, CEA Saclay, France*

[b]*INSTM, Unit of Trieste, Dipartimento di Scienze Farmaceutiche, Università di Trieste, Italy*

13.1 Introduction

Functionalization of Carbon Nanotubes (CNTs) has been pursued by several groups and has led to a completely new class of CNT hybrids, commonly called functionalized carbon nanotubes (*f*-CNTs). *f*-CNTs offer the invaluable opportunity to combine the outstanding properties of CNT with those of other classes of materials. This chapter deals with a basic introduction to CNT functionalization, followed by an extensive description of *f*-CNT applications.

13.2 Functionalization of Carbon Nanotubes

Carbon nanotubes (CNTs) are cylinder-shaped macromolecules with a radius as small as a few nanometers and with length typically reaching the micrometer or the millimeter scale. Mainly three techniques are used to produce carbon nanotubes: arc discharge, laser ablation or gas-phase catalytic growth from carbon monoxide or other

Chemistry of Nanocarbons Edited by Takeshi Akasaka, Fred Wudl and Shigeru Nagase

carbon sources. These fabrication techniques give rise to raw materials which contain carbon nanotubes mixed with amorphous carbon and catalytic metal particles as impurities. Consequently, an effective purification of the nanotubes is required before their further processing.

During the last decade, several methods have been described to purify carbon nanotubes and many of these methods are based on strong oxidative treatments: sonication/heating in HNO_3 or H_2SO_4/HNO_3 mixtures or treatment with H_2O_2 or piranha solutions (H_2SO_4/H_2O_2). Additional HCl or NaOH treatment are sometimes performed to remove metal catalyst or oxidized amorphous carbon, respectively [1–4]. These methods permit the purification of the nanotube but they also cause the formation of carboxylic functions at the nanotube edges and at the sidewalls. Historically, oxidation of nanotubes followed by functionalization of the carboxylic groups has represented a powerful pathway for solubilization and modifications of carbon nanotubes [5]. Functionalization of carbon nanotubes is not limited to the chemistry of carboxylic acids, as more elaborated methods have been developed to link organic moieties directly at the nanotube sidewalls. In this section, we will give a short overview of carbon nanotube functionalizations (Figure 13.1), for a more extensive review see Chapters 1 and 12 or references 6–7 [6, 7].

A. A widely used method for nanotube functionalization consists in the oxidation of the nanotubes and the introduction of alcohol or amine moieties *via* activation of the carboxylic groups with carbodiimides or oxalyl or thionyl chloride. Many groups have been introduced like alkyl chains [5, 8, 9], porphyrins [10], tetrathiafulvalene [11], quantum dots [12, 13], polymers [14, 15], and bioactive molecules [16–18].
B. Recently, Hirsch has reported the direct addition of nitrene derivatives produced by thermal decomposition of alkyl azidoformates to the nanotubes sidewalls. This reaction was first applied for small addends (where R = ethyl and *tert*-butyl) [19] and then extended to more complex substituents such as long alkyl chains, aromatic groups, dendrimers, crown ethers, and oligoethylene glycol units [20].
C. CNTs are able to react with radicals. Khabashesku *et al.* described the reaction of SWNT with organic peroxides to produce alkyl or 3-carboxypropyl functionalized SWNTs [21, 22]. The functionalization of nanotubes with radicals was also reported by the group of Billups [23]. In this case, thermal decomposition of benzoyl peroxides produced radicals as initiators for the formation of alkyl radicals from various alkyl iodides. The same group demonstrated the functionalization of nanotubes via the reductive alkylation of SWNTs using lithium and alkyl halides in liquid ammonia [24].
D. The addition of aryl diazonium on carbon nanotubes is a popular reaction for nanotube functionalization. Initially, Tour *et al.* described the synthesis of several nanotube conjugates *via* electrochemical reduction and grafting of different aryl diazonium salts onto the nanotube sidewalls [25]. Subsequently, they extended this method to diazonium derivatives, chemically generated in situ [26] and then in solvent-free conditions [27] or in water using sodium dodecylsulfate (SDS) as surfactant [28].
E. 1,3-dipolar cycloaddition reactions have been successfully applied for sidewall functionalization. In the initial reports, we described the synthesis of several derivatives containing solubilizing alkyl or ethylene glycol chains [29, 30]. The group of Langa described the functionalization of CNTs with 2,5-diarylpyrazoline or isoxazoline

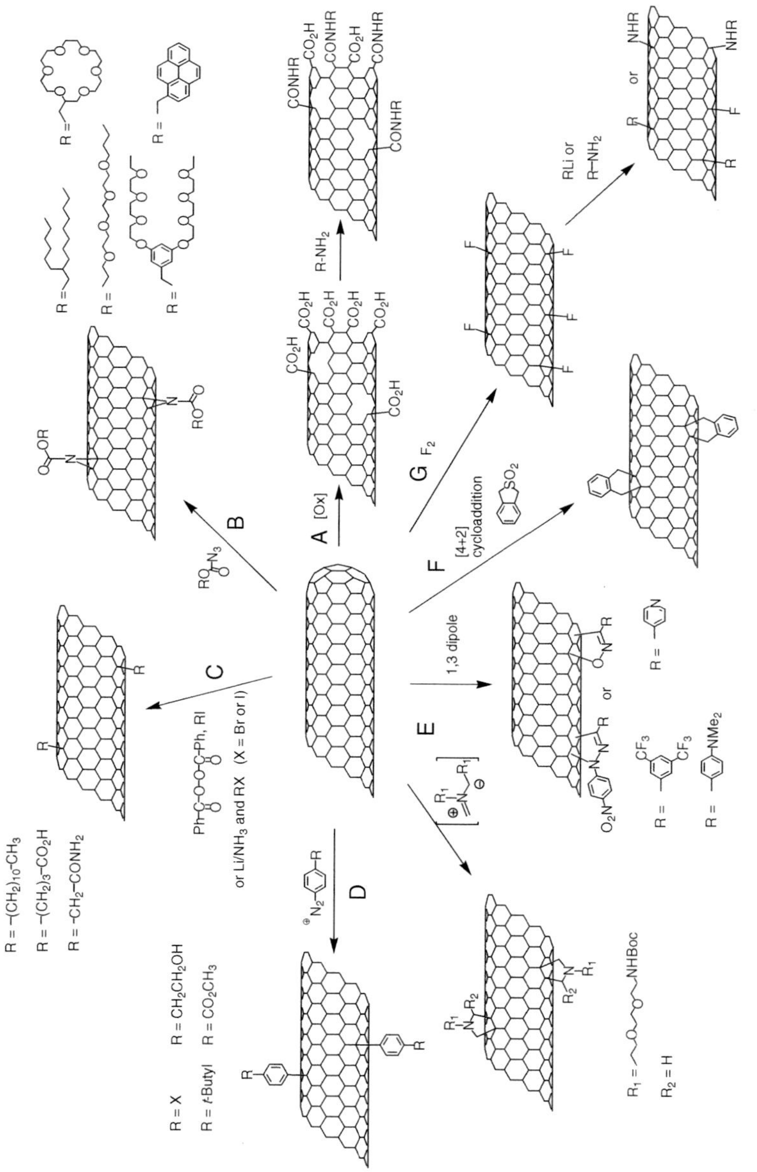

Figure 13.1 *Example of covalent functionalization of SWNT*

rings [31]. Recently, another example of 1,3-dipolar cycloaddition was reported by Swager using a zwitterion formed by reaction of dimethylaminopyridine (DMAP) with dimethyl acetylenedicarboxylate [32, 33].

F. The Diels-Alder reaction was also described by Langa and co-workers [9]. The nanotubes were treated in the presence of *o*-quinodimethane (generated in situ from the corresponding sultine derivative) under microwave irradiation. Another example of [4 + 2] cycloaddition was reported by Mioskowski, namely, the reaction of SWNTs with an electron rich diene in the presence of $Co(CO)_6$ [34].

G. 'Fluoronanotubes' were obtained by direct fluorination of SWNTs with F_2 gas; the stoichiometry of the addition was to found to be near to 1 fluorine for 2 carbon atoms [35]. The fluorination of the nanotubes drastically enhanced the reactivity of sidewalls and subsequent derivatization is possible. Billups demonstrated that alkyllithium reagents may be used to attach alkyl groups on the sidewalls of fluoronanotubes [36]. Khabashesku found that amino derivatives are able to substitute fluorine atoms on nanotubes [37]. Barron reported the functionalization of fluorinated nanotubes using the [4 + 2] cycloaddition [38].

13.3 Properties and Applications

13.3.1 Electron Transfer Properties and Photovoltaic Applications

The first step toward the realization of energy conversion devices is the design of structures in which electron donor and electron acceptor moieties are in close proximity and can interact with each others. The electrical conductivity, morphology, and good chemical stability of SWNTs are promising features that stimulate their integration into photovoltaic systems. It has been demonstrated that carbon nanotubes readily accept charges, which can be transported under nearly ideal conditions along the tubular SWNT axis [39, 40]. In these systems, carbon nanotubes act as electron acceptors and many electron donors (discrete molecules like porphyrins, phthalocyanines, tetrathiafulvalenes (TTF) or conjugated polymers like polythiophenes or poly-*meta*-phenylenevinylenes) have been linked to nanotubes either by covalent or noncovalent coupling.

13.3.1.1 Covalent Approach

The covalent linkage of porphyrins to SWNT was achieved through an esterification reaction between the carboxylic groups of SWNTs and porphyrins containing hydroxyl groups (Figure 13.2) [10]. Photophysical investigations of the SWNT-H_2P conjugates showed that the quenching of the fluorescence of the porphyrins was strongly dependent on the length of the spacer between the nanotube and the chromophore. In the case of **1**, where the spacer is a CH_2 moiety, no fluorescence quenching was observed, while in **2**, where the spacer contains a six-carbon chain, the fluorescence of the porphyrin was approximately 70 % of that of the reference compound. This result could be attributed to the flexibility of the alkyl chain which allows for a better interaction between the nanotube and the porphyrin.

The photophysical properties of a family of SWNT–TTF derivatives were investigated as a function of the linker between the SWNTs and the TTF moieties and also as a function of the nature of the tetrathiafulvalene unit (simple TTF or π-extended TTF) (Figure 13.2) [11].

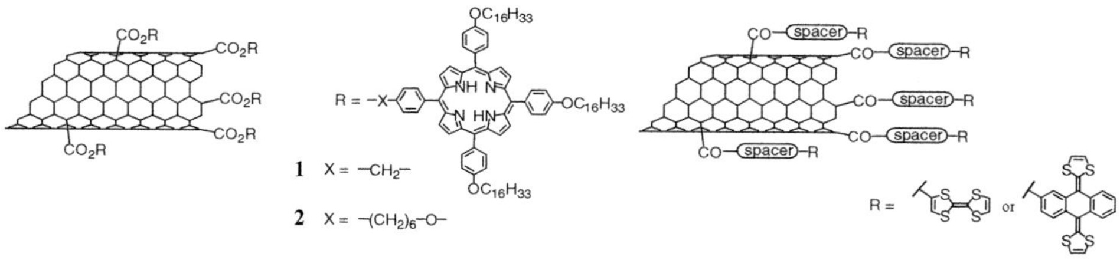

Figure 13.2 *Example of SWNT functionalized with electro-active species (i.e. porphyrin and tetrathiafulvalene (TTF) moieties)*

The SWNT-TTF conjugates gave rise to photoinduced electron transfer, while the lifetime of the charge separated states (typically on the order of several hundreds of nanoseconds) depends on the length of the spacer between the two electroactive moieties (longer lifetimes when the length increases) and on the nature of the TTF unit (longer lifetimes with π-extended TTF, probably due to a better charge delocalization).

More recently, photoactive films made of SWNT–porphyrin were prepared and used to fabricate photoelectrochemical devices on nanostructured SnO_2 electrodes [41]. The films exhibited a photon-to-current efficiency of about 4.9% under a bias voltage of 0.08 V. According to the authors, the electron injection from the excited states of the SWNTs to the conduction band of the SnO_2 electrode is responsible for the photocurrent generation. Despite the efficient quenching of the porphyrin-excited singlet state by the SWNTs in the porphyrin-linked SWNTs, the photocurrent action spectra revealed that the excitation of the porphyrin moieties makes no contribution to the photocurrent generation. The evolution of an exciplex between the porphyrin-excited singlet state and the SWNTs and the subsequent rapid decay to the ground state without generating the charge-separated state was proposed to explain the unusual photoelectrochemical behavior.

The modification of a self-assembled monolayer (SAM of cysteamine/2-thioethane-sulfonic acid on gold surface) with short oxidized carbon nanotubes was reported. SWNTs were linked to the amino surface on the first extremity while CdS nanoparticles were attached to the second extremity [12]. The photocurrent generated by the functionalized electrode in the presence of triethanolamine as a sacrificial electron donor was found to be as high as 830 nA under irradiation at 390 nm. The authors estimated a quantum efficiency for the photon-to-electron conversion of about 25%.

In an earlier work, the covalent attachment of ferrocene onto nanotube sidewalls was achieved (Figure 13.3) [42]. The photoexcitation of SWNT–Fc with visible light led to electron transfer that yields a long-lived $SWNT^{\bullet -}$-$Fc^{\bullet +}$ species.

Based on the same approach, nanohybrids containing phthalocyanine (Pc) and porphyrin derivatives were obtained (Figure 13.3) [43, 44]. For all nanoconjugates a strong communication (i.e. photoinduced electron transfer) between the nanotube and the macrocycle subunits has been observed. Phthalocyanines are of particular interest since they are synthetic porphyrin analogues, exhibiting particularly intense absorption characteristics in the red spectral region, where porphyrins fail to absorb appreciably.

The previous examples demonstrate that the combination of CNTs with photoactive molecules is of particularly high interest. However, fabrication of nanotube-based molecular assemblies is still limited because of the difficulty to incorporate highly engineered

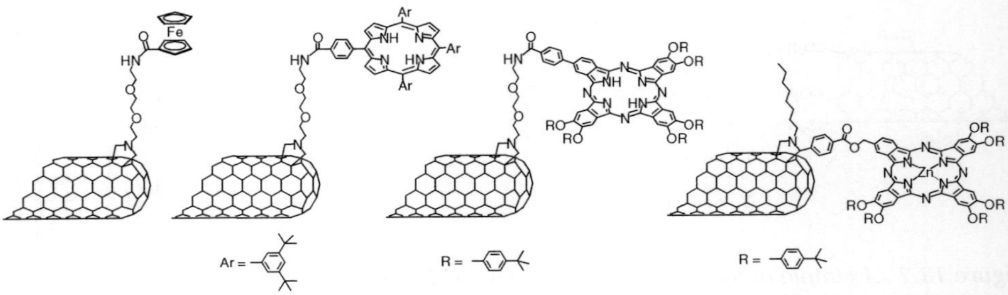

Figure 13.3 *Examples of functionalized SWNT containing ferrocene, porphyrin or phthalocyanine derivatives*

molecules on the nanotube surfaces. This problematic issue can have mainly two origins: incompatibility between the functionality on the molecules and the conditions required for nanotube functionalization and/or the fact that nanotube functionalization requires a large excess of reagent which is difficult or impossible to recycle. Therefore, there is a real need for simple and versatile procedures which allow the introduction of new functionalities onto the nanotube surfaces. Recently, we investigated the functionalization of SWNTs with Zn-phthalocyanine derivatives via 'click chemistry' and demonstrated that this concept can be used for the realization of photovoltaic cells [45]. The term 'click chemistry' [46] defines a series of chemical reactions clean, versatile, specific, easy to realize and exhibiting simple purification processes (absence of by-products). Among the large collection of organic reactions, the Cu-catalyzed variant of the *Huisgen* cycloaddition [47] (1,3-dipolar cycloaddition between azide and acetylene derivatives) represents the most effective example of the 'click chemistry' [48–51]. It has been demonstrated that the emerging field of 'click chemistry' can bring very elegant solutions to easily achieve nanotube-based functional materials [45, 52, 53]. In our recent experiments, we described the functionalization of single-wall carbon nanotubes (SWNTs) with 4-(2-trimethylsilyl)ethynylaniline and the subsequent attachment of a zinc-phthalocyanine (ZnPc) derivative using the reliable *Huisgen* 1,3-dipolar cycloaddition (Figure 13.4) [45]. The SWNT-ZnPc nanoconjugate was fully characterized and a photoinduced communication between the two photoactive components (i.e. SWNT and ZnPc) was identified. Such beneficial features led us to incorporate the SWNT-ZnPc hybrid as photoactive material in an ITO photoanode in a photoelectrochemical cell (Figure 13.4).

While it is important to improve the reaction on nanotubes to facilitate the incorporation of active molecules, it is also interesting to increase the number of these molecules on the sidewalls without altering the nanotube properties. Indeed it is generally admitted that extensive covalent functionalization of SWNTs sidewalls disrupts the conjugated π-system of the tubes – affecting their optical and electronic properties [54]. In this context, functionalization of nanotubes with polymers or dendrimers represents a particularly promising strategy. Dendrimers are regular hyperbranched macromolecules; at high generations, they possess globular structure with a large density of functional groups at the periphery [55–59]. In order to increase significantly the number of light harvesting chromophores on the nanotubes, a polyamidoamine (PAMAM) dendrimer [60] was built on the nanotube sidewalls that was further functionalized with tetraphenylporphyrins [61].

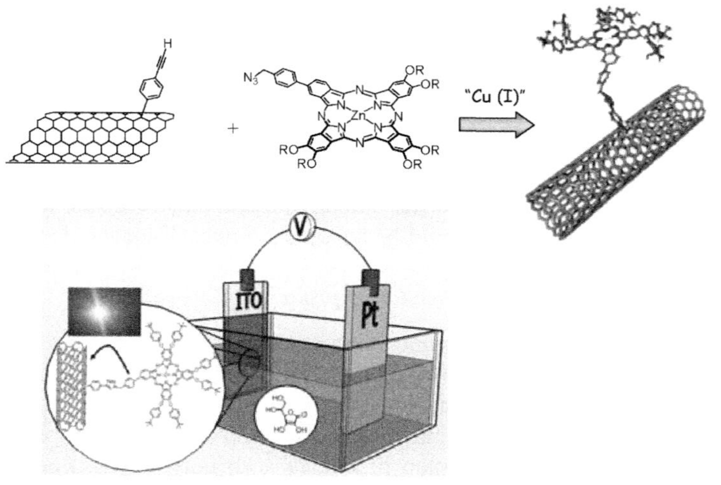

Figure 13.4 Functionalization of SWNTs with phthalocyanine via 'Click Chemistry' and schematic representation of the photoelectrochemical cell

The focal point of the dendrimer is the amino group of the functionalized nanotubes, while on average, two porphyrins units per dendron were estimated (Figure 13.5).

In response to visible light irradiation, the SWNT-$(H_2P)_x$ nanoconjugate gave rise to fast charge separation evolving from the photoexcited H_2P chromophores, the oxidized H_2P chromophore was identified through its fingerprint absorption in the 550–800 nm range, while the signature of the reduced SWNT appeared in the 850–1400 nm range.

Recently, dendrimers attached onto nanotubes were used as templates for the synthesis of metallic or semiconducting nanoparticles. The synthesis of Ag particles in PAMAM dendrimers linked to MWNT was reported [62], while, in another contribution, the fabrication of CdS nanoparticles in dendrimers was described [63]. These examples demonstrate that dendrimers can play several roles: amplifiers of functional groups on

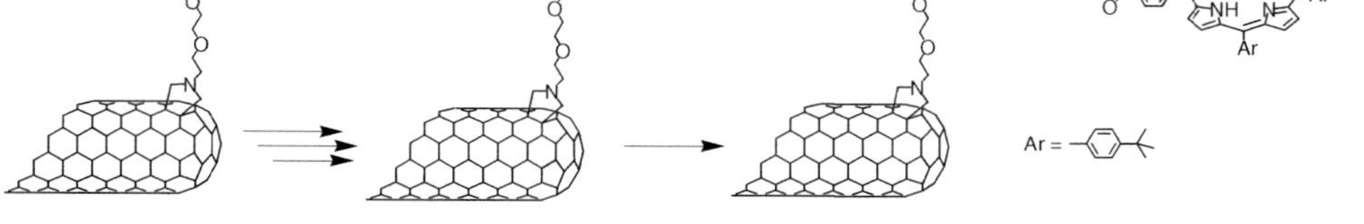

Figure 13.5 SWNT bearing PAMAM dendrimers and functionalized with porphyrins

nanotube surface and templates for nanoparticle growth. The latter would be a very elegant way to combine the properties of nanotubes (electrical conductivity, aspect ratio, etc.) with those of nanoparticles (catalysis, luminescence, etc.).

In general, covalent linkage of photo/electro-active groups to the SWNT sidewalls has allowed to obtain model compounds for studying electron and/or energy transfer processes in solution. However, such nanoconjugates remain difficult to synthesize and the degradation of the nanotube properties due to the oxidative treatments or to the insertion of sp^3 carbon in the conjugated π-system could be a serious drawback for the device efficiency.

13.3.1.2 Hybrid Covalent/Supramolecular Approach

The functionalization of carbon nanotubes with isoxazolines containing pyridyl pendant groups was reported recently [33]. In this construction, pyridyl moieties form axial complexes with zinc porphyrins. Upon photoexcitation, energy transfer between the singlet excited state of the porphyrin and the nanotube was observed in the complexes.

As discussed earlier, functionalization of SWNT with polymers can be of particular interest to introduce a large number of repetitive units on nanotubes without inducing the transformation ($sp^2 \rightarrow sp^3$) of too many carbon atoms of the framework. Several studies have been reported, for example the functionalization of SWNTs with poly(sodium 4-styrenesulfonate) to form SWNT-PSS^{n-} [64]. The negative charges on the polymer were used to form an electrostatic complex with a positively charged porphyrin (H_2P^{8+} and ZnP^{8+}) (Figure 13.6a) [65, 66]. The complementary approach was also explored: a positively-charged polymer has been grafted on the nanotube sidewall via free-radical polymerization of (vinylbenzyl)trimethylammonium chloride. The poly[(vinylbenzyl)trimethylammonium]-nanotube conjugate was then complexed with negatively-charged porphyrins (Figure 13.6b) [67]. In another example, polyvinylpyridines were attached to SWNTs and zinc porphyrins were complexed axially to the pyridine moieties (Figure 13.6c) [68].

The polymers covalently linked to the sidewalls ensure the solubility of the nanotubes and the interaction with the choromophores. Since only a limited number of polymer chains are attached on the nanotubes, the absorption spectra showed that the electronic fine structures of the SWNT are retained in the Vis-NIR region. The complexation of SWNT-polymer conjugates with porphyrin derivatives was followed by absorption and fluorescence spectroscopy. In the complexes, photoexcitation of the porphyrin chromophore led to a rapid and efficient intrahybrid charge transfer. The lifetime of the charge-separated radical ion pair was found to be on the order of several microseconds (11 μs for SWNT-PSS^{n-}/H_2P^{8+}, 2.2 μs for SWNT-$PVBTA^{n+}$/ZnP^{8-} and 3.8 μs for SWNT-PVP/ZnP). The properties of the SWNT-PSS^{n-}/Zn_2P^{8+} and of the SWNT-$PVBTA^{n+}$-ZnP^{8-} hybrids were studied by transient absorption and by photoelectrochemical measurements. The photocurrent measurements gave an internal photon-to-current efficiency (IPCE) of about 1% for SWNT-PSS^{n-}/Zn_2P^{8+} and about 3.8% when a potential of 0.5 V was applied.

13.3.1.3 Supramolecular Approach

For applications where the electronic properties of SWNT are important, the most promising approach remains the pure supramolecular functionalization (i.e. noncovalent association of nanotubes with electron donors). Interaction of nanotubes with functionalized surfactant

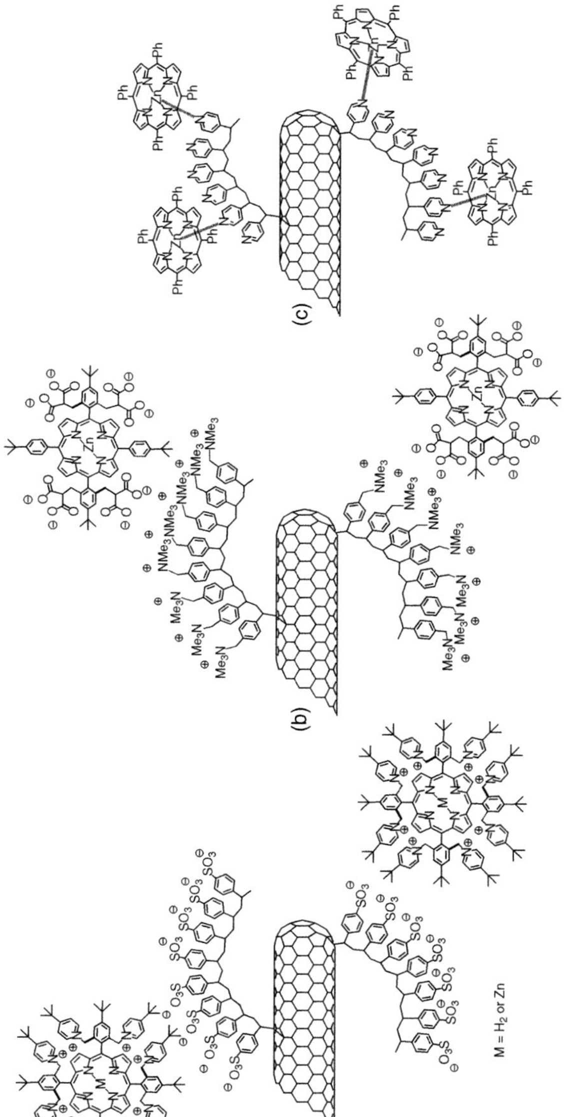

Figure 13.6 *SWNT functionalized with polystyrenesulfonate (PSS^{n-}), poly[(vinylbenzyl)trimethylammonium] ($PVBTA^{n+}$) or polyvinylpyridine (PVP) complexed with porphyrin derivatives*

is so far the easiest method to disperse nanotubes in water or in organic solvents and many examples have been reported during the last five years [69–73].

13.3.1.3.1 WRAPPING OF MOLECULAR ENTITIES AROUND SWNT

The dispersion of SWNTs by wrapping with a photoactive polymer was reported [74]. The polymer is in fact a copolymer of methyl methacrylate and porphyrin-modified methacrylic acid (Figure 13.7). The photophysical properties of the wrapped SWNTs were determined by means of transient absorption spectroscopy; the complex gave rise to a photoinduced electron transfer from the porphyrin moiety to the nanotube. A similar approach was described for the complexation of SWNTs with porphyrin-rich units [75]. Here a macromolecule containing 16 porphyrin pendant groups was used to disperse the nanotubes. The authors suggested that a diameter-selective dispersion was accomplished through non-covalent complexation of the nanotubes with the flexible porphyrin polypeptide. In addition, photoexcitation of the supramolecular complex afforded the long-lived charge-separated species. An interesting feature of these approaches is that the control of the chromophore quantity incorporated in the copolymer can allow for a fine tuning of the properties of the resulting nanohybrid.

Photovoltaic devices were fabricated very simply by mixing SWNTs with poly-3-octylthiophene (P3OT) in chloroform. The photoactive films were deposited by drop or spin casting from a solution on indium-tin oxide ITO on quartz substrates followed by evaporation of aluminium [76]. Al/SWNT-P3OT/ITO diodes with a low nanotube concentration ($<1\%$) showed photovoltaic behavior, with an open circuit voltage of 0.7–0.9 V. The short circuit current was increased by two orders of magnitude as compared with the pristine polymer diodes. It was proposed that the main reason for this increase was the photoinduced electron transfer at the polymer/nanotube interface. SWNT/conductive polymer composites

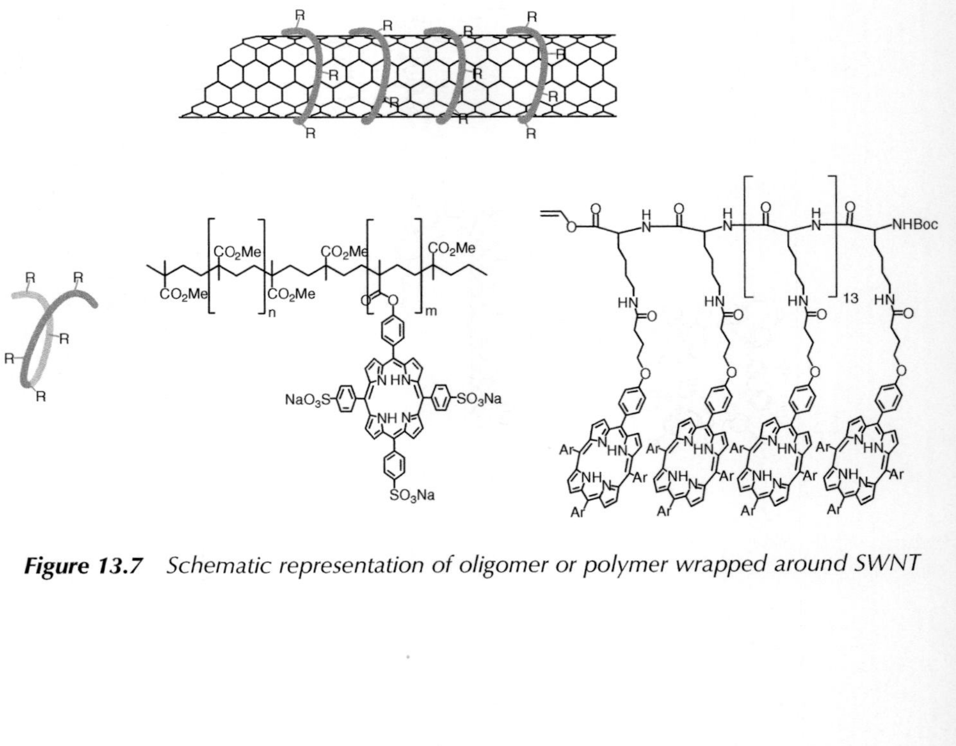

Figure 13.7 Schematic representation of oligomer or polymer wrapped around SWNT

may represent an alternative class of organic semiconducting materials promising for organic photovoltaic cells with improved performance.

13.3.1.3.2 COMPLEXATION BY π-STACKING

Large π-conjugated systems exhibit strong interactions with the nanotube sidewalls. In a family of aromatic compounds containing a polar head, polyaromatic compounds like anthracene and even better pyrene were shown to be able to give stable suspensions of SWNTs [73]. Aromatic macrocycles such as porphyrins or phthalocyanines are also suitable for nanotube dispersion. In particular, several groups demonstrated that monomeric [77–80] or polymeric porphyrins [81] as well as fused porphyrins [82] could stick to the nanotube surface by π-stacking interactions. However, despite of the simplicity and the versatility of these systems, their photovoltaic properties have not been tested.

The dispersion of SWNT with a pyrene derivative bearing an imidazole ring was achieved. The imidazolyl moiety was used for axial complexation of zinc porphyrin (ZnP) and naphthalocyanine (ZnNc) derivatives (Figure 13.8) [83]. Photophysical measurements showed efficient fluorescence quenching of the donor ZnP and ZnNc entities in the nanohybrids and revealed that the photoexcitation of the chromophores results in one-electron oxidation of the donor unit with a simultaneous one-electron reduction of SWNT. The experiments were also conducted in the presence of electron and hole mediators (hexyl-viologen dication and 1-benzyl-1,4-dihydronicotinamide respectively). Accumulation of the radical cation ($HV^{\bullet+}$) was observed in high yields, which provided additional proof for the occurrence of photoinduced charge separation. The same type of experiments were also carried out with porphyrins bearing 18-crown-6 substituents and complexed to nanotubes via pyrene ammonium salt anchors [84].

Moreover, pyrene-containing TTF and π-extended TTF were synthesized and combined with nanotubes via π-stacking interactions (Figure 13.9) [85, 86]. Photoexcitation of the SWNT-exTTF nanohybrid allowed for the first time a complete characterization of the radical ion pair state, especially in light of injecting electrons into the conduction band of SWNTs. These electrons, injected from photoexcited exTTF, shift the transitions that are associated with the van Hove singularities to lower energies.

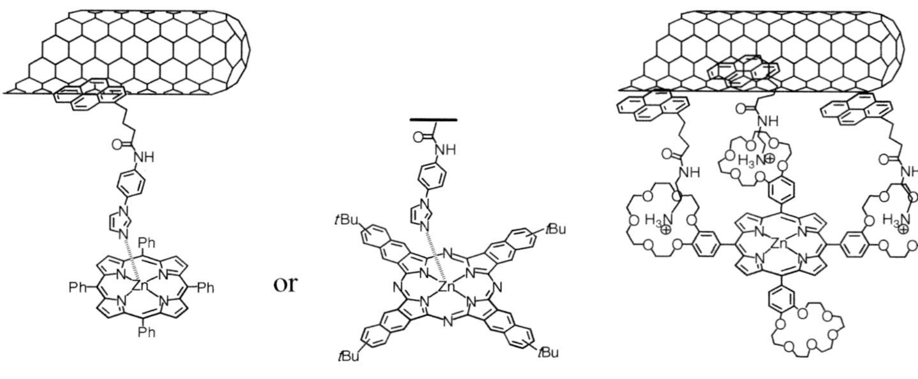

Figure 13.8 *SWNT-Pyrene supramolecular assemblies complexed with porphyrins or naphthalocyanines*

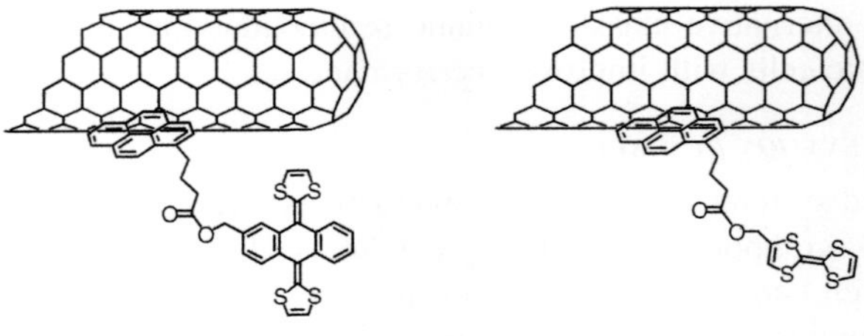

Figure 13.9 *SWNT functionalized by π-stacking with pyrene-TTF derivatives*

Amphiphilic pyrene derivatives are known to disperse carbon nanotubes through π-π interactions [72]. Using 1-(trimethylammonium acetyl)pyrene bromide (pyrene$^+$), CNTs were dispersed in aqueous media and were combined with negatively charged chromophores [87–90]. The interactions between pyrenes and nanotubes were investigated by absorption and emission spectroscopy. Immobilization of pyrene on SWNTs caused a slight red-shift (i.e. 1–2 nm) of the π–π^* transitions of pyrene$^+$ indicating electronic communication between the two components of the system. To prepare donor-acceptor complexes, the trimethylammonium group of pyrene$^+$ was used as an electrostatic anchor to bind anionic porphyrin (H_2P^{8-} and ZnP^{8-}) or polythiophene derivatives (Figure 13.10). In the SWNT/pyrene/porphyrin composite systems, fluorescence and transient absorption studies in solutions showed rapid intrahybrid electron transfer, creating intrinsically long-lived radical ion pairs. Following the initial charge separation event, the spectroscopic features of the oxidized donors disappear with time. Through analysis at several wavelengths, it was possible to obtain lifetimes for the newly formed ion-pair state of about 0.65 μs and 0.4 μs for H_2P^{8-} and ZnP^{8-} respectively [88].

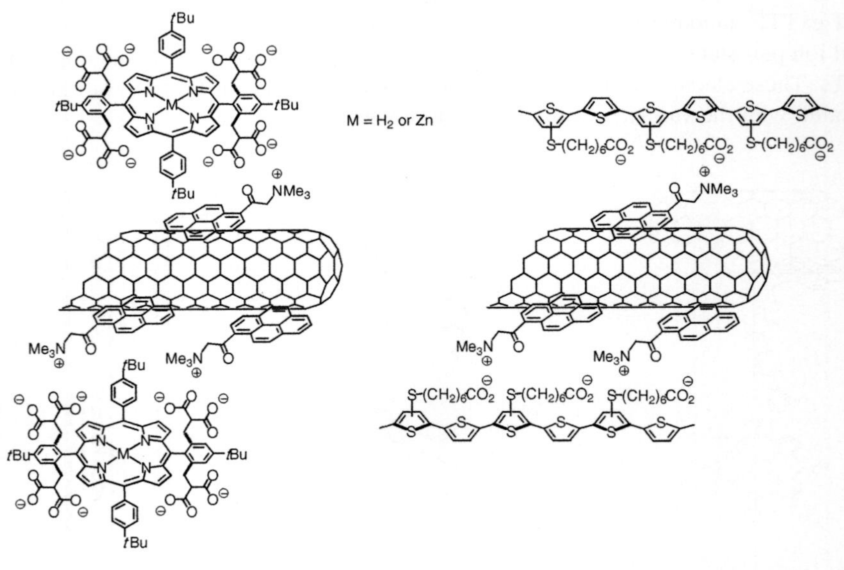

Figure 13.10 *Example of supramolecular donor/acceptor assemblies formed by electrostatic interactions*

The favorable charge separation features that result from the combination of SWNT with porphyrins in SWNT/pyrene$^+$/MP^{8-} (M = H_2 or Zn) are promising for the construction of photoactive electrode surfaces. Using electrostatically driven layer-by-layer (LBL) assembly technique, semitransparent ITO electrodes have been realized from SWNT/pyrene$^+$/MP^{8-} and SWNT/pyrene$^+$/polythiophene^{n-}. The ITO electrodes were first coated by poly (diallyl dimethylammonium) chloride (PDDA^{n+}) and sodium poly(styrene-4-sulfonate) (PSS^{n-}); the hydrophobic interactions between the surface and the polymer chains ensure the stability of the modified electrode on which the nanotubes will be deposited. After deposition of a layer of SWNT/pyrene$^+$ on PDDA^{n+}/PSS^{n-} coated ITO respectively, the layer of negatively charged porphyrin or polythiophene was deposited. The process was repeated to obtain electrodes containing up to 15 layers of SWNT/pyrene$^+$/MP^{8-} (or polythiophene^{n-}) [66, 89, 90].

The photoelectrochemical cells were finally constructed using a Pt electrode connected to the modified ITO electrode. An example of a cell is given in Figure 13.11. Upon illumination, electron transfer from the porphyrins to the nanotubes occurs. The electrons are then injected into the ITO layer then travelling to the Pt electrode. The oxidized porphyrins are converted to their ground state through the reduction via sodium ascorbate, which serves as a sacrificial electron donor. These systems gave rise to promising monochromatic internal photoconversion efficiencies (IPCE) of up to 8.5% [66].

Recently, a similar approach has been described: ITO electrodes were modified with carbon nanotubes and connected to ruthenium complexes via viologen derivatives [91]. The formation of the donor-acceptor hybrids (SWNT/viologen/Ru complexes) was ensured by electrostatic interactions: interaction of carboxylate groups on SWNTs and Ru complexes with the ammonium groups of the viologen derivative. Photovoltaic properties of the ITO-modified electrode were measured in a three-electrode glass cell containing a

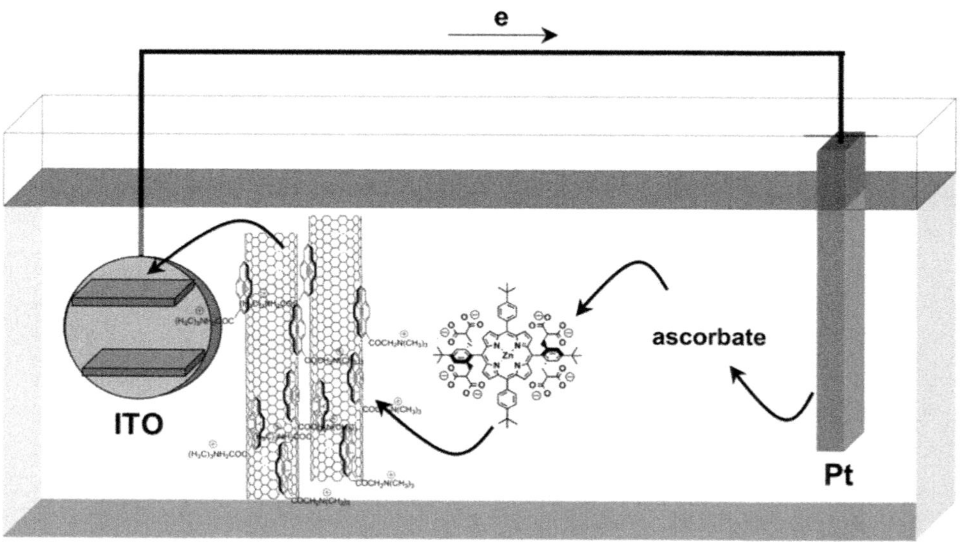

Figure 13.11 *Schematic representation of an electrochemical photovoltaic cell containing carbon nanotubes, positively charged pyrenes and negatively charged porphyrins*

reference electrode (Ag/AgCl) and a platinum counter electrode in a solution of I^-/I_3^- in acetonitrile. The device showed a photocurrent of 10.3 nA/cm^2 under white light illumination (100 mW/cm^2).

13.3.2 Functionalized Carbon Nanotubes for Electrical Measurements and Field Effect Transistors

SWNTs can be seen as sheets of graphene rolled up to form hollow tubes. Depending on the orientation of the tube axis with respect to the hexagonal lattice, the structure of a nanotube can be completely specified through its chiral vector, which is denoted by the chiral indices (n, m). Nanotubes in which $n = m$ are metallic and quasi metallic (with a tiny band gap) if n-m is divisible by 3. All other tubes are semiconducting with band gaps of the order of 0.5 eV. The electronic properties of SWNTs depend on the geometry of the tube: carbon nanotubes are to date the only material known to have this unique property [92]. The use of carbon nanotubes for producing field effect transistors (FETs) has been extensively studied, but for such applications, only semiconducting tubes are suitable [93].

The first examples of carbon nanotube field effect transistors (CNT-FETs) (Figure 13.12) were reported in 1998 simultaneously by Avouris [94] and Dekker [95]. The transistors were made from laser-ablation nanotubes dispersed by sonication and deposited on a Si/SiO_2 surface patterned with noble metal electrodes. Since SWNTs are made by rolled graphene sheets, all the atoms constituting the nanotubes are located on the surface and there are no constituent atoms inside as it is the case for bulk materials. All the atoms of a nanotube are in close contact with the environment, making nanotubes incredibly sensitive materials.

Recently, Blanchet *et al.* demonstrated on CNT-FET that the covalent functionalization of SWNTs with fluorinated olefins can drastically increase the on/off ratio (and so improve the transistor characteristics) without affecting too much the carrier mobility (at least for low functionalization degrees) [96]. As it has been already observed for diazonium additions to SWNTs [97], the fluorinated olefins reacted preferentially on metallic nanotubes *via* a [2 + 2] cycloaddition. For low concentrations of olefin (less than 0.01 moles of olefin per mole of SWNTs), mainly metallic nanotubes were functionalized while the semiconducting ones were unaffected. For high concentrations of olefin the semiconducting tubes reacted as well and the properties of the devices were, in this case, affected. Cabana and Martel also explored the influence of covalent functionalization on the properties of CNT-FET [98]. In

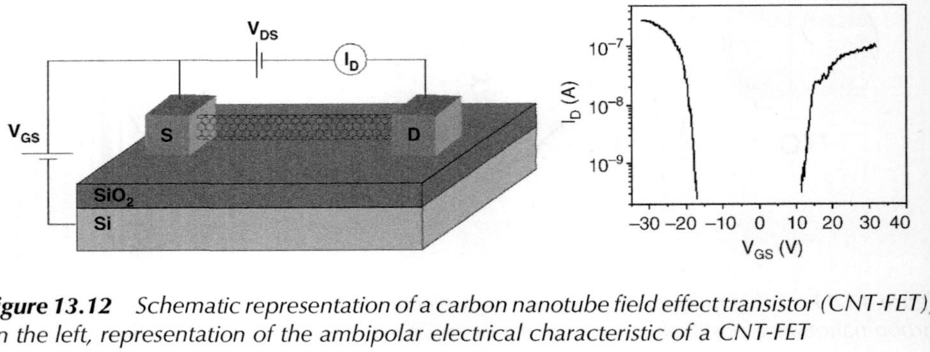

Figure 13.12 Schematic representation of a carbon nanotube field effect transistor (CNT-FET); on the left, representation of the ambipolar electrical characteristic of a CNT-FET

particular, they studied the reversibility of the diazonium addition reaction by cycling functionalization and defunctionalization processes. They demonstrated that the reaction was not fully reversible and led to an accumulation of defects onto the SWNT sidewall during the reaction/annealing cycles which degrade the properties of the devices.

In 2000, chemical sensors based on individual single-walled carbon nanotubes FET were described [99]. The electrical characteristics of the transistors changed upon exposure to gaseous molecules such as NO_2 or NH_3 in argon flow. The nanotube sensors exhibited a fast response and a high sensitivity for 1% NH_3 and 200 ppm NO_2. The reversibility was achieved by slow recovery under ambient conditions or by heating to high temperatures. At the same time, the effect of oxygen on CNT-FET was demonstrated [100]. By comparing the characteristics of CNT-FETs in vacuo and in the presence of oxygen, it was observed that exposure to air or oxygen influences the electrical resistance of the SWNT: the resistance decreased when the devices were exposed to oxygen. In these two examples, one can see that an adsorbed gas strongly influences the behaviour of the CNT-FETs, thus demonstrating that CNTs are extremely sensitive to their chemical environment. However, this extreme sensitivity is countered by a lack of selectivity. In 2003, the fabrication of CNT-FET arrays was reported for the detection of gas molecules [101]. Functionalization of the nanotubes with polymers was used to impart high sensitivity and selectivity to the sensors. Polyethyleneimine coating afforded n-type nanotube devices capable of detecting NO_2 at less than 1 ppb concentrations while being insensitive to NH_3. Coating nanotubes with nafion (a sulfonated tetrafluorethylene copolymer) allowed for the selective detection of NH_3. CNT-FETs functionalized non covalently with peptide-modified polymers were also tested for the selective detection of heavy metals [102]. They allowed for the detection of metal ions with concentrations in the pico- to micromolar range using several different peptides.

Another way was explored for ammonia detection [103, 104]. The approach was based on covalent functionalization of SWNTs with poly(*m*-aminobenzenesulfonic acid) (PABS). The polyaniline-based polymer linked to nanotubes was found to be sensitive to NH_3 by deprotonation of the sulfonic groups of the PABS side chains. The devices fabricated with SWNT-PABS showed an increase of resistance during exposure to ammonia compared to pristine nanotubes. The SWNT-PABS sensors rapidly recovered their resistance when NH_3 was replaced with nitrogen and this system allowed detection of NH_3 at concentrations as low as 5 ppm.

The use of carbon nanotubes as gas sensors went beyond the research step. Nanomix (www.nano.com) commercialized a detection platform based on carbon nanotube networks. The principle is based, as in the other cases, on the changes in the electronic characteristics of the device as it interacts with the analyte. The carbon nanotube networks were functionalized with different recognition agents to induce the proper performance characteristics such as specificity, sensitivity, etc [105–109]. Notably, the coating of CNT networks by metal nanoparticles for gas detection was described [109]. In this work, the differences in catalytic activity of 18 metals for detection of H_2, CH_4, CO, and H_2S gas were examined. The electronic response of metal-decorated CNT-FET devices to all four combustible gases was similar and resulted in a decrease in the device conductance. Furthermore, a sensor array was fabricated by site-selective electroplating of Pd, Pt, Rh, and Au metals on isolated SWNT networks located on a single chip (Figure 13.13). The resulting electronic sensor array was exposed to a randomized series of toxic/combustible gas and the electronic responses of all sensor elements

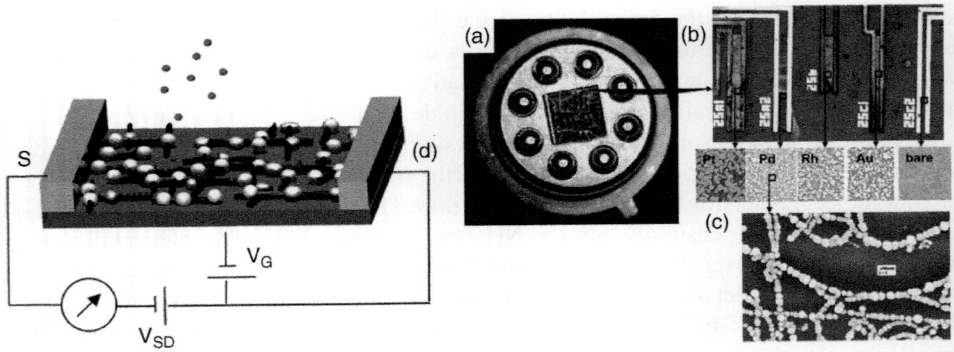

Figure 13.13 Gas sensors based on CNT-FETs functionalized with metal nanoparticles. Adapted from A. Star et al., J. Phys. Chem. B, **110**, 21014–21020 (2006), with permission from American Chemical Society

were recorded and analyzed using statistical analysis tools allowing the determination of the specific response of each element.

In the previous series of examples it has been possible to highlight the high sensitivity of CNT towards the detection of chemicals through direct interaction between the analyte and the nanotube or through interaction the analyte and a receptor in close interaction with the nanotube. If the receptor is light sensitive, its light-induced transformation can change the environment of the nanotube which can detect the event electrically.

Recently, the use of functionalized CNT-FETs as light detectors was reported [110]. The nanotube transistors were functionalized noncovalently with a zinc porphyrin derivative by drop casting of a solution of porphyrin onto the SWNT network. Upon illumination, the response of the device was a shift of the threshold voltage toward positive voltages, indicating hole doping of the SWNTs. The direction of the threshold voltage shift indicates a photoinduced electron transfer from the nanotubes to the porphyrins.

In a similar approach, photochromic molecules were used to switch the conductance of a single-walled carbon nanotube transistor. Spiropyrans [111] are well known photo-switchable molecules: the spiro form (colorless) in which the two aromatic parts of the molecule are separated by a spiro sp^3 carbon can open under UV irradiation leading to a completely conjugated zwitterionic molecule (called merocyanine). CNT-FETs were functionalized with spiropyran molecules and the influence of the irradiation on the transistor characteristics was studied [112]. These authors used an alkyl chain or a pyrene moiety as an anchor to hold the photoswitchable spiropyrans in proximity to the tube surface (Figure 13.14). Under UV irradiation, the conductance of the photosensitive device decreased significantly while the threshold voltage did not change appreciably and after visible irradiation, the initial characteristics were restored. Therefore, the decrease of conductivity is due to the isomerization of the spiropyrans and the authors explained this by the fact that the charge-separated state of the merocyanine introduces scattering sites for the carriers by creating localized dipole fields around the tubes. These sites then scatter charge which flows in the nearby SWNT channel and thereby lowers the mobility in the devices. Another possibility may be that the nearby phenoxide ion quenches the p-type carriers in the tubes and behaves like a charge trap.

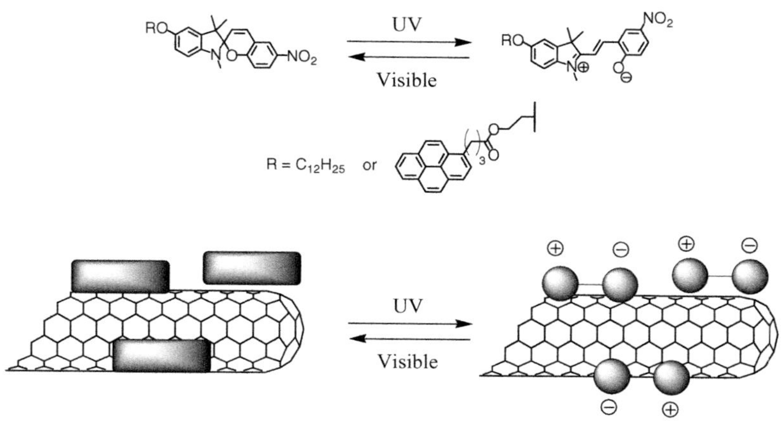

Figure 13.14 *SWNTs functionalized by π-stacking with photoactive switches*

Light induced isomerization of azobenzene-based chromophores was also used to control the CNT-FET characteristics [113]. First the nanotube-based transistors were fabricated and then the SWNTs were functionalized with an azobenzene derivative bearing a pyrene subunit using π-stacking interactions. Upon UV illumination, the conjugated chromophore gave rise to *cis-trans* isomerization leading to charge redistribution near the nanotube. This charge redistribution changed the local electrostatic environment, shifting the threshold voltage and increasing the conductivity of the nanotube transistor. The conductance change was reversible and repeatable over long periods of time. The same group recently demonstrated that these devices could be used as nanoscale color detectors [114]. The authors designed some azobenzene derivatives with specific absorption in the visible range that they attached onto SWNTs. The measurements suggested that upon illumination, the chromophores isomerized from the ground state trans configuration to the excited state cis configuration. The isomerization was accompanied by a large change in dipole moment which was detected by the nanotubes by changing its electrostatic environment.

In 2005, a method to connect molecules into gaps in nanotubes and to study the conductance through the junction was described [115]. To this aim, a gap in carbon nanotubes grown on surface was opened via oxidative cutting and the molecules under study were introduced by covalent coupling on carboxylic functions. This technique gave rise to SWNTs electrodes separated by gaps of ≤ 10 nanometers bridged with a series of molecules. It is important to note that the reconnected SWNTs recovered their original general electrical behavior (either metallic or semiconducting). The nanotube gaps were functionalized with several π-conjugated molecules like benzoxazole derivatives, oligothiophenylethynylenes, terpyridine complexes or oligoanilines (Figure 13.15a). A series of protonation and deprotonation experiments were performed on the oligoaniline-based devices and the result was that the protonated form was more conductive than the neutral form. These devices provided a local probe for monitoring pH on the basis of one or only a few molecules. Molecular switches based on photoisomerizable diarylethene derivatives were also introduced in the nanotube gaps (Figure 13.15b) [116]. As in the case of spiropyran-merocyanine systems, diarylethene derivatives [117] possess open and closed forms; however, in the case of diarylethene derivatives, the open form is nonconjugated

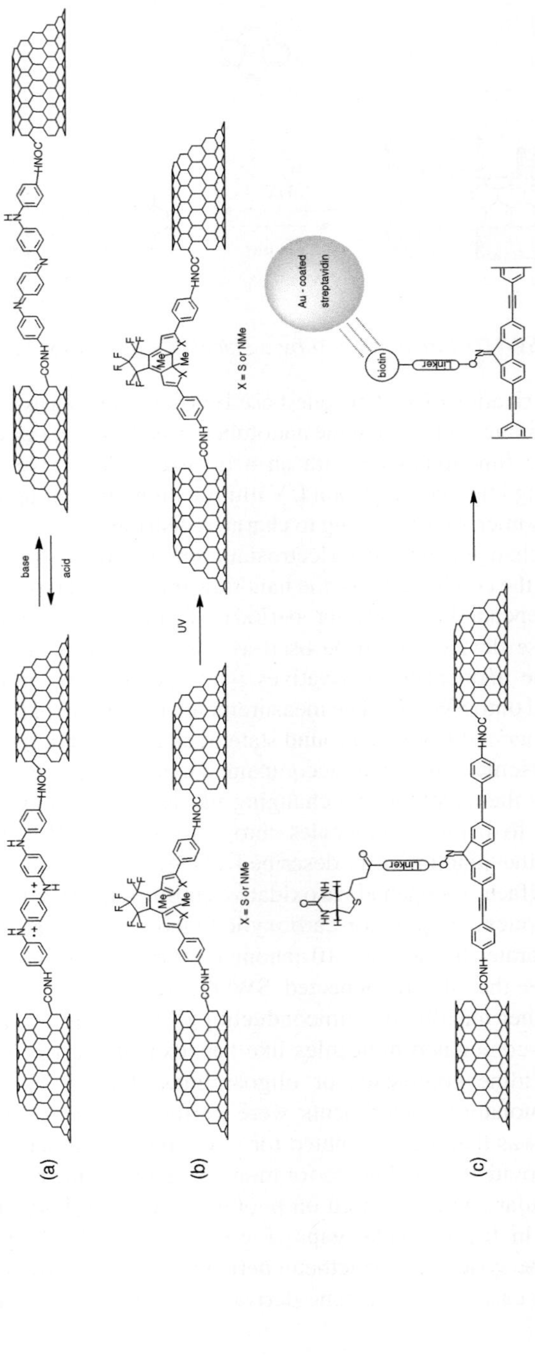

Figure 13.15 *Representation of nanotube gaps functionalized with active molecules*

while the closed form is conjugated. It was found that under UV irradiation (ring closure), the conductance of the devices increased up to 25-fold when thiophene-based switches were tested. However, since these structures did not permit a proof of the reversibility of the isomerization, the reversibility was demonstrated using pyrrole-based switches.

In another study, cut SWNTs were first functionalized with diaminofluorenone derivatives and then biotin probes were attached. These probes were linked to gold nanoparticles coated with streptavidin through formation of noncovalent complexes (Figure 13.15c) [118]. Each step of the chemical functionalization and biological assembly was detected electrically at the single event level. The formation of the biotinylated derivatives gave rise to a decrease in the ON-state resistance and threshold voltage of the devices. When coupling with streptavidin gold nanoparticles, large changes in the resistance were noticed. Because these devices are able to sense individual binding events, this approach makes possible the formation of ultrasensitive and real-time measurements of individual events. Very recently, the same group attached DNA between single-walled carbon nanotube electrodes and measured their electrical properties [119]. Several DNA sequences were measured, well-matched duplex DNA in the gap between the electrodes exhibited a resistance on the order of a few MΩ. However, a single mismatch in the DNA sequence led to a dramatic increase of the resistance of the devices.

In 2004, a CNT-FET was combined with a photosensitive polymer to fabricate optoelectronic memory devices [120]. The authors demonstrated that the electrical characteristics of CNT-FETs coated with poly{(*m*-phenylenevinylene)-*co*-[(2,5-dioctyloxy-*p*-phenylene)-vinylene]} (PmPV) or poly-3-octylthiophene (P3OT) changed upon illumination and required a long time, after the illumination was stopped, to recover their original conductance. However, the origin of this effect remained mainly unexplained until 2006, when the group of Bourgoin fabricated the same kind of CNT-FETs based on one or a few nanotubes and covered them with P3OT (Figure 13.16) [121]. They proposed a mechanism for the memory effect based on the trapping of photogenerated electrons at the nanotube/gate dielectric interface. The optical gating mechanism can be understood under the assumption that when the photoexcitation is turned on, a large quantity of electron-hole pairs are generated throughout the whole polymer film. Figure 13.16b shows that the CNT-FET gets settled in its ON-state regardless of V_{GS}. The p-type transistor induces the depletion of holes and accumulation of electrons at the polymer/SiO_2 interface. Because of their density and proximity, these trapped electrons define the nanotube conductance more efficiently than the back gate. When the illumination is stopped and due to the positive gate voltage, the electrons remain trapped and keep the transistor ON. The device is brought back to its initial state by applying a short negative V_{GS} pulse (–4 V).

13.3.3 Biosensors

Electronic detection of biomolecules using carbon nanotubes appears to be a field of intense research and several reviews have recently listed the principal efforts made in this direction [122–125]. It is well known that nanotube devices are very sensitive to their environment and then CNT-FETs have been developed and used for biosensing and biodetection. In this section, we will mainly focus on sensors based on nanotube field effect transistors but we will also give the principle and few examples of nanotube-based electrochemical biosensors

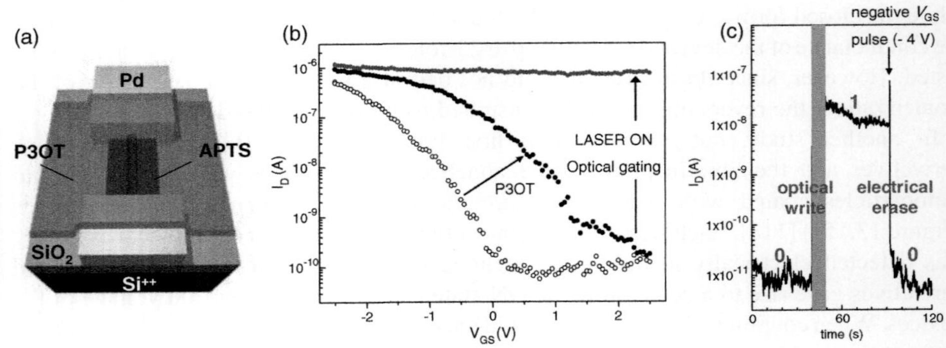

Figure 13.16 (a) Representation of the optical gated CNT-FET; (b) characteristics of the naked transistor in the dark (open black circles), coated with P3OT in the dark (filled black circles), and upon illumination ($\lambda = 457$ nm, gray circles); (c) principle of the writing and erasing of the memory device. The band in blue represents the light pulse use to write the electric information. From J. Borghetti et al., Adv. Mater., **18**, 2535–40 (2006), Copyright Wiley-VCH Verlag GmbH & Co. KGaA. Reproduced with permission from John Wiley and Sons

13.3.3.1 CNT-FET Based Biosensors

In 2002, the interactions between streptavidin (SA) and SWNTs were investigated. It was demonstrated that proteins could link through nonspecific binding (NSB) with nanotubes via hydrophobic interactions [126]. To prevent NSB, the authors functionalized nanotubes by co-adsorption of triton and poly(ethylene glycol) on the sidewalls; this polymer coating did not allow the fixation of streptavidin. On the contrary, specific binding of SA onto SWNTs was achieved by co-functionalization of nanotubes with biotin and protein-resistant polymers. A similar approach was used to fabricate nanotube-based biosensors capable of the selective detection of biological objects in solution [127]. In this work, several biological targets like biotin, staphylococcal protein A (SpA) and U1A antigen (a 33 kDa protein) were covalently attached to Tween 20 surfactant. These assemblies were immobilized on the nanotube surface for specific recognition respectively with streptavidin, immunoglobulin G (IgG) and 10E3 monoclonal antibodies (Figure 13.17a–c). The binding process of the biological objects with their respective targets immobilized on CNTs was followed by quartz crystal microbalance (QCM) analysis and CNT-FET electrical resistance measurements. For the microbalance measurements, the immunosensing system was assembled on a QCM crystal, whereas the electrical measurements were conducted on the CNTs bridging two microelectrodes. For example in the case of SpA-IgG system, the QCM frequency decreased upon the specific binding of the IgG whereas little perturbation was observed for proteins that did not interact specifically with SpA (Figure 13.17d). In separate experiments, it was demonstrated that CNT-FET coated with the SpA-Tween conjugate exhibited specific detection with an appreciable conductance change upon exposure to IgG but not to unrelated proteins (Figure 13.17e). Thus, specific interactions between antibodies and antigens can be probed by using nanotubes directly as electronic transducers.

Another example of Tween-functionalized CNT-FET was reported recently [128]. The nanotube-based sensor was designed to selectively detect thrombin (a coagulation protein)

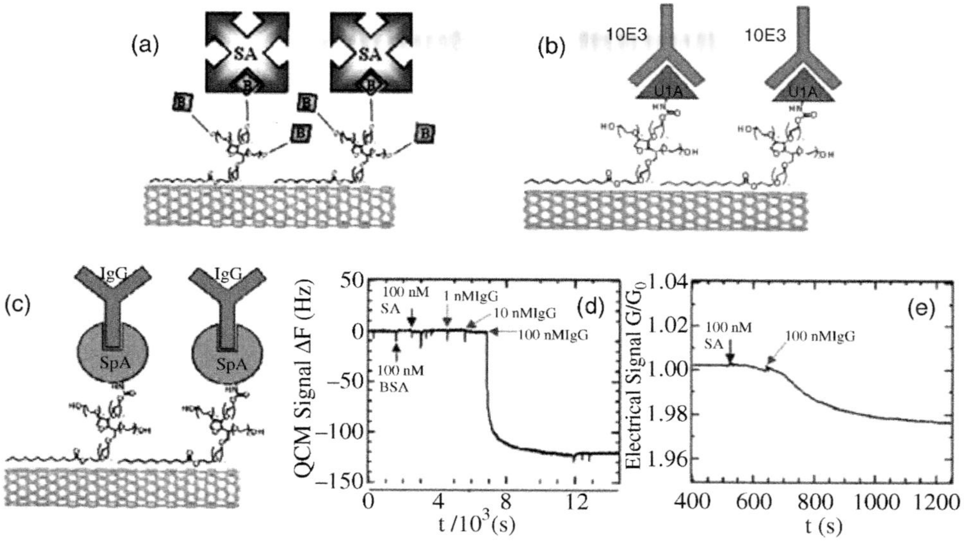

Figure 13.17 *(a–c) Schematic representation of SWNTs non-covalently functionalized with bioactive species (biotins, U1A antigen and staphylococcal protein A). The bioactive moieties attached to Tween 20 surfactant ensured the recognition properties of the system and allowed for specific detection; (d–e): examples of QCM and electrical characteristic curves are given after introduction of immunoglobulin G (IgG). Adapted from R. J. Chen et al., Proc. Natl. Acad. Sci. USA,* ***100****, 4984–9 (2003), Copyright (2003) National Academy of Sciences, USA*

through selective interaction of the protein with a specific DNA sequence. The fabrication of the sensor was based on the modification of activated Tween 20 adsorbed on the sidewalls of the CNT transistor with a 3′-amino-modified single stranded DNA. The electrical transfer characteristics of the CNT-FET were measured at each process stage. The immobilization of the ss-DNA caused a rightward shift in the gate-threshold voltage, presumably due to the negatively charged DNA backbone. Upon addition of thrombin, a sharp decrease in conductance of the device was observed. The sensitivity of the device increased strongly up to a protein concentration of about 100 nM and then became saturated around a concentration of 300 nM. The lowest detection limit of the sensor reported in this work was around 10 nM.

Steptavidin exhibits a strong affinity for biotin; the dissociation constant of the complex is on the order of $\sim 10^{-15}$, ranking among one of the stronger known noncovalent interactions. This explains why the biotin-streptavidin complex has been extensively used in nanotechnology. The fabrication of a CNT-FET sensitive to SA using a biotin functionalized carbon nanotube bridging two microelectrodes was reported [105]. The CNT-FET was coated with a mixture of polyethyleneimine/polyethyleneglycol (PEI/PEG) and the amino groups present on PEI were allowed to react with biotin *N*-hydroxy-succinimidyl ester to permit the specific binding of SA on the device. The characteristics of the transistor showed significant changes upon streptavidin exposure. The control experiment, realized on coated CNT-FET but in the absence of biotin on PEI, permitted a demonstration that SA did not bind to the polymer layer. AFM was used to show the SA binding: the biotinylated device was exposed to streptavidin labeled with gold nanoparticles. The presence of the nanoparticle on the images confirmed the presence of SA.

The realization of individual CNT-FET functionalized with glucose oxidase (GOx) bearing pyrene moieties through π-π interactions was described [129]. Controlled immobilization of GOx onto the sidewalls of a semiconducting SWNT resulted in the decrease of the nanotube conductance. The conductivity of the GOx-coated SWNTs exhibited a strong dependence on pH and showed an increase in conductance upon addition of glucose, suggesting their potential use as a sensor for enzymatic activity.

In 2005, the selective detection of a prostate specific antigen (PSA), which is an oncological marker for the presence of prostate cancer, was reported using both n-type In_2O_3nanowire (NW) and p-type carbon nanotube transistors [130]. The originality of this approach is the complementary detection of PSA using n- and p-type devices. To ensure the selective binding of PSA, the devices were functionalized with an anti-PSA monoclonal antibody. In the case of CNT-FETs, the SWNT surface was first functionalized with 1-pyrenebutanoic acid succinimidyl ester followed by treatment with the PSA antibody solution; for the nanowires, 3-phosphonopropionic acid was anchored on the In_2O_3 surface after which the antibodies were introduced after the activation of the carboxylic groups. Upon addition of PSA, an enhancement of the conductance for nanowire devices and a reduction of conductance for CNT-FET were observed. The gate dependence of both NW and SWNT devices changed and the threshold voltage of the NW device was shifted toward a more negative value. The influence of PSA resulted in an n-doping of the devices. The real-time detection measurements showed that upon exposure to PSA, the NW device showed an increase in conductance for protein concentration of 0.14 nM while the SWNT device exhibited a decrease in conductance for protein concentrations of 1.4 nM. Therefore, the sensitivity of the In_2O_3-based sensor was found to be better than the one for a carbon nanotube-based sensor.

13.3.3.2 Electrochemical Sensors

Carbon materials have been widely used as components in electrochemical biosensors for decades and notably in these last years, carbon nanotubes have attracted particular attention inside the scientific community [124, 131, 132]. The outstanding ability of CNTs to accept and transport charges makes them very promising materials for their incorporation in electrochemical sensing devices. Simple modification of glassy carbon or metal electrodes with unfunctionalized carbon nanotubes have been reported to improve the characteristic of the electrodes [132]. However, to improve solubility and specificity of nanotubes in the device, their chemical functionalization seems much more appropriate. For example, in order to detect DNA hybridization, amino-terminated ss-DNA were covalently linked to oxidized multi-walled carbon nanotubes (MWNTs) on a gold substrate [133] or on a glassy carbon electrode [134, 135]. In general, fabrication of electrochemical sensors requires the use of MWNTs because these nanotubes exhibit metallic character and they are easier to manipulate as compared with SWNTs. Only a few examples reported the use of modified SWNT for electrochemical sensing application.

The realization of an electrochemical glucose sensor was reported, based on the association of GOx on carbon nanotube modified electrodes [136]. SWNT arrays were fabricated on a gold electrode by covalent linkage of shortened carbon nanotubes on a

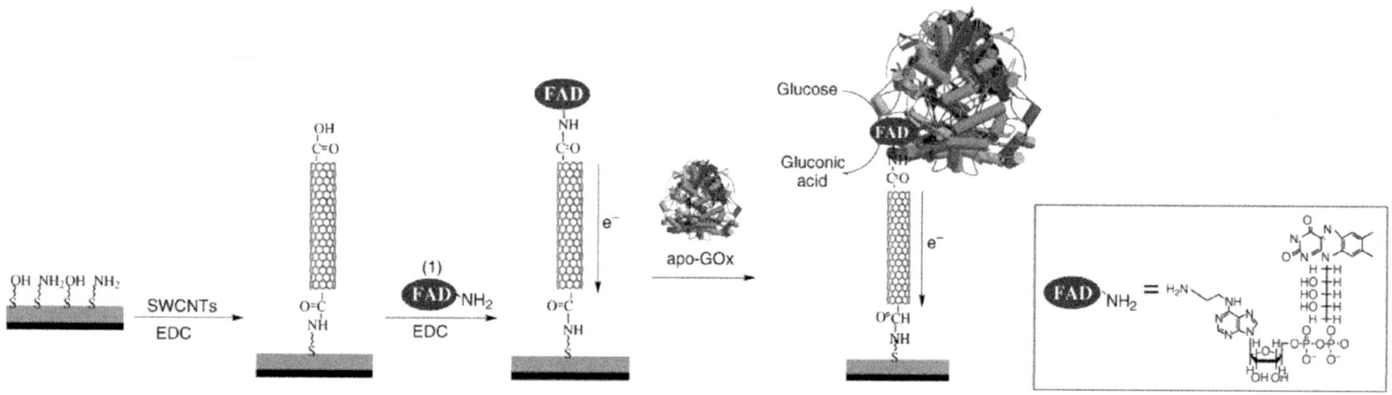

Figure 13.18 *Electrochemical sensor obtained by immobilization of glucose oxidase on SWNT vertically aligned on a gold electrode. From F. Patolsky et al., Angew. Chem., Int. Ed.,* ***43****, 2113–17 (2004), Copyright Wiley-VCH Verlag GmbH & Co. KGaA. Reproduced with permission from John Wiley and Sons*

cystamine/thioethanol mixed self-assembled monolayer. An amino derivative of the flavine adenine dinucleotide cofactor (FAD) was attached to the second extremity of the nanotubes and then GOx was reconstituted on the surface of the SWNT array. The nanotubes perpendicularly oriented to the surface played the role of electron acceptor and charge carrier from the reactive center to the electrode (Figure 13.18). QCM experiments were performed to estimate the surface coverage of the SWNTs and cyclic voltammetry of FAD modified nanotube electrode was used to prove that the FAD units were electrically connected to the surface. Coulometric assay of the FAD redox waves and microgravimetric QCM experiments indicated an average surface coverage of about $1.5{\cdot}10^{-10}\,\mathrm{mol\,cm^{-2}}$. Finally, the binding of GOx on the FAD cofactor was supported by AFM. Upon addition of glucose, an increase of the electrocatalytic anodic current was observed as the concentration of glucose increased and it was shown that the electron transfer rate was strongly dependent of the nanotube lengths: very short nanotube wires on surface (i.e. 25 nm) improved the electrical communication between proteins and electrode.

In a similar way, SWNTs functionalized with ferrocene were tested as amperometric glucose sensor [137]. For this purpose, a SWNT-Fc derivative (see Figure 13.3) was co-immobilized with glucose oxidase within a thin polypyrrole film adsorbed onto the glassy carbon electrode surface. The SWNT-Fc/GOx/polypyrrole films were examined for their catalytic properties with respect to glucose oxidation. For the detection, the modified electrode potential was held at 0.5 V which corresponds to oxidation of ferrocenyl moieties, and the anodic current was monitored while adding subsequent amounts of glucose. After each addition, an anodic current step was observed, reaching a stationary value within 10 s; the glucose sensitivity of the composite film was found to be about $0.3\,\mathrm{mA\,M^{-1}\,cm^{-2}}$. In contrast, no response was obtained with pure SWNT-Fc/polypyrrole films, *i.e.* in the absence of the redox protein.

13.4 Conclusion

The already rich variety of CNT applications can be further improved when these carbon cylinders are functionalized. The main reason is that *f*-CNT become more versatile while the new molecular pieces can be combined with the CNT properties. The result is a wide variety of interesting hybrid materials that can be used in a high number of applications. This field is in current expansion, as new methodologies for the functionalization of CNT are produced continuously and new applications are reported with improved performances.

References

[1] J. Liu, A. G. Rinzler, H. Dai, J. H. Hafner, R. K. Bradley, P. J. Boul, A. Lu, T. Iverson, K. Shelimov, C. B. Huffman, F. Rodriguez-Marcias, Y.-S. Shon, T. R. Lee, D. T. Colbert and R. E. Smalley, Fullerene Pipes, *Science*, **280**, 1253–1256 (1998).
[2] K. J. Ziegler, Z. Gu, H. Peng, E. L. Flor, R. H. Hauge and R. E. Smalley, Controlled Oxidative Cutting of Single-Walled Carbon Nanotubes, *J. Am. Chem. Soc.*, **127**, 1541–1547 (2005).
[3] L. Capes, E. Valentin, S. Esnouf, A. Ribayrol, O. Jost, A. Filoramo and J.-N. Patillon, High yield non destructive purification of single wall carbon nanotubes monitored by EPR measurements, *Proc. of the 2002 2nd IEEE Conference on Nanotechnology*, 439–442 (2002).

[4] C. G. Salzmann, S. A. Llewellyn, G. Tobias, M. A.H. Ward, Y. Huh and M. L.H. Green, The Role of Carboxylated Carbonaceous Fragments in the Functionalization and Spectroscopy of a Single-Walled Carbon-Nanotube Material, *Adv. Mater.*, **19**, 883–887 (2007).

[5] J. Chen, M. A. Hamon, H. Hu, Y. Chen, A. M. Rao, P. C. Eklund and R. C. Haddon, Solution Properties of Single-Walled Carbon Nanotubes, *Science*, **282**, 95–98 (1998).

[6] D. Tasis, N. Tagmatarchis, A. Bianco and M. Prato, Chemistry of Carbon Nanotubes, *Chem. Rev.*, **106**, 1105–1136 (2006).

[7] P. Singh, S. Campidelli, S. Giordani, D. Bonifazi, A. Bianco and M. Prato, Organic functionalisation and characterisation of single-walled carbon nanotubes, *Chem. Soc. Rev.*, **38**, 2214–2230 (2009).

[8] D. Bonifazi, C. Nacci, R. Marega, S. Campidelli, G. Ceballos, S. Modesti, M. Meneghetti and M. Prato, Microscopic and Spectroscopic Characterization of Paintbrush-like Single-walled Carbon Nanotubes, *Nano Lett.*, **6**, 1408–1414 (2006).

[9] J. L. Delgado, P. de la Cruz, F. Langa, A. Urbina, J. Casado and J. T. López Navarrete, Microwave-assisted sidewall functionalization of single-wall carbon nanotubes by Diels–Alder cycloaddition, *Chem. Commun.*, 1734–1735 (2004).

[10] H. Li, R. B. Martin, B. A. Harruff, R. A. Carino, L. F. Allard and Y.-P. Sun, Single-Walled Carbon Nanotubes Tethered with Porphyrins: Synthesis and Photophysical Properties, *Adv. Mater.*, **16**, 896–900 (2004).

[11] M. A. Herranz, N. Martín, S. Campidelli, M. Prato, G. Brehm and D. M. Guldi, Control over Electron Transfer in Tetrathiafulvalene-Modified Single-Walled Carbon Nanotubes, *Angew. Chem., Int. Ed.*, **45**, 4478–4482 (2006).

[12] L. Sheeney-Haj-Ichia, B. Basnar and I. Willner, Efficient Generation of Photocurrents by Using CdS/Carbon Nanotube Assemblies on Electrodes, *Angew. Chem., Int. Ed.*, **44**, 78–83 (2005).

[13] S. Banerjee and S. S. Wong, Synthesis and Characterization of Carbon Nanotube–Nanocrystal Heterostructures, *Nano Lett.*, **2**, 195–200 (2002).

[14] B. Zhao, H. Hu and R. C. Haddon, Synthesis and Properties of a Water-Soluble Single-Walled Carbon Nanotube-Poly(*m*-aminobenzene sulfonic acid) Graft Copolymer, *Adv. Funct. Mater.*, **14**, 71–76 (2004).

[15] M. D'Este, M. De Nardi and E. Menna, A Co-Functionalization Approach to Soluble and Functional Single-Walled Carbon Nanotubes, *Eur. J. Org. Chem.*, 2517–2522 (2006).

[16] N. Wong Shi Kam, T. C. Jessop, P. A. Wender and H. Dai, Nanotube Molecular Transporters: Internalization of Carbon Nanotube-Protein Conjugates into Mammalian Cells, *J. Am. Chem. Soc.*, **126**, 6850–6851 (2004).

[17] F. Pompeo and D. E. Resasco, Water Solubilization of Single-Walled Carbon Nanotubes by Functionalization with Glucosamine, *Nano Lett.*, **2**, 369–373 (2002).

[18] K. A. Williams, P. T.M. Veenhuizen, B. G. de la Torre, R. Eritja and C. Dekker, Carbon nanotubes with DNA recognition, *Nature*, **420**, 761–1761 (2002).

[19] M. Holzinger, O. Vostrowsky, A. Hirsch, F. Hennrich, M. Kappes, R. Weis and F. Jellen, Sidewall Functionalization of Carbon Nanotubes, *Angew. Chem., Int. Ed.*, **40**, 4002–4005 (2001).

[20] M. Holzinger, J. Abraham, P. Whelan, R. Graupner, L. Ley, F. Hennrich, M. Kappes and A. Hirsch, Functionalization of Single-Walled Carbon Nanotubes with (R-)Oxycarbonyl Nitrenes, *J. Am. Chem. Soc.*, **125**, 8566–8580 (2003).

[21] H. Peng, P. Reverby, V. N. Khabashesku and J. L. Margrave, Sidewall functionalization of single-walled carbon nanotubes with organic peroxides, *Chem. Commun.*, 362–363 (2003).

[22] H. Peng, L. B. Alemany, J. L. Margrave and V. N. Khabashesku, Sidewall Carboxylic Acid Functionalization of Single-Walled Carbon Nanotubes, *J. Am. Chem. Soc.*, **125**, 15174–15182 (2003).

[23] Y. Ying, R. K. Saini, F. Liang, A. K. Sadana and W. E. Billups, Functionalization of Carbon Nanotubes by Free Radicals, *Org. Lett.*, **5**, 1471–1473 (2003).

[24] F. Liang, A. K. Sadana, A. Peera, J. Chattopadhyay, Z. Gu, R. H. Hauge and W. E. Billups, A Convenient Route to Functionalized Carbon Nanotubes, *Nano Lett.*, **4**, 1257–1260 (2004).

[25] J. Bahr, J. Yang, D. V. Kosynkin, M. J. Bronikowski, R. E. Smalley and J. M. Tour, Functionalization of Carbon Nanotubes by Electrochemical Reduction of Aryl Diazonium Salts: A Bucky Paper Electrode, *J. Am. Chem. Soc.*, **123**, 6536–6542 (2001).

[26] J. Bahr and J. M. Tour, Highly Functionalized Carbon Nanotubes Using in Situ Generated Diazonium Compounds, *Chem. Mater.*, **13**, 3823–3824 (2001).

[27] C. A. Dyke and J. M. Tour, Solvent-Free Functionalization of Carbon Nanotubes, *J. Am. Chem. Soc.*, **125**, 1156–1157 (2003).

[28] C. A. Dyke and J. M. Tour, Unbundled and Highly Functionalized Carbon Nanotubes from Aqueous Reactions, *Nano Lett.*, **3**, 1215–1218 (2003).

[29] V. Georgakilas, K. Kordatos, M. Prato, D. M. Guldi, M. Holzinger and A. Hirsch, Organic Functionalization of Carbon Nanotubes, *J. Am. Chem. Soc.*, **124**, 760–761 (2002).

[30] V. Georgakilas, N. Tagmatarchis, D. Pantarotto, A. Bianco, J.-P. Briand and M. Prato, Amino acid functionalisation of water soluble carbon nanotubes, *Chem. Commun.*, 3050–3051 (2002).

[31] M. Alvaro, P. Atienzar, P. de la Cruz, J. L. Delgado, H. Garcia and F. Langa, Sidewall Functionalization of Single-Walled Carbon Nanotubes with Nitrile Imines. Electron Transfer from the Substituent to the Carbon Nanotube, *J. Phys. Chem. B*, **108**, 12691–12697 (2004).

[32] W. Zhang and T. M. Swager, Functionalization of Single-Walled Carbon Nanotubes and Fullerenes via a Dimethyl Acetylenedicarboxylate-4-Dimethylaminopyridine Zwitterion Approach, *J. Am. Chem. Soc.*, **129**, 7714–7715 (2007).

[33] M. Alvaro, P. Atienzar, P. de la Cruz, J. L. Delgado, V. Troiani, H. Garcia, F. Langa, A. Palkar and L. Echegoyen, Synthesis, Photochemistry, and Electrochemistry of Single-Wall Carbon Nanotubes with Pendent Pyridyl Groups and of Their Metal Complexes with Zinc Porphyrin. Comparison with Pyridyl-Bearing Fullerenes, *J. Am. Chem. Soc.*, **128**, 6626–6635 (2006).

[34] C. Ménard-Moyon, F. Dumas, E. Doris and C. Mioskowski, Functionalization of Single-Wall Carbon Nanotubes by Tandem High-Pressure/Cr(CO)6 Activation of Diels-Alder Cycloaddition, *J. Am. Chem. Soc.*, **128**, 14764–14765 (2006).

[35] V. N. Khabashesku, W. E. Billups and J. L. Margrave, Fluorination of Single-Wall Carbon Nanotubes and Subsequent Derivatization Reactions, *Acc. Chem. Res.*, **35**, 1087–1095 (2002).

[36] R. K. Saini, I. W. Chiang, H. Peng, R. E. Smalley, W. E. Billups, R. H. Hauge and J. L. Margrave, Covalent Sidewall Functionalization of Single Wall Carbon Nanotubes, *J. Am. Chem. Soc.*, **125**, 3617–3621 (2003).

[37] J. L. Stevens, A. Y. Huang, H. Peng, I. W. Chiang, V. N. Khabashesku and J. L. Margrave, Sidewall Amino-Functionalization of Single-Walled Carbon Nanotubes through Fluorination and Subsequent Reactions with Terminal Diamines, *Nano Lett.*, **3**, 331–336 (2003).

[38] L. Zhang, J. Yang, C. L. Edwards, L. B. Alemany, V. N. Khabashesku and A. R. Barron, Diels-Alder addition to fluorinated single walled carbon nanotubes, *Chem. Commun.*, 3265–3267 (2005).

[39] A. Javey, J. Guo, Q. Wang, M. Lundstrom and H. Dai, Ballistic carbon nanotube field-effect transistors, *Nature*, **424**, 654–657 (2003).

[40] S. Latil, S. Roche and J.-C. Charlier, Electronic Transport in Carbon Nanotubes with Random Coverage of Physisorbed Molecules, *Nano Lett.*, **5**, 2216–2219 (2005).

[41] T. Umeyama, M. Fujita, N. Tezuka, N. Kadota, Y. Matano, K. Yoshida and H. Imahori, Electrophoretic Deposition of Single-Walled Carbon Nanotubes Covalently Modified with Bulky Porphyrins on Nanostructured SnO2 Electrodes for Photoelectrochemical Devices, *J. Phys. Chem. C*, **111**, 11484–11493 (2007).

[42] D. M. Guldi, M. Marcaccio, D. Paolucci, F. Paolucci, N. Tagmatarchis, D. Tasis, E. Vázquez and M. Prato, Single-Wall Carbon Nanotube–Ferrocene Nanohybrids: Observing Intramolecular Electron Transfer in Functionalized SWNTs, *Angew. Chem., Int. Ed.*, **42**, 4206–4209 (2003).

[43] B. Ballesteros, S. Campidelli, G. de la Torre, C. Ehli, D. M. Guldi, M. Prato and T. Torres, Synthesis, characterization and photophysical properties of a SWNT-phthalocyanine hybrid, *Chem. Commun.*, 2950–2952 (2007).

[44] C. Ehli, S. Campidelli, F. G. Brunetti, M. Prato and D. M. Guldi, Single-wall carbon nanotube porphyrin nanoconjugates, *J. Porphyr. Phthalocyanines*, **11**, 442–447 (2007).

[45] S. Campidelli, B. Ballesteros, A. Filoramo, D. Díaz-Díaz, G. de la Torre, T. Torres, G. M.A. Rahman, C. Ehli, D. Kiessling, F. Werner, V. Sgobba, D. M. Guldi, C. Cioffi, M. Prato and J.-P. Bourgoin, Facile Decoration of Functionalized Single-Wall Carbon Nanotubes with Phthalocyanines via 'Click Chemistry', *J. Am. Chem. Soc.*, **130**, 11503–11509 (2008).
[46] H. C. Kolb, M. G. Finn and K. B. Sharpless, Click Chemistry: Diverse Chemical Function from a Few Good Reactions, *Angew. Chem., Int. Ed.*, **40**, 2004–2021 (2001).
[47] R. Huisgen, Cycloadditions – Definition, Classification, and Characterization, *Angew. Chem., Int. Ed. Engl.*, **7**, 321–328 (1968).
[48] V. D. Bock, H. Hiemstra and J. H. van Maarseveen, Cu(I)-Catalyzed Alkyne-Azide 'Click' Cycloadditions from a Mechanistic and Synthetic Perspective, *Eur. J. Org. Chem.*, 51–68 (2006).
[49] J.-F. Lutz, 1.3-Dipolar Cycloadditions of Azides and Alkynes: A Universal Ligation Tool in Polymer and Materials Science, *Angew. Chem., Int. Ed.*, **46**, 1018–1025 (2007).
[50] M. Meldal and C. W. Tornøe, Cu-Catalyzed Azide-Alkyne Cycloaddition, *Chem. Rev.*, **108**, 2952–3015 (2008).
[51] W. H. Binder and R. Sachsenhofer, 'Click' Chemistry in Polymer and Material Science: An Update, *Macromol. Rapid. Commun.*, **29**, 952–981 (2008).
[52] H. Li, F. Cheng, A. M. Duft and A. Adronov, Functionalization of Single-Walled Carbon Nanotubes with Well-Defined Polystyrene by 'Click' Coupling, *J. Am. Chem. Soc.*, **127**, 14518–14524 (2005).
[53] Z. Guo, L. Liang, J.-J. Liang, Y.-F. Ma, X.-Y. Yang, D.-M. Ren, Y.-S. Chen and J.-Y. Zheng, Covalently β-cyclodextrin modified single-walled carbon nanotubes: a novel artificial receptor synthesized by 'click' chemistry, *J. Nanopart. Res.*, **10**, 1077–1083 (2008).
[54] J. Bahr and J. M. Tour, Covalent chemistry of single-wall carbon nanotubes, *J. Mater. Chem.*, **12**, 1952–1958 (2002).
[55] F. Vögtle, *Dendrimers II Architecture, Nanostructure and Supramolecular Chemistry*; Springer Berlin, Germany (2000).
[56] J.-P. Majoral and A.-M. Caminade, Dendrimers Containing Heteroatoms (Si, P, B, Ge, or Bi), *Chem. Rev.*, **99**, 845–880 (1999).
[57] A. W. Bosman, H. M. Janssen and E. W. Meijer, About Dendrimers: Structure, Physical Properties, and Applications, *Chem. Rev.*, **99**, 1665–1668 (1999).
[58] G. R. Newkome, E. He and C. N. Moorefield, Suprasupermolecules with Novel Properties: Metallodendrimers, *Chem. Rev.*, **99**, 1689–1746 (1999).
[59] S. M. Grayson and J. M.J. Fréchet, Convergent Dendrons and Dendrimers: from Synthesis to Applications, *Chem. Rev.*, **101**, 3819–3867 (2001).
[60] D. A. Tomalia, H. Baker, J. R. Dewald, M. Hall, G. Kallos, S. Martin, J. Roeck, J. Ryder and P. Smith, A New Class of Polymers: Starburst-Dendritic Macromolecules, *Polym. J.*, **17**, 117–132 (1985).
[61] S. Campidelli, C. Sooambar, E. Lozano-Diz, C. Ehli, D. M. Guldi and M. Prato, Dendrimer-Functionalized Single-Wall Carbon Nanotubes: Synthesis, Characterization, and Photoinduced Electron Transfer, *J. Am. Chem. Soc.*, **128**, 12544–12552 (2006).
[62] L. Tao, G. Chen, G. Mantovani, S. York and D. M. Haddleton, Modification of multi-wall carbon nanotube surfaces with poly(amidoamine) dendrons: Synthesis and metal templating, *Chem. Commun.*, 4949–4951 (2006).
[63] S.-H. Hwang, C. N. Moorefield, P. Wang, K.-U. Jeong, S. Z.D. Cheng, K. K. Kotta and G. R. Newkome, Dendron-Tethered and Templated CdS Quantum Dots on Single-Walled Carbon Nanotubes, *J. Am. Chem. Soc.*, **128**, 7505–7509 (2006).
[64] S. Qin, D. Qin, W. T. Ford, J. E. Herrera, D. E. Resasco, S. M. Bachilo and R. B. Weisman, Solubilization and Purification of Single-Wall Carbon Nanotubes in Water by in Situ Radical Polymerization of Sodium 4-Styrenesulfonate, *Macromolecules*, **37**, 3965–3967 (2004).
[65] D. M. Guldi, G. M.A. Rahman, J. Ramey, M. Marcaccio, D. Paolucci, F. Paolucci, S. Qin, W. T. Ford, D. Balbinot, N. Jux, N. Tagmatarchis and M. Prato, Donor–acceptor nanoensembles of soluble carbon nanotubes, *Chem. Commun.*, 2034–2035 (2004).

[66] D. M. Guldi, G. M.A. Rahman, M. Prato, N. Jux, S. Qin and W. T. Ford, Single-Wall Carbon Nanotubes as Integrative Building Blocks for Solar-Energy Conversion, *Angew. Chem., Int. Ed.*, **44**, 2015–2018 (2005).
[67] G. M.A. Rahman, A. Troeger, V. Sgobba, D. M. Guldi, N. Jux, M. N. Tchoul, W. T. Ford, A. Mateo-Alonso and M. Prato, Improving Photocurrent Generation: Supramolecularly and Covalently Functionalized Single-Wall Carbon Nanotubes–Polymer/Porphyrin Donor-Acceptor Nanohybrids, *Chem. Eur. J.*, **14**, 8837–8846 (2008).
[68] D. M. Guldi, G. M.A. Rahman, S. Qin, M. Tchoul, W. T. Ford, M. Marcaccio, D. Paolucci, F. Paolucci, S. Campidelli and M. Prato, Versatile Coordination Chemistry towards Multifunctional Carbon Nanotube Nanohybrids, *Chem. Eur. J.*, **12**, 2152–2161 (2006).
[69] M. J. O'Connell, P. Boul, L. M. Ericson, C. Huffman, Y. Wang, E. Haroz, C. Kuper, J. Tour, K. D. Ausman and R. E. Smalley, Reversible water-solubilization of single-walled carbon nanotubes by polymer wrapping, *Chem. Phys. Lett.*, **342**, 265–271 (2001).
[70] A. Star, J. F. Stoddart, D. Steuerman, M. Diehl, A. Boukai, E. W. Wong, X. Yang, S.-W. Chung, H. Choi and J. R. Heath, Preparation and Properties of Polymer-Wrapped Single-Walled Carbon Nanotubes, *Angew. Chem., Int. Ed.*, **40**, 1721–1725 (2001).
[71] A. Star, D. W. Steuerman, J. R. Heath and J. F. Stoddart, Starched Carbon Nanotubes, *Angew. Chem., Int. Ed.*, **41**, 2508–2512 (2002).
[72] N. Nakashima, Y. Tomonari and H. Murakami, Water-Soluble Single-Walled Carbon Nanotubes via Noncovalent Sidewall-Functionalization with a Pyrene-Carrying Ammonium Ion, *Chem. Lett.*, 638–639 (2002).
[73] Y. Tomonari, H. Murakami and N. Nakashima, Solubilization of Single-Walled Carbon Nanotubes by using Polycyclic Aromatic Ammonium Amphiphiles in Water-Strategy for the Design of High-Performance Solubilizers, *Chem. Eur. J.*, **12**, 4027–4034 (2006).
[74] D. M. Guldi, H. Taieb, G. M.A. Rahman, N. Tagmatarchis and M. Prato, Novel Photoactive Single-Walled Carbon Nanotube-Porphyrin Polymer Wraps: Efficient and Long-Lived Intracomplex Charge Separation, *Adv. Mater.*, **17**, 871–875 (2005).
[75] K. Saito, V. Troiani, H. Qiu, N. Solladié, T. Sakata, H. Mori, M. Ohama and S. Fukuzumi, Nondestructive Formation of Supramolecular Nanohybrids of Single-Walled Carbon Nanotubes with Flexible Porphyrinic Polypeptides, *J. Phys. Chem. C*, **111**, 1194–1199 (2007).
[76] E. Kymakis and G. A.J. Amaratunga, Single-wall carbon nanotube/conjugated polymer photovoltaic devices, *Appl. Phys. Lett.*, **80**, 112–114 (2002).
[77] H. Murakami, T. Nomura and N. Nakashima, Noncovalent porphyrin-functionalized single-walled carbon nanotubes in solution and the formation of porphyrin–nanotube nanocomposites, *Chem. Phys. Lett.*, **378**, 481–485 (2003).
[78] J. Chen and C. P. Collier, Noncovalent Functionalization of Single-Walled Carbon Nanotubes with Water-Soluble Porphyrins, *J. Phys. Chem. B*, **109**, 7605–7609 (2005).
[79] G. M.A. Rahman, D. M. Guldi, S. Campidelli and M. Prato, Electronically interacting single wall carbon nanotube–porphyrin nanohybrids, *J. Mater. Chem.*, **16**, 62–65 (2006).
[80] D. R. Kauffman, O. Kuzmych and A. Star, Interactions between Single-Walled Carbon Nanotubes and Tetraphenyl Metalloporphyrins: Correlation between Spectroscopic and FET Measurements, *J. Phys. Chem. C*, **111**, 3539–3543 (2007).
[81] F. Cheng and A. Adronov, Noncovalent Functionalization and Solubilization of Carbon Nanotubes by Using a Conjugated Zn–Porphyrin Polymer, *Chem. Eur. J.*, **12**, 5053–5059 (2006).
[82] F. Cheng, S. Zhang, A. Adronov, L. Echegoyen and F. Diederich, Triply Fused ZnII-Porphyrin Oligomers: Synthesis, Properties, and Supramolecular Interactions with Single-Walled Carbon Nanotubes (SWNTs), *Chem. Eur. J.*, **12**, 6062–6070 (2006).
[83] R. Chitta, A. D. Sandanayaka, A. L. Schumacher, L. D'Souza, Y. Araki, O. Ito and F. D'Souza, Donor-Acceptor Nanohybrids of Zinc Naphthalocyanine or Zinc Porphyrin Noncovalently Linked to Single-Wall Carbon Nanotubes for Photoinduced Electron Transfer, *J. Phys. Chem. C*, **111**, 6947–6955 (2007).
[84] F. D'Souza, R. Chitta, A. S.D. Sandanayaka, N. K. Subbaiyan, L. D'Souza, Y. Araki and O. Ito, Self-Assembled Single-Walled Carbon Nanotube: Zinc–Porphyrin Hybrids through

Ammonium Ion–Crown Ether Interaction: Construction and Electron Transfer, *Chem. Eur. J.*, **13**, 8277–8284 (2007).
[85] M. A. Herranz, C. Ehli, S. Campidelli, M. Gutiérrez, G. L. Hug, K. Ohkubo, S. Fukuzumi, M. Prato, N. Martín and D. M. Guldi, Spectroscopic Characterization of Photolytically Generated Radical Ion Pairs in Single-Wall Carbon Nanotubes Bearing Surface-Immobilized Tetrathiafulvalenes, *J. Am. Chem. Soc.*, **130**, 66–73 (2008).
[86] C. Ehli, D. M. Guldi, M. A. Herranz, N. Martín, S. Campidelli and M. Prato, Pyrene-tetrathiafulvalene supramolecular assembly with different types of carbon nanotubes, *J. Mater. Chem.*, **18**, 1498–1503 (2008).
[87] D. M. Guldi, G. M.A. Rahman, N. Jux, N. Tagmatarchis and M. Prato, Integrating Single-Wall Carbon Nanotubes into Donor-Acceptor Nanohybrids, *Angew. Chem., Int. Ed.*, **43**, 5526–5530 (2004).
[88] C. Ehli, G. M.A. Rahman, N. Jux, D. Balbinot, D. M. Guldi, F. Paolucci, M. Marcaccio, D. Paolucci, M. Melle-Franco, F. Zerbetto, S. Campidelli and M. Prato, Interactions in Single Wall Carbon Nanotubes/Pyrene/Porphyrin Nanohybrids, *J. Am. Chem. Soc.*, **128**, 11222–11231 (2006).
[89] V. Sgobba, G. M.A. Rahman, D. M. Guldi, N. Jux, S. Campidelli and M. Prato, Supramolecular Assemblies of Different Carbon Nanotubes for Photoconversion Processes, *Adv. Mater.*, **18**, 2264–2269 (2006).
[90] G. M.A. Rahman, D. M. Guldi, R. Cagnoli, A. Mucci, L. Schenetti, L. Vaccari and M. Prato, Combining Single Wall Carbon Nanotubes and Photoactive Polymers for Photoconversion, *J. Am. Chem. Soc.*, **127**, 10051–10057 (2005).
[91] W. Lee, J. Lee, S.-H. Lee, J. Chang, W. Yi and S.-H. Han, Improved Photocurrent in Ru(2,2′-bipyridine-4,4′-dicarboxylic acid)-2(NCS)2/di(3-aminopropyl) viologen/Single-Walled Carbon Nanotubes/Indium Tin Oxide System: Suppression of Recombination Reaction by Use of Single-Walled Carbon Nanotubes, *J. Phys. Chem. C*, **111**, 9110–9115 (2007).
[92] R. Saito, M. Fujita, G. Dresselhaus and M. S. Dresselhaus, Electronic structure of chiral graphene tubules, *Appl. Phys. Lett.*, **60**, 2204–2206 (1992).
[93] M. Freitag, Carbon nanotube electronics and devices; In *Carbon Nanotubes: Properties and Applicatons*; M. J. O'Connell, ed. CRC Press: 2006; pp. 83–117.
[94] R. Martel, T. Schmidt, H. R. Shea, T. Hertel and P. Avouris, Single- and multi-wall carbon nanotube field-effect transistors, *Appl. Phys. Lett.*, **73**, 2447–2449 (1998).
[95] S. J. Tans, A. R.M. Verschueren and C. Dekker, Room-temperature transistor based on a single carbon nanotube, *Nature*, **393**, 49–52 (1998).
[96] M. Kanungo, H. Hu, G. G. Malliaras and G. B. Blanchet, Suppression of Metallic Conductivity of Single-Walled Carbon Nanotubes by Cycloaddition Reactions, *Science*, **323**, 234–237 (2009).
[97] M. S. Strano, C. A. Dyke, M. L. Ursey, P. W. Barone, M. J. Allen, H. W. Shan, C. Kittrell, R. H. Hauge, J. M. Tour and R. E. Smalley, Electronic Structure Control of Single-Walled Carbon Nanotube Functionalization, *Science*, **301**, 1519–1522 (2003).
[98] J. Cabana and R. Martel, Probing the Reversibility of Sidewall Functionalization Using Carbon Nanotube Transistors, *J. Am. Chem. Soc.*, **129**, 2244–2245 (2007).
[99] J. Kong, N. R. Franklin, C. Zhou, M. G. Chapline, S. Peng, K. Cho and H. Dai, Nanotube Molecular Wires as Chemical Sensors, *Science*, **287**, 622–625 (2000).
[100] P. G. Collins, K. Bradley, M. Ishigami and A. Zettl, Extreme Oxygen Sensitivity of Electronic Properties of Carbon Nanotubes, *Science*, **287**, 1801–1804 (2000).
[101] P. Qi, O. Vermesh, M. Grecu, Q. Wang, H. Dai, S. Peng and K. J. Cho, Toward Large Arrays of Multiplex Functionalized Carbon Nanotube Sensors for Highly Sensitive and Selective Molecular Detection, *Nano Lett.*, **3**, 347–351 (2003).
[102] E. S. Forzani, X. Li, P. Zhang, N. Tao and R. Zhang, Tuning the Chemical Selectivity of SWNT-FETs for Detection of Heavy-Metal Ions, *Small*, **2**, 1283–1291 (2006).
[103] E. Bekyarova, M. Davis, T. Burch, M. E. Itkis, B. Zhao, S. Sunshine and R. C. Haddon, Chemically Functionalized Single-Walled Carbon Nanotubes as Ammonia Sensors, *J. Phys. Chem. B*, **108**, 19717–19720 (2004).

[104] E. Bekyarova, I. Kalinina, M. E. Itkis, L. Beer, N. Cabrera and R. C. Haddon, Mechanism of Ammonia Detection by Chemically Functionalized Single-Walled Carbon Nanotubes: In Situ Electrical and Optical Study of Gas Analyte Detection, *J. Am. Chem. Soc.*, **129**, 10700–10706 (2007).

[105] A. Star, J. C.P. Gabriel, K. Bradley and G. Grüner, Electronic Detection of Specific Protein Binding Using Nanotube FET Devices, *Nano Lett.*, **3**, 459–463 (2003).

[106] A. Star, T.-R. Han, V. Joshi, D. Steuerman and J. R. Stetter, Sensing with Nafion Coated Carbon Nanotube Field-Effect Transistors, *Electroanalysis*, **16**, 108–112 (2004).

[107] A. Star, T.-R. Han, V. Joshi, J. C.P. Gabriel and G. Grüner, Nanoelectronic Carbon Dioxide Sensors, *Adv. Mater.*, **16**, 2049–2052 (2004).

[108] A. Star, E. Tu, J. Niemann, J. C.P. Gabriel, C. S. Joiner and C. Valcke, Label-free detection of DNA hybridization using carbon nanotube network field-effect transistors, *Proc. Natl. Acad. Sci. USA*, **103**, 921–926 (2006).

[109] A. Star, V. Joshi, S. Skarupo, D. Thomas and J. C. P. Gabriel, Gas Sensor Array Based on Metal-Decorated Carbon Nanotubes, *J. Phys. Chem. B*, **110**, 21014–21020 (2006).

[110] D. S. Hecht, R. J.A. Ramirez, M. Briman, E. Artukovic, K. S. Chichak, J. F. Stoddart and G. Grüner, Bioinspired Detection of Light Using a Porphyrin-Sensitized Single-Wall Nanotube Field Effect Transistor, *Nano Lett.*, **6**, 2031–2036 (2006).

[111] G. Berkovic, V. Krongauz and V. Weiss, Spiropyrans and Spirooxazines for Memories and Switches, *Chem. Rev.*, **100**, 1741–1753 (2000).

[112] X. Guo, L. Huang, S. O'Brien, P. Kim and C. Nuckolls, Directing and Sensing Changes in Molecular Conformation on Individual Carbon Nanotube Field Effect Transistors, *J. Am. Chem. Soc.*, **127**, 15045–15047 (2005).

[113] J. M. Simmons, I. In, V. E. Campbell, T. J. Mark, F. Léonard, P. Gopalan and M. A. Eriksson, Optically Modulated Conduction in Chromophore-Functionalized Single-Wall Carbon Nanotubes, *Phys. Rev. Lett.*, **98**, 086802-1-4 (2007).

[114] X. Zhou, T. Zifer, B. M. Wong, K. L. Krafcik, F. Léonard and A. L. Vance, Color Detection Using Chromophore-Nanotube Hybrid Devices, *Nano Lett.*, **9**, 1028–1033 (2009).

[115] X. Guo, J. P. Small, J. E. Klare, Y. Wang, M. S. Purewal, I. W. Tam, B. H. Hong, R. Caldwell, L. Huang, S. O'Brien, J. Yan, R. Breslow, S. J. Wind, J. Hone, P. Kim and C. Nuckolls, Covalently Bridging Gaps in Single-Walled Carbon Nanotubes with Conducting Molecules, *Science*, **311**, 356–359 (2006).

[116] A. C. Whalley, M. L. Steigerwald, X. Guo and C. Nuckolls, Reversible Switching in Molecular Electronic Devices, *J. Am. Chem. Soc.*, **129**, 12590–12591 (2007).

[117] M. Irie, Diarylethenes for Memories and Switches, *Chem. Rev.*, **100**, 1685–1716 (2000).

[118] X. Guo, A. C. Whalley, J. E. Klare, L. Huang, S. O'Brien, M. L. Steigerwald and C. Nuckolls, Single-Molecule Devices as Scaffolding for Multicomponent Nanostructure Assembly, *Nano Lett.*, **7**, 1119–1122 (2007).

[119] X. Guo, A. A. Gorodetsky, J. Hone, J. K. Barton and C. Nuckolls, Conductivity of a single DNA duplex bridging a carbon nanotube gap, *Nat. Nanotechnol.*, **3**, 163–167 (2008).

[120] A. Star, Y. Lu, K. Bradley and G. Grüner, Nanotube Optoelectronic Memory Devices, *Nano Lett.*, **4**, 1587–1591 (2004).

[121] J. Borghetti, V. Derycke, S. Lenfant, P. Chenevier, A. Filoramo, M. Goffman, D. Vuillaume and J.-P. Bourgoin, Optoelectronic Switch and Memory Devices Based on Polymer-Functionalized Carbon Nanotube Transistors, *Adv. Mater.*, **18**, 2535–2540 (2006).

[122] E. Katz and I. Willner, Biomolecule-Functionalized Carbon Nanotubes: Applications in Nanobioelectronics, *ChemPhysChem.*, **5**, 1084–1104 (2004).

[123] G. Grüner, Carbon nanotube transistors for biosensing applications, *Anal. Bioanal. Chem.*, **384**, 322–325 (2006).

[124] K. Balasubramanian and M. Burghard, Biosensors based on carbon nanotubes, *Anal. Bioanal. Chem.*, **385**, 452–468 (2006).

[125] B. L. Allen, P. D. Kichambare and A. Star, Carbon Nanotube Field-Effect-Transistor-Based Biosensors, *Adv. Mater.*, **19**, 1439–1451 (2007).

[126] M. Shim, N. Wong Shi Kam, R. J. Chen, Y. Li and H. Dai, Functionalization of Carbon Nanotubes for Biocompatibility and Biomolecular Recognition, *Nano Lett.*, **2**, 285–288 (2002).

[127] R. J. Chen, S. Bangsaruntip, K. A. Drouvalakis, N. Wong Shi Kam, M. Shim, Y. Li, W. Kim, P. J. Utz and H. Dai, Noncovalent functionalization of carbon nanotubes for highly specific electronic biosensors, *Proc. Natl. Acad. Sci. USA*, **100**, 4984–4989 (2003).

[128] H.-M. So, K. Won, Y. H. Kim, B.-K. Kim, B. H. Ryu, P. S. Na, H. Kim and J.-O. Lee, Single-Walled Carbon Nanotube Biosensors Using Aptamers as Molecular Recognition Elements, *J. Am. Chem. Soc.*, **127**, 11906–11907 (2005).

[129] K. Besteman, J.-O. Lee, F. G.M. Wiertz, H. A. Heering and C. Dekker, Enzyme-Coated Carbon Nanotubes as Single-Molecule Biosensors, *Nano Lett.*, **3**, 727–730 (2003).

[130] C. Li, M. Curreli, H. Lin, B. Lei, F. N. Ishikawa, R. Datar, R. J. Cote, M. E. Thompson and C. Zhou, Complementary Detection of Prostate-Specific Antigen Using In2O3 Nanowires and Carbon Nanotubes, *J. Am. Chem. Soc.*, **127**, 12484–12485 (2005).

[131] J. Wang, Carbon-Nanotube Based Electrochemical Biosensors: A Review, *Electroanalysis*, **17**, 7–17 (2005).

[132] G. A. Rivas, M. D. Rubianes, M. C. Rodŕguez, N. F. Ferreyra, G. L. Luque, M. L. Pedano, S. A. Miscoria and C. Parrado, Carbon nanotubes for electrochemical biosensing, *Talanta*, **74**, 291–307 (2007).

[133] J. Li, H. T. Ng, A. Cassell, W. Fan, H. Chen, Q. Ye, J. Koehne, J. Han and M. Meyyappan, Carbon Nanotube Nanoelectrode Array for Ultrasensitive DNA Detection, *Nano Lett.*, **3**, 597–602 (2003).

[134] H. Cai, X. Cao, Y. Jiang, P. He and Y. Fang, Carbon nanotube-enhanced electrochemical DNA biosensor for DNA hybridization detection, *Anal. Bioanal. Chem.*, **375**, 287–293 (2004).

[135] S. G. Wang, R. Wang, P. J. Sellin and Q. Zhang, DNA biosensors based on self-assembled carbon nanotubes, *Biochem. Biophys. Res. Commun.*, **325**, 1433–1437 (2004).

[136] F. Patolsky, Y. Weizmann and I. Willner, Long-Range Electrical Contacting of Redox Enzymes by SWCNT Connectors, *Angew. Chem., Int. Ed.*, **43**, 2113–2117 (2004).

[137] A. Callegari, S. Cosnier, M. Marcaccio, D. Paolucci, F. Paolucci, V. Georgakilas, N. Tagmatarchis, E. Vázquez and M. Prato, Functionalised single wall carbon nanotubes/polypyrrole composites for the preparation of amperometric glucose biosensors, *J. Mater. Chem.*, **14**, 807–810 (2003).

14

Dispersion and Separation of Single-walled Carbon Nanotubes

Yutaka Maeda,[a] Takeshi Akasaka,[b] Jing Lu[c] and Shigeru Nagase[d]

[a]*Department of Chemistry, Tokyo Gakugei University, Koganei, Tokyo Japan; PRESTO, Japan Science and Technology Agency, Chiyoda, Tokyo, Japan*
[b]*Tsukuba Advanced Research Alliance, University of Tsukuba, Tsukuba, Ibaraki, Japan*
[c]*Mesoscopic Physics Laboratory, Department of Physics, Peking University, Beijing, P. R. China*
[d]*Department of Theoretical and Computational Molecular Science, Institute for Molecular Science, Okazaki, Japan*

14.1 Introduction

Single-walled carbon nanotubes (SWNs), hollow cylindrical tubes with diameters of 0.4–4 nm, have excellent mechanical and electrical properties suggesting many potential applications [1–4]. The form of SWNTs can be visualized as carbon tubes formed through the rolling of one graphene sheet seamlessly. The SWNT structure can be specified completely through its chiral vector, which is donated by the chiral index (n, m) [5]. According to their structures, SWNTs are classifiable into three categories: armchair, zigzag, and chiral tubes. In terms of their electronic structures, SWNTs are classifiable into two categories: metallic (n-$m = 3k$, where k is an integer) and semiconducting (n-$m \neq 3k$) tubes. Metallic SWNTs can function as nanometer-sized conductors and transparent conductive films; semiconducting SWNTs can serve as field effect transistors and saturable absorbers. Widespread application of SWNTs requires the use of SWNTs of a uniform electronic type. However, SWNTs are typically grown as bundles of metallic and semiconducting tubes. Selective syntheses of SWNTs according to their electronic structure

Chemistry of Nanocarbons Edited by Takeshi Akasaka, Fred Wudl and Shigeru Nagase

remain a big challenge. In order to overcome this limitation, one has to exfoliate SWNT bundles and separate metallic from semiconducting SWNTs. This review describes the research about dispersion of SWNTs in organic solvent and separation of SWNTs using amine.

14.2 Dispersion of SWNTs

14.2.1 Dispersion of SWNTs Using Amine

Single-walled carbon nanotubes (SWNTs) have excellent mechanical and electrical properties that have suggested numerous promising applications. Nevertheless, practical applications of SWNTs have been hindered by their poor dispersibility. Therefore, dispersion of bundled SWNTs to individual ones in organic solvents is an important scientific goal, which makes homogeneous chemical reactions possible. Noncovalent bond formation of SWNTs with polymers [6] and π-conjugated compounds [7, 8] has been suggested for dispersion of bundled SWNTs in nonaqueous solution without changing their structure and properties.

From microscopic observations, Choi *et al.* have reported that amines untangle SWNTs in a nonaqueous solution [9, 10]. In amidation reactions, the SWNT dispersibility depends on the amount of amines [11–14]. To provide insight into the dispersion efficiencies, series of amines with different substituents were used as SWNT dispersants (Figure 14.1) [15]. The dispersion efficiencies obtained by measuring the optical absorption intensity of the dispersion solution of SWNTs [16] at 1310 nm are presented in Table 14.1 [17, 18]. Dispersibility decreases in order of a primary, secondary, and tertiary amine, suggesting that the interaction between SWNTs and amines is sensitive to steric hindrance around a nitrogen atom. As presented in Table 14.1, the interaction between SWNTs and amines is

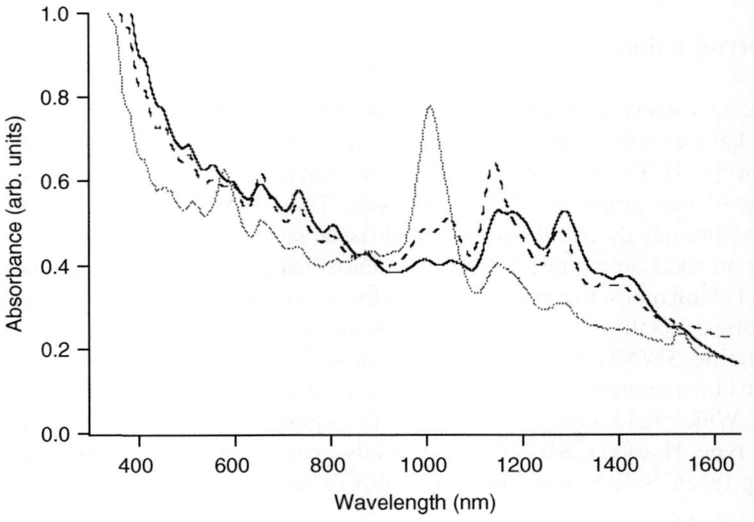

Figure 14.1 Absorption spectra of SWNTs dispersed in octylamine-THF solution. Solid line: HiPco SWNTs. Dotted line: CoMoCAT SWNTs. Dashed line: ACCVD SWNTs

Table 14.1 Absorption intensity ($\lambda_{1310\ nm}$) of SWNTs in THF solution with amine

compounds	$\lambda_{1310\ nm}$	compounds	$\lambda_{1310\ nm}$
N,N-ethylenediamine	12.4	propylamine	5.8
1,4-butanediamine	10.5	1methylpropylamine	5.0
ethylenediamine	10.2	dipropylamine	4.6
DBU	10	cyclohexylamine	3.9
N,N,N′,N′-ethylenediamine	7.6	octadecylamine	3.6
octylamine	7.4	tripropylamine	3.2
N-methyl-propylamine	7.0	methylpiperidine	2.1
dodecylamine	6.7	pyridine	1.6
aminoethanol	6.7	aniline	1
pentaethylenehexamine	6.2	DMF	1
poperidine	6.2	propionamide	1
isopropylamine	6.2	none	1

correlated with the basicity of the amines. The most likely mechanism is that the amine nitrogen interacts strongly with the SWNT surface. The binding energy between amines and SWNTs is estimated to be considerable [19].

The observed near-infrared fluorescence from an amine-THF solution of SWNTs displays distinct emission transitions of several different semiconducting SWNTs. Figure 14.2 portrays contour plots of fluorescence intensities for SWNTs in an amine-THF solution as a function of the wavelengths of excitation and resultant emission. These features are characteristic of individually dispersed SWNTs solutions, which are also found recently with surfactants after sonication treatment in aqueous solution [20, 21]. The fluorescence peaks of SWNTs in amine-THF solution are shifted to red-region and broadened compared to those of SWNTs in SDS-D_2O solution. The SWNT-amine interaction might include a charge-transfer character.

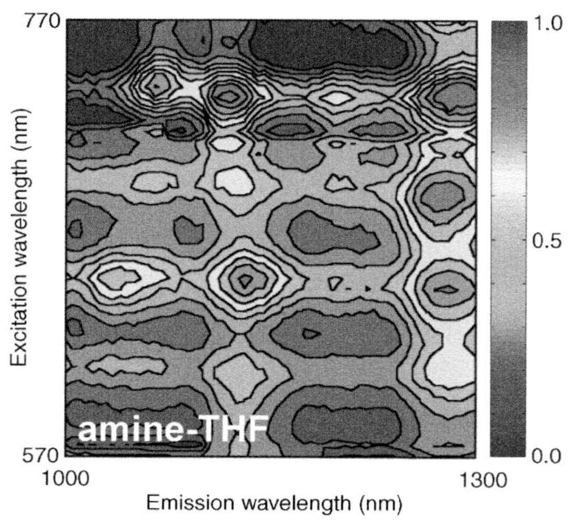

Figure 14.2 Contour plots of fluorescence intensities for SWNTs in octylamine-THF solution

Atomic force microscopic (AFM) measurements show that SWNTs in a THF/octylamine solution have a length distribution of 300–700 nm, with tube diameters of 0.8–4 nm. These diameters are close to the 0.9–1.3 nm expected for SWNTs.

The effective amine-assisted dispersion method was applied to peapods. Peapods [22] (SWNTs encapsulating fullerenes) are currently of great interest as a new form of SWNT-based material that might be applicable for nanometer-sized devices [23, 24]. The absorption bands corresponding to the van Hove transition of semiconducting tubes of C_{60}@Metro-SWNTs (1500–1750 nm) and La@C_{82}@Metro-SWNTs (1500–2200 nm) change in comparison with that of Metro-SWNTs [25]. Theoretical [26–29] and experimental [23, 24, 30] studies show that the structure and electronic properties of SWNTs are changed markedly upon encapsulating fullerenes and endohedral metallofullerenes. In this context, the difference in the absorption spectra of peapods is explainable by the structural deformation of SWNTs and charge transfer between SWNTs and C_{60} or La@C_{82}.

14.2.2 Dispersion of SWNTs Using C_{60} Derivatives

Several groups have reported that endohedral interaction between SWNTs and fullerenes make the band-gap and field effect transistor (FET) properties of SWNTs tunable. This was confirmed based on observations using scanning tunneling microscopy [23, 24] and FET measurements [30]. Exohedral interaction between SWNTs and fullerene has also been reported. Using a high-resolution transmission electron microscope (HRTEM), Liu *et al.* observed C_{60} derivatives stabilizing the SWNT surface [31].

Fullerodendrons having dendritic poly(amidoamine) substituents assist dispersion of SWNTs into D_2O and tetrahydrofuran via noncovalent functionalization (Figure 14.3) [32]. Bundled SWNTs were dispersed in THF by sonication in the presence of fullerodendron (0.1 mM). According to z-scan analysis of AFM, tube diameters of 2.3–9.6 nm were observed. Considering that SWNTs have a diameter distribution of 0.9–1.3 nm, SWNTs

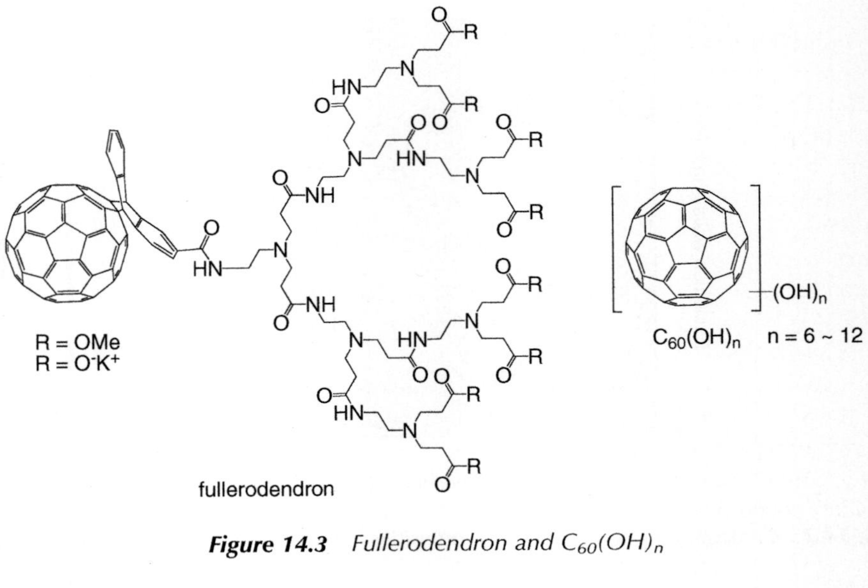

Figure 14.3 Fullerodendron and $C_{60}(OH)_n$

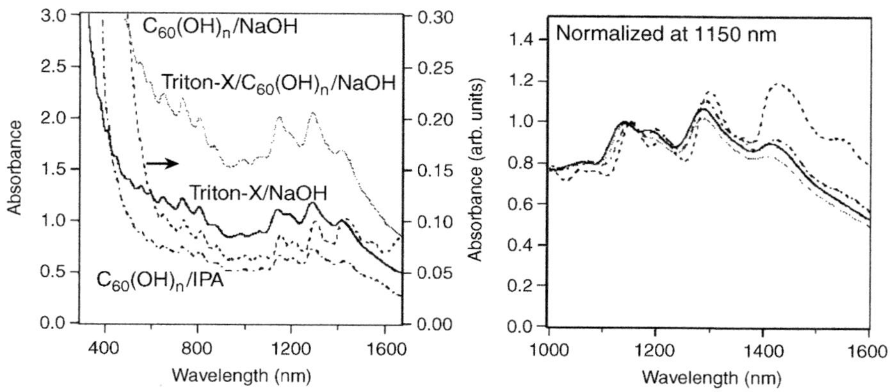

Figure 14.4 *Absorption spectra of SWNTs*

in a D_2O/fullerodendron solution have a polymer-wrapped and less bundled structure. Furthermore, such a picture is supported by HRTEM results at a higher magnification, which shows soft materials on the surface of small bundle of SWNTs. Comparable TEM images, which revealed formation of a monolayer of fullerene on the outside of the SWNTs under supercritical conditions, were observed by Britz *et al.* [33].

Dispersion of SWNTs in an aqueous and nonaqueous solution using amphiphilic C_{60} derivatives was also achieved (Figure 14.3) [34]. Actually, $C_{60}(OH)_n$ is soluble in an aqueous alkaline solution and isopropyl alcohol (IPA). The SWNTs were dispersed in D_2O containing 1 wt% NaOH and 0.1 mg/ml $C_{60}(OH)_n$ and in IPA containing 0.1 mg/ml $C_{60}(OH)_n$. The absorption intensity of SWNTs dispersed in D_2O containing 1 wt% NaOH, 1 wt% Triton-X, and 0.1 mg/ml $C_{60}(OH)_n$ increased compared to that dispersed in D_2O containing NaOH and Triton-X (Figure 14.4), which suggests that dispersibility of SWNTs is increased by the addition of $C_{60}(OH)_n$ and Triton-X. However, the stability of the dispersion in IPA and in D_2O, each of which contains $C_{60}(OH)_n$, is lower than that of the dispersion in the same solvents containing Triton-X and a mixture of Triton-X and $C_{60}(OH)_n$. The low stability of the dispersion in IPA and D_2O containing $C_{60}(OH)_n$ might result from the small molecular size and large curvature of $C_{60}(OH)_n$. The interaction between SWNTs and $C_{60}(OH)_n$ might be weak compared to the interaction between SWNTs and Triton-X. A SWNT-$C_{60}(OH)_n$ complex and its components might exist in equilibrium, so that SWNT aggregation can occur.

Raman spectra of SWNT films prepared from dispersion were measured under excitation at 514.5 nm (Figure 14.5) [35]. The characteristic peaks, the radial breathing mode (RBM), disordered carbon mode (D band), tangential Raman mode (G-band), and G′ mode were observed. It is particularly interesting that a large and broad Breit–Wigner–Fano (BWF) line at the lower energy side of the G-band, which is characteristic of metallic SWNTs, became sharp in the SWNT films that were obtained from dispersion in IPA containing $C_{60}(OH)_n$. A shift of the G′ mode was also observed in the same sample. However, SWNT films prepared from dispersion in IPA, which did not contain $C_{60}(OH)_n$, showed no spectral change. Electrochemical studies and solvent effect on SWNTs revealed that the changing of G-band and G′ mode is indicative of a charge transfer

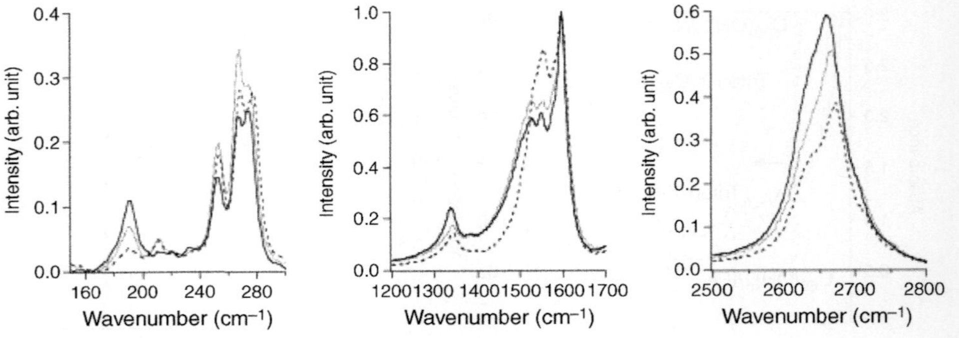

Figure 14.5 Raman spectra of SWNTs (Film). Solid line: Triton-X/NaOH/D_2O. Dotted line: $C_{60}(OH)_n$/NaOH/D_2O. Dashed line: $C_{60}(OH)_n$/IPA

interaction between SWNTs and adsorbent [36, 37]. It is also reported that charge transfer of SWNTs induces the intensity loss of absorption band from large-diameter SWNTs [38–40]. The intensity loss of S11 band was not observed in dispersion in IPA containing $C_{60}(OH)_n$ compared to other dispersions. These results suggest that a weak electronic interaction exists between SWNTs and $C_{60}(OH)_n$. On the other hand, a small spectral change was observed in SWNT films prepared from an aqueous NaOH solution containing $C_{60}(OH)_n$. This difference might result from the effect of NaOH. Further studies are necessary to elucidate the phenomena observed in the Raman spectra.

Figure 14.6 portrays contour plots for the PL intensity of SWNTs dispersions [20, 21]. The peak positions are similar to those reported previously for dispersion of SWNTs in D_2O. These peaks can be assigned to the emission from the first interband (S11) of (6,5), (7,5), (7,6), (8,3), (8,4), (8,6), (8,7), (9,4), (9,5), (10,2), (10,3), and (10,5) SWNTs. The PL intensity of semiconducting SWNTs having a large diameter increased when $C_{60}(OH)_n$ was present. The absorption spectra, normalized based on the peak intensity at 1150 nm in S11 bands of the SWNTs dispersions, are depicted in Figure 14.4. The intensity of semiconducting SWNTs having a large diameter increased when SWNTs were dispersed in D_2O containing $C_{60}(OH)_n$ and NaOH. This might be a reason for the increase in the PL intensity of semiconducting SWNTs. On the other hand, no significant difference was found in the absorption intensity of semiconducting SWNTs dispersed in an aqueous NaOH solution containing Triton-X and that containing Triton-X and $C_{60}(OH)_n$. Reportedly, exfoliation of SWNTs improves the PL intensity because it prevents quenching of semiconducting SWNTs by metallic SWNTs, which indicates that large-diameter SWNTs are exfoliated by $C_{60}(OH)_n$.

The vis-NIR and PL spectra strongly indicate that SWNTs are dispersed and exfoliated by $C_{60}(OH)_n$. Nakashima *et al.* reported that pyrene derivatives with an ammonium group selectively disperse large-diameter semiconducting SWNTs [41]. They described the contribution of a π–π interaction between the pyrene group and SWNTs towards the dispersion of SWNTs. The cation-π interaction between the ammonium moiety and SWNTs also played a minor part in the dispersion process of SWNTs. Here, the π–π interaction between SWNTs and $C_{60}(OH)_n$ might be the dominant factor for dispersion; weak electronic interaction between SWNTs and $C_{60}(OH)_n$ also facilitated dispersion.

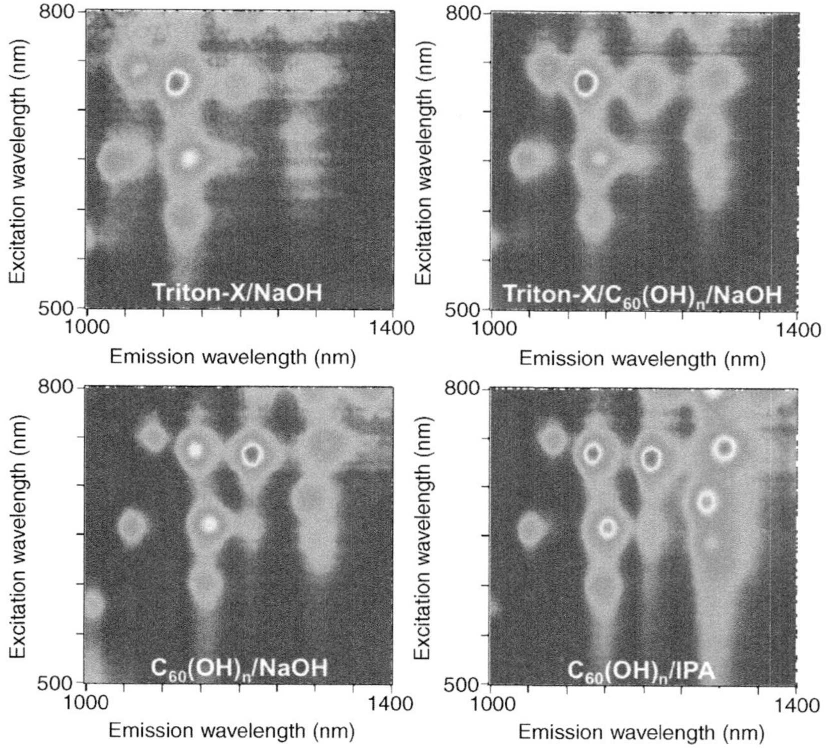

Figure 14.6 *Contour plots of fluorescence intensities for SWNTs*

14.2.3 Dispersion of SWNTs in Organic Solvents

Smalley and co-workers reported that dispersibility of SWNTs in several organic solvent and *o*-dichlorobenzene (ODCB) is a suitable organic solvent for dispersing SWNTs [42]. Geckeler *et al.* reported that the dispersion of individual SWNTs can be achieved using a combination of ultrasonication and ultracentrifugation in ODCB [43]. The dispersion and exfoliation of SWNTs in an organic solvent enable complex formation and functionalization of SWNTs with organic materials under homogeneous conditions.

It is noteworthy that the intensity of the absorption bands corresponding to S_{11} decreased in ODCB compared to those in amine-THF or pyrene-THF solution [15, 44, 45]. The S11 band of SWNTs was clearly observed after removal of ODCB, which suggests that the interaction between SWNTs and ODCB is reversible. These phenomena resemble those occurring during the acid treatment of SWNTs. Reportedly, protons on SWNTs oxidize the SWNTs [46, 47]. Interaction between SWNTs and an acid increases the solubility of SWNTs. This interaction is expected not to involve covalent bond formation. Moreover, it might be important for stable and high concentrated dispersion of SWNTs in ODCB.

Recently, Shin *et al.* reported a strong correlation between the sheet resistance and electronic structures of SWNTs treated with organic solvents [48]. They described that

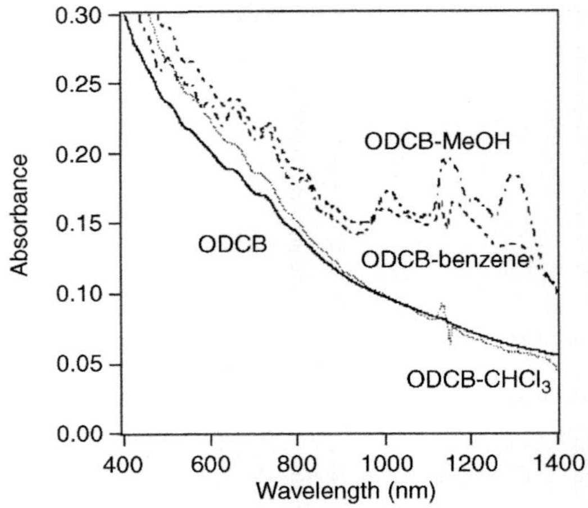

Figure 14.7 *Absorption spectra of SWNTs*

the sheet resistance of SWNT films treated with organic solvents decreases concomitantly with decreased metallic intensity of the G-band. They assumed that the electronic structure of SWNTs can be tuned systematically through appropriate selection of the backbones of solvent molecules and electron-donating and electron-withdrawing groups.

The characteristic absorption spectra of SWNTs were observed in ODCB-MeOH, ODCB-benzene, and ODCB-$CHCl_3$ solutions (Figure 14.7). The characteristic absorption bands of SWNTs were not recovered completely in mixed solutions diluted with benzene. When chloroform was added to the ODCB dispersion, the characteristic absorption bands were not recovered, either. These results indicate that the aromatic ring and chloro substituents, electron withdrawing groups, are important for the interactions of SWNTs and ODCB. The SWNT films were prepared on a membrane filter by filtration of these dispersions in mixed solvents. The intensity of the BWF line at the lower energy side of the G-band (514.5 nm) and relative intensity of RBM toward the G-band (514.5 and 633 nm) decreased in the following order: ODCB-MeOH, ODCB-benzene, ODCB, and ODCB-$CHCl_3$ (Figure 14.8). Corio *et al.* reported that the G′ band at ca. 2600 cm^{-1} was sensitive to the carrier density of the SWNT π valence bond [36]. The decreasing order of the carrier density is the same as that described above. The order can be explained reasonably in terms of the degree of *p*-type doping. The volume resistance of SWNT films, which was measured at room temperature in air using four-point probe conductivity measurement, decreased in the following order: ODCB-MeOH ($84 \times 10^{-2}\,\Omega\cdot cm$), ODCB-benzene ($82 \times 10^{-2}\,\Omega\cdot cm$), ODCB ($26 \times 10^{-2}\,\Omega\cdot cm$), and ODCB-$CHCl_3$ ($12 \times 10^{-2}\,\Omega\cdot cm$). These results show that ODCB is useful not only to disperse SWNTs, but also to control their electronic properties easily when other organic solvents are added under homogeneous conditions. Mickelson reported that sidewall functionalization of SWNTs increases the resistance of SWNTs [49]. The spectroscopic and resistance change of SWNTs show that the interaction between SWNTs and ODCB causes *p*-doping of SWNTs.

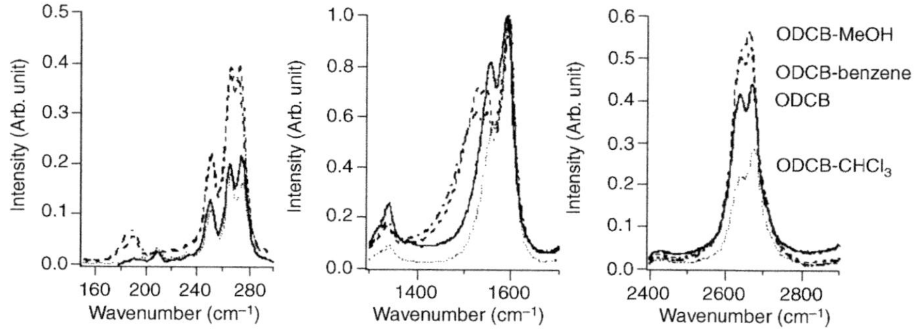

Figure 14.8 *Raman spectra of SWNTs*

14.3 Purification and Separation of SWNTs Using Amine

14.3.1 Purification and Separation of SWNTs Prepared by CVD Methods

Typically, SWNTs are grown as bundles of metallic and semiconducting tubes, thereby hindering their widespread application. Therefore, it is technologically important to separate metallic and semiconducting SWNTs in high yields. Several methods have been investigated to separate metallic from semiconducting tubes [50–74]. Among them, dispersion and centrifugation process are simple methodologies for separation of metallic and semiconducting SWNTs.

The physical ground of the amine-assisted method is that metallic SWNTs are more strongly adsorbed by amines than semiconducting SWNTs are. The adsorption energy of NH_2CH_3 on a SWNT is defined as

$$E_a = E(\text{SWNT}) + E(\text{NH}_2\text{CH}_3) - E(\text{SWNT} + \text{NH}_2\text{CH}_3).$$

The calculated adsorption energies [64] from the density functional theory (DFT) within the local density approximation (LDA) are presented in Table 14.2 together with the optimized adsorption configurations. In fact, NH_2CH_3 tends to adsorb SWNT through the interaction of the H atoms (rather than the N lone pair). The most noteworthy entry in Table 14.2 is that the metallic (7,7) SWNT is more strongly adsorbed by NH_2CH_3

Table 14.2 *Adsorption energy (Ea) of NH_2CH_3 on the (13,0) and (7,7) SWNTs [64]*

	mode (I)	mode (II)	mode (III)	mode (IV)	mode (V)	mode (VI)
(7,7)	–0.11	–0.18	–0.17	–0.17	-	–0.18
(13,0)	–0.04	–0.11	–0.07	–0.07	–0.08	–0.08
	mode (I)	mode (II)	mode (III)	mode (IV)	mode (V)	mode (VI)

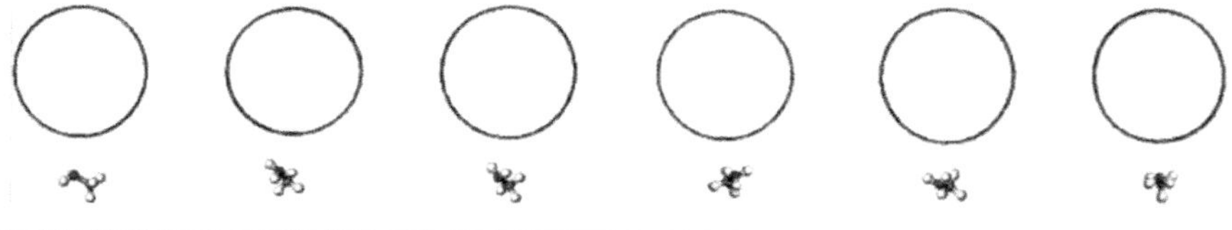

than the semiconducting (13,0) SWNT, irrespective of the adsorption mode. The adsorption energies between NH_2CH_3 and the (7,7) SWNT are small (–0.04 – –0.18 eV), suggesting that NH_2CH_3 is easily removable and also has no strong effect on the electronic structure of the (7,7) SWNT. In stark contrast, the electronic structures of SWNTs are altered considerably by covalent functionalization [75–77].

The chemical vapor deposition (CVD) method attracts broad attention for one of possible low-cost and large-scale production. The CVD method is advantageous because appropriate catalysts can be used and many of experimental parameters such as temperature and atmosphere are adjustable. The first step towards separation is to disperse SWNT bundles. The SWNTs produced using the HiPco method were dispersed in a THF solution containing 5 M propylamine and then centrifuged (45,620 × *g*, 12 h, labeled as SWNTs-P5) [64–67]. Three regions are identifiable in the absorption spectrum of SWNTs: the first interband transitions for metallic SWNTs, M_{11} (400–650 nm), and the first and second interband transitions for semiconducting SWNTs: S_{11} (900–1600 nm) and S_{22} (550–900 nm), respectively. Remarkably, SWNTs-P5 has stronger absorption peaks in the metallic M_{11} band. Weaker absorption peaks are detected in the semiconducting S_{11} and S_{22} bands than SWNTs-O1 (treated in 1 M octylamine solution), which is indicative of the enrichment of metallic SWNTs in the supernatant (Figure 14.9). The selective decay of semiconducting absorption bands and the enhancement of metallic absorption bands in SWNTs-P5 demonstrate that the dispersion – centrifugal separation process is effective for separation of SWNTs according to their electronic properties. Raman spectroscopy is a powerful tool for characterization of SWNTs. Using it, their diameter and electronic properties can be estimated. From a detailed study of Raman spectra of SWNTs, Kataura *et al.* proposed that the RBM peaks appear in the range around 260 cm^{-1} and 180 cm^{-1} when metallic SWNTs were excited respectively at 514.5 and 633 nm, although the peaks appear at 200–260 cm^{-1} when semiconducting SWNTs were excited, respectively, at 514.5 and

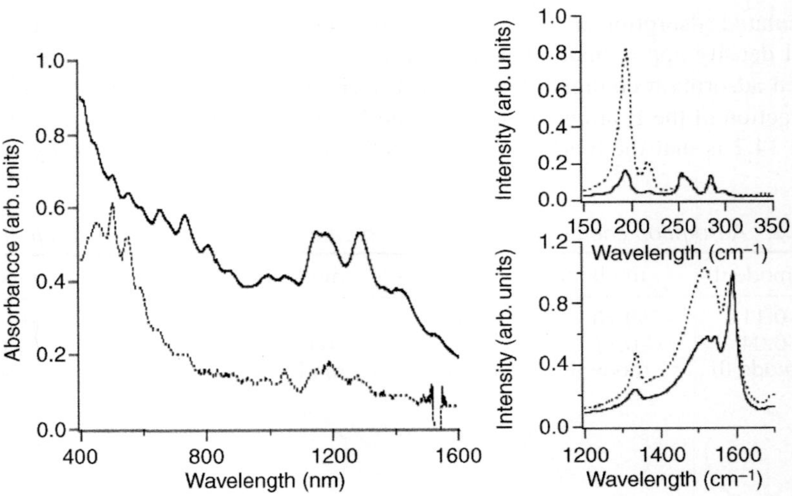

Figure 14.9 *Absorption and Raman spectra of SWNTs. Solid line: SWNTs-O1. Dotted line: SWNTs-P5*

633 nm (Figure 14.9) [17]. The strong peaks assigned to metallic SWNTs in SWNTs-P5 provide additional evidence for the metal-semiconducting selective separation using our simple extraction method, which is further supported by the much broader G-band of SWNTs-P5: a characteristic of metallic SWNTs [78].

CoMoCAT and ACCVD SWNTs have a diameter and helicity distribution different from those of HiPco SWNT. Not only HiPco SWNTs, but also SWNTs prepared by other CVD methods, such as CoMoCAT [65] and alcohol catalytic CVD (ACCVD) SWNTs [66] are effectively dispersed by amine. Depending on the dispersion and centrifugation condition, metallic SWNTs is separated from CoMoCAT and ACCVD SWNTs dispersion. In the case of ACCVD SWNTs, metal catalyst is removed through dispersion and centrifugation process. It thus suggests that the interaction between amine and SWNTs chiefly depends on the electronic structure of SWNTs, nearly irrespective of the diameter and chirality of SWNTs.

14.3.2 Purification and Separation of Metallic SWNTs Prepared by Arc-Discharged Method

Among the various SWNT synthesis methods, the arc discharge method is widely used because, at low cost, it yields gram quantities of SWNTs having a large diameter and small band gap. This method is advantageous because it yields crystalline SWNTs of uniform diameter. The SWNTs, however, generate considerable amounts of carbonaceous impurities, such as amorphous carbon, fullerene, and graphite. Carbonaceous impurities are usually removed through oxidative treatment in the liquid phase (wet chemical oxidation) and thermal treatment (air oxidation) [79, 80]. Reportedly, the diameter, electronic properties, and chemical treatment strongly affect the covalent and nonbonding interactions of SWNTs. It would be of great interest to elucidate whether the interaction between SWNTs and amine depends on their diameter, electronic properties, or chemical treatment.

Thermal gravimetric analysis (TGA) of AP-grade SWNTs (ArcNTs) produced using the arc discharge method and oxidized SWNTs (PArcNTs) obtained by air oxidation of ArcNTs were conducted under atmospheric conditions. Two components were observed in the thermal analysis of ArcNTs at around 370 °C and 430 °C. The disappearance of the peak at 370 °C for PArcNTs is evidence of the removal of amorphous carbons by air oxidation. Supernatant solutions of ArcNTs and ParcNTs were obtained after sonication of 10 mg of ArcNTs and PArcNTs in a THF solution (10 ml) containing 1M octylamine, with subsequent centrifugation (supernatant: labeled respectively as ArcNTs-O and PArcNTs-O). Three characteristic absorption bands for SWNTs are observed approximately at 1800, 1000, and 700 nm [81]. The first two bands are attributed to electronic transitions between the first (S_{11}) and second (S_{22}) pairs of van Hove singularities in semiconducting SWNTs. The other is attributed to the first pair (M_{11}) of singularities in metallic SWNTs. The weak characteristic absorption intensity of ArcNTs-O suggests that these dispersions contain many carbonaceous impurities. The vis-NIR spectrum of PArcNTs-O shows a fine structure in the S_{11}, S_{22}, and M_{11} regions, presumably because of the absence of impurities such as amorphous carbon. Increased characteristic absorption intensities and resolution improvement have already been documented and correlated to the dispersibility and purity of SWNTs [13, 14]. Although carbonaceous impurities were observed in ArcNTs and PArcNTs, the SEM image of PArcNTs-O showed highly pure SWNTs (Figure 14.10).

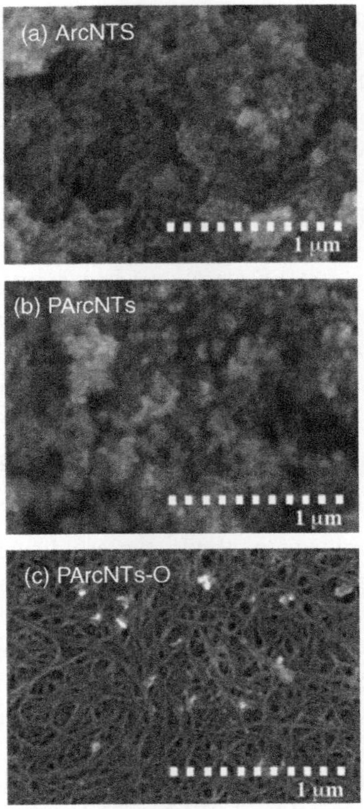

Figure 14.10 *SEM images of ArcNTs, PArcNTs, and PArcNTs-O*

The dispersion and separation efficiencies of SWNTs depend strongly on the structure and concentration of amine used. Therefore, propylamine was used for dispersion of PArc in THF solution [64–67]. It is worthwhile to separate metallic SWNTs and large diameter and narrow band gap semiconducting SWNTs using intermolecular interaction with an amine. Figure 14.11 portrays vis-NIR spectra of the supernatant of PArcNTs treated using a dispersion–centrifugation process with 3 M propylamine (PArcNTs-P) and PArcNTs-O. Actually, PArcNTs-P has stronger absorption peaks in the metallic M_{11} band and weaker absorption peaks in the semiconducting S_{22} bands than PArcNTs-O does. Their absorption spectra exhibit fine structures assigned to metallic SWNTs (M_{11}) and semiconducting SWNTs (S_{22}). Intensity of the M_{11} band was stronger than that of S_{22} band in PArcNTs-P. This result reveals a selective separation of metallic SWNTs and a large diameter and narrow band gap semiconducting SWNTs using this simple extraction method. Raman spectra of ArcNTs, PArcNTs, PArcNTs-O, and PArcNTs-P were obtained using excitation wavelengths of 514.5 and 633 nm. The increase of the integrated G/D ratio in the Raman peaks of PArcNTs-P compared to that of ArcNTs indicates the improvement in purification of SWNTs [35]. The efficiency of separation of metallic SWNTs and semiconducting SWNTs is estimated based on the area in the RBM peak. Unfortunately, those of metallic and semiconducting SWNTs were not observed simultaneously because the metallic and semiconducting SWNTs, with their small diameter and large band gap distribution, can not

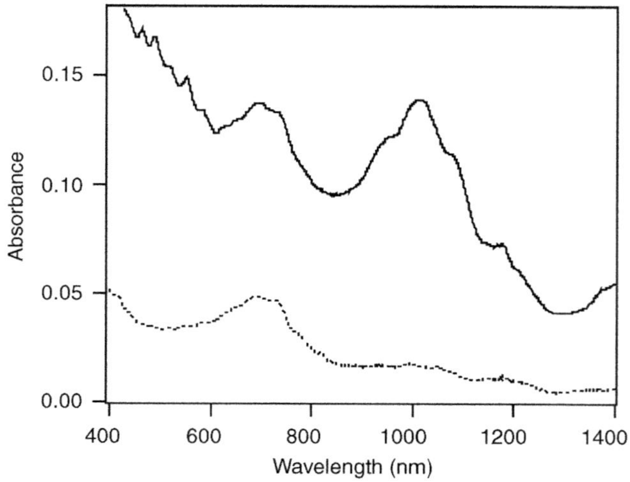

Figure 14.11 *Absorption spectra of SWNTs. Solid line: PArcNTs-O. Dotted line: PArcNTs-P*

be excited by irradiation of the same wavelength. An increase of the peak attributable to Breit–Wigner–Fano resonance was observed when the PArcNTs-P was excited at 633 nm [78, 82]. These results demonstrate the increased contents of metallic SWNTs in PArcNTs-P.

It is particularly interesting that different electronic types have been enriched using an amine as a dispersant to separate SWNTs. Metallic SWNTs were enriched in a supernatant from as-prepared SWNT, or SWNTs treated by air oxidization, whereas semiconducting SWNTs were concentrated in a filtrate from oxidized SWNTs purified through wet chemical oxidation [64–69]. Lu and coworkers studied the adsorption energy change of an amine onto SWNTs with the hole doping concentration using the DFT method [83]. The results showed that, although an amine is adsorbed more strongly onto neutral metallic SWNTs than onto the neutral semiconducting SWNTs, it is adsorbed more strongly onto semiconducting SWNTs than onto metallic SWNTs when the hole concentration exceeds 3.6×10^{-3} $|e|$ C atom^{-1} (Figure 14.12). This dramatic change of selectivity in adsorption of an amine onto SWNTs caused by hole-doping is a plausible explanation for the observed different electronic types in enriched SWNTs using amine as a dispersant if the hole doping level in SWNTs treated by air oxidization is markedly lower than that in SWNTs treated by wet chemical oxidation. Actually, Wiltshire and coworkers reported that wet chemical oxidation introduces many more carboxyl groups than air oxidation does [84]. Furthermore, Barros and coworkers described, based on detailed analyses of Raman and infrared spectroscopy, that the SWNTs act as a donor by donating electrons to the carboxyl groups [85].

14.3.3 Preparation of SWNTs and Metallic SWNTs Films

Recently, SWNTs have been anticipated as candidate materials for preparation of transparent and conductive thin films that complement indium – tin oxide (ITO) [86–91]. For ITO coatings on plastic, surface resistivity of 4 Ω/sq at 78% transmittance has been reported [91]. Development of a thin film of carbon nanotubes consisting of a ubiquitous element has been

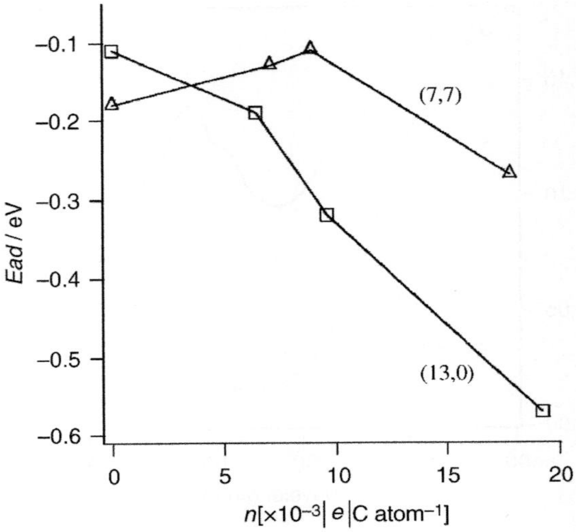

Figure 14.12 Adsorption energy of NH_2CH_3 on the (13,0) and (7,7) SWNTs as a function of hole concentration (n)

demanded because of limited production and reserves of indium. Additionally, carbon nanotubes present the advantage of being flexible. For instance, when applied in a touch panel, high durability is expected. Roth and coworkers reported the relation between transmittance and conductivity of carbon nanotubes at room temperature [87]. They found that SWNTs are more suitable for transparent conductive materials than multi-walled carbon nanotubes (MWNTs) because of the considerably larger diameter of MWNTs, which increases light absorption but not conductivity. Recently, a difference in sheet resistance of SWNTs based on the synthetic method of SWNTs was reported [89]. The factor of sheet resistance of SWNTs remains unclear. Dimensions and defect density of SWNTs, resistive impurity, and ease of exfoliation of bundled SWNTs might affect the sheet resistance. It is useful to use metallic and semiconducting SWNTs separately according to applications. For transparent and conductive thin films, metallic SWNTs are more suitable than a mixture of metallic and semiconducting SWNTs because the conductivity of metallic SWNTs is expected to be higher than that of the mixture. Methods for preparation of SWNT thin films by filtration [86], spin coating [92], Langmuir–Blodgett deposition [93], and dip coating [94] have been reported.

The surface morphology of metallic SWNTs on PET films prepared from metallic SWNTs dispersion in amine-THF solution by air spray method was examined using SEM and AFM (Figure 14.13). The SEM and AFM images show a dense and homogeneous network with no noticeable impurities [95]. These results suggest that this method is effective to prepare uniform thin films of SWNTs. Surface resistivity of SWNT films on substrates was measured using a four-point probe conductivity measurement at room temperature in air. The transmittance of SWNT films (%T = [SWNTs on PET film] − [PET film]; Transmittance of PET film: ca. 86%T) was determined based on visible light spectra in the range of 400–800 nm. The resistivity of metallic SWNTs (9.0 kΩ/sq) was one 24th that of SWNTs (215 kΩ/sq) at transmittance of 97.1 and 96.6% (Figure 14.14). The resistivity of

Figure 14.13 *SEM image of metallic SWNTs on PET*

metallic SWNTs (690 Ω/sq) is one twelfth that of SWNTs (8.9 kΩ/sq) at the transmittance of 81.4 and 80.0%. A similar tendency was observed when the metallic SWNT and SWNT films were prepared on quartz glass. The resistivity of metallic SWNTs is 800 Ω/sq at the transmittance 80.7%; that of SWNTs is 8.6 kΩ/sq at the transmittance of 78.2%. These films' thicknesses were estimated respectively using a surface profiler as about 28 and 30 nm, suggesting that metallic SWNTs are more suitable for use as transparent and conductive thin films than as a mixture of metallic and semiconducting SWNTs, especially at high transmittance. The sheet resistance was reduced after HCl treatment for 30 min. The changes of the sheet resistances from 690 Ω/sq to 330 Ω/sq in metallic SWNTs and from 8.9 kΩ/sq to 2.8 kΩ/sq in SWNTs were observed respectively at transmittances of 82.1 and

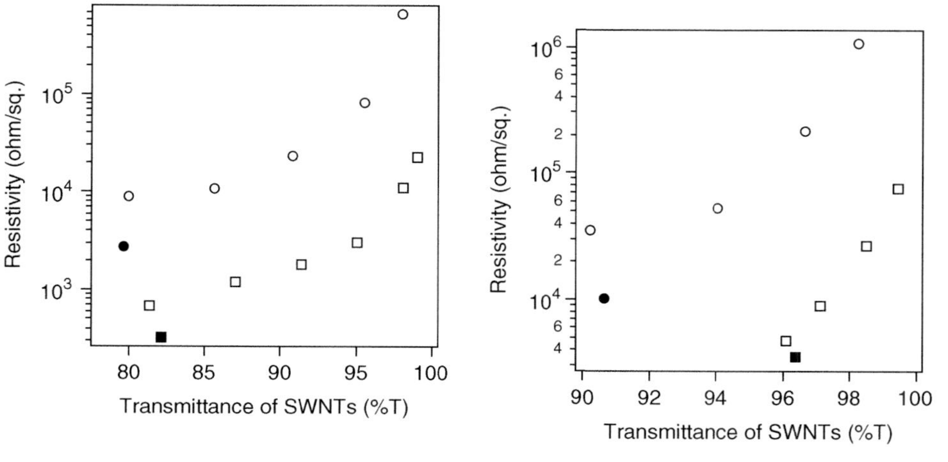

Figure 14.14 *Transmittance vs. resistivity plots for sprayed SWNT layers on PET films: metallic SWNTs (□); SWNTs (○); after HCl treatment, (■) and (●)*

79.6%. Geng *et al.* reported that HNO_3 treatment of SWNT films prepared by SDS dispersion increases the conductivity because of SDS removal [96]. The decrease of the sheet resistivity of SWNT films by HCl washing might result from removal of the adsorbed amine and *p*-type doping effect [19].

Metallic SWNTs are more suitable materials for use in transparent and conductive thin films than a mixture of metallic and semiconducting SWNTs. Particularly, higher conductivity was achieved in a concentrated sample of metallic SWNTs at higher transmittance than in the mixture. Practical use of SWNT thin films demands development of large-scale synthesis methods of SWNTs and improvement of separation methods of metallic and semiconducting SWNTs.

14.4 Conclusion

Because of their excellent mechanical and electrical properties, SWNTs have been studied extensively. It is often necessary to use SWNTs with uniform electronic type in their applications. To date, it has been difficult to selectively synthesize metallic SWNT or semiconducting SWNTs. Various postsynthetic separation approaches have been developed. It is expected that the application of SWNTs will develop rapidly once a simple, efficient, and inexpensive method for separation of metallic and semiconducting SWNTs is established.

References

[1] S. Iijima and T. Ichihashi, *Nature* **363**, 603 (1993).

[2] D. S. Bethune, C. H. Klang, M. S. de Vries, G. Gorman, R. Savoy, J. Vazquez, and R. Beyers, *Nature* **363**, 605 (1993).

[3] S. Niyogi, M. A. Hamon, H. Hu, B. Zhao, P. Bhowmik, R. Sen, M. E. Itkis, and R. C. Haddon, *Acc. Chem. Res.* **35**, 1105 (2002).

[4] R. H. Baughman, A. A. Zakhidov, and W. A. de Herr, *Science* **297**, 787 (2002).

[5] R. Saito, M. Fujita, G. Dresselhaus, and M. S. Dresselhaus, *Appl. Phys. Lett.* **60**, 2204 (1992).

[6] M. J. O'Connell, P. Boul, L. M. Ericson, C. Huffman, Y. Wang, E. Haroz, C. Kuper, K. Tour, D. K. Ausman, and R. E. Smalley, *Chem. Phys. Lett.* **342**, 265 (2001).

[7] R. J. Chen, Y. Zhang, D. Wang, and H. Dai, *J. Am. Chem. Soc.* **123**, 3838 (2001).

[8] J. Chen, H. Liu, W. A. Weimer, M. D. Halls, D. H. Waldeck, and G. C. Walker, *J. Am. Chem. Soc.* **124**, 9034 (2002).

[9] N. Choi, M. Kimura, H. Kataura, S. Suzuki, Y. Achiba, W. Mizutani, and H. Tokumoto, *Jpn. J. Appl. Phys.* **41**, 6264 (2002).

[10] C. A. Furtado, U. J. Kim, H. R. Gutierrez, L. Pan, E. C. Dickey, and P. C. Eklund, *J. Am. Chem. Soc.* **126**, 6095 (2004).

[11] Y. Maeda, Y. Lian, T. Wakahara, M. Kako, T. Akasaka, N. Choi, H. Tokumoto, S. Kazaoui, and N. Minami, *ITE Lett.* **4**, 798 (2003).

[12] Y. Lian, Y. Maeda, Y. T. Wakahara, T. Akasaka, S. Kazaoui, N. Minami, N. Choi, and H. Tokumoto, *J. Phys. Chem. B* **107**, 12082 (2003).

[13] Y. Maeda, S. Kimura, T. Hasegawa, Y. Lian, T. Wakahara, T. Akasaka, H. Abe, N. Choi, T. Shimizu, H. Tokumoto, S. Kazaoui, and N. Minami, *ITE Lett.*, **5**, 263 (2004).

[14] Y. Lian, Y. Maeda, T. Wakahara, T. Akasaka, S. Kazaoui, N. Minami, T. Shimizu, N. Choi, and H. Tokumoto, *J. Phys. Chem. B* **108**, 8848 (2004).

[15] Y. Maeda, S. Kimura, Y. Hirashima, M. Kanda, Y. Lian, T. Wakahara, T. Akasaka, T. Hasegawa, H. Tokumoto, T. Shimizu, H. Kataura, Y. Miyauchi, and S. Maruyama, *J. Phys. Chem. B*, **108**, 18395 (2004).
[16] P. Nikolaev, M. J. Bronikowski, R. K. Bradley, F. Rohmund, D. T. Colbert, K. A. Smith, and R. E. Smalley, *Chem. Phys. Lett.*, **313**, 91 (1999).
[17] H. Kataura, Y. Kumazawa, Y. Maniwa, I. Umezu, S. Suzuki, Y. Ohtsuka, and Y. Achiba, *Synth. Met.*, **103**, 2555 (1999).
[18] I. W. Chiang, B. E. Brinson, A. Y. Huang, P. A. Willis, M. J. Bronikowski, J. L. Margrave, R. E. Smalley, and R. H. Hauge, *J. Phys. Chem. B*, **105**, 8297 (2001).
[19] J. Kong and H. Dai, *J. Phys. Chem. B* **105**, 2890 (2001).
[20] M. J. O'Connell, S. M. Bachilo, C. B. Huffman, V. C. Moore, M. S. Strano, E. H. Haroz, K. L. Rialon, P. J. Boul, W. H. Noon, C. Kittrell, J. Ma, R. H. Hauge, R. B. Weisman, and R. E. Smalley, *Science*, **297**, 593 (2002).
[21] S. M. Bachilo, M. S. Strano, C. Kittrell, R. H. Hauge, R. E. Smalley, and R. B. Weisman, *Science* **298**, 2361 (2002).
[22] B. W. Smith, M. Monthiouz, and D. E. Luzzi, *Nature* **396**, 323 (1998).
[23] J. Lee, H. Kim, S. -J. Kahng, G. Kim, Y. -W. Son, J. Ihm, H. Kato, Z. W. Wang, T. Okazaki, H. Shinohara, and Y. Kuk, *Nature* **415**, 1005 (2002).
[24] D. J. Hornbaker, S. –J. Kahng, S. Misra, B. W. Smith, A. T. Johnson, E. J. Mele, D. E. Luzzi, and A. Yazdani, *Science* **295**, 828 (2002).
[25] H. Kataura, Y. Maniwa, T. Kodama, K. Kikuchi, K. Hirahara, K. Suenaga, S. Iijima, S. Suzuki, Y. Achiba, and W. Kraetschmer, *Synth. Met.*, **121**, 1195 (2001).
[26] L. Yang and J. Han, *Phys. Rev. Lett.* **85**, 0031–9007 (2000).
[27] M. Otani, S. Okada, and A. Oshiyama, *Phys. Rev. B* **68**, 125424 (2003).
[28] Y. Chow, S. Han, G. Kim, H. Lee, and J. Ihm, *Phys. Rev. Lett.* **90**, 106402–106411 (2003).
[29] J. Lu, S. Nagase, S. Zhang, and L. Peng, *Phys. Rev. B* **68**, 121402 (2003).
[30] T. Shimada, T. Okazaki, R. Taniguchi, T. Sugai, H. Shinohara, K. Suenaga, Y. Ohno, S. Mizuno, S. Kishimoto, and T. Mizutani, *Appl. Phys. Lett.* **81**, 4067 (2002).
[31] Z. Liu, K. Suenaga, and S. Iijima, *J. Am. Chem. Soc.*, **129**, 6666 (2007).
[32] Y. Takaguchi, M. Tamura, Y. Sako, Y. Yanagimoto, S. Tsuboi, T. Uchida, K. Shimamura, S. Kimura, T. Wakahara, Y. Maeda, and T. Akasaka, *Chem. Lett.*, **34**, 1608 (2005).
[33] D. A. Britz, A. N. Khlobystov, J. Wang, A. S. O'Neil, M. Poliakoff, A. Ardavan, and G. A. D. Briggs, *Chem. Commun.* 176 (2004).
[34] Y. Maeda, T. Kato, J. Higo, T. Hasegawa, T. Kitano, T. Tsuchiya, T. Akasaka, T. Okazaki, J. Lu, and S. Nagase, Nano. **3**, 455 (2008).
[35] A. M. Rao, E. Richter, S. Bandow, B. Chase, P. C. Eklund, K. A. Williams, S. Fang, K. R. Subbaswamy, M. Menon, A. Thess, R. E. Smalley, G. Dresselhaus, and M. S. Dresselhaus, *Science* **275**, 187 (1997).
[36] P. Corio, P. S. Santos, P. S. Brar, G. G. Samsonidze, S. G. Chou, and M. S. Dresselhaus, *Chem. Phys. Lett.* **370**, 675 (2003).
[37] H. -J. Shin, S. M. Kim, S. -M. Yoon, A. Benayad, K. K. Kim, S. J. Kim, H. K. Park, J. -Y. Choi, and Y. H. Lee, *J. Am. Chem. Soc.* **130**, 2062 (2008).
[38] A. Kukovecz, T. Pichler, R. Pfeiffer, and H. Kuzmany, *Chem. Commun.* 1730 (2002).
[39] M. J. O'Connell, E. E. Eibergen, and S. K. Doorn, *Nature Mat.* **4**, 412 (2005).
[40] J. Zhou, Y. Maeda, J. Lu, A. Tashiro, T. Hasegawa, G. Luo, L. Wang, L. Lai, T. Akasaka, S. Nagase, Z. Gao, R. Qin, W. N. Mei, G. Li, and D. Yu, *Small* **5**, 244 (2009).
[41] Y. Tomonari, H. Murakami, and N. Nakashima, *Chem. Eur. J.* **12**, 4027 (2006).
[42] J. L. Bahr, E. T. Mickelson, M. J. Bronikowski, R. E. Smalley, and J. M. Tour, *Chem. Commun.* 193 (2001).
[43] D. S. Kim, D. Nepal, and K. E. Geckeler, *Small* **1**, 1117 (2005).
[44] R. J. Chen, Y. Zhang, D. Wang, and H. Dai, *J. Am. Chem. Soc.* **123** (2001).
[45] N. Nakashima, Y. Tomonari, and H. Murakami, *Chem. Lett.* **638** (2002).
[46] M. S. Strano, C. B. Huffman, V. C. Moore, M. J. O'Connell, E. H. Haros, J. Hubbard, M. Miller, K. Pialon, C. Kittrell, S. Ramesh, R. H. Hauge, and R. E. Smalley, *J. Phys. Chem. B* **107**, 6979 (2003).

[47] S. Ramesh, L. M. Ericson, V. A. Davis, R. K. Saini, C. Kittrell, M. Pasquali, W. E. Billups, W. W. Adams, R. H. Hauge, and R. E. Smalley, *J. Phys. Chem. B* **108**, 8794 (2004).
[48] H. J. Shin, S. M. Kim, S. M. Yoon, A. Benayad, K. K. Kim, S. J. Kim, H. K. Park, J. Y. Choi, and Y. H. Lee, *J. Am. Chem. Soc.* **130**, 2062 (2008).
[49] E. T. Mickelson, C. B. Huffman, A. G. Rinzler, R. E. Smalley, R. H. Hauge, and J. L. Margrave, *Chem. Phys. Lett.* **296**, 188 (1998).
[50] P. Collins, M. Arnold, and P. Avouris, *Science* **292**, 706 (2001).
[51] H. Huang, R. Maruyama, K. Noda, H. Kajiura, and K. Kadono, *J. Phys. Chem. B* **110**, 7316 (2006).
[52] Y. Zhang, Y. Zhang, X. Xian, J. Zhang, and Z. Liu, *J. Phys. Chem. C* **112**, 3849 (2008).
[53] G. Zhang, P. Qi, X. Wang, Y. Lu, X. Li, R. Tu, S. Bangsaruntip, D. Mann, L. Zhang, and H. Dai, *Science* **314**, 974 (2006).
[54] Y. Miyata, Y. Maniwa, and H. Kataura, *J. Phys. Chem. B* **110**, 25 (2006).
[55] R. Krupke, F. Hennrich, H. Löhneysen, and M. Kappes, *Science* **301**, 344 (2003).
[56] T. Tanaka, H. Jin, Y. Miyata, and H. Kataura, *Appl. Phys. Exp.* **1**, 114001 (2008).
[57] M. Zheng, A. Jagota, E. Semke, B. A. Diner, R. Mclean, S. Lustig, R. Richardson, and N. Tassi, *Nat. Mat.* **2**, 338 (2003).
[58] M. Zheng, A. Jagota, M. S. Strano, A. P. Santos, P. Barone, S. Grace Chou, B. A. Diner, M. S. Dresselhaus, R. S. Mclean, G. Bibiana Onoa, G. G. Samzonidze, E. D. Semke, M. Usrey, and D. J. Walls, *Science* **302**, 1545 (2003).
[59] M. S. Strano, C. A. Dyke, M. L. Usrey, P. W. Barone, M. J. Allen, H. Shan, C. Kittrell, R. H. Hauge, J. M. Tour, and R. E. Smalley, *Science* **301**, 1519 (2003).
[60] C. A. Dyke, M. P. Stewart, and J. M. Tour, *J. Am. Chem. Soc.* **127**, 4497 (2005).
[61] S. Toyota, Y. Yamaguchi, M. Hiwatashi, Y. Tomonari, H, Murakami, and N. Nakashima, *Chem Asian J.* **2**, 145 (2007).
[62] Z. Chen, X. Du, M. Du, C. D. Rancken, H. Cheng, A. G. Rinzler, *Nano Lett.* **3**, 1245 (2003).
[63] H. Li, B. Zhou, Y. Lin, L. Gu, W. Wang, K. A. S. Fernando, S. Kumar, L. F. Allard, and Y. Sun, *J. Am. Chem. Soc.* **126**, 1014 (2004).
[64] Y. Maeda, S. Kimura, M. Kanda, Y. Hirashima, T. Hasegawa, T. Wakahara, Y. Lian, T. Nakahodo, T. Tsuchiya, T. Akasaka, J. Lu, X. Zhang, Z. Gao, Y. Yu, S. Nagase, S. Kazaoui, N. Minami, T. Shimizu, H. Tokumoto, and R. Saito, *J. Am. Chem. Soc.* **127**, 10287 (2005)
[65] Y. Maeda, M. Kanda, M. Hashimoto, T. Hasegawa, S. Kimura, Y. Lian, T. Wakahara, T. Akasaka, S. Kazaoui, N. Minami, T. Okazaki, Y. Hayamizu, K. Hata, J. Lu, and S, Nagase, *J. Am. Chem. Soc.* **128**, 12239 (2006)
[66] Y. Maeda, M. Hashimoto, T. Hasegawa, M. Kanda, T. Tsuchiya, T. Wakahara, T. Akasaka, H. Miyauchi, S. Maruyama, J. Lu, and S, Nagase, *Nano*. **2**, 221 (2007)
[67] Y. Maeda, Y. Takano, A. Sagara, M. Hashimoto, M. Kanda, S. Kimura, Y. Lian, T. Nakahodo, T. Tsuchiya, T. Wakahara, T. Akasaka, T. Hasegawa, S. Kazaoui, N. Minami, J. Lu, and S. Nagase, *Carbon* **46**, 1563 (2008)
[68] D. Chattopadhyay, I. Galeska, and F. Papadimitrakopoulos, *J. Am. Chem. Soc.* **125**, 3370 (2003)
[69] G. G. Samsonidze, S. G. Chou, A. P. Santos, V. W. Brar, G. Dresselhaus, M. S. Dresselhaus, A. Selbst, A. K. Swan, M. S. Ünlü, B. B. Goldberg, D. Chattopadhyay, S. N. Kim, and F. Papadimitrakopoulos, *Appl. Phys. Lett.* **85**, 1006 (2004).
[70] S. N. Kim, Z. Luo, and F. Papadimitrakopoulos, *Nano Lett.* **5**, 2500 (2005).
[71] A. Nish, J. Hwang, J. Doig, and R. J. Nicholas, *Nat. Nanotech.* **2**, 640 (2007).
[72] F. Chen, B. Wang, Y. Chen, and L. Li, *Nano Lett.* **7**, 3013 (2007).
[73] M. S. Arnold, A. A. Green, J. F. S. I. Stupp, and M. C. Hersam, *Nano Lett.* **5**, 713 (2005).
[74] M. S. Arnold, A. A. Green, F. J. F. Hulvat, S. I. Stupp, and M. C. Hersam, *Nat. Tech.* **1**, 60 (2006).
[75] K. Kamaras, M. E. Itkis, H. Hu, B. Zhao, and R. C. Haddon, *Science* **301**, 1501 (2003).
[76] J. Zhao, K. Park, J. Han, and J. P. Lu, *J. Phys. Chem. B* **108**, 4227 (2004).
[77] V. Barone, J. Heyd, and G. E. Scuseria, *Chem. Phys. Lett.* **389**, 289 (2004).
[78] M. Pimenta, A. Marucci, S. A. Empedocles, M. G. Bawendi, E. B. Hanlon, A. M. Rao, P. C. Eklund, R. E. Smalley, G. Dresselhaus, and M. S. Dresselhaus, *Phys. Rev B* **58**, 16016 (1998).

[79] S. Bandow, A. M. Rao, K. A. Williams, A. Thess, R. E. Smalley, and C. P. Eklund, *J. Phys. Chem. B* **101**, 8839 (1997).
[80] W. Holzer, A. Penzlpfer, S. H. Gong, A. Bleyer, and D. D. Bradley, *Adv. Mater.* **8**, 974 (1996).
[81] J. Chen, M. A. Hamon, H. Hu, Y. Chen, A. M. Rao, P. C. Eklund, and R. C. Haddon, *Science* **282**, 95 (1998).
[82] M. S. Dresselhaus, M. S. Dresselhaus, A. Jorio, A. Souza Filho, and R. Saito, *Carbon* **40**, 2043 (2002).
[83] J. Lu, L. Lai, G. Luo, J. Zhou, R. Qin, D. Wang, L. Wang, W. N. Mei, G. Li, Z. Gao, S. Nagase, Y. Maeda, T. Akasaka, and D. Yu, *Small* **3**, 1566 (2007).
[84] E. B. Barros, A. G. S. Filho, V. Lemos, J. M. Filho, S. B. Fagan, M. H. Herbst, J. M. Rosolen, C. A. Luengo, and J. G. Huber, *Carbon* **43**, 2495 (2005).
[85] U. J. Kim, C. A. Furtado, X. Liu, G. Chen, and P. C. Eklund, *J. Am. Chem. Soc.* **127**, 15437 (2005).
[86] Z. Wu, Z. Chen, X. Du, J. M. Logan, J. Sippel, M. Nikolou, K. Kamaras, J. R. Reynolds, D. B. Tanner, A. F. Hebard, and A. G. Rinzler, *Science* **305**, 1273 (2004).
[87] M. Kaempgen, G. S. Duesberg, and S. Roth, *Appl. Surface Science* **252**, 425 (2005).
[88] N. Saran, K. Parikh, D. S. Suh, E. Muñoz, H. Kolla, and S. K. Manohar, *J. Am. Chem. Soc.* **126**, 4462 (2004).
[89] D. Zhang, K. Ryu, X. Liu, E. Polikarpov, J. Ly, M. E. Tompson, and C. Zhou, *Nano Lett* **6**, 1880 (2006)
[90] C. M. Aguirre, S. Auvray, S. Pigeon, R. Izquierdo, P. Desjardins, and R. Martel, *Appl. Phys. Lett.* **88**, 183104 (2006).
[91] J. Herrero, and C. Guillen, *Vacuum* **67**, 611 (2002).
[92] M. Y. Lay, J. P. Novak, and E. S. Snow, *Nano Lett.* **4**, 603 (2004).
[93] Y. Kim, N. Minami, W. Zhu, S. Kazaoui, R. Azumi, and M. Matsumoto, *Jpn. J. Appl. Phys.* **42**, 7629 (2003).
[94] H. Shimoda, S. J. Oh, H. Z. Geng, R. J. Walker, X. B. Zhang, L. E. McNeil, and O. Zhou, *Adv. Mater.* **14**, 899 (2002).
[95] Y. Maeda, M. Hashimoto, S. Kaneko, M. Kanda, T. Hasegawa, T, Tsuchiya, T. Akasaka, Y. Naitoh, T. Shimizu, H. Tokumoto, J. Lu, and S. Nagase, *J. Mater. Chem.* **18**, 4189 (2008).
[96] Z. Geng, K. K. Kim, K. P. Then, Y. S. Lee, Y. Chang, and Y. H. Lee, *J. Am. Chem. Soc.* **129**, 7758 (2007).

15

Molecular Encapsulations into Interior Spaces of Carbon Nanotubes and Nanohorns

T. Okazaki, S. Iijima and M. Yudasaka

Nanotube Research Center, Meijo University, Japan

15.1 Introduction

Single-wall carbon nanotubes (SWCNTs) are single-graphene tubules with diameters of about 1 nm and length of micrometer to millimeter orders, which was discovered in 1993 [1]. Materials encapsulated inside SWCNTs were first reported in 1998, which shows a linear array of C_{60} molecules inside SWCNTs [2] This discovery of C_{60}@SWCNT was accidental, and C_{60} entrance mechanism was unclear at that time. Later on, C_{60}@SWCNT was found to be prepared easily by putting SWCNTs in the C_{60} vapor [3, 4]. This means that the C_{60} molecules in the vapor directly enter inside SWCNTs from the open ends, or the C_{60} molecules adsorb on the SWCNT surface, migrate, and reach the holes to enter inside SWCNTs. The theoretical studies suggest that the former model is likely [5].

Inside SWCNTs, not only the alignments, but also motions of molecules are considerably restricted. This effect enables the observation of the various molecules individually [6–9] and their chemical changes [10–13] with transmission electron microscope otherwise difficult because they cannot be isolated and fixed. An example of such interesting result is cis-trans isomerization of retinal encapsulated inside SWCNTs [6].

Chemistry of Nanocarbons Edited by Takeshi Akasaka, Fred Wudl and Shigeru Nagase

The inner spaces of SWCNTs are also useful as a one-dimensional reaction sites: the C_{60} molecules inside SWCNT coalesced to form second SWCNTs by heating up to 1000 °C, resulting in the double-wall carbon nanotube creation [14].

The electronic structures of SWCNTs are effectively affected by the encapsulated molecules. Recent discovery of photoluminescence (PL) from SWCNTs allows us to investigate the molecular encapsulated effects on the electronic states of each (n, m) SWCNTs because the emission and excitation spectra show characteristic peaks depending on the molecular structure of SWCNTs [15]. In Section 15.2, we review the C_{60} encapsulation effects on the electronic states of SWCNTs by the photoluminescence excitation (PLE) spectroscopy.

Materials storage inside SWNH are potentially useful in various applications. Large quantity methane adsorption is the primary one, and molecule-size depending adsorption through different-size holes is unique in the carbonaceous materials. Recently studies on the drug delivery applications have presented the high potentiality.

Thus, apparently, the molecules incorporation inside SWCNTs and SWNHs are useful to study the unique properties and applications of the incorporated molecules. We introduce a part of those in the followings.

15.2 SWCNT Nanopeapods

15.2.1 Synthesis Methods

A bulk production method of nanopeapods has now been achieved by several methods. The most standard way is to heat the end-opened SWCNTs together with fullerenes in an evacuated sealed glass tube [3]. During the heat treatment, fullerenes are encapsulated into the hollow space of SWCNTs by thermal kinetic energy. To obtain high-purity and high-yield nanopeapods, one should prepare high-purity end-opened SWCNTs. For example, now our research group has purified the as-prepared SWCNTs by using a combination of the acid and alkali treatments [16]. First, the as-prepared SWCNTs are heated at ~350 °C for 30 minutes in air to remove the most of amorphous carbon and other carbon materials which coats catalyst metal particles. The obtained SWCNTs were treated in methanol solution of sodium hydroxide (0.2 g/100 ml) for 30 minute and washed by ethanol for several times. Then the remaining metal particles are washed by hydrochloric acid and heated at 600 °C for 2 hours in vacuum.

To open the cap of SWCNTs, the purified SWCNTs were heated at 500 °C for 30 minutes in air. The treated SWCNTs and fullerenes are sealed under vacuum ($\sim 1 \times 10^{-4}$ Pa) in quartz tube and heated at 600 °C for 24 hours. The obtained nanopeapods are washed with toluene to remove the fullerenes adsorbed on the outside of the walls. After the filtration, a dark, paper-like sheet, so-called buckypaper is obtained (Figure 15.1).

Figure 15.2 shows typical transmission electron microscope (TEM) images of the produced nanopeapods. The TEM images clearly show that C_{60} and $Er@C_{82}$ fullerenes are highly packed inside the SWCNTs.

Besides the gas phase doping, nanopeapods can also be made through liquid phase (see also Section 15.3.2) [17]. The method involves immersing open-ended SWNTs in a saturated solution of the filling molecules. The solution method is especially advantageous in bio-medical use of nano-peapods because it can be applied for thermally unstable molecules such as large organic and biological molecules.

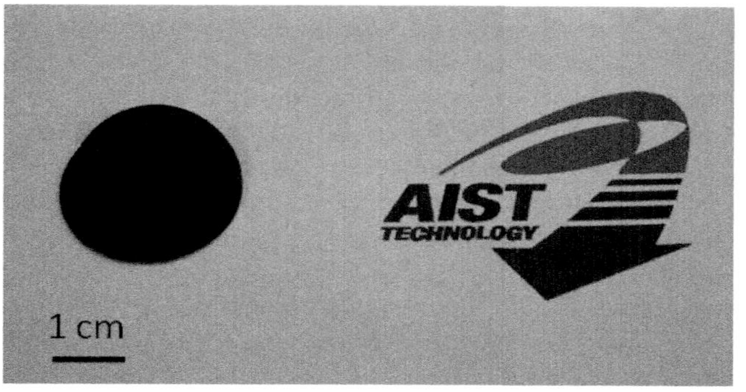

Figure 15.1 *Picture of nanopeapods buckypaper*

15.2.2 Electronic Structures of C_{60} Nanopeapods

15.2.2.1 Photoluminescence Spectroscopy

Fullerene encapsulation effects on the electronic structures of SWCNTs were firstly investigated by the scanning tunneling spectroscopy at low temperature. For example, a down shift in the first van Hove band gap of ~60 meV was observed for SWCNTs with d_t~1.3 nm upon C_{60} encapsulation [18]. Such a band gap modulation leads to the substantial

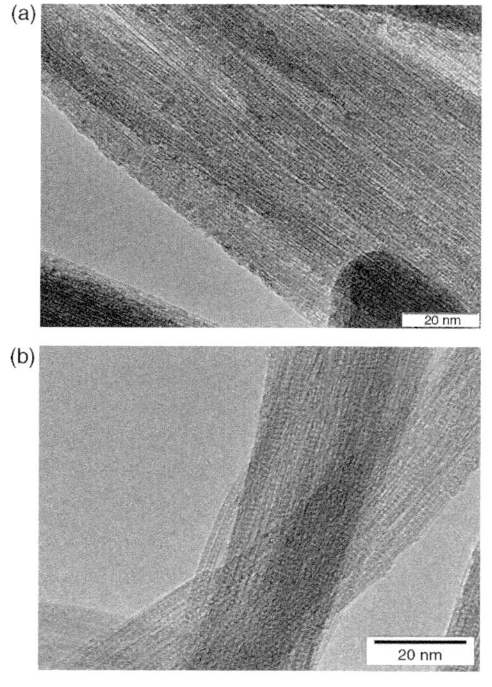

Figure 15.2 *Typical TEM images of (a) C_{60} nanopeapods and (b) Er@C_{82} nanopeapods, respectively*

change in transport properties of SWCNTs. Actually, the *p*-type field-effect transistor (FET) behavior of SWCNTs was changed to be ambipolar by insertion of fullerenes [19].

The detailed information about band gap modification of SWCNTs upon fullerene insertions can be obtained by photoluminescence excitation (PLE) spectroscopy [16, 20]. The origin of the observed photoluminescence (PL) peak can be reliably assigned to SWCNTs with specific chiral indexes (*n, m*) because the emission and excitation spectra show characteristic peaks depending on the molecular structure of SWCNTs [15]. Consequently, the PL method can provide rich information about electronic properties for individual (*n, m*) nanopeapods at a resolution of a few meV.

Figure 15.3 shows a 2D PL contour plot of the unfilled SWCNTs in SDBS-D_2O as a function of emission (λ_{11}) and excitation (λ_{22}) wavelengths [16, 20]. Here two SWCNTs having different diameter distributions are used; one is synthesized by pulsed-laser vaporization (PLV) method [20] (Figure 15.3a) and the other by arc-discharging method [16] (Figure 15.3b). The PL maxima (spots) on the map are clearly seen in the second interband (E_{22}) excitation region ($\lambda_{22} = 800$–1060 nm) and the first interbands (E_{11}) emission region ($\lambda_{11} = 1400$–1800 nm) of SWNTs with 1.2–1.6 nm in diameter, which can be assigned to the specific (*n, m*) SWNTs by using the empirical relations of Weisman *et al* [21]. It is noted that all peak positions were slightly and uniformly red-shifted after the tube opening (~9 meV in E_{11} and ~15 meV in E_{22}). This can be explained by the presence of the solvents in the interior space of the open-ended SWCNTs. Actually, it has been known that water molecules are encapsulated inside SWCNTs [22]. The filling water solvents should cause a reduction of the electron-electron repulsion and the exciton binding energy by dielectric screen effects [23], which results in the observed red-shifts.

Overall features of PL behavior of SWCNTs drastically change upon C_{60} encapsulations. Figure 15.3c,d show the 2D PL contour maps of PLV-SWCNTs and arc-SWCNTs, respectively, after C_{60} encapsulations. The PL peaks in Figure 15.3c can be divided into two groups (the white line in Figure 15.3c). One is the peaks whose positions are the same as those of unfilled SWCNTs. The origin of these peaks are thus easily assigned to the (11, 6), (15, 1), (10, 8), (12, 5) and (11, 7) tubes, respectively (Figure 15.3c). The characteristic feature of this group is that the tube diameters are less than ~1.25 nm. On the other hand, the original PL peaks of larger diameter tubes with $d_t > 1.25$ nm such as (13, 5), (12, 7) and (10, 9) tubes almost completely disappears in Figure 15.3c. Alternatively, new PL peaks appear at different positions to those of the unfilled SWCNTs. The threshold diameter of ~1.25 nm closely matches the theoretically predicted value of ~1.28 nm for unstrained filling of C_{60} [24]. This agreement strongly suggests that C_{60} is preferentially encapsulated into SWNTs with $d_t > 1.25$ nm and that the filling causes spectral shifts for those nanotubes.

The observed new PL peaks were assigned to each (*n, m*) nanopeapods by examining many samples that has a different chirality distribution and a different filling yield [16, 20]. The results are shown in Figure 15.3C, D.

15.2.2.2 Mechanisms for the Optical Band Gap Modification

In order to investigate the detail mechanisms of band gap modification, the energy differences in E_{11} and E_{22} between C_{60} nanopeapods and SWCNTs ($\Delta E_{ii} = E_{ii}^{\text{nanopeapods}} - E_{ii}^{\text{SWCNTs}}$, $i = 1, 2$) are plotted as a function of tube diameter and a chiral angle (θ) (Figure 15.4A–D). Apparently, the diameter dependencies of ΔE_{11} and ΔE_{22} are different

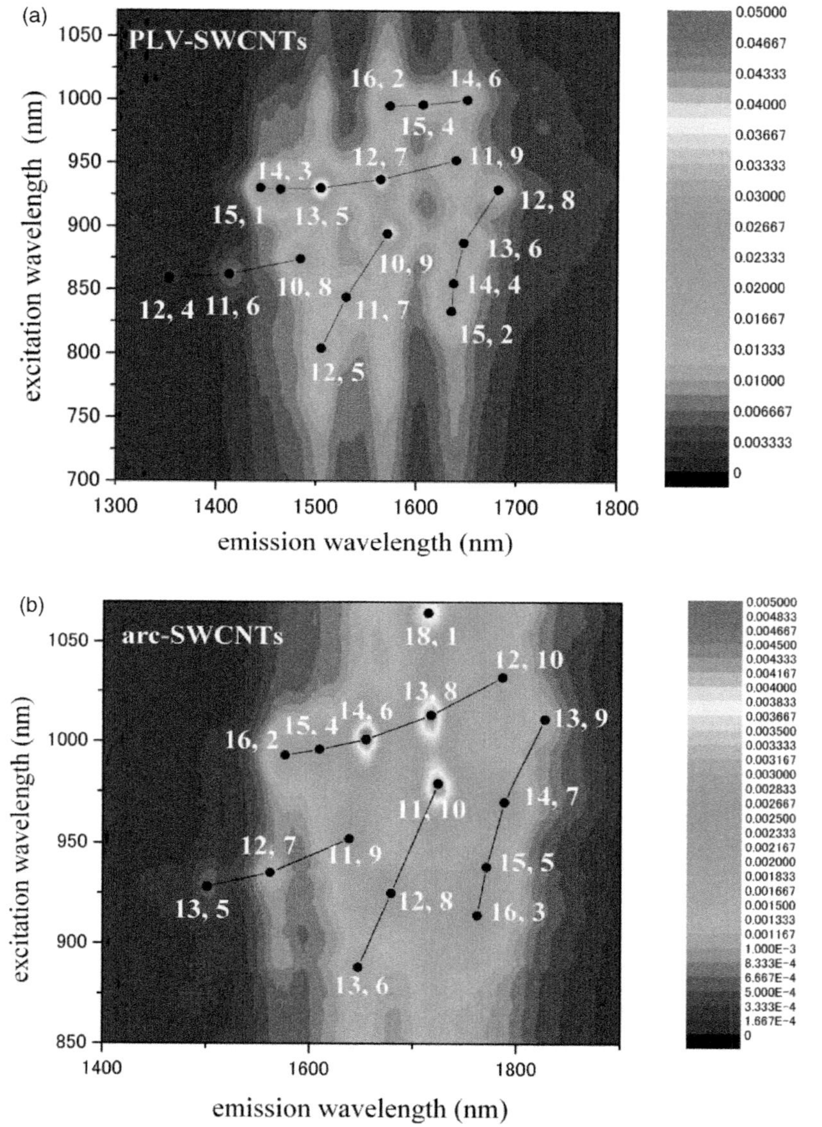

Figure 15.3 *LE maps of (a) PLV-SWCNTs, (b) arc-SWCNTs, (c) PLV-nanopeapods and (d) arc-nanopeapods, respectively*

between the '$2n + m$' family types (type I (mod($2n + m$, 3) = 1) and type II (mod($2n + m$, 3) = 2), respectively). For example, ΔE_{11} for type I tubes show positive values in a smaller diameter regime. As increasing tube diameter, it exponentially decreases towards –0.02 eV and approaches to zero. On the contrary, the ΔE_{11} for type II tubes increase towards 0.02 eV as d_t increase, and then gradually decrease again (Figure 15.4a). As is the case of ΔE_{11},

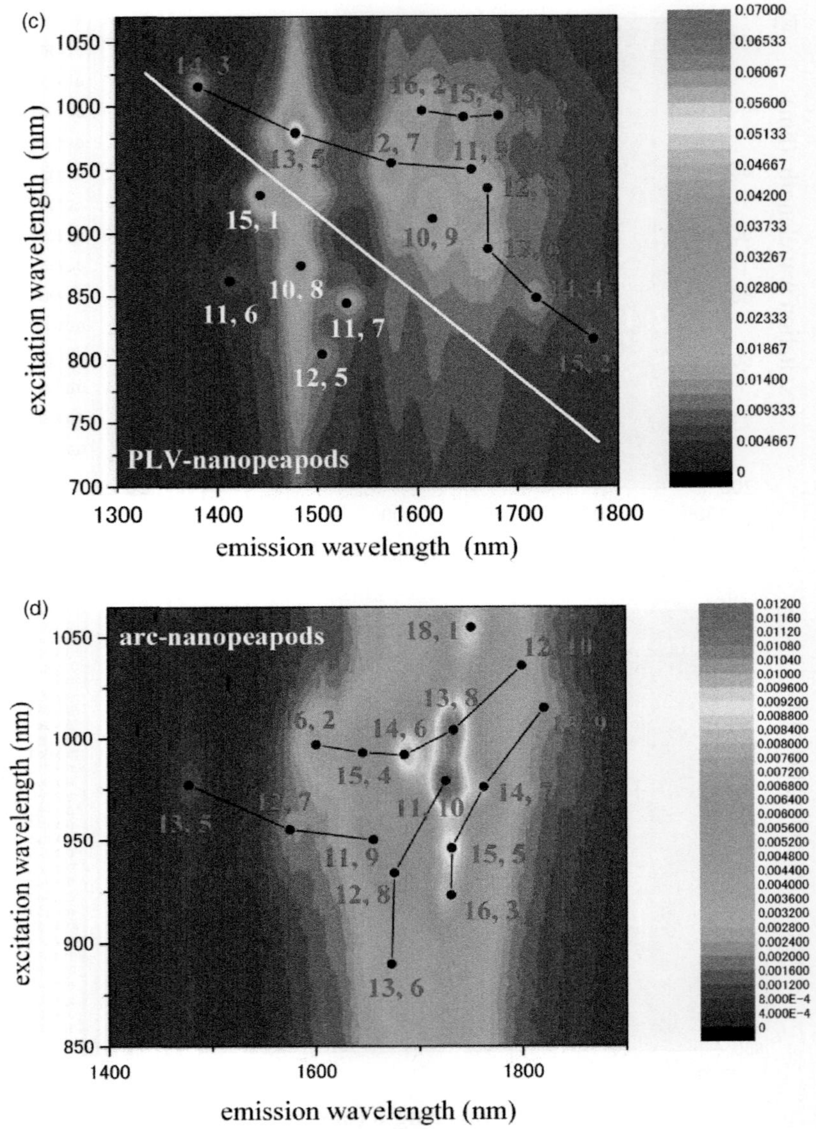

Figure 15.3 *(Continued)*

ΔE_{22} also shows clear family-type dependence (Figure 15.4b). Namely, ΔE_{22} for type I tubes increase as increasing tube diameter and then decrease at $d_t \sim 1.4$ nm, whereas ΔE_{22} for type II tubes exhibit a totally opposite trend (Figure 15.4b).

Such a strong dependence on '$2n + m$' family type is a characteristic feature of the strain-induced spectral shift [25]. Theoretical calculations predicted that the band gaps of E_{11} and E_{22} change towards opposite directions each other upon the structural deformation of

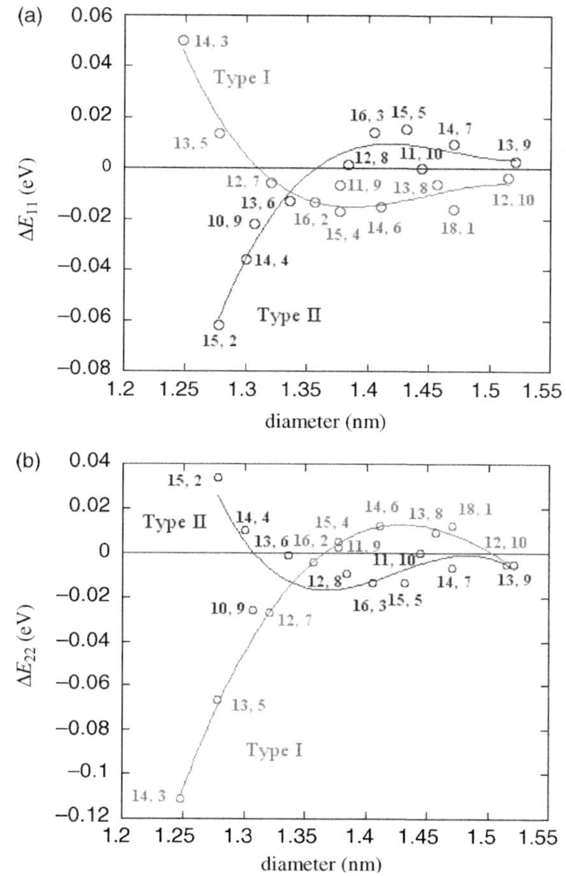

Figure 15.4 *Differences in optical transition energies in E_{11} and E_{22} (ΔE_{11} and ΔE_{22}, respectively) between C_{60} nanopeapods and SWNTs as a function of (a, b) a tube diameter and (c, d) chiral angle*

SWCNTs caused by the mechanical strain [25]. Furthermore, the strain-induced band gap shifts for type I tubes are entirely contrary to those of type II tubes. For example, E_{11} of type I tubes increase under a radial expansion and/or a compressive strain along the tube, whereas that of type II tubes decrease [25]. The observed ΔE_{11} and ΔE_{22} for smaller diameter tubes ($d_t < \sim 1.32$ nm) correspond to this situation (Figure 15.4a,b). The tube diameter expansion due to the fullerene encapsulation is responsible to the observed PL peak shifts.

As increasing the tube diameter, the ΔE_{11} and ΔE_{22} approach to zero line and change their signs (Figure 15.4a,b). Okada and coworkers predicted that C_{60} molecules can enter the interior space of SWCNTs without friction at $d_t > \sim 1.3$ nm [24, 26, 27]. The diameters of the SWCNTs are unchanged upon fullerene insertion within the calculation accuracy (~ 0.001 nm). In this diameter region, efficient coupling (hybridization) between the π states of C_{60} and the nearly free electron (NFE) states of SWCNTs occurs [24, 26, 27]. Electrons are transferred from π orbitals of SWCNTs and C_{60} to the space between them,

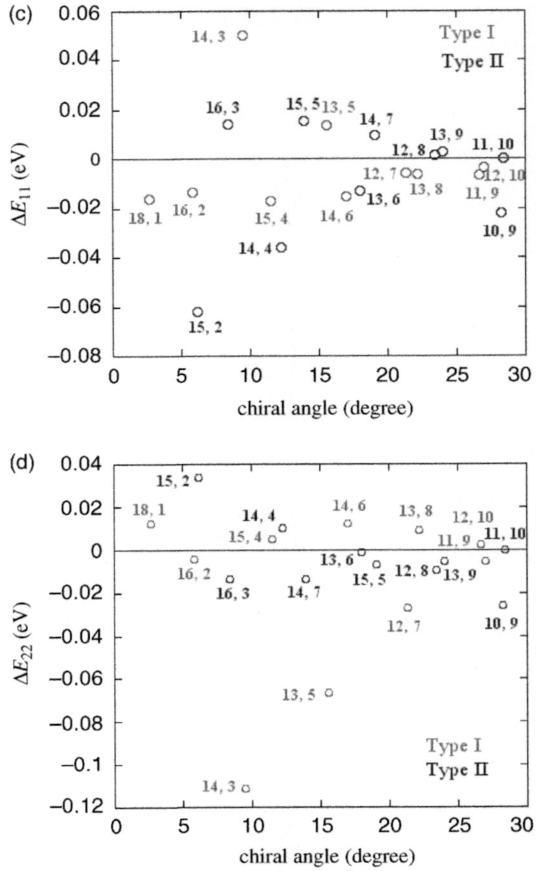

Figure 15.4 *(Continued)*

so that π electron clouds around the nanotube walls expand to the inner and the outer spaces [24, 26, 27]. Indeed electron density in the vicinity of the tube wall is found to decrease due to the expansion of the electron clouds for C_{60}@(11, 11) nanopeapods ($\sim$0.05e) ($d_t = 1.48$ nm) [27]. The expansion of the electron clouds results in the increase of the resonance integral between the neighboring atoms. Because the π states of SWCNTs and C_{60} radially-overlapped in nanopeapods, the resonance integral increases around the circumference of the tube rather than along tube axis. Such an increase of the resonance integral should cause the same effect as the decrease of the C-C bond length and the corresponding reduction of the effective tube diameter. As a result, the band gap of SWCNTs changes in a family type dependent manner ($\Delta E_{11} < 0$ and $\Delta E_{22} > 0$ for type I, and $\Delta E_{11} > 0$ and $\Delta E_{22} < 0$ for type II) even though the positions of the carbon atoms are unchanged.

The PL shifts for smaller diameter tubes can be ascribed to the mechanical strain to SWCNTs by the encapsulated C_{60} [20]. The local strain upon C_{60} encapsulation may disappear at 1.30∼1.35 nm because the PL shift changes their signs around it (Figure 15.4a, b). Surprisingly, this value matches well with the nearest neighboring distance between C_{60} and SWCNTs obtained by a simple calculation. The intermolecular distance between C_{60} molecules inside SWCNTs was found to be 0.97 nm by electron diffraction methods [28]. It is well known that the interlayer distance between two graphenes is 0.34 nm in graphite [29]. Hence, the nearest distance from the center of C_{60} to the tube wall can be calculated to be 0.66 (=(0.97 + 0.34)/2) nm, which suggests that C_{60} require the tube diameter of larger than 1.31 (=0.97 + 0.34) nm to enter the interior space without friction. Excellent agreement about the threshold diameters strongly suggests the van der Waals nature of the interaction between C_{60} and SWCNTs.

On the other hand, the observed PL spectral shifts do not correlate with the chiral angle of SWCNTs (Figure 15.4c,d). All SWCNTs that show large ΔE_{11} and ΔE_{22} are smaller diameter tubes such as (14, 3), (15, 2) and (13, 5).

15.2.2.3 *Exciton Effects*

It is generally accepted that the PL behaviors of SWCNTs are dominated by the Coulomb interaction between the produced electron-hole pairs (excitons). For example, although the one electron theory predicts that the energy ratio between E_{22} and E_{11} (E_{22}/E_{11}) must approach two as tube diameter increases, experimental results show that it approaches a value smaller than two. This 'ratio problem' can be explained by the strong exciton effects in SWCNTs [30]. Figure 15.5 shows the obtained E_{22}/E_{11} values of C_{60} nanopeapods and SWCNTs as a function of tube diameter. The experimental error for E_{22}/E_{11} is less than 0.01. Interestingly, both E_{22}/E_{11} values approach almost same values (∼1.75, solid line). This strongly suggests the exciton effects are very similar between SWCNTs and nanopeapods.

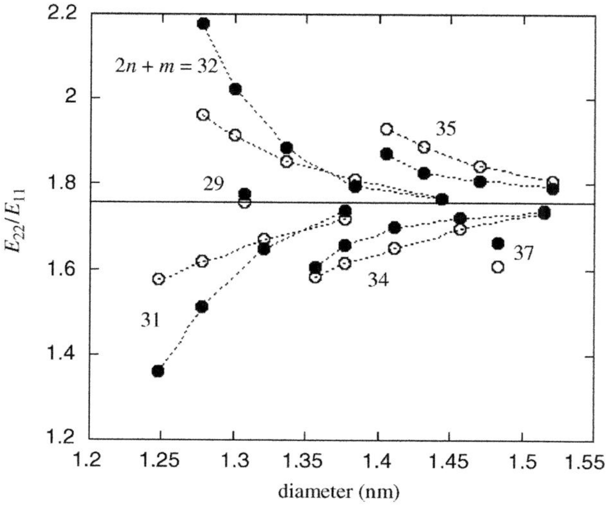

Figure 15.5 *E_{22}/E_{11} for SWCNTs (open circle) and nanopeapods (solid circle) as a function of a tube diameter*

15.3 Material Incorporation and Release in/from SWNH

15.3.1 Structure of SWNH and SWNHox

SWNHs are formed by CO_2 laser ablation of graphite [31, 32]. A single-wall carbon nanohorn is a single-graphene tubule with diameters of 5 nm and length of 40–50 nm (Figure 15.6) [31]. About 2000 of SWNHs assemble and form a spherical aggregate (Figure 15.6) [31]. The aggregate has not been separated into individual SWNHs.

As-grown SWNHs are closed tubules, therefore the holes have to be created to use the inner hollow spaces. The holes are opened by oxidation using oxygen gas [32, 33], carbon dioxide gas [34], or oxidative acid solutions [35–37]. The holes of hole-opened SWNH (SWNHox) are visible with transmission electron microscopy (Figure 15.7a) [38]. The hole size histogram showed that the tip holes had smaller sizes than sidewall holes (Figure 15.7b).

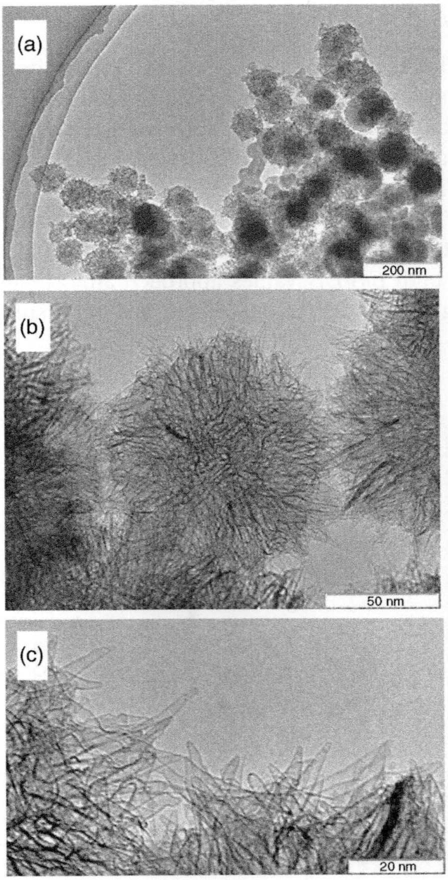

Figure 15.6 Transmission electron microscopy images of single-wall carbon nanohorns. Spherical aggregates with dahlia-like-shapes (a, b) and the surface of the aggregate (c). Scale bars: 200 nm (a), 50 nm (b), and 20 nm (c)

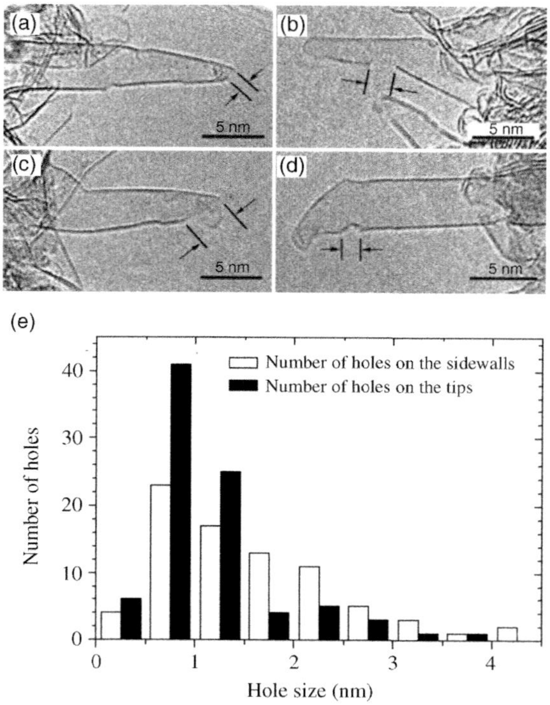

Figure 15.7 *Transmission electron microscopy images of SWNHox. Holes are and their widths are shown with bars and arrows (a~d). Histograms of holes at the tips (Black bars) and side walls (blank bars) (e) [38]*

The number and size of holes are controlled with the oxidation conditions as shown in Figure 15.8 [38]. In Figure 15.8a, the specific surface area of SWNHox increased with the oxidation temperature in oxygen atmosphere. Here, the specific surface area was estimated through nitrogen adsorption quantities measured at 77 K [39]. Figure 15.8b shows SWNHox obtained by 'slow combustion' adsorbed m-xylene, and the adsorption quantities increased with the oxidation temperature [33]. Here, the 'slow combustion' and 'm-xylene adsorption' is explained briefly. In the 'slow combustion', temperature was increased to 400–600 °C with an increase rate of 1 °C/min in dry air [33]. This is an excellent hole-opening method in terms of small-amount generation of carbonaceous dusts. We (M.Y. and S.I.) often adopt the m-xylene adsorption quantity instead of nitrogen adsorption quantity as a measure for the specific surface area or pore volume of SWNHox because the m-xylene adsorption is measured in a short time by thermo gravimetry, and the correspondence between the nitrogen adsorption quantity and m-xylene adsorption quantity was good (Table 15.1) [33].

15.3.2 Liquid Phase Incorporation at Room Temperature

As described in Section 15.2.1, incorporations of fullerene inside SWCNTs are often carried out in gas phase at high temperatures. Since we (M.Y. and S.I.) are interested in the

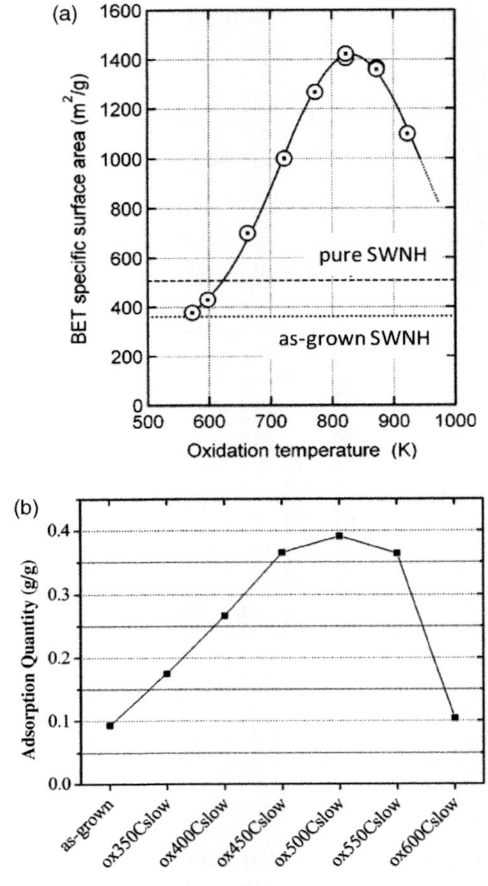

Figure 15.8 *BET specific surface area of SWNHox treated in oxygen gas at various temperatures (a) [39]. Specific surface of 'Pure SWNH' was estimated by removing those of impurities. Quantities of m-xylene adsorbed by SWNHox oxidized by 'Slow combustion' method at various temperatures in dry air [33]*

application of SWNHox as drug delivery systems, we studied the incorporation of materials in liquid phase at room temperatures [39].

In the liquid phase incorporation, affinity balances among guest molecules, solvents, and SWNHox are the key [39]. When the affinity between the guest molecule and SWNHox is higher than those between the guest molecules and solvents or the solvents and SWNHox, the guest molecules are incorporated inside SWNHox by mixing the guest materials, SWNHox, and solvent. If the affinity between the gust material and SWNHox is weak, the incorporation uses the condition of super-saturation. This is shown using two examples in the followings.

In the preparation of cisplatin (cis-diaminedichloroplatinum (II), $Pt(NH_3)_2Cl_2$, CDDP) incorporated SWNHox (CDDP@SWNHox), SWNHox was dispersed in aqueous solution of cisplatin, and left until the water was evaporated [40]. The obtained black powders were

Table 15.1 *Specific surface area and pore volumes of SWNHox estimated from the nitrogen adsorption isotherms measured at 77 K. The values for the SWNHox obtained by the quick combustions (upper table) and slow combustions (lower table) are shown [14]*

T(target) (°C)	BET surface area[a] ($m^2\ g^{-1}$)	pore volume[b] ($cm^{-3}\ g^{-1}$)
	Quick Combustion[c] (O_2 keeping period = 10 min)	
as-grown	320	0.21
350	550	0.35
400	990	0.63
450	1200	0.75
500	1300	0.82
550	1400	0.92
580	1270	0.85
	Slow Combustion[d] (air; Rate(rise) = 1 °C/min; keeping period = 0 min)	
400	1120	0.71
450	1300	0.81
500	1450	0.92
550	1360	0.89
600	540	0.41

[a] Calculated in the pressure region from $P/P_0 = 0.001–0.1$.
[b] Estimated from adsorption amount at $P/P_0 = 0.7$.
[c] SWNHox were treated by quick combustion performed at various values of *T*(target) in 100% oxygen gas for 10 min.
[d] SWNHox were treated by slow combustion with various values of *T*(target) while keeping Rate(rise) at 1 °C/min in dry air. The *T*(target) holding period was 0 min.

CDDP@SWNHox as shown with transmission electron microscope (TEM) images and elemental mapping in Figure 15.9. The quantity of CDDP in CDDP@SWNHox was about 50% as estimated from inductively-coupled plasma atomic emission spectroscopy measurements [40].

The C_{60} was incorporated inside SWNHox by several methods (Figure 15.10) [38, 41]. The large quantity of C_{60}@$SWNH_{ox}$ was obtained by immersing SWNHox in C_{60}-toluene solution, and left until the toluene was evaporated [41]. The TEM image of the obtained C_{60}@SWNHox showed that there were C_{60} inside SWNHox (Figure 15.10). The content of C_{60} in C_{60}@SWNHox was controlled by the initial C_{60} quantity, reaching about 0.2 g/g (C_{60}/SWNHox) at maximum. Evidence for the absence of C_{60} outside of SWNHox were given by X-ray diffraction analysis not showing the C_{60} crystal diffractions and Raman spectra exhibiting peaks of C_{60} molecular vibration at 1466 cm^{-1}. In the Raman spectra, the C_{60} peak was not observed when it was outside of SWNHox, while the peak was clearly appeared when it was incorporated. This phenomenon was explained that laser irradiation on the sample in the Raman spectrum measurements in air, the oxygen-mediated laser decomposition occurred quickly on C_{60} located outside SWNHox, while it went slowly on C_{60} located inside SWNHox [41].

15.3.3 Adsorption Sites of SWNHox

Nitrogen adsorption isotherm measured at 77 K revealed that there were three adsorption sites in SWNHox, interstitial space among SWNHox tubules, wall surface of inside SWNHox, and near-center inside SWNHox. Their pore volume ratio was about 1:2:2 [42]. The nitrogen adsorption stability was high in the interstitial sites, followed inner surface and inner center. Similar results were obtained when temperature profiles of

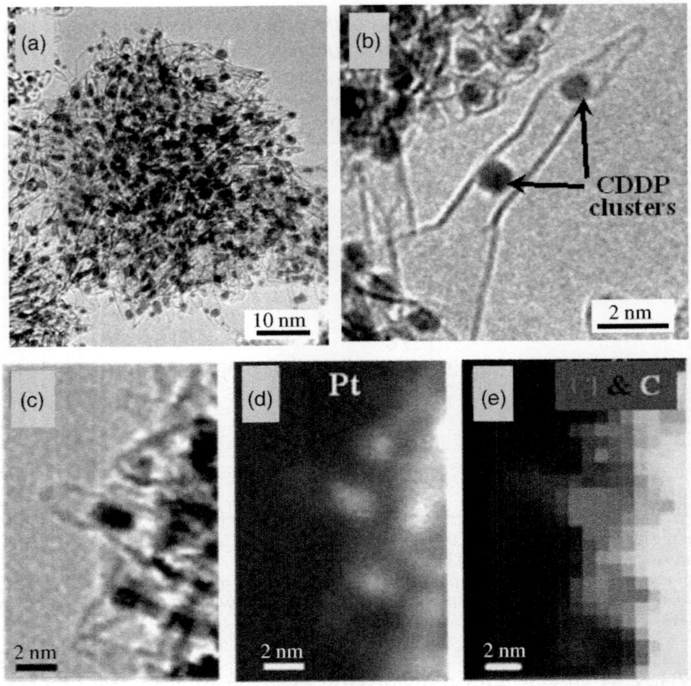

Figure 15.9 *Transmission electron microscopy images of cisplatin (CDDP) incorporated SWNHox (a, b). Black particles are CDDP clusters. Arrows in (b) indicate the CDDP clusters. Scanning transmission electron microscopy image (c), Z-contrast image (d), and mappings of chlorine and carbon measured by electron energy loss spectroscopy [40]*

m-xylene desorption quantity was analyzed, additionally showing that there are the fourth adsoprtion sites, which was assigned to the inside tip of SWNHox (Figure 15.7b) [43]. The D_2/H_2 separation was possible using SWNHox, and it is inferred that this separation proceeded at the inside of tips [44]. In the recent report, it has been revealed that there are five sites near the center of aggregate [45]. When Gd_2O_3 cluster was used as a prove for searching the adsorption sites, large Gd_2O_3 particles with sizes of 10–20 nm were found near the center of the aggregate, indicating that there were caves in the aggregate. This cave has not been found in studying the adsorption and desorption of small molecules, perhaps because the cave size is so large and does not serve as an energetically deep adsorption site.

The release rates of materials are slower from the deeper potential sites if material-wall interactions are mainly van der Waals type as in the case of nitrogen adsorption or m-xylene adsorption. However, there are several another factors that influence the release of materials incorporated inside SWNHox, which are shown below.

15.3.4 Release of Materials from inside SWNHox

When the release of C_{60} from inside SWNHox was examined in solvents, the release rate depended on the solvent (toluene, ethanol, or toluene-ethanol mix-solvents)

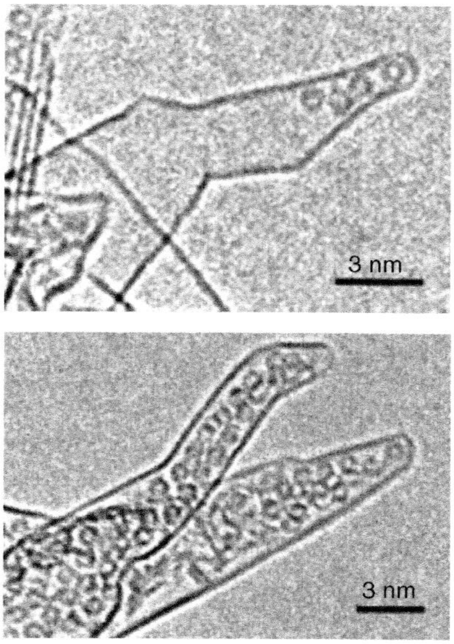

Figure 15.10 *Transmission electron microscopy images of C_{60}-incorporated SWNHox [38]*

(Figure 15.11) [41]. Release was fast and release quantity was large in toluene, and both decreased with the increase of ethanol content in the mix-solvents. Since the release was strongly influenced by the solubility of C_{60} in the solvents, the temporal changes of the released C_{60} quantity was well simulated by assuming only two sites, energetically shallow and deep, inside SWNHox [41].

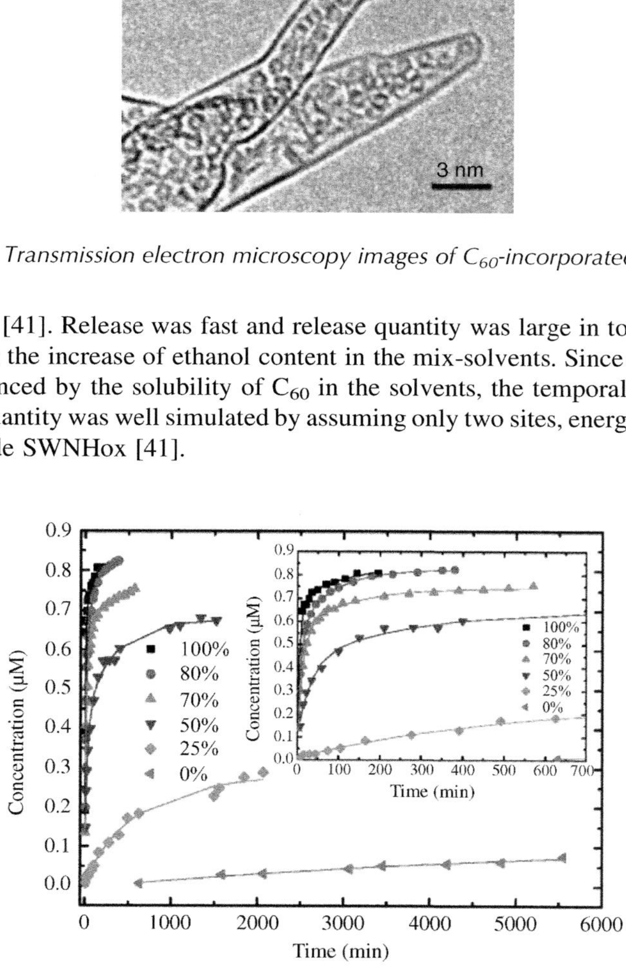

Figure 15.11 *Quantities of C_{60} released from C_{60}@SWNHox into toluene, ethanol, or their mixtures increased with the immersion periods. (Concentrations (%) of toluenes in the toluene-ethnaol mixtures are indicated in graphs.) [41]*

Curious phenomena were found concerning the entrance and exit of C_{60} inside SWNHox. The C_{60} went out of SWNHox when C_{60}@SWNHox was exposed to ethanol vapor or liquid ethanol, and the C_{60} re-entered inside SWNHox by exposing to toluene vapor [46]. These exit and entrance were repeatable. The exit of C_{60} by ethanol was mediated by toluene existing with C_{60} inside SWNHox. If the toluene was removed from inside C_{60}@SWNHox, the exit of C_{60} with ethanol was not observed [46].

In the study of CDDP@SWNHox for the possible application to cancer therapy, the release quantities of CDDP from CDDP@SWNHox were measured in water, phosphate buffer saline, and culture medium. The CDDP was released quickly at the beginning and followed by the slow release, taking about three days for the almost complete release [40]. Referring to the C_{60}@SWNHox cases described above, CDDP should be released more quickly in a good solvent of water. The slow release of CDDP from SWNHox in water might be caused by the CDDP cluster formation: The cluster could have the reduced electrical-dipole-moments, by which CDDP could gain certain stabilization energy inside SWNHox. Similar effect of the cluster was found in the water adsorption inside SWNHox. Adsorption of water inside SWNHox is explained by the water pentamer formation, which corresponds to the hysteresis observed in adsorption-desorption isotherms [47].

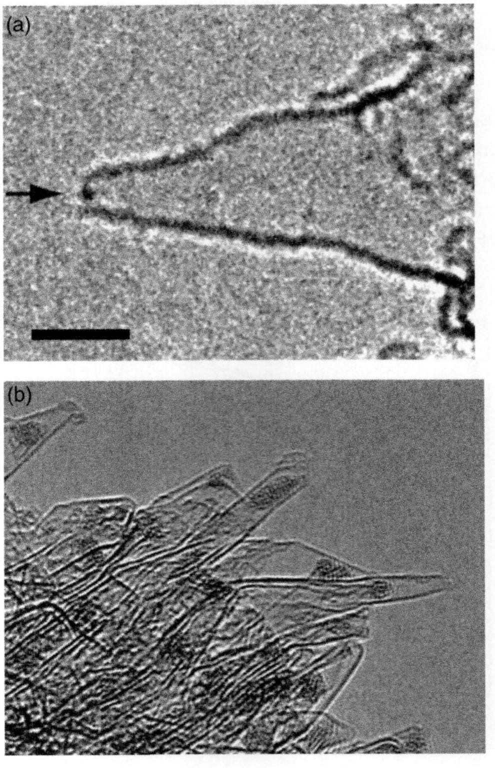

Figure 15.12 Transmission electron microscopy images of a Gd acetate molecule (an arrow) at a hole of tips of SWNHox (a). Clusters of Gd acetates (an arrow) are located inside SWNHox (b) [49]

15.3.5 Plug

The release described in Section 15.3.4 was the passive one, while here the active control of release using plugs located at holes is introduced. Gd acetate was found to attach to the edges of holes of SWNHox or made clusters at or near the holes (Figure 15.12), and it was suggested that they hindered the incorporation of C_{60} inside SWNHox [48]. This plug effect of Gd acetate was precisely studied by the release of C_{60} incorporated inside SWNHox [49]. It was found that the C_{60} of C_{60}@SWNHox was not released in toluene when the Gd acetate plugs was put at the holes. After removal of the Gd acetate plugs from the SWNHox by washing with water, the C_{60} inside SWNHox was released in toluene almost completely [48].

The study also indicated that the Gd acetate attached to the hole edges (Figure 15.12a) more effectively confined C_{60} inside SWNHox than the Gd acetate clusters located at or near the holes (Figure 15.12b) [49].

15.4 Summary

The interior spaces of carbon nanotubes and nanohorns have almost same size for many organic molecules so that the effective interaction between the guest molecules and the host materials is expected. In fact, as seen in this chapter, the ability to modify the electronic properties of SWCNTs can be realized by encapsulating fullerenes. Such hybrid structures are also useful for drug-delivery systems. Now we can not only encapsulate molecules into SWNHs, but also release them at will. One of the great challenges for nanotechnology is the creation of novel functional materials through molecular level control of material composition. We believe that carbon nanotubes and nanohorns will lead to a new class of multifunctional materials by utilizing their 'empty' spaces.

References

[1] S. Iijima and T. Ichihashi, Ingle-shell carbon nanotubes of 1-nm diameter, *Nature*, **363**, 603–605 (1993).

[2] B. W. Smith, M. Monthioux and D. E. Luzzi, Encapsulated C_{60} in carbon nanotubes, *Nature*, **396**, 323 (1998).

[3] K. Hirahara, K. Suenaga, S. Bandow, H. Kato, T. Okazaki, H. Shinohara and S. Iijima, One dimensional metallofullerenes crystal generated inside single-walled carbon nanotube, *Phys. Rev. Lett.*, **85**, 5384 (2000).

[4] H. Kataura, Y. Maniwa, T. Kodama, K. Kikuchi, K. Hirahara, K. Suenaga, S. Iijima, S. Suzuki, Y. Achiba and W. Krätschmer, High-yield fullerene encapsulation in single-wall carbon nanotubes *Synth. Met.*, **121**, 1195 (2001).

[5] S. Berber, Y.-K. Kwon, and D. Tománek, Microscopic formation machanism of nanotube peapods, *Phys. Rev. Lett.*, **88**, 185502 (2002).

[6] Z. Liu, K. Yanagi, K. Suenaga, H. Kataura and S. Iijima, 'Imaging the dynamic behaviour of individual retinal chromophores confined inside carbon nanotubes', *Nat. Nanotech.*, **2**, 422–425 (2007).

[7] L. Guan, K. Suenaga, Z. Shi, Z. Gu and S. Iijima, 'Polymorphic structures of iodine and their phase transition in confined nanospace', *Nano Lett.*, **7**, 1532–1535 (2007).

[8] M. Koshino, T. Tanaka, N. Solin, K. Suenaga, H. Isobe and E. Nakamura, 'Imaging of single organic molecules in motion', *Science*, **316**, 853 (2007).

[9] Z. Liu, M. Koshino, K. Suenaga, A. Mrzel, H. Kataura and S. Iijima, 'Transmission electron microscopy imaging of individual functional groups of fullerene derivatives', *Phys. Rev. Lett.*, **96**, 088304–088308 (2006).

[10] Z. Liu, K. Suenaga and S. Iijima, Imaging the structure of an individual C_{60} fullerene molecule and its deformation process using HRTEM with atomic sensitivity, *J. Am. Chem. Soc.*, **129**, 6666–6667 (2007).

[11] H. Shiozawa, T. Pichler, A. Gruneis, R. Pfeiffer, H. Kuzmany, Z. Liu, K. Suenaga and H. Kataura, Catalytic reaction inside a single-wall carbon nanotube, *Adv. Mater.*, **20**, 1443–1449 (2008).

[12] R. Pfeiffer, M. Holzweber, H. Peterlik, H. Kuzmany, Z. Liu, K. Suenaga and H. Kataura, Dynamics of carbon nanotube growth from fullerenes, *Nano Lett.*, **7**, 2428–2434 (2007).

[13] L. Guan, K. Suenaga, T. Okazaki, Z. Shi, Z. Gu and S. Iijima, 'Coalescence of C_{60} molecules assisted by doped iodine inside carbon nanotubes', *J. Am. Chem. Soc.*, **129**, 8954–8955 (2007).

[14] S. Bandow, M. Takizawa, K. Hirahara, M. Yudasaka and S. Iijima, 'Raman scattering study of double-wall carbon nanotubes derived from the chains of fullerenes in single-wall carbon nanotubes', *Chem. Phys. Lett.*, **337**, 48–54 (2001).

[15] R. B. Weisman, in *Applied Physics of Carbon Nanotubes: Fundamentals of Theory, Optics and Transport Devices*, S. V. Rotkin, S. Subramoney (eds) (Springer: Berlin, 2005), pp. 183–202.

[16] S. Okubo, T. Okazaki, N. Kishi, S.-K. Joung, T. Nakanishi, S. Okada and S. Iijima, Diameter-selective interaction between single-walled carbon nanotubes and encapsulated fullerenes, *J. Phys. Chem. C*, **113**, 571–575 (2009).

[17] M. Yudasaka, K. Ajima, K. Suenaga, T. Ichihashi, A. Hashimoto and S. Iijima, Nano-extraction and nano-condensation for C_{60} incorporation into single-wall carbon nanotubes in liquid phases, *Chem. Phys. Lett.*, **380**, 42 (2003).

[18] D. J. Hornbaker, S.-J. Kahng, S. Misra, B. W. Smith, A. T. Johnson, E. J. Mele, D. E. Luzzi and A. Yazdani, Mapping the one-dimensional electronic states of nanotube peapod structures, *Science*, **295**, 828–831 (2002).

[19] T. Shimada, T. Okazaki, R. Taniguchi, T. Sugai, H. Shinohara, K. Suenaga, Y. Ohno, S. Mizuno, S. Kishimoto and T. Mizutani, Ambipolar field-effect transistor behavior of Gd@C_{82} metallo-fullerene peapods, *Appl. Phys. Lett.*, **81**, 4067 (2002).

[20] T. Okazaki, S. Okubo, T. Nakanishi, S.-K. Joung, T. Saito, M. Otani, S. Okada, S. Bandow and S. Iijima, Optical band gap modification of single-walled carbon nanotubes by encapsulated fullerenes, *J. Am. Chem. Soc.*, **130**, 4122–4128 (2008).

[21] R. B. Weisman and S. M. Bachilo, Dependence of optical transition energies on structure for single-walled carbon nanotubes in aqueous suspension: an empirical Kataura plot, *Nano Lett.*, **3**, 1235–1238 (2003).

[22] Y. Maniwa, H. Kataura, M. Abe, S. Suzuki, Y. Achiba, H. Kira and K. Matsuda, K. Phase transition in confined water inside carbon nanotubes, *J. Phys. Soc. Jpn.*, **71**, 2863–2866 (2002).

[23] T. Ando, Excitons in carbon nanotubes revisited: dependence on diameter, Aharonov-Bohm flux, and strain, *J. Phys. Soc. Jpn.*, **73**, 3351–3363 (2004).

[24] S. Okada, S. Saito and A. Oshiyama, Energetics and electronic structures of encapsulated C_{60} in a carbon nanotube, *Phys. Rev. Lett.*, **86**, 3835–3838 (2001).

[25] L. Yang and J. Han, Electronic structure of deformed carbon nanotubes, *Phys. Rev. Lett.*, **85**, 154–157 (2000).

[26] M. Otani, S. Okada and A. Oshiyama, Energetics and electronic structures of one-dimensional fullerene chains encapsulated in zigzag nanotubes, *Phys. Rev. B*, **68**, 125424 (2003).

[27] S. Okada, Radial-breathing mode frequencies for nanotubes encapsulating fullerenes, *Chem. Phys. Lett.*, **438**, 59–62 (2007).

[28] K. Hirahara, S. Bandow, K. Suenaga, H. Kato, T. Okazaki, H. Shinohara and S. Iijima, Electron diffraction study of one-dimensional crystals of fullerenes, *Phys. Rev. B*, **64**, 115420 (2001).

[29] H. A. R. Palser, Interlayer interactions in graphite and carbon nanotubes, *Phys. Chem. Chem. Phys.*, **1**, 4459 (1999).

[30] C. L. Kane and E. J. Mele, Ratio problem in single carbon nanotube fluorescence spectroscopy, *Phys. Rev. Lett.*, **90**, 207401 (2003).
[31] S. Iijima, M. Yudasaka, R. Yamada, S. Bandow, K. Suenaga, F. Kokai and K. Takahashi, Nanoaggregates of single-walled graphitic carbon nano-horns, *Chem. Phys. Lett.*, **309**, 165 (1999).
[32] T. Azami, D. Kasuya, R. Yuge, M. Yudasaka, S. Iijima, T. Yoshitake and Y. Kubo, Large-scale production of single-wall carbon nanohorns with high purity, *J. Phys. Chem. C*, **112**, 1330–1334.
[33] J. Fan, M. Yudasaka, J. Miyawaki, K. Ajima, K. Murata and S. Iijima, Control of hole opening in single-wall carbon nanotubes and single-wall carbon nanohorns using oxygen, *J. Phys. Chem. B*, **110**, 1587–1591 (2006).
[34] E. Bekyarova, K. Kaneko, M. Yudasaka, D. Kasuya, S. Iijima, A. Huirobro and F. R.-Reinoso, Controlled opening of single-wall carbon nanohorns by heat treatment in carbon dioxide, *J. Phys. Chem. B*, **107**, 4479–4484 (2003).
[35] C.-M. Yang, D. Kasuya, M. Yudasaka, S. Iijima and K. Kaneko, Microporosity development of single-wall carbon nanohorn with chemically induced coalescence of the assembly structure *J. Phys. Chem. B*, **108**, 17775–17782 (2004).
[36] C.-M. Yang, H. Noguchi, K. Murata, M. Yudasaka, A. Hashimoto, S. Iijima and K. Kaneko, Highly ultramicroporous single-walled carbon nanohorn assemblies, *Adv. Mater.*, **17**, 866–870 (2005).
[37] M. Zhang, M. Yudasaka, K. Ajima, J. Miyawaki and S. Iijima, Light-assisted oxidation of single-wall carbon nanohorns for abundant creation of oxygenated groups that enables chemical modifications with proteins to enhance biocompatibility, *ACSNano*, **1**, 265–272 (2007).
[38] K. Ajima, M. Yudasaka, K. Suenaga, D. Kasuya, T. Azami and S. Iijima, Materials storage mechanism in porous nanocarbons, *Adv. Mater.*, **16**, 397–401 (2004).
[39] S. Utsumi, J. Miyawaki, H. Tanaka, Y. Hattori, T. Ito, N. Ichikuni, H. Kanoh, M. Yudasaka, S. Iijima and K. Kaneko, Opening mechanism of internal nanoporosity of single-wall carbon nanohorn, *J. Phys. Chem. B*, **109**, 14319 (2005).
[40] K. Ajima, T. Murakami, Y. Mizoguchi, K. Tsuchida, T. Ichihashi, S. Iijima and M. Yudasaka, Enhancement of in vivo anticancer effects of cisplatin by incorporation inside single-wall carbon nanohorns, *ACS Nano*, **2**, 2057–2064 (2008).
[41] R. Yuge, M. Yudasaka, J. Miyawaki, Y. Kubo, T. Ichihashi, H. Imai, E. Nakamura, H. Isobe, H. Yorimitsu and S. Iijima, Controlling the incorporation and release of C_{60} in nanometer-scale hollow spaces inside single-wall carbon nanohorns, *J. Phys. Chem. B*, **109**, 17861–17867 (2005).
[42] K. Murata, K. Kaneko, W. Steele, F. Kokai, K. Takahashi, D. Kasuya, K. Hirahara, M. Yudasaka and S. Iijima, Molecular potential structures of heat-treated single-wall carbon nanohorn assemblies *J. Phys. Chem. B*, **105**, 10210–10216 (2001).
[43] M. Yudasaka, J. Fan, J. Miyawaki and S. Iijima, Studies on the adsorption of organic materials inside thick carbon nanotubes, *J. Phys. Chem. B*, **109**, 8909–8913 (2005).
[44] H. Tanaka, H. Kanoh, M. Yudasaka, S. Iijima and K. Knaeko, Quantum effects on hydrogen isotope adsorption on single-wall carbon nanohorns, *J. Am. Chem. Soc.*, **127**, 7511–7516 (2005).
[45] R. Yuge, T. Ichihashi, J. Miyaswaki, T. Yoshitake, S. Iijima and M. Yudasaka, Hidden caves in an aggregate of single-wall carbon nanohorns found by using Gd_2O_3 probes, *J. Phys. Chem. C*, in press.
[46] J. Miyawaki, M. Yudasaka, R. Yuge and S. Iijima, Organic-vapor-induced repeatable entrance and exit of C_{60} into/from single-wall carbon nanohorns at room temperature, *J. Phys. Chem. C*, **111**, 9719–9722 (2007).
[47] E. Bekyarova, Y. Hanzawa, K. Kaneko, J. S-Albero, A. S. -Escribano, F. R. -Reinoso, D. Kauysa, M. Yudasaka and S. Iijima, Cluster-mediated filling of water vapor in intra-tube and interstitial nanospaces of single-wall carbon nanohorns, *Chem. Phys. Lett.*, **366**, 463–468 (2002).
[48] R. Yuge, M. Yudasaka, J. Miyawaki, Y. Kubo, H. Isobe, H. Yorimitsu, E. Nakamura and S. Iijima, Plugging and unplugging of holes of single-wall carbon nanohorns, *J. Phys. Chem. C*, **111**, 7348–7351 (2007).
[49] A. Hashimoto, H. Yorimitsu, K. Ajima, K. Suenaga, H. Isobe, J. Miyawaki, M. Yudasaka, S. Iijima and E. Nakamura, Selective deposition of a gadolinium(III) cluster in a hole opening of single-wall carbon nanohorn, *Proc. Natl. Acad. Sci.*, **101**, 8527–8530 (2004).

16

Carbon Nanotube for Imaging of Single Molecules in Motion

Eiichi Nakamura

Nakamura Functional Carbon Cluster Project, ERATO, Japan Science and Technology Agency and Department of Chemistry, The University of Tokyo, Japan

16.1 Introduction

In this chapter, we describe a new experimental tool for analysis of the motional behavior of a single small organic molecule, which has been fixed loosely onto a single wall carbon nanotube. The method allows us for the first time to observe the motions and the interactions of small organic molecules at a near atomic resolution with the aid of transmission electron microscopy. The results are illustrated for the study on the small organic molecules such as **1–4** illustrated in Figure 16.1.

Although chemists often draw a single molecule on paper and use a molecular model to discuss chemistry in question, the present knowledge of chemistry has been constructed on the basis of experimental data based on the study of molecular ensembles. In this sense, chemistry has been the molecular sociology. With our new capacity to study the motion of individual small organic molecules, we will be able to understand chemistry on the basis of the studies of individual molecules–a new deal of chemistry, the molecular psychology.

Seeing something invisible by naked eyes was a dream of human being, which was for the first time fulfilled by the invention of microscopes and telescopes around 1600. Galileo Galilei contributed much to astronomy with the telescope that he improved. Robert Hooke on the other hand explored smaller world with his microscope, identified 'cells' in cork and studied small insects, and popularized the wonder of the small world through his book

Chemistry of Nanocarbons Edited by Takeshi Akasaka, Fred Wudl and Shigeru Nagase

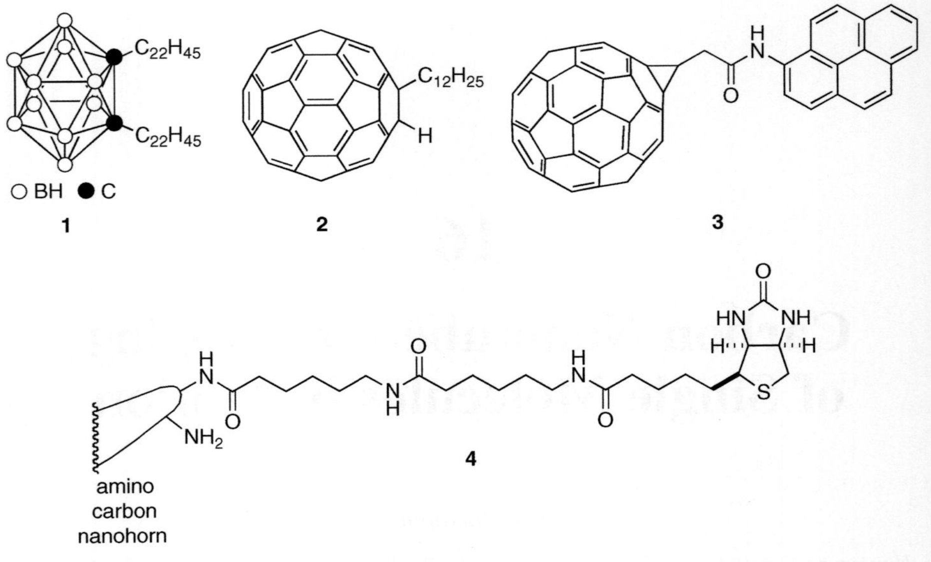

Figure 16.1 *The organic molecules observed by TEM in this study*

Micrographia published in 1665. Recent improvement of optical microscope is such that nanometer objects may be visualized so that one can study the motion of proteins. Recent advances on scanning probe microscopy give us sub-nanometer structural information on the molecules fixed on a flat substrate. Yet, it gives only static, top views of the molecules.

16.2 Electron Microscopic Observation of Small Molecules

Chemists have been dreaming of seeing molecules in motion as if we see molecular models. In principle, state-of-the-art transmission electron microscopy should be able to do the job. I stepped into this area in the late 1990s through collaboration with Prof. S. Iijima on the subject to performing controlled organic synthesis on the platform of carbon nanotubes. After tantalizing experience in the chemical synthesis and in the detection of organic molecules attached to the surface of a carbon nanotube, I realized that it is impossible to detect the molecules even with the best TEM technique. Instead, we discovered that one can observe a single gadolinium atom attached to an open oxidized hole (Figure 16.2) [1]. High-resolution TEM images of these objects were taken by the talented scientist, Dr. K. Suenaga in the Iijima group.

After the gadolinium and the related works having been finished [2], the dream of the imaging of organic molecules recurred to my mind, and I started a collaborative work with Dr. Suenaga on the imaging of small organic molecules attached to a carbon nanotube. The organic chemistry part was taken care of by Prof. H. Isobe of my university research group.

TEM should have long been able to provide the images of single organic molecules, but it had not. The TEM specialist concluded that organic molecules are too unstable to see by TEM. Retrospectively, there were at least several theses that discouraged people to attempt

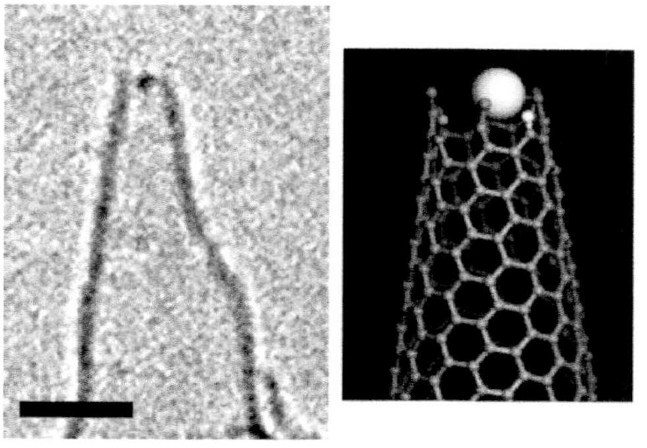

Figure 16.2 A gadolinium atom attached to an oxidized open of a tapered carbon nanotube (black dot; scale bar = 2 nm) and its molecular model

single molecule observation by TEM. First, TEM requiring the specimen molecule to be placed in vacuum with sub-nm precision, methods were lacking to hold a small molecule in a position suitable for TEM observation. Second, in vacuum, single organic molecules were considered to vibrate too quickly to be observed by such a slow time scale method as TEM. Third, an accepted wisdom was established in the TEM society: organic molecules give too weak contrast and they are too unstable under the TEM observation conditions. Indeed, in the TEM studies of crystalline organic molecules, high-energy electrons hit numerous molecules in the solid and either destroy the molecules or disturb the crystal structure. Finally, sample charging was suspected to hamper high resolution imaging. We noted however that there was no direct proof that a single molecule isolated in vacuum cannot be seen.

We discovered that the imaging of small organic molecules in motion does not suffer from any of these expected problems. Single wall carbon nanotube was found to serve a robust, conductive and essentially transparent substrate for holding single molecules, and the single molecules to move slowly in the time scale of seconds – the time scale of the TEM observation. Single organic molecules in vacuum were found to be stable against thousand times higher electron dose than that known to disturb observation of the molecules in its crystalline state. Therefore we can study the molecular motions at least for a few minutes with a commonly available TEM instrument placed at room temperature.

16.3 TEM Imaging of Alkyl Carborane Molecules

After a year of fruitless period, it occurred to us that we need to design special molecules to prove the feasibility of our dream experiments. Thus, we synthesized ortho-carborane **1** and related compounds bearing one or two long alkyl chains (Figure 16.1). The molecule **1** about 0.8 nm in diameter and 3.1 nm in length for R = $C_{22}H_{45}$. A real joy of being synthetic chemists is our capability to synthesize any molecules that we wish to make. We thus synthesized 10^{20} molecules (i.e., sub g scale) each of the four compounds, encapsulated them into carbon nanotubes, and submitted the sample for TEM observation. The end caps of the carbon nanotubes were removed by oxidation prior to the encapsulation procedure.

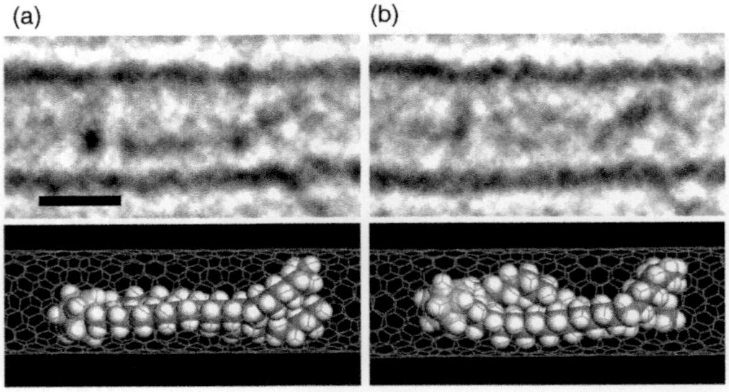

Figure 16.3 Time-lapse images of a dialkylcarborane and its molecular model at time 4.2 sec (a) and 6.3 sec (b). Scale bar = 1 nm

The carborane **1** and the carbon nanotubes were mixed together and heated in vacuum to allow the molecules to enter into the interior on the tubes. The nanotubes containing the molecules were taken onto the TEM grid and analyzed by a rather ordinary 120 kV TEM at room temperature. The spatial resolution of our TEM (0.23 nm) and the time resolution (0.5 sec) are just good enough to the purpose of the study, suggesting further improvement being possible soon. It took us several months to convince ourselves that the objects we saw in the nanotubes are the alkylcarboranes that we put in. And, a surprising bonus came out: We discovered two single molecules of the C_{22} carborane slowly changing their conformation [3]. Two pictures from the time-lapse images are shown in Figure 16.3a. The images were taken in an interval of 2.1 sec with an exposure time of 0.5 sec. One can clearly see two carbon chains and the carborane head at 6.3 sec (Figure 16.3b) for the corresponding molecular model), while one can see only one overlapping chains at 4.2 sec (Figure 16.3a). The 0.5 sec exposure time is sometimes too slow that the image is blurred. Conversely, if the image is clear and well defined, the molecule stays in one position for 0.5 sec. We found that the conformational change of the hydrocarbon chain does not occur as a continuous motion. This is expected by the theory of conformational theory but no one has seen experimentally such a ratchet motion.

Dr. Suenaga carried out independent studies on imaging of organic molecules and found that the structure of fulleropyrrolidine [4] and the cis-trans isomerism of a trisubstituted olefin [5] can be identified by TEM.

16.4 Alkyl Chain Passing through a Hole

Molecular transport through a nanosized pore in a film, membrane, or wall structure is an event of fundamental importance in a number of physical, chemical, and biological processes [1–6]. However, there has been a lack of experimental methods which can provide such information as the structure and the orientation of the molecules during the passage through a pore. We made a serendipitous discovery that the TEM study of **2** in a nanotube provides such information–the way an alkyl chain interacts with a hole in the wall of a carbon nanotube [6].

To obtain the image shown in Figure 16.4, we performed experiments on the molecule **2** shown in Figure 16.1 in a nanotube placed on a specimen stage kept at 4 K on a TEM instrument designed by Dr. Suenaga. Figure 16.4 shows the alkyl chain of **2** moving out through a hole in

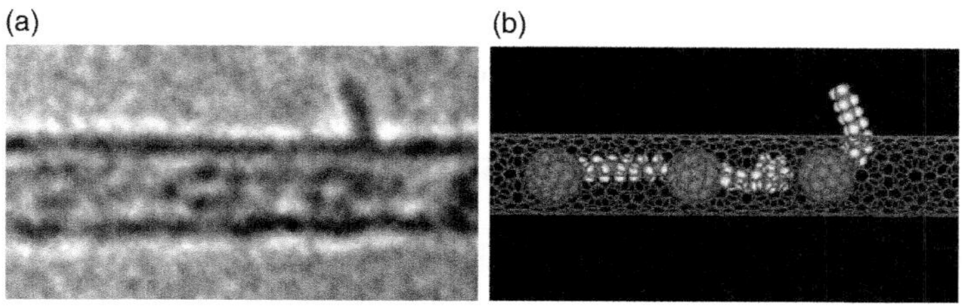

Figure 16.4 The alkyl group in alkylfullerene **2** egressing through a hole. (a) TEM image at 4K. (b) Its 3D model

the tube in its stretched conformation. We saw that the chain vibrated slowly in its linear conformation, and then was drawn back into the nanotube interior. Interestingly enough, the observed translation and the conformational changes of the molecule at 4K occurred at approximately the same timescale of seconds as those at 293 K (i.e. the image of the alkyl chain recorded as clearly as other images in this chapter taken on an instrument operating at 4 K).

16.5 3D Structural Information on Pyrene Amide Molecule

The above discovery demonstrated that TEM provides not only the static image of the hydrocarbons and organoboranes but also their structural changes. What chemically more interesting functional groups can be imaged? Being an important constituent of proteins, amide groups are of great interest for chemists. We employed an amide molecule **3** containing a fullerene moiety and a pyrene moiety (Figure 16.5a). The fullerene part is used to maximize the affinity of the molecule to the nanotube interior and the pyrene part is used to monitor the orientation of the amide [7].

Unexpectedly, we could identify not only the presence but also the orientation of the amide carbonyl and the pyrene groups (Figure 16.5b vs. 5c). We see here that carbonyl and the pyrene groups are oriented parallel to the electron beam. This conformation agrees with what the textbook conformational analysis based says; that is, the pyrene ring and the amide carbonyl group share the same plane of symmetry (and slow bond rotation around the nitrogen atom). An additional bonus: the molecule rotates by itself and allows us to see the

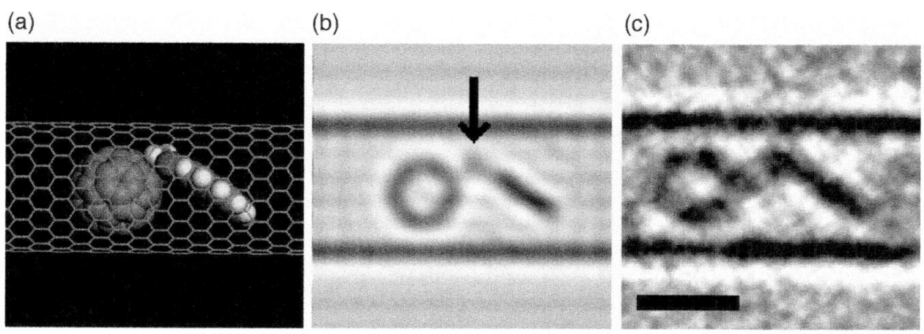

Figure 16.5 Amide molecule **3** bearing fullerene and pyrene. (a) Molecular model. (b) TEM simulation. (c) TEM image. Scale bar = 1 nm. The arrows indicates the carbonyl group oriented parallel to the electron beam. With the permission of CSJ

molecule from various directions and hence to obtain the 3D information of the sample molecule (i.e. the pyrene being flat and square).

16.6 Complex Molecule 4 Fixed outside of Nanotube

In the preceding studies, we loosely fixed the specimen molecules in the interior of a nanotube. This approach inherently limits the size of the specimen molecule and its motion. An obviously more general approach is to fix the specimen molecule on the exterior of a nanotube by a covalent bond. One concern was whether or not the molecule in the vacuum space vibrates too fast to be observed by TEM.

Thus, we first synthesize a polyaminated carbon nanohorn by the synthetic procedure and bonded a biotin diamide molecule **5** through amide bond formation (Figure 16.6) [8]. Low-magnification search of the molecules revealed the presence of many molecules that are attached predominantly to the cap or curved regions of the nanohorn (Figure 16.7a) and are taking various mobile conformations. Analysis of each frame of the motion pictures of single biotin triamide molecules allows us to assign a possible conformation of the molecule to each image (Figure 16.7b,c) [9].

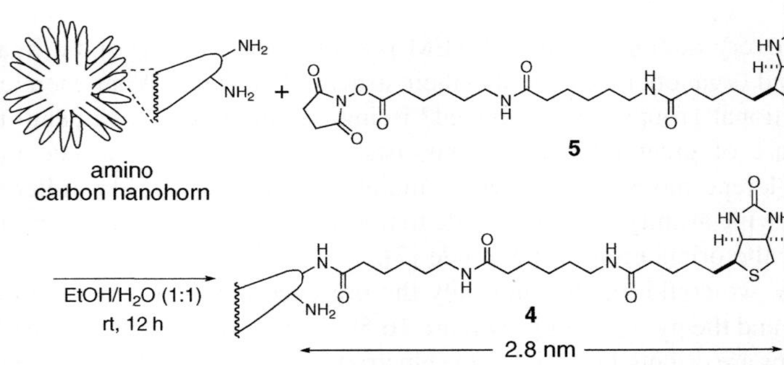

Figure 16.6 A biotin molecule attached an amino carbon nanohorn

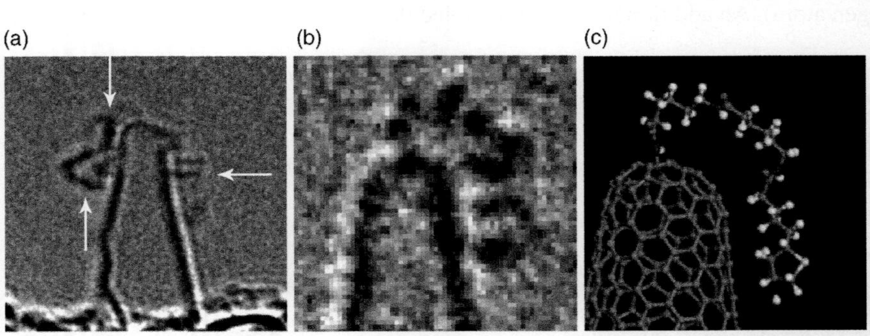

Figure 16.7 TEM image of biotin molecules attached an amino carbon nanohorn (a) A few molecules indicated by arrows. (b) TEM image of one molecule. (c) The corresponding molecular model. With the permission of ACS

16.7 Conclusion

Several new discoveries were made in this research.

First, we have demonstrated that small organic molecules maintains their structure under the TEM observation conditions. We found little sign of deleterious effects of sample charging either. After all, the scattering cross section of the interaction between an electron and a carbon atom is very small, and an organic molecule isolated in vacuum is much less prone to decompose than molecules packed in solid state –Egg of Columbus.

Second, the method provides a new method to study in situ how a single organic molecule changes its structure and how it interacts with other objects. We also observed a sign of chemical reaction such as decomposition and conformational changes that suggest breaking and forming of hydrogen bonds. The motions that we observed were generally very slow (i. e. in the timescale of seconds), and were ratchet motions rather than continuous motions (i.e. the images are generally well defined but sometimes blurred because of motion during the 0.5 sec exposure time). Of particular interest is that the observed speed of motions at 4 K was qualitatively the same as that in the experiments carried out on a TEM instrument kept at room temperature. These facts appear to pose us something fundamental as to the origin of the motion; for instance, the observed motions at 4K and perhaps even at room temperature under vacuum reflect the conformational change due to tunneling instead of thermal effects (caused either by electrons or external heat).

Third, the method represents an ultimate microanalytical method that allows one to determined the molecular structures by using a single molecule. Even with the present 0.2 nm resolution of the instrument, we can readily differentiate all four molecules in Figure 16.1; that is, given the information that the TEM image in Figure 16.5 is one of the four compounds in Figure 16.1, you can immediately tell that it is pyrene amide **3**. Forthcoming improvement of the time and spatial resolutions of TEM might soon make obsolete the conventional spectroscopic and crystallographic method of structural determination. The history of science tells us that a solution will come, once the problem is identified.

Fourth, the TEM observation of the motions of single molecules suggests us to bridge the quantum mechanical world of molecular science to the classical world. The translation and conformational change of the molecules attached to a graphene sheet must be strongly related to the friction and lubrication phenomena involving graphite surface and a hydrocarbon lubricant; a typical classical mechanical world or the world of mechanical engineering. After all, isn't it rather curious that the quantum substance like molecules behave as if they are molecular models which is a classical mechanical substance?

Acknowledgements

This work is the result of fruitful collaboration with Dr. Kazutomo Suenaga under the Nakamura Functional Carbon Cluster Project (ERATO) administrated by Japan Science and Technology Agency, and could not have been achieved without intellectual contributions from the co-workers whose names are listed in the references. We also thank JEOL Co. for continuous technical support, and for MEXT for financial support (KAKENHI #18655012 to E.N.).

References

[1] A. Hashimoto, H. Yorimitsu, K. Ajima, K. Suenaga, H. Isobe, J. Miyawaki, M. Yudasaka, S. Iijima and E. Nakamura, Selective deposition of a one-atom to a multi-atom gadolinium(III) cluster in a hole opening of single-wall carbon nanohorn, *Proc. Natl. Acad. Sci.*, **101**, 8527–8530 (2004).

[2] J. Miyawaki, M. Yudasaka, H. Imai, H. Yorimitsu, H. Isobe, E. Nakamura and S. Iijima, In vivo magnetic resonance imaging of carbon nanotubes by labeling with magnetite nanoparticles, *Adv. Mater.*, **10**, 1010–1014 (2006).

[3] M. Koshino, T. Tanaka, N. Solin, K. Suenaga, H. Isobe and E. Nakamura, Imaging of single organic molecules in motion, *Science*, **316**, 853 (2007).

[4] Z. Liu, M. Koshino, K. Suenaga, A. Mrzel, H. Kataura and S. Iijima, Transmission electron microscopy imaging of individual functional groups of fullerene derivatives, *Phys. Rev. Lett.*, **96**, 088304 (2006).

[5] Z. Liu, K. Yanagi, K. Suenaga, H. Kataura and S. Iijima, Imaging the dynamic behaviour of individual retinal chromophores confined inside carbon nanotubes, *Nat. Nanotechnol.*, **2**, 422–425 (2007).

[6] M. Koshino, N. Solin, T. Tanaka, H. Isobe and E. Nakamura, Imaging the passage of a single hydrocarbon chain through a nanopore, *Nat. Nanotechnol.*, **3**, 595–597 (2008).

[7] N. Solin, M. Koshino, T. Tanaka, S. Takenaga, H. Kataura, H. Isobe and E. Nakamura, Imaging of aromatic amide in motion, *Chem. Lett.*, **36**, 1208–1209 (2007).

[8] The Avidin-Biotin complex in bioanalytical applications, M. Wilchek and E. A. Bayer, *Anal. Biochem.*, **171**, 1–32 (1988).

[9] E. Nakamura, M. Koshino, Y. Niimi, K. Harano, Y. Nakamura and H. Isobe, Imaging of conformational change of biotinylated triamide molecules covalently bonded to carbon nanotube surface, *J. Am. Chem. Soc.*, **130**, 7808–7809 (2008).

17
Chemistry of Single-Nano Diamond Particles

Eiji Ōsawa

Nanocarbon Research Institute, Ltd, Asama Research Extension Centre, Faculty of Textile Science and Technology, Shinshu University, Ueda, Nagano, Japan

17.1 Introduction

The known nanocarbon species have been almost exclusively dominated by graphitic or sp^2-hybridized carbon atoms including fullerenes and carbon nanotubes, as are most of the chapters in this book. However, there is no a priori reason that sp^3-hybridized carbon atoms cannot form nanocarbon. This chapter deals with the first entry of nanocarbon from the sp^3-carbon family. Its general name is *detonation nanodiamond*, and naturally *the world's smallest diamond* [1]. DN was accidentally discovered in 1963 by V. V. Danilenko and his collaborators [2] when they were tracing the shock synthesis of diamond, which was reported in 1961 by DeCarli and Jamieson [3]. In the shock synthesis, graphite was irradiated with shock wave generated by detonation of explosives to induce phase transition of graphitic carbon to diamond crystals. While DeCarli did not describe about the kind of explosive they used, Danilenko used a common military explosive called Composition B which is 1:1 mixture of TNT and cyclonite (**1**) and observed that the yield of diamond increased with the amount of explosives used, even if the amount of graphite was decreased. He continued to decrease graphite and eventually found that diamond still formed even if no graphite was added at all. Fortunately the components of Composition B are both oxygen-imbalanced, thus produced small amounts of soot upon explosion. It was in this soot that he found diamond. Then he correctly deduced how diamond was formed; the explosion

Chemistry of Nanocarbons Edited by Takeshi Akasaka, Fred Wudl and Shigeru Nagase

produces carbon atoms as the result of incomplete combustion, which are then exposed to the high-temperature high-pressure zone of shock wave to grow into diamond crystals. These observations and deductions led to remarkable discovery of DN (Table 17.1). The difference between shock and detonation synthesis is that in the former diamond carbon atoms come from added graphite, while in the latter originate from explosive molecules.

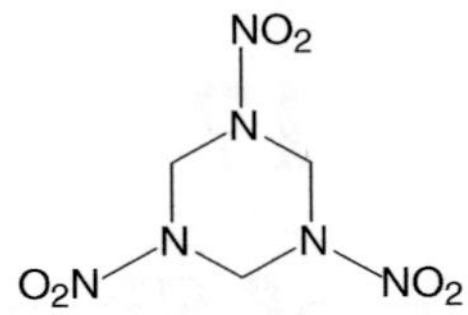

1

Unfortunately no official document on the discovery and subsequent developmental works on DN in the former Soviet Union have ever been released from the military regime which had classified these works as top military secret. Flow of information within the regime seems to have been pretty well controlled, as can be indicated by a surprising story that DN was discovered four times under the same project [1]. Nevertheless, the production process by detonation appears to have been well-studied and the advanced technology well-inherited, based on the fact that immediately after Perestroika, a large factory was quickly built in Bijsk, Siberia, to produce DN in semi-continuous system in the early 1990s.

However, apparently they have not advanced well in the analysis of DN and overlooked the very unusual and tight aggregation that had occurred among primary particles during the production of DN. It is likely that they misinterpreted the TEM image where the primary particles are clearly visible, no equipment to measure particle-size were available and the product behaved in some cases as if well-dispersed. The last point contains an interesting clue to further developments, but we are not ready to discuss it here.

They began marketing the unusually tight aggregates under the very attractive name of *ultra-dispersed diamond* (UDD), explicitly claiming that the particle-size ranges between 4 and 5 nm. In retrospect, the name aroused a lot of expectation and then disappointment among scientists. Soon UDD was ignored by the scientific community at large, although UDD found certain special markets and still produced in small scale (about 8 tons per year) in a few CIS countries, Ukraine and China [4].

According to our own examination, UDD is a fine, dark grey to grey powder of thermal conglomerates resulting from intense drying process. When dilute aqueous suspension of the powder was subjected to sonication with powerful ultrasonic generator, the thermal aggregates can be destroyed to core aggregates having wide size-distribution between 60 and 200 nm. Dynamic light scattering (DLS) analysis of particle-size distribution failed to detect any sign of primary particles below 10 nm above the detection limit of 0.1% [1].

In early 2002, the unusually tight aggregation was finally recognized and within the same year dispersed by beads-milling into primary particles [5–7]. This means that the primary crystals of detonation nanodiamond were isolated 39 years after Danilenko's discovery. At that time, we could not identify the force acting in the stubborn core aggregates, hence gave a special terminology of *agglutination* to describe this special form of aggregation. As we learned only recently [8], the force that binds the primary particles together to form the assemblies was suggested to be of totally unprecedented type (see below)!

Table 17.1 *Four major diamond syntheses*

Method (raw material)	Discoverers/break-thru (year)	Form of diamond (av size)	Country of production (developing)	Major applications	Ref
hydrostatic HPHT (graphite + Fe + cat)	GE(1955)	single crystalline particles (av 50 μm)	China	polishing, lapping	[a]
shock(explosive, Cu powder, graphite)	DeCarli- Jamieson (1961)	polycrystalline particles (av 50 μm)	China	polishing, lapping	[a]
detonation (TNT+ cyclo-nite/water)	Danilenko (1963)	Agglutinate powder (60–200 nm)	CIS, Ukraine, China	HDD finishing, plating additive	[2]
	NCRI (2003)	primary particles (4–5 nm)	Japan	drug carrier, CVD seeding, lubricant, plating additive	[1, 10]
CVD($CH_4 + H_2$, $C_{60} + Ar$)	Eversole (1950)/NIMS (1982)	polycrystalline films (thickness up to μm)	(Japan, US)	Low-friction coating, window material, electrodes	[b]

[a] R. M. Hazen, *The Diamond Makers*, Cambridge University Press, Cambridge, 1999, p. 244.
[b] K. Kobashi, *Diamond Films*, Elsevier, Amsterdam, 2005, p. 336.

We thought it desirable to clearly distinguish dispersed primary particles from UDD and not to confuse scientists any more by using ambiguous name DN. For this reason, we call UDD here as 'agglutinates', and named our particles as Single-Nano Buckydiamond [9]. For convenience the long naming is abbreviated to SNBD throughout this chapter.

By now the production technology as well as the science of SNBD has advanced well enough for extensive developmental drive to start. At this point, it is considered highly desirable to introduce the researchers of nanoparticles about enormous potential of SNBD. The purpose of this chapter is to present the latest development in our chemical approach to SNBD, which we believe to grow as a general-purpose nanocarbon material shortly [10].

Various forms of SNBD are shown in Figure 17.1. Aqueous colloidal solutions are intensive black in color but surprisingly stable at least for three years with no sign of flocculation or precipitation according to our experience. SNBD is well 'soluble' in a few other organic solvents like DMSO and ethylene glycol as well [11]. Evaporation of water from the colloidal solution gives first black soft gel, then hard gel with more intense black color, which can be pulverized in a mortar to brownish and dry-looking powder. When the hard gel holds 20–30% of water, it can be readily redispersed by simply stirring with water, irradiating with supersonic wave and removing small particles by centrifugation to give a dark brown colored, transparent sol. Hard gel loses its water slowly on exposure to air.

TEM photograph (Figure 17.1) gives direct evidence for the single-nano size of SNBD particles, but does not give information on dispersity. Although their shape is not clear, partial images of edges, facets and apices are seen (marked with arrows, Figure 17.1). In

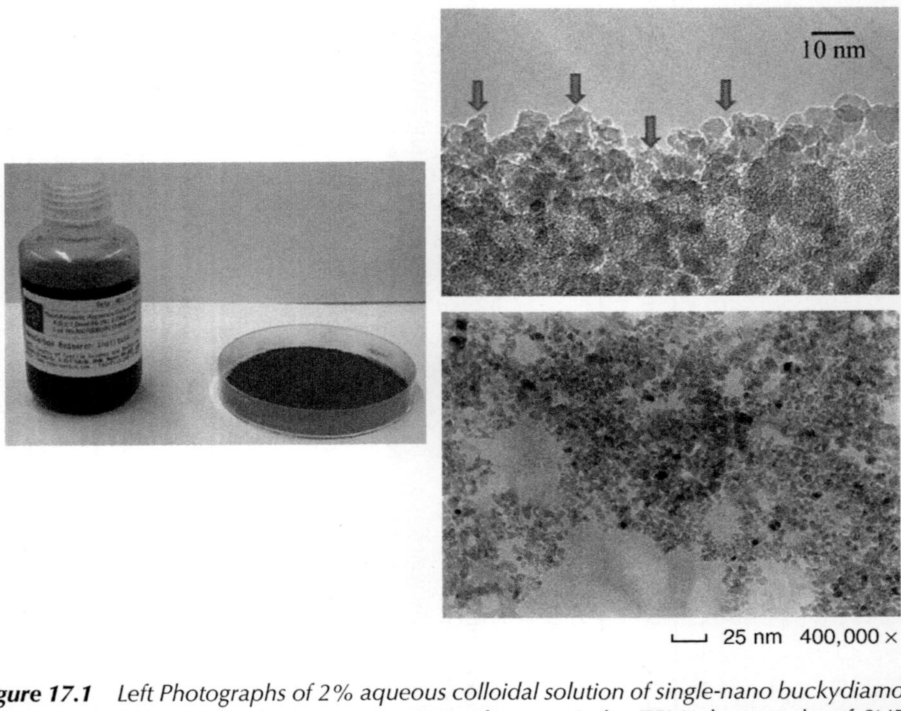

Figure 17.1 *Left Photographs of 2% aqueous colloidal solution of single-nano buckydiamond (SNBD) and hard hydrogel containing 21.7% of water. Right: TEM photographs of SNBD, re-aggregated while drying. Arrows in the top indicate partial images of apexes and edges*

view of very short growth period and sudden termination of crystallization process, we cannot hope for well-defined shapes but should duly expect irregular polyhedra. Improved TEM techniques like high-vacuum aberration-correction and direct mounting of samples on conductive grid provided clearer images suggesting polyhedral origin [12, 13].

17.2 Geometrical Structure

Let us start this section with a brief explanation of the meaning of SNBD. About the size, we currently adopt an average size of 4.8 ± 0.7 nm obtained by simply averaging results from three different methods (Table 17.2). Among the methods included, XRD and apparently DLS as well measure the coherent scattering region of crystal, whereas TEM images are processed by visually defining contrasted periphery of particles, all by spherical approximation. Almost perfect agreements among these sizes indicate that the graphitic shells are thin, probably only one or two layers at most (see below). Good agreements among samples detonated in separate countries suggest that the detonation involves identical crystal growth process as long as the same explosive composition is used. The name 'buckydiamond' was first coined by Raty and Galli [14], and then later elaborated by Barnard [8, 15, 16] as a stable form of extremely small diamond crystal models up to 2 nm in diameter. We anticipate that our DN may be the largest stable buckydiamond.

Information on the internal structure of SNBD is contributed heavily by computation. Barnard and Sternberg [16] performed systematic Self-Consistent Charges Density Functional Theory of Tight Binding (SCC-DFTB) calculations using four polyhedral models of various sizes up to about 1700 carbon atoms. Here are mentioned representative results obtained on two tetradecahedral models, truncated octahedra (Figures 17.2A) and cuboctahedra (Figure 17.2B). Starting from all-diamond structures (Figures 17.2C and 17.2D) they obtained energy-minimum structures, both partially graphitized on [111] facets and propagated to some depths below (Figures 17.2E and 17.2F). The extent of graphitization strongly depends on the model and size; in octahedral model most eminently but in cube least evident, and the smaller the more extensive in all models. In contrast, [100] facets are resistant to the transition and diamond carbons are retained on and below the [100] facets.

Table 17.2 *Size of single-nano buckydiamond (SNBD) by independent methods*

Method	Size (std dev), nm	Origin of sample
TEM[a]	4.8 (0.8)	China[d]
DLS[b]	5.0 (0.8)	Japan[e]
XRD[c]	4.5 (0.5)	Russia[f]
Av	4.8 (0.7)	

[a]Particle-size distribution in the TEM image shown in the lower right of Fig. 1 was analyzed by using a semi-automatic analysis software Mac-View, ver. 3.5 (2005), from Mountec Co., Tokyo.
[b]Dynamic light scattering analysis of particle-size distribution performed on a Particle Analyzer for Concentrated Systems, type FPAR1000 equipped with an automatic sampler FP3000. Poor reproducibility in the DLS measurements was overcome by taking non-uniform concentration dependence of scattering intensities in poly-disperse system into account. See, E. Ōsawa, Remarks on the particle-size determination of NanoAmando by dynamic light scattering, *NCRI Technical Bull.* No. 2, *http://nano-carbon.jp*, (2007).
[c]Based on X-ray diffraction intensities, M. V. Baidakova, V. I. Siklitsky and A. Ya. Vul, *Chaos, Solitons & Fractals*, **10**, 2153–2163 (1999).
[d]Agglutinates manufactured by Guangzhou Guangda Electromechanical Co., Guangzhou.
[e]Agglutinates manufactured by Nippon Kayaku Co., Tokyo.
[f]Agglutinates manufactured by Diamond Center, St. Petersburg.

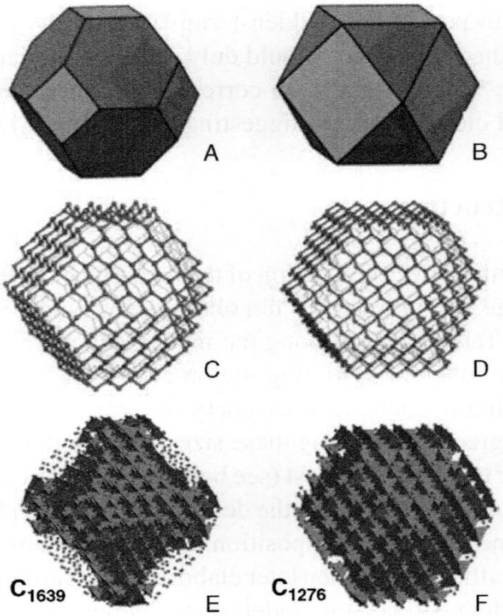

Figure 17.2 *Left column corresponds to truncated octahedral model (A, C, E), whereas right column contains cuboctahedral model (B, D, F). The first line illustrates the polyhedral models (A and B), the second line all-diamond starting structure before geometry optimization (C and D), and the third line the energy-minimum structures (E and F). In E and F, graphitic shells are located over and below the [1 1 1] facets and its positions of carbon nuclei denoted by black dots in order to demonstrate perspective image, while the position of carbon nuclei in the diamond core are given by 3D-tetrahedral models. All the shapes are seen from the same direction and height*

This feature makes internal structures of SNBD quite complicated, the cross-sections of which are shown in Figure 17.3 for a hypothetical seamless sphere (A) for clarity and a truncated octahedral model (B) as a more realistic illustration. Observation of several characteristic extrusions in the TEM image (Figure 17.1, arrows) prompts us to suggest a

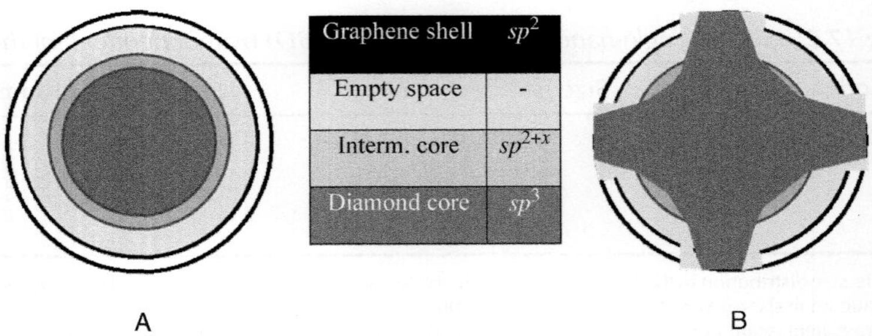

Figure 17.3 *Cross-sectional illustrations of core-shell structure in SNBD particle. A: Seamless and oversimplified spherical model, B: Handwritten truncated octahedral model. Graphene shell is considered one or at most two layers. Diamond core and intermediate core are connected. There is an empty shell-shaped space between shell and core. In the more realistic polyhedral model, the top of diamond core is supposed to expose itself at six [100] facets*

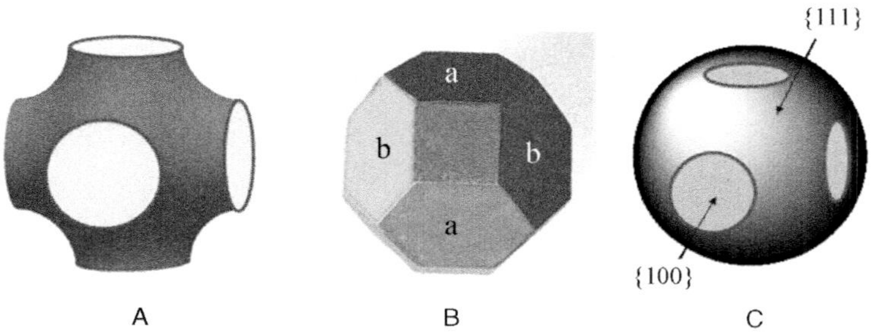

Figure 17.4 Three most likely candidates for the surface shape of SNBD based on truncated octahedral model. A: Concave expression identical with Schwarzite P-type surface with negative Gaussian curvature. B: The original truncated octahedral model with 14 planar facets. C: Convex spherical expression

faint possibility of P-Schwarzite shape with negative Gaussian curvature [17] (Figure 17.4A) from a truncated octahedral model (Figure 17.4B). However, graphitization at the unstable [111]/[111] edges in the latter model will readily induce Stone-Wales rearrangements [[17]a] to produce pentagonal rings and lead to fullerene-like but holey shell with positive Gaussian curvatures (Figure 17.4C). Its cross-sectional view is given in Figure 17.3B.

As mentioned above, we think the graphitic surface of SNBD will be very thin based on TEM observation, but SCC-DFTB calculations suggest significantly higher proportion of graphitic carbons (56% for truncated octahedral model shown in Figure 17.2E, and 37% for cuboctahedral, Figure 17.2F). The discrepancy probably arose from the strong size-dependence of the phase transition. We have long wished to bleach SNBD into colorless particles having simpler surface and internal structures with higher diamond content, but we are inclined to take computational results on the energy-minimum core-shell structure. While our attempts to bleaching are still continuing and promising results are being obtained, we decided at least for the time-being to concentrate on exploiting applications using the core-shell SNBD particles (see below).

17.3 Electronic Structure

Surprising high electron mobility within SNBD particle was borne out of Barnard's SCC-DFTB calculations [8, 16], which to our knowledge has never been expected before. As shown in Figure 17.5, the optimized truncated octahedral models show strong and diverse electrostatic potential fields, generally localized on the facets. All other models used in Barnard's calculations show more or less similar trends on the surface charges. Here again the effects are strongly size- and model-dependent. Concentrating on Figure 17.5 for the moment, we notice that the charges are always strongly positive on [100] facets and even flow to the neighboring [111] facets over the edges. The overflow occurs at the two opposing edges of square [100] facets more evidently than the other pair of edges, thus creating two kinds of [111] facets regarding the charge distribution, $[111]_a$ and $[111]_b$ (inset of Figure 17.5). As a consequence, $[111]_b$ acquires complex distribution of charges within the facet, while $[111]_a$ maintains essentially uniform negative potential except for the near-edge

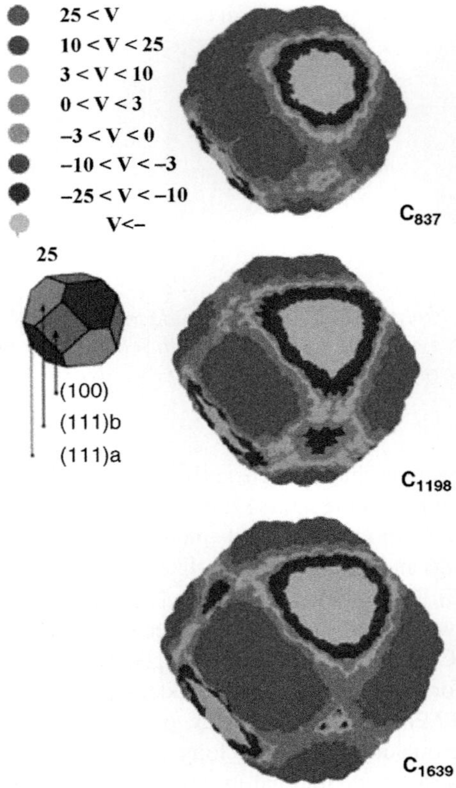

Figure 17.5 *Distribution of electrostatic potential fields on the facets of three smaller truncated octahedral models of SNBD particle obtained by SCC-DFTB geometry-optimization (ref. [16])*

region where negative charge diminishes and in some place even turned itself into positively charged.

The surface electrostatic features thus revealed are truly intriguing. First of all, such a multi-pole is difficult to detect experimentally, hence represents another unique capability of computational method that works for problems difficult to solve by other means like the mathematically rigorous determination of transition states. Unlike *van der Waals* interaction between planar planes, interactions between charged facets could be strongly attractive or repulsive depending on the charge, and should profoundly affect the inter-particle interactions in nanoparticles like aggregation. Secondly, charged facets should themselves be active sites for solvents and reagents, e.g. surface reactions and ligand exchange equilibria. As there will be a plenty of crystalline core-shell nanoparticles to appear in the future, the Coulombic interactions involving crystal facets will be a prevalent interparticle interaction force. It should be noted that the single-nano particles have especially large specific surfaces areas.

Barnard analyzed factors affecting the Coulombic interparticle interactions by adjusting interplanar distance and rotations around the common normal axes passing through the centers of approaching facets [8]. She found, for truncated octahedral models, very strong attractive interactions to occur between $[100]/[111]_b$ and $[111]_a/[111]_b$ interparticle pairs with calculated attractive energies of –2.46 and –3.28 eV, respectively, at a distance of 1.78 Å under certain rotation angle, and proposed that such forces should be the source of enigmatic *agglutination* in the crude product of DN. This is highly reasonable interpretation as there will be plenty of time for the SNBD particles to find the optimum interfacial orientations and distances with neighboring SNBD particles at high temperatures after the shock wave is passed and the diamond crystal growth terminated. She called such a strong attraction as *coherent interfacial Coulombic interaction* [8].

In more general case like re-aggregation of dispersed nanoparticles, still considerably strong Coulombic interactions could occur among particles but will never reach coherent orientation and distance under mild conditions. Such *incoherent interfacial Coulombic interactions* could be significant and frequent cause of interparticle events under ordinary conditions. In chemistry, we used to analyze intermolecular interactions in terms of atom-atom mode, but in nanoscience of crystalline nanoparticles, inter-particle interactions represented by coherent and incoherent facet-facet Coulombic interactions will be more useful.

Before analyzing the results of more 'accurate' calculations which are being executed at the moment, we can qualitatively rationalize the self-polarization in small polyhedral models with the help of orbital interaction theory [9]. Our starting point is to take advantage of the contrasting electronegativity features of core and shell in SNBD structure. Let us first take a spherical seamless model (Figure 17.3A) to start the analysis. The cross-sectional view illustrates that shell and core portions are separated by an empty space, which is created by delamination of the last giant fullerenic shell from the top of core; hence shell and core comprise separate electronic systems (Figure 17.6, left and right). The diamond-graphene phase transition not only produces a shell-shaped space but also leaves skeletal strain at the top of core, thus producing an intermediate layer consisting of sp^{2+x} $(0 < x < 1)$ hybridized carbon atoms, but the intermediate layers and diamond core comprise a united electronic system.

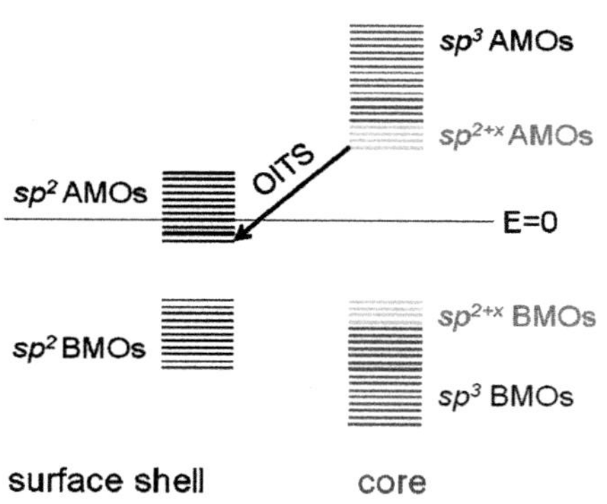

Figure 17.6 *Electronic structure of a SNBD particle consists of two independent molecular orbital domains, surface shell and core, but respective orbitals may interact through narrow space (OITS) between the domains. AMO = anti-bonding molecular orbitals, BMO = bonding molecular orbitals. Electron migration takes place from electropositive diamond core to electronegative fullerenic surface shell (arrow) to produce self-polarization and a multi-pole particle*

Thus, schematic representation of the molecular orbitals in SNBD consists of two separate systems, where shell portion is noted for high electronegativity due to the lower anti-bonding molecular orbital deeply penetrating into bonding energy region (left side of Figure 17.6), while core is characterized by the opposite trend in electronegativity, which is even 'negative', due to very wide band gap between high-energy anti-bonding MOs and low energy bonding MOs (right side of Figure 17.6). As the diamond surface is transformed into spherical fullerene-like shell one by one, a large number of free spins are left on the surface of intermediate core, and enter into the lowest vacant orbital of the same intermediate core, which will find a low-energy vacant orbital with matched phase in the shell portion of close vicinity. Because of this small geometrical separation between these two anti-bonding molecular orbitals (less than 0.35 nm), orbital interactions will take place through this space with large energy gain (arrow in Figure 17.6). The transferred electrons will stay in the low-energy sp^2 AMOs producing negative charge on the surface of [111] facets. On the other hand, positive holes created by the electron migration in the sp^{2+x} AMOs of core will migrate by orbital interactions through bonds (OITB) into the core BMOs with matched phase, or hyperconjugation, eventually reaching [100] facets and producing high positive charge densities.

As a whole, free electrons formed as the result of diamond-graphite phase transition migrate spontaneously from the top of core to shell taking advantage of the large difference in electronegativity between shell and core through orbital interactions through space and bonds. Positive holes created by the transition also will migrate in the core. Thus, we came to an inevitable conclusion that *diamond could polarize in the core-shell structure of SNBD*. If we remove graphitic shell from SNBD, the resulting free electrons in the surface of pure diamond will have to be stabilized by bonding with surface substituents like hydrogen. These qualitative explanations of self-polarization in SNBD await verification by more advanced computation.

17.4 Properties

17.4.1 Tight Hydration

In the past the behaviors of nanodiamond particles like the high stability of SNBD colloid in water have traditionally been interpreted as endowed by hydroxyl and carboxyl groups, believed to have been generated on the particle surface during the purification process of the crude detonation product. We have a different view on the origin of polarity in SNBD. First, we strongly doubt the presence of carboxyl groups, at least in amounts that will significantly influence the polarity of large SNBD particles, based on our continued failure in titrating this group. We reported that, under high pH, the carboxylate groups are decarboxylated at high temperatures up to 300 °C, a condition often created to remove trace of acid after oxidation of soot during the production process [11]. The C-OH groups have been proved to be absent by careful in situ IR experiments by Ji and his coworkers [18] by heating samples up to 200 °C under inert atmosphere in the IR spectrometer. Intense OH vibrational peaks that appear with overwhelming intensities in the conventional IR spectra of agglutinates and SNBD alike disguised previous workers for a long time, but should be assigned to water strongly adsorbed onto the particle surface.

In the past we had often noticed difficulties in completely drying the agglutinates, and hence now impose a severe condition of heating at 200 °C under 15 hPa for 10 hours for the determination of water contents in the SNBD gel. The dried sample is extremely hygroscopic and it is impossible to perform elementary analysis in conventional equipments. The reported elementary composition generally includes 4–10% of oxygen, which has been generally attributed to surface oxygen-containing functional groups formed by oxidation of surface carbon atoms [7]. Here again, we must use perfectly dried sample for this kind of interpretation to be valid.

Instead of oxygen-containing polar groups, we can now think of multi-pole nature of SNBD surface as a source of polarity; namely, surface charges, both positive and negative, exposed as the result of de-agglutination during beads milling will be instantly attract surrounding water molecules. As the molecule is bipolar, the hydration will be extensive and complete over the whole exposed surface of primary particles.

Experimentally a hint for the tight hydration of SNBD had been obtained by Korobov and his group [19, 20]; when they were studying dynamic scanning calorimetry of the solid gel obtained by removing most of bulk water from the aqueous SNBD colloidal solution, they detected a large endothermic peak at −8 °C in the warming process. This low-freezing layer must have been formed on the particle surface by strongly oriented absorption of bulk water. By carefully reducing the bulk water content until the ice peak disappeared, we obtained the mass ratio of the adsorbed water to the mass of a SNBD particle to be 0.47, which would give a layer thickness of 1 nm. However, if we include this water layer into effective size of particles, the resulting size becomes 0.4 nm too large than that derived by Kelvin equation, which relates effective droplet size to its vapor pressure. Hence we concluded that the water molecules in the low-freezing layer can exchange with bulk water, and cannot be included in the effective size of colloidal particle, but there is one more layer of water directly surrounding the particle surface which never freezes and is more strongly held than the low-freezing layer (Figure 17.7).

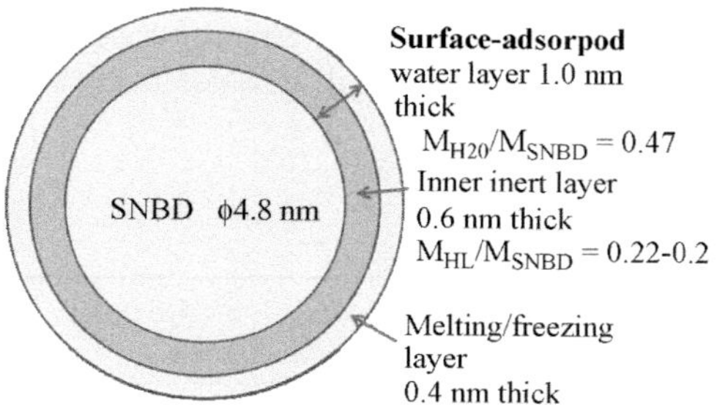

Figure 17.7 Double hydration shell on the surface of SNBD particles in aqueous colloid as deduced from DSC analysis of the gel. The inner inert layer is supposed to consist of water molecules strongly oriented and adsorbed onto the particle which will not freeze at all and exchange with bulk water. A part of them could well penetrate into the inside of particle through the space between shell and core. It is difficult to remove the inner layer by conventional drying. The outer melting/freezing layer freezes at −8 °C, can exchange with bulk water and can be removed readily under usual drying conditions

We assume that it is this *inner inert layer* of water which prevents colloidal particles of SNBD from aggregating each other and gives so stable colloid. Then one would ask, what will happen if water is removed by heating under vacuum or exposing long time to air at room temperature. The answer is that very tight aggregates are formed. These are incoherent interfacial Coulombic aggregates but certainly not agglutinates; hence it is possible to redisperse the aggregates in water by repeating intense sonication and centrifugal removal of stubborn insoluble particulates. Nevertheless, aggregation of single-nano particles of any kind is not easy to disperse, and it is recommended to use colloidal solution as often as possible.

17.4.2 Gels

Gels of SNBD are more useful and convenient for transportation and storage than the dried aggregates of SNBD. Colloidal solution of SNBD starts to coagulate into soft gel when the concentration exceeds about 10%. Gellation never takes place with suspension of agglutinates. Gels of SNBD colloid are the network of particles and considered to start when the particles began to feel mutual attraction. How close do the SNBD particles approach each other in the start of gel formation? Simple arithmetic using basic number of SNBD particle (Table 17.3) gives an average centre-to-centre distance of 10.6 nm between neighboring particles in 10 wt% SNBD colloid. Taking average particle size of 4.8 nm and the thickness of hydration layer of 1.0 nm into account, the average closest distance between a pair of particles is 3.8 nm, slightly smaller than the particle size itself.

Further removal of water decreases the interparticle distance to smaller and smaller. Jellylike body is turned into wet and black solid, then breaks up into smaller masses and finally becomes dry-looking powder. At this point, the water content reaches to about 50 to 40% of the weight of SNBD. Air-drying for a day or two leaves black powder which is crushable into smaller pieces with the tip of spatula. Color of the solid is real black up to this point. The coarse powder can be ground in a mortar with a ceramics pestle into brown colored fine powder containing 15–25% of water. The remaining water can be considered to correspond to the inner inert layer (Figure 17.7), and is probably enough to keep much of the particle surface from directly touching each other. In fact the hard gel thus prepared can be redispersed in water fairly readily with the help of intense sonication.

Table 17.3 *Basic numbers of SNBD particles*

Selected numbers for one SNBD particle	Diameter $D(=2r)$	4.8 nm	100 nm[d]
	Volume $v(=4\pi r^3/3)$	5.79×10^{-20} cm^3	5.23×10^{-16} cm^3
	Weight $w(=\rho v)$[a]	1.74×10^{-19} g	1.64×10^{-15} g
	PW[b] $(=wN)$[c]	104 000	9.88×10^{8}
	Number of C atoms	8660	8.23×10^{7}
Number of particles in 1 g of SNBD $n(=1/w)$		5.75×10^{18}	6.10×10^{14}

[a] ρ = sp. gr. of diamond = 2.99 g/cm^3 (preliminary results).
[b] 'Particulate weight', a concept corresponding to 'molecular weight'.
[c] N = Avogadro number.
[d] A model of agglutinated product of detonation nanodiamond.

17.4.3 Number Effect

Due to the tenacious hydration, it is difficult to determine fundamental physical constants of SNBD particles like density, elementary composition, and dielectric constant which need to be performed on perfectly dried samples under rigorous exclusion of atmospheric moisture. For this reason, none of the fundamental properties of SNBD has been measured at the moment. Thermogravimetric analyses will be carried out when we have finalized standard procedure for the preparation of SNBD.

Nevertheless, we recently found a new attribute of SNBD which we have so far neglected but could be unique and generally useful for the smaller nanoparticles. The new attribute is the unexpectedly large number of particles in unit volume or unit weight. I noticed the impact of overwhelmingly large number associated with nanoparticles, especially single-nano particles, when I spotted a word '*nanodiamond*' in an advertisement for mechanical pencils. Naturally I was aroused and found a major pencil company claiming to have improved writing performance of their new brand of mechanical pencils by *dispersing 400 million particles of nanodiamond into one piece of lead* with 0.5 mm in diameter and 60 mm in length [[20]a]. *Four hundred million* particles sounded to me an unbelievably large number, and I even suspected whether it was at all possible to disperse such a large number of particles in a tiny stick of lead. However, the question was quickly solved by simple arithmetic. The number density of the added particles is given by $400 \times 10^6/0.0118 = 3.39 \times 10^{10}/\text{cm}^3$ (the volume of one lead is equal to $0.0118\,\text{cm}^3$). This volume density can be approximately translated into a linear density of 3230/cm, or 310 nm for the average center-to-center distance between the nearest particles. As a nanoparticle is defined to have diameters between 1 and 100 nm, there is a plenty of room to pack 400 million particles in a lead, even if the largest nanodiamond is used. In other word, this advertised density sounds great but actually is not really large for nanoparticles.

Then it occurred to me that, if the number of spherical nanoparticles were a decisive factor for the writing performance of lead, how much SNBD is needed to pack 400 million particles of them into a lead? It turned out that we need only 0.5 mg of SNBD per 1 kg of graphite to achieve the advertised volume density. Now I am really impressed by the enormous number of particles that tiny amounts of SNBD contain. Basic figures necessary to count the number of SNBD particles are summarised in Table 17.3. In the last line of this table the number of particles contained in 1 g of SNBD is given, which is 5 billion squared! To be more specific, the number is in the order of a quintillion (10^{18}).

Such large numbers are beyond our comprehension. We could have some idea of the largeness up to millions (10^6) and billions (10^9); for example populations of large cities and countries are of these orders of magnitude. Although it is difficult to grasp the immensity of numbers beyond a trillion (10^{12}), we should pay attention to the effects of great numbers when dealing with nanoparticles, of which there should be plenty. A few cases will be presented below.

17.5 Applications

It would be good to find good applications for new and interesting materials like SNBD particles. In fact it is difficult to cite any known material that surpasses the overall performance than SNBD, which not only inherits numerous outstanding properties of bulk diamond, but also acquired many new advantages due to its unique size and internal structure. It is then perfectly natural to expect that many new and interesting applications will be found for SNBD soon.

Already a variety of leads for grand applications of SNBD particles are being developed that cover wide and diverse fields of technology [1]. The developments have so far been slow because we could not understand the properties and behaviors of SNBD well enough. Now that we have an excellent model, thanks to the help from theoretical calculations, we may duly expect accelerated developments, for example for the following promising cases:

- A new platform for drug carrier based on SNBD gel, and characterized by inactivated delivery and slow release from within the sick cell [21–24].
- High-density seeding by SNBD particles for homoepitaxial CVD growth of polycrystalline diamond film [25–27].
- Nanocomposites using dispersed SNBD particles as reinforcing additive including electrolytic plating [28–34].
- Cell imaging with intensely fluorescing SNBD particles [35].
- A new type of lubrication and polishing system consisting of SNBD particles in oil and water [36, 37].
- Hard masking for dense vertical diamond nanowires [38–40].

These applications take advantage of high dispersity, super-hardness, highly charged crystal facets, high crystallinity, processability of nanocomposites and gigantic particle numbers. As the space given for the chapter has already exceeded the limit, we will mention below only our latest finding, namely a new lubrication system based on SNBD, which seems particularly important for environmental protection.

17.5.1 Lubrication Water

All the movements on earth, including physical activities of living creatures, use friction, which is fine-tuned by lubrication. Artificial lubricants have been, traditionally and exclusively, oil. However, disposal of the used lubrication oil has become so damaging to environment, hence costly, that alternative lubricators are being sought. Environmentally most benign lubricator is water, but water has been considered unsuitable as lubricant because it has too low viscosity to form liquid film in boundary lubrication. However, we found aqueous colloidal solution of SNBD acts as an entirely new type of lubricant. We envisioned the following mechanism:

1. SNBD particles are supposed to play a role of hard spacer with low rolling friction to prevent direct contacts between the shearing planes under the boundary condition.
2. Such spacer particles must be available abundantly any time in any microscopic sections of true contact area. This rare requirement is fulfilled by dilute colloidal solutions of SNBD.
3. It is desirable that each SNBD particle is wrapped with soft liquid film in order to reduce the pressure at the point of contact between the particle and the shearing plane. The inner inert layer of hydration shell (Figure 17.7) will be a good candidate for such a soft surface film.

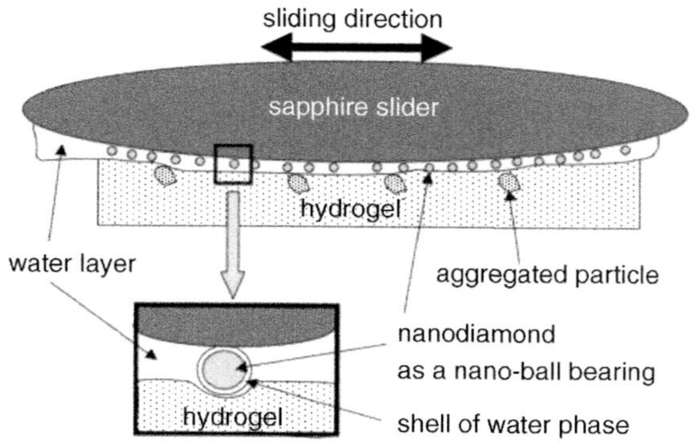

Figure 17.8 *Illustration of an experimental setup for the SNBD aqueous colloid lubrication. Test piece [hydrogel of poly(acryl amide)] is immersed in 1 and 5% SNBD aq colloid. A sapphire slider ball of 2 mm in diameter was reciprocated over a distance of 1.5 mm under a load of 10 mN at a frequency of 10 Hz*

Our first experiments were carried out by dipping hydrogel of poly(acryl amide) into 1 and 5% aqueous colloid of SNBD, and a sapphire slider was placed on the surface of hydrogel and reciprocated under application of a fixed load (Figure 17.8). Results are quite encouraging; friction coefficient of 0.02 was reached in 20 seconds (Figure 17.9) and this level of friction was maintained at least for eight hours of continuous operation. Similar experiments were repeated using Si-wafer as a work surface to obtain comparable coefficients. In this case no serious scars by the slider ball were visible on the wafer surface. The same friction coefficient level obtained by using 1 and 5% colloidal solution indicates that the concentration may have been too high [37].

We estimated the number of SNBD nanospacers in the true contact areas using published data for the shearing pair of polished soft-steel surfaces in boundary contact (Figure 17.10

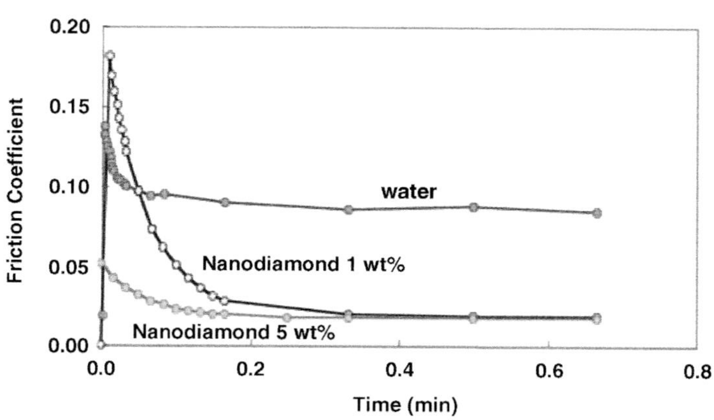

Figure 17.9 *The first result of a new lubrication system consisting of a large number of SNBD spacer balls dispersed in water. Changes in friction coefficient in the initial stage of lubrication test carried out in a set-up shown in Figure 17.8*

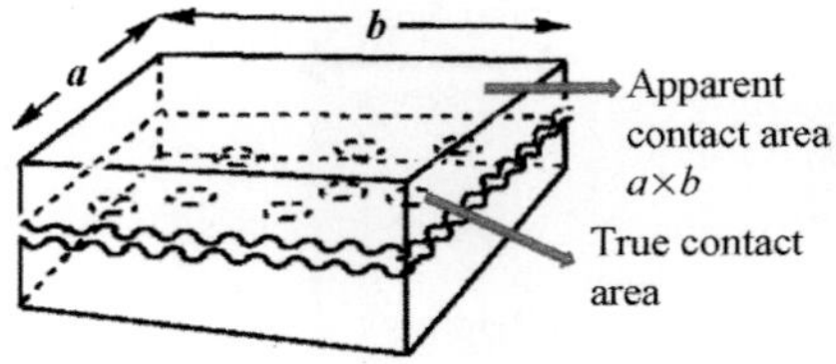

Figure 17.10 Illustration of a pair of soft-steel planes in shearing motion against each other. Whereas apparent contact area is given by nominal area a × b, true contact area under a given pressure is much smaller as shown in Table 17.4

and Table 17.4) [41]. As shown in the last column of Table 17.4, more than enough SNBD particles exist at each true contact area in the order of billion. Hence our interpretation on the identical results for 1 and 5% colloid seems to be justified; we used far too concentrated solution. For lubrication to work in our spacer mechanism, the colloid concentration could be much, much lower. It is likely that too many particles do not increase the lubrication effect in our mechanism.

In fact, we have once observed a mysterious result before in lubrication oil. When SNBD was dispersed in poly-α-olefin oil together with small amounts of glycerin mono-oleate, a well-known additive to protect boundary contact of diamond particles with work surface, and a surfactant, very good lubrication effect was observed for a shearing pair of polished SUS surfaces. The friction coefficient did not increase even if the contents of SNBD particles were reduced to more than tens of thousand times as small. Now we recognize the hidden characteristics of nanoparticles: the number of nanoparticles in conventional percentile concentrations is much greater than what we know with fine micro- and larger particles [42].

17.6 Recollection and Perspectives

We have quickly gone through the history of single-nano diamond of detonation origin: discovery, misfortune in the first few ten years of developments, misconception about the

Table 17.4 Area and number of true contact points and number of SNBD spacers present at a true contact point between a pair of shearing soft-steel surface (Figure 17.10) polished to a surface roughness of 20 nm. Concentration of SNBD is 1 wt%[a]

Load W kgf	Area of TCP[a] A mm^2 [2]	A/S	Number of TCP[a]	SNBD particles per one TCP[a] $\times 10^9$ [7]
500	5	1/400	35	403
100	1	1/2000	22	50.6
20	0.2	1/10000	9	4.14
5	0.05	1/40000	5	0.575
2	0.02	1/100000	3	0.138

[a]Apparent contact area = 2000 mm^2, thickness of liquid film filling the contact area = average roughness × 2 = 40×10^{-6} mm, volume of liquid film in apparent contact area = 8×10^{-2} mm^3, number of SNBD particles in the volume = $8 \times 10^{-2} \times 5.75 \times 10^{13} = 4.6 \times 10^{12}$.

dispersity, experimental breakthrough that led to the isolation of primary particles, the second stagnation caused by compound reasons and arrival of relief from theoretical modeling. Where do we go from here? It seems worthwhile to go back to Table 17.1 and see what happened with the other methods of diamond synthesis. First we note an interesting fact that all the four major methods of diamond synthesis were discovered within a short span of time between 1950 and 1963. Perhaps this was the period when the desire of scientists for artificial synthesis of diamond grew to an unprecedented height.

Among these, the HPHT and shock methods, both for microdiamond particles, have simultaneously retreated from the fore stage a few years ago. Technology of microdiamond syntheses was almost perfected, production cost went down to respectable levels, and market grew slowly but steadily. Why did the founding manufacturers stop? From the beginning to the end, strict secrecy prevailed in this industry of artificial diamond. Accurate statistics on the yearly production have never been disclosed. Likewise, there is no official reason for the retreat announced; hence the following description is our guess.

Although diamond is widely recognized as the singularly outstanding material having so many '*the highest on earth*' properties [43–45], it has one deadly defect when applied to industrial purposes, which is the total lack of processability. This is the result of several factors including high hardness, high tensile modulus, too high heat-conductivity, and too small specific heat. A great deal of efforts must have been poured into finding a way to anneal diamond, but apparently no good method could be found so far. They could find only a disappointingly small market as polishers. Diamond tools turned out to be a better business but the technology of diamond coating remains low. A bitter lesson we received from microdiamond is that '*The first form is the final form.*'

The other two methods, CVD and detonation, are still under developmental stages for too long time, but the chance of overcoming the above lesson is quite good. For example, if we can prepare high-quality ultra-thin single-crystal diamond films by CVD, they may be deformable at least to some extent. Our SNBD is clearly more convincing. Throughout this chapter, we described only those aspects which were acquired by having single-nano size. Fortunately significant events happened by the reduction of crystal size from micron to nano. Actually the magnitude of size reduction was from 50 μm (Table 17.1) to 5 nm, or by a factor of 10^{-5}. As a result, the significance of particle surface increased exponentially and spontaneous phase transition took place on the surface, leading to spontaneous electron polarization. Most of the application leads mentioned above for SNBD came from the polarized structure, hence cannot be transferred to microdiamonds.

Among the prospective applications of SNBD, the most significant in view of the microdiamond lesson is nanocomposites (see above). We can perfectly overcome the problem of nonprocessability in microdiamonds by dispersing small amounts of SNBD particles in matrices including polymers, elements, metals, alloys, ceramics, glasses and others. The only problem is that we must yet to develop a general method to disperse SNBD in solid matrices. We are presently most concerned with this problem, but the first sign of success appeared recently, wherein we could improve Vickers microhardness of electrolytically plated chromium film by 36% by incorporation of only 0.1 wt% of SNBD [33].

For chemists, SNBD offers wide opportunities remaining almost unexplored, namely derivatization. Introduction of appropriate functional groups in appropriate numbers on the particle surface will produce significant changes in the properties of particle. Homogeneous reactions can be conducted in water and a few organic solvents, but we must be sure that the reagent molecules reached the reaction site through the solvation shell without being hydrolyzed. Chemistry of SNBD is virtually unknown at present.

Finally, small comments are due about the potential toxicity, health risk and environmental effects of SNBD particles in view of increasing social concern towards nanoparticles. Careful evaluation of cytotoxicity and genotoxicity has been carried out by Professor L. Dai and his group and SNBD proved to be one of the safest materials ever known among nanoparticles [46].

Acknowledgements

We are grateful to the donors of research funds including NEDO, Chiba Prefecture, Chiba Bank, Futaba Corporation, Nippon Kayaku Co., and other sources.

References

[1] E. Ōsawa, Monodisperse single-nano diamond particulates, *Pure & Appl. Chem.*, **80**, 1365–1379 (2008).

[2] V. V. Danilenko, On the history of detonation nanodiamond discovery, *Phys. Solid State*, **46**, 595–599 (2004).

[3] P. S. DeCarli and J. C. Jamieson, Formation of diamond by explosive shock, *Science*, **133**, 1821–1822 (1961).

[4] Private communications from Professors A. Ya. Vul' and V. Danilenko.

[5] E. Ōsawa, Nanodiamond agglutinates: characterization and disintegration, First International Symposium on Detonation Nanodiamonds: Technology, Properties and Applications, July 7–9, 2003, St Petersburg, Russia.

[6] See also: E. Ōsawa, Disintegration and purification of crude aggregates of detonation nanodiamond – a few remarks on nano methodology, in *Synthesis, Properties and Applications of Ultrananocrystalline Diamond, Proceedings of the NATO Advanced Research Workshop on Ultrananocrystalline Diamond, June 7–10, 2004, St. Petersburg, Russia*, D. Gruen, A. Ya. Vul', O. Shenderova (Eds), NATO Science Series, Series II: Mathematics, Physics and Chemistry, Vol. **192**, Springer, Dortrecht, (2005). pp. 231–240.

[7] A. Krüger, F. Kataoka, M. Ozawa, T. Fujino, Y. Suzuki, A. E. Aleksenskii, A. Ya. Vul and E. Ōsawa, Unusually tight aggregation in detonation nanodiamond: identification and disintegration, *Carbon*, **43**, 1722–1730 (2005).

[8] A. S. Barnard, Self-assembly in nanodiamond agglutinates, *J. Mater. Chem.*, **18**, 4038–4041 (2008).

[9] First used in: E. Ōsawa, D. Ho, H. Huang, M. V. Korobov and N. N. Rozhkova, Consequences of strong and diverse electrostatic potential field on the surface of detonation nanodiamond particles, *Diam. Rel. Mater.*, **18**, 904–908 (2009).

[10] See also: E. Ōsawa, Single-nano buckydiamond particles-synthetic strategies, characterization methodologies and emerging applications, *in Nanodiamonds: Applications in Biology and Nanoscale Medicine*, D. Ho (Ed.) Springer Science+Business Media, Norwell, Chapt. 1, pp. 1–33.

[11] M. Ozawa, M. Inakuma, M. Takahashi, F. Kataoka, A. Krüger and E. Ōsawa, Preparation and behaviors of brownish clear nanodiamond colloids, *Adv. Mater.*, **19**, 1201–1206 (2007).

[12] K. Iakoubovskii, K. Mitsuichi and K. Furuya, High-resolution electron microscopy of detonation nanodiamond, *Nanotechnol.*, **19**, 155705 (5pp) (2008).

[13] S. Hu, J. Sun, X. Du, F. Tian and L. Jiang, The formation of multiply twinning structure and photoluminescence of well-dispersed nanodiamonds produced by pulsed-laser irradiation, *Diam. Rel. Mater.*, **17**, 142–146 (2008).

[14] J.-Y. Raty and G. Galli, Ultradispersity of diamond at the nanoscale, *Nature Mater.*, **2**, 792–795 (2003).
[15] A. S. Barnard, S. P. Russo and I. K. Snook, Visualization of hybridization in nanocarbon systems, *J. Comput. Theor. Nanosci.*, **2**, 68–74 (2005).
[16] A. S. Barnard and M. Sternberg, Crystallinity and surface electrostatics of diamond nanocrystals, *J. Mater. Chem.*, **17**, 4811–4819 (2007).
[17] K. Makino, S. Tejima, I. Minami, H. Nakamura and E. Ōsawa, Large-scale simulation on the formation of nanocarbon structures (in Japanese), *RIST News*, **46**, 28–37 (2009). a E. Ōsawa, H. Ueno, M. Yoshida, Z. Slanina, X. Zhao, M. Nishiyama and H. Saito, Combined Topological and Energy Analysis of Annealing Process in Fullerene Formation. Stone-Wales Interconversion Pathways among IPR Isomers of Higher Fullerenes, *J. Chem. Soc., Perkin Trans.*, **2**, 943–950 (1998).
[18] S. Ji, T. Jiang, K. Xu and S. Lee, FTIR study of adsorption of water on ultradisperse diamond powder surface, *Appl. Surf. Sci.*, **133**, 231–238 (1998).
[19] M. V. Korobov, M. M. Efremova, N. V. Avramenko, N. I. Ivanova, N. N. Rozhkova and E. Ōsawa, Solvation structure on single-nano buckydiamond in gel and dried powder by differential scanning calorimetry and nitrogen adsorption, *J. Phys. Chem. C*.
[20] M. V. Korobov, N. V. Avramenko, A. G. Bogachev, N. N. Rozhkova and E. Ōsawa, Nanophase of water in nanodiamond gel, *J. Phys. Chem. C.*, **111**, 7330–7334 (2007).a http://uniball.com/home/index.html.
[21] R. Lam, M. Chen, E. Pierstorff, H. Huang, E. Ōsawa and D. Ho, Nanodiamond-embedded microfilm devices for localized chemotherapeutic elution, *ACS NANO*, **2**, 2095–2102 (2008).
[22] M. Chen, E. D. Pierstorff, R. Lam, S. Li, R. Chatterton, S. Khan, E. Ōsawa and D. Ho, Nanodiamond-mediated delivery of water-insoluble therapheutics, *ACS Nano*, **3** 10.1021/nn900480m (2009).
[23] H. Huang, E. Pierstorff, E. Ōsawa and D. Ho, Protein-mediated assembly of nanodiamond hydrogels into biocompatible and biofunctional multilayer nanofilm, *ACS Nano*, **2**, 203–212 (2008).
[24] H. Huang, E. Pierstorff, E. Ōsawa and D. Ho, Active nanodiamond hydrogels for chemotherapeutic delivery, *Nano Lett.*, **17**, 3305–3314 (2007).
[25] J. C. Arnault, S. Saada, M. Nesladek, O. A. Williams, K. Haenen, P. Bergonzo and E. Ōsawa, Diamond nanoseeding on silicon: stability under H_2 MPCVD exposures and early stages of growth, *Diam. Rel. Mater.*, 17, doi: 10.1016/j.diamond.2008.01.008 (2008).
[26] O. A. Williams, M. Nesladek, M. Daenen, S. Michaelson, O. Ternyak, A. Hoffman, E. Ōsawa, K. Haenen, R. B. Jackman and D. M. Gruen, Growth and electronic properties of nano-diamond, *Diam. Rel. Mater.*, **17**, 1080–1088 (2008).
[27] O. A. Williams, O. Douheret, M. Daenen, K. Haenen, E. Ōsawa and M. Takahashi, Enhanced diamond nucleation on monodispersed nanocrystalline diamond, *Chem. Phys. Lett.*, **445**, 255–258 (2007).
[28] J. B. Correia, V. Livramento, N. Shohoji, E. Tresso, K. Yamamoto, T. Taguchi, K. Hanada and E. Ōsawa, Bulk copper-nanodiamond nanocomposites; processing and properties, *Mater. Sci. Forum*, **587–8**, 443–447 (2008).
[29] K. Hanada, K. Yamamoto, T. Taguchi, E. Ōsawa, M. Inakuma, V. Livramento, J. B. Correia and N. Shohoji, Further studies on copper nanocomposite with dispersed single-digit-nanodiamond particles, *Diam. Rel. Mater.*, **16**, 2054–2057 (2007).
[30] D. H. Wang, L.-S. Tan, H. Huang, L. Dai and E. Ōsawa, *In-situ* nanocomposite synthesis: arylcarbonylation and grafting of primary diamond nanoparticles with a poly(ether-ketone) in polyphosphoric acid, *Macromol.*, **42**, 114–124 (2009).
[31] K. Yamamoto, T. Taguchi, K. Hanada, E. Ōsawa, M. Inakuma, V. Livramento, J. B. Correia and N. Shohoji, TEM studies of nanocarbons and nanodiamonds (ND): mechanical milling of ND and Cu, *Diam. & Rel. Mater.*, **16**, 2058–2062 (2007).
[32] V. Livramento, J. B. Correia, N. Shohoji and E. Ōsawa, Nanodiamond as an effective component for nanocopper alloy, *Diamond & Rel. Mater.*, **16**, 202–204 (2007).

[33] G. K. Burkat, V. Yu. Dolmatov, E. Ōsawa and E. A. Orlova, Investigation of properties of chrome-nanodiamond coatings based on detonation nanodiamonds (DND) from different producers, IWFACT09, St. Petersburg, Russia, July 2009.

[34] D. Nunes, V. Livramento, J. B. Correia, K. Hanada, P. A. Carvalho, R. Mateus, N. Shohoji, H. Fernandes, C. Silva, E. Alves and E. Ōsawa, Consolidation of Cu-nDiamond nanocomposites: hot extrusion vs spark plasma sintering, The Vth International Materials Symposium, Materiais 2009, Technical University of Lisbon, Lisboa, Portugal, Apr. 5–8, 2009.

[35] B. R. Smith, D. Inglis, B. Sandnes, J. R. Rabeau, A. V. Zvyagin, D. Gruber, C. Noble, R. Vogel, E. Ōsawa and T. Plakhotnik, Lighting up 5-nm monocrystalline nanodiamond with luminescent nitrogen-vacancy defect centers, *Small* doi.org/10.1002/smll.20081802.

[36] W. M. Lin, T. Kato, H. Ohmori and E. Ōsawa, Study on tribo-fabrication in polishing by nanodiamond colloid, *Key Engineeing Mater.*, **404**, 131–136 (2009).

[37] S. Mori, A. Kanno, H. Nanao, I. Minami and E. Ōsawa, Tribological performance of nanodiamond for water lubrication, *Proceedings of the 3rd International Symposium on Detonation Nanodiamonds: Technology, Properties and Applications*, July 1–4, 2008, St Petersburg, Russia, pp. 21–8, Ioffe Physico-Technical Institute.

[38] N. Yang, H. Uetsuka, E. Ōsawa and C. E. Nebel, Vertically aligned diamond nanowires for DNA sensing, *Angew. Chem. Int. Ed.*, **47**, 5183–5185 (2008).

[39] N. Yang, H. Uetsuka, E. Ōsawa and C. E. Nebel, Vertically aligned nanowires from boron-doped diamond, *Nano Lett.*, **8**, 3572–3576 (2008).

[40] C. Nebel, N. Yang, H. Uetsuka, E. Ōsawa and O. A. Williams, Detection of nucleic acids using diamond, *New Diamonds* (2008) in press.

[41] T. Bowden and D. Tabor, *Friction and Lubrication in Solids*, Japanese edition, Maruzen, Tokyo 1961, p. 27.

[42] Y. Mabuchi and E. Ōsawa, manuscript in preparation.

[43] M. A. Prelas, G. Popovici and L. K. Bigelow (eds) *Handbook of Industrial Diamonds and Diamond Films*, Marcel Dekker Inc., New York, 1998.

[44] J. E. Field, (ed), *The Properties of Natural and Synthetic Diamond*, Academic Press, San Diego, 1992.

[45] G. E. Harlow (ed), *The Nature of Diamonds*, Cambridge Univ. Press, Cambridge, 1998.

[46] A. M. Schrand, J. Johnson, L. Dai, S. M. Hussain, J. J. Schlager, L. Zhu, Y. Hong and E. Ōsawa, Cytotoxicity and genotoxity of carbon nanomaterials, Schrand, A. M.; Johnson, J.; Dai, L.; Hussain, S. M.; Schlager, J. J.; Zhu, L.; Hong, Y.; Ōsawa, E. in *Safety of Nanoparticles: From Manufacturing to Medical Applications*, Webster, T. J. (Ed.), Springer Science+Business Media, New York, (2008). Chapter 8, pp. 159–188.

18

Properties of π-electrons in Graphene Nanoribbons and Nanographenes

De-en Jiang[a], *Xingfa Gao*[b], *Shigeru Nagase*[b] *and Zhongfang Chen*[c]

[a]*Chemical Science Division, Oak Ridge National Laboratory, MS6201, Oak Ridge, TN, USA*
[b]*Department of Theoretical and Computational Molecular Science, Institute for Molecular Science, Myodaiji, Okazaki, Japan*
[c]*Department of Chemistry, Institute for Functional Nanomaterials, University of Puerto Rico, Rio Piedras Campus , San Juan, PR, USA*

18.1 Introduction

In his Nobel lecture in 1996, the late Richard Smalley remarked: 'Carbon has this genius of making a chemically stable two-dimensional, one-atom-thick-membrane in a three-dimensional world. And that, I believe, is going to be very important in the future of chemistry and technology in general.' Looking back, this statement was so insightful that it predicted the exploding interest in graphene research today. From graphite to fullerenes to nanotubes, carbon displays the versatility of sp^2 hybridization coupled with the right valence. The ability of carbon atoms to adopt sp^2 hybridization dictates many structural and physical properties of carbon materials, for example, the stability of a graphene sheet.

Even not long ago, it was thought that a strictly two-dimensional crystal was physically impossible. The successful isolation of graphene, a single layer of graphite, in 2004 [1–3] refreshed our minds. The simple top-down approach to graphene preparation by mechanical exfoliation (repeated peeling) enabled a plethora of experimental studies to probe the unique properties of electrons in the two-dimensional honeycomb lattice [4–8]. Graphene's many exciting unusual properties [9–11] are bringing revolutions to many advanced materials we

Chemistry of Nanocarbons Edited by Takeshi Akasaka, Fred Wudl and Shigeru Nagase

desire. For example, graphene is the strongest material ever measured [12], chemically stable and inert, and conducts electricity better than any other known material at room temperature [13]. These properties endow graphene-based materials, among others, many applications for ultra-strong lightweight materials used in space shuttles to improve fuel efficiency, for transparent conductive electrodes in solar cells [14] and various optoelectronical devices such as liquid crystal displays and electric-field-activated optical modulators [15], and electrical energy storage in batteries and supercapacitors to enable production of renewable energy at a large scale.

Exciting news keeps coming like waves about the large-scale preparation of graphenes. Among others, Kim *et al.* [16] produced large, centimetre-scale areas of high-quality graphene films (Figure 18.1) with excellent electronic properties on arbitrary substrates using chemical vapor deposition (CVD) method; Pan *et al.* [17, 18] prepared high-quality large continuous pieces of graphene (up to a few square millimeters in size) on crystals by a simple annealing technique.

Nanographenes confine their π-electrons in a nanometer-sized space defined by their σ-bond network. The π-electrons will respond to both the confinement and the shape of the peripheries. This confinement effect manifests itself in the HOMO-LUMO gap, while the edge effect can be evidenced by the spin ordering along the edge. Nanographenes have different shapes and topologies, such as one-dimensional nanoribbons and zero-dimensional polycyclic aromatic hydrocarbons (PAHs). Nanographene is a playground where chemists, physicists, and materials scientists meet and an intersection where the top-down approach from macroscopic graphene meets the bottom-up approach from small molecular building blocks. Applications of nanographenes are rapidly expanding. More sophisticated applications demand novel properties, and more members of the nanographene family will see their potentials explored in the coming years.

This chapter reviews the recent progress in understanding the properties of nanographenes, especially nanoribbons. Section 18.2 focuses on the zigzag edge effects. Section 18.3 examines the electronic and magnetic properties of nanographenes in both one-dimension (i.e. nanoribbons) and zero-dimension (i.e. PAHs). In Section 18.4, we give a brief outlook for nanographene research.

Figure 18.1 *Graphene films on the polydimethylsiloxane (PDMS) substrates are transparent and flexible. Reprinted figure with permission from: K. S. Kim, Y. Zhao, H. Jang, S. Y. Lee, J. M. Kim, K. S. Kim, J. H. Ahn, P. Kim, J. Y. Choi, B. H. Hong, Nature,* ***457****, 706 (2009). Copyright (2009) by the Nature Publishing Group*

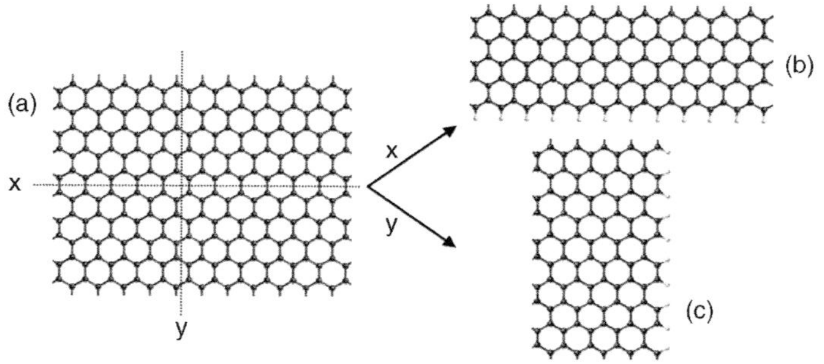

Figure 18.2 *Cutting through an infinite graphene sheet (a) to obtain a semi-infinite sheet with a hydrogen-terminated zigzag (b) or armchair (c) edge. Carbon atoms are shown in black and hydrogen atoms are in gray. Reprinted figure with permission from: D. E. Jiang, B. G. Sumpter, S. Dai, J. Chem. Phys.,* ***126****, 134701 (2007). Copyright (2007) by the American Institute of Physics*

18.2 Edge Effects in Graphene Nanoribbons and Nanographenes

The importance of an edge to a graphene sheet parallels that of a surface to a crystal. Cutting through an infinite graphene sheet (Figure 18.2a), one first breaks C–C σ bonds and then obtains two semi-infinite graphene sheets, each with a one-dimensional edge. The dangling σ bonds at the edges can be saturated with hydrogen (so-called hydrogenated or hydrogen-terminated edges) and all the carbon atoms remain sp^2 hybridized. Depending on the cutting direction, two unique types of edges can be obtained: zigzag (Figure 18.2b) and armchair (Figure 18.2c). The cutting also introduces a boundary at the edge to the previously fully delocalized π-electron system.

The edge geometry makes a huge difference in the π-electron structure at the edge. By constructing an analytical solution to the π-electron wavefunction, Nakada *et al.* [19] showed that the zigzag edge in a semi-infinite graphene sheet gives rise to a degenerate flat band near the Fermi level for the k vector between $2\pi/3$ and π. For $k = \pi$, the wavefunction is completely localized at the edge sites, leading to a so-called localized state at the zigzag edge. This flat-band feature and its corresponding localized state are unique to the zigzag edge (they are completely absent from the armchair edge). Using the DV-Xα method with the linear combination of atomic orbitals (LCAO) bases, Kobayashi [20] in 1993 predicted the existence of such a flat band and localized state on a zigzag-edged vicinal graphite surface. Independently, Klein [21] analytically examined the band structure of the simple Hückel model for several graphene ribbons with zigzag edges, but the flat bands were predicted to be in $0 < k < 2\pi/3$ with a small gap at the Fermi level. This description was subsequently modified by the same author [22]. More thorough theoretical investigations were presented by Fujita, Nakada, and others [19, 23–34]. Encouragingly, with modern scanning tunneling spectroscopy, the zigzag edge was recently observed on highly oriented pyrolytic graphite surfaces and the localized state at the zigzag edge was confirmed [35–37].

The semi-infinite graphene sheet can be cut again parallel to the edge, to generate a graphene ribbon with two edges (Figure 18.3). In 1996, Fujita and coworkers theoretically introduced graphene nanoribbons, and found the peculiar behavior of zigzag edges [19, 23]. If the ribbon width is within the nanometer range, the effect of edge atoms is more

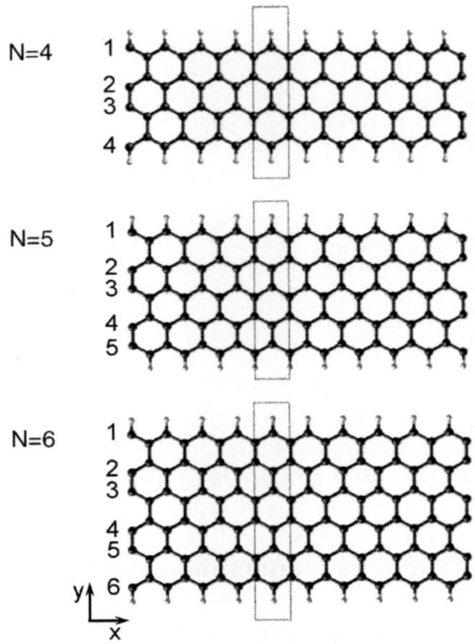

Figure 18.3 Zigzag-edged graphene nanoribbons with different width. The rectangular boxes indicate the unit cell. The ribbons are infinitely repeated along the x axis. Reprinted figure with permission from: D. E. Jiang, B. G. Sumpter, S. Dai, J. Chem. Phys., **126**, 134701 (2007). Copyright (2007) by the American Institute of Physics

pronounced [19] and the two edges can interact with each other. Using the Hubbard model with the unrestricted Hartree-Fock approximation, Fujita *et al.* [23] revealed weak ferromagnetism along one edge and antiferromagnetism between two edges (one edge spin-up, the other spin-down) on a zigzag-edged graphene nanoribbon (ZGNR). Here what is remarkable is that the magnetism arises from a system made of only sp^2 carbon *without* σ*-dangling bonds.*

The magnetism of ZGNR has been used to explain observed magnetic properties in nanographite materials [38, 39]. Very recently, Son *et al.* predicted half metallicity in ZGNRs from first principles [40]. The half metallicity is caused by opposite responses of energy bands to the external electric field for the up and down spins.

In honor of Mitsutaka Fujita's original contribution of proposing the importance of graphene edge state and introducing graphene nanoribbons, the mono-hydrogenated zigzag edge is often named as 'Fujita edge'.

Before experimental verification, ZGNRs with reasonable length that can preserve magnetic and electronic properties predicted for infinitely long ribbons need to be synthesized. Chemists have made great stride in the bottom-up synthesis of nanographenes [41] since the publication of Clar's two volumes on polycyclic hydrocarbons [42]. However, the synthesis of rectangular nanographenes with consecutive zigzag edges remains a challenge [43, 44]. One class of nanographenes is polycyclic aromatic hydrocarbons (PAHs). Figure 18.4 defines a rectangular PAH [X,Y]: X and Y represent the numbers of fused rings in the zigzag and armchair edges, respectively [45]. From small

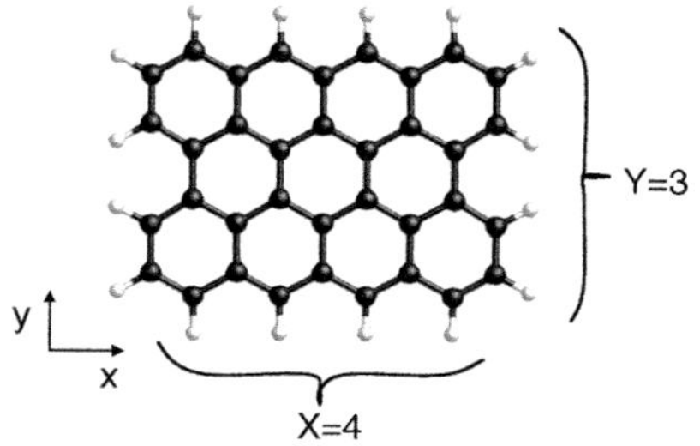

Figure 18.4 *Defining a rectangular polycyclic aromatic hydrocarbon, PAH [X,Y]. Here X = 4 and Y = 3 indicate the lengths of zigzag and armchair edges, respectively. Carbon atoms are in black and H atoms in white. Reprinted figure with permission from: D. E. Jiang, B. G. Sumpter, S. Dai, J. Chem. Phys.,* ***127****, 124703 (2007). Copyright (2007) by the American Institute of Physics*

PAHs to infinitely long ZGNRs, one wonders: (1) how small can a PAH be but still have a magnetic ground state and (2) how do the magnetic properties of rectangular PAHs change with sizes and how do they compare with infinitely long ribbons? Answering these questions will help understand the magnetism of nanographenes and nanographites and also guide synthesis towards a small enough PAH that still behaves electronically like a long ribbon.

As emphasized above, the edge matters for nanographenes. Zigzag edges tend to cause localized edge states, spin ordering, and instability, where armchair edges are more stable and in most cases can be understood as a double bond. As a consequence, most PAHs synthesized have armchair peripheries, while PAHs with consecutive zigzag edges remain challenging to synthesize. The zigzag edge's instability can be understood from Clar's empirical but highly useful sextet rules [42]. According to Clar's rules, a PAH is most stable when it has the greatest number of aromatic sextets. For example, **2** (phenanthrene) is more stable than the isomer **1** (anthracene) because **2** has two sextets, while **1** has only one. This again relates to the HOMO-LUMO gap. The computed HOMO-LUMO gap at the B3LYP/6-31G* level of theory is 3.6 eV for **1** and 4.7 eV for **2**, indicating that the HOMO-LUMO gap correlates to the number of sextets and the stability. We will relate to Clar's sextet rules in several occasions later when discussing the stability of specific PAHs with zigzag edges.

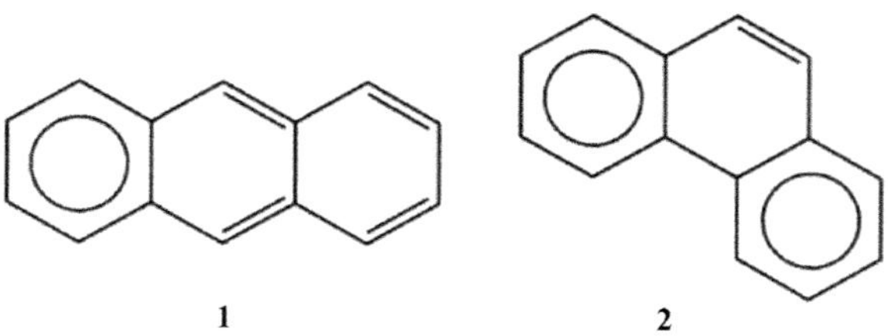

1 **2**

For nanographene ribbons, the majority of published research is theoretical and computational studies. This reflects both the unusual difficulties in synthesis and the high reactivity of such strips. Considering the better stability of armchair edges, nanographene ribbons with armchair edges are more feasible as synthesis targets, and recent experimental successes confirmed this notion [38,41]. Zigzag-edged oligomers are being actively pursued. The most famous example is acenes, the linearly fused benzene rings. Crystals of substituted heptacene compounds have been obtained recently [46, 47], and applications of these low gap materials in molecular electronics are expected in the near future [48]. Organic chemists are also actively working on synthesizing even higher acenes and other unique zigzag-edged PAHs. This is an area that computational studies can provide useful information for experimental exploration.

18.3 Electronic and Magnetic Properties of Graphene Nanoribbons and Nanographenes

From two-dimensional macroscopic graphene to one-dimensional graphene nanoribbons or zero-dimensional nanographenes, properties of graphene change. Two factors determine the electronic structure of nanographenes: edge effect and quantum confinement. These two factors are not necessarily independent. The edge effect manifests itself most obviously in zigzag-edged nanographenes. We discuss graphene nanoribbons and nanographenes in Section 18.3.1 and Section 18.3.2 respectively.

18.3.1 Graphene Nanoribbons

Edge effects become significant when a graphene layer is shrunk to nanometer scale for electronic devices such as transistors and logic gates [49]. Zigzag edges are more interesting to us because they lead to edge states, while armchair edges behave more 'normally'. We discuss zigzag edges first, and then armchair edges.

18.3.1.1 Zigzag-edged Graphene Nanoribbons

The properties of zigzag edges have been reported as earlier as in 1993 when two theoretical papers predicted localized electronic states at the zigzag edges [20, 21]. Later, detailed theoretical studies predicted an antiferromagnetic (AFM) ground state for zigzag-edged graphene nanoribbons (ZGNR) [19, 23]. Recently, unique physical and chemical properties of ZGNRs from first principles density functional theory (DFT) calculations have been reported [40, 50]. In the meantime, the localized electronic states predicted for ZGNRs were confirmed by scanning tunneling microscopy and spectroscopy [37, 51], but the AFM phase for a single ZGNR predicted by theory remains to be confirmed. In reality, no ZGNR has been made to the best of our knowledge. Rectangular nanographenes, as finite-sized ZGNRs, should serve as an ideal target for organic synthesis (see Section 18.3.2).

18.3.1.1.1 ELECTRONIC STRUCTURE

The early calculations of ZGNR's electronic structure from the tight-binding method produced an interesting fact that valence and conduction bands become flat and degenerate for part of the one-dimensional Brillourin Zone [19, 23]. This degenerate flat band is absent in armchair-edged graphene nanoribbons (AGNRs) and contrasts with graphene whose valence and conduction bands intersect at only one point. In the density of states (DOS) plot, this flat region in the band structure results in a sharp peak at the Fermi level. Analytical solutions of approximate wavefunctions of π electrons in ZGNRs shows that the degenerate flat band corresponds to π electrons localized at the zigzag edges, leading to so-called 'edge

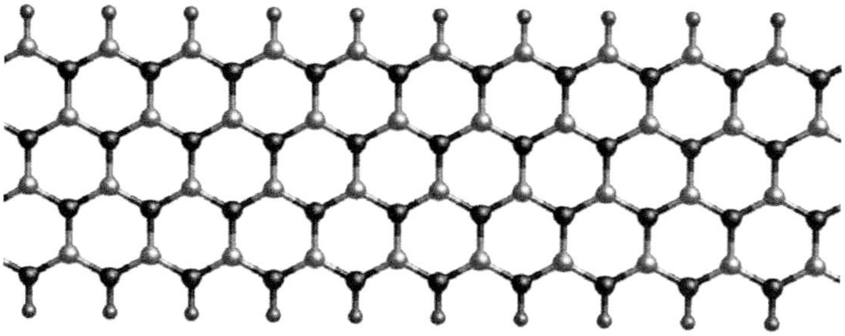

Figure 18.5 The bipartite lattice (A and B sublattices in dark and light gray, respectively) of a zigzag-edged graphene nanoribbon, inherited from the honeycomb lattice of graphene which consists of two interpenetrating hexagonal lattices. Small gray balls are terminal H atoms

states' [19, 23]. Fujita *et al.* [23] also examined possible spin ordering in ZGNRs by using the Hubbard model within the UHF approximation. They found that for even a small on-site Coulomb repulsion, the antiferromagnetic (AFM) state becomes the ground state, causing spin up and spin down electrons to localize at opposite edges. They attributed this behavior to the bipartite lattice of ZGNRs (Figure 18.5) and electron-electron interaction. Both the edge states and their associated spin ordering have profound consequences for their electronic and chemical properties.

These pioneering works in ZGNRs were followed by many theoretical studies. Table 18.1 summarizes some first principles DFT results for magnetic phases of ZGNRs with different widths [50]. The AFM phase is indeed predicted to be the ground state and the edge carbon atom has a local magnetic moment about 0.14 μ_B. Figure 18.6 plots the local density of states (LDOS) for the nonmagnetic (NM), ferromagnetic (FM) [52], and AFM magnetic phases of 4-ZGNR. For the NM phase, the Fermi level bisects the sharp peak, which leads to instability and is subject to Stoner magnetism [53]. Consequently, the two spin states shift in the opposite direction relative to the Fermi level. The finite DOS at the Fermi level for the FM phase indicates that the phase is metallic, whereas a small gap opens up for the AFM phase, indicating that the phase is semiconducting.

18.3.1.1.2 HALF-METALLICITY

A recent exciting finding in the computational studies of ZGNRs is that when an in-plane external electric field (perpendicular to the main axis of the ribbon) is applied, the spin up

Table 18.1 Relative energies of different magnetic phases for zigzag-edged graphene nanoribbons, and local magnetic moment on the edge carbon of the antiferromagnetic phases ($M_{edge\text{-}AFM}$)[a]

Width (N)	E_{NM} (meV/cell)[a]	E_{FM} (meV/cell)[a]	E_{AFM} (meV/cell)[b]	$M_{edge\text{-}AFM}$ (μ_B)
4	0	−37	−51	0.139
5	0	−50	−62	0.143
6	0	−60	−72	0.145

[a]Reprinted table with permission from: D. E. Jiang, B. G. Sumpter, S. Dai, *J. Chem. Phys.*, **126**, 134701 (2007). Copyright (2007) by the American Institute of Physics.
[b]The unit cells are shown in Figure 18.3.

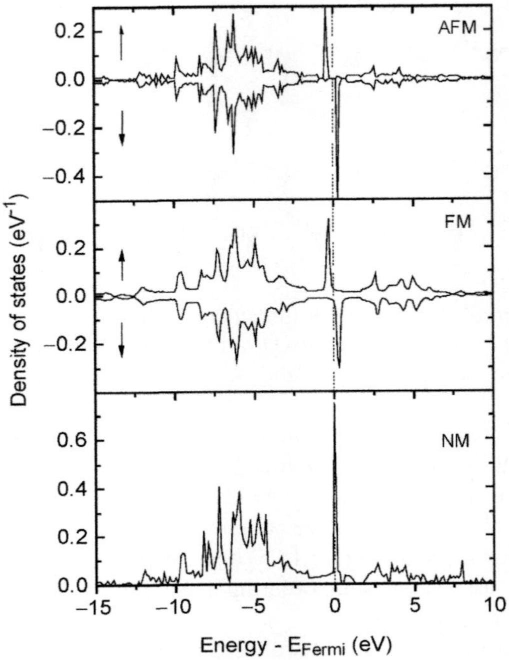

Figure 18.6 *Local density of states plots for the p-orbital of an edge carbon atom in the nonmagnetic (NM), ferromagnetic (FM), and antiferromagnetic (AFM) phases of a zigzag-edged graphene nanoribbon (N=4). Arrows indicate the direction of spin polarization. Reprinted figure with permission from: D. E. Jiang, B. G. Sumpter, S. Dai, J. Chem. Phys.,* ***126****, 134701 (2007). Copyright (2007) by the American Institute of Physics*

and spin down electrons in the AFM phase (Figure 18.6) can respond oppositely to the field; when certain field strength is reached, one spin channel becomes metallic while the other becomes insulating, resulting in half metallicity [40]. This study employed DFT calculations in the local density approximation (LDA) for electron exchange and correlation, respectively. However, Rudberg *et al.* [54] showed that when a hybrid functional is used, electric-field induced half metallicity disappears for a finite-sized ZGNR. They found that although one spin channel indeed decreases its band gap with the field strength initially, the band gap for this spin reaches a minimum and then increases until the ribbon becomes nonmagnetic. The hybrid functional results showed that ZGNRs are half-semiconductors within a window of electric field strength. Later, a hybrid functional study [55] of infinitely long ZGNRs showed that ZGNRs can still be a half metal at certain field strength, indicating that Rudberg *et al.*'s conclusion is probably due to the finite-size effect.

Motivated by these findings, Dalosto and Levine employed generalized-gradient approximation (GGA) method and studied the influence of the electric field produced by a polar ad-molecule, namely $NH_3(CH)_6CO_2$ and $NH_3(CH)_{10}CO_2$, to the band gaps of ZNGRs, and found that the presence of polar ad-molecules affects the band gap in the same way external electric field does [56]. More interestingly, Kan *et al.* showed that half metallicity can be realized in ZGNRs by tuning the terminal groups at the two zigzag edges, forgoing an electric field, for example, by terminating a ZGNR by NO_2 groups at one edge and by CH_3 groups on the other side [57]. The half-metallicity of graphene nanoribbons opens up an opportunity for spintronics devices.

18.3.1.1.3 CHEMICAL REACTIVITY

Spin-polarized π-electrons are predicted to be localized on the zigzag carbon atoms of ZGNRs, which from a chemical view, prompts us to think of them as a 'partial radical'. That is, these ZGNRs have unpaired π-electrons distributed mainly on the two edges, but on average each edge carbon atom has only ~0.14 unpaired electrons (estimated from the local magnetic moment on the edge carbon; see Table 18.1). Due to the partial radical character, these edge carbon atoms should offer special chemical reactivity, comparing with non-edge zigzag-ribbon carbon atoms and armchair carbon atoms, which show little or no radical character.

Jiang *et al.* [50, 58] examined the reaction of the zigzag edge with common radicals. The bond dissociation energy (BDE) [59] is calculated for the newly formed C-H bond in which C is an edge carbon. Comparing with other $C(sp^3)$-H BDEs (4.553 eV for C_2H_5-H and 4.315 eV for cyclo-C_6H_{11}-H) [60], the C-H bond formed at the zigzag edge has a strength of ~60% of the C-H bond between a molecular carbon radical and H (Table 18.2). This observation indicates that a 'partial radical' concept is useful to characterize the chemical reactivity of the zigzag edge. Of course, a question arises accordingly: why the edge carbon has a partial charge of ~0.14 e but the edge C-H bond has a strength of ~60% that of a common C-H bond? This question brings up another aspect of the chemical reactivity at the zigzag edge. Although the localized state at the zigzag edge offers only a partial amount of the π-electron density on a per edge carbon basis, these partial electrons are not confined to those edge carbons, but can act collectively when interacting with another radical. So here the 'localization' (or 'localized' state) at the edge sites is meant to be with respect to the inner sites.

In addition to hydrogen, Jiang *et al.* [50] also examined other common radicals, the BDEs are also listed in Table 18.2. Like the C-H bond, the C-OH and C-CH_3 bonds formed at the edge have a BDE of 50–70% of C_2H_5-OH and C_2H_5-CH_3 bonds. For the halogens, the BDE of edge-X decreases from F to I and follows the same trend as that of C_2H_5-X. This trend can be attributed to the decreasing electronegativity from F to I. Being the most electronegative, F has the greatest BDE. Moreover, the BDE ratio of edge-X to C_2H_5-X increases dramatically from Cl to F, indicating the extraordinary ability of F to pull electrons from the GNR's zigzag edge.

Table 18.2 *Comparison of bond dissociation energy (BDE, in eV) of zigzag edge-X bonds with experimental BDE of C_2H_5-X*[a,b]

Radical: X	H	OH	CH_3	F	Cl	Br	I
BDE (Edge-X)	2.86	2.76	2.22	3.71	2.18	1.65	1.18
BDE (C_2H_5-X)[c]	4.358	4.055	3.838	4.904	3.651	3.036	2.420

[a]Reprinted table with permission from: D. E. Jiang, B. G. Sumpter, S. Dai, *J. Chem. Phys.*, **126**, 134701 (2007). Copyright (2007) by the American Institute of Physics.
[b]The coverage is at 1/6 X/edge-C (see Figure 18.7); zero-point-energy corrections not included.
[c]Experimental values, from Ref. [59].

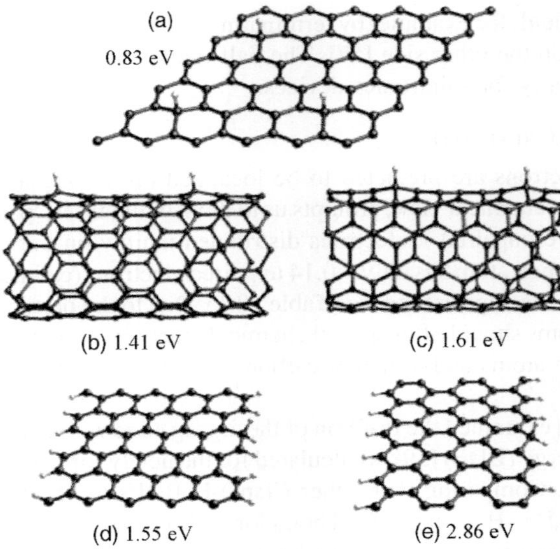

Figure 18.7 *Reaction of a hydrogen atom with (a) a graphene sheet, (b) (10,0) carbon nanotube, (c) (5,5) carbon nanotube, (d) a graphene nanoribbon's armchair edge, and (e) a graphene nanoribbon's zigzag edge. Numbers give C-H bond dissociation energy. Reprinted figure with permission from: D. E. Jiang, B. G. Sumpter, S. Dai, J. Chem. Phys.,* ***126****, 134701 (2007). Copyright (2007) by the American Institute of Physics*

The above discussion has demonstrated the 'partial radical' nature of the GNR's zigzag edges. Jiang *et al.* [50] also examined how these zigzag edge carbon atoms differ from other sp^2-carbon atoms by calculating the BDEs for the C–H bonds formed on a graphene sheet, a metallic nanotube, a semiconducting nanotube, and a graphene nanoribbon with armchair edges (Figure 18.7 displays the optimized structures and BDEs).The graphene sheet has the lowest reactivity toward a hydrogen radical, whereas the zigzag edges have the highest reactivity, and the nanotubes and the armchair edge are in between.

A single graphene sheet is a zero-gap semiconductor (see LDOS in Figure 18.8), and due to its stable π-electron system and flat structure, the interaction between a graphene sheet and an isolated radical ($S = 1/2$) [61] has been demonstrated to be weak (usually, BDE < 1 eV) [58, 62]. The (5,5) carbon nanotube is metallic with low DOS at the Fermi level, whereas the (10,0) carbon nanotube is a small-gap semiconductor (Figure 18.8). The two tubes' C–H BDEs are 0.6–0.8 eV higher than that of graphene, mainly due to the tubes' curvature. Our C-H BDEs for the (10,0) and (5,5) tubes agree very well with a previous periodic DFT-GGA study using all-electron Gaussian basis sets [63]. The slightly greater BDE for the (5,5) tube compared to the (10,0) tube may result from its metallic character. The band gap in the armchair-edged GNR (AGNR) depends on the ribbon width [64]. The AGNR (Figure 18.6d), like the semiconducting (10,0) nanotube, has a small band gap (Figure 18.8), and their C–H BDEs are also similar. The zigzag-edged GNR has a significantly higher C–H BDE than the armchair-edged GNR, nanotubes, or a graphene sheet. The ZGNR's unique electronic structure, as clearly shown in their LDOS (Figure 18.6), has a substantial peak near the Fermi level, which directly leads to its stronger bonding to hydrogen.

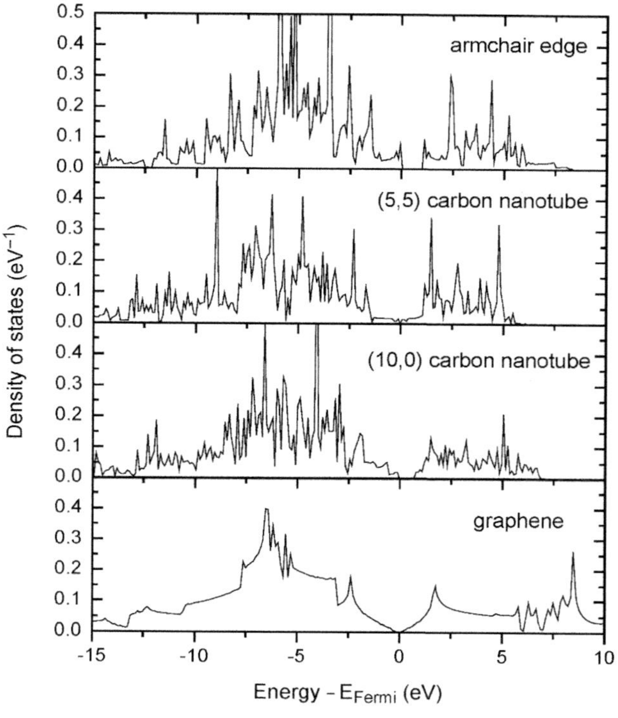

Figure 18.8 *Local density of states plots for the p-orbital of an carbon atom in a graphene sheet, (10,0)carbon nanotube, (5,5)carbon nanotube, and an armchair edge. Reprinted figure with permission from: D. E. Jiang, B. G. Sumpter, S. Dai, J. Chem. Phys.,* ***126****, 134701 (2007). Copyright (2007) by the American Institute of Physics*

18.3.1.2 Armchair-edged Graphene Nanoribbons

Armchair edge is more stable than the zigzag edge for graphene nanoribbons, and the π electrons tend to form localized double bonds on the armchair edge. In their pioneering study of graphene nanoribbons, Nakada *et al.* [19] also examined AGNRs with the tight-binding method. For an AGNR defined in Figure 18.9 with a width index N, they found that the ribbon is metallic for $N = 3M-1$ (M is positive integer) and nonmetallic for other N. For the metallic AGNRs, the valence and conduction bands touch at one point, similar to that of graphene. For nonmetallic AGNRs, the direct gap at the Γ point decreases with N, approaching the behavior of graphene. No magnetic phase was found for AGNRs. Later, DFT-LDA calculations [65] showed that the band gap in AGNRs is nonzero for all N and follows the order of $3M-1 < 3M < 3M+1$. For the three families of AGNRs, an empirical relation of band gap versus M has been proposed to fit the computed data. GW calculations showed significant self-energy corrections for AGNRs (and also for ZGNRs), which increases the DFT-LDA band gaps by 0.5–3.0 eV for ribbons of width 2.4–0.4 nm [65].

The most exciting recent development in AGNR research is the experimental realization of an AGNR about 10 nm in length [66]. Müllen and coworkers used Suzuki-Miyaura polymerization and intramolecular Scholl reaction and synthesized polymers of 9-AGNR of

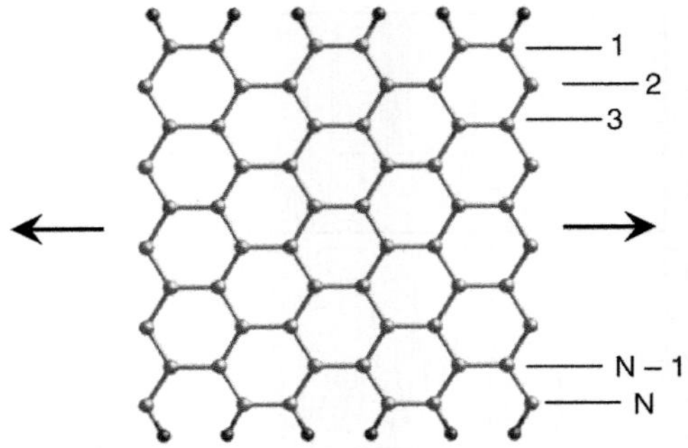

Figure 18.9 N-AGNR: Armchair-edged graphene nanoribbon with a width index N. Carbon, light gray; H, dark gray. Arrows indicate the repeating directions of the ribbon. Reprinted figure with permission from: K. Nakada, M. Fujita, G. Dresselhaus, M. S. Dresselhaus, Phys. Rev. B, **54**, 17954 (1996). Copyright (1996) by the American Physics Society

8–12 nm in length (**3**). **3** is found to dissolve in common organic solvents such as THF. The UV-VIS absorption of **3** in THF shows a maximum at 485 nm. Drop casting of **3** in THF on a silica substrate yielded rod-shaped microcrystallines of 100 nm in diameter and 5 μm in length. TEM revealed π stacking of the ribbons in the rods. **3** has great potential in device applications.

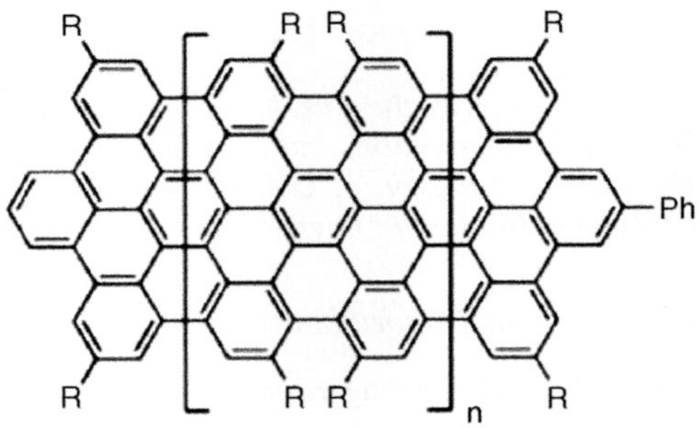

up to 12 nm

3

The chemical reactivity of AGNRs' π-electrons at edges is expected to be between those of a double bond and a benzene ring. This estimate is based on two considerations: (1) armchair edges can conveniently adopt a double bond configuration when necessary; (2) armchair edges facilitate formation of more Clar's sextets. Therefore, chemistry of a C–C double bond or a benzene ring can be translated to armchair edges, which we will not discuss further here.

18.3.2 Nanographenes

18.3.2.1 Nanographenes with Two Symmetric Zigzag Sides

From discussions of electronic structure of ZGNRs above, we know that the bipartite lattice of ZGNRs leads to an AFM ground state with one edge spin up and the other spin down. For

Table 18.3 Relative energies (in meV) of antiferromagnetic (AFM), ferromagnetic (FM), and nonmagnetic (NM) phases of PAHs[a]

PAH	[3,3]	[3,5]	[3,7]	[3,9]	[4,3]	[5,3]	[5,5]	[7,5]	[9,5]
$E_{FM}-E_{NM}$	0	102	−88	−173	43	−202	−277	−260	−502
$E_{AFM}-E_{NM}$	0	−18	−113	−180	−52	−230	−290	−349	−548

[a]Reprinted table with permission from: D. E. Jiang, B. G. Sumpter, S. Dai, *J. Chem. Phys.*, **127**, 124703 (2007). Copyright (2007) by the American Institute of Physics.

a nanographene with such two symmetric zigzag edges, there ought to be a critical size beyond which a closed-shell NM ground state yields to an AFM open-shell ground state. We discuss this critical size in three related nanographenes: rectangular PAHs, acenes, and cyclacenes (and short zigzag nanotubes).

18.3.2.1.1 RECTANGULAR PAHS

Using DFT-GGA, Jiang *et al.* [67] examined the energetics of three magnetic phases (AFM, FM, and NM) for different sizes of PAHs as defined in Figure 18.3. Table 18.3 displays the data for PAHs [3,3] to [9,5]. PAH [3,3] (chemical name, bisanthene) does not have a magnetic phase; initial guesses for AFM and FM phases both converged to the NM phase. This is also the case for PAH [2,3] (perylene) and PAH [4,1] (tetracene) (results not shown in Table 18.3). As PAH [X,Y] increases either side of the rectangle greater than 3, a stable AFM phase appears for all PAHs considered here. So does an FM phase. The AFM phase is found to be the most stable for PAHs larger than [3,3]. The FM phase is energetically in between the AFM and NM phases for most larger PAHs, except for [4,3] and [3,5] whose NM phase is more stable than the FM phase. The closeness in energy between AFM and NM phases for PAHs [4,3] and [3,5] indicates that they are in a transition between a nonmagnetic PAH [3,3] and larger PAHs that have more stable AFM phases and metastable FM phases. A good indication of how PAH [X,Y] approaches infinitely long zigzag ribbons in terms of stability is to compare the energetic difference between the AFM and NM phases (normalized to the dimension of the zigzag edge, X) of a PAH with that of an infinitely long ribbon of the same width (Y). Figure 18.10 shows that the normalized energetic difference enlarges greatly from PAH [3,5] to PAH [5,5]. The change is relatively small from PAH [5,5] to PAH [9,5] and the magnitude is only slightly smaller than that of the infinite ribbon.

Although PAH [4,3] has an energetically unfavorable FM phase and a slightly more stable AFM phase, the distribution of its spin densities is prototypical for rectangular PAHs considered in the present work. Figure 18.11 plots the isosurfaces of spin density magnetization ($\rho_\uparrow - \rho_\downarrow$) for the AFM and FM phases of PAH [4,3]. The magnetization mainly localizes at the periphery of the PAH structure and concentrates at the middle of the two zigzag edges. In the AFM phase, the two zigzag edges have opposite spins. Another feature is that some minor magnetization also appears at the armchair edges.

Recent theoretical studies have predicted many fascinating properties associated with ZGNRs, such as half-metallicity [40]. However, these predictions are mainly based on infinitely long ribbons. Before experimental verification of these predictions, ZGNRs that have reasonable length and similar magnetic and electronic properties need to be synthesized. Our results indicate that PAHs such as [5,5] preserve the properties of infinitely long ribbons of the same width. Therefore, these molecules should be good candidates for testing

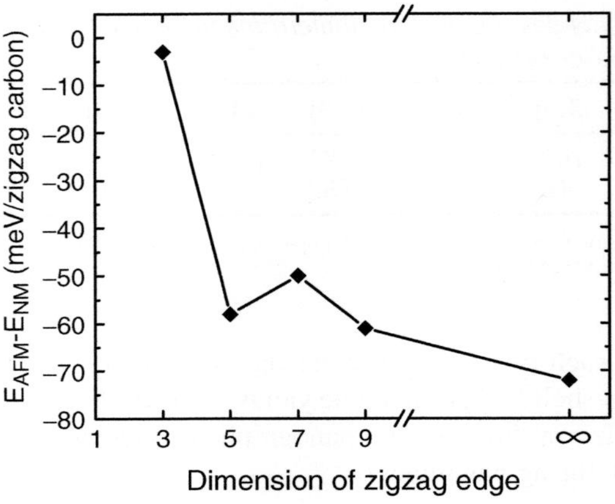

Figure 18.10 *Change of normalized energetic difference between the AFM and NM phases of PAH [X,5] with the dimension of the zigzag edge, X. Corresponding value for an infinite long ribbon is also plotted. Reprinted figure with permission from: D. E. Jiang, B. G. Sumpter, S. Dai, J. Chem. Phys.,* ***127****, 124703 (2007). Copyright (2007) by the American Institute of Physics*

predictions such as half-metallicity. Although synthetic chemistry of PAHs have made tremendous progress, synthesis of PAHs such as [4,3] remains a challenge due to their instability and low solubility in common solvents [43, 44].

The critical size in the rectangular nanographenes has also been examined by several other groups [68, 69]. Gao *et al.* [68] examined rectangular-shaped PAHs with slightly different armchair sides and a different naming system, as shown in Figure 18.12a. They systematically examined the HOMO-LUMO gap for the closed-shell singlet (that is, the NM state) as a function of zigzag and armchair dimensions, n_a and n_z. The energy gap decreases with both n_a and n_z, indicating a decreasing kinetic stability as the total number of benzenoid rings contained in carbon nanographene (CNG) increases (see Figure 18.12b). The inverse dependence of gap on n_z, however, is remarkably stronger than that on n_a. CNGs [*n*,1] and [1,*n*] are known as polyphenanthrene and acene, whose electronic structures and stabilities have stimulated great research interest [70–73]. The gap of CNG [1,*n*] reduces from 6.8 to below 1.0 eV rapidly as *n* increases from 1 to 10; contrarily, CNG [*n*,1] has a much slower

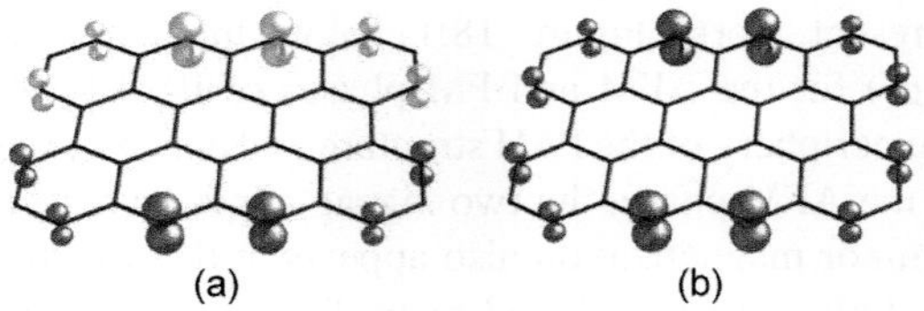

Figure 18.11 *Isosurfaces of spin density magnetization for the AFM (a) and FM (b) phases of PAH [4,3]. Dark and light isosurfaces are 0.075 and −0.075 e/Å^3, respectively. Only C–C bonds are shown (C–H bonds not shown). Reprinted figure with permission from: D. E. Jiang, B. G. Sumpter, S. Dai, J. Chem. Phys.,* ***127****, 124703 (2007). Copyright (2007) by the American Institute of Physics*

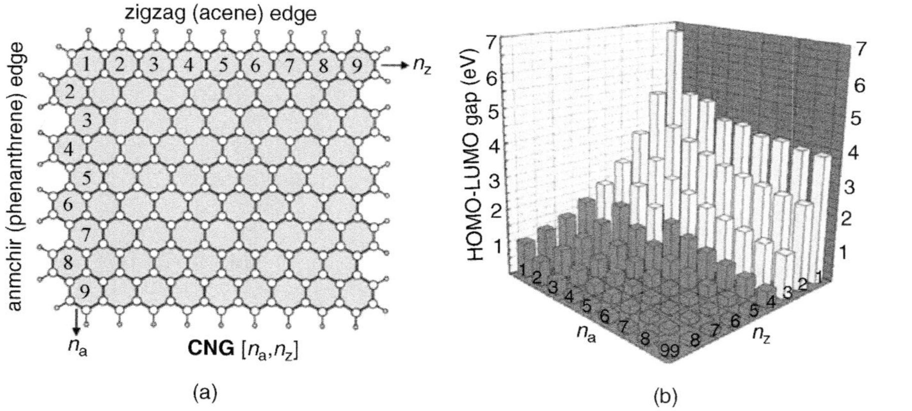

Figure 18.12 *(a) Rectangular shaped carbon nanographene (CNG) with zigzag and armchair edges; (b) HOMO-LUMO gaps of CNGs at the lowest-energy closed-shell singlet states (RB3LYP/6-31G*//RB3LYP/6-31G*) (The species with closed-shell singlet ground states and open-shell singlet ground states are represented by white and dark gray bars, respectively). Reprinted figure with permission from: X. F. Gao, Z. Zhou, Y. L. Zhao, S. Nagase, S. B. Zhang, Z. F. Chen, J. Phys. Chem. C,* ***112****, 12677 (2008). Copyright (2008) by the American Chemical Society*

decreasing tendency. We discuss electronic structure of acenes more closely in the next section (Section 18.3.2.1.2).

The small HOMO-LUMO gap of a large CNG hints that a single-determinant wavefunction might no longer appropriately describe the ground state of the CNG and that other types of wavefunctions might have lower energies. Figure 18.12b illustrates whether the restricted Kohn-Sham (RKS) solution is the ground state for a CNG: all CNGs with $n_z \leq 3$ have closed-shell singlet ground states as shown as the white bars; however, all those with $n_z \geq 6$ have open-shell singlet ground states as shown as the dark bars; the ground states of those with $n_z = 4$, 5 and 6 are determined by the values of both n_a and n_a. Figure 18.13 plots the normalized energy difference of E_{cs} and E_{os} (energy difference divided by the number of benzene rings) which shows ($E_{CS}-E_{OS}$) increases as the size of CNG increases. These results suggest that, among the computationally economic single-determinant wavefunction methods, using an unrestricted broken spin-symmetry (UBS) ansatz is essential to properly calculate the ground state properties for a large CNG.

18.3.2.1.2 ACENES

Acenes (**4**), as a chain of linearly fused benzene rings, are a special class of rectangular PAHs with two symmetric zigzag edges and occupy a peculiar position in PAH research. On the one hand, pentacene is widely used in molecular electronics; on the other hand, octacene or higher acenes have not been synthesized. In between, hexacene and heptacene compounds have been successfully synthesized and characterized by single-crystal X-ray diffraction recently [46, 47]. We show that this current situation indeed has to do with the two zigzag edges of acenes and the critical size.

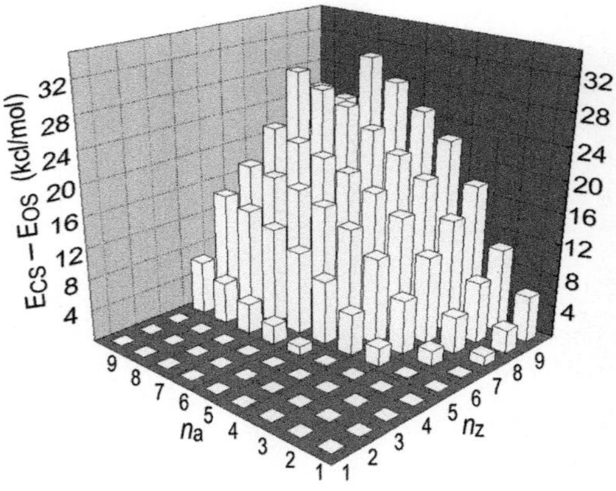

Figure 18.13 *The normalized energy difference between closed-shell singlet and open-shell singlet states of CNGs. Reprinted figure with permission from: X. F. Gao, Z. Zhou, Y. L. Zhao, S. Nagase, S. B. Zhang, Z. F. Chen, J. Phys. Chem. C,* ***112****, 12677 (2008). Copyright (2008) by the American Chemical Society*

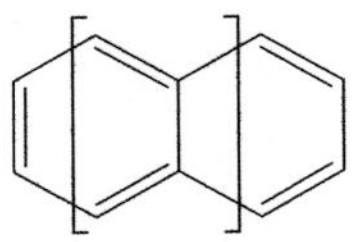

4

Several theoretical studies have attempted to understand the electronic structures of higher acenes and explain their high reactivity. Houk and co-workers predicted a triplet ground state for acenes with n (the number of fused benzene rings) >8 [70], using unrestricted B3LYP (UB3LYP) for the triplet but restricted B3LYP for the singlet (both with 6-31G* basis). Later, Bendikov *et al.* showed that the ground state is an open-shell singlet for n > 6, by using UB3LYP (with 6-31G* basis) also for the singlet, and claimed that this open-shell singlet represents a diradical [74]. Because a diradical is a molecule with two unpaired electrons [75], their claim of an open-shell singlet diradical implies that there are one up and one down spins in acenes with n > 6. In a recent study, dos Santos [76] reported a higher spin ground state, a quintet, for n = 20–23, also based on UB3LYP/6-31G*. More recently, Chan and coworkers used a density matrix renormalization group (DMRG) algorithm and studied acenes with n = 2–12. They found that the ground states for longer acenes are polyradical singlets [77].

Jiang and Dai studied the electronic structure of higher acenes for n up to 40 with the spin-polarized DFT-GGA method [73]. Figure 18.14 shows how the relative energetics changes with the size of the oligomer from 5 to 20: the AFM state emerges as a ground state when n is greater than 7. At n = 7, the AFM and NM phases are very close in energy. When n is smaller than 7, the NM phase is the ground state and the AFM state is unstable and relaxed to the NM state. The AFM state has high chemical reactivity because unpaired electrons are piled up at the edges and yield partial radical characters at the edge carbon atoms [67]. Therefore, the appearance of the AFM phase can be used an indicator of acenes' chemical reactivity.

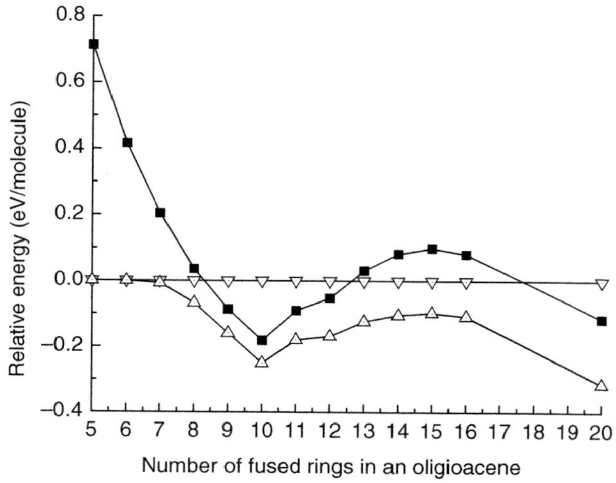

Figure 18.14 *Energetic differences between the AFM (open-shell singlet), FM (triplet), and NM (closed-shell singlet) states of an oligoacene vs. the number of fused rings in the oligoacene. Reprinted figure with permission from: Jiang, D. E., Dai, S. J. Phys. Chem. A,* ***112****, 332 (2008). Copyright (2008) by the American Chemical Society*

The heptacene-containing single crystals have been obtained only recently [46, 47], while the parent unsubstituted heptacene was synthesized by photochemical routes in a poly (methacrylate) matrix very recently [78, 79]. The acenes with more than 7 fused rings have not been synthesized. All these facts support the use of the AFM state as a reactivity indicator.

The number of unpaired spin-up electrons for several n-acenes by integrating their spin density magnetization is shown in Table 18.4. The total number of unpaired spin-up electrons increase with the size of the acene and is greater than one even for n = 10, thus, the diradical description is inaccurate for higher acenes. This result is in agreement with the

Table 18.4 *Sum of spin-up moments of spin density magnetization ($\rho_\uparrow - \rho_\downarrow$) ($M_{up\text{-}AFM}$) and averaged magnetic moment per benzene ring ($M_{edge\text{-}AFM}$) for the antiferromagnetic phase of polyacene and n-acenes*[a]

System	$M_{up\text{-}AFM}$ (μ_B)[b]	$M_{edge\text{-}AFM}$ (μ_B)
Polyacene	0.226	0.226
7-acene	0.96	0.137
10-acene	2.030	0.203
15-acene	2.839	0.189
20-acene	4.103	0.205
40-acene	8.315	0.208

[a]Reprinted table with permission from: Jiang, D. E., Dai, S., *J. Phys. Chem. A*, **112**, 332 (2008). Copyright (2008) by the American Chemical Society.
[b]Per unit cell for polyacene and per molecule for acenes.

density-matrix renormalization group (DMRG) study by Chan and coworkers [77]. In addition, the average spin-up moments per benzene ring slowly increase with n from 10 to 20 to 40, and approach that of an infinite polyacene. Therefore, our results provide a consistent picture of how finite-sized n-acene approaches infinite polyacene with n.

18.3.2.1.3 CIRCUMACENES

PAH [X,3] is also called periacenes. Circumacenes and periacenes as zigzag-edged nanographenes differ only in that circumacenes have one extra benzene ring on each of the two armchair sides. Using DFT-GGA, Jiang and Dai showed that this slight difference in the boundary shape dramatically affects the critical size at which the open-shell AFM state supersedes the closed-shell NM state as the ground state [80]. Structures of circumacenes (CAs) and periacenes (PAs) are compared in Figure 18.15. For the infinite limit ($n \to \infty$), both n-CAs and n-PAs should converge to the zigzag-edged graphene nanoribbons, which have been shown to have an AFM ground state. At the other end of the limit ($n = 1$), 1-CA (coronene) and 1-PA (perylene) are closed-shell molecules with large HOMO-LUMO gaps.

The appearance of the AFM state with size can be attributed to the decreased HOMO-LUMO gap for the closed-shell state. As the HOMO-LUMO gap approaches zero, the near-degeneracy at the Fermi level coupled with electron-electron interaction renders the AFM state more stable [23]. Figure 18.16 shows how the HOMO-LUMO gap changes for the closed-shell state of PAs and CAs with the molecular size. The gap for PAs drops more quickly with n from 1.87 eV for 1-PA to 0.08 eV for 4-PA, while the gap for CAs decreases rather smoothly from 2.86 eV for 1-CA to 0.22 eV for 7-CA.

Because 3-PA has an AFM ground state and is the critical size for PAs, we can use its HOMO-LUMO gap as a criterion for the appearance of the AFM ground state. According to 3-PA's 0.35-eV HOMO-LUMO gap, we judge that 6-CA should be the critical size for CAs (see the dashed horizontal line in Figure 18.16). This simple determination from the gap criterion is supported by the computed energetic differences between the AFM state and

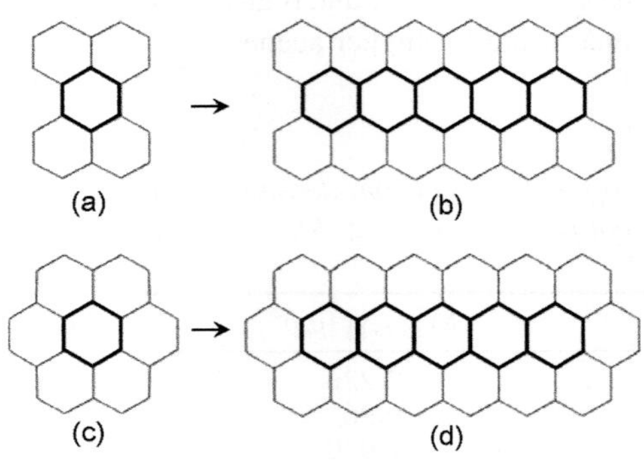

Figure 18.15 *Structures of n-periacenes (n-PAs) and n-circumacenes (n-CAs): from 1-PA (a) to 5-PA (b), and from 1-CA (c) to 5-CA (d). Here n is the number of benzene rings in the middle (highlighted in thicker lines). All carbons are in sp^2 hybridization, and C-H bonds are not shown. Reprinted figure with permission from: Chem. Phys. Lett., **466**, 72 (2008). Copyright (2008) by Elsevier (including Academic Press)*

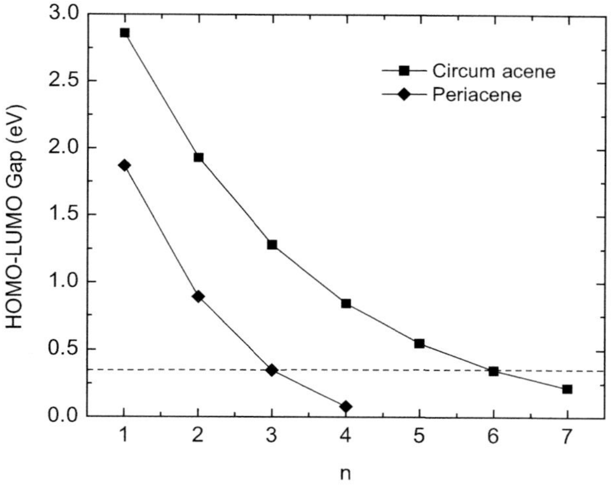

Figure 18.16 HOMO-LUMO gap versus n for the closed-shell nonmagnetic state of n-circumacenes and n-periacenes. Reprinted figure with permission from: Chem. Phys. Lett., **466**, 72 (2008). Copyright (2008) by Elsevier (including Academic Press)

the closed-shell nonmagnetic (NM) state (Table 18.5): while 5-CA has a NM ground state, 6-CA has an AFM ground state and is indeed the critical size for CAs.

From n-PA to n-CA, the increased HOMO-LUMO gap indicates increased stability. This increased stability can be explained by Clar's sextet rule; namely, stability increases with the number of aromatic sextets [42]. The extra two rings on the sides of CAs effectively increase the number of sextets by two (for example, see sextets in 3-CA and 3-PA in Figure 18.17), thereby rendering the corresponding CAs more stable. However, when n is large enough (>5), the effect of side boundaries becomes less important and the zigzag edges dominate the electronic structure, leading to an AFM ground state.

The critical size of CAs is the rather large 6-CA. Therefore, n-CAs with $n < 6$ have a closed-shell ground state which may make them more likely to be synthesized. Although 1-CA (coronene) and 2-CA (ovalene) are well-known stable compounds, 3-CA's synthesis has been reported only by Diederich's group [81], and 4-CA and 5-CA's syntheses have not been achieved. Our results indicate that both 4-CA and 5-CA have a closed-shell ground

Table 18.5 Energetic difference between the open-shell antiferromagnetic (AFM) and closed-shell nonmagnetic (NM) states of n-circumacene and the largest local magnetic moment on the edge carbon atoms (M_{edge}) of the AFM state[a]

n	5	6	7
$E_{AFM}-E_{NM}$ (meV)	0^b	–34	–105
M_{edge} (μ_B)	0^b	0.12	0.14

[a]Reprinted figure with permission from: *Chem. Phys. Lett.*, **466**, 72 (2008). Copyright (2008) by Elsevier (including Academic Press).

[b]For $n=5$, $E_{AFM}-E_{NM}=0$ and $M_{edge}=0$ mean that the AFM state is unstable and relaxed to the NM state.

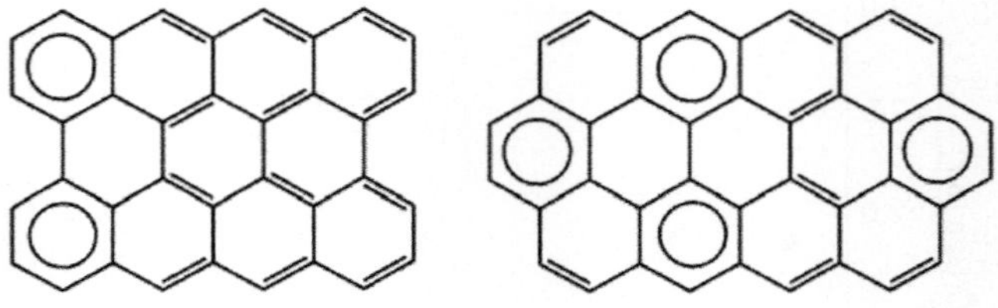

Figure 18.17 Clar's sextets (circles) in 3-periacene (a) and 3-circumacene (b). Reprinted figure with permission from: Chem. Phys. Lett., **466**, 72 (2008). Copyright (2008) by Elsevier (including Academic Press).

state which should make them potential synthesis targets. Their small HOMO-LUMO gaps may be found useful in molecular electronic applications [41].

18.3.2.1.4 CYCLACENES

Joining the two ends of acenes together, one obtains cyclacenes (**5**). Unlike their planar counterpart, no cyclacene has been synthesized. There have been several notable attempts to synthesize these 'molecular belts'; Stoddart and coworkers' efforts to synthesize [12] cyclacene (consisting of 12 fused benzene rings) realized a hydrogenated derivative, but its dehydrogenation did not yield the cyclacene [82–84]. Cory *et al.* attempted to synthesize [8] cyclacene by a stereospecific double Diels-Alder macroannulation [85, 86]. A belt of eight fused six-membered rings was obtained, but further reaction to produce [8]cyclacene did not succeed.

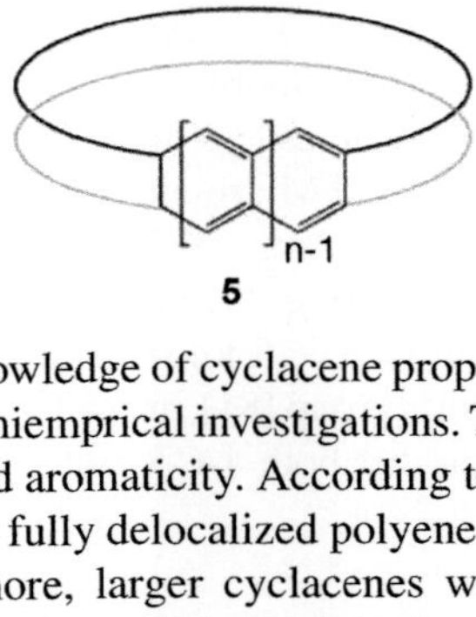

5

Consequently, the current knowledge of cyclacene properties stems from several density functional theory (DFT) and semiemprical investigations. The DFT studies [70, 87] focused on their energies, structures, and aromaticity. According to these computational investigations, cyclacenes consist of two fully delocalized polyene (trannulene) ribbons, connected by long C–C bonds. Furthermore, larger cyclacenes were predicted to possess triplet electronic ground states, but these results were based on a spin-restricted ansatz for the singlet states. By employing an UBS wavefunction, Chen *et al.* [88] demonstrated that [n] cyclacenes, where n is 6 or larger, as well as short zigzag (n,0) tubes have open-shell singlet ground states. Figure 18.18 plots the changes in energies of the triplet and the closed-shell (CS) singlet relative to the open-shell (OS) singlet with the number of fused benzene rings in [n]cyclacenes. Clearly the ground state is the OS singlet when n is greater than 5. When n is 4 or 5, the OS singlet relaxes to the CS singlet ground state. For n = 3, the D_{3h} cyclacene structure is unstable (geometry optimization gives a different structure, which is not a local minimum).

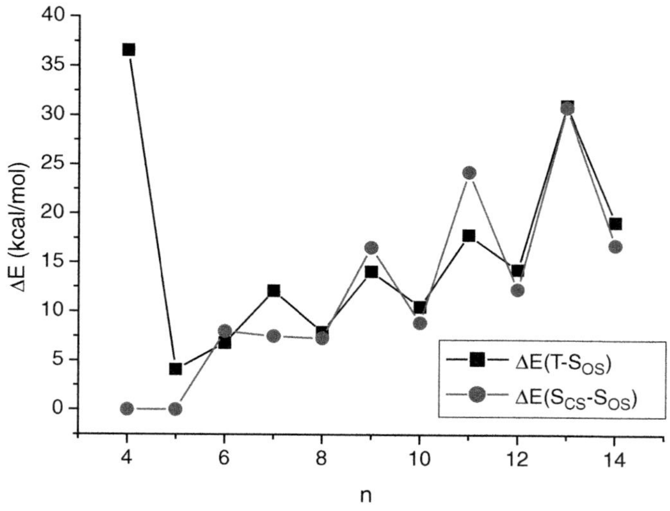

Figure 18.18 *The relative energies of triplet [$\Delta E(S_T\text{-}S_{OS})$] and restricted closed-shell (CS) singlet states [$\Delta E(E_{CS}\text{-}S_{OS})$] to the UBS open-shell (OS) singets of [n]cyclacenes at the B3LYP/6-31G* level of theory. Reprinted figure with permission from: Z. Chen, D. E. Jiang, X. Lu, H. F. Bettinger, S. Dai, P. v. R. Schleyer, K. N. Houk, Org. Lett.,* **9**, *5449 (2007). Copyright (2007) by the American Chemical Society*

Because of their radical character, the open-shell singlets should be highly reactive and unstable. Along with strain, this may help explain why previous attempts to synthesize [8] and [12] cyclacenes did not succeed. Unfortunately, smaller cyclacenes with closed-shell singlet ground state are too highly strained to be good targets for synthesis. The belt structure of [5]cyclacene (Figure 18.19) results in ca. 70 kcal/mol strain. Although [4]cyclacene also has a CS singlet ground state, its strain (ca. 120 kcal/mol) is even larger. The expected high reactivity of the radicaloid singlet ground states and the strain energies complicate cyclacene synthesis. Hopefully suitably kinetically stabilized members of this series of fascinating hydrocarbons can be achieved.

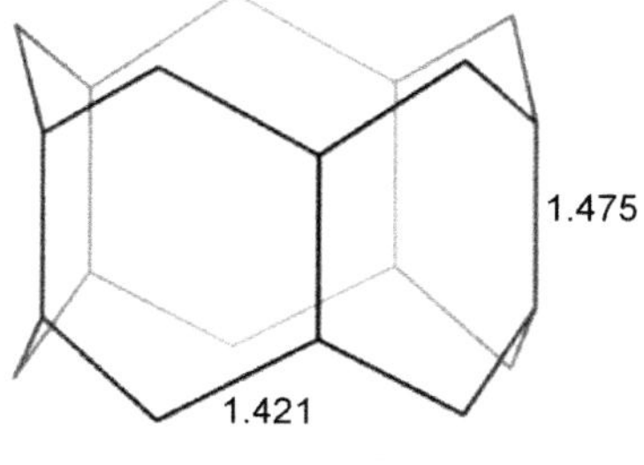

Figure 18.19 *B3LYP/6-31G* optimized geometry for [5]cyclacene. C-CH and C-C bond lengths are shown. Reprinted figure with permission from: Z. Chen, D. E. Jiang, X. Lu, H. F. Bettinger, S. Dai, P. v. R. Schleyer, K. N. Houk, Org. Lett.,* **9**, *5449 (2007). Copyright (2007) by the American Chemical Society*

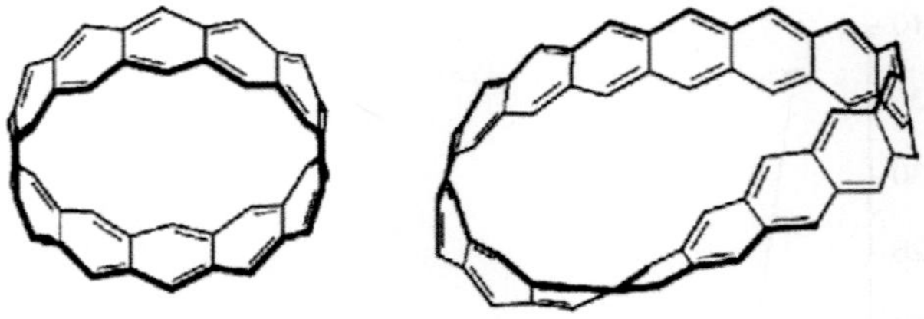

Figure 18.20 A Hückel (or regular) cyclacene (left) and its Möbius counterpart (right). Reprinted figure with permission from: R. Herges, Chem. Rev., **106**, 4820 (2006). Copyright (2006) by the American Chemical Society

18.3.2.2 Möbius Ribbons of Nanographene

From the above discussion we know that two symmetric zigzag edges will yield an AFM ground state for nanographenes beyond a certain size with one edge spin up and the other spin down. But Möbius graphene nanoribbons have only one edge topologically. So how would π-electrons respond to such a boundary shape? Using the Hubbard model with the unrestricted Hartree-Fock approximation, Wakabayashi and Harigaya [21] found that the total magnetization of the Möbius strip is zero for even N (ribbon width) and nonzero for odd N. In a later study, Harigaya *et al.* [89] found a helical spin state for the Möbius strip whose total magnetization is also zero for even N. Because odd N Möbius strips have odd number of edge carbon atoms, it is not surprising that their total magnetization is nonzero. For even N, however, one would expect that the one-edged Möbius ribbons should favor one spin over the other and therefore have a nonzero total magnetization, such as for Möbius cyclacenes ($N = 2$).

Figure 18.20 displays cyclacene in the Hückel (two separate edges) and Möbius (one edge) topologies [90, 91]. As demonstrated from DFT-GGA results for various spin states for cyclacene and Möbius cyclacene with a length of $L = 15$ (see Table 1 in Ref. [92]), the open-shell singlet is the most stable for the cyclacene, and the triplet's energy is higher than the closed-shell singlet state. However, the triplet becomes the ground state for the Möbius cyclacene, followed by the open-shell singlet. Higher spin states are less stable than the closed-shell singlet for both cyclacene and Möbius cyclacene. Figure 18.21 shows spin magnetization density for the open-shell singlet and triplet states of the Möbius cyclacene. For

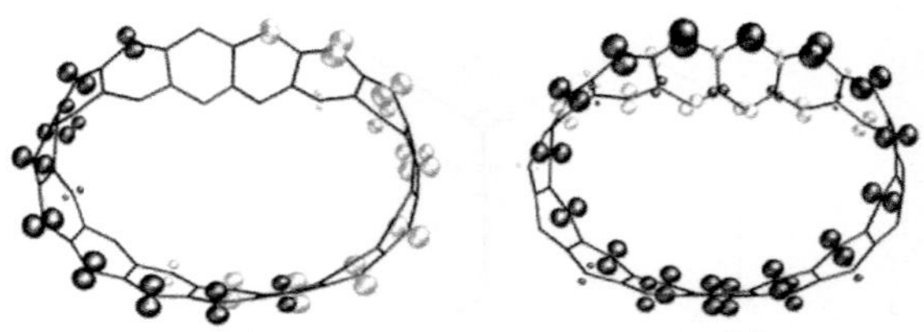

Figure 18.21 Isosurfaces of spin density magnetization ($\rho_{\uparrow} - \rho_{\downarrow}$) for (a) the AFM and (b) triplet states of Möbius [15]cyclacene. Dark and light isosurfaces are 0.05 and -0.05 e/Å^3, respectively. Reprinted figure with permission from: D. E. Jiang, S. Dai, J. Phys. Chem. C, **112**, 5348 (2008). Copyright (2008) by the American Chemical Society

the open-shell singlet, the spin up and spin down electrons are divided into two domains around the ribbon and the domain wall is near the top of the Figure 18.20a. At the bottom of the Figure 18.20a, part of the ribbon has spin up at one edge and spin down across the ribbon. For the triplet state, the spin up electrons dominate the edge and minor down spins appear at the inner edge (at the upper part of Figure 18.20b, where there is a significant twist in the ribbon).

18.3.2.3 Other-Shaped Nanographenes

Triangular zigzag-edged nanographenes have odd number of π electrons, so they are naturally metallic and magnetic as a radical [93, 94]. The most famous example is the phenalenyl radical (**6**). As the side of the triangle increases, the imbalance between the number of carbon atoms in the two sublattices ($N_z = N_a - N_b$) of the honeycomb graphene also becomes greater. It follows from Lieb's theorem that the total spin of the nanographene, S, equals to $N_z/2$ [95]. This has been confirmed by both tight-binding and DFT calculations [96].

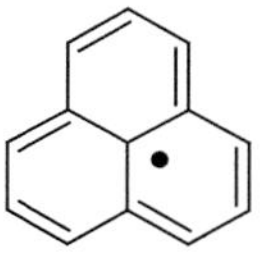

6

Parallelogram-shaped nanographenes with four zigzag edges can be constructed as in Figure 18.22. Their N_z is zero. Tight-binding calculations showed that the number of orbital levels near the Fermi level increases with dimensions of the nanographene [93]. Although spin-polarized DFT calculations have not been performed on these nanographenes, a critical size for a transition from a nonmagnetic ground state to an AFM ground state is expected, as for the rectangular nanographenes.

More than 20 years ago, by means of Hückel molecular orbital (HMO) calculations, Stein and Brown showed that large hexagonal zigzag-edged nanographenes display a peak in the DOS at the Fermi level [97]. This peak of course is due to the famous edge state. Recent tight-binding and DFT calculations of a series of hexagonal zigzag-edged nanographenes concluded that the AFM phase becomes the ground state beyond a critical size [96].

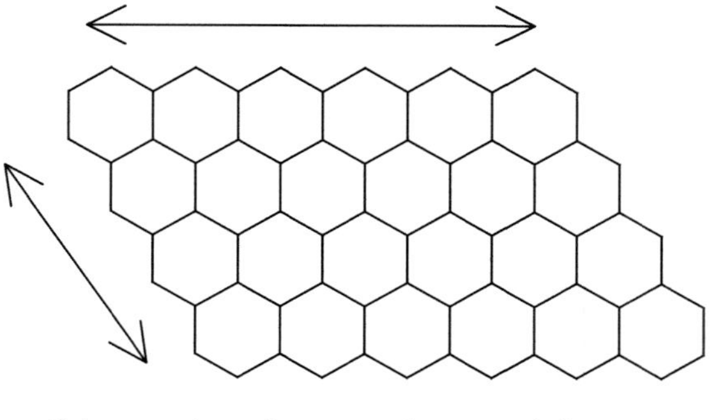

Figure 18.22 *A parallelogram-shaped nanographenes with four consecutive zigzag edges*

Armchair-edged nanographenes are more stable than their zigzag counterparts and are nonmagnetic. Their properties approach that of graphene with increasing sizes. Both HMO and tight-binding calculations of triangular and hexagonal armchair-edged nanographenes showed this trend [93, 97]. As a result, armchair-edged nanographenes have been primary synthesis targets, and many successful examples have been recently reviewed by Müllen and coworkers [41].

18.4 Outlook

Though 'the "chemistry part" of graphene story has only just begun' [98], it has shown promising abilities in conquering the bottleneck problems in graphene production, solubilization and functionalization. However, further efforts are still desired to seek productive, economic and environmentally friendly methods of making high quality graphene. Improving the efficiency of the available methods, such as chemical vapor deposition, chemical reduction of exfoliated graphite oxides and exfoliation of graphite intercalated compounds, or developing novels synthetic method would be extremely interesting. Graphene has unique structural and electronic properties that are potentially useful for electronic devices. The structural and electronic designs of graphene-based electronic devices by utilizing these properties should be highly rewarding. Since the magnetic properties of graphenes dramatically depend on the kinds of edge and their macroscopic arrangements, to controllably generate, manufacture and protect these structures would be necessary. The design of suitable chemical reaction to these ends would be challenging problems. Chemical modifictions of graphenes with appropriate addends to achieve different structures and properties are promising. The very recent achievement of hydrogenating graphenes to 'graphanes', and thus converting the highly conductive graphene from a semimetal into an insulator, is a shining example [99]. Modern computational techniques are providing deeper insights to the unusual properties of nanographenes, and playing an increasingly important role in designing graphene-based devices. We can foresee an even closer interplay between theory and experiments in the fascinating graphene research.

Acknowledgement

This work was supported in the USA by the Office of Basic Energy Sciences, Department of Energy, under Contract No. DE-AC05-00OR22725 with UT-Battelle, LLC, by NSF Grant CHE-0716718 and the Institute for Functional Nanomaterials (NSF Grant 0701525), and in Japan by by the Grand-in-Aid for Scientific Research on Priority Area and Next Generation Super Computing Project (Nanoscience Program).

References

[1] K. S. Novoselov, A. K. Geim, S. V. Morozov, D. Jiang, Y. Zhang, S. V. Dubonos, I. V. Grigorieva and A. A. Firsov, Electric Field Effect in Atomically Thin Carbon Films, *Science* **306**, 666–669 (2004).

[2] K. S. Novoselov, D. Jiang, F. Schedin, T. J. Booth, V. V. Khotkevich, S. V. Morozov and A. K. Geim, Two-Dimensional Atomic Crystals, *Proc. Natl. Acad. Sci. U. S. A.* **102**, 10451–10453 (2005).
[3] J. C. Meyer, A. K. Geim, M. I. Katsnelson, K. S. Novoselov, T. J. Booth and S. Roth, The Structure of Suspended Graphene Sheets, *Nature* **446**, 60–63 (2007).
[4] K. S. Novoselov, Z. Jiang, Y. Zhang, S. V. Morozov, H. L. Stormer, U. Zeitler, J. C. Maan, G. S. Boebinger, P. Kim and A. K. Geim, Room-Temperature Quantum Hall Effect in Graphene, *Science* **315**, 1379–11379 (2007).
[5] K. S. Novoselov, E. McCann, S. V. Morozov, V. I. Fal'ko, M. I. Katsnelson, U. Zeitler, D. Jiang, F. Schedin and A. K. Geim, Unconventional Quantum Hall Effect and Berry's Phase of 2_π in Bilayer Graphene, *Nat. Phys.* **2**, 177–180 (2006).
[6] S. V. Morozov, K. S. Novoselov, M. I. Katsnelson, F. Schedin, L. A. Ponomarenko, D. Jiang and A. K. Geim, Strong Suppression of Weak Localization in Graphene, *Phys. Rev. Lett.* **97**, 016801 (2006).
[7] M. I. Katsnelson, K. S. Novoselov and A. K. Geim, Chiral Tunnelling and the Klein Paradox in Graphene, *Nat. Phys.* **2**, 620–625 (2006).
[8] K. S. Novoselov, A. K. Geim, S. V. Morozov, D. Jiang, M. I. Katsnelson, I. V. Grigorieva, S. V. Dubonos and A. A. Firsov, Two-Dimensional Gas of Massless Dirac Fermions in Graphene, *Nature* **438**, 197–200 (2006).
[9] A. K. Geim and K. S. Novoselov, The Rise of Graphene, *Nat. Mater.* **6**, 183–191 (2007).
[10] D. Li and R. B. Kaner, Materials Science-Graphene-Based Materials, *Science* **320**, 1170–1171 (2007).
[11] J. A. Rogers, Electronic Materials-Making Graphene for Macroelectronics, *Nat. Nanotechnol.* **3**, 254–255 (2008).
[12] X. Wei Lee, J. W. Kysar and J. Hone, Measurement of the Elastic Properties and Intrinsic Strength of Monolayer Graphene, *Science* **321**, 385–388 (2008).
[13] L. A. Ponomarenko, F. Schedin, M. I. Katsnelson, R. Yang, E. W. Hill, K. S. Novoselov and A. K. Geim, Chaotic Dirac Billiard in Graphene Quantum Dots, *Science* **320**, 356–358 (2008); highlighted by Westervelt, R. M. Graphene Nanoelectronics, *Science* **320**, 324–5 (2008).
[14] X. Wang, L. Zhi and K. Müllen, Transparent, Conductive Graphene Electrodes for Dye-Sensitized Solar Cells, *Nano Lett.* **8**, 323–327 (2008).
[15] P. Blake, P. D. Brimicombe, R. R. Nair, T. J. Booth, D. Jiang, F. Schedin, L. A. Ponomarenko, S. V. Morozov, H. F. Gleeson, E. W. Hill, A. K. Geim and K. S. Novoselov, Graphene-Based Liquid Crystal Device, *Nano Lett.* **8**, 1704–1708 (2008).
[16] K. S. Kim, Y. Zhao, H. Jang, S. Y. Lee, J. M. Kim, K. S. Kim, J. H. Ahn, P. Kim, J. Y. Choi and B. H. Hong, Large-Scale Pattern Growth of Graphene Films for Stretchable Transparent Electrodes, *Nature* **457**, 706–710 (2009). highlighted by L. Venema, *Nature* **457**, 274 (2009), and Segal, M. Graphene: on the Grow, *Nat. Nanotechnol.*, doi:10.1038/nnano.2009.13 (2009).
[17] Y. Pan, H. Zhang, D. Shi, J. Sun, S. Du, F. Liu and H. Gao, Highly Ordered, Millimeter-Scale, Continuous, Single-Crystalline Graphene Monolayer Formed on Ru (0001), *Adv. Mater.* doi: 10.1002/adma200800761. highlighted by T. Reid, Graphene: Making it big, *Nature China*, doi:10.1038/nchina.2009.11 (Single-Crystalline Graphene Monolayer Formed).
[18] D. E. Jiang, M. H. Du and S. Dai, First Principles Study of the Graphene/Ru(0001) Interface, *J. Chem. Phys.* **130**, 074705 (2009).
[19] K. Nakada, M. Fujita, G. Dresselhaus and M. S. Dresselhaus, Edge State in Graphene Ribbons: Nanometer Size Effect and Edge Shape Dependence, *Phys. Rev. B* **54**, 17954–17961 (1996) for online abstract, see http://prola.aps.org/abstract/PRB/v54/i24/p17954_1
[20] K. Kobayashi, Electronic Structure of a Stepped Graphite Surface, *Phys. Rev. B* **48**, 1757–1760 (1993).
[21] D. J. Klein, Graphitic Polymer Strips with Edge States, *Chem. Phys. Lett.* **217**, 261–265 (1994).
[22] D. J. Klein and L. Bytautas, Graphitic Edges and Unpaired π-Electron Spins, *J. Phys. Chem. A* **103**, 5196–5210 (1999).

[23] M. Fujita, K. Wakabayashi, K. Nakada and K. Kusakabe, Peculiar Localized State at Zigzag Graphite Edge, *J. Phys. Soc. Jpn.* **65**, 1920–1923 (1999).
[24] M. Fujita, M. Igami and K. Nakada, Lattice Distortion in Nanographite Ribbons, *J. Phys. Soc. Jpn.* **66**, 1864–1867 (1997).
[25] K. Nakada, M. Igami and M. Fujita, Electron-Electron Interaction in Nanographite Ribbons, *J. Phys. Soc. Jpn.* **67**, 2388–2394 (1998).
[26] K. Wakabayashi, M. Fujita, H. Ajiki and M. Sigrist, Electronic and Magnetic Properties of Nanographite Ribbons, *Phys. Rev. B* **59**, 8271–8282 (1999).
[27] T. Kawai, Y. Miyamoto, O. Sugino and Y. Koga, Graphitic Ribbons without Hydrogen-Termination: Electronic Structures and Stabilities, *Phys. Rev. B* **62**, R16349–R16352 (2000).
[28] K. Harigaya, New Type of Antiferromagnetic State in Stacked Nanographite, *Chem. Phys. Lett.* **340**, 123–128 (2001).
[29] K. Wakabayashi and K. Harigaya, Magnetic Structure of Nano-Graphite Möbius Ribbon, *J. Phys. Soc. Jpn.* **72**, 998–1001 (2003).
[30] A. Yamashiro, Y. Shimoi, K. Harigaya and K. Wakabayashi, Spin- and Charge-Polarized States in Nanographene Ribbons with Zigzag Edges, *Phys. Rev. B* **68**, 193410 (2003).
[31] K. I. Sasaki, S. Murakami and R. Saito, Gauge Field for Edge State in Graphene, *J. Phys. Soc. Jpn.* **75**, 74713 (2006).
[32] K. Sasaki, S. Murakami and R. Saito, Stabilization Mechanism of Edge States in Graphene, *Appl. Phys. Lett.* **88**, 113110 (2006).
[33] T. Enoki, Y. Kobayashi and K. I. Fukui, Electronic Structures of Graphene Edges and Nanographene, *Int. Rev. Phys. Chem.* **26**, 609–645 (2007).
[34] K. I. Sasaki and R. Saito, Magnetism as a Mass Term of the Edge States in Graphene, *J. Phys. Soc. Jpn.* **77**, 054703 (2008).
[35] Z. Klusek, Z. Waqar, E. A. Denisov, T. N. Kompaniets, I. V. Makarenko, A. N. Titkov and A. S. Bhatti, Observations of Local Electron States on the Edges of the Circular Pits on Hydrogen-Etched Graphite Surface by Scanning Tunneling Spectroscopy, *Appl. Surf. Sci.* **161**, 508–514 (2000).
[36] Y. Kobayashi, K. Fukui, T. Enoki and K. Kusakabe, Edge State on Hydrogen-Terminated Graphite Edges Investigated by Scanning Tunneling Microscopy, *Phys. Rev. B* **73**, 125415 (2006).
[37] Y. Niimi, T. Matsui, H. Kambara, K. Tagami, M. Tsukada and H. Fukuyama, Scanning Tunneling Microscopy and Spectroscopy of the Electronic Local Density of States of Graphite Surfaces near Monoatomic Step Edges, *Phys. Rev. B* **73**, 85421 (2006).
[38] Y. Shibayama, H. Sato, T. Enoki and M. Endo, Disordered Magnetism at the Metal-Insulator Threshold in Nano-Graphite-Based Carbon Materials, *Phys. Rev. Lett* **84**, 1744–1747 (2000).
[39] T. Enoki and Y. Kobayashi, Magnetic Nanographite: an Approach to Molecular Magnetism, *J. Mater. Chem* **15**, 3999–4002 (2005).
[40] Y.-W. Son, M. L. Cohen and S. G. Louie, Half-Metallic Graphene Nanoribbons, *Nature* **444**, 347–349 (2006).
[41] J. S. Wu, W. Pisula and K. Müllen, Graphenes as Potential Material for Electronics, *Chem. Rev.* **107**, 718–747 (2007).
[42] E. Clar, *Polycyclic Hydrocarbons*, Vol. **1, 2**, Academic Press, London (1964).
[43] Z. H. Wang, E. Tomovic, M. Kastler, R. Pretsch, F. Negri, V. Enkelmann and K. Müllen, Graphitic Molecules with Partial 'Zig/Zag' Periphery, *J. Am. Chem. Soc* **126**, 7794–7795 (2004).
[44] M. Kastler, J. Schmidt, W. Pisula, D. Sebastiani and K. Müllen, From Armchair to Zigzag Peripheries in Nanographenes, *J. Am. Chem. Soc* **128**, 9526–9534 (2006).
[45] K. May, S. Dapprich, F. Furche, B. V. Unterreiner and R. Ahlrichs, Structures, C–H and C–CH3 Bond Energies at Borders of Polycyclic Aromatic Hydrocarbons, *Phys. Chem. Chem. Phys* **2**, 5084–5088 (2000).
[46] M. M. Payne, S. R. Parkin and J. E. Anthony, Functionalized Higher Acenes: Hexacene and Heptacene, *J. Am. Chem. Soc.* **127**, 8028–8029 (2005).

[47] D. Chun, Y. Cheng and F. Wudl, The Most Stable and Fully Characterized Functionalized Heptacene, *Angew. Chem. Int. Ed.* **47**, 8380–8385 (2008).
[48] J. E. Anthony, The Larger Acenes: Versatile Organic Semiconductors, *Angew. Chem. Int. Ed.* **47**, 452–483 (2008).
[49] V. Barone, O. Hod and G. E. Scuseria, Electronic Structure and Stability of Semiconducting Graphene Nanoribbons, *Nano Lett.* **6**, 2748–2754 (2006).
[50] D. E. Jiang, B. G. Sumpter and S. Dai, Unique Chemical Reactivity of a Graphene Nanoribbon's Zigzag Edge, *J. Chem. Phys.* **126**, 134701 (2007).
[51] Y. Kobayashi, K. Fukui, T. Enoki, K. Kusakabe and Y. Kaburagi, Observation of Zigzag and Armchair Edges of Graphite Using Scanning Tunneling Microscopy and Spectroscopy, *Phys. Rev. B* **71**, 193406 (2005).
[52] Here we use the term 'ferromagnetic' in a rather loose sense, just to indicate that the spins at the two edges are all up. Strictly speaking, this phase should be called ferrimagnetic because some of the carbon atoms between the two edges are spin down.
[53] C. Herring, *Magnetism* **4** (Eds.: G. T. Rado, H. Suhl), Academic Press, New York, (1966).
[54] E. Rudberg, P. Salek and Y. Luo, Nonlocal Exchange Interaction Removes Half-Metallicity in Graphene Nanoribbons, *Nano Lett.* **7**, 2211–2213 (2007).
[55] E. J. Kan, Z. Y. Li, J. L. Yang and J. G. Hou, Will Zigzag Graphene Nanoribbon Turn to Half Metal under Electric Field?, *Appl. Phys. Lett.* **91**, 243116 (2007).
[56] S. D. Dalosto and Z. H. Levine, Controlling the Band Gap in Zigzag Graphene Nanoribbons with an Electric Field Induced by a Polar Molecule, *J. Phys. Chem. C.* **112**, 8196–8199 (2008).
[57] E. J. Kan, Z. Y. Li, J. L. Yang and J. G. Hou, Half-Metallicity in Edge-Modified Zigzag Graphene Nanoribbons, *J. Am. Chem. Soc.* **130**, 4224–4225 (2008).
[58] D. E. Jiang, B. G. Sumpter and S. Dai, How Do Aryl Groups Attach to a Graphene Sheet?, *J. Phys. Chem. B* **110**, 23628–23632 (2006).
[59] Zero-point-energy [ZPE] was not included in our calculations of BDEs. Including ZPE corrections was estimated to decrease the calculated BDEs by $\sim$0.2 eV for C-H bonds and by a much less amount for other bonds in the present work.
[60] Y. R. Luo, Handbook of Bond Dissociation Energies in Organic Compounds, CRC Press, Boca Raton, FL (2002).
[61] Here S is the radical's total spin moment.
[62] X. W. Sha and B. Jackson, First-Principles Study of the Structural and Energetic Properties of H Atoms on a Graphite (0001) Surface, *Surf. Sci.* **496**, 318–330 (2002).
[63] V. Barone, J. Heyd and G. E. Scuseria, Interaction of Atomic Hydrogen with Single-Walled Carbon Nanotubes: A Density Functional Theory Study, *J. Chem. Phys.* **120**, 7169–7173 (2004).
[64] Y.-W. Son, M. L. Cohen and S. G. Louie, Energy Gaps in Graphene Nanoribbons, *Phys. Rev. Lett.* **97**, 216803 (2006).
[65] L. Yang, C. H. Park, Y. W. Son, M. L. Cohen and S. G. Louie, Quasiparticle Energies and Band Gaps in Graphene Nanoribbons, *Phys. Rev. Lett.* **99**, 186801 (2007).
[66] X. Y. Yang, X. Dou, A. Rouhanipour, L. J. Zhi, H. J. Rader and K. Müllen, Two-Dimensional Graphene Nanoribbons, *J. Am. Chem. Soc.* **130**, 4216–4217 (2008).
[67] D. E. Jiang, B. G. Sumpter and S. Dai, First Principles Study of Magnetism in Nanographenes, *J. Chem. Phys.* **127**, 124703 (2007).
[68] X. F. Gao, Z. Zhou, Y. L. Zhao, S. Nagase, S. B. Zhang and Z. F. Chen, Comparative Study of Carbon and BN Nanographenes: Ground Electronic States and Energy Gap Engineering, *J. Phys. Chem. C* **112**, 12677–12682 (2008).
[69] O. Hod, V. Barone and G. E. Scuseria, Half-Metallic Graphene Nanodots: A Comprehensive First-Principles Theoretical Study, *Phys. Rev. B* **77**, 035411–035416 (2008).
[70] K. N. Houk, P. S. Lee and M. Nendel, Polyacene and Cyclacene Geometries and Electronic Structures: Bond Equalization, Vanishing Band Gaps, and Triplet Ground States Contrast with Polyacetylene, *J. Org. Chem.* **66**, 5517–5521 (2001).
[71] M. Bendikov, H. M. Duong, K. Starkey, K. N. Houk, E. A. Carter and F. Wudl, Oligoacenes: Theoretical Prediction of Open-Shell Singlet Diradical Ground States, *J. Am. Chem. Soc.* **126**, 7416–7417 (2004). erratum *J. Am. Chem. Soc.* **126**, 10493 (2004).

[72] M. C. dos Santos, Electronic Properties of Acenes: Oligomer to Polymer Structure, *Phys. Rev. B* **74**, 045426 (2006).

[73] D. E. Jiang and S. Dai, Electronic Ground State of Higher Acenes, *J. Phys. Chem. A* **112**, 332–335 (2008).

[74] M. Bendikov, H. M. Duong, K. Starkey, K. N. Houk, E. A. Carter and F. Wudl, Oligoacenes: Theoretical Prediction of Open-Shell Singlet Diradical Ground States, *J. Am. Chem. Soc.* **126**, 7416–7417 (2004). Erratum in *J. Am. Chem. Soc.* **126**, 10493–10493 (2004).

[75] L. Salem and C. Rowland, The Electronic Properties of Diradicals, *Angew. Chem. Int. Ed.* **11**, 92–111 (1972).

[76] M. C. dos Santos, Electronic Properties of Acenes: Oligomer to Polymer Structure, *Phys. Rev. B* **74**, 045426 (2006).

[77] J. Hachmann, J. J. Dorando, M. Avilés and G. K.-L. Chan, The Radical Character of the Acenes: A Density Matrix Renormalization Group Study, *J. Chem. Phys.* **127**, 134309 (2007).

[78] R. Mondal, B. K. Shah and D. C. Neckers, Photogeneration of Heptacene in a Polymer Matrix, *J. Am. Chem. Soc.* **128**, 9612–9613 (2006).

[79] H. F. Bettinger and R. Mondal, D. C. Stable Photoinduced Charge Separation in Heptacene, Neckers, *Chem. Commun.* 5209–5211 (2007).

[80] D. E. Jiang and S. Dai, Circumacenes versus Periacenes: HOMO–LUMO Gap and Transition from Nonmagnetic to Magnetic Ground State with Size, *Chem. Phys. Lett.* **466**, 72–75 (2008).

[81] R. D. Broene and F. Diederich, The Synthesis of Circumanthracene, *Tetrahedron Lett.* **32**, 5227–5230 (1991).

[82] F. H. Kohnke, A. M. Z. Slawin, J. F. Stoddart and D. J. Williams, The Larger Acenes: Versatile Organic Semiconductors, *Angew. Chem. Int. Ed.* **26**, 892–894 (1987).

[83] P. R. Ashton, N. S. Isaacs, F. H. Kohnke, A. M. Z. Slawin, C. M. Spencer, J. F. Stoddart and D. J. Williams, Towards the Making of [12]Collarene, *Angew. Chem. Int. Ed.* **27**, 966–969 (1988).

[84] P. R. Ashton, G. R. Brown, N. S. Isaacs, D. Giuffrida, F. H. Kohnke, J. P. Mathias, A. M. Z. Slawin, D. R. Smith, J. F. Stoddart and D. J. Williams, Molecular LEGO. 1. Substrate-Directed Synthesis via Stereoregular Diels-Alder Oligomerizations, *J. Am. Chem. Soc.* **114**, 6330–6353 (1992).

[85] R. M. Cory and C. L. McPhail, Transformations of a Macrocyclic Cyclophane Belt into Advanced [8]Cyclacene and [8]Cyclacene Triquinone Precursors, *Tetrahedron Lett.* **37**, 1987–1990 (1996).

[86] R. M. Cory, C. L. McPhail, A. J. Dikmans and J. J. Vittal, Macrocyclic Cyclophane Belts via Double Diels-Alder Cycloadditions: Macroannulation of Bisdienes by Bisdienophiles. Synthesis of a Key Precursor to an [8]Cyclacene, *Tetrahedron Lett.* **37**, 1983–1986 (1996).

[87] H. S. Choi and K. S. Kim, Structures, Magnetic Properties, and Aromaticity of Cyclacenes, *Angew. Chem. Int. Ed.* **38**, 2256–2258 (1999).

[88] Z. Chen, D. E. Jiang, X. Lu, H. F. Bettinger, S. Dai, P. v. R. Schleyer and K. N. Houk, Open-Shell Singlet Character of Cyclacenes and Short Zigzag Nanotubes, *Org. Lett.* **9**, 5449–5452 (2007).

[89] K. Harigaya, A. Yamashiro, Y. Shimoi and K. Wakabayashi, Theoretical Investigation on Electronic Properties of Topological Materials: Möbius Nanographite and Conjugated Polymers, *Synth. Met.* **152**, 261–264 (2005).

[90] R. Herges, Topology in Chemistry: Designing Möbius Molecules, *Chem. Rev.* **106**, 4820–4842 (2006).

[91] H. S. Rzepa, Möbius Aromaticity and Delocalization, *Chem. Rev.* **105**, 3697–3715 (2005).

[92] D. E. Jiang and S. Dai, Spin States of Zigzag-Edged Möbius Graphene Nanoribbons from First Principles, *J. Phys. Chem. C* **112**, 5348–5351 (2008).

[93] M. Ezawa, Metallic graphene nanodisks: Electronic and Magnetic Properties, *Phys. Rev. B* **76**, 245415–245416 (2007).

[94] M. Ezawa, Graphene Nanoribbon and Graphene Nanodisk, *Physica E* **40**, 1421–1423 (2008).

[95] E. H. Lieb, Two Theorems on the Hubbard Model, *Phys. Rev. Lett.* **62**, 1201–1204 (1989).

[96] J. Fernandez-Rossier and J. J. Palacios, Magnetism in Graphene Nanoislands, *Phys. Rev. Lett.* **99**, 177204 (2007).

[97] S. E. Stein and R. L. Brown, .pi.-Electron Properties of Large Condensed Polyaromatic Hydrocarbons, *J. Am. Chem. Soc.* **109**, 3721–3729 (1987).
[98] R. Ruoff, Graphene: Calling All Chemists, *Nat. Nanotechnol.* **3**, 10–11 (2008).
[99] D. C. Elias, R. R. Nair, T. M. G. Mohiuddin, S. V. Morozov, P. Blake, M. P. Halsall, A. C. Ferrari, D. W. Boukhvalov, M. I. Katsnelson, A. K. Geim and K. S. Novoselov, Control of Graphene's Properties by Reversible Hydrogenation: Evidence for Graphane, *Science* **323**, 610–613 (2009); highlighted by Armstrong, G.; Graphene: Here comes graphane? *Nature Chemistry*, doi:10.1038/nchem.116 (2009).

19

Carbon Nano Onions

Luis Echegoyen[a], *Angy Ortiz*[a], *Manuel N. Chaur*[a] *and Amit J. Palkar*[b]

[a]*Department of Chemistry, Clemson University, Clemson, SC USA*

[b]*ConocoPhillips Company, Ponca City, Oklahoma USA*

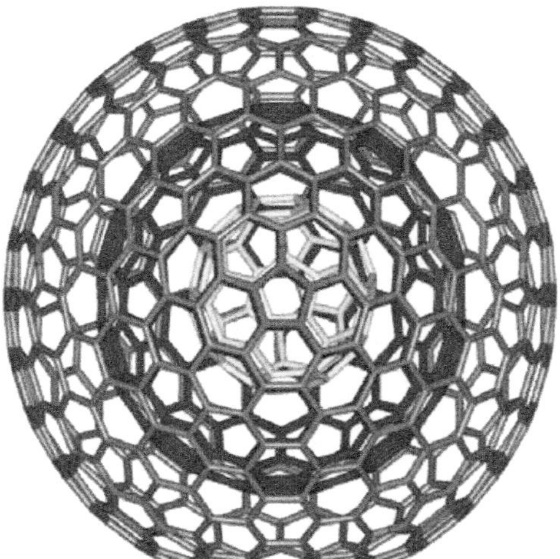

Chemistry of Nanocarbons Edited by Takeshi Akasaka, Fred Wudl and Shigeru Nagase

19.1 Introduction

The laser desorption experiment of Smalley, Curl and Kroto in 1985 led to the discovery of a new allotropic form of carbon known as [60]fullerene or C_{60} [1]. This breakthrough prompted a search for previously unobserved allotropes of carbon that can be obtained synthetically. In 1990, Huffman and Krätchmer along with their co-workers demonstrated simple electric arc discharge methods for large scale synthesis of these compounds [2]. A wide family of nanometric graphitic systems were synthesized later by making slight modifications to the electric arc experiment (nanotubes [3, 4] nanoparticles [5], metal-filled nanoparticles [6]), thus resulting in a wealth of revolutionary discoveries. As high resolution transmission electron microscopy [TEM] was the only reliable technique for obtaining structural details of these particles, most of them were subjected to this analysis.

Ugarte, while studying such graphitic particles using HRTEM observed the curling and closure of these particles to form spherical multilayered onion-like structures which he described as carbon nano onions (Figure 19.1) [7, 8]. These carbon nano onions show a perfectly spherical shell-inside structures with each shell separated from the adjacent ones by a distance approximately equal to the distance between two [220] graphitic planes (0.334 nm). However, as it is obvious from this process of synthesis, this method produced only minuscule amounts of carbon nano onions and thus was not useful for further exploration of their properties. Surprisingly, the first breakthrough for obtaining carbon nano onions in bulk came not through a modification of electric arcing experiments, but from annealing of nanodiamonds. Nanodiamonds are products formed during an inert atmosphere detonation of explosives such as TNT and RDX [9]. The kinetic products of such violent conditions are typically 2–5 nm diamond particles showing all sp^3 diamond framework. Kuznetsov *et al.* observed that the annealing of these nanodiamonds at temperatures higher than 800 °C lead to graphitization of the diamonds to ultimately form carbon nano onions [10]. The only inconvenient requirement of this synthetic process was

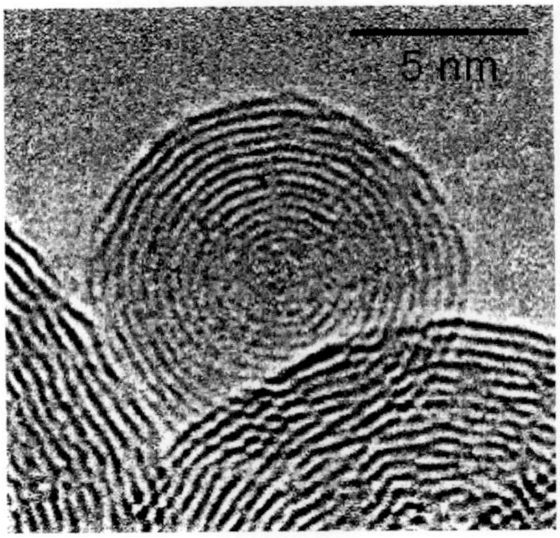

Figure 19.1 *HREM image of a quasi-spherical onion-like graphitic generated by electron irradiation (dark lines represent graphitic shells, and distance between layers is 0.34 nm) [scale estimated from the figure description]. Reference [7]*

the need to use vacuum at high temperatures which required specialized equipment. Some other preparation methods have also been described in the literature such as carbon-ion implantation at high temperature into copper substrates [11], thermal treatment of laser pyrolysis carbon blacks [12, 13] and co-sputtering [14]. However, most of these are low yield processes and have not been pursued further. Nevertheless, the method annealing of nanodiamonds has been pursued by some research groups around the world and this has led to the further study of carbon nano onions and their physical properties.

19.2 Physical Properties of Carbon Nano Onions Obtained from Annealing

19.2.1 Annealing Process

A detailed analysis of the effect of annealing temperature on the fate of carbon nano onions has been reported by Kuznetsov *et al.* [15], Obratsova *et al.* [16] and Tomita *et al.* [17] The initial structure of a nanodiamond shows regularly arranged [111] lattice planes with some amorphous carbon layers (Figure 19.2a). Tomita *et al.* have shown that the annealing of nanodiamonds in a vacuum up to 800 °C does not affect the structure of the nanodiamonds. Heating the nanodiamonds above 900 °C results in the initial thermal desorption of noncarbonaceous material which may be present in the sample. The thermal desorption data indicates that the last noncarbon contaminant (hydrogen) leaves the nanodiamond surface at approximately 850 °C [18, 19]. It has been suggested that at the higher annealing temperatures, a rearrangement of the surface carbons of the nanodiamond particles takes place that results in the minimization of the surface energy [16]. Upon heating to higher temperatures, a transformation of the diamond phase to a graphitic phase begins [17].

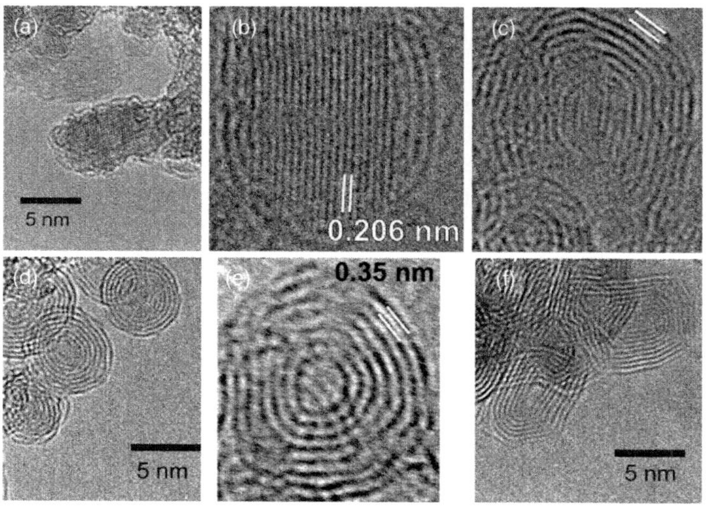

Figure 19.2 *HRTEM images of (a) nanodiamond, (b) 827 °C (c) 1400 °C (d) 1700 °C (e) 1700 °C (high resolution image) (f) > 1900 °C. (a), (d), (f) from reference [17] and (b), (c), (e) from reference [15]*

Initially, above 900 °C the outer surface of the nanodiamond undergoes graphitization (Figure 19.2b). However, a small amount of nanodiamond crystals remains at the core of the particle. Upon further annealing at 1100 °C the diamond cores from the smaller particles completely disappear to form graphitic shells while the larger particles still have some residual core (Figure 19.2c). Further heating of nanodiamonds at 1500 °C transforms the nanodiamonds completely into graphitic structures (Figure 19.2d). The major product is carbon nano onions having six to seven shells (Figure 19.2e). At temperatures of 1500–1800 °C the existence of the separate onions becomes energetically less favorable than the formation of joint graphitic layers of neighboring onions. This results in the appearance of extended multi-shell graphitic cages with hollow centers as well as graphite-like ribbons with parallel graphitic planes (Figure 19.2f). At the highest annealing temperatures (2100 °C) nanodiamonds transform into a faceted multi-shelled structure.

Kuznetsov *et al.* have proposed a mechanism for the formation of carbon nano onions using what has been described as a zipper type mechanism. It has been proposed that out of every three [111] diamond planes, the center plane migrates into the bulk of the diamond crystal. The atoms of the central diamond layer distribute equally in the formation of two graphitic sheets. This is the only way in which one can obtain the same lengths for the two growing graphitic sheets. Figure 19.3 illustrates the process, which resembles the opening of a zipper. Each of the marked groups of atoms contributes to the formation of an additional row of graphite sixfold rings. The process probably proceeds via successive insertion of two carbon atoms per two sixfold ring of a growing sheet with the intermediate formation and subsequent reconstruction of eightfold rings [15].

19.3 Raman Spectroscopy of Carbon Nano Onions Prepared by Annealing Nanodiamonds

Raman spectroscopy is a very powerful tool for exploring the behavior of various forms of carbon. Most graphitic nano particles show characteristic D and G Raman bands. The D band has been attributed to the disorder induced in the structure due to presence of sp^3 carbons. The G band corresponds to the E_{2g} mode of the sp^2 carbon framework and it is the only band present in highly oriented pyrolitic graphite (HOPG) and is observed at $1582\,cm^{-1}$. The width of the *G* band is related to the disorder between the sheets of sp^2 hybridized carbon (Figure 19.4) [17]. The effect of heating the nanodiamond has been examined by Raman spectroscopy. The spectrum of a crude nanodiamond powder shows a

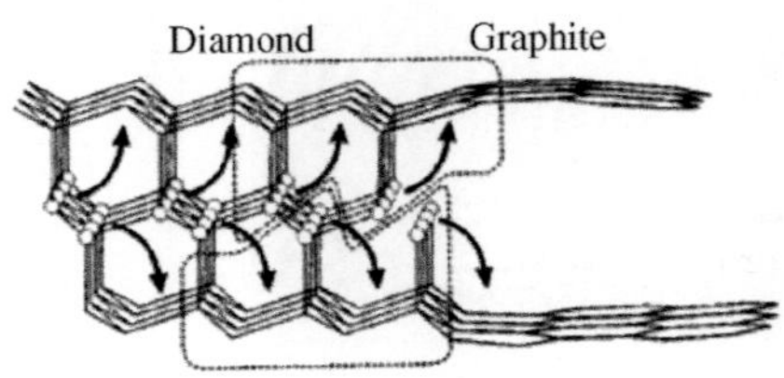

Figure 19.3 *Proposed mechanism of transformation of nanodiamonds into carbon nano onions. From reference [15]*

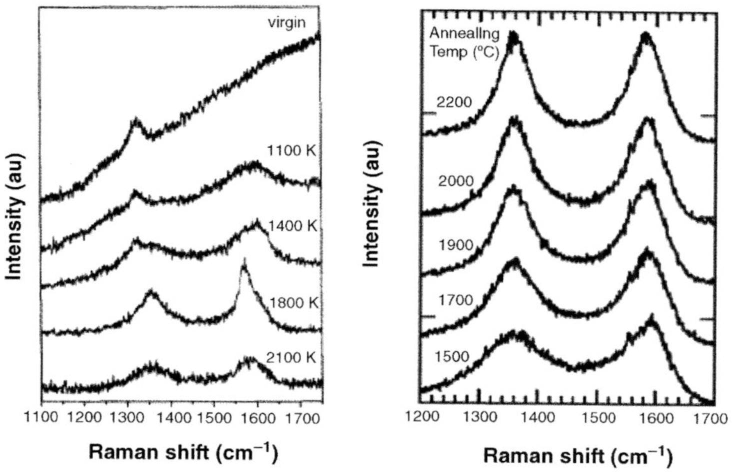

Figure 19.4 *Evolution of the Raman spectra of the nanodiamonds with increasing temperature, reproduced from reference [16] and [17]*

relatively narrow single Raman band at 1323 cm^{-1} over a strong photoluminescence background. This band is a characteristic signal of the nanodiamonds. Due to crystal size effects, the position is shifted toward lower wavenumbers when compared with that of a single diamond crystal (1332 cm^{-1}) [16].

The Raman spectrum of the nanodiamonds annealed at temperatures around 900 °C show two wide Raman bands, the D band at 1350 cm^{-1} and the G band at 1600 cm^{-1}. As a consequence it has been proposed that the two bands arise from the carbon phase formed on the surface of a diamond particle to stabilize the sp^3-bonds left dangling after the thermal desorption of the foreign impurities (Figure 19.4) [16]. At 1100 °C the weak peaks from the diamond core almost completely disappear and the D band corresponding to the defective carbon nano onions being formed becomes even more pronounced. At temperatures around 1500 °C the Raman spectrum of this sample shows peaks of disordered carbon ('D' peak at 1350 cm^{-1} with' and 'G' peak at 1584 cm^{-1} with a distinctive side band at 1572 cm^{-1}. This has been attributed to the Raman spectrum of carbon nano onions obtained from annealing of nanodiamonds by Obratsova *et al.* (Figure 19.4) [16]. At temperatures between 1500–1800° onions form joint graphitic layers with their neighboring onions and the Raman spectra are similar to those of glassy carbon. At the highest annealing temperatures nanodiamonds transform into a faceted multi-shelled structure with the Raman bands becoming even more pronounced.

19.3.1 X-Ray Diffraction Studies

The diffraction pattern of the crude nanodiamond powder compared to those of the carbon nano onions synthesized at various temperatures is shown in Figure 19.5. The diffraction pattern of nanodiamond samples (Figure 19.5a) has prominent peaks at angles of 43.34°, 74.68°, 90.36° and 118.37° which correspond to the normal peaks for the [111], [220], [311] and [400] diamond planes, respectively. In addition, a small peak centered at 26°

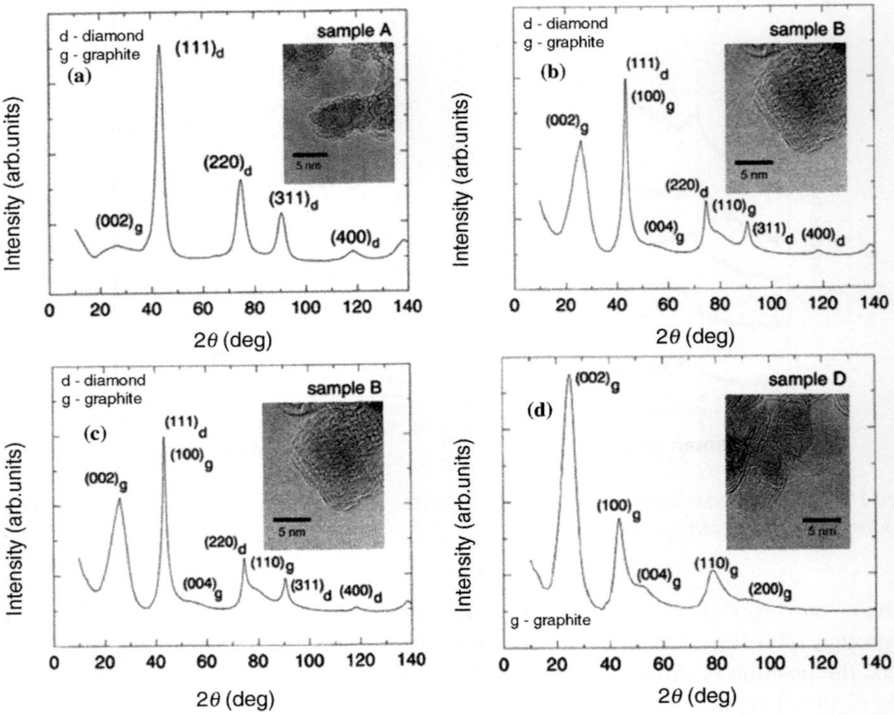

Figure 19.5 X-ray diffraction intensity of (a) nanodiamond (b) after heating at 1400 °C (c) after heating at 1700 °C (d) after heating at 2000 °C from reference [20]

corresponding to the [002] peak of graphite is also observed, suggesting that there may be some contribution from surface sp^2 carbon fragments. Upon annealing at 1400 °C [Figure 19.5b] a significant narrowing of the diamond peaks followed by a significant increase in the intensity of the [002] graphitic peaks and the appearance of the [004] and [110] graphitic peaks is observed. Further annealing at 1700 °C results in an almost complete disappearance of the diamond [111] peak as well as a significant reduction of the second, third and fourth diamond peaks. This indicates that much of the sp^3 coordination has been effectively removed at higher temperatures [Figure 19.5c]. Further heating transforms the spectrum completely into a graphitic one, with well pronounced [002], [100], [004], [110] and [200] peaks [Figure 19.5d] [20].

19.3.2 Electrical Resistivity Studies

Nanodiamond samples heated up to 900 °C do not show any significant graphitization and therefore the resistivity of the samples is very high (10^9 Ω). As graphitization proceeds, the resistivity drops significantly [Figure 19.6]. In the case of nanodiamonds samples treated at temperatures above 1300 °C, the resistivity drops to values between 0.2–0.5 Ω. Increasing the annealing temperature continues to reduce the resistivity; however, at 1500 °C the resistivity shows an abnormal increase as compared to the values at both 1300 °C and

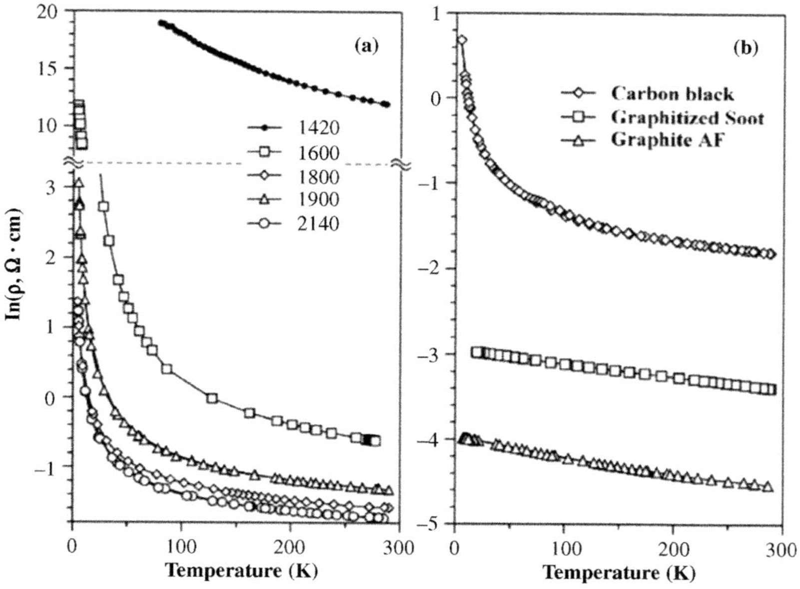

Figure 19.6 *Temperature dependence of resistivity of nanodiamond and the corresponding annealing products. From reference [21]*

1600 °C. This has been postulated to be the effect of the formation of perfectly spherical onions, which have the lowest density of defects of the graphitic structure. At 1600 °C the graphitic rings become unstable and undergo reorganization by forming joint graphitic layers with the neighboring onions, resulting in decreasing resistivity of the samples [21].

19.4 Electron Paramagnetic Resonance Spectroscopy

Unlike other carbon based nanoparticles, nanodiamonds and carbon nano onions show strong EPR signals. Figure 19.7 shows the evolution of the EPR spectrum as the nanodiamonds are annealed to form carbon nano onions as a function of temperature. Nanodiamonds show a narrow EPR line (line width of 8 G) with a g value of 2.0024. The proximity of the g value to that of the free electron spin (g value 2.0023) indicates that the signal does not originate from any metallic impurities but most likely from carbon based spins. As the annealing temperature increases, the line width continues to increase with a simultaneous decrease of the g value. At temperatures between 1500 and 1700 °C the g value changes to 2.0020 with the signal showing Curie- type behavior [17]. At temperatures above 2000 °C the signal becomes extremely broad with a line width of approximately 60 G. Tomita *et al.* have theorized that the carbon based free radicals are present on the surface of the carbon nano onions which was later confirmed by Echegoyen *et al.* [22]. The spin concentration of diamond nanoparticles was estimated using $CuSO_4.5H_2O$ as a reference at 3.5×10^{20} spins/g, while the concentration of the localized spins in spherical onions was

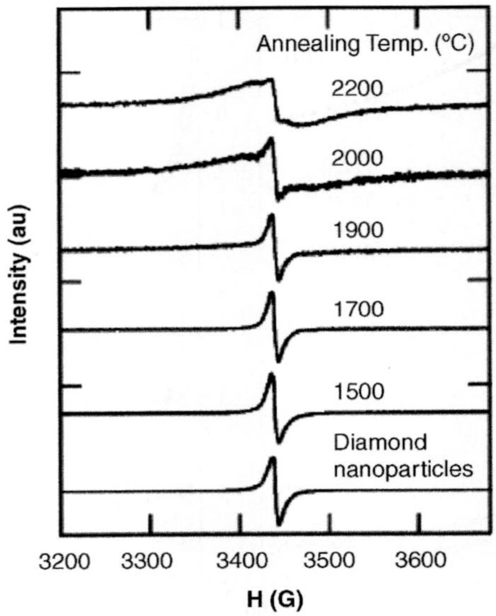

Figure 19.7 *Evolution of EPR spectra during transformation of nanodiamonds to carbon nano onions from reference [17]*

3.9×10^{19} spins/g. This value corresponds to about 10 dangling bonds per spherical onion, when the number of carbon atoms in a spherical onion is assumed to be 12 000 carbon atoms [17].

19.5 Carbon Nano Onions Prepared from Arcing Graphite Underwater

In 2001 Sano and co-workers reported the arc discharge synthesis of carbon nano onions underwater [23, 24]. In this method, two high purity graphite rods were electrically arced under water by applying a bias potential of 16–17 V between them and maintaining a constant current of 30 A. During the synthesis, carbon nano onions were obtained as a floating powder on the surface of the water while other subproducts such as multi-walled carbon nanotubes precipitated to the bottom of the container (see Figure 19.8). Although it immediately may be presumed that the density of carbon nano onions is less than that of water, this is not the case. The density of carbon nano onions is far greater than water (1.64 g/cm^3) therefore the floating carbon onions on the surface of water is presumed to be due to the formation of large van der Waals crystals as described in the next section [24].

The arc discharge method offers an alternative to vacuum processes for the synthesis of larger carbon nano onions in reasonable quantities. In addition, the arc discharge method is considerably cheaper and simpler than other methods. Carbon nano onions obtained by this method exhibit diameters between 15 and 25 nm (20–30 shells). Xu and co-workers reported later that the mean carbon nano onion size is 40 nm, considerably larger than carbon nano onions obtained from nanodiamonds [25]. The UV-vis characteristic of these carbon nano onions exhibit remarkable similarities with the interstellar dust ultraviolet

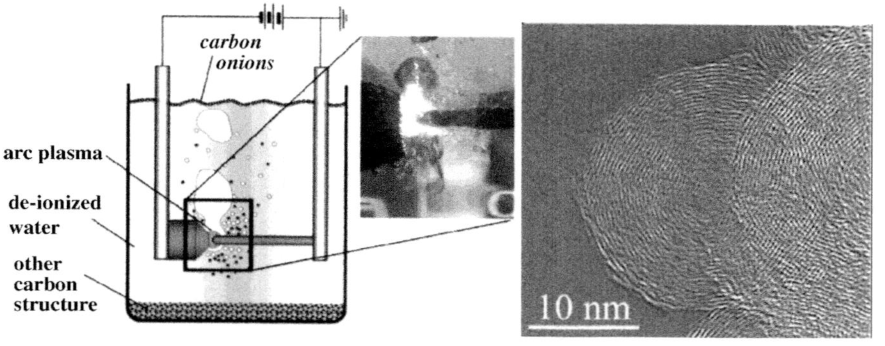

Figure 19.8 Scheme of the apparatus used for the arc discharge synthesis of CNOs. Reproduced with permission from ref. [24]

absorption spectrum which has been widely studied and therefore the presence of carbon nano onions in dense interstellar clouds has been proposed [26].

19.5.1 Mechanism of Formation

Figure 19.9 shows a pictorial representation of the model used by Sano to explain the formation of carbon nano onions under water [24]. Sano proposed that during an arcing process a gradient of temperature exists which can be divided into multiple zones. At the interface between the two electrodes the temperature can reach up to 4000 K and the vaporized carbon exists as chaotic ions and radicals along with the water which is electrolyzed into oxygen and hydrogen gases. These gases expand and form bubbles around the electrodes the carbon atom radicals and ions are expelled away from the electrodes. The upon reaching zone 1 there is a rapid cooling due to the dissipation of heat in the water. However, as the expelled particles are travelling in a directional fashion this promotes their combination, allowing the formation of multiwalled nanotubes in this zone. However as particles travel away from the arc into zone 2 which is at even lower temperature they lose their directionality and some of them coalesce to form the carbon nano onions in the much cooler zone ($\sim$373 K). Once formed, the carbon nano onions cluster together forming large van der Waals crystals, which accounts for their flotation on top of the water surface [24].

19.5.2 Properties of Carbon Nano Onions Obtained from Arc Discharge

Although carbon nano onions obtained from arcing have not been studied as extensively as the ones obtained from annealing nanodiamonds some of their properties are known. The BET surface area of the carbon nano onions determined by nitrogen gas adsorption has been found to be 984.3 m^2/g, which is significantly larger than that reported for single and multiwalled carbon nanotubes (SWNTs and MWNTs) [24]. The Raman spectroscopy of these carbon nano onions shows the D (1344 cm^{-1}) and the G band (1569–1577 cm^{-1}), respectively. The ratio of the intensities of the D and G peaks is often used to estimate the degree of perfection of graphene planes. For example, the absence of the D peak in HOPG implies perfect graphene planes. Thus, a very weak D peak and a strong G peak from carbon nano onions signify that only slight imperfections are present in the graphene planes of the

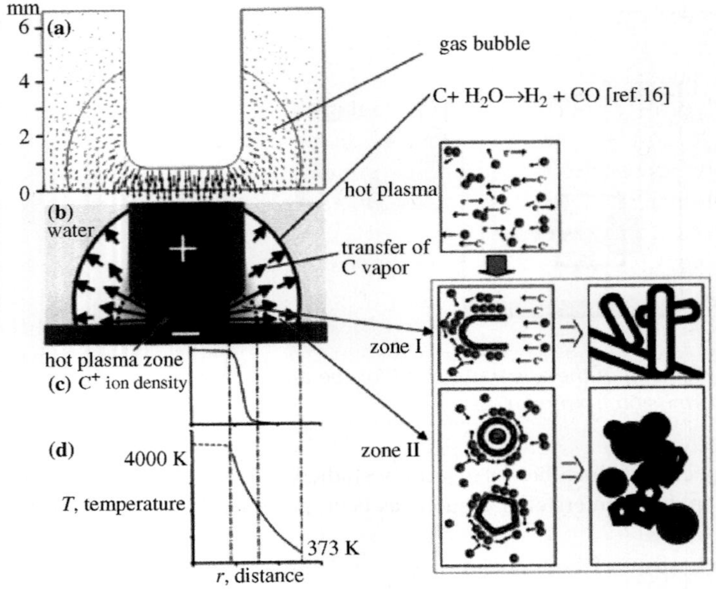

Figure 19.9 *Suggested mechanism for the formation of carbon nano onions produced in water. (a) Direction of the electric field formed between the two electrodes. (b) Direction of thermal expansion from the plasma to water. (c) Ion density distribution as a function the distance from the plasma. (d) Gradient of temperature and formation of different nanoparticles. Figure reproduced from reference [24]*

onions [27]. Upon revisiting the Raman spectra of carbon nano onions obtained from annealing of nanodiamonds (Figure 19.4) it can be seen that the case is reversed and the D band is as strong as the G band thus indicating that the former are much less defective than the latter.

Other methods of synthesis of carbon nano onions have appeared in the literature since the discovery of the arc synthesis. As mentioned before, graphitization of nanodiamonds to obtain carbon nano onions can be achieved by annealing nanodiamond particles above 900 °C and the same methodology was recently applied to acetylene carbon blacks using an iron catalyst to generate carbon nano onions with diameters between 40 to 100 nm [28]. In addition, Iijima and co-workers found that electron irradiation of nanodiamonds can also lead to the production of carbon nano onions [29]. Several years later, Roddatis and co-workers reported a series of TEM images showing the complete transformation from nanodiamonds to carbon nano onions during electron irradiation [30]. Contrary to Iijima's results, in this case the graphitization occurred starting at the outer shells, not from the core. Gubarevich and co-workers reported the deposition of carbon nano onions from the detonation of nanodiamond powders by an electromagnetically accelerated plasma spraying (EMAPS) [31]. Synthesis of carbon nano onions by laser irradiation has been achieved using different methodologies such as irradiation of amorphous silicon carbide [32], irradiated gold nanoparticles [33], laser ablation of Ni-doped or Ni-Co-doped graphite targets [34], electron irradiation of nanodiamonds particles [35] and laser irradiation of carbon black under water [36]. Interestingly the latter method results in hydrophilic carbon nano onions.

In 2005 Liu and co-workers reported the synthesis of carbon nano onions by thermal reduction of glycerin with magnesium [37]. In this method, glycerin and magnesium were heated at 650 °C during 12 hours resulting in the formation of carbon nano onions with diameters ranging between 60 and 90 nm. Two years later Zhao and coworkers reported the assembly of aromatic molecules such as polymers which lead to the synthesis of carbon nano onions with diameters of 60–80 nm [38]. Another alternative for the production of carbon nano onions is chemical vapor deposition (CVD) over a catalyst. This method was first introduced in 2001 by Chen and co-workers and it allows the synthesis of large quantities of carbon nano onions by the deposition of methane over the surface of a Co catalyst. The average size of these carbon nano onions ranges from a few to several tens of nanometers [39]. Modifications of this method include changing the catalysts to Nickel supported on aluminum or Si(100) suface [40, 41]. Two other methods for the synthesis of carbon nano onions include catalytic synthesis using counterflow diffusion flames and heavy shocking of polycrystalline silicon carbide powder [42, 43].

Among all the methods published, annealing of nanodiamonds and arc discharge under water of graphite electrodes offer optimal amounts of pure CNOs although their sizes differ considerably. Current research has therefore been directed towards the selective synthesis of small CNOs with regular size, since smaller CNOs feature larger surface areas and therefore are more reactive towards chemical functionalization. This was demonstrated by Echegoyen et al. who studied the differences between carbon nano onions obtained from arcing [A-CNOs] and from annealing of nanodiamonds [N-CNOs] [22].

19.6 Reactivity of Carbon Nano Onions (CNOs)

In the last couple of years different reactions have been reported with CNOs, mainly designed to increase the solubility and potential applications. As already mentioned, small CNOs (N-CNOs) with 6–8 shells (5 nm) are obtained from nanodiamond annealing [23] while large CNOs (A-CNOs) with 20–30 shells (15–25 nm) are obtained from arc discharge [10]. The different functionalization reactions that have been performed on these CNOs are described below.

19.6.1 1,3-Dipolar Cycloaddition Reaction

In 2003, Prato and co-workers reported the first chemical functionalization reaction on A-CNOs by the 1,3 dipolar cyloaddition reaction to form a pyrrolidine adduct **1** (Figure 19.10). A mixture of the α-amino acid (((2-(2-methoxyethoxy)ethoxy) ethylamino)acetic acid), paraformaldehyde and the raw soot generated by arc discharge in toluene was refluxed for 4 days. After purification the TEM images showed onion like structures with varying sizes from 60–300 nm with surprisingly large and explained (4 nm) distances between the internal shells and UV-*vis* spectral bands at 230 and 265 nm attributed to π-plasmons of sp^2-like carbon atoms [44]. No other characterization techniques were discussed and no explanation was presented for the large interplanar distances observed. Later, in 2006, Echegoyen and co-workers reported pyrrolidine-derivatized A-CNO derivative **2a** and **2b** (Figure 19.10)

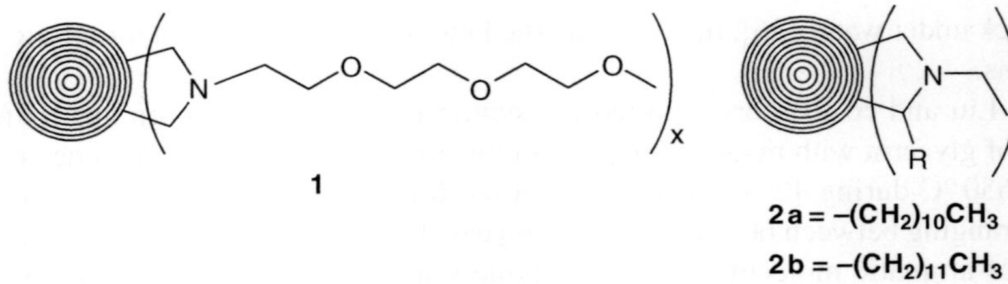

Figure 19.10 *Pyrrolidine derivatives of A-CNOs. Reference [44, 45]*

characterized by means of TEM, TGA and Raman spectroscopy, using onions having approximately 30 shells instead of onions with 32 shells (10.5 nm, 0.33 nm inter-shell). The solubility of the A-CNO was increased with the use of the aldehyde tridecanal. The pyrrolidine-A-CNOs derivatives **2a** and **2b** were electrophoretically deposited on ITO electrodes in order to obtain a sharp Raman spectrum and four bands were obtained at 1311, 1580 cm^{-1} (*D* and *G* bands, respectively), 1609 and 2618 cm^{-1} thus proving functionalized products contains onions [45].

19.6.2 Amidation Reactions

In 2006, Echegoyen and co-workers reported the addition of polyethylene glycol PEG_{1500N} and 1-Octadecylamine (ODA) to A-CNOs via amidation. The defective surface sites were oxidized with 3 M HNO_3 for 48 hours in order to produce COOH groups which could be later coupled with other reagents. In the case of PEGylation the PEG_{1500N} was added to the oxidized A-CNOs solution and heated at 140 °C for 19 days to prepare A-CNOs **3** (Figure 19.11) which are water soluble derivatives. The Raman spectrum showed the absence of the typical *D* and *G* bands (1320 and 1580 cm^{-1}, respectively). The ^1H NMR spectrum showed the disappearance of two weak broad peaks at 2.28 and 2.68 ppm ($-CH_2$ terminal groups of PEG_{1500N}) as well as the three ^{13}C NMR signals of the PEG_{1500N} carbons closest to the A-CNOs cage. This was attributed to the proximity of the protons and carbon atoms to the A-CNOs surface, which results in longer correlation and consequently shorter relaxation times. In the case of amidation with ODA a solid-state reaction was done with purified A-CNOs, and soluble derivatives **4** (Figure 19.11) were purified by extraction with chloroform. A weight gain of around 40% was observed using TGA experiments. Additionally, the absence of the ^1H NMR and ^{13}C NMR signals for the atoms close to the A-CNOs surface was also noted [45].

In 2008, Echegoyen *et al.* introduced the first amidation reaction of N-CNO (**5**, Figure 19.11). The oxidation was achieved by refluxing the N-CNOs in 3 M HNO_3

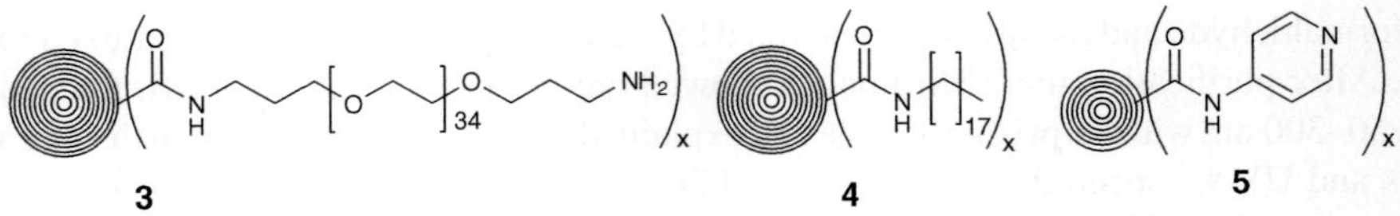

Figure 19.11 *Amides derivatives of A-CNOs (**3** and **4**) and N-CNO (**5**). Reference [45] and [46]*

for 48 hours. The resulting product showed an increase in the ratio of *D* to *G* bands in the Raman spectrum, indicating the increase in the number of defects on the surface [46]. Amidation of these oxidized onions was done by heating to 170 °C during 24 hours with 4-aminopyridine. The product (**5**, Figure 19.11) was soluble in water to the extent of 2 mg/mL. The product was characterized by means of ^{1}H and ^{13}C NMR, UV-*vis* spectroscopy, TEM, Raman spectroscopy and TGA analysis. Two bands were observed in the Raman spectrum at 1316 and 1596 cm^{-1} (*D* and *G* bands, respectively) [46]. The group also reported the first solution UV-vis spectrum of carbon nano onions which showed an absorbance maximum at 262 nm [46]. It should be noted that Tomita *et al.* using modeling studies have predicted that the absorption maxima for isolated carbon nano onions in water would be observed at around 256.4 nm [47]. Agreement of the experimental absorption maximum to the theoretical value indicates the deaggregation of the carbon nano onions into individual particles in solution [47]. TGA experiments showed a weight gain of 43% by TGA analysis which corresponded to approximately one aminopyridine group per 120 carbons. They also reported electrochemical and ^{1}H NMR studies of the complexes formed between the pyridyl (Py) groups of the N-CNOs **5** (Figure 19.11) and the Zn atom of ZnTPP [46]. Upon complexation, the reduction potentials were shifted anodically by about 170 mV and the oxidation potentials for zinc porphyrin by about 60 mV [47]. In the case of NMR titration the aromatic protons next to the N atom were shifted upfield upon addition of ZnTPP, indicating the binding of the ligand to the functionalized N-CNOs [46].

19.6.3 [2+1] Cycloaddition Reactions

In 2007, Echegoyen *et al.* reported the first [2 + 1] Bingel-Hirsch cyclopropanation reaction of N-CNOs [22]. The N-CNOs were reacted with dodecyl malonate ester, carbon tetrabromide and 1,8-diazabicyclo[5.4.0]undec-7-ene (DBU) in *o*-dichlorobenzene (ODCB) for 24 h. The ATR-IR (attenuated total reflection infrared) spectrum of the solid product showed a broad band at around 1705 cm^{-1} that can be attributed to the C = O stretching mode. The TGA analysis in air showed the removal of the malonate groups at around 250 °C, as well as the subsequent oxidative degradation of the N-CNOs (a, Figure 19.12). The Raman spectrum showed an increase of the *D* band (1307 cm^{-1}) which was obtained as a result of formation of additional *sp*^{3}-carbons. Interestingly A-CNOs were not functionalized under the same reaction conditions (b, Figure 19.12) [22].

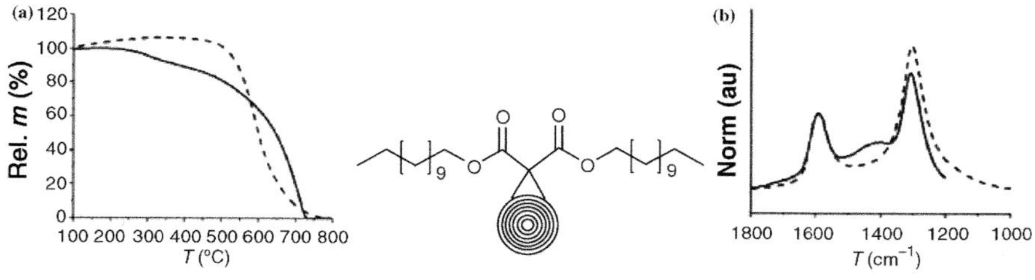

Figure 19.12 (a) TGA plots of cyclopropanated A-CNOs (----) and N-CNOs (–) in air. (b) Raman spectra of crude N-CNOs (–) and N-CNOs after cyclopropanation (----). Reference [22]

Recently, a [2 + 1] cycloaddition reaction of nitrenes with CNOs was reported by Gao and co-workers [48]. After thermal decomposition of the organic azide (2-azidoethanol and azidoethyl 2-bromo-2-methyl propano-ate) the generated nitrenes underwent a [2 + 1] cycloaddition on the CNOs, giving rise to hydroxylated (CNO-OH, **7**, Scheme 19.1) and brominated CNOs (CNO-Br, **9**, Scheme 19.1), respectively. A weight gain of 8.8% for CNO-OH **7** and 10.2% CNO-Br **9** was obtained by TGA analysis. The differences in the binding energies of CNO, CNO-OH **7** and CNO-Br **9** were analyzed by XPS (X-ray photoelectron spectroscopy) to obtain evidence of the covalent bond formation of the hydroxyl and bromine onto the CNOs surfaces. The C1s XPS peak of CNO-OH and CNO-Br are shifted significantly due to the formation of sp^3-C on the surface of CNOs. An increase in the atomic ratios O/C and N/C (0.07 and ~0.001 for CNOs, respectively) of CNO-OH (0.21 and 0.023, respectively) and CNO-Br (0.24 and 0.027, respectively) was observed due to the functionalization of the surface [48]. Furthermore, only a peak for Br3d5 at 69.4 eV was observed for CNO-Br, confirming the presence of the Br element. These two new derivatives (CNO-OH **7** and CNO-Br **9**) were used as starting materials to prepare polymers **8** and **10** (Scheme 19.1) by in-situ ring-opening polymerization (ROP) of ε-caprolactone and atom transfer radical polymerization (ATRP) of styrene, respectively [48].

19.6.4 Free-Radical Addition Reactions

N-CNO and A-CNOs were treated with benzoyl peroxide as phenyl radical precursors. After purification of the products, a 25% weight gain was observed for N-CNOs but no weight gain was observed for A-CNOs, indicating their considerable lower reactivity under these conditions. The N-CNO derivative **12** (Figure 19.14) was no longer EPR active, which confirmed that the unpaired spins are localized on the surface (Figure 19.13) [22]. The phenylated CNO compounds **12** (Figure 19.14) were sulfonated with oleum (30% SO_3 in H_2SO_4) and refluxed for 2 hours followed by treatment with NaOH [22]. The purified

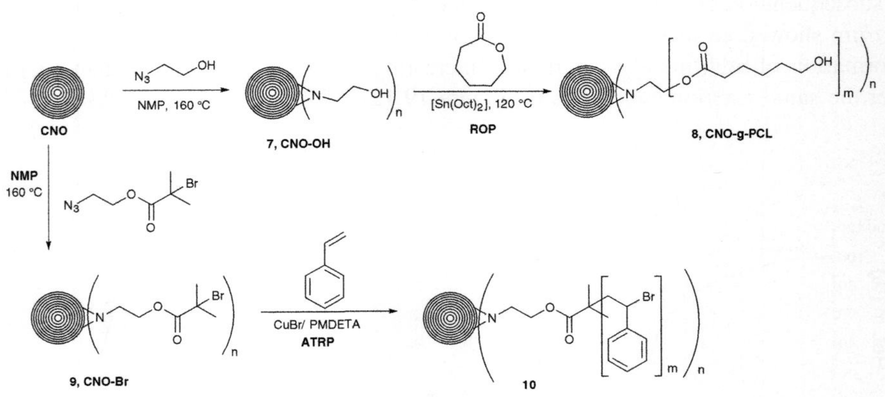

Scheme 19.1 *Facile functionalization of CNOs by direct [2 + 1] cycloaddition of nitrenes to prepare hydroxyl CNOs (**7**, CNO-OH) and brominated CNOs (**9**, CNO-Br), and grafting polymer from the functional CNOs by in-situring-opening polymerization (ROP) of e-caprolactone and atom transfer radical polymerization (ATRP) of styrene. Reference [48]*

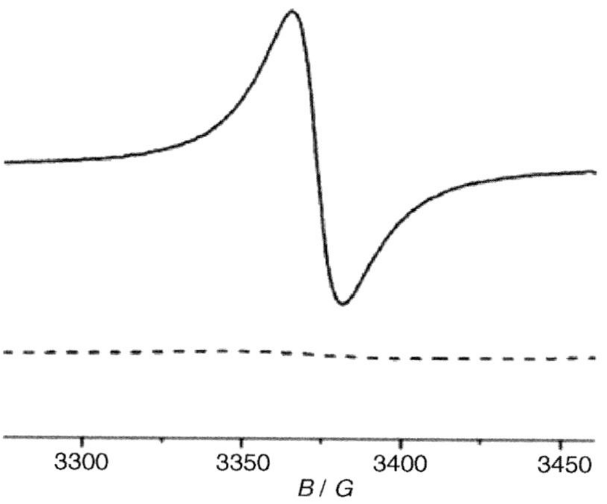

Figure 19.13 Powder X-band EPR spectra of N-CNOs before (–) and after (---) derivatization with phenyl radicals. Reference [22]

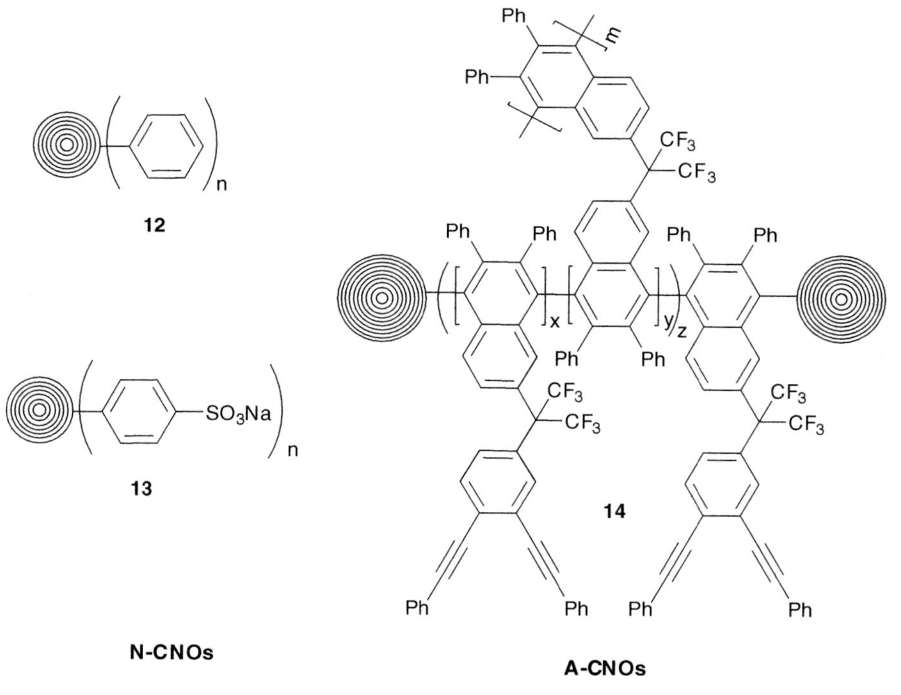

Figure 19.14 N-CNOs derivatives (**12** and **13**) and A-CNOs derivative (**14**) via free-radical reaction. References [22] and [50]

residual solid **13** (Figure 19.14) was highly dispersible in water and ethanol and resulted in a solution that was stable for months [22].

A direct fluorination of large CNOs (50–100 nm) was reported by Khabashesku's group using a custom-built reactor fluorination apparatus equipped with a flow of F_2, H_2 and He at variable temperatures between 350 and 480 °C. The yields of fluorinated CNOs (CNO-F, **11**, Scheme 19.2) were $C_{10.1}F$, $C_{3.3}F$, and $C_{2.3}F$ at 350, 410 and 480 °C, respectively. The CNO-Fs **11** (Scheme 19.2) were characterized by means of FTIR, Raman, UV-*vis*, XPS, SEM, TEM, X-ray diffraction and TGA analysis. The spherical onion-like shape was retained in the CNO-F **11** according to SEM and TEM images [49]. CNOs were recovered by a 'healing' process, via defluorination of CNO-F **11** by hydrazine treatment (Scheme 19.2) [49]. Finally, Smith Jr. and co-workers reported the direct functionalization of A-CNOs via the addition of *bis-o*-diynyl arene (BODA) monomers by formation of a thermal diradical [50]. A suspension of A-CNOs in N-methyl-2-pyrrolidinone was added to a twofold excess of BODA and refluxed for four days. The soluble A-CNO-BODA-copolymers **14** (Figure 19.14) were characterized by TEM, GPC, TGA, Raman spectroscopy, and XPS [50].

19.7 Potential Applications of CNOs

- *Nonlinear optical limiting*: In 2002, Koudoumas and co-workers studied the nonlinear optical responses of polydispersed and ultradispersed diamond powders and N-CNO structures in water suspension. The N-CNOs exhibited a nonlinear absorption after excitation with 532 nm, 10 ns laser pulses with a very efficient optical limiting action compared to the diamond powder suspensions [51].
- *Catalytic applications*: Schlögl *et al.* showed that CNOs can be used as catalysts for oxidative dehydrogenation reactions (Figure 19.15) because their graphite network has a high degree of curvature and an absence of inner particle porosity, just as with other carbon catalytic materials currently used [52].
- *Solid lubricants*: It is well known that WS_2 and MoS_2 onion-like structures have excellent wear behavior due to the absence of dangling bonds which decreases their reactivity with O_2 and water. Also, their spherical structure gives them the ability to roll, thus inducing a low friction coefficient. As expected CNOs show similar characteristics. Cabioc'h and co-workers investigated the optical and tribological properties of thin films

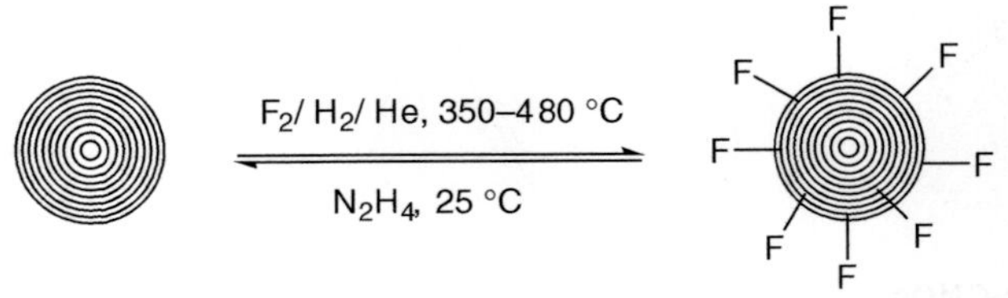

11, CNO-F

Scheme 19.2 *Functionalization of CNOs by direct fluorination, reference [49]*

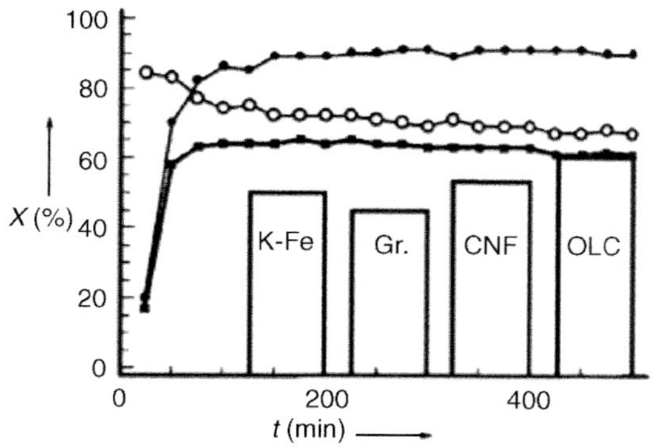

Figure 19.15 *Performance of CNOs (OLC) in the oxidative dehydrogenation of ethylbenzene at 790 K with time on stream. X = ethylbenzene conversion (•), selectivity to styrene (○), and styrene yield (■). Steady-state styrene yields for graphite (Gr.), carbon nanofilaments (CNF), and the industrial K-Fe catalyst are also given. Data are given relative to the catalyst mass. Reference [52]*

of CNOs (3–15 nm) and composite CNOs/silver deposited on two different substrates; iron steel and silicon. Only the CNO/silver composite showed that wear behavior of the silver thin film was improved due to the CNOs, but the friction coefficient (0.2) remained high [53]. In 2004, Street and co-workers reported the tribological properties of suspended CNOs on Krytox 143AB under ultrahigh vacuum (1.3. 10^{-6} Pa) and in air (relative humidity 50–60%). The CNOs did not improve the lifetime or change the friction coefficient of this oil under vacuum conditions, but the lifetime was improved with a low friction coefficient (0.04–0.05) under air conditions [54]. Also Hirata and co-workers reported a tribological study of N-CNO using a range of sizes (5 to 10 nm), both under vacuum (1.3×10^{-3} Pa) and in air (relative humidity 55%) (Figure 19.16). It was discovered that under both conditions, N-CNOs have a lower friction coefficient (<0.1) and lower wear compared to graphite powder. Interestingly, the larger N-CNOs had worse tribological properties than the smaller N-CNOs, which was attributed to the fact that N-CNOs have more defects, thus increasing the bonding with counter materials [55]. Another possible explanation is that the improvement of the tribological properties of the oils by the presence of N-CNOs could be due to the formation graphene nanosheets or a kind of DLC tribofilm by the friction-induced proposed by Joly-Pottuz and co-wokers in 2008 [56–58].

Although carbon nano onions exhibit interesting properties similar to those of carbon nanotubes and fullerenes they have yet to be studied in more detail. These novel materials can now be synthesized in bulk and exhibit interesting properties. For example, TGA analysis indicates the weight gain of the CNOs, Raman spectroscopy shows how the *D* and *G* bands changes with the new functional groups, NMR spectroscopy shows how the peaks of the addend changes in the presence of the CNOs, and so on. The different reactivity between A-CNOs and N-CNOs has been established [22], where mild condition reactions are needed for N-CNO functionalization compared to strong conditions for A-CNOs, and in some cases

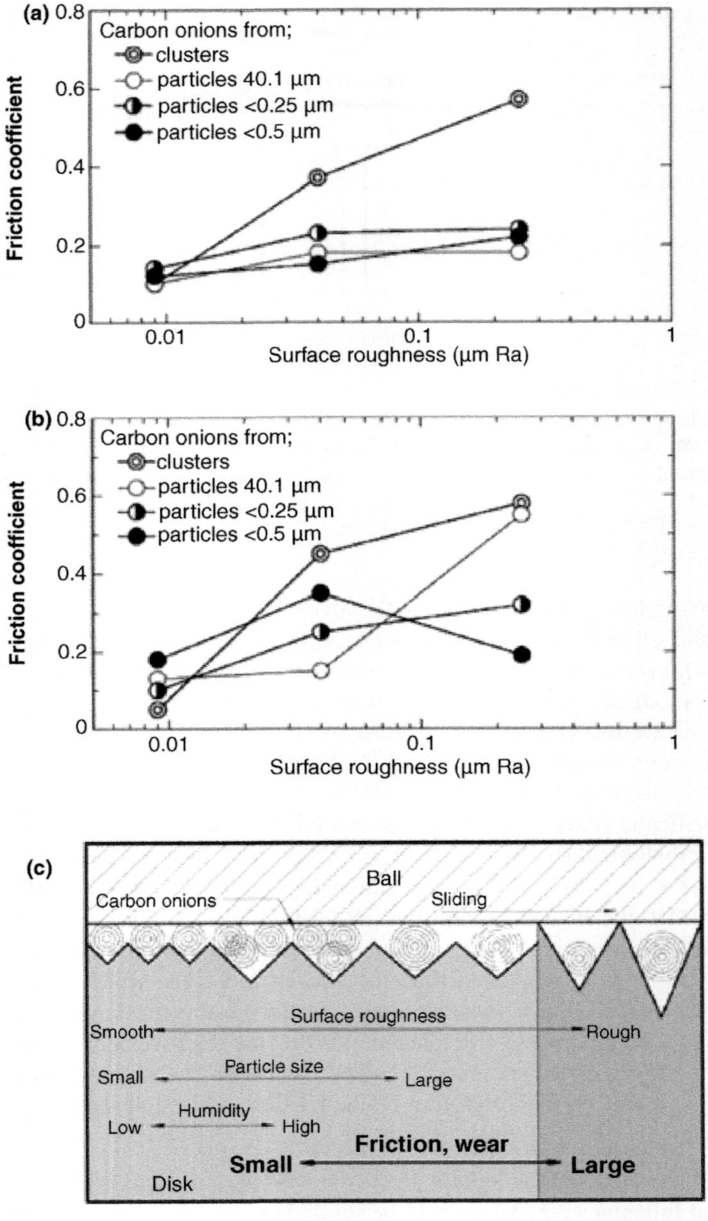

Figure 19.16 *Friction coefficient as a function of surface roughness for silicon discs when various CNOs sizes were employed in (a) air; and (b) vacuum. (c) Schematic illustration describing tribological behavior when CNOs reside between sliding surfaces. Reference [55]*

no reaction was observed for A-CNOs. How these materials find use in potentially useful applications still remains as a wide open field for future development. Just as practical applications for the fullerenes have been slow in coming, CNOs will probably find interesting and unique applications in due time.

Acknowledgements

Financial support from the National Science Foundation (Grant CHE-0509989 to A.J.A. and L.E.) and (Grant DMR-0809129 to S.C. and LE) is greatly appreciated. This material is based on work supported by the National Science Foundation while LE was working there. Any opinion, findings and conclusions or recommendations expressed in this material are those of the authors and do not necessarily reflect the views of the NSF.

References

[1] H.W. Kroto, J.R. Heath, S.C. Obrien, R.F. Curl and R.E. Smalley, C-60 – Buckminsterfullerene. *Nature*, **318**, 162–163 (1985).
[2] W. Kratschmer, L.D. Lamb, K. Fostiropoulos and D.R. Huffman, Solid C_{60}: a new form of carbon. *Nature*, **347**, 354–358 (1990).
[3] S. Iijima, Helical microtubules of graphitic carbon. *Nature*, **354**, 56–58 (1991).
[4] T.W. Ebbesen and P.M. Ajayan, Large-scale synthesis of carbon nanotubes. *Nature*, **358**, 220–222 (1992).
[5] Y. Saito, T. Yoshikawa, M. Inagaki, M. Tomita and T. Hayashi, Growth and structure of graphitic tubules and polyhedral particles in arc-discharge. *Chem. Phys. Lett.*, **204**, 277–282 (1993).
[6] R.S. Ruoff, D.C. Lorents, B. Chan, R. Malhotra and S. Subramoney, Single crystal metals encapsulated in carbon nanoparticles. *Science*, **259**, (5093) 346–348 (1993).
[7] D. Ugarte, Onion-Like graphitic particles. *Carbon*, **33**, 989–993 (1995).
[8] D. Ugarte, Curling and closure of graphitic networks under electron-beam irradiation. *Nature*, **359**, 707–709 (1992).
[9] V.L. Kuznetsov, A.L. Chuvilin, E.M. Moroz, V.N. Kolomiichuk, S.K. Shaikhutdinov, Y.V. Butenko and I.Y. Malkov, Effect of explosion conditions on the structure of detonation soots – ultradisperse diamond and onion carbon. *Carbon*, **32**, 873–882 (1994).
[10] V.L. Kuznetsov, A.L. Chuvilin, Y.V. Butenko, I.Y. Malkov and V.M. Titov, Onion-Like carbon from ultra-disperse diamond. *Chem. Phys. Lett.*, **222**, 343–348 (1994).
[11] V.Z. Mordkovich, The observation of large concentric shell fullerenes and fullerene-like nanoparticles in laser pyrolysis carbon blacks. *Chem. Mater.*, **12**, 2813–2818 (2002).
[12] V.Z. Mordkovich, Multishell and giant fullerenes in laser pyrolysis carbon blacks. *Mol. Mater.*, **13**, 371–376 (2000).
[13] V.Z. Mordkovich, A.G. Umnov, T. Inoshita and M. Endo, The observation of multiwall fullerenes in thermally treated laser pyrolysis carbon blacks. *Carbon*, **37**, 1855–1858 (1999).
[14] O. Mamezaki, H. Adachi, S. Tomita, M. Fujii and S. Hayashi, Thin films of carbon nanocapsules and onion-like graphitic particles prepared by the cosputtering method. *Jpn. J. Appl. Phys. Part 1*, **39**, 6680–6683 (2000).
[15] V.L. Kuznetsov, I.L. Zilberberg, Y.V. Butenko, A.L. Chuvilin and B. Segall, Theoretical study of the formation of closed curved graphite-like structures during annealing of diamond surface. *J. Appl. Phys.*, **86**, 863–870 (1999).
[16] E.D. Obraztsova, M. Fujii, S. Hayashi, V.L. Kuznetsov and Y.V. Butenko, Raman identification of onion-like carbon. *Carbon*, **36**, 821–826 (1998).
[17] S. Tomita, T. Sakurai, H. Ohta, M. Fujii and S. Hayashi, Structure and electronic properties of carbon onions. *J. Chem. Phys.*, **114**, 7477–7482 (2001).

[18] V.L. Kuznetsov, A.L. Chuvilin, Y.V. Butenko, I.Y. Malkov, A.K. Gutakovskii, S.V. Stankus and S.P. Krdirulin, Science and technology of fullerene materials. *MRS Symposium. Procceedings.*, **359**, 105 (1995).

[19] Y.V. Butenko, V.L. Kuznetsov, E.A. Paukshtis, A.I. Stadnichenko, I.N. Mazov, S.I. Moseenkov, A.I. Boronin and S.V. Kosheev, The thermal stability of nanodiamond surface groups and onset of nanodiamond graphitization. *Fullerenes Nanotubes and Carbon Nanostructures*, **14**, 557–564 (2006).

[20] S. Tomita, A. Burian, J.C. Dore, D. LeBolloch, M. Fujii and S. Hayashi, Diamond nanoparticles to carbon onions transformation: X-ray diffraction studies. *Carbon*, **40**, 1469–1474 (2002).

[21] V.L. Kuznetsov, Y.V. Butenko, A.L. Chuvilin, A.I. Romanenko and A.V. Okotrub, Electrical resistivity of graphitized ultra-disperse diamond and onion-like carbon. *Chem. Phys. Lett.*, **336**, 397–404 (2001).

[22] A. Palkar, F. Melin, C.M. Cardona, B. Elliot, A.K. Naskar, D.D. Edie, A. Kumar and L. Echegoyen, Reactivity differences between carbon nano onions (CNOs) prepared by different methods. *Chem. Asian J.*, **2**, 625–633 (2007).

[23] N. Sano, H. Wang, M. Chhowalla, I. Alexandrou and G.A.J. Amaratunga, Synthesis of carbon 'onions' in water. *Nature*, **414**, 506–507 (2001).

[24] N. Sano, H. Wang, I. Alexandrou, M. Chhowalla, K.B.K. Teo and G.A.J. Amaratunga, Properties of carbon onions produced by an arc discharge in water. *J. Appl. Phys.*, **92**, 2783–2788 (2002)

[25] J. Guo, X. Wang, Y. Yao, X. Yang, X. Liu and B. Xu, Structure of Nanocarbons Prepared by Arc Discharge in Water. *Mater. Chem. Phys.*, **105**, 175–178 (2007).

[26] M. Chhowalla, H. Wang, N. Sano, K.B.K. Teo, S.B. Lee and G.A.J. Amaratunga, Carbon onions: carriers of the 217.5 nm interstellar absorption feature. *Phys. Rev. Lett.*, **90**, 155504-1/155504-4 (2003).

[27] D. Roy, M. Chhowalla, H. Wang, N. Sano, I. Alexandrou, T.W. Clyne and G.A.J. Amaratunga, Characterisation of carbon nano-onions using raman spectroscopy. *Chem. Phys. Lett.*, **373**, 52–56 (2003).

[28] W. Lian, H. Song, X. Chen, L. Li, J. Huo, M. Zhao and G. Wang, The transformation of acetylene black into onion-like hollow carbon nanoparticles at 100 °C using iron catalyst. *Carbon*, **46**, 525–530 (2008).

[29] L.C. Qin and S. Iijima, Onion-like graphitic particles produced from diamond. *Chem. Phys. Lett.*, **262**, 252–258 (1996).

[30] V.V. Roddatis, V.L. Kuznetsov, Y.V. Butenko, D.S. Su and R. Schlogl, Onion-like graphitic particles produced from diamond. *Phys. Chem. Chem. Phys.*, **4**, 1964–1967 (2002).

[31] A.V. Gubarevich, J. Kitamura, S. Usuba, H. Yokoi, Y. Kakudate and O. Odawara, Onion-like carbon deposition by plasma spraying of nanodiamonds. *Carbon*, **41**, 2601–2606 (2003).

[32] T. Gorelik, S. Urban, F. Falk, U. Kaiser and U. Glatzel, Carbon onions produced by laser irradiation of amorphous silicon carbide. *Chem. Phys. Lett.*, **373**, 642–645 (2003).

[33] H.E. Troiani, A. Camacho-Bragado, V. Armendariz, J.L.G. Torresday and M.J. Yacaman, Synthesis of carbon onions by gold nanoparticles and electron irradiation. *Chem. Mater.*, **15**, 1029–1031 (2003).

[34] G. Radhakrishnan, P.M. Adams and L.S. Bernstein, Plasma characterization and room temperature growth of carbon nanotubes and nano-onions by excimer laser ablation. *Appl. Surf. Sci.*, **253**, 7651–7655 (2007).

[35] J. Hiraki, H. Mori, E. Taguchi, H. Yasuda, H. Kinoshita and N. Ohmae, Transformation of diamond nanoparticles into onion-like carbon by electron irradiation studied directly inside an ultrahigh-vacuum transmission electron microscope. *Appl. Phys. Lett.*, **86**, 22310111–22310113 (2005).

[36] S. Hu, P. Bai, F. Tian, S. Cao and J. Sun, Hydrophilic carbon onions synthesized by millisecond pulsed laser irradiation. *Carbon*, **47**, 876–883 (2009).

[37] J. Du, Z. Liu, Z. Li, B. Han, Z. Sun and Y. Huang, Carbon onions synthesized via thermal reduction of glycerin with magnesium. *Mater. Chem. Phys.*, **93**, 178–180 (2005).

[38] Y. Yan, H. Yang, F. Zhang, B. Tu and D. Zhao, Low-temperature solution synthesis of carbon nanoparticles, onions and nanoropes by the assembly of aromatic molecules. *Carbon*, **45**, 2209–2216 (2007).
[39] X.H. Chen, F.M. Deng, J.X. Wang, H.S. Yang, G.T. Wu, X.B. Zhang, J.C. Peng and W.Z. Li, New method of carbon onion growth by radio-frequency plasma-enhanced chemical vapor deposition. *Chem. Phys. Lett.*, **336**, 201–204 (2001).
[40] C. He, N. Zhao, X. Du, C. Shi, J. Ding, J. Li and Y. Li, Low-temperature synthesis of carbon onions by chemical deposition using a nickel catalyst supported on aluminum. *Scripta Mater.*, **54**, 689–693 (2006).
[41] F. Shoji, Z. Feng, A. Kono and T. Nagai, Spherical carbon particle growth in a methane plasma. *Appl. Phys. Lett.*, **89**, 171504 (2006).
[42] S.S. Hou, D.H. Chung and T.H. Lin, High-yield synthesis of carbon nano-onions in counterflow diffusion flames. *Carbon*, **47**, 938–947 (2009).
[43] T. Kobayashi, T. Sekine and H. He,Chem. Mater., Formation of carbon onion from heavily shocked SiC,. *Chem. Mater.*, **15**, 2681–2683 (2003).
[44] V. Georgakilas, D.M. Guldi, R. Signorini, R. Bozio and M. Prato, Organic functionalization and optical properties of carbon onions. *J. Am. Chem. Soc.*, **125**, 14268–14269 (2003).
[45] A.S. Rettenbacher, B. Elliott, J.S. Hudson, A. Amirkhanian and L. Echegoyen, Preparation and functionalization of multilayer fullerenes (carbon nano-onions). *Chem. Eur. J.*, **12**, 376–387 (2006).
[46] A. Palkar, A. Kumbhar, A.J. Athans and L. Echegoyen, Pyridyl-functionalized and water-soluble carbon nano onions: first supramolecular complexes of carbon nano onions. *Chem. Mater.*, **20**, 1685–1687 (2008).
[47] S. Tomita, M. Fujii and S. Hayashi, Optical extinction properties of carbon onions prepared from diamond nanoparticles. *Phys. Rev. B*, **66**, 245424-1/245424-7 (2002).
[48] L. Zhou, C. Gao, D. Zhu, W. Xu, F.F. Chen, A. Palkar, L. Echegoyen and E.S-W. Kong, Facile functionalization of multilayer fullerenes (carbon nano-onions) by nitrene chemistry and 'grafting from' strategy. *Chem. Eur. J.*, **15**, 1389–1396 (2009).
[49] Y. Liu, R.L Vander Wal and V.N. Khabashesku, Functionalization of carbon nano-onions by direct fluorination. *Chem. Mater.*, **19**, 778–786 (2007).
[50] A.S. Rettenbacher, M.W. Perpall, L. Echegoyen, J. Hudson and D.W. Smith, Radical addition of a conjugated polymer to multilayer fullerenes (carbon nano-onions). *Chem. Mater.*, **19**, 1411–1417 (2007).
[51] E. Koudoumas, O. Kokkinaki, M. Konstantaki, S. Couris, S. Korovin, P. Detkov, V. Kuznetsov, S. Pimenov and V. Pustovoi, Onion-like carbon and diamond nanoparticles for optical limiting. *Chem. Phys. Lett.*, **357**, 336–340 (2002).
[52] N. Keller, N.I. Maksimova, V.V. Roddatis, M. Schur, G. Mestl, Y.V. Butenko, V.L. Kuznetsov and R. Schlögl, The catalytic use of onion-like carbon materials for styrene synthesis by oxidative dehydrogenation of ethylbenzene. *Angew. Chem. Int. Ed.*, **41**, 1885–1888 (2002).
[53] T. Cabioc'h, E. Thune, J.P. Rivière, S. Camelio, J.C. Girard, P. Guérin, M. Jaouen, L. Henrard and Ph. Lambin, structure and properties of carbon onion layers deposited onto various substrates. *J. Appl. Phys.*, **91**, 1560–1567 (2002).
[54] K.W. Street, M. Marchetti, R.L. Vander Wal, and A.J. Tomasek, Evaluation of the tribological behavior of nano-onions in Krytox 143AB. *Tribol. Lett.*, **16**, 143–149 (2004).
[55] A. Hirata, M. Igarashi and T. Kaito, Study on solid lubricant properties of carbon onions produced by heat treatment of diamond clusters or particles. *Tribol. Int.*, **37**, 899–905 (2004).
[56] L. Joly-Pottuz N. Matsumoto, H. Kinoshita, B. Vacher, M. Belin, G. Montagnac, J.M. Martin and N. Ohmae, Diamond-derived carbon onions as lubricant additives. *Tribol. Int.*, **41**, 69–78 (2008).
[57] L. Joly-Pottuz and N. Ohmae, Carbon-based Nanolubricants, in *Nanolubricants*, Wiley, Great Britain, 2008.
[58] L. Joly-Pottuz, B. Vacher, N. Ohmae, J.M. Martin and T. Epicier, Anti-wear and friction reducing mechanisms of carbon nano-onions as lubricant additives. *Tribol. Lett.*, **30**, 69–80 (2008).

Index

Page numbers in *italics* refer to figures; page numbers in **bold** refer to tables.

Chemistry of Nanocarbons Edited by Takeshi Akasaka, Fred Wudl and Shigeru Nagase

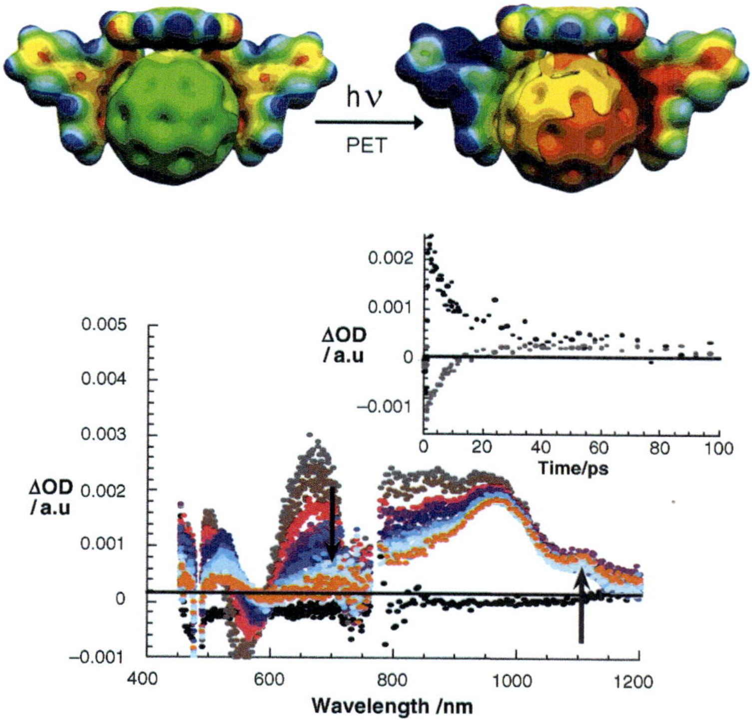

Figure 2.9 *Top: B3LYP/6-31G** electrostatic potential maps calculated for MTW•C_{60} in the ground electronic state and in the photoinduced charge-separated state. Bottom: differential absorption spectra (visible and near-infrared) obtained upon femtosecond flash photolysis (484 nm) of a MTW•C_{60} mixture (MTW: 2.5×10^{-5}; C_{60}: 2.5×10^{-3}) in benzonitrile. Inset – time-absorption profile at 670 nm (open circles) and 580 nm (filled circles), reflecting the charge separation and charge recombination dynamics*

Figure 6.2 *X-ray crystallographic structures of metal-penta(organo)[60]fullerene complexes*

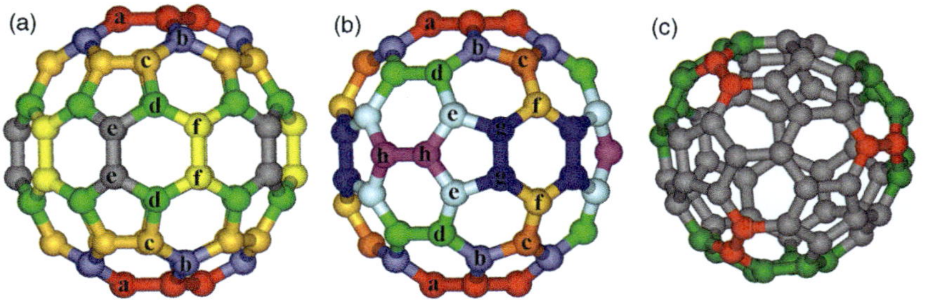

Figure 9.4 *(a) D_{5h}-C_{80} fullerene cage b) $D_{3h}(5)$-C_{78} fullerene cage and c) $D_3(6140)$-C_{68} fullerene cage. The bonds highlighted in red have been suggested as the preferred site for addition. The atoms highlighted in green correspond to the pentalene patches where the metal atoms of the cluster are bound*

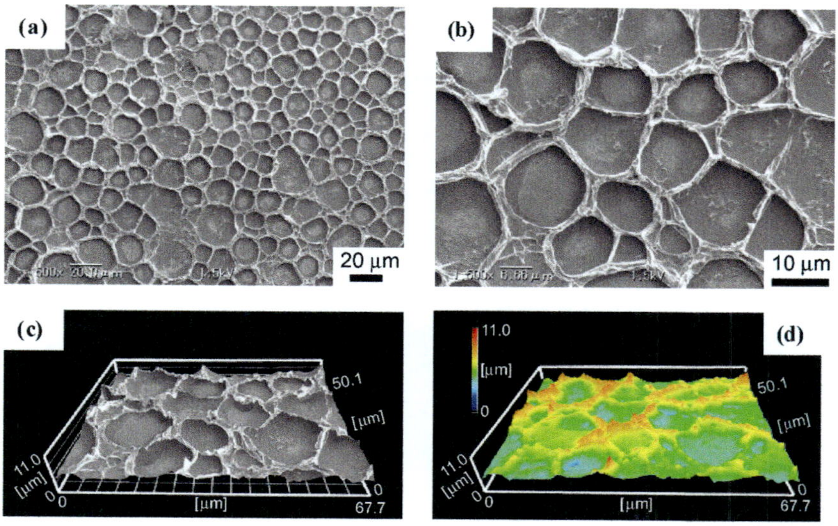

Figure 12.17 *SEM images of Complex **1** on a glass substrate after the ion-exchange*

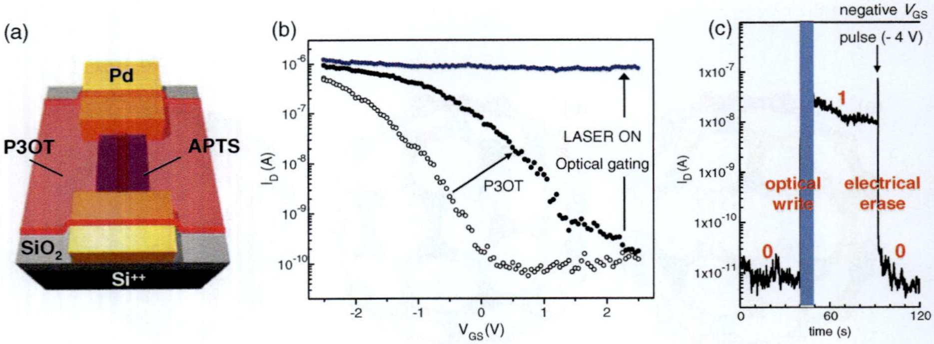

Figure 13.16 (a) Representation of the optical gated CNT-FET; (b) characteristics of the naked transistor in the dark (open black circles), coated with P3OT in the dark (filled black circles), and upon illumination ($\lambda = 457$ nm, blue circles); (c) principle of the writing and erasing of the memory device. The band in blue represents the light pulse use to write the electric information. From J. Borghetti et al., Adv. Mater., **18**, 2535–40 (2006), Copyright Wiley-VCH Verlag GmbH & Co. KGaA. Reproduced with permission from John Wiley and Sons

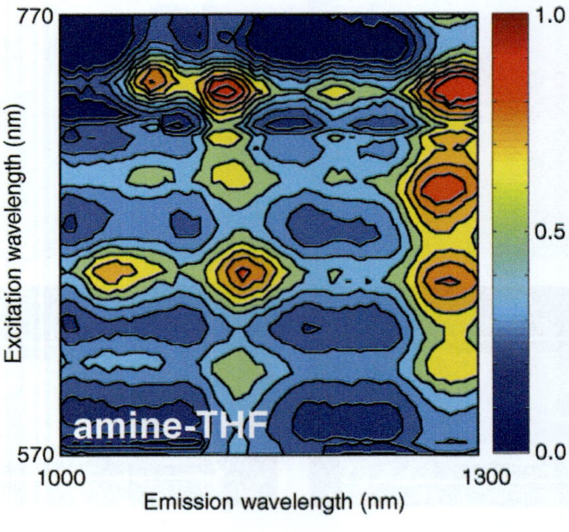

Figure 14.2 Contour plots of fluorescence intensities for SWNTs in octylamine-THF solution

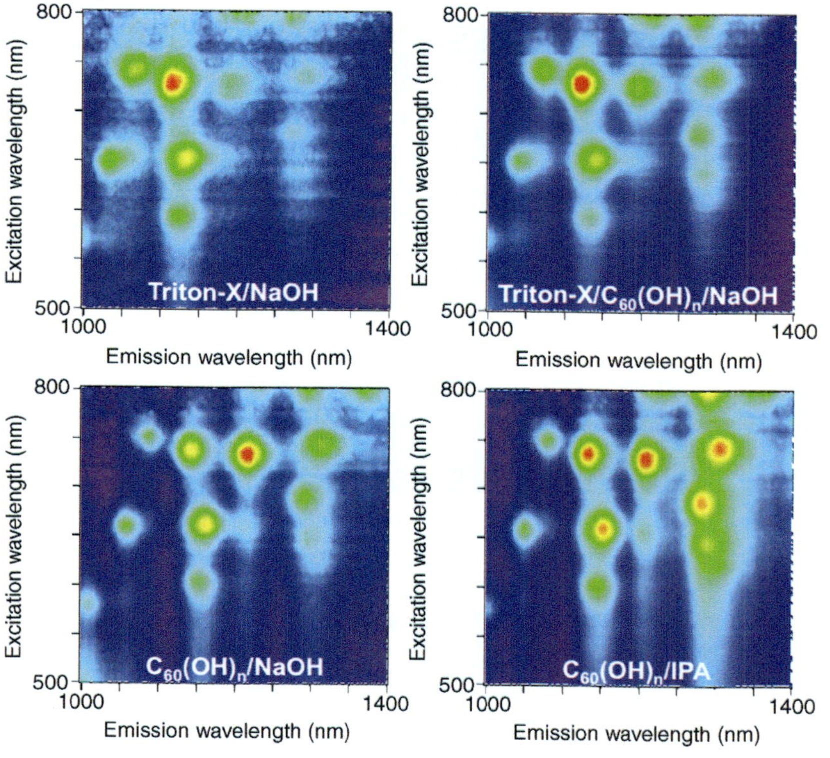

Figure 14.6 *Contour plots of fluorescence intensities for SWNTs*

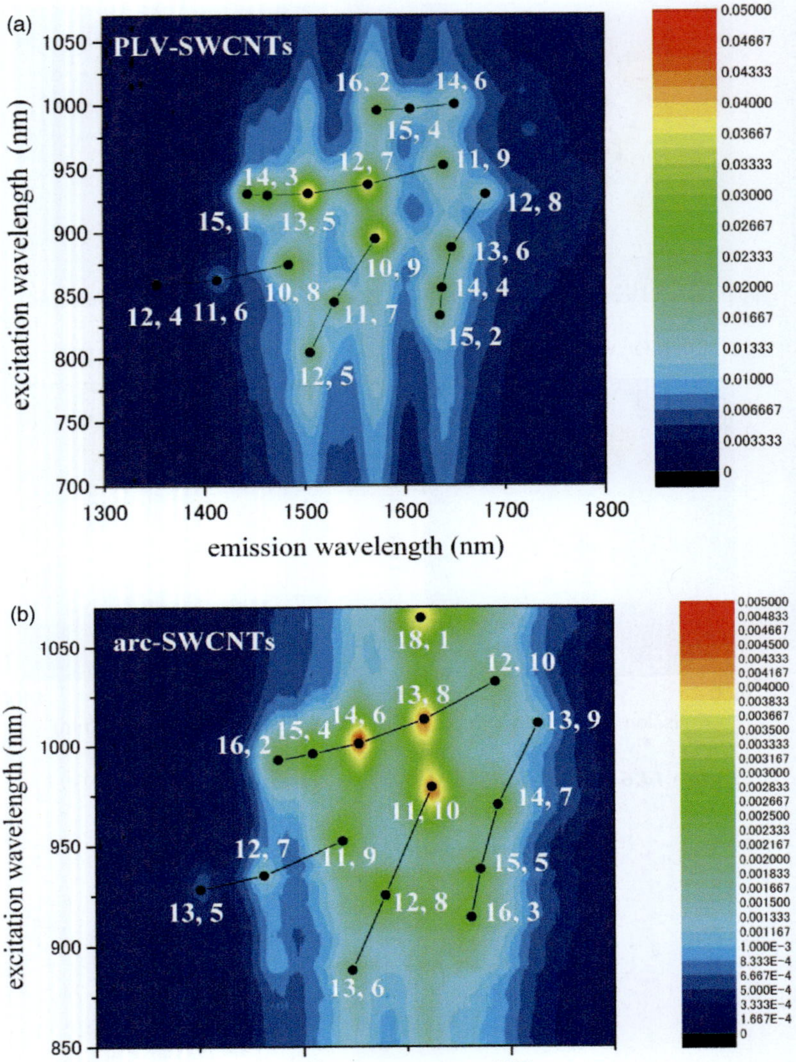

Figure 15.3 *LE maps of (a) PLV-SWCNTs, (b) arc-SWCNTs, (c) PLV-nanopeapods and (d) arc-nanopeapods, respectively*

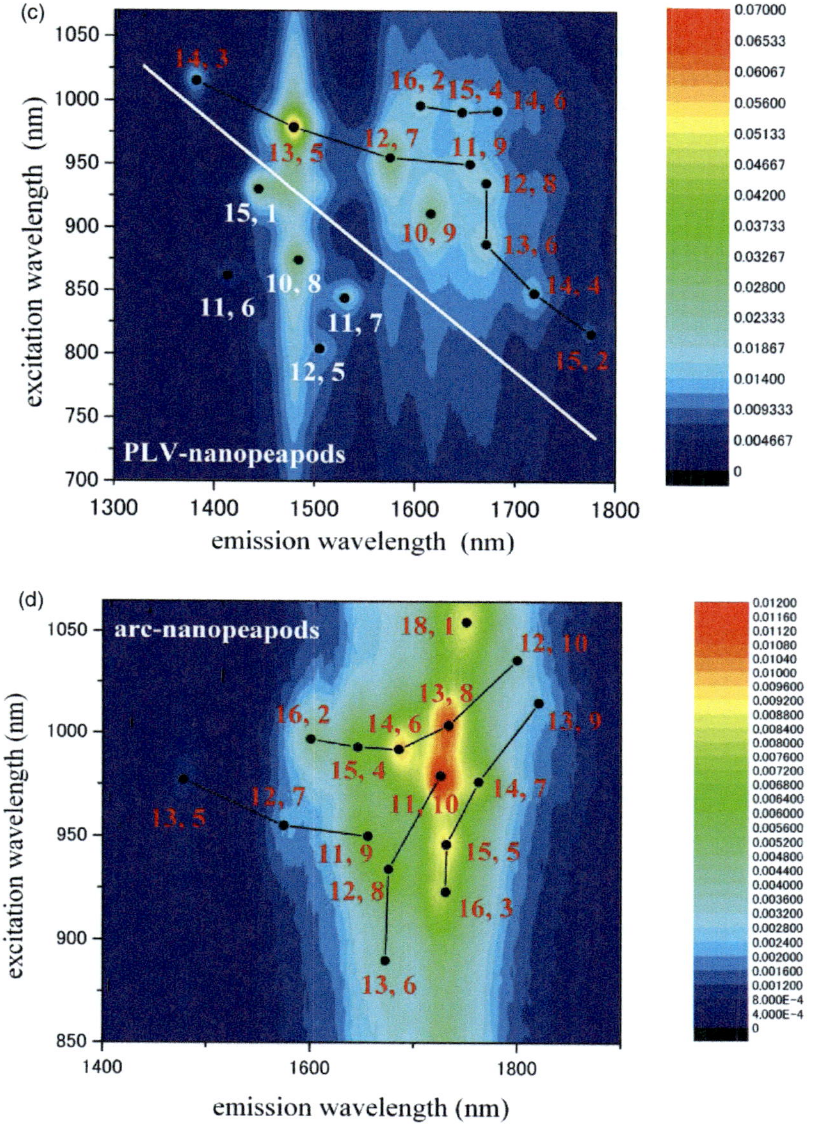

Figure 15.3 *(Continued)*

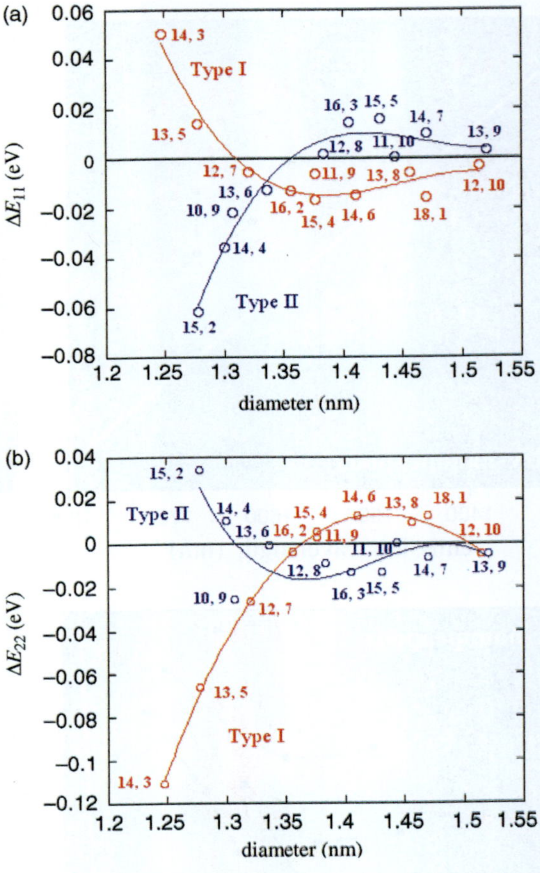

Figure 15.4 *Differences in optical transition energies in E_{11} and E_{22} (ΔE_{11} and ΔE_{22}, respectively) between C_{60} nanopeapods and SWNTs as a function of (a, b) a tube diameter and (c, d) chiral angle*

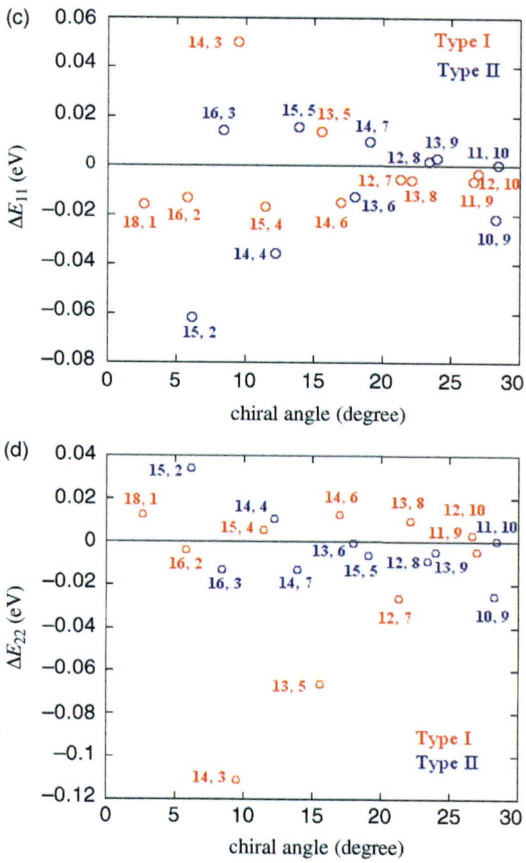

Figure 15.4 *(Continued)*

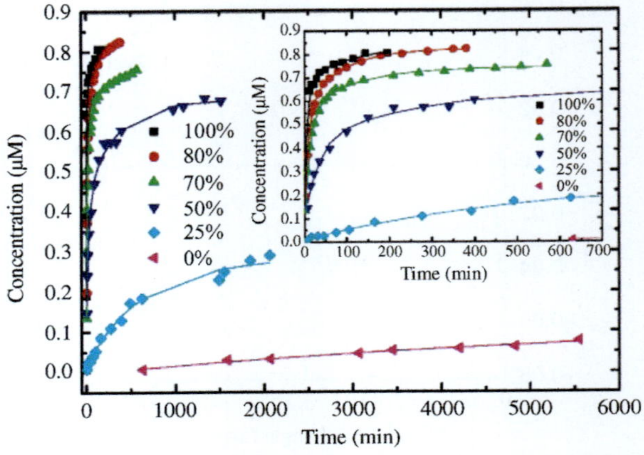

Figure 15.11 *Quantities of C_{60} released from C_{60}@SWNHox into toluene, ethanol, or their mixtures increased with the immersion periods. (Concentrations (%) of toluenes in the toluene-ethnaol mixtures are indicated in graphs.) [41]*

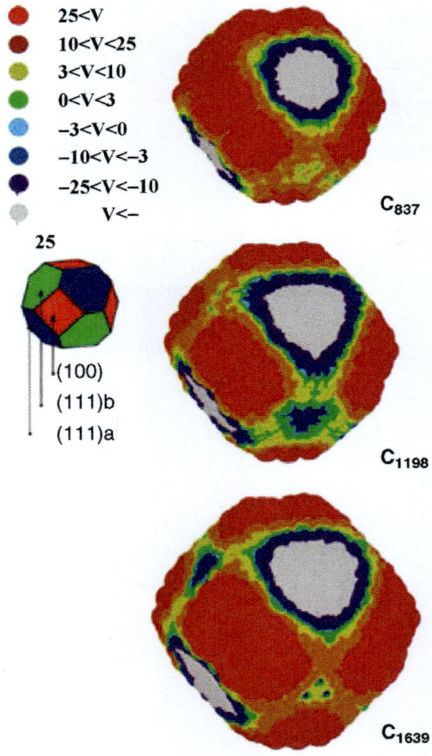

Figure 17.5 *Distribution of electrostatic potential fields on the facets of three smaller truncated octahedral models of SNBD particle obtained by SCC-DFTB geometry-optimization (ref. [16])*